Lecture Notes in Computer Science   12823

More information about this subseries at http://www.springer.com/series/7412

ghlights of the conference included the keynote talks given by Masaki
recipient of the IAPR/ICDAR Outstanding Achievements Award, and
ustaty, recipient of the IAPR/ICDAR Young Investigator Award, as well as
uished keynote speakers Prem Natarajan, vice president at Amazon, who
on "OCR: A Journey through Advances in the Science, Engineering, and
tion of AI/ML", and Beta Megyesi, professor of computational linguistics at
University, who elaborated on "Cracking Ciphers with 'AI-in-the-loop':
ion and Decryption in a Cross-Disciplinary Field".
of 340 publications were submitted to the main conference, which was held
ulieu convention center ... 8–10, 2021. Based on the reviews,
... esentation and 142 papers
... or the ICDAR-IJDAR
... was integrated in
... al consortium
... using on
... ition,

Josep Lladós · Daniel Lopresti ·
Seiichi Uchida (Eds.)

Document Analysis and Recognition – ICDAR 2021

16th International Conference
Lausanne, Switzerland, September 5–10, 2021
Proceedings, Part III

Springer

Editors
Josep Lladós 🅾
Universitat Autònoma de Barcelona
Barcelona, Spain

Daniel Lopresti 🅾
Lehigh University
Bethlehem, PA, USA

Seiichi Uchida 🅾
Kyushu University
Fukuoka-shi, Japan

Foreword

Our warmest welcome to the proceedings of ICDAR 2021, the 16th IAPR Int
Conference on Document Analysis and Recognition, which was held in Switz
the first time. Organizing an international conference of significant size d
COVID-19 pandemic, with the goal of welcoming at least some of the pa
physically, is similar to navigating a rowboat across the ocean during a st
tunately, we were able to wor
amount of flexibility and
the Beaulieu con
the internation
have support
setup

ISSN 0302-9743   ISSN 1611-3349   (electronic)
Lecture Notes in Computer Science
ISBN 978-3-030-86333-3   ISBN 978-3-030-86334-0   (eBook)
https://doi.org/10.1007/978-3-030-86334-0

LNCS Sublibrary: SL6 – Image Processing, Computer Vision, Pattern Recognition, and Graphics

This Springer imprint is published by the registered company Springer Nature Switzerland AG
The registered company address is: Gewerbestrasse 11, 6330 Cham, Switzerland

Foreword

Our warmest welcome to the proceedings of ICDAR 2021, the 16th IAPR International Conference on Document Analysis and Recognition, which was held in Switzerland for the first time. Organizing an international conference of significant size during the COVID-19 pandemic, with the goal of welcoming at least some of the participants physically, is similar to navigating a rowboat across the ocean during a storm. Fortunately, we were able to work together with partners who have shown a tremendous amount of flexibility and patience including, in particular, our local partners, namely the Beaulieu convention center in Lausanne, EPFL, and Lausanne Tourisme, and also the international ICDAR advisory board and IAPR-TC 10/11 leadership teams who have supported us not only with excellent advice but also financially, encouraging us to setup a hybrid format for the conference.

We were not a hundred percent sure if we would see each other in Lausanne but we remained confident, together with almost half of the attendees who registered for on-site participation. We relied on the hybridization support of a motivated team from the Lule University of Technology during the pre-conference, and professional support from Imavox during the main conference, to ensure a smooth connection between the physical and the virtual world. Indeed, our welcome is extended especially to all our colleagues who were not able to travel to Switzerland this year. We hope you had an exciting virtual conference week, and look forward to seeing you in person again at another event of the active DAR community.

With ICDAR 2021, we stepped into the shoes of a longstanding conference series, which is the premier international event for scientists and practitioners involved in document analysis and recognition, a field of growing importance in the current age of digital transitions. The conference is endorsed by IAPR-TC 10/11 and celebrates its 30th anniversary this year with the 16th edition. The very first ICDAR conference was held in St. Malo, France in 1991, followed by Tsukuba, Japan (1993), Montreal, Canada (1995), Ulm, Germany (1997), Bangalore, India (1999), Seattle, USA (2001), Edinburgh, UK (2003), Seoul, South Korea (2005), Curitiba, Brazil (2007), Barcelona, Spain (2009), Beijing, China (2011), Washington DC, USA (2013), Nancy, France (2015), Kyoto, Japan (2017), and Syndey, Australia (2019).

The attentive reader may have remarked that this list of cities includes several venues for the Olympic Games. This year the conference was be hosted in Lausanne, which is the headquarters of the International Olympic Committee. Not unlike the athletes who were recently competing in Tokyo, Japan, the researchers profited from a healthy spirit of competition, aimed at advancing our knowledge on how a machine can understand written communication. Indeed, following the tradition from previous years, 13 scientific competitions were held in conjunction with ICDAR 2021 including, for the first time, three so-called "long-term" competitions addressing wider challenges that may continue over the next few years.

Other highlights of the conference included the keynote talks given by Masaki Nakagawa, recipient of the IAPR/ICDAR Outstanding Achievements Award, and Mickaël Coustaty, recipient of the IAPR/ICDAR Young Investigator Award, as well as our distinguished keynote speakers Prem Natarajan, vice president at Amazon, who gave a talk on "OCR: A Journey through Advances in the Science, Engineering, and Productization of AI/ML", and Beta Megyesi, professor of computational linguistics at Uppsala University, who elaborated on "Cracking Ciphers with 'AI-in-the-loop': Transcription and Decryption in a Cross-Disciplinary Field".

A total of 340 publications were submitted to the main conference, which was held at the Beaulieu convention center during September 8–10, 2021. Based on the reviews, our Program Committee chairs accepted 40 papers for oral presentation and 142 papers for poster presentation. In addition, nine articles accepted for the ICDAR-IJDAR journal track were presented orally at the conference and a workshop was integrated in a poster session. Furthermore, 12 workshops, 2 tutorials, and the doctoral consortium were held during the pre-conference at EPFL during September 5–7, 2021, focusing on specific aspects of document analysis and recognition, such as graphics recognition, camera-based document analysis, and historical documents.

The conference would not have been possible without hundreds of hours of work done by volunteers in the organizing committee. First of all we would like to express our deepest gratitude to our Program Committee chairs, Joseph Lladós, Dan Lopresti, and Seiichi Uchida, who oversaw a comprehensive reviewing process and designed the intriguing technical program of the main conference. We are also very grateful for all the hours invested by the members of the Program Committee to deliver high-quality peer reviews. Furthermore, we would like to highlight the excellent contribution by our publication chairs, Liangrui Peng, Fouad Slimane, and Oussama Zayene, who negotiated a great online visibility of the conference proceedings with Springer and ensured flawless camera-ready versions of all publications. Many thanks also to our chairs and organizers of the workshops, competitions, tutorials, and the doctoral consortium for setting up such an inspiring environment around the main conference. Finally, we are thankful for the support we have received from the sponsorship chairs, from our valued sponsors, and from our local organization chairs, which enabled us to put in the extra effort required for a hybrid conference setup.

Our main motivation for organizing ICDAR 2021 was to give practitioners in the DAR community a chance to showcase their research, both at this conference and its satellite events. Thank you to all the authors for submitting and presenting your outstanding work. We sincerely hope that you enjoyed the conference and the exchange with your colleagues, be it on-site or online.

September 2021

Andreas Fischer
Rolf Ingold
Marcus Liwicki

Editors
Josep Lladós 🆔
Universitat Autònoma de Barcelona
Barcelona, Spain

Daniel Lopresti 🆔
Lehigh University
Bethlehem, PA, USA

Seiichi Uchida 🆔
Kyushu University
Fukuoka-shi, Japan

ISSN 0302-9743   ISSN 1611-3349   (electronic)
Lecture Notes in Computer Science
ISBN 978-3-030-86333-3   ISBN 978-3-030-86334-0   (eBook)
https://doi.org/10.1007/978-3-030-86334-0

LNCS Sublibrary: SL6 – Image Processing, Computer Vision, Pattern Recognition, and Graphics

This Springer imprint is published by the registered company Springer Nature Switzerland AG
The registered company address is: Gewerbestrasse 11, 6330 Cham, Switzerland

Josep Lladós · Daniel Lopresti ·
Seiichi Uchida (Eds.)

Document Analysis and Recognition – ICDAR 2021

16th International Conference
Lausanne, Switzerland, September 5–10, 2021
Proceedings, Part III

 Springer

Preface

It gives us great pleasure to welcome you to the proceedings of the 16th International Conference on Document Analysis and Recognition (ICDAR 2021). ICDAR brings together practitioners and theoreticians, industry researchers and academics, representing a range of disciplines with interests in the latest developments in the field of document analysis and recognition. The last ICDAR conference was held in Sydney, Australia, in September 2019. A few months later the COVID-19 pandemic locked down the world, and the Document Analysis and Recognition (DAR) events under the umbrella of IAPR had to be held in virtual format (DAS 2020 in Wuhan, China, and ICFHR 2020 in Dortmund, Germany). ICDAR 2021 was held in Lausanne, Switzerland, in a hybrid mode. Thus, it offered the opportunity to resume normality, and show that the scientific community in DAR has kept active during this long period.

Despite the difficulties of COVID-19, ICDAR 2021 managed to achieve an impressive number of submissions. The conference received 340 paper submissions, of which 182 were accepted for publication (54%) and, of those, 40 were selected as oral presentations (12%) and 142 as posters (42%). Among the accepted papers, 112 had a student as the first author (62%), and 41 were identified as coming from industry (23%). In addition, a special track was organized in connection with a Special Issue of the International Journal on Document Analysis and Recognition (IJDAR). The Special Issue received 32 submissions that underwent the full journal review and revision process. The nine accepted papers were published in IJDAR and the authors were invited to present their work in the special track at ICDAR.

The review model was double blind, i.e. the authors did not know the name of the reviewers and vice versa. A plagiarism filter was applied to each paper as an added measure of scientific integrity. Each paper received at least three reviews, totaling more than 1,000 reviews. We recruited 30 Senior Program Committee (SPC) members and 200 reviewers. The SPC members were selected based on their expertise in the area, considering that they had served in similar roles in past DAR events. We also included some younger researchers who are rising leaders in the field.

In the final program, authors from 47 different countries were represented, with China, India, France, the USA, Japan, Germany, and Spain at the top of the list. The most popular topics for accepted papers, in order, included text and symbol recognition, document image processing, document analysis systems, handwriting recognition, historical document analysis, extracting document semantics, and scene text detection and recognition. With the aim of establishing ties with other communities within the concept of reading systems at large, we broadened the scope, accepting papers on topics like natural language processing, multimedia documents, and sketch understanding.

The final program consisted of ten oral sessions, two poster sessions, three keynotes, one of them given by the recipient of the ICDAR Outstanding Achievements Award, and two panel sessions. We offer our deepest thanks to all who contributed their time and effort to make ICDAR 2021 a first-rate event for the community. This year's ICDAR had a large number of interesting satellite events as well: workshops, tutorials, competitions, and the doctoral consortium. We would also like to express our sincere thanks to the keynote speakers, Prem Natarajan and Beta Megyesi.

Finally, we would like to thank all the people who spent time and effort to make this impressive program: the authors of the papers, the SPC members, the reviewers, and the ICDAR organizing committee as well as the local arrangements team.

September 2021

Josep Lladós
Daniel Lopresti
Seiichi Uchida

Organization

Organizing Committee

General Chairs

Andreas Fischer	University of Applied Sciences and Arts Western Switzerland, Switzerland
Rolf Ingold	University of Fribourg, Switzerland
Marcus Liwicki	Luleå University of Technology, Sweden

Program Committee Chairs

Josep Lladós	Computer Vision Center, Spain
Daniel Lopresti	Lehigh University, USA
Seiichi Uchida	Kyushu University, Japan

Workshop Chairs

Elisa H. Barney Smith	Boise State University, USA
Umapada Pal	Indian Statistical Institute, India

Competition Chairs

Harold Mouchère	University of Nantes, France
Foteini Simistira	Luleå University of Technology, Sweden

Tutorial Chairs

Véronique Eglin	Institut National des Sciences Appliquées, France
Alicia Fornés	Computer Vision Center, Spain

Doctoral Consortium Chairs

Jean-Christophe Burie	La Rochelle University, France
Nibal Nayef	MyScript, France

Publication Chairs

Liangrui Peng	Tsinghua University, China
Fouad Slimane	University of Fribourg, Switzerland
Oussama Zayene	University of Applied Sciences and Arts Western Switzerland, Switzerland

Sponsorship Chairs

David Doermann	University at Buffalo, USA
Koichi Kise	Osaka Prefecture University, Japan
Jean-Marc Ogier	University of La Rochelle, France

Local Organization Chairs

Jean Hennebert   University of Applied Sciences and Arts Western
                             Switzerland, Switzerland
Anna Scius-Bertrand   University of Applied Sciences and Arts Western
                             Switzerland, Switzerland
Sabine Süsstrunk   École Polytechnique Fédérale de Lausanne,
                             Switzerland

Industrial Liaison

Aurélie Lemaitre   University of Rennes, France

Social Media Manager

Linda Studer   University of Fribourg, Switzerland

Program Committee

Senior Program Committee Members

Apostolos Antonacopoulos   University of Salford, UK
Xiang Bai   Huazhong University of Science and Technology,
                             China
Michael Blumenstein   University of Technology Sydney, Australia
Jean-Christophe Burie   University of La Rochelle, France
Mickaël Coustaty   University of La Rochelle, France
Bertrand Coüasnon   University of Rennes, France
Andreas Dengel   DFKI, Germany
Gernot Fink   TU Dortmund University, Germany
Basilis Gatos   Demokritos, Greece
Nicholas Howe   Smith College, USA
Masakazu Iwamura   Osaka Prefecture University, Japan
C. V. Javahar   IIIT Hyderabad, India
Lianwen Jin   South China University of Technology, China
Dimosthenis Karatzas   Computer Vision Center, Spain
Laurence Likforman-Sulem   Télécom ParisTech, France
Cheng-Lin Liu   Chinese Academy of Sciences, China
Angelo Marcelli   University of Salerno, Italy
Simone Marinai   University of Florence, Italy
Wataru Ohyama   Saitama Institute of Technology, Japan
Luiz Oliveira   Federal University of Parana, Brazil
Liangrui Peng   Tsinghua University, China
Ashok Popat   Google Research, USA
Partha Pratim Roy   Indian Institute of Technology Roorkee, India
Marçal Rusiñol   Computer Vision Center, Spain
Robert Sablatnig   Vienna University of Technology, Austria
Marc-Peter Schambach   Siemens, Germany

Srirangaraj Setlur	University at Buffalo, USA
Faisal Shafait	National University of Sciences and Technology, India
Nicole Vincent	Paris Descartes University, France
Jerod Weinman	Grinnell College, USA
Richard Zanibbi	Rochester Institute of Technology, USA

Program Committee Members

Sébastien Adam
Irfan Ahmad
Sheraz Ahmed
Younes Akbari
Musab Al-Ghadi
Alireza Alaei
Eric Anquetil
Srikar Appalaraju
Elisa H. Barney Smith
Abdel Belaid
Mohammed Faouzi Benzeghiba
Anurag Bhardwaj
Ujjwal Bhattacharya
Alceu Britto
Jorge Calvo-Zaragoza
Chee Kheng Ch'Ng
Sukalpa Chanda
Bidyut B. Chaudhuri
Jin Chen
Youssouf Chherawala
Hojin Cho
Nam Ik Cho
Vincent Christlein
Christian Clausner
Florence Cloppet
Donatello Conte
Kenny Davila
Claudio De Stefano
Sounak Dey
Moises Diaz
David Doermann
Antoine Doucet
Fadoua Drira
Jun Du
Véronique Eglin
Jihad El-Sana
Jonathan Fabrizio

Nadir Farah
Rafael Ferreira Mello
Miguel Ferrer
Julian Fierrez
Francesco Fontanella
Alicia Fornés
Volkmar Frinken
Yasuhisa Fujii
Akio Fujiyoshi
Liangcai Gao
Utpal Garain
C. Lee Giles
Romain Giot
Lluis Gomez
Petra Gomez-Krämer
Emilio Granell
Mehdi Hamdani
Gaurav Harit
Ehtesham Hassan
Anders Hast
Sheng He
Jean Hennebert
Pierre Héroux
Laurent Heutte
Nina S. T. Hirata
Tin Kam Ho
Kaizhu Huang
Qiang Huo
Donato Impedovo
Reeve Ingle
Brian Kenji Iwana
Motoi Iwata
Antonio Jimeno
Slim Kanoun
Vassilis Katsouros
Ergina Kavallieratou
Klara Kedem

Christopher Kermorvant
Khurram Khurshid
Soo-Hyung Kim
Koichi Kise
Florian Kleber
Pramod Kompalli
Alessandro Lameiras Koerich
Bart Lamiroy
Anh Le Duc
Frank Lebourgeois
Gurpreet Lehal
Byron Leite Dantas Bezerra
Aurélie Lemaitre
Haifeng Li
Zhouhui Lian
Minghui Liao
Rafael Lins
Wenyin Liu
Lu Liu
Georgios Louloudis
Yue Lu
Xiaoqing Lu
Muhammad Muzzamil Luqman
Sriganesh Madhvanath
Muhammad Imran Malik
R. Manmatha
Volker Märgner
Daniel Martín-Albo
Carlos David Martinez Hinarejos
Minesh Mathew
Maroua Mehri
Carlos Mello
Tomo Miyazaki
Momina Moetesum
Harold Mouchère
Masaki Nakagawa
Nibal Nayef
Atul Negi
Clemens Neudecker
Cuong Tuan Nguyen
Hung Tuan Nguyen
Journet Nicholas
Jean-Marc Ogier
Shinichiro Omachi
Umapada Pal
Shivakumara Palaiahnakote

Thierry Paquet
Swapan Kr. Parui
Antonio Parziale
Antonio Pertusa
Giuseppe Pirlo
Réjean Plamondon
Stefan Pletschacher
Utkarsh Porwal
Vincent Poulain D'Andecy
Ioannis Pratikakis
Joan Puigcerver
Siyang Qin
Irina Rabaev
Jean-Yves Ramel
Oriol Ramos Terrades
Romain Raveaux
Frédéric Rayar
Ana Rebelo
Pau Riba
Kaspar Riesen
Christophe Rigaud
Syed Tahseen Raza Rizvi
Leonard Rothacker
Javad Sadri
Rajkumar Saini
Joan Andreu Sanchez
K. C. Santosh
Rosa Senatore
Amina Serir
Mathias Seuret
Badarinath Shantharam
Imran Siddiqi
Nicolas Sidère
Foteini Simistira Liwicki
Steven Simske
Volker Sorge
Nikolaos Stamatopoulos
Bela Stantic
H. Siegfried Stiehl
Daniel Stoekl Ben Ezra
Tonghua Su
Tong Sun
Yipeng Sun
Jun Sun
Suresh Sundaram
Salvatore Tabbone

Kazem Taghva
Ryohei Tanaka
Christopher Tensmeyer
Kengo Terasawa
Ruben Tolosana
Alejandro Toselli
Cao De Tran
Szilard Vajda
Ernest Valveny
Marie Vans
Eduardo Vellasques
Ruben Vera-Rodriguez
Christian Viard-Gaudin
Mauricio Villegas
Qiu-Feng Wang

Da-Han Wang
Curtis Wigington
Liang Wu
Mingkun Yang
Xu-Cheng Yin
Fei Yin
Guangwei Zhang
Heng Zhang
Xu-Yao Zhang
Yuchen Zheng
Guoqiang Zhong
Yu Zhou
Anna Zhu
Majid Ziaratban

Contents – Part III

Document Analysis Systems

Office Automation

Signature Verification

Document Forensics and Provenance Analysis

Pen-Based Document Analysis

Graphics Recognition

Extracting Document Semantics

MiikeMineStamps: A Long-Tailed Dataset of Japanese Stamps via Active Learning

Paola A. Buitrago[1,2], Evgeny Toropov[3(✉)], Rajanie Prabha[1,2], Julian Uran[1,2], and Raja Adal[4]

[1] Pittsburgh Supercomputing Center, Pittsburgh, PA 15203, USA
[2] Carnegie Mellon University, Pittsburgh, PA 15203, USA
[3] DeepMap Inc., East Palo Alto, CA 94303, USA
etoropov@nvidia.com
[4] University of Pittsburgh, Pittsburgh, PA 15260, USA

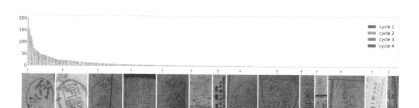

Fig. 1. Long tail distribution of stamps in the MiikeMineStamps dataset and stamp samples from selected classes. The labeling was performed in cycles across documents from different time periods.

Abstract. Mining existing image datasets with rich information can help advance knowledge across domains in the humanities and social sciences. In the past, the extraction of this information was often prohibitively expensive and labor-intensive. AI can provide an alternative, making it possible to speed up the labeling and mining of large and specialized datasets via a human-in-the-loop method of active learning (AL). Although AL methods are helpful for certain scenarios, they present limitations when the set of classes is not known before labeling (i.e. an open-ended set) and the distribution of objects across classes is highly unbalanced (i.e. a long-tailed distribution). To address these limitations in object detection scenarios we propose a multi-step approach consisting of 1) object detection of a generic "object" class, and 2) image classification with an open class set and a long tail distribution. We apply our approach to recognizing stamps in a large compendium of historical documents from the Japanese company Mitsui Mi'ike Mine, one of the largest business archives in modern Japan that spans half a century, includes tens of thousands of documents, and has been widely used by labor historians, business historians, and others. To test our approach we produce and make publicly available the novel and expert-curated MiikeMineStamps dataset. This unique dataset consists of 5056 images

© Springer Nature Switzerland AG 2021
J. Lladós et al. (Eds.): ICDAR 2021, LNCS 12823, pp. 3–19, 2021.
https://doi.org/10.1007/978-3-030-86334-0_1

of 405 different Japanese stamps, which to the best of our knowledge is the first published dataset of historical Japanese stamps. We hope that the MiikeMineStamps dataset will become a useful tool to further explore the application of AI methods to the study of historical documents in Japan and throughout the world of Chinese characters, as well as serve as a benchmark for image classification algorithms with an open-ended and highly unbalanced class set.

Keywords: Active learning · Object detection · Long tail · Open set · Stamp · Japanese · Historical · Dataset

1   Introduction

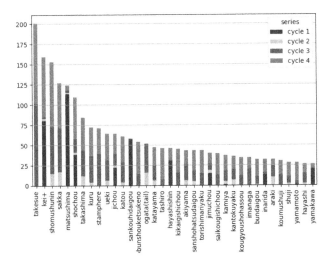

Fig. 2. The most frequent stamps by class as collected across active learning cycles.

Mining existing image datasets with rich information can help advance knowledge across various domains in the humanities, social sciences, and beyond. In East Asia, stamps often take the place of signatures. When opening a bank account or completing a contract, instead of signing one's name, it is common to stamp it onto the account application or contract. Stamps are therefore the primary instrument for verifying one's identity, but they have also been used for a number of other purposes. It is not uncommon for businesses and government offices to stamp the date onto a document, along with the name of the company or branch office, or the status of the document, such as "approved" or "top secret". Documents emanating from East Asian government bureaucracies or businesses often feature multiple stamps on a single page. Mining these stamps opens unprecedented scenarios, making it possible to transform a document archive into a rich dataset that can reveal individual names, information flows, and interpersonal networks.

The Mitsui Mi'ike Mine archive is probably the most complete business archive for the study of modern Japan available today. Its uniqueness lies in its size, more than 30,000 pages, and its span, half a century ranging from 1889 to 1940. Without the aid of machine learning, mining the tens of thousands of stamps in this archive would require an expensive team of research assistants trained in reading the frequently stylized and hard-to-read Chinese characters that are used in East Asian stamps. The research assistants would need to open a photograph of each document, input a document identifier in a spreadsheet, and then work on recognizing the stamps that appear on that document. Since every document has, on average, several stamps, this would have to be repeated tens of thousands of times, requiring thousands of hours of work. However, the cost of such work would be secondary to the real challenge of finding, hiring, and training such a team of expert assistants.

During the past decade, machine learning has been widely used and applied to discovering and automating such tasks. The most promising type of algorithms falls under the supervised learning category [31]. These algorithms depend on the availability of large volumes of labeled data, making it possible to learn "by example". Producing the much-needed labeled data traditionally requires an expensive and heavily involved process which can be prohibitive. The labeling challenge is particularly significant in domains that involve specialized knowledge. Active learning (AL), which is concerned with optimally selecting the next data samples to label based on feedback from prior iterations, has become a useful approach to making labeling possible while making a reasonable investment in time and effort. Until now, most AL research has been applied to classification rather than object detection. For AL in detection, however, the main area of focus is defining ideal criteria that make it possible to select ideal next candidates for labeling.

A significant limitation of the existing AL approaches for detection is that they do not consider open-class (i.e. undefined number of classes), long-tail data distributions (i.e. a large number of classes and few samples for a significant portion of them). Datasets exhibiting these characteristics are common and particularly challenging.

To facilitate the labeling work in scenarios like the one described here, we propose a method that leverages active learning concepts and popular algorithms in the area tuned to the application. The method relies on the following elements:

1. Break the task into two parts: detect generic "objects" and classify them.
2. Use a classification model to manage open-class, long-tail distributions.

We illustrate this method by applying it to the Mitsui Mi'ike Mine catalogue of historical documents, whose characteristics make it ideal for this type of work:

– Stamps share similar visual features. Previously unseen classes of stamps can still be identified by a generic stamp detector.
– The stamp class set is not known in advance.
– The long-tail distribution limits the accuracy of off-the-shelf object detectors.

In this work, we use AL to crop out and annotate stamps from the historical documents and produce the resulting MiikeMineStamps dataset. This unique dataset contains 5056 images of 405 different Japanese stamps enriched with relevant domain-specific metadata. Figure 1 presents examples of the stamp images and the distribution of the stamp classes.

The contribution of this paper is therefore twofold. First, we introduce MiikeMineStamps, a unique dataset of stamps from Japanese historical documents, and second, we also present the application of a known AL approach for the object detection of datasets with open class and long-tail distributions. We trust that this dataset will become a useful tool to further study Japanese historical documents, as well as serve as a benchmark for image classification algorithms with highly unbalanced class sets.

2 Related Work

2.1 Kuzushiji and Stamps in Japanese Historic Documents

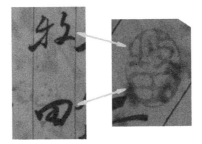

Fig. 3. Handwritten and stamped Chinese characters. The last name Makita looks very different when handwritten in cursive (left) and imprinted as a stamp (right).

Much of the leading-edge research in the recognition of the Chinese characters, which are used in Japan, China, Korea, and a few other parts of East Asia, has focused on the recognition of handwritten cursive script. Today, most Japanese is printed or handwritten in easy-to-read block or semi-cursive characters. Until the beginning of the twentieth century, however, most documents were either printed with woodblocks or handwritten with a brush using a cursive script known as *kuzushiji*. Not only does the kuzushiji cursive script link multiple characters, making it difficult to know where a character begins and where it ends, but there was no standard way for writing each character. Learning how to read a character meant learning three, four, or more ways in which it could be written. Since reading the kuzushiji cursive script requires special training, only trained archivists and historians are able to read it. Recently, however, the Center for Open Data in the Humanities (CODH) in Tokyo published a revolutionary machine-learning model known as Kuronet [4], which made it possible to read cursive kuzushiji script with an F-score in the range of 80% to 90% for most woodblock-printed books and with lower and sufficient accuracy for handwritten documents. This model has elicited enormous interest from archivists and historians in Japan and internationally.

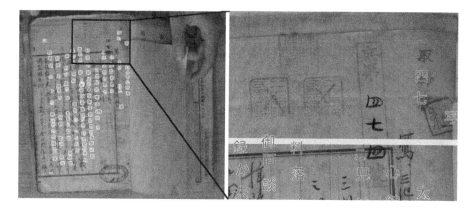

Fig. 4. The result of using Kuronet Kuzushiji Recognition Service on a document with stamps. Kuronet [4] successfully recognizes semi-cursive, but stamps are not detected at all and when they are, they are incorrectly recognized.

Machine learning models for reading kuzushiji cursive characters such as Kuronet, however, are not capable of recognizing the innumerable stamps that populate Japanese bureaucratic documents, as well as documents from China, Korea, and other parts of East Asia, from several thousand years ago to today. The scripts used to make stamps are stylized in ways that are very different from the kuzushiji cursive writing and use multiple, often archaic, fonts (Fig. 3). They can also combine multiple characters on a single stamp or can be used to simultaneously stamp two conforming copies of a document, such as a letter or a contract, so that the top half of the stamp appears on one document and the bottom half of the stamp on another. Figure 4 shows how Kuronet, a model created to recognize cursive handwritten or woodblock printed documents, is incapable of recognizing stamps. Although Kuronet successfully recognized portions of the semi-cursive writing, it did not recognize any of the stamps or even detect most of them. This is not surprising if we consider that Kuronet was never trained to recognize stamps.

As a result, a different model and a different dataset are needed for recognizing stamps in historical documents. The labeled dataset of stamps MiikeMineStamps together with the AI model to distinguish them fills this gap.

2.2   Active Learning

With the wide adoption of data-hungry deep learning methods, the need for large labeled datasets encouraged the development of Active Learning (AL) methods. AL aims to efficiently label large datasets in order to reduce the annotation cost. In AL, a small subset of data is annotated first, then an acquisition function selects the next batch to be annotated. A machine learning model trained on previously collected data helps the annotator by producing machine-generated labels, which the annotator verifies or corrects. The process repeats until the whole dataset has been labeled.

Until recently, AL research in computer vision focused primarily on image classification [2, 8, 14, 25, 26, 33, 37]. Only a few recent works explore AL in context of object detection [1, 12, 21, 22]. All these works focus on the optimal selection of the acquisition function.

A few works consider the class imbalance when applying AL. In [21], the authors address the class imbalance for the task of object detection in aerial images. The long tail distribution of classes is also explored in [9] in the image classification scenario. Most other AL approaches consider datasets with a small number of well-balanced classes, such as DOTA [35] with 15 classes, CityPersons [38] with 30 classes, PASCAL2007 [6] with 7 classes, BDD100K [36] with 10 classes, or CIFAR100 [15] with 100 classes.

In our dataset, the number of classes is not predefined during the cycles of AL (i.e. the open class set problem) reaching 405 by the final iteration. On every cycle, most classes contain only a handful of instances, making the dataset highly unbalanced (Fig. 2) and meaning that the object detectors used in existing AL work simply can not be bootstrapped.

2.3   Image Classification with Unbalanced Data

Over the past few years, convolutional neural networks (CNNs) have excelled on image classification tasks. These classic CNN architectures, however, only perform well on well-balanced academic datasets, such as ImageNet [5], CIFAR-100 [15], COCO [17], Caltech-256 [10], CelebA [18], VisualGenome [13], and others. Most of these datasets rarely capture the state of the real world in which highly skewed, unbalanced data prevails.

As a result, multiple few-shot learning algorithms [34] were introduced. Matching Networks [32], Prototypical Networks [27], and Model-Agnostic Meta-Learning [7] are some cutting-edge research papers that aim at solving the image classification problem with very few images or instances per class. In 2019, "Large-Scale Long-Tailed Recognition in an Open World" (OLTR) [19] was presented. It addressed the long-tail and open-set nature of real-world datasets. We compare three of the aforementioned models for the task of classifying stamps in our MiikeMineStamps dataset.

3   Methodology

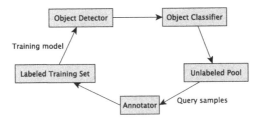

Fig. 5. The proposed AL pipeline. Arrows indicate the flow of information.

The proposed method (Fig. 5) follows the general principle of active learning. On every cycle, a machine learning model first predicts bounding boxes and object class for all unlabeled images in the dataset. Then we pick a subset of images based on an adjustable criterion. In our experiments, we favor images with a large number of objects that have high uncertainty. The images and the predictions are passed over to human experts to verify the labels and correct them if necessary. The ML model is then retrained on all the verified data available, and the cycle is considered complete.

In the case of the open class set, we do not know object classes beforehand and cannot train an object detector model that looks for a specific set of classes. Instead, we propose a two-step approach. First, an object detector model finds instances of the generic "stamp" class, then an image classification model is used to recognize a specific class in cropped out images of "stamps". This approach provides the advantage of transferring the difficulty of dealing with open class sets and long-tail distribution from the detection to the classification setup, where there are more tools to manage it.

We apply the detection algorithm on the images to extract stamps and resize these stamps to 80×80 pixels. Then, the cropped stamps are individually passed to the image classifier. While any off-the-shelf object detector architecture can be taken for the "stamp" detection step, the image classification model must be able to handle the open class set and the long tail challenges. We assume the number of instances per class varies from one to several hundred. Furthermore, we assume the existence of previously unseen classes. We compared three image classification models: FaceNet [24], Prototypical Networks [27], and OLTR [19]. While FaceNet and Prototypical Networks produce reasonable results, these models do not address the long-tail class distribution, and their performance falls behind OLTR. We direct the reader to the respective work for the details of the architecture.

4   Experimental Results

4.1   Mitsui Mi'ike Mine Documents

We use the presented two-step active learning approach on a compendium of historical documents from the Mitsui Mi'ike Mine company. In this company, like in many other Japanese companies from this era, when a letter or other document crossed someone's desk, it usually incurred a stamp, either to inscribe the name of the manager who approved it or to label it in some other way. Recognizing stamps across this archive will make it possible to trace the circulation of documents within this company. Considering that the full archive consists of more than thirty-two thousand pages and each document usually features multiple stamps, the automatic detection and classification of stamps is of considerable advantage.

Figure 6 (left) shows examples of the ground truth annotations in a page of a document. Documents were photographed as colored images with resolution 6000×4000 pixels. We collected annotations for the total of 677 images that have 5056 stamps. In Sect. 6, we present a dataset that consists of images of stamps, cropped from the original archive, annotated with stamp names and other metadata. The original image archive is not published in order to preserve the privacy of employees.

The active learning workflow follows a pipeline proposed in Sect. 3. Below, we describe the detection and classification components in detail.

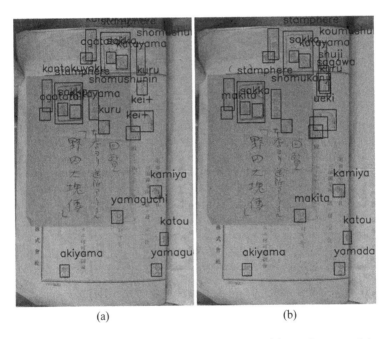

(a)   (b)

Fig. 6. (a): an example image with ground truth labels; (b): predictions of detector + classifier.

4.2   Detection

Given that the method is agnostic to the specific choice of a detector, we used a well-known RetinaNet [16] detector with ResNet-50 backbone, pretrained on COCO.

Hyperparameters were chosen via 5-fold cross-validation separately for every cycle. For the last cycle, the learning rate was set to $lr = 0.0001$ and batch size to $batch = 4$.

Table 1. Object detection average precision (@IoU = 0.5) across active learning cycles (%). Numbers on the main diagonal are the average "test" result in 5-fold cross-validation. Models trained on data from the 1st, 2nd, 3rd, and 4th cycle produce exceedingly better results when evaluated on the 4th cycle (in bold).

Trained on	Tested on			
	Cycle 1	Cycle 2	Cycle 3	Cycle 4
Cycle 1	89.3		31.6	**44.7**
Cycles 1–2		86.8	63.2	**55.3**
Cycles 1–3			83.5	**75.0**
Cycles 1–4				**84.3**

Table 1 tracks the performance of the object detector across AL cycles. The numbers on the main diagonal, i.e. trained and tested on the same cycle, are obtained from training with cross-validation. The three models trained on the first three cycles (in bold) perform increasingly better when trained on more data, showing that the active learning is gathering useful training data.

Figure 7 presents the precision-recall curves of detectors trained on cycles 1, 1–2 and 1–3, and evaluated on the last cycle 4. The detector performance can be seen to be steadily improving.

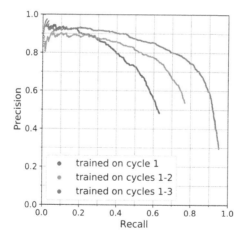

Fig. 7. Object detector trained on different cycles and evaluated on cycle 4.

4.3   Classification

Once stamps are detected via an object detector, the next task is to classify them. Figure 1 shows the high imbalance across classes. In fact, many classes have only a handful of examples. To overcome that, we picked classification models that are capable of working with long-tailed and open-set datasets. We evaluate (1) FaceNet [24], (2) Prototypical Networks [27], and (3) OLTR [19] image classification models. We now describe the experiments with each of them.

FaceNet is a popular architecture designed to work with a high number of classes but few instances per class. We split all data from cycle 1 with classes having more than 2 instances into the train, validation, and test sets. The remaining classes with 2 or fewer instances were combined into a class called "other", which was added to the test set. This gave us a total of 29 classes with 507 images in the train-val set and 30 classes and 135 images in the test set. We used Inception ResNet v1 as the backbone model for the FaceNet model with the softmax loss. We trained the model for 200 epochs on images of stamps resized 160×160 and generated embeddings in the 512-dimensional space. SVM was chosen as the last layer of FaceNet owing to its popularity [24, 28]. It proved to be a better choice over the Random Forest classifier as per our experiments. After applying the SVM classifier in this embedding space, we achieved the test accuracy of 63% with RBF (radial basis function) kernel. The triplet loss failed to work because of the bias in the selection of the triplets. Randomly selected triplets do not lead to model convergence, and using the hardest triplets results in the model getting stuck in local minima. Additionally, adjusting class weights proved to have no effect on such a long-tailed dataset.

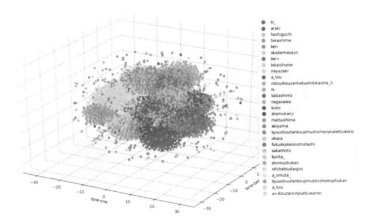

Fig. 8. Prototypical network: t-SNE on the test set (29 classes)

Owing to these limitations of the FaceNet model, we explored a few-shot learning architecture, specifically Prototypical Networks that is additionally tolerant to the long-tail data distribution. We trained this model with 5-shot, 5-query examples per class, and achieved the test accuracy of 69% for cycle 1 and 76% on the combined cycles 1 and 2 respectively. The t-SNE plot for the cycle 1 test set is shown in Fig. 8. Through this figure, we aim to illustrate that some classes form well-defined clusters in the t-SNE space, while other classes are highly diffuse. It graphically represents the class imbalance in the dataset.

The motivation behind exploring OLTR model was because of its ability to handle the open-set property of the dataset. This model promised successful results based on similarly distributed datasets, and henceforth will be used for future cycles of our dataset. All classes from cycle 1–3 with less than 3 image instances were moved to the open set (novel set). We used ResNet-10 [11] as our backbone and trained it with feature dimension 512 on 200 classes (many-shot, median-shot, low-shot combined). One important aspect of our work is reducing the labeling effort for subsequent cycles. To this end, subsequent cycles are automatically annotated with top-3 class predictions, given these predictions are above a certain confidence threshold. The expert can either choose from them or input their own class. Accordingly, we report top-3 and top-5 accuracy of 71.15% and 78.30% on the test set respectively. As expected, classes from the "many-shot" set perform better than classes from the "median-shot" set by 15%, which in turn perform better than classes from the "low-shot" set by another 15%. For the open-set (novel classes), we achieved 64% accuracy with the confidence threshold of 0.4. In order to assert robustness, we did 5-fold cross-validation for all models. A few examples of correct predictions by the OLTR model are presented in the top row of Fig. 10. The first stamp in the bottom row was incorrectly predicted to belong to either "kodama" or "sakka" classes, instead of the correct "kurihara" class. The last four stamps in the second row of the same figure illustrate the visual similarity between these three classes, which led to the incorrect prediction.

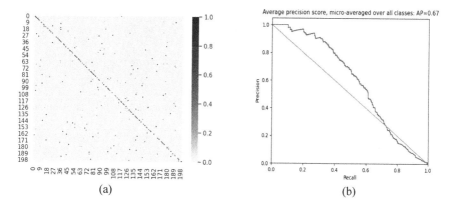

(a)   (b)

Fig. 9. OLTR model. Classification results for 200 random classes from cycle 3. (a) confusion matrix (b) precision-recall curve.

To sum up, OLTR performs best on our dataset as compared to FaceNet or Prototypical Networks, but the challenge of achieving high classification confidence (> 50%) still exists. Figure 9(a) shows the confusion matrix for the test set of 200 classes and Fig. 9(b) shows the precision-recall curve showing that the threshold lies close to 0.4.

The OLTR model trained on all the available stamps is released together with the MiikeMineStamps dataset.

5   Technical Details

Each cycle of AL includes the manual labeling process that requires a domain expert to inspect and annotate labels for hundreds of images. A labeling tool was required for streamlining the annotation process and making it as fast and accurate as possible. Our research identified critical requirements for a labeling tool: web-based, open-source, and/or free of charge for the relevant volume of data, the ability to specify labels dynamically in the interface as opposed to choosing from a given set, the compatibility of the label files format, the ability to export, the support for uploading new or modified labels, and the usability of the interface.

As a result of comparing 17 different tools, the well-known LabelMe Annotation Tool [23] was chosen for this project. The comparison is released together with the code. The authors hope that it will be useful for future AL researchers.

Furthermore, active learning with thousands of objects presents the challenge of tracking changes in the datasets. As one example, the manual cleaning step after each labeling cycle included (1) expanding the bounding box around each stamp, (2) tiling stamps of the same class into one "collage" image, (3) exporting

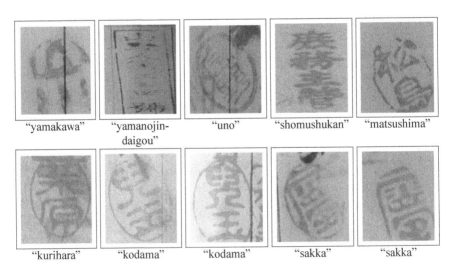

Fig. 10. 1st row: Examples of correct classification. 2nd row: The first stamp "kurihara" was incorrectly classified as "kodama" or "sakka". The last four stamps belong to the "kodama" and "sakka" classes. The visual similarity between these three classes explains the model's incorrect prediction of the first stamp.

to LabelMe format, (4) importing the cleaned results from LabelMe, (5) back-projecting stamps from collages back into their original images, and (6) shrinking bounding boxes back to their original size. This pipeline as well as other work on managing datasets, including filtering, splitting and merging, visualization and querying, was performed using the Shuffler toolbox [29].

The project code is available at https://github.com/pscedu/ml4docs.

6   MiikeMineStamps Dataset

In this section, we describe the published MiikeMineStamps dataset.

Once the annotation process via AL was completed, the annotated stamps were cropped out of the original documents, resulting in 5056 images from 405 stamp classes. The average dimensions of a stamp are 167×257 pixels, but both width and height vary significantly from 27 pixels to 1200 pixels (Fig. 11b).

The distribution of the number of stamps by class is very unbalanced. The most common class, "takesue", has 201 images, at the same time, 158 classes are represented by a single instance. Two stamps with the same letters but different shapes belong to the same class. The published dataset contains 14 such classes.

Additionally, the date on each original document was transferred to the stamps, which allowed us to track the flow of individual stamps over decades. Figure 12 illustrates this distribution for a small subset of stamps, while the full information is available in the published dataset.

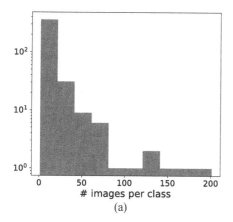
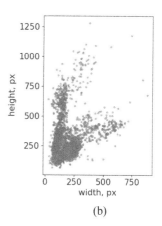

(a)　　　　　　　　　　　(b)

Fig. 11. (a) histogram of the number of stamp images per class; (b) distribution of stamps sizes.

The dataset introduced in this paper is publicly available under a Creative Commons Attribution 4.0 International license. The data is available for free to researchers for non-commercial use. This dataset includes the stamp images and labels. Additionally, we attach the information about the position of each stamp relative to its page, and other useful details, such as the year of the source document. The dataset DOI is https://doi.org/10.1184/R1/14604768. More information on the dataset and how to retrieve it can be found at https://kukuruza.github.io/MiikeMineStamps/. The original images of the historic documents are not publicly available as they may contain sensitive and personally identifiable information.

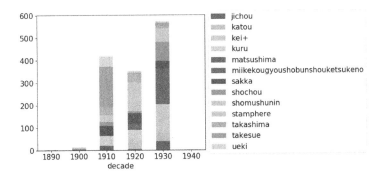

Fig. 12. Distribution of the most frequent classes across decades.

7   Discussion and Conclusion

The dataset of cropped stamps is interesting as it provides a completely different perspective on an archive. It instantly shows, for example, which stamps are most commonly used, providing clues as to who might be the gatekeepers of the organization. The benefits of this dataset increases considerably when stamps are classified into a series of classes and matched to the individual document(s) on which they appear. This will make it possible to identify all of the documents that came across the desk of an individual. Even more interesting is that stamps can show the way in which documents circulate in a company or government office. A memorandum will often circulate across the desk of multiple individuals, departments, or branches. At each location, it will usually incur a stamp that attests that someone has seen and approved it. Mining stamps on a large scale opens the door to tracing the circulation not only of one such document but of thousands of them. It helps to answer numerous questions in archives that feature a large number of stamps, not only in this archive of the Mitsui Mi'ike Mine but in most institutional archives in East Asia. For example, how does the circulation of documents change when a family-owned company becomes a joint-stock company? How does the circulation of documents in a ministry of foreign affairs change during wartime? Do the gatekeepers change? How is censorship implemented? What is the decision-making process in times of crisis? And more broadly, how do different bureaucratic decision-making processes lead to different outcomes? The answers to these questions are of interest to historians, political scientists, sociologists, anthropologists, media scholars, and researchers interested in the study of business management, among other fields.

The recent Large-Scale Long-Tailed Recognition in an Open World paper [19] presents long-tailed versions of three well-known datasets: ImageNet-LT, Places-LT, and MS1M-LT. In this work, we collected a naturally long-tailed dataset in the domain of documents, that we called MiikeMineStamps, which can serve as a benchmark for OLTR problems.

Acknowledgements. This work used the Extreme Science and Engineering Discovery Environment (XSEDE) which is supported by National Science Foundation grant number ACI-1548562. Specifically, it used the Bridges and Bridges-2 systems, which is supported by NSF award number ACI-1445606 and ACI-1928147, at the Pittsburgh Supercomputing Center (PSC) [3, 20, 30]. The work was made possible through the XSEDE Extended Collaborative Support Service (ECSS) program.

We are grateful to the Mitsui Archives for giving us permission to reproduce their documents and publish the stamps.

Finally, this work would not have been possible without the expert labeling and assistance of Ms. Mieko Ueda.

References

1. Aghdam, H.H., González-García, A., van de Weijer, J., López, A.M.: Active learning for deep detection neural networks. In: ICCV, pp. 3671–3679 (2019)
2. Beluch, W.H., Genewein, T., Nurnberger, A., Kohler, J.M.: The power of ensembles for active learning in image classification. In: CVPR, pp. 9368–9377 (2018). https://doi.org/10.1109/CVPR.2018.00976
3. Buitrago, P.A., Nystrom, N.A.: Neocortex and bridges-2: a high performance AI+HPC ecosystem for science, discovery, and societal good. In: Nesmachnow, S., Castro, H., Tchernykh, A. (eds.) High Performance Computing, pp. 205–219. Springer International Publishing, Cham (2021)
4. Clanuwat, T., Lamb, A., Kitamoto, A.: KuroNet: pre-modern Japanese Kuzushiji character recognition with deep learning. In: ICDAR, pp. 607–614 (2019)
5. Deng, J., Dong, W., Socher, R., Li, L.J., Li, K., Fei-Fei, L.: ImageNet: a large-scale hierarchical image database. In: CVPR (2009)
6. Everingham, M., Van Gool, L., Williams, C.K.I., Winn, J., Zisserman, A.: The PASCAL visual object classes (VOC) challenge. IJCV **88**(2), 303–338 (2010)
7. Finn, C., Abbeel, P., Levine, S.: Model-agnostic meta-learning for fast adaptation of deep networks. In: Precup, D., Teh, Y.W. (eds.) ICML, vol. 70, pp. 1126–1135 (2017)
8. Gal, Y., Islam, R., Ghahramani, Z.: Deep Bayesian active learning with image data. ICML **70**, 1183–1192 (2017)
9. Geifman, Y., El-Yaniv, R.: Deep active learning over the long tail (2017)
10. Griffin, G., Holub, A., Perona, P.: Caltech-256 object category dataset. CalTech Report, March 2007
11. He, K., Zhang, X., Ren, S., Sun, J.: Deep residual learning for image recognition. In: 2016 IEEE Conference on Computer Vision and Pattern Recognition (CVPR), pp. 770–778 (2016). https://doi.org/10.1109/CVPR.2016.90
12. Kao, C.-C., Lee, T.-Y., Sen, P., Liu, M.-Y.: Localization-aware active learning for object detection. In: Jawahar, C.V., Li, H., Mori, G., Schindler, K. (eds.) ACCV 2018. LNCS, vol. 11366, pp. 506–522. Springer, Cham (2019). https://doi.org/10.1007/978-3-030-20876-9_32

13. Krishna, R., et al.: The visual genome dataset v1.0 + v1.2 images. https://visualgenome.org/
14. Krishnamurthy, A., Agarwal, A., Huang, T.K., Daume, H., III., Langford, J.: Active learning for cost-sensitive classification. JMLR **20**(65), 1–50 (2019)
15. Krizhevsky, A., Nair, V., Hinton, G.: CIFAR-100 (Canadian Institute for Advanced Research)
16. Lin, T., Goyal, P., Girshick, R., He, K., Dollár, P.: Focal loss for dense object detection. In: ICCV, pp. 2999–3007 (2017)
17. Lin, T.-Y., Maire, M., Belongie, S., Hays, J., Perona, P., Ramanan, D., Dollár, P., Zitnick, C.L.: Microsoft COCO: common objects in context. In: Fleet, D., Pajdla, T., Schiele, B., Tuytelaars, T. (eds.) ECCV 2014. LNCS, vol. 8693, pp. 740–755. Springer, Cham (2014). https://doi.org/10.1007/978-3-319-10602-1_48
18. Liu, Z., Luo, P., Wang, X., Tang, X.: Deep learning face attributes in the wild. In: ICCV (2015)
19. Liu, Z., Miao, Z., Zhan, X., Wang, J., Gong, B., Yu, S.X.: Large-scale long-tailed recognition in an open world. In: CVPR (2019)
20. Nystrom, N.A., Levine, M.J., Roskies, R.Z., Scott, J.R.: Bridges: a uniquely flexible HPC resource for new communities and data analytics. In: XSEDE 2015: Scientific Advancements Enabled by Enhanced Cyberinfrastructure (2015). https://doi.org/10.1145/2792745.2792775
21. Qu, Z., Du, J., Cao, Y., Guan, Q., Zhao, P.: Deep active learning for remote sensing object detection (2020)
22. Roy, S., Unmesh, A., Namboodiri, V.: Deep active learning for object detection. In: BMVC (2019)
23. Russell, B., Torralba, A., Murphy, K., Freeman, W.: LabelMe: a database and web-based tool for image annotation. Int. J. Comput. Vis. **77**, 157–173 (2008)
24. Schroff, F., Kalenichenko, D., Philbin, J.: FaceNet: a unified embedding for face recognition and clustering. CoRR abs/1503.03832 (2015)
25. Sener, O., Savarese, S.: Active learning for convolutional neural networks: a core-set approach. In: ICLR (2018)
26. Sinha, S., Ebrahimi, S., Darrell, T.: Variational adversarial active learning. In: ICCV, pp. 5971–5980 (2019). https://doi.org/10.1109/ICCV.2019.00607
27. Snell, J., Swersky, K., Zemel, R.: Prototypical networks for few-shot learning. NIPS **30**, 4077–4087 (2017)
28. Taigman, Y., Yang, M., Ranzato, M., Wolf, L.: Deepface: closing the gap to human-level performance in face verification. In: CVPR, pp. 1701–1708 (2014). https://doi.org/10.1109/CVPR.2014.220
29. Toropov, E., Buitrago, P.A., Moura, J.M.F.: Shuffler: A large scale data management tool for machine learning in computer vision. In: PEARC (2019)
30. Towns, J., Cockerill, T., Dahan, M., Foster, I., Gaither, K., Grimshaw, A., Hazlewood, V., Lathrop, S., Lifka, D., Peterson, G.D., Roskies, R., Scott, J., Wilkins-Diehr, N.: XSEDE: accelerating scientific discovery. Comput. Sci. Eng. **16**(05), 62–74 (2014). https://doi.org/10.1109/MCSE.2014.80
31. Villalonga, G., Lopez, A.M.: Co-training for on-board deep object detection (2020)
32. Vinyals, O., Blundell, C., Lillicrap, T., Kavukcuoglu, k., Wierstra, D.: Matching networks for one shot learning. In: Lee, D., Sugiyama, M., Luxburg, U., Guyon, I., Garnett, R. (eds.) NIPS, vol. 29, pp. 3630–3638 (2016)
33. Wang, K., Zhang, D., Li, Y., Zhang, R., Lin, L.: Cost-effective active learning for deep image classification. IEEE Trans. Circ. Syst. Video Technol. **27**(12), 2591–2600 (2017). https://doi.org/10.1109/TCSVT.2016.2589879

34. Wang, Y., Yao, Q., Kwok, J., Ni, L.: Few-shot learning: a survey. arXiv preprint arXiv:1904.05046 (2019)
35. Xia, G., et al.: DOTA: a large-scale dataset for object detection in aerial images. In: CVPR, pp. 3974–3983 (2018). https://doi.org/10.1109/CVPR.2018.00418
36. Xu, H., Gao, Y., Yu, F., Darrell, T.: End-to-end learning of driving models from large-scale video datasets. In: CVPR, pp. 3530–3538 (2017)
37. Yoo, D., Kweon, I.S.: Learning loss for active learning. In: CVPR, pp. 93–102 (2019). https://doi.org/10.1109/CVPR.2019.00018
38. Zhang, S., Benenson, R., Schiele, B.: CityPersons: a diverse dataset for pedestrian detection. In: CVPR, pp. 4457–4465 (2017). https://doi.org/10.1109/CVPR.2017.474

Deep Learning for Document Layout Generation: A First Reproducible Quantitative Evaluation and a Baseline Model

Romain Carletto[✉], Hubert Cardot, and Nicolas Ragot

Université de Tours, LIFAT, 6300 Tours, EA, France
{romain.carletto,hubert.cardot,nicolas.ragot}@univ-tours.fr

Abstract. Deep generative models have been recently experimented in automated document layout generation, which led to significant qualitative results, assessed through user studies and displayed visuals. However, no reproducible quantitative evaluation has been settled in these works, which prevents scientific comparison of upcoming models with previous models. In this context, we propose a fully reproducible evaluation method and an original and efficient baseline model. Our evaluation protocol is meticulously defined in this work, and backed with an open source code available on this link: https://github.com/romain-rsr/quant_eval_for_document_layout_generation/tree/master.

Keywords: Document layout generation · Quantitative evaluation · Generative adversarial network.

1 Introduction

1.1 Document Layout Generation

For decades, developments in information and communication technologies leveraged interest in automated document layout generation. This application usually consists in automatically laying out elements on a canvas of given dimensions. It can take as inputs a random vector and additional optional features such as elements categories, reading order of the elements, element contents (texts or images) and geometric constraints such as aspect ratios or areas of the elements. While the outputs can take many forms, it is usually a list of bounding boxes, one for each element, with for each of these boxes its category, its dimensions and positions on the canvas. While former automated attempts in this field faced lack of both functionality and flexibility, recent solutions based on deep generative models reached interesting possibilities and provided encouraging visual and user study results. However, quantitative evaluations of these later solutions show important deficiencies, which prevents any scientific comparison with upcoming works.

© Springer Nature Switzerland AG 2021
J. Lladós et al. (Eds.): ICDAR 2021, LNCS 12823, pp. 20–35, 2021.
https://doi.org/10.1007/978-3-030-86334-0_2

1.2   A Tricky Quantitative Evaluation

Quantitative evaluation of generative models has always been a delicate matter, and is even more complicated in the specialized field of document layout generation. Real world layout guidelines are mainly implicit, even to layout designers themselves, and while metrics from image generation field are still being adapted, standard metrics from document layout community do not apply. As an example, the IoU score which is commonly used in layout analysis has limited utility in layout generation, where different positions for a same element can be of similar quality.

While in related works interesting quantitative evaluation methods have emerged and are progressively converging to a common standard, none of these works, after a meticulous study, have provided a reproducible and reliable definition of a quantitative evaluation method. In this context, the comparison between a new model and previous works relies mainly on subjective interpretations of visual results, which are of high interest but not sufficient to match hard sciences principles. Therefore, this paper aims to propose, on main document layout datasets, a first reproducible quantitative evaluation of any document layout generation model, as well as an original baseline model.

2   Related Works

2.1   From Explicit Methods to Deep Learning Methods

Former works in automated document layout generation focused on turning layout guidelines into explicit and static quantitative rules. Earliest works were based on templates [1,2] or on interactive tools combining basic layout rules [3] and showed poor possibilities. Later works were based on geometrical objective functions to be optimized [4–6] faced qualitative biases and a strong lack of flexibility. To tackle these various shortcomings, recent researches focused on learning methods where guidelines are turned in a dynamic and data-driven way into implicit quantitative rules projected in a multidimensional feature space. More specifically, most recent researches have been based on deep learning generative models, which can learn very complex rules from simple objective functions, and are already successful in other generative tasks such as image or video generation.

In [7], a Generative Adversarial Network (GAN) generates layouts from only a random vector as input. It can generate layouts of documents, layouts of pixels subset (extracted from MNIST handwritten digits images), clipart scenes and geometric tangrams. In [8], authors extend [7] previous work to propose two main applications: layout generation and layout adjustment. Layout generation is split into three sub-applications. Image layout generation takes geometric constraints as input and generates several layouts propositions containing only the product image. Attribute-guided layout generation takes these first layout propositions as input and add additional elements, according to these elements attributes, to produce different propositions of complete layouts. Finally, grouping and ranking application allows to select the best layout among different layout propositions, through overlap, alignment and discriminator scores.

The model presented in [9] is based on graph neural network and Variational Auto-Encoders (VAE) and generates a layout iteratively, element after element. It takes as inputs a set of elements, the order of elements to be iteratively laid out, and optional user-specified constraints. They experimented their models on generating application page layouts, magazine page layouts and private web advertising layouts.

In [12], authors merge GAN and VAE architectures to propose a model that encodes element attributes at different steps: when encoding a real layout into a latent vector, when generating a layout from a latent vector and when classifying a layout to be real or generated. These attributes can thus be learned in association with real training layouts during the training step and added as inputs, aside with a random vector, when using the model. They contain particularly rich and diversified information, such as image contents, text and label values than can be both continuous (such as aspect ratios) or discrete. Author architecture also allows the users to add optional soft constraints as inputs, such as reserved area for specific types of elements. Experiments are focused on generating magazine page layouts.

While not exactly focusing on document layout generation, [13] proposes a model based on VAE to generate real scene layouts. Generation steps are divided into several sub-steps: a first VAE generates a list of elements, then another VAE generates bounding boxes from it. The model is applied on a dataset adapted from MNIST, where handwritten digits are laid out on a black screen, and is also applied on COCO dataset, containing real life scenes with for each scene a labeled bounding box for every important object and person in the scene. Other moderately related but yet interesting works, [14] and [15], consider a potential sub-task of layout generation and generate bounding boxes from graphs in which strong relational and geometrical constraints are already indicated, e.g. which element must be at the right of which element, or which element has to be bigger than which element.

2.2   Deficiencies in Existing Quantitative Evaluation Methods

Previous works usually display visual results and often provide user studies of their experiments, but while this information can draw attention on the presented models, it can't allow any objective comparison with upcoming methods. In parallel, quantitative evaluation methods are also presented in the course of these works, but are not defined in a valid and reproducible way.

While some authors [12] do not provide any quantitative evaluation at all, others laid incomplete groundwork to define a valid quantitative metric. In [7], absolute alignment and overlap scores are provided for generated document layouts but the dataset on which these scores have been obtained, referred only as "the document layout dataset", has not been made available. The exact same problem is found in [8] where alignment and overlap scores are given for a private advertising layout dataset, not available to the scientific community.

In [9], authors propose an alignment evaluation on two public datasets but while our understanding of their alignment evaluation method has been confirmed by the authors, our results differ from theirs by orders of magnitude when we apply this method on the ground truth layouts of both evaluated datasets. Aside from the alignment evaluation method, [9] proposes an evaluation based on the Fréchet Inception Distance (FID). As described in their works, this method relies on the feature distribution of the penultimate layer of the discriminator: the distribution obtained on real layouts is compared with the distribution obtained on generated layouts and produces a score ranging from zero to positive infinity, with lower values indicating better performances. Given a fixed discriminator architecture with fixed feature parameters and fixed hyperparameters, this evaluation method allows for a quantitative comparison between two different generative models. But in [9], these parameters are missing.

3 Experimental Protocol

3.1 Datasets

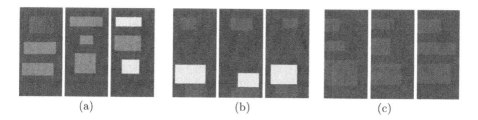

(a)   (b)   (c)

Fig. 1. Layouts from the first synthetic dataset. (a) layouts have been produced through general rules only, while (b) and (c) layouts have been produced by two different specific rules.

Synthetic Dataset. As explained in introduction, real world layout guidelines are mainly implicit which makes it difficult to quantitatively evaluate the quality of a layout. In this context, [10] proposes two synthetic datasets. The first dataset contains 100,000 synthetic document layouts, in which elements have fictitious semantic categories and are laid out following a combination of arbitrary and fictitious layout rules, that are both hard to learn and easy to quantitatively evaluate. A second dataset contains 100,000 other synthetic document layouts, in which elements have this time similar categories as in the web advertising industry: Product Image, Text, Call-To-Action (CTA) and Logo. In this second dataset, elements are laid out according to basic rules and distributions of the web advertising industry (e.g. logos are always on the top or on the bottom of the layout, in a majority of layouts images are bigger than CTAs and logos, ...). These layout rules are thus both realistic and easy to quantitatively evaluate. Any learning document generation model can thus be trained on these two synthetic datasets and then be evaluated in an exact and quantitative way, through reusing the explicit rules that were used in the first place to produce each of these datasets.

In the first dataset, while general rules alone have been used to create the majority of the layouts, additional specific rules have been used when creating certain layouts, as seen in Fig. 1. The use of these additional rules and their precedence on general rules is triggered by specific sequences of categories within the list of elements in each layout to be generated. Furthermore, some of the specific rules are deliberately in contradiction with the general layout rules to challenge the ability of a layout generation model to discern patterns within a complex and implicit combination of layout rules, similarly to what is expected from the model in a real case document generation application. The datasets are available at the following link: https://github.com/romain-rsr/synth_datasets_for_web_advertising_layout/tree/master

RICO. This public dataset [11] contains 65,538 ground truth layouts extracted from various application pages. Numerous works have experimented on this dataset for its quality and for the high quantity of layouts it contains. Elements within the layouts can be nested and can present very different sizes and aspect ratios. RICO dataset can be found at this link: http://interactionmining.org/rico

Magazine. Magazine dataset [12] contains 3,919 magazine page layouts, that can be more similar to advertising layouts than RICO application page layouts. This is of great interest since there is no public dataset available for advertising layouts but high industrial application in this field for sophisticated automated layout. The dataset is available at this link: https://xtqiao.com/projects/content_aware_layout/

3.2   Baseline Model

Architecture. Our model is a GAN with both generator and discriminator based on a fully connected residual block architectures. Instead of the usual residual block containing two layers, residual blocks within our model are tangled and contain one layer each. As seen in Fig. 2, both generator and discriminator contain 5 layers assembled in tangled residual blocks, each containing 100 neurons except the last layer. In the generator, the number of neurons in the last layer is equal to the product of the number of elements to be laid out and the number of feature for each element. In discriminator, the last layer contain only one neuron to output the cagory of the input layout: real or fake. In both generator and discriminator, each layer is activated by a ReLU activation function except the last layer activated by a sigmoid function.

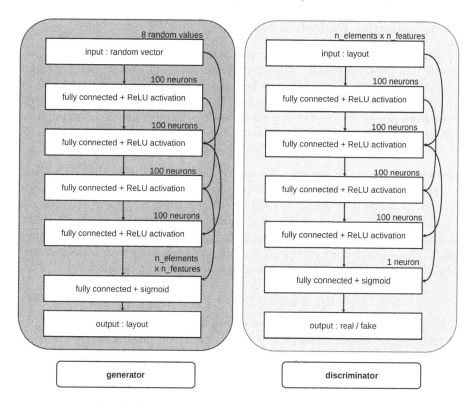

Fig. 2. Our generator and discriminator architectures.

Training. Our experimental protocol can be applied on layouts of any number of elements. Yet, layouts have first to be grouped according to the number of elements they contain so the experiments are run on each group separately, with the number of elements per layout being given as an input parameter of our model and evaluation process. For greater simplicity in the analysis of our experiments, the presented results focus on layouts with three elements.

Model's performance is monitored through the binary cross-entropy loss function during training, and the model is trained through Adam optimizer, which extends rmsprop optimizer on one hand, by adapting its learning rate to each parameter, and extends adagrad optimizer on an other hand, by adapting its learning rate to first and second moment (respectively mean and uncentered variance) of recent gradient magnitudes. In each experiment, we select a subset of the complete dataset and split it in train, validation and test sets, respectively of 60%, 20% and 20% of the subset.

3.3 Quantitative Evaluation Metrics

Our quantitative evaluation is based on four metrics: the Fréchet Inception Distance (FID), the Comparative Alignment Score, the Comparative Overlap Score and the Comparative Diversity Score. Some of them are inspired and revised version of incomplete yet interesting quantitative evaluations described in related works. Both pseudo-code and ready-to-use code of these evaluations are publicly available in our git, aside with real and generated evaluated samples.

Fréchet Inception Distance. This metric, initially defined in [16], is applied as specified in [17] to the output feature distribution within the discriminator penultimate layer. The mean and the covariance matrix of these features are first computed when applying the discriminator on generated layouts, then the same operation is carried out with real layouts. Finally, mean μ_X and covariance matrix $\sum_X$, obtained on generated layouts, are compared with mean μ_Y and covariance matrix $\sum_Y$, obtained on real layouts, following equation:

$$d^2(F,G) = |\mu_X - \mu_Y| + tr(\textstyle\sum_X + \sum_Y - 2(\sum_X \sum_Y)^{1/2}) \tag{1}$$

The resulting score is an absolute value which must be as low as possible. As specified earlier, this evaluation is of interest only if fixed architecture, feature parameters and hyperparameters are settled for the discriminator used in the evaluation, so that when comparing two models, the shift in the FID score comes only from the shift in the similarity between real and generated layouts. Also, using a discriminator of a specific existing work in layout generation would give an unfair advantage to this work when comparing new models with it, so a standard and non specialised discriminator architecture has to be used instead. Therefore, we used the pre-classification layers of the open source inception v3 discriminator, whom fixed parameters are indicated in our git.

Absolute Alignment Score. In related works, alignment score is often mentioned, as an absolute measure applied on generated layouts only, and measures the closest possible alignment of elements on each layout, according to one of the possible vertical alignment axes (element lefts, centers or rights). Here is the [9] definition of this absolute alignment score :

$$alignment_{gen} = \frac{1}{N} \sum_k \sum_i \min_{j,i\neq j}\{min(l(e_i^k, e_j^k), m(e_i^k, e_j^k), r(e_i^k, e_j^k)\}) \tag{2}$$

where N is the total number of generated layouts, e_i^k and e_j^k are the i_{th} and j_{th} elements of the k_{th} generated layout and where l, m and r are respectively the distances between lefts, centers and rights of two considered elements.

Comparative Metric. Absolute score can be interpreted very differently according to the type of evaluated documents. On RICO application pages, for example, lower alignment score could be preferred by designers while on advertising layouts, greater misalignment can be a quality factor so absolute alignment score has no general objective value. Therefore, our evaluation goes further and compares the absolute alignment score obtained on generated layouts with the absolute alignment score obtained on real layouts. While simply computing the ratio between two absolute scores would have been a straightforward comparison, it is not adapted if the score used as the divider is equal to zero. To overcome this problem and obtain similar score scales as in the FID evaluation, the following comparative metric has been defined to compare absolute scores $score_{gen}$ and $score_{real}$, obtained respectively on generated and real layouts:

$$comp(score_{gen}, score_{real}) = |\log \frac{score_{gen} + 1e{-}10}{score_{real} + 1e{-}10}| \tag{3}$$

Comparative Alignment Score. We apply the comparative metric on the couple of absolute alignment scores obtained on real and generated layouts to get the comparative alignment score, which is finally a robust and objective indication of how similar generated layouts are to the real layouts, independently of the subjective interpretation of what is a good absolute alignment value within each dataset:

$$alignment_{comp} = comp(alignment_{gen}, alignment_{real}) \tag{4}$$

Absolute Overlap Score. In parallel to the alignment measure, absolute overlap score is used in several related works and measures the ratio of overlapping areas over the total canvas area:

$$overlap_{gen} = \frac{1}{N} \sum_{k} \sum_{i} \sum_{j,j<i} \frac{intersection(area_i^k, area_j^k)}{area_c^k} \tag{5}$$

where N is the number of generated layouts, where $area_c^k$ is the total canvas area of the k_{th} generated layout (in most datasets this value is constant) and where $area_i^k$ and $area_j^k$ are the respective areas of i_{th} and j_{th} elements of the k_{th} generated layout

Comparative Overlap Score. Absolute overlap score encounters the same interpretation concerns as the absolute alignment score: RICO layouts, for example, present nested elements, which are fully overlapping, while in other types of document layouts, overlaps are unacceptable. Therefore, as for the alignment evaluation, our final overlap score is obtained by applying the comparative metric to the absolute overlap scores obtained on generated and real layouts, and hence benefits of the same objectiveness and generalisation as our alignment comparative score:

$$overlap_{comp} = comp(overlap_{gen}, overlap_{real}) \tag{6}$$

Comparative Diversity Score. Our evaluation method additionally incorporates the comparative diversity score, which compares standard deviations of real and generated layouts features. As an example, a standard deviation is computed for the left position of the first element, over all generated layouts, and one standard deviation is computed for each pair of element rank - element feature. The minimum value of these standard deviations is then computed for the generated layouts. The same operation is applied on real layouts, then real and generated obtained minimums are compared through the comparative metric.

$$diversity_{comp} = comp(\min_{i,j}\{\sigma_{gen}^{ij}\}, \min_{i,j}\{\sigma_{real}^{ij}\}) \tag{7}$$

where σ_{gen}^{ij} (respectively σ_{real}^{ij}) is the standard deviation of the j_{th} feature of the i_{th} element over all generated layouts (respectively over all real layouts). Lefts and widths standard deviations are normalised on layouts width while tops and heights standard deviations are normalised on layouts height.

Applying a Same Comparative Metric on Different Datasets. Note that comparative alignment score is neither penalizing or rewarding alignment, it is only penalizing differences between the absolute alignment score obtained on generated layouts, and the one obtained on real layouts. The same reasoning applies for comparative overlap score and for comparative diversity score. Therefore these comparative scores can be easily compared from different datasets, even with high variation of any given property between and within those datasets (such as the number of layout elements).

Independence Between Training Metrics and Comparative Metrics. Comparative metrics are used only during evaluation and are not used at all during training, so that evaluation scores remain independent of the training process. The only metric used during training is the binary cross-entropy loss function, which is agnostic to our evaluation metrics.

3.4   Baseline Evaluation Results

As specified earlier in the related work section, there is no reproducible quantitative baseline for document layout generation, which make impossible for us to compare our model to previous work. As an example, [9] unconstrained document layout generator achieves an FID score of 143.51 on RICO dataset while our model achieves an FID score of 66.96 on the same dataset. While these results could show that our model allow a significant performance gain, this actually cannot be asserted since the discriminator architecture and parameters used for the FID computation in [9] is not available.

Table 1. Quantitative evaluation results

Quantitative metric	Synthetic dataset I	RICO dataset	Magazine dataset
Fréchet inception distance	33.86	66.96	90.15
Comparative alignment score	0.25	0.29	0.32
Comparative overlap score	12.32	17.69	$3.59e^{-4}$
Comparative diversity score	19.83	0.65	1.37

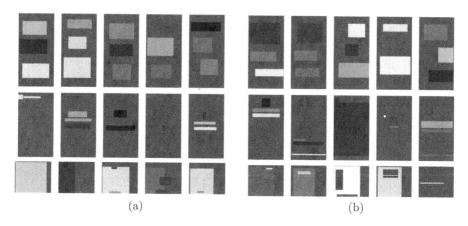

       (a)   (b)

Fig. 3. Generated (a) and real (b) layouts from different datasets. First row layouts are from the first synthetic dataset, second row layouts are from RICO dataset and third row layouts are from Magazine dataset.

Therefore, we propose our own quantitative baseline results, indicated in Table 1. These results are also available in our git along with the evaluation functions and the real and generated layouts that have been used to obtain them. While scores are globally of the same order of magnitude from one dataset to another, some score discrepancies remain noteworthy. FID is particularly lower on synthetic dataset, where size and location ranges are smaller and where there is no nested element. On this dataset, diversity score is also much higher, which is due to first element top position being constant on each real layout. Therefore, even a short deviation in generated layouts corresponding feature is highly penalized. Finally, due to a high number of nested elements, real layouts overlap score is not as tight in Magazine as in the other datasets. Therefore, the absolute difference in overlap score between generated and real layouts is less sensitive and penalized in this dataset, which explains such a low comparative overlap score for Magazine Dataset.

In order to put into perspective these quantitative results, generated layout representations have been randomly selected and are displayed along with real layout representations in Fig. 3. We can see on results obtained on synthetic dataset that only one layout, among all displayed generated layouts, contains an overlapping error. On results obtained on RICO dataset, we see strong similarity between generated and real layout patterns, e.g. a very large bounding box covering the majority of the canvas with much smaller, horizontally aligned bounding boxes above it. We see in both generated and real RICO layouts that elements are generally extremely close to each other without overlapping, which shows that our model matches industrial precision standards.

3.5   Additional Results: Application of the Quantitative Evaluation Metrics on Specific Examples

Table 2. Absolute and comparative metrics: intensive training vs. poor training

Intensive training on Synth II	Alignment	Overlap	Diversity
Absolute score on real layouts	0.112	0	0
Absolute score on generated layouts	0.116	0	0.003
Comparative score	0.031	0	17.166
Poor training on Synth II	Alignment	Overlap	Diversity
Absolute score on real layouts	0.112	0	0
Absolute score on generated layouts	0.088	1.15e-05	0.022
Comparative score	0.238	11.653	19.230
Intensive training on Rico	Alignment	Overlap	Diversity
Absolute score on real layouts	0.120	0.274	0.234
Absolute score on generated layouts	0.095	0.317	0.174
Comparative score	0.236	0.146	0.293
Poor training on Rico	Alignment	Overlap	Diversity
Absolute score on real layouts	0.120	0.274	0.234
Absolute score on generated layouts	0.019	0.157	0.014
Comparative score	1.830	0.558	2.843

In order to assert the resilience and the versatility of our quantitative evaluation metrics, an additional set of experiments focused on more particular examples. More specifically, these experiments aim to verify that the accuracy and the consistency of our metrics remain proportional to the level of training of the evaluated model and remain independent of datasets properties such as the number of element per layout or the degree of alignment, overlap and diversity within each dataset.

Evaluation Scores after Intensive Training and after Poor Training. We evaluated the consistency of our metrics with respect to the level of training by comparing the obtained results after 100 and 10.000 training epochs. This protocol was first ran on the second synthetic dataset, then ran again on Rico

Table 3. Absolute and comparative results on datasets with high differences in terms of alignment, overlap and diversity

Highly aligned real layouts (from Rico)	Metric	Score
Absolute score on real layouts	Alignment	0.0
Absolute score on generated layouts	Alignment	0.00023
Comparative score	Alignment	14.657
Poorly aligned real layouts (from Rico)	Metric	Score
Absolute score on real layouts	Alignment	0.365
Absolute score on generated layouts	Alignment	0.340
Comparative score	Alignment	0.070
Highly overlapping real layouts (from Rico)	Metric	Score
Absolute score on real layouts	Overlap	0.414
Absolute score on generated layouts	Overlap	0.696
Comparative score	Overlap	0.518
Poorly overlapping real layouts (from Rico)	Metric	Score
Absolute score on real layouts	Overlap	0
Absolute score on generated layouts	Overlap	0.001
Comparative score	Overlap	16.228
Highly diversified real layouts (from Synth II)	Metric	Score
Absolute score on real layouts	Diversity	0.005
Absolute score on generated layouts	Diversity	4.39e-04
Comparative score	Diversity	2.396
Poorly diversified real layouts (from Synth II)	Metric	Score
Absolute score on real layouts	Diversity	0.002
Absolute score on generated layouts	Diversity	2e-04
Comparative score	Dsiversity	2.516

dataset. Results of Table 2 show that on both datasets and on each property (alignment, overlap and diversity), better training leads to better comparative scores. As expected, the consistency of these comparative scores is contrasting with the divergence of some absolute scores. As an example, on the second synthetic dataset, more training leads to a higher absolute overlap score while on Rico, more training leads to a lower absolute overlap score. While the interpretation of these two absolute scores relies on the subjective understanding of their related datasets, a lower comparative score systematically implies a better performance, on any dataset.

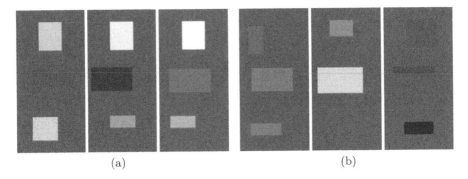

Fig. 4. Poorly diversified (a) and highly diversified real layouts (b) from the second synthetic dataset

Evaluating Datasets with High Differences in Terms of Alignment, Overlap and Diversity. The results in Table 3 have been obtained by experimenting the same model on specifically selected layouts of a same dataset, showing opposite values on a given property (alignment, overlap or diversity). As seen in the section focusing on highly and poorly diversified layouts, we see that comparative diversity scores for both data subsets are in the same order of magnitude, which is consistent with the fact that the same model has been experimented on both subsets of layouts, leading to similar performances. Additionally, the sections focusing on highly aligned and poorly overlapping layouts present very high comparative scores (which implies lower performance). These scores show the ability of the comparative metric to encompass the critical difference between a property equal to zero in real layouts (reflecting a hard constraint) and a very low non-zero value for the same property in generated layouts (which are then missing the hard constraint), by heavily penalizing the relative comparative scores.

Evaluating Layouts with Different Numbers of Elements. Table 4 shows that the comparative metrics are consistent over layouts of different number of elements. Except the sections where generated layouts do not comply with hard constraints, comparative scores remain in the same order of magnitude, independently of the number of elements. Moreover, in the last section of the table we show that one comparative score can easily be applied to two sets of layouts, each containing layouts with a distinct number of elements.

Table 4. Applying absolute and comparative metrics on layouts with different numbers of elements

Layouts with 3 elements (from Synth II)	Alignment	Overlap	Diversity
Absolute score on real layouts	0.023	0	0
Absolute score on generated layouts	0.024	0	0.001
Comparative score	0.031	0	15.715
Layouts with 5 elements (from Synth II)	Alignment	Overlap	Diversity
Absolute score on real layouts	0.043	0	0
Absolute score on generated layouts	0.042	1.56e-05	2.14e-05
Comparative score	0.017	11.960	12.272
Layouts with 7 elements (from Synth II)	Alignment	Overlap	Diversity
Absolute score on real layouts	0.043	0	0
Absolute score on generated layouts	0.048	2.85e-05	2.38e-04
Comparative score	0.115	12.562	14.686
3-elements versus 7-elements layouts	Alignment	Overlap	Diversity
Absolute score on real layouts (3 elements)	0.023	0	0
Absolute score on real layouts (7 elements)	0.041	0	0
Comparative score	0.575	0	0

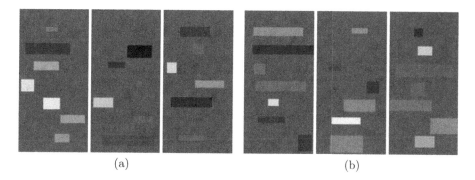

(a)   (b)

Fig. 5. Generated (a) and real layouts (b) with seven elements, from the second synthetic dataset

4 Conclusion

4.1 Contributions

In a context where recent publications on automated document layout generation show off impressive model architectures and promising user studies, we aimed at setting a sorely missing quantitative basis for scientific comparison and cooperation in the field of document layout generation. We thus propose a first baseline and made our quantitative evaluation method fully reproducible, backing it with a turn-key git containing both data and evaluation metrics that led to the presented results. The model we propose is based on an original yet easy to implement adaptation of the residual block concept and shows satisfying results, in both quantitative and visual aspects.

4.2 Future Works

Now that a first reproducible quantitative evaluation is settled, it would be interesting to monitor quantitative performance shifts when adding related work modules to our baseline model, or when adding functionalities such as attribute-guided and constrained layout generation.

Another interesting approach would be to add a background image processing module to our layout generation model, since background and foreground graphical balance are a central problem when generating sophisticated graphical layouts.

Finally, a critical step to achieve in automated document layout generation would be to go beyond the bounding box model and consider generating layouts in a pixel-wise dimension. This could also have very interesting applications for image generation, allowing users to add specific constraints or attributes on reserved areas of an image to be generated.

Acknowledgements. This work is financed by Centre Val de Loire Region, in France, and by Madmix Digital, a creative studio based in Paris and New-York, who helped us to identify and scientifically match the major challenges of document layout generation.

References

1. Myers, B.A.: User interface software tools. ACM Trans. Comput. Hum. Interact. (1994)
2. Lok, S., Feiner, S., Ngai, G.: Evaluation of visual balance for automated layout. In: 9th International Conference on Intelligent User Interfaces (2004)
3. Feiner, S., Nagy, S., Van Dam, A.: An experimental system for creating and presenting interactive graphical documents. ACM Trans. Graph. (1982)
4. Merell, P., Schkufza, E., Li, Z., Agrawala, M., Koltun, V.: Interactive furniture layout using interior design guidelines. SIGGRAPH (2011)
5. Lin, X.: Active layout engine: algorithms and applications in variable data printing. Comput. Aided Des. (2005)
6. Purvis, L., Harrington, S., O'Sullivan, B., Freuder, E.: Creating personalized documents: an optimization approach. In: ACM Symposium on Document Engineering (2003)
7. Li, J., Yang, J., Hertzmann, A., Zhang, J., Xu, T.: LayoutGAN: generating graphic layouts with wireframe discriminators. In: ICLR (2019)
8. Li, J., Yang, J., Zhang, J., Liu, C., Wang, C., Xu, T.: Attribute-conditioned layout GAN for automatic graphic design. IEEE Trans. Visual. Comput. Graph. (2020)
9. Lee, H..-Y.., et al.: Neural design network: graphic layout generation with constraints. In: Vedaldi, A., Bischof, H., Brox, T., Frahm, J.-M. (eds.) ECCV 2020. LNCS, vol. 12348, pp. 491–506. Springer, Cham (2020). https://doi.org/10.1007/978-3-030-58580-8_29

10. Carletto, R., Cardot, H., Ragot, N.: Automatic generation of web advertising layouts: a synthetic dataset and a deep learning baseline model. In: ICPRS (2021)
11. Deka, B., et al.: Rico: a mobile app dataset for building data-driven design applications. In: ACM Symposium on User Interface Software and Technology (2017)
12. Zheng, X., Qiao, X., Cao, Y., LAU, R.W.H.: Content-aware generative modeling of graphic design layouts. ACM Trans. Graph. (2019)
13. Jyothi, A.A., Durand, T., He, J., Sigal, L., Mori, G.; LayoutVAE: stochastic scene layout generation from a label set. In: ICCV (2019)
14. Nauata, N., Chang, K.-H., Cheng, C.-Y., Mori, G., Furukawa, Y.: House-GAN: relational generative adversarial networks for graph-constrained house layout generation. In: Vedaldi, A., Bischof, H., Brox, T., Frahm, J.-M. (eds.) ECCV 2020. LNCS, vol. 12346, pp. 162–177. Springer, Cham (2020). https://doi.org/10.1007/978-3-030-58452-8_10
15. Schroeder, B., Tripathi, S., Tang, H.: Triplet-aware scene graph embeddings. In: Scene Graph Representation Learning Workshop at ICCV (2019)
16. Dowson, D.C., Landau, B.V.: The fréchet distance between multivariate normal distributions. J. Multivar. Anal. (1982)
17. Heusel, M., Ramsauer, H., Unterthiner, T., Nessler, B., Hochreiter, S.: GANs trained by a two time-scale update rule converge to a local nash equilibrium. NIPS (2017)

Text and Symbol Recognition

MRD: A Memory Relation Decoder for Online Handwritten Mathematical Expression Recognition

Jiaming Wang[1], Qing Wang[1], Jun Du[1,2(✉)], Jianshu Zhang[1], Bin Wang[3], and Bo Ren[3]

[1] University of Science and Technology of China, Hefei, Anhui, China
{jmwang66,xysszjs}@mail.ustc.edu.cn, {qingwang2,jundu}@ustc.edu.cn
[2] Guangdong Artificial Intelligence and Digital Economy Laboratory (Pazhou Lab), Guangzhou, China
[3] Youtu Lab, Tencent, Hefei, Anhui, China
{bingolwang,timren}@tencent.com

Abstract. Recently, attention based encoder-decoder methods have been widely used in online handwritten mathematical expression recognition, which achieve significant improvements compared to traditional methods. The encoder-decoder methods usually employ string decoders to generate the recognition result, which are not well matched for tree-structured languages like math expression. A novel sequential relation decoder (SRD) was introduced to recognize the online mathematical expression as a math tree, which can be decomposed into a subtree sequence and each subtree consists of a relation node and two symbol nodes (related symbol node and primary symbol node). However, the alignments between these two symbol nodes were implemented by spatial attention probabilities, leading to incorrect recognition if spatial attention is not accurate. In this paper, we propose a memory relation decoder (MRD), equipped with a memory based attention model to determine the correspondence between two symbol nodes. Specifically, at each decoding step, this memory based attention finds the corresponding primary symbol node in the memory and treats it as the related symbol node, which actually achieves the alignments between two symbol nodes in an explicit manner. Besides, we propose to introduce global visual information while calculating attention probabilities to help alleviate the ambiguous problems in online handwritten mathematical expression recognition. Evaluated on a benchmark published by CROHME competition, the proposed approach can substantially outperform previous encoder-decoder methods.

Keywords: Handwritten mathematical expression recognition · Encoder-decoder · Tree structure · Memory based attention

© Springer Nature Switzerland AG 2021
J. Lladós et al. (Eds.): ICDAR 2021, LNCS 12823, pp. 39–54, 2021.
https://doi.org/10.1007/978-3-030-86334-0_3

1   Introduction

Handwritten mathematical expression recognition (HMER) plays an essential role in electronic technology documents, machine scoring and many other applications. As mathematical expression is a complicated two-dimensional structure (inherent tree structure) [3,6,15], HMER usually meets more challenges than Chinese text recognition or other sequence recognition problems, which are usually written in one direction and the alignments between input and output are monotonic, i.e., the correspondence between the input and output shares the same order.

The main problems of HMER can be roughly divided into two branches [7], namely symbol recognition and structural analysis. Symbol recognition denotes grouping strokes which belong to the same symbol and then determines the class of each symbol. Structural analysis denotes generating the most likely math tree based on the symbol recognition. Traditional methods usually solve these two problems separately or jointly, namely sequential or global methods. While contextual information is not fully utilized and symbol recognition errors will be inherited afterwards to structural analysis in sequential methods [1,22], global methods [2,4] seem to be more suitable as they optimize symbol recognition and structural analysis concurrently, but previous sequential methods usually outperform traditional global methods.

Recently, several researches [20,23,26] proposed a global way to recognize a mathematical expression as a LaTeX string instead of a math tree since the LaTeX string and the math tree are actually one to one correspondence and can be converted into each other equivalently. As deep learning came into prominence, attention based encoder-decoder methods were widely used in sequence to sequence learning, such as machine translation [11,13,18], speech recognition [5,8] and so on. Online handwritten mathematical expression recognition can also be treated as a sequence to sequence problem and attention based encoder-decoder methods [19,23] can be employed to generate a LaTeX string as the recognition result. These encoder-decoder methods can usually achieve better performance than traditional methods due to their powerful modeling capabilities and free of pre-defined grammar or symbol segmentation.

However, using LaTeX strings as the recognition results of handwritten mathematical expressions will meet several problems [24,25]. Therefore, [24] proposed a sequential relation decoder (SRD), which obtained recognition results in math tree formats using encoder-decoder methods and can be trained in an end-to-end manner. Specifically, SRD decomposed the complete math tree into a subtree sequence. At each step, SRD can generate a subtree, containing a relation node and two symbol nodes (first generated a primary symbol node and then a related symbol node based on the obtained primary symbol node). Besides, SRD employed spatial attention probabilities to acquire alignments between primary symbol nodes and related symbol nodes. [25] proposed a tree decoder, which employed a memory based attention model to achieve the alignments between primary symbol nodes and related symbol nodes instead.

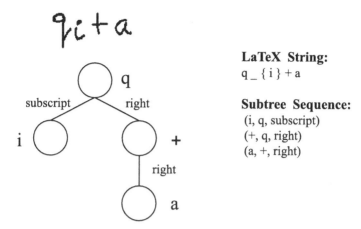

LaTeX String:
q _ { i } + a

Subtree Sequence:
(i, q, subscript)
(+, q, right)
(a, +, right)

Fig. 1. An example of handwritten mathematical expression, which can be represented as a LaTeX string or a subtree sequence.

In this work, we propose a memory relation decoder (MRD) for online handwritten mathematical expression recognition, which also recognizes the mathematical expression as a tree structure. As shown in Fig. 1, the complete math tree can be decomposed into a subtree sequence and MRD can generate a subtree at each decoding step and finally all the subtrees can be utilized to compose the complete tree. More specifically, MRD first generates a related symbol node using a predicted related GRU, a related GRU and a related attention model. Then a primary symbol node, following depth-first order, is generated by a predicted primary GRU, a primary GRU and a primary attention model. Moreover, we propose to insert a global visual feature into original features to help alleviate ambiguous problems in online handwritten mathematical expression recognition. Based on two obtained symbol nodes, a relation node can be predicted, indicating the attribute between related and primary symbol nodes. Unlike SRD, we employ an improved version of memory based attention model [25] to achieve alignments between related symbol nodes and primary symbol nodes, which additionally exploits related and primary context vectors. This memory based attention model actually determines which primary symbol node that the related symbol node should be corresponded to at each decoding step in an explicit manner. Furthermore, two attention guiders, namely related attention guider and primary attention guider are employed to help guide the learning of related attention and primary attention.

The main contributions of this paper can be summarized as:

- A memory relation decoder (MRD) is proposed for online handwritten mathematical expression recognition, which significantly outperforms both previous string decoders and tree-structured decoders.
- We introduce global visual information inserted in attention models to help alleviate the ambiguous problems.

– We demonstrate the effectiveness of memory based attention and global visual information through complete experimental analysis.

2   The Proposed Approach

In this section, we elaborate the overall system for online handwritten mathematical expression recognition, which consists of an encoder and a memory relation decoder (MRD). The encoder employs a stacked RNNs to extract high-level features from handwriting traces. Then, the memory relation decoder is introduced to generate a subtree at each decoding step t, including a related symbol node, a primary symbol node and a relation node. The two symbol nodes are determined by related decoder and primary decoder, respectively. Then the relation node can be determined by both related symbol node and primary symbol node. We employ a memory based attention model to implement the alignments between related symbol nodes and primary symbol nodes in an explicit manner, which is necessary to reconstruct the complete tree by the generated subtree sequence. Besides, we introduce a global visual feature to alleviate ambiguous problems and two attention guiders to help guide the learning of related attention and primary attention.

2.1   Encoder

The raw data of online handwritten mathematical expression recognition is handwriting traces collected during the writing procedure. Following [20], we first normalize the traces and then obtain an 8-dimensional feature vector for each point i as follows:

$$\mathbf{x}_i = [x_i, y_i, \Delta x_i, \Delta y_i, \Delta' x_i, \Delta' y_i, \text{strokeFlag1}, \text{strokeFlag2}] \tag{1}$$

where $\Delta x_i = x_{i+1} - x_i$, $\Delta y_i = y_{i+1} - y_i$, $\Delta' x_i = x_{i+2} - x_i$, $\Delta' y_i = y_{i+2} - y_i$. The last two terms indicate the pen status, which record whether the point is the last one of a stroke, i.e., $[1,0]$ means pen-down while $[0,1]$ means pen-up. Then, this sequence of 8-dimensional feature vectors is considered as the input of the encoder.

As shown in Fig. 2, to capture contextual information from input, we employ a stack of recurrent neural networks (RNN) with gated recurrent units (GRU). Besides, we actually adopt bidirectional GRU instead of unidirectional GRU as both past and future contextual information are useful for recognition. The bidirectional GRU will scan the input forwards and backwards with two separate GRU layers and concatenate their hidden states. The final output of stacked GRUs is an annotation sequence of variable length, which is referred as $\mathbf{A} = \{\mathbf{a}_1, \mathbf{a}_2, \cdots, \mathbf{a}_L\}$ and $\mathbf{a}_i \in \mathbb{R}^{D_1}$.

Besides, we believe that the attention models can benefit from global visual information as it can help alleviate ambiguous problems in online handwritten mathematical expression recognition. Therefore, we first convert handwriting

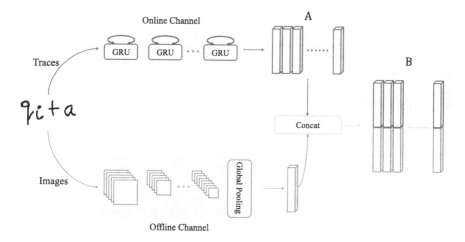

Fig. 2. The architecture of the encoder.

traces into a static image [20] and then employ an additional convolution neural networks (CNN) with dense blocks [12] to extract visual features, which is a tensor with size $H \times W \times D_2$. A global average pooling layer is built on top of CNN to obtain the global visual feature, $\mathbf{fea} \in \mathbb{R}^{D_2}$. We combine this global visual feature with each hidden state as follows:

$$\mathbf{B} = \{\mathbf{b}_1, \mathbf{b}_1, \cdots, \mathbf{b}_L\} \qquad \mathbf{b}_i = \mathrm{Concat}\,(\mathbf{a}_i, \mathbf{fea}) \tag{2}$$

where $\mathbf{b}_i \in \mathbb{R}^D$ and $D = D_1 + D_2$. Overall, the encoder can extract an annotation sequence A used to compute context vectors and an annotation sequence B used in attention models. The implementation details of the encoder can be seen in Sect. 3.1.

2.2   Memory Relation Decoder

As shown in Fig. 1, the target of memory relation decoder (MRD) is to generate a complete math tree for recognition, which can be decomposed into a variable length subtree sequence:

$$\mathbf{Y} = \{(\mathbf{y}_1^{\mathrm{r}}, \mathbf{y}_1^{\mathrm{p}}, \mathbf{y}_1^{\mathrm{re}}), (\mathbf{y}_2^{\mathrm{r}}, \mathbf{y}_2^{\mathrm{p}}, \mathbf{y}_2^{\mathrm{re}}), \cdots, (\mathbf{y}_T^{\mathrm{r}}, \mathbf{y}_T^{\mathrm{p}}, \mathbf{y}_T^{\mathrm{re}})\} \tag{3}$$

where T denotes the total number of subtrees and each subtree t contains a related symbol node $\mathbf{y}_t^{\mathrm{r}}$, a primary symbol node $\mathbf{y}_t^{\mathrm{p}}$ and a relation node $\mathbf{y}_t^{\mathrm{re}}$. These subtrees can be generated by several unidirectional GRUs step by step, using two annotation sequences $\mathbf{A}$ and $\mathbf{B}$ extracted from the encoder.

Note that there are three rules to confirm that the predicted subtree sequence can reconstruct the complete tree: (i) the subtree sequence is serialized by traversing the complete tree following a depth-first order. (ii) every primary symbol node must have a corresponding related symbol node and only occur once. (iii) each related symbol node must be selected from existing primary symbol nodes.

Related Decoder. As shown in Fig. 3, to decode the related symbol node, we employ two GRUs, namely predicted related GRU and related GRU with a related attention model, which can be represented as follows:

$$\hat{\mathbf{s}}_t^r = \text{PRGRU}\left(\mathbf{y}_{t-1}^p, \mathbf{s}_{t-1}^p\right) \tag{4}$$

$$\mathbf{c}_t^r = f_{\text{ratt}}\left(\hat{\mathbf{s}}_t^r, \mathbf{A}, \mathbf{B}\right) \tag{5}$$

$$\mathbf{s}_t^r = \text{RGRU}\left(\mathbf{c}_t^r, \hat{\mathbf{s}}_t^r\right) \tag{6}$$

where PRGRU and RGRU denote predicted related GRU and related GRU, respectively. f_{ratt} denotes related attention model, considering handwriting information and global visual information at the same time, which is designed as:

$$\mathbf{F}^r = \mathbf{Q}^r * \sum\nolimits_{\tau=1}^{t-1} \boldsymbol{\alpha}_\tau^r \tag{7}$$

$$e_{tj}^r = \boldsymbol{\nu}_r^T \tanh\left(\mathbf{W}_{\text{att}}^r \hat{\mathbf{s}}_t^r + \mathbf{U}_{\text{att}}^r \mathbf{b}_j + \mathbf{U}_F^r \mathbf{f}_j^r\right) \tag{8}$$

$$\alpha_{tj}^r = \frac{\exp\left(e_{tj}^r\right)}{\sum_k \exp\left(e_{tk}^r\right)} \tag{9}$$

$$\mathbf{c}_t^r = \sum\nolimits_{j=1}^L \alpha_{tj}^r \mathbf{a}_j \tag{10}$$

where $*$ denotes a convolution layer and $\mathbf{f}_j^r$ denotes the j-th element of F, which is utilized as a coverage vector to help alleviate the lack of coverage in the standard attention model. α_{tj}^r denotes the related attention probability of j-th element at decoding step t while $\mathbf{c}_t^r$ denotes related context vector at decoding step t. $\mathbf{a}_j$ and $\mathbf{b}_j$ are j-th elements of $\mathbf{A}$ and $\mathbf{B}$, respectively. Note that we employ $\mathbf{B}$ to compute attention probabilities as global visual information can help solve ambiguous problems in online handwritten mathematical expression recognition. Instead, we only adopt original features $\mathbf{A}$ to compute related context vector because these features are enough when accurate trace points are selected by attention model. Another reason is that global visual information is unsuitable to be considered at each decoding step t as only local visual information corresponded to current predicted symbol is needed.

Primary Decoder. After obtaining related symbol node $\mathbf{y}_t^r$ and related decoder hidden state $\mathbf{s}_t^r$, we can generate primary context vector $\mathbf{c}_t^r$ and primary decoder hidden state $\mathbf{s}_t^p$. As shown in Fig. 3, the architecture of primary decoder is similar with related decoder, which consists of predicted primary GRU and primary GRU:

$$\hat{\mathbf{s}}_t^p = \text{PPGRU}\left(\mathbf{y}_t^r, \mathbf{s}_t^r\right) \tag{11}$$

$$\mathbf{c}_t^p = f_{\text{patt}}\left(\hat{\mathbf{s}}_t^p, \mathbf{A}, \mathbf{B}\right) \tag{12}$$

$$\mathbf{s}_t^p = \text{PGRU}\left(\mathbf{c}_t^p, \hat{\mathbf{s}}_t^p\right) \tag{13}$$

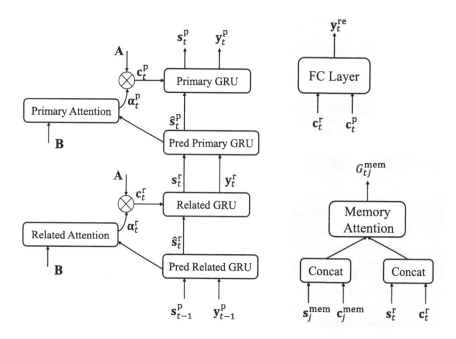

Fig. 3. The architecture of the decoder, which consists of a related decoder and a primary decoder. The right part illustrates the prediction of relation node and the memory based attention model.

where PPGRU and PGRU denote predicted primary GRU and primary GRU, respectively. f_{patt} has the same structure with f_{ratt} but the parameters are not shared. Besides, different from related decoder, we will additionally compute the probability of each primary symbol node $\mathbf{y}_t^{\text{p}}$ by feeding the concatenation of related symbol node $\mathbf{y}_t^{\text{r}}$, primary decoder hidden state $\mathbf{s}_t^{\text{p}}$ and primary context vector $\mathbf{c}_t^{\text{p}}$ into a fully connected layer with a softmax activation function:

$$p\left(\mathbf{y}_t^{\text{P}}\right) = \text{softmax}\left(\mathbf{W}_{\text{out}}^{\text{P}}\left(\mathbf{y}_t^{\text{r}}, \mathbf{s}_t^{\text{p}}, \mathbf{c}_t^{\text{P}}\right)\right) \tag{14}$$

Then the classification loss of primary symbol node, namely the training loss of primary decoder part is defined as:

$$\mathcal{L}_{\text{p}} = -\sum_t \log p\left(w_t^{\text{p}}\right) \tag{15}$$

where w_t^{p} denotes the ground-truth primary symbol node at decoding step t.

In addition, to generate a subtree at each decoding step t, we still need to compute the relation node, which describes the attribute between related symbol node and primary symbol node, such as right, above, superscript and so on. This relation node is computed by feeding the concatenation of related context vector $\mathbf{c}_t^{\text{r}}$ and primary context vector $\mathbf{c}_t^{\text{p}}$ into a fully connected layer with a softmax activation function:

$$p^{\text{re}}\left(\mathbf{y}_t^{\text{re}}\right) = \text{softmax}\left(\mathbf{W}_{\text{out}}^{\text{re}}\left(\mathbf{c}_t^{\text{r}}, \mathbf{c}_t^{\text{P}}\right)\right) \tag{16}$$

Then we define the training loss of the relation part as:

$$\mathcal{L}_{\text{re}} = -\sum_t \log p\left(v_t\right) \tag{17}$$

where v_t denotes the ground-truth relation node at decoding step t.

Memory Based Attention. In the above sections, we have introduced how to compute the probability of primary symbol node $\mathbf{y}_t^{\text{p}}$ at each decoding step t, which can be used to determine $\mathbf{y}_t^{\text{p}}$ during the testing stage. Nevertheless, we do not determine related symbol nodes in this way as there is no explicit order for related symbol nodes while primary symbol nodes always follow the depth-first order. In contrast, we adopt a memory based attention model to help determine related symbol node $\mathbf{y}_t^{\text{r}}$ at each decoding step t.

Specifically, we can get the related decoder hidden state $\mathbf{s}_t^{\text{r}}$, related context vector $\mathbf{c}_t^{\text{r}}$, primary decoder hidden state $\mathbf{s}_t^{\text{p}}$, primary context vector $\mathbf{c}_t^{\text{p}}$ and primary symbol node $\mathbf{y}_t^{\text{p}}$ at each decoding step t. During decoding, we append the concatenation of primary decoder state $\mathbf{s}_t^{\text{p}}$ and primary context vector $\mathbf{c}_t^{\text{p}}$ into the key memory and append the primary symbol node $\mathbf{y}_t^{\text{p}}$ into the value memory.

To determine the related symbol node at decoding step t, as shown in Fig. 3, we employ a memory based attention using the concatenation of related decoder hidden state $\mathbf{s}_t^{\text{r}}$ and related context vector $\mathbf{c}_t^{\text{r}}$ as query and the concatenation of primary decoder hidden state $\mathbf{s}_t^{\text{p}}$ and primary context vector $\mathbf{c}_t^{\text{p}}$ as key. Then, the attention probabilities can be computed as:

$$\mathbf{G}_{tj}^{\text{mem}} = \sigma\left(\nu_{\text{mem}}^T\left(\tanh\left(\mathbf{W}_{\text{mem}}\mathbf{z}_t^{\text{r}} + \mathbf{U}_{\text{mem}}\mathbf{z}_j^{\text{mem}}\right)\right)\right) \tag{18}$$

where $\mathbf{z}_t^{\text{r}}$ denotes the concatenation of the related decoder state $\mathbf{s}_t^{\text{r}}$ and related context vector $\mathbf{c}_t^{\text{r}}$, $\mathbf{z}_j^{\text{mem}}$ denotes the j-th element in the key memory.

In the training stage, we define the training loss of related decoder part as a binary classification loss:

$$\mathcal{L}_{\text{r}} = -\sum_t \sum_j \left[\bar{\mathbf{G}}_{tj}^{\text{mem}} \log\left(\mathbf{G}_{tj}^{\text{mem}}\right) + \left(1 - \bar{\mathbf{G}}_{tj}^{\text{mem}}\right) \log\left(1 - \mathbf{G}_{tj}^{\text{mem}}\right)\right] \tag{19}$$

where $\bar{\mathbf{G}}_{tj}^{\text{mem}}$ denotes the ground-truth of the alignment between related symbol node $\mathbf{y}_t^{\text{r}}$ and primary symbol node $\mathbf{y}_j^{\text{p}}$. In other words, $\bar{\mathbf{G}}_{tj}^{\text{mem}}$ is 1 when t-th related symbol node is aligned to the j-th element of the memory, otherwise 0.

In the testing stage, we choose $\mathbf{y}_{\hat{j}}^{\text{p}}$, $\hat{j} = \operatorname{argmax}\left(\mathbf{G}_{tj}^{\text{mem}}\right)$ in the value memory as the related symbol node at decoding step t.

Symbol Node Attention Guider. The alignment accuracy between input and output provided by attention models are important for recognition. However, how to train attention properly remains challenging. Therefore, we employ two symbol node attention guiders, namely related attention guider and primary attention guider, to help guide the learning of related attention and primary attention models. These two guiders can be implemented as the oracle alignment information can be acquired in the training stage in online handwritten mathematical expression recognition, namely which annotation features should be aligned when decoding each related symbol node $\mathbf{y}_t^{\mathrm{r}}$ and each primary symbol node $\mathbf{y}_t^{\mathrm{p}}$. We design these two symbol node attention guiders as:

$$\mathcal{L}_{\mathrm{rali}} = -\sum_t \sum_j \left[\bar{\alpha}_{tj}^{\mathrm{r}} \log \left(\alpha_{tj}^{\mathrm{r}} \right) + \left(1 - \bar{\alpha}_{tj}^{\mathrm{r}} \right) \log \left(1 - \alpha_{tj}^{\mathrm{r}} \right) \right] \tag{20}$$

$$\mathcal{L}_{\mathrm{pali}} = -\sum_t \sum_j \left[\bar{\alpha}_{tj}^{\mathrm{p}} \log \left(\alpha_{tj}^{\mathrm{p}} \right) + \left(1 - \bar{\alpha}_{tj}^{\mathrm{p}} \right) \log \left(1 - \alpha_{tj}^{\mathrm{p}} \right) \right] \tag{21}$$

where $\bar{\alpha}_{tj}^{\mathrm{r}}$ and $\bar{\alpha}_{tj}^{\mathrm{p}}$ denote the ground-truth alignments of related attention and primary attention models. These two guiders will be regarded as additional losses of the total training loss, namely related alignment loss and primary alignment loss, respectively.

3 Experiments

In this section, we design a set of experiments to evaluate the effectiveness of the proposed method on CROHME benchmark [14,17], which is currently the most widely used dataset for online handwritten mathematical expression recognition. We use CROHME 2014 training set as our training set, which consists of 8836 handwritten mathematical expressions and CROHME 2014 testing set as our testing set, which has 986 handwritten mathematical expressions. There are totally 101 math symbol classes and 6 math relations (above, below, right, inside, superscript (sup), subscript (sub)). To prove the generalization and robustness, we also evaluate our proposed method on CROHME 2016 and CROHME 2019 testing sets, which consist of 1147 expressions and 1199 expressions, respectively.

3.1 Training and Testing Details

Training. The overall model can be trained in an end-to-end manner and the training target is to minimize the weighted summation of the related decoder loss, primary decoder loss, relation loss, related alignment loss and primary alignment loss, which can be represented as follows:

$$O = \lambda_1 \mathcal{L}_{\mathrm{r}} + \lambda_2 \mathcal{L}_{\mathrm{p}} + \lambda_3 \mathcal{L}_{\mathrm{re}} + \lambda_4 \mathcal{L}_{\mathrm{rali}} + \lambda_5 \mathcal{L}_{\mathrm{pali}} \tag{22}$$

We set $\lambda_1 = \lambda_2 = \lambda_3 = 1$ as we believe that the prediction of the related symbol node, primary symbol node and relation node are equally important. Besides, we set $\lambda_4 = \lambda_5 = 0.1$ for alignment losses, which can be regarded as the regularization losses. For encoder, we employ 4 stacked GRU layers to extract high-level features. Each GRU layer is bidirectional and has 256 forward and 256 backward GRU units. There are two pooling layers of factor 2 on the top 2 GRU layers. Besides, the CNN is the same as DenseNet-99 in [19], which consists of three dense blocks and each block has 16 3×3 and 16 1×1 convolution layers. As for decoder, PRGRU, RGRU, PPGRU, PGRU are all unidirectional GRU layers and each layer has 256 GRU units. The attention dimension of related attention and primary attention models are both 512. The kernel size of coverage model is

7×1 and the number of output channel is 256. The embedding dimension is set to 256. We train our model by AdaDelta algorithm and the corresponding hyper-parameters are set as $\rho = 0.95$, $\epsilon = 10^{-8}$. All the experiments are implemented with Pytorch and on a single NVIDIA Tesla 1080Ti 11G GPU.

Testing. In the testing stage, we expect to obtain the most likely subtree sequence. As we do not have the ground-truth related symbol nodes and primary symbol nodes during testing, we employ a hierarchical version of beam search algorithm [9] of beam size 3. Specifically, at each decoding step, 3 most likely previous primary symbol nodes (i.e., 3 hypotheses) are maintained to compute the current related symbol nodes. Then each hypothesis is expanded with 3 most likely current related symbol nodes and these current related symbol nodes are utilized to compute the current primary symbol nodes. In total, we have $3 \times 3 = 9$ hypotheses kept and then choose 3 beams according to the combined likelihood of related symbol node and primary symbol node, which are used for the next step decoding.

3.2 Recognition Performance

Table 1. Performance Comparison on CROHME 2014 testing set (in %). ExpRate denotes the percentage of predicted mathematical expressions matching the ground truth. ≤ 1 s. error and ≤ 2 s. error denote the expression recognition accuracies with at most one and two errors. StruRate only focuses on whether the structure is correctly recognized and ignores symbol recognition errors.

System	ExpRate	≤ 1 s. error	≤ 2 s. error	StruRate
I	37.2	44.2	47.3	–
II	25.7	33.2	35.9	–
III	26.1	33.9	38.5	–
WYGIWYS [10]	35.9	–	–	–
PAL [21]	39.7	–	–	–
WAP [26]	48.4	66.1	70.2	70.1
TAP [23]	48.5	63.3	67.3	67.2
TAP + WAP + LM [23]	61.2	75.5	77.7	–
SRD [24]	50.6	57.9	62.1	–
TD [25]	49.1	64.2	67.8	68.6
MRD1	**55.2**	**70.0**	**73.4**	**73.3**
MRD2	**55.8**	**72.0**	**75.3**	**75.3**

In this section, we first compare the proposed MRD based encoder-decoder system with other state-of-the-arts, including traditional methods, string decoder based encoder-decoder methods and tree-structured decoder based encoder-decoder methods on CROHME 2014 testing set. As shown in Table 1, we list the best 3 systems in CROHME 2014 competition [16], which only used official datasets. We also list the recognition performance of 5 string decoder based encoder-decoder systems and the details can be seen in [10,21,23,26]. SRD and TD are tree-structured decoder based encoder-decoder systems [24,25]. MRD1 denotes the MRD based encoder-decoder system without global visual feature while MRD2 uses global visual feature. Note that although TAP + WAP + LM can achieve a high result, it actually ensembled three TAP, three WAP and three GRU-based language models, which is not fairly comparable. Apart from expression recognition rate (ExpRate), we also adopt those with at most one, two object-level errors (≤ 1 s. error, ≤ 2 s. error) and structural recognition rate (StruRate) as additional metrics to further conduct the effectiveness of the proposed methods.

It is obvious that tree-structured decoder based encoder-decoder systems can outperform string decoder based encoder-decoder systems. Furthermore, MRD1 can achieve a significant improvement compared with SRD/TD and the ExpRate improvement is more than 5%, which demonstrates that our memory based attention model can implement more accurate alignments between related symbol nodes and primary symbol nodes and accordingly improve performance. Besides, MRD2 can still improve the performance compared with MRD1, proving the necessary of the global visual information. The improvements for ≤ 1 s. error, ≤ 2 s. error and StruRate are more significant and further conduct the effectiveness of the proposed MRD.

Table 2. Performance comparison on CROHME 2016 and CROHME 2019 testing sets (in %).

Dataset	System	ExpRate	≤ 1 s. error	≤ 2 s. error	StruRate
CROHME16	Tokyo	43.9	50.9	53.7	61.6
	São Paulo	33.4	43.5	49.2	57.0
	Nantes	13.3	21.0	28.3	21.5
	WAP	46.8	64.6	65.5	66.2
	TAP	44.8	59.7	62.8	63.1
	TAP + WAP + LM	57.0	72.3	75.6	–
	SRD	46.6	–	–	–
	TD	48.5	62.3	65.3	65.9
	MRD1	**51.3**	**65.9**	**68.9**	**69.2**
	MRD2	**52.5**	**68.4**	**71.5**	**71.7**
CROHME19	WAP	48.1	63.5	67.2	68.0
	TAP	44.2	58.8	62.7	63.6
	SRD	45.9	–	–	–
	TD	51.4	66.1	69.1	69.8
	MRD1	**52.3**	**67.3**	**70.2**	**70.8**
	MRD2	**53.6**	**68.9**	**72.1**	**72.3**

To confirm the generalization of the proposed MRD, we also compare MRD with competition systems and other state-of-the-art systems on both CROHME 2016 [17] and CROHME 2019 [14] testing sets in Table 2. The systems Tokyo, São Paulo and Nantes denote the best 3 systems of all submitted systems in CROHME 2016 competition using only official dataset and we do not list the results of submitted systems in CROHME 2019 competition as they all use additional training sets or other strategies such as ensemble. It is obvious that MRD1 can still achieve better performance compared to both string decoder based encoder-decoder systems and other tree-structured decoder based encoder-decoder systems. Similarly, MRD2 can further outperform MRD1 and the improvement is larger on these two testing sets with more ambiguous problems.

3.3   Visualization Analysis

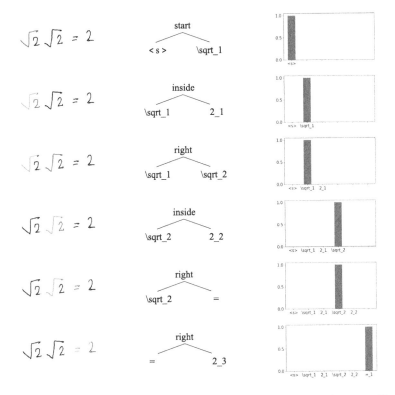

Fig. 4. An example of how MRD generates a complete tree step by step. For each step, from left to right, we show the attention visualization of related attention and primary attention, the predicted subtree and the attention visualization of memory based attention. Memory based attention actually achieves the alignments between related symbol nodes and primary symbol nodes in an explicit manner, illustrated in the right part of the figure (Color figure online).

In Sect. 3.2, we have demonstrated that the proposed MRD can outperform both previous string decoder and tree-structured decoder. In this section, we further show how MRD achieves to generate a complete tree as the recognition result for online handwritten mathematical recognition. As shown in Fig. 4, we show the attention visualization results of related attention, primary attention and memory based attention at each decoding step.

Specifically, each line in Fig. 4 denotes a decoding step, which has three parts. The left part shows the related attention result in green color and the primary attention result in red color. The middle part shows the corresponding subtree, including the related symbol node, primary symbol node and relation node. For example, the related symbol node, primary symbol node and relation node of the subtree in the second line are "\sqrt", "2" and "inside", respectively. Note that we use "\sqrt_1" instead of "\sqrt", "2_1" instead of "2" to distinguish other same symbols in this expression. The right part shows the result of memory based attention. As described in Sect. 2.2, at each decoding step, memory based attention is designed to determine which primary symbol node that the related symbol node should be corresponded to. We take the third line as an example. At this decoding step, there are already three primary symbol nodes in the memory, which are appended in the previous steps. Then, the memory based attention will compute a probability over these symbols, which is represented as the vertical coordinate. Obviously, the probability of primary symbol node, "\sqrt_1" is the largest. Therefore, we select "\sqrt_1" as the related symbol node at this step, indicating both the symbol class and the alignment between the related symbol node and primary symbol node. This memory based attention is very accurate and the probability distribution is very close to the ground-truth probability distribution (the value is nearly 1 or 0).

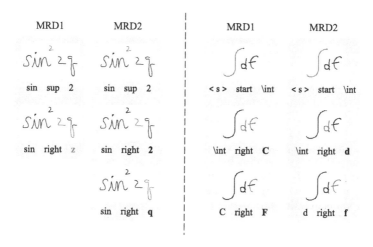

Fig. 5. Two examples to show the effectiveness of global visual information, which helps generate more accurate attention results. The incorrect recognition results are shown in blue color (Color figure online).

Furthermore, compared with MRD1, MRD2 equipped with global visual information can acquire more accurate attention results. As shown in Fig. 5, we show two examples that MRD2 can correctly recognize while MRD1 not. Note that we only show the primary attention results in red color as related symbol nodes are actually selected from primary symbol nodes. In the left example, MRD1 incorrectly recognizes "2 q" as "z" as MRD1 attends both "2" and "q" simultaneously. Therefore, the redundant parts make MRD1 misidentify "2" as "z" and the one step attention omission makes MRD1 miss "q". However, MRD2 can acquire more accurate attention with global visual information and recognize correctly. The similar observation can be seen in the right example.

4 Conclusion

In this study, we propose a memory relation decoder (MRD) for online handwritten mathematical expression recognition. To alleviate the ambiguous problems, we further introduce global visual information, which can help generate more accurate attention results. The proposed MRD can achieve significant improvements compared to string decoder based encoder-decoder methods and other tree-structured decoder base encoder-decoder methods on a benchmark published by CROHME competition, including CROHME 2014, 2016 and 2019 testing sets. Through attention visualization, we show how the proposed MRD implements the alignments between related symbol nodes and primary symbol nodes in an explicit manner and how the global visual information helps achieve better spatial attention results, which can both improve the recognition performance. In the future, we aim to investigate an approach utilizing both string and tree-structured decoders to further improve the recognition performance.

Acknowledgement. This work was supported in part by the MOE-Microsoft Key Laboratory of USTC, and Youtu Lab of Tencent.

References

1. Álvaro, F., Sánchez, J.A., Benedí, J.M.: Recognition of on-line handwritten mathematical expressions using 2D stochastic context-free grammars and hidden Markov models. Pattern Recogn. Lett. **35**, 58–67 (2014)
2. Alvaro, F., Sánchez, J.A., Benedí, J.M.: An integrated grammar-based approach for mathematical expression recognition. Pattern Recogn. **51**, 135–147 (2016)
3. Anderson, R.H.: Syntax-directed recognition of hand-printed two-dimensional mathematics. In: Symposium on Interactive Systems for Experimental Applied Mathematics: Proceedings of the Association for Computing Machinery Inc., Symposium, pp. 436–459 (1967)
4. Awal, A.M., Mouchère, H., Viard-Gaudin, C.: A global learning approach for an online handwritten mathematical expression recognition system. Pattern Recogn. Lett. **35**, 68–77 (2014)
5. Bahdanau, D., Chorowski, J., Serdyuk, D., Brakel, P., Bengio, Y.: End-to-end attention-based large vocabulary speech recognition. In: International Conference on Acoustics, Speech and Signal Processing, pp. 4945–4949 (2016)

6. Belaid, A., Haton, J.P.: A syntactic approach for handwritten mathematical formula recognition. IEEE Trans. Pattern Anal. Mach. Intell. **1**, 105–111 (1984)
7. Chan, K.F., Yeung, D.Y.: Mathematical expression recognition: a survey. Int. J. Doc. Anal. Recogn. **3**(1), 3–15 (2000)
8. Chan, W., Jaitly, N., Le, Q., Vinyals, O.: Listen, attend and spell: a neural network for large vocabulary conversational speech recognition. In: International Conference on Acoustics, Speech and Signal Processing, pp. 4960–4964 (2016)
9. Cho, K.: Natural language understanding with distributed representation. arXiv preprint arXiv:1511.07916 (2015)
10. Deng, Y., Kanervisto, A., Ling, J., Rush, A.M.: Image-to-markup generation with coarse-to-fine attention. In: International Conference on Machine Learning, pp. 980–989 (2017)
11. He, T., et al.: Layer-wise coordination between encoder and decoder for neural machine translation. In: Advances in Neural Information Processing Systems, pp. 7944–7954 (2018)
12. Huang, G., Liu, Z., Van Der Maaten, L., Weinberger, K.Q.: Densely connected convolutional networks. In: IEEE Conference on Computer Vision and Pattern Recognition, pp. 4700–4708 (2017)
13. Huang, P.Y., Liu, F., Shiang, S.R., Oh, J., Dyer, C.: Attention-based multimodal neural machine translation. In: Conference on Machine Translation, vol. 2, pp. 639–645 (2016)
14. Mahdavi, M., Zanibbi, R., Mouchere, H., Viard-Gaudin, C., Garain, U.: ICDAR 2019 CROHME+ TFD: competition on recognition of handwritten mathematical expressions and typeset formula detection. In: 2019 International Conference on Document Analysis and Recognition (ICDAR), pp. 1533–1538. IEEE (2019)
15. Miller, E.G., Viola, P.A.: Ambiguity and constraint in mathematical expression recognition. In: AAAI, pp. 784–791 (1998)
16. Mouchere, H., Viard-Gaudin, C., Zanibbi, R., Garain, U.: ICFHR 2014 competition on recognition of on-line handwritten mathematical expressions (CROHME 2014). In: International Conference on Frontiers in Handwriting Recognition, pp. 791–796 (2014)
17. Mouchère, H., Viard-Gaudin, C., Zanibbi, R., Garain, U.: ICFHR 2016 CROHME: competition on recognition of online handwritten mathematical expressions. In: 2016 15th International Conference on Frontiers in Handwriting Recognition (ICFHR), pp. 607–612. IEEE (2016)
18. Vaswani, A., et al.: Attention is all you need. In: Advances in Neural Information Processing Systems, pp. 5998–6008 (2017)
19. Wang, J., Du, J., Zhang, J.: Stroke constrained attention network for online handwritten mathematical expression recognition. arXiv preprint arXiv:2002.08670 (2020)
20. Wang, J., Du, J., Zhang, J., Wang, Z.R.: Multi-modal attention network for handwritten mathematical expression recognition. In: International Conference on Document Analysis and Recognition, pp. 1181–1186 (2019)
21. Wu, J.-W., Yin, F., Zhang, Y.-M., Zhang, X.-Y., Liu, C.-L.: Image-to-markup generation via paired adversarial learning. In: Berlingerio, M., Bonchi, F., Gärtner, T., Hurley, N., Ifrim, G. (eds.) ECML PKDD 2018. LNCS (LNAI), vol. 11051, pp. 18–34. Springer, Cham (2019). https://doi.org/10.1007/978-3-030-10925-7_2
22. Zanibbi, R., Blostein, D., Cordy, J.R.: Recognizing mathematical expressions using tree transformation. IEEE Trans. Pattern Anal. Mach. Intell. **24**(11), 1455–1467 (2002)

23. Zhang, J., Du, J., Dai, L.: Track, Attend and Parse (TAP): an end-to-end frame-work for online handwritten mathematical expression recognition. IEEE Trans. Multimedia **21**(1), 221–233 (2019)
24. Zhang, J., Du, J., Yang, Y., Song, Y.Z., Dai, L.: SRD: a tree structure based decoder for online handwritten mathematical expression recognition. IEEE Trans. Multimedia (2020)
25. Zhang, J., Du, J., Yang, Y., Song, Y.Z., Wei, S., Dai, L.: A tree-structured decoder for image-to-markup generation. In: International Conference on Machine Learning, pp. 11076–11085. PMLR (2020)
26. Zhang, J., et al.: Watch, attend and parse: an end-to-end neural network based approach to handwritten mathematical expression recognition. Pattern Recogn. **71**, 196–206 (2017)

Full Page Handwriting Recognition via Image to Sequence Extraction

Sumeet S. Singh$^{(\boxtimes)}$ ⓘ and Sergey Karayev

Turnitin, 2101 Webster Street #1800, Oakland, CA 94612, USA
{ssingh,skarayev}@turnitin.com

Abstract. We present a Neural Network based Handwritten Text Recognition (HTR) model architecture that can be trained to recognize full pages of handwritten or printed text without image segmentation. Being based on Image to Sequence architecture, it can extract text present in an image and then sequence it correctly without imposing any constraints regarding orientation, layout and size of text and non-text. Further, it can also be trained to generate auxiliary markup related to formatting, layout and content. We use character level vocabulary, thereby enabling language and terminology of any subject. The model achieves a new state-of-art in paragraph level recognition on the IAM dataset. When evaluated on scans of real world handwritten free form test answers - beset with curved and slanted lines, drawings, tables, math, chemistry and other symbols - it performs better than all commercially available HTR cloud APIs. It is deployed in production as part of a commercial web application.

1 Overview

With the advancement of Deep Neural Networks, the field of HTR has progressed by leaps and bounds. Neural Networks have enabled algorithm developers to increasingly rely on features and algorithms learned from data rather than hand crafted ones. This makes it easier than ever before to create HTR models for new datasets and languages. That said, HTR models still embody domain-specific inductive biases, assumptions and heuristics in their architectures regarding the layout, size and orientation of text. We aim to address some of these limitations in this work.

1.1 Challenges of Real World Full Page HTR

Our target data - *Free Form Answers* dataset - was derived from scans of test paper submissions of STEM subjects, from school level all the way up to post-graduate courses. It consists of images containing possibly multiple blurbs of handwritten text, math equations, tables, drawings, diagrams, side-notes, scratched out text and text inserted using an arrow/circumflex and other artifacts, all put together with no reliable layout (Figs. 1 and 2). Length of transcription ranges from 0 to 1100 characters averaging around 160.

© Springer Nature Switzerland AG 2021
J. Lladós et al. (Eds.): ICDAR 2021, LNCS 12823, pp. 55–69, 2021.
https://doi.org/10.1007/978-3-030-86334-0_4

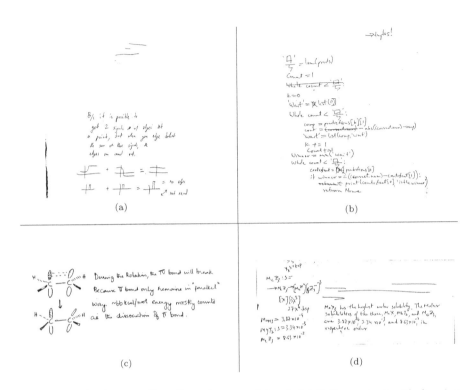

Fig. 1. Samples from our Free Form Answers dataset. (a) Full page text with drawing. (b) Full page computer source code. (c) Diagrams and text with embedded math. (d) Math and text regions, embedded math and stray artifacts.

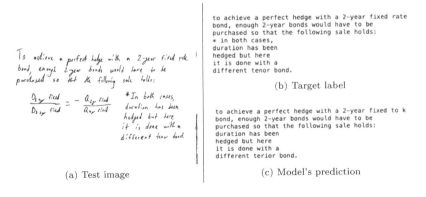

Fig. 2. Test image with two text and one math regions. The model successfully skips the math region and transcribes the two text regions, with some mistakes.

Problem Framework. In order to get a handle on this seemingly random structure, we define each independent blurb as a *region* and view each image as being composed of one or more regions. The regions are further classified as text, math, drawing, tables and deleted (scratched) text. Non textual regions and deleted text (hereafter *untranscribed regions*) are optionally demarcated with special *auxiliary tags* but left untranscribed otherwise. Text regions range from a few characters to multiple paragraphs and are possibly interspersed or embedded with untranscribed regions. Text was transcribed in the order it was meant to be read[1] and line ends, empty lines, spaces and indentations are also faithfully transcribed. These can be programmatically removed later if so desired. Since the regions can be randomly situated on the image, we define a predictable region sequencing order *the natural reading order* as the order in which somebody would read the answer aloud. This order is implicitly captured in the transcription and the model must learn to reproduce it.

With the above framework in mind, the problem becomes: 1) Implicitly identify and classify regions on the image 2) Extract text from each text region and emit it in natural reading order 3) Ignore/skip unrecognized artifacts and 4) Produce auxiliary markup. We call the above formulation *extractive*, since it seeks to extract the identified and desirable content but ignore the unknown and undesirable. This formulation is generic enough to cover simple to complex scenarios but also teaches the model to skip over artifacts that were not encountered in the dataset.

1.2 Limitations of Existing Architectures

Most state-of-the-art HTR models rely on a prior image segmentation method to cleanly isolate pieces of text (words, lines or paragraphs). There are several problems with this approach.

First, image segmentation in such models is a separate system, usually based on hand-crafted features and heuristics which do not hold up when the data changes significantly. Some methods go even further and correct segmented text for slant, curve, and height, using the same problematic features and heuristics.

Second, and more importantly, clean segmentation of units of text is not even possible in many cases of real world handwriting. For instance, observe that lines are curved or interspersed with non-textual symbols and artifacts in Figs. 1 and 2. Further discussion of such limitations can be found in the literature [3,4].

Third, stitching a complete transcription from the individually transcribed text regions introduces yet another system, with its own potentials for errors, and brittleness to changing the data distribution (for example, right-to-left languages would require a different stitching system).

Lastly, formatting and indentation tends to get lost in this three-step process because text segments are usually concatenated using space or newline as a separator. This would be unacceptable when transcribing Python language source code, for instance.

---

[1] This becomes relevant when text is not horizontal or when inserted using a circumflex or arrow.

In our end-to-end model all of the above steps are implicit and learned from data. Adapting the model to a new dataset or adding new capabilities is just a matter of retraining or fine-tuning with different labels or different data.

1.3   Model Overview

Image to Sequence Architecture. Our model (Fig. 3) belongs to the lineage of attention-based sequence-to-sequence and tensor-to-tensor architectures [29, 33]. It consists of a ResNet [11] *encoder* and a Transformer [28] *decoder*. It has a *Image to Sequence architecture* i.e., it learns to map an image to a sequence of tokens. The model signals end of output sequence via a special <EOS> token, so there is no limit on sequence length.[2]

Formatting and Auxiliary Markup. We trained the model to extract text from text regions only, and to either skip the untranscribed regions or produce special markup tags (<END-OF-REGION>, <MATH>, <DELETED-TEXT>, <TABLE> AND <DRAWING>) for them. This requires the model to identify the untranscribed regions, which are almost always *unique per sample*. It generalizes reasonably well in either of these tasks. Similarly, because whitespace is preserved in our ground truth transcription, the model learns to faithfully replicate multiple empty lines and indentation spaces, which is important for computer source code for example.

Token Vocabulary. In order to cater to a variety of subject matter terminology and proper nouns, we chose to use character level token vocabularies in our base configuration. That said, we experimented with different types of vocabularies, for example character vs hybrid word/character, or lowercase vs mixed case. Model performance did not differ significantly between these variations. Therefore, we are confident that more complex transcription tasks – such as transcribing math [6,26] and complex layouts such as tables – should be possible given appropriately transcribed training data and a character-level vocabulary.

Layout Agnostic Architecture. Since there are no assumptions regarding the layout of text, the model works equally well on single-column vs two-column text, and easily learns to produce a column separator tag (Table 2). The multistage HTR models described in Subsect. 1.2 on the other hand, would concatenate adjacent lines from the two columns, unless a separate code module was added to recognize columns and modifications made to the rest of the code.

Performance. The model outperforms the best commercially available API on our proprietary Free Form Answers dataset (Table 2) and state of art full page recognition [4] and full paragraph recognition [3] models on the academic IAM dataset (Table 1).

---

[2] Except a limit set at prediction to prevent an endless loop.

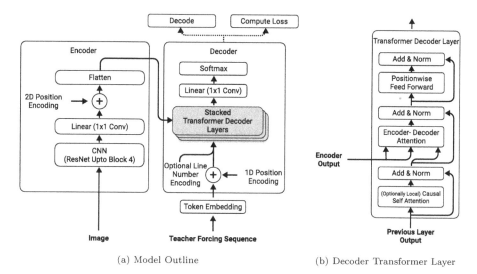

(a) Model Outline   (b) Decoder Transformer Layer

Fig. 3. Model Schematics. Left: CNN Encoder and Transformer Decoder. Right: The Transformer Layer is identical to [28] except that Self Attention maybe optionally Local. Teacher Forcing Sequence refers to the ground-truth shifted right during training or the predicted tokens when predicting.

2   Related Work

The first work to use Neural Networks for offline HTR was by Lecun et al. [15] which recognized single digits. The next evolution was by Graves and Schmidhuber [10] who used an MDLSTM model and CTC loss and decoding [9] to recognize Tunisian town names in Arabic. Their model was designed to recognize a single line of text, oriented horizontally. Pham et al. [22] refined the MDLSTM + CTC model by adding dropout, but it remained a single-line recognition model dependent on prior line segmentation and normalization. Voigtlaender et al. [30] further increased performance by adding deslanting preprocessing (which is entirely dataset specific) and using a bigger network.

Bluche et al. [4] developed a model with the same vision as ours: full-page recognition with no layout or size assumptions. However, they subsequently abandoned this approach citing prohibitive memory requirements, unavailability of GPU acceleration for training of MDLSTM and intractable inference time. Follow-up work by the same authors [2,3] saw a come-back to the encoder-only + CTC approach but with a scaled-back MDLSTM attention model that could isolate individual lines of text, in effect performing automatic line segmentation and enabling the model to recognize paragraphs. While far superior to single-line recognition models, this approach still hard-codes the assumption that lines of text stretch horizontally from left to right, and fill the entire image. Therefore, it can't handle arbitrary text layouts and instead, relies on paragraph segmentation. This approach also does not output a variable-length sequence rather a fixed number of lines T (some of which may be empty), T being baked into the model during training. Finally, since the predicted lines are joined together using a fixed separator it is unlikely that the model can faithfully reproduce empty lines and indentation.

Further on, embracing the trend towards more parallel architectures, Messina [2] and Puigcerver [23] replaced the MDLSTM encoder with CNN, but continued to rely on CTC.

Another recent trend in deep learning has been the move away from encoder-only towards encoder-decoder sequence-to-sequence architectures. These architectures decouple the sequence length of the output from the input thereby eliminating the need for CTC loss and decoding. Cross-entropy loss and greedy/beam-search decoding which are less compute-intensive are employed instead. Ly et al. [16] applied such architecture to recognizing very short passages of Japanese text (a handful of short vertical lines). Similarly Wang et al. [31] applied a bespoke sequence-to-sequence architecture for recognizing single lines.

It is well established now that Transformers can completely replace LSTMs in function and are more parameter efficient, accurate and enable longer sequences. Embracing this trend Kang et al. [13] published the most 'modern' architecture to date employing CNN and Transformers somewhat like ours. However since it collapses the vertical dimension of the image feature map, this model is designed to recognize single lines only. It also employs a transformer encoder thereby making it larger than our model. To our knowledge, ours is the only work other than [4] that attempts full page handwriting recognition.

3   Model Architecture

Our Neural Network architecture is shown in Fig. 3. It is an encoder-decoder architecture, using ResNet [11] for encoding the image, and Transformer [28] for decoding the encoded representation into text. We refer you to [26–28, 33] for a background on neural image-to-sequence and sequence-to-sequence models. This section will fill in the remaining details necessary to reproduce our model architecture.

We use the term *base configuration* to refer to our most frequently used model configuration, and all configuration parameters we list hereafter describe this configuration unless stated otherwise.

Encoder. The encoder uses a CNN to extract a 2D feature-map from the input image. It uses the ResNet architecture without its last two layers: the average-pool and linear projection. The feature-map is then projected to match the Transformer's hidden-size d_{model}, then a 2D positional encoding added and finally flattened into a 1D sequence. 2D positional encoding is a fixed sinusoidal encoding as in [28], but using the first $d_{model}/2$ channels to encode the Y coordinate and the rest to encode the X coordinate (Eq. 1) (similar to [20]). Output I of the *Flatten* layer is made available to all Transformer decoder layers, as is standard.

$$PE(y, 2i) = sin(y/10000^{2i/d_{model}})$$
$$PE(y, 2i + 1) = cos(y/10000^{2i/d_{model}})$$
$$PE(x, d_{model}/2 + 2i) = sin(x/10000^{2i/d_{model}}) \quad (1)$$
$$PE(x, d_{model}/2 + 2i + 1) = cos(x/10000^{2i/d_{model}})$$
$$i \in [0, d_{model}/4)$$

Decoder. The decoder is a Transformer stack with non-causal attention to the encoder output (its layers can attend to the encoder's entire output) and causal self-attention (it can only attend to past positions of its text input). As is standard, training is done with *teacher forcing*, which means that the ground truth text input is shifted one off from the output.

In total, the base configuration has 27.8 million parameters (6.3M decoder, 21.4M ResNet).

The input vectors are enhanced with 1D position encoding, as is standard. Additionally, we concatenate a *line number encoding* (*lne*) - the scaled text line number (l) that the token lies on - to it (Fig. 3a). Assuming a maximum of 100 text lines, $l \in [1, 100]$ and $lne = l/100$. We added *lne* to the model in order to address line level errors i.e., missing or duplicated lines but it was applied only in Table 1. We haven't yet officially concluded on its impact on model performance and mention it here only for completeness sake.

In order to improve memory and computation requirements of our model, we implemented a localized form of causal self-attention by limiting the attention span to 50 (configurable) past positions. This is similar to Sliding Window Attention of [1]) or a 1D version of Local Self Attention of [20]). We hypothesized that a look back of 50 characters should be enough to satisfy the language modeling needs of our task, while the limited attention span should help training converge faster, both assumptions being validated by experiment. Final model performance however, was not impacted by it. That said, a thorough ablation study was not performed. Practically though, it allowed us to use larger mini-batches by about 12%.

Objective Function. For each step t the model outputs the token probability distribution p_t over the vocabulary set $\{1, \ldots, V\}$ (Eq. 2). This distribution is conditioned upon the tokens generated thus far $y_{<t}$ and I (Eq. 3). Probability of the entire token sequence is therefore given by Eq. 4.

$$p_t \quad : \quad \{1,\dots,V\} \to [0,1] \quad ; \ \boldsymbol{Y}_t \sim \boldsymbol{p}_t \tag{2}$$

$$\boldsymbol{p}_t(\boldsymbol{y}_t) \quad := \quad \mathbb{P}(\boldsymbol{Y}_t{=}\boldsymbol{y}_t|\boldsymbol{y}_{<t}, \boldsymbol{I}) \tag{3}$$

$$\mathbb{P}(\boldsymbol{y}|\boldsymbol{I}) \quad = \quad \prod_{t=1}^{\tau} \boldsymbol{p}_t(\boldsymbol{y}_t) \tag{4}$$

As is typical with sequence generators, the training objective here is to maximize probability of the target sequence $\boldsymbol{y}^{GT}$. We use the standard per-word cross-entropy objective (Eq. 5), modified slightly for the mini-batch (Eq. 6). We did not use any regularization objective, relying instead on dropout, data-augmentations and synthetic data to provide regularization.

$$\mathcal{L}_{seq} \quad = \quad -\frac{1}{\tau}\sum_t \ln\left(\boldsymbol{p}_t(\boldsymbol{y}_t^{GT})\right) \qquad ; \ \tau \equiv \text{ sequence length} \tag{5}$$

$$\mathcal{L}_{batch} \quad = \quad -\frac{1}{n}\sum_{batch}\sum_t \ln\left(\boldsymbol{p}_t(\boldsymbol{y}_t^{GT})\right) \quad ; \ n \equiv \# \text{ of tokens in batch} \tag{6}$$

The final Linear layer of the decoder (Fig. 3a) is a 1×1 convolution function that produces logits which are then normalized by softmax to produce $\boldsymbol{p}_t$.

Combination of Vision and NLP. One of the strengths of our architecture is in the combination of Vision and Language models. CNNs such as ResNet are considered best for processing image data. And Transformers are considered best for Language Modeling (LM) and Natural Language Understanding (NLU) tasks [7,24,25], possessing properties that are very useful in dealing with noisy and incomplete text that often occurs in real handwriting. Having both the visual feature map and a language model, the model can do a much better job than one relying on visual features alone.

Inference. We use simple greedy decoding, which picks the highest probability token at each step. Beam search decoding [8] did not yield any accuracy improvement indicating that the model is quite opinionated/confident.

4   Training Configuration and Procedure

The base configuration uses grayscale images scaled down to 140–150 dots per inch. Higher resolutions yielded slightly better accuracy at the cost of compute and memory. We use the 34-layer configuration of ResNet, but have also successfully trained the 18-layer and 50-layer configurations; larger models tending to do better in general as expected.

The following is the base configuration of the Transformer stack:

- N (number of layers) = 6
- d_{model} = 260
- h (number of heads) = 4
- d_{ff} (inner-layer of positionwise feed-forward network) = 1024
- Activation function inside feed-forward layer = GELU [12]
- dropout = 0.5

The model was implemented in PyTorch [21], and training was carried out using 8 NVIDIA 2080Ti GPUs. For full page datasets a mini-batch size of 56 combined with a gradient accumulation factor of 2 was used, yielding an effective batch-size of 112. Single-line datasets had batch sizes as high as 200, but were adjusted downwards when using higher angles of image rotation. ADAM optimizer [14] was employed with a fixed learning rate (α) of 0.0002, $\beta_1 = 0.9$ and $\beta_2 = 0.999$.

While all images in a batch must have the same size; we also set all batches to have the same image size, padding smaller images as needed. This helps during training because any impending GPU OOM errors surface quickly at the beginning of the run. It also makes the validation/test results agnostic of the batching scheme since the images will always be the same size regardless of how they are grouped. Smaller images within a batch are centered during validation and testing. Padding color can be either the max of 4 corner pixels or simply 0 (black), the choice having no impact on model accuracy.

The base configuration vocabulary consists of all lowercase ASCII printable characters, including space and newline.

We observed that model performance increases with increased layer sizes and also image resolution. It also tends to improve monotonically with training time. Therefore, we trained our models for as long as possible the longest being 11 days on full pages (~47M samples) and 8 days (~102M samples) on single lines. Typical training length though is roughly 24M total training samples. Model state is saved periodically during training. At the end of training, the checkpoint that performs best on validation is selected for final testing.

5   Data

Data Sources. The following is a comprehensive set of all data sources, although each experiment used only a subset:

- IAM: The IAM dataset [17] with the RWTH Aachen University split [19], which is a widely used benchmark for handwriting recognition.
- WIKITEXT: The WikiText-103 [18] dataset was used to generate synthetic images (explained later).
- FREE FORM ANSWERS: Our proprietary dataset with about 13K samples described in Subsect. 1.1.
- ANSWERS2: Another proprietary dataset of about 16K segmented words and lines extracted from handwritten test answers. However since we train full pages, we stitched back neighboring words and lines using heuristics.

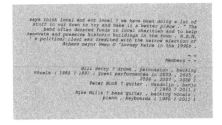 

Fig. 4. Synthetic samples generated from WikiText-103 Dataset [18]. For the two column sample (b), the model – as per training – predicts the left column first, then a column separator tag <col> and then the right column.

- NAMES: Yet another proprietary dataset of about 22K handwritten names and identifiers.

Proprietary datasets were only used for results reported in Table 2.

Synthetic Data Augmentation. Since the IAM data set has word-level segmentation, we generate synthetic IAM samples by stitching together images of random spans of words or lines. This made it possible to significantly augment the IAM paragraph dataset beyond the mere 747 training forms available in the RWTH Aachen split. Without this augmentation the model would not generalize at all.

We also generate synthetic images on the fly from the WikiText data by picking random text spans and rendering them into single-column and/or two-column layout, using 34 fonts in multiple sizes for a total of 114 combinations (Fig. 4). The WikiText data is over 530 million characters long from which over 583 billion unique strings of lengths ranging from 1 to 1100 may be created. Multiplying this with 114 (fonts) yields an epoch size of 66.46 trillion which we would never get through in any of our runs. The dataset thus provides us with a continuous stream of unique samples which builds the implicit language model. Furthermore this trick can be used to 'teach' the model new language and terminology by using an appropriate seed text file. Addition of this dataset reduces the error rate on IAM paragraphs by only about 0.4% (Table 1.) on the IAM dataset but significantly improves the validation loss - about 30%.

Additionally, we generate empty images of varying backgrounds on the fly. Without this data, the model generates text even on blank images – which provides evidence of an underlying language model working a little overzealously.

Image Augmentation. The following transformations were applied to individual training images: 1) Image scale, 2) rotation, 3) brightness, 4) background color of synthetic images, 5) contrast, 6) perspective and 7) Gaussian noise. At the batch level the images were randomly placed within the larger batch image size during training but centered during validation and testing.

Data Sampling. We found that the model generalized best when prior biases were eliminated from the training data. Therefore the datasets are sampled on the fly for every mini-batch with a configurable distribution. We also do not group images by size or sequence length rather we sample them uniformly. Further, parameters for synthetic data generation (e.g. the text span to render or the font to render it in) and image augmentation (e.g. scale of image, angle of rotation, background color) are also sampled uniformly.

Inspite of this sampling scheme, one bias does enter into the training data: padding around text. This is so because most images of a batch are smaller than the batch image size and therefore get padded. This causes the model to perform poorly in inference unless the image was sufficient padding. The optimal padding amount becomes a new hyperparameter that requires tuning after training. We circumvented this problem by padding all images to the same (large) size both during training and inference. This scheme though has the downside of consuming the highest amount of encoder compute regardless of the original image size. Therefore the first scheme is preferable for production.

6   Results

Table 1 shows character error rates (Levenstein edit distance divided by length of ground truth) of paragraph and single line level tests on the IAM dataset. FPHR refers to our model trained with only the IAM dataset whereas FPHR + Aug refers to our model trained with the IAM plus WikiText based synthetic dataset[3]. FPHR trained at ~145 DPI outperforms previous state of art in the paragraph level test[4] and FPHR+Aug improves error rate by a further 0.4%. When trained on single lines only, performance is similar to full page but short of the specialized state-of-the-art single-line models. This is because the model does equally well on different text lengths and number of lines provided they were uniformly distributed during training i.e., performance tracks the training data distribution. Notice that our corpus level CERs are lower than the averages indicating better performance on longer text. This is not a characteristic of the model, rather that of the training data which had fewer short sequences. We believe that the observed performance (6+% test and 4+% validation CER) is the upper limit of this system and task configuration i.e., the model architecture, its size, the dataset and its split. Increasing the model size and/or image resolution does improve performance slightly (albiet by less than 1%) as does increasing the amount of training data.

When trained on all datasets and evaluated on Free Form Answers, our model gets a error rate of 7.6% vs the best available cloud API's 14.4% (Table 2).

---

[3] We view synthetic WikiText based data as an augmentation method since it does not rely on proprietary data or method.

[4] Results from [2] are not included because it was trained on a lot more than IAM data and 30% of it was proprietary.

Table 1. Comparison on the IAM dataset with and without closed lexicon decoding (LM). Figures in brackets are corpus level scores. *Model requires paragraph segmentation. ‡FPHR trained with single lines only. SLS = Shredding Line Segmentation.

Test type	Model	Mean test CER w/o LM	Mean test CER w/ LM
Paragraph level	Bluche et al. [4] (150 dpi)	16.2%	
	FPHR (~ 145 dpi)	**6.7** (6.5)%	
	FPHR + Aug (~ 145 dpi)	**6.3** (6.1)%	
Paragraph level*	Bluche [3] (150 dpi)	10.1%	6.5%
	Bluche [3] (300 dpi)	7.9%	**5.5**%
	SLS + MDLSTM + CTC [4] (150 dpi)	11.1%	
	SLS + MDLSTM + CTC [4] (300 dpi)	7.5%	
Single line	Puigcerver [23]	5.8%	4.4%
	Wigington et al. [32]	6.4%	
	Wang et al. [31]	6.4%	
	Kang et al. [13]	**4.7**%	
	FPHR + Aug‡ (~ 145 dpi)	6.5 (5.9)%	

Table 2. Character Error Rates on Free Form Answers and multi column synthetic datasets. *FPHR trained on one and two col. WikiText synthetic data.

Test data set	Best cloud API	FPHR
Free form answers	14.4%	**7.6**%
Wikitext (1 column)	1.4%	0.008%*
Wikitext (2 column)	57%	0.012%*

Favoring the cloud models, we removed auxiliary markup, line indentations and lower cased all predicted and ground truth text for this comparison[5].

The model has no trouble transcribing two column layout; its performance on two-column and single-column data are comparable (Table 2) providing evidence of its adaptability to different text layouts. The Cloud API on the other hand does well on single-column data, but falters with two-column text precisely because it concatenates adjacent lines across columns.

Inferencing takes an average of 4.6 s on a single CPU thread for a set of images averaging 2500 × 2200 pixels, 456 chars and 11.65 lines without model compression i.e., model pruning, distillation or quantization.

---

[5] We evaluated Microsoft, Google and Mathpix cloud APIs. Microsoft performed the best and its results are reported here. This is not intended to be a comparison of models, rather a practical data point that can be used to make build-vs-buy decisions.

7 Conclusion

We have presented a "modern" neural network architecture that can be trained to perform full page handwriting recognition without image segmentation, delivers state of art accuracy and which is also small and fast enough to be deployed into commercial production. It adapts reasonably to different vocabularies, text layouts and auxiliary tasks and is therefore fit to serve a variety of handwriting and printed text recognition scenarios. The model is also quite easy to replicate using open source 'off the shelf' modules making it all the more compelling. Although the overall Full Page HTR problem is not solved yet, we believe it takes us one step forward from [4] and [3].

7.1 Limitations and Future Work

Although the presented framework encompasses multiple tasks, available datasets are usually heavily biased towards one or two thereby masking the model's performance on outlier tasks. For e.g., there's usually only one transcribed text region per sample in the Free Form dataset which makes the model tend to transcribe only one (main) text region while skipping others. On the other hand when the dataset is balanced e.g., with one and two column synthetic text, it performs well on both layouts. That said, this aspect needs to be explored more thoroughly and hopefully with standardized datasets and tasks so that the research community can iterate over it.

Second, we have only trained with text up to 1100 characters long and averaging 360 characters. Should there be a need to transcribe longer lengths of text say 10K characters, then some more work becomes necessary in order to deal with longer sequence lengths. Possible solutions include the use of multicharacter vocabularies and sparse Transformers such as [1,5].

Other desirable improvements to the model include 1) reducing its sensitivity to image padding. 2) Reducing the encoder's size, which currently stands at almost 22 million parameters. 3) Separating vision models for significantly different visual data (e.g., Synthetic v/s Free Response Answers) so that they may both contribute to the language model but not interfere with each other's vision models.

We believe that the Full Page HTR problem cannot be considered "solved" until the error rate has been brought down to less than 1%. Therefore, more work remains for the community.

Acknowledgements. We would like to thank Saurabh Bipin Chandra for implementing the fast inference path $(O(N^2))$ of the Transformer decoder, which was lacking in PyTorch.

References

1. Beltagy, I., Peters, M.E., Cohan, A.: Longformer: the long-document transformer (2020)

2. Bluche, T., Messina, R.: Gated convolutional recurrent neural networks for multi-lingual handwriting recognition. In: 2017 14th IAPR International Conference on Document Analysis and Recognition (ICDAR), vol. 1, pp. 646–651 (2017)
3. Bluche, T.: Joint line segmentation and transcription for end-to-end handwritten paragraph recognition. arXiv:1604.08352 (2016)
4. Bluche, T., Louradour, J., Messina, R.O.: Scan, attend and read: end-to-end handwritten paragraph recognition with MDLSTM attention. CoRR arxiv:1604.03286 (2016)
5. Dai, Z., Yang, Z., Yang, Y., Carbonell, J., Le, Q.V., Salakhutdinov, R.: Transformer-XL: attentive language models beyond a fixed-length context (2019)
6. Deng, Y., Kanervisto, A., Ling, J., Rush, A.M.: Image-to-markup generation with coarse-to-fine attention. In: ICML (2017)
7. Devlin, J., Chang, M.W., Lee, K., Toutanova, K.: BERT: pre-training of deep bidirectional transformers for language understanding (2019)
8. Graves, A.: Supervised sequence labelling with recurrent neural networks. In: Studies in Computational Intelligence (2008)
9. Graves, A., Fernández, S., Gomez, F., Schmidhuber, J.: Connectionist temporal classification: labelling unsegmented sequence data with recurrent neural networks. In: ICML 2006 (2006)
10. Graves, A., Schmidhuber, J.: Offline handwriting recognition with multidimensional recurrent neural networks. In: NIPS (2008)
11. He, K., Zhang, X., Ren, S., Sun, J.: Deep residual learning for image recognition. CoRR arxiv:1512.03385 (2015)
12. Hendrycks, D., Gimpel, K.: Bridging nonlinearities and stochastic regularizers with gaussian error linear units. CoRR arxiv:abs/1606.08415 (2016)
13. Kang, L., Riba, P., Rusiñol, M., Fornés, A., Villegas, M.: Pay attention to what you read: non-recurrent handwritten text-line recognition (2020)
14. Kingma, D.P., Ba, J.: Adam: a method for stochastic optimization (2017)
15. Lecun, Y., Bottou, L., Bengio, Y., Haffner, P.: Gradient-based learning applied to document recognition. Proc. IEEE **86**(11), 2278–2324 (1998)
16. Ly, N.T., Nguyen, C.T., Nakagawa, M.: An attention-based end-to-end model for multiple text lines recognition in japanese historical documents. In: 2019 International Conference on Document Analysis and Recognition (ICDAR), pp. 629–634 (2019). https://doi.org/10.1109/ICDAR.2019.00106
17. Marti, U.V., Bunke, H.: The IAM-database: an English sentence database for offline handwriting recognition. Int. J. Doc. Anal. Recogn. **5**, 39–46 (2002). https://doi.org/10.1007/s100320200071
18. Merity, S., Xiong, C., Bradbury, J., Socher, R.: Pointer sentinel mixture models. CoRR arxiv:abs/1609.07843 (2016)
19. Open SLR: Aachen data splits (train, test, val) for the IAM dataset. https://www.openslr.org/56/. Identifier: SLR56
20. Parmar, N., et al.: Image transformer. Shazeer (2018)
21. Paszke, A., et al.: PyTorch: an imperative style, high-performance deep learning library. In: Wallach, H., Larochelle, H., Beygelzimer, A., d' Alché-Buc, F., Fox, E., Garnett, R. (eds.) Advances in Neural Information Processing Systems vol. 32, pp. 8024–8035. Curran Associates Inc. (2019). http://papers.neurips.cc/paper/9015-pytorch-an-imperative-style-high-performance-deep-learning-library.pdf
22. Pham, V., Kermorvant, C., Louradour, J.: Dropout improves recurrent neural networks for handwriting recognition. CoRR arxiv:1312.4569 (2013)

23. Puigcerver, J.: Are multidimensional recurrent layers really necessary for handwritten text recognition? In: 2017 14th IAPR International Conference on Document Analysis and Recognition (ICDAR), vol. 1, pp. 67–72 (2017)
24. Radford, A.: Improving language understanding by generative pre-training (2018)
25. Raffel, C., et al.: Exploring the limits of transfer learning with a unified text-to-text transformer (2020)
26. Singh, S.S.: Teaching machines to code: neural markup generation with visual attention. CoRR arxiv:1802.05415 (2018)
27. Sutskever, I., Vinyals, O., Le, Q.V.: Sequence to sequence learning with neural networks. CoRR arxiv:1409.3215 (2014)
28. Vaswani, A., et al.: Attention is all you need. CoRR arxiv:1706.03762 (2017)
29. Vaswani, A., et al.: Tensor2Tensor for neural machine translation. CoRR arxiv:1803.07416 (2018)
30. Voigtlaender, P., Doetsch, P., Ney, H.: Handwriting recognition with large multidimensional long short-term memory recurrent neural networks. In: 2016 15th International Conference on Frontiers in Handwriting Recognition (ICFHR), pp. 228–233 (2016)
31. Wang, T., et al.: Decoupled attention network for text recognition (2019)
32. Wigington, C., Tensmeyer, C., Davis, B., Barrett, W., Price, B., Cohen, S.: Start, follow, read: end-to-end full-page handwriting recognition. In: Proceedings of the European Conference on Computer Vision (ECCV), September 2018
33. Xu, K.,et al.: Show, attend and tell: neural image caption generation with visual attention. In: ICML (2015)

SPAN: A Simple Predict & Align Network for Handwritten Paragraph Recognition

Denis Coquenet[1,2,3(✉)] ⓘ, Clément Chatelain[1,4] ⓘ, and Thierry Paquet[1,2] ⓘ

[1] LITIS Laboratory - EA 4108, Rouen, France
{denis.coquenet,clement.chatelain,thierry.paquet}@litislab.eu
[2] Rouen University, Rouen, France
[3] Normandie University, Mont-Saint-Aignan, France
[4] INSA of Rouen, Saint-Étienne-du-Rouvray, France

Abstract. Unconstrained handwriting recognition is an essential task in document analysis. It is usually carried out in two steps. First, the document is segmented into text lines. Second, an Optical Character Recognition model is applied on these line images. We propose the Simple Predict & Align Network: an end-to-end recurrence-free Fully Convolutional Network performing OCR at paragraph level without any prior segmentation stage. The framework is as simple as the one used for the recognition of isolated lines and we achieve competitive results on three popular datasets: RIMES, IAM and READ 2016. The proposed model does not require any dataset adaptation and can be trained without line breaks in the transcription labels. Our code and trained model weights are available at https://github.com/FactoDeepLearning/SPAN.

Keywords: Handwritten paragraph recognition · Fully convolutional network · Recurrence-free model

1 Introduction

Offline handwritten text recognition consists in recognizing the text from a scanned document. This task is usually performed in two steps, by two different neural networks. In a first step, the document image is cut into text regions: this is the segmentation step. Then, Optical Character Recognition (OCR) is applied on each text region images. Following the advances of the recognition process over time, segmentation was performed on larger and larger entities, from the character in the early ages, to text lines more recently, gradually decreasing the amount of segmentation labels required to train the system.

As a matter of fact, producing segmentation labels by hand, in addition to transcription labels, is costly. Moreover, the use of a two-step process requires to clearly define what a line should be in a non-latent pivot format, *i.e.* a line of text, to generate target labels. However, the definition of a text line raises several questions that prevent to optimize its detection in order to maximize the recognition performance: is a text line a bounding box, a polygon, a set of pixels or a baseline?

J. Lladós et al. (Eds.): ICDAR 2021, LNCS 12823, pp. 70–84, 2021.
https://doi.org/10.1007/978-3-030-86334-0_5

How should it be measured? Which loss should it be trained with? Given all these open questions, we claim that segmentation and recognition should be trained in an end-to-end fashion using a latent space between both stages. Indeed, this allows to circumvent any text line definition, while leveraging the annotation needs.

In this paper, we propose the Simple Predict & Align Network (SPAN), an end-to-end recurrence-free Fully Convolutional Network (FCN) model free of those issues, further reducing the needs for labels in two ways. First, the proposed model performs OCR at paragraph level, so it does not need line-level segmentation labels. Second, it does not even require line breaks in the transcription labels. The proposed model totally circumvents the line segmentation problem using a very straightforward approach. The input paragraph image is analysed in a 2D fashion using a classical fully convolutional architecture, leading to a 2D latent space that is reshaped into a 1D sequential latent space. Finally, the CTC loss is simply used to align the 1D character prediction sequence with the paragraph transcription.

This paper is organized as follows. Related Works are presented in Sect. 2. The proposed SPAN architecture is described in Sect. 3. Section 4 presents the experimental environment and the results. We draw conclusion in Sect. 5.

2   Related Works

In the literature, multi-line text recognition is mainly carried out in two steps. First, a text region (line/word) segmentation is performed [12,16,17]; then, an OCR is applied on the extracted text regions images [7,8,14,23] thanks to the Connectionist Temporal Classification (CTC) [10]. As shown in [9], deep neural networks perform well on both task separately but, when put together, errors in the segmentation stage leads to errors in the OCR stage, leading to higher Character Error Rate (CER).

Recently, one can notice a trend towards the use of unified models. We can classify them into two categories: those performing a text region segmentation prior to the recognition and those without explicit segmentation.

2.1   Segmentation-Based Approaches

Segmentation-based approaches, by definition, require line or word segmentation labels, in addition to the associated transcription label; so the line breaks must be annotated.

Among these approaches, [3,4,6] are based on object-detection methods: a Region Proposal Network (RPN), followed by a non-maximal suppression process and Region Of Interest (ROI), generates line or word bounding boxes. An OCR is then applied on these bounding boxes.

Other approaches are based on predicting the start-of-line coordinates. While in [15] the line is considered horizontal, in [20,21], lines are normalized, recurrently predicting coordinates. Finally, an OCR is applied on these lines.

2.2 Segmentation-Free Approaches

Since they do not explicitly segment the input image, segmentation-free approaches do not require any segmentation labels; the models can be trained using transcription labels only.

In [1,9], the proposed models incorporate an attention mechanism to recurrently generate line features, performing a kind of implicit line segmentation. Indeed, an encoder generates features from the input image; then, the attention process sequentially selects features to focus on. Finally, a decoder predicts characters from these features. [2] proposed a similar approach with an implicit character segmentation.

To our knowledge, only two other works have proposed segmentation-free approaches for multi-line text recognition. [18] focuses on the loss to tackle the two-dimensional aspect of the task, providing a Multi-Dimensional Connectionist Classification (MDCC). Ground truth transcription labels are converted to a two-dimensional model, using a Conditional Random Field (CRF). This model enables to jump from one line to the next one, adding a new line separator label in addition to the standard CTC blank label. In [22], the model is trained to unfold the input multi-line text image into a sequence of lines, thus forming a single large line. Thus, the task is reduced to a one-dimensional problem and the model can be trained with the standard CTC loss.

[18] and [2] are part of the first works proposed for multi-line text recognition, but they remain below the state of the art. While [1] requires pretraining on line-level images, [7] requires line breaks in the transcription labels. [22] is the only model that can be trained from scratch, without any segmentation labels nor line breaks in the transcription labels; but, as a counterpart, it requires some hyperparameters to be adapted for each dataset. Indeed, it requires input images of fixed sizes and includes intermediate bilinear interpolations with fixed dimensions which are specific to each dataset.

In this work, we propose an end-to-end model trained in the same conditions, *i.e.* with paragraph transcription as the only used label, without any line breaks. Instead of unfolding the input image as in [22], we propose to train a model to both predict and align characters so as to get vertical separation between lines, preserving the two-dimensional nature of the task. Contrary to the work presented in [22], the proposed model is able to handle variable size input images, making it flexible enough to be used on multiple datasets without modifying any hyperparameter.

3 SPAN Architecture

We propose an end-to-end model to perform the optical character recognition of paragraphs. We wanted to keep the original shapes of the input images in order to preserve both their ratio and their details as well as to be flexible enough to adapt to a large variety of datasets. To this end, we use a Fully Convolutional Network as the encoder to analyse the 2D paragraph images. An implicit line segmentation is performed by reducing the vertical axis through row concatenation, reshaping the 2D latent space into a 1D latent space, acting as a collapse operator for this dimension. The training process is based on the standard CTC loss that aligns the label sequence with the data in the 1D latent space without any need for line breaks in the annotation. Figure 1a shows an overview of the model architecture: it consists of an FCN encoder, which extracts the features, followed by a convolutional layer, which predicts the character probabilities. Finally, the rows of predictions are concatenated to obtain one single large row of predictions. This brings us back to a one-dimensional sequence alignment problem which is handled with the standard CTC loss.

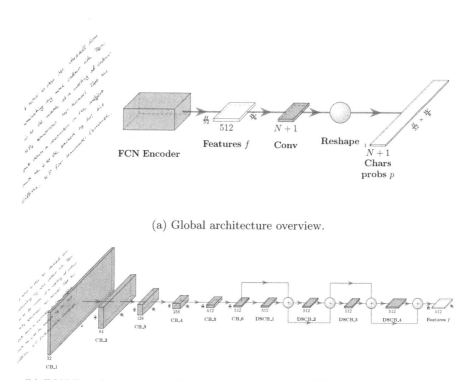

(a) Global architecture overview.

(b) FCN Encoder overview. CB: Convolution Block, DSCB: Depthwise Separable Convolution Block.

Fig. 1. Model visualization. (a) presents an overview of the architecture and (b) focuses on the encoder.

3.1 Encoder

The purpose of the encoder is to extract features from the input images. It employs some convolutions with stride in order to reduce the memory consumption: it takes an input image $X \in \mathbb{R}^{H \times W \times C}$ and outputs some feature maps $f \in \mathbb{R}^{\frac{H}{32} \times \frac{W}{8} \times 512}$ where H, W and C are respectively the height, the width and the number of channels (C = 1 for a grayscale image, C = 3 for a RGB image). The encoder architecture is depicted in Fig. 1b. It corresponds to the encoder proposed in [9], to which the number of channels has been modified going from 16-256 up to 32-512. It is made up of a succession of Convolution Blocks (CB) and Depthwise Convolution Blocks (DSCB).

CB is defined as two convolutional layers followed by instance normalization and a third convolutional layer. This third convolutional layer has a stride of 1×1 for CB_1, 2×2 for CB_2 to CB_4 and 2×1 for CB_5 to CB_6. The strides are chosen to progressively decrease the dimensions. It is a trade-off between memory consumption and performance.

DSCB follows the same structure as CB but the convolutional layers are superseded by Depthwise Separable Convolutions [5] in order to reduce the number of parameters at stake. Moreover, the third DSC has always a stride of 1×1. This enables to introduce residual connections with element-wise sum operator between the DSCB.

For both blocks, convolutional layers have a 3×3 kernel, 1×1 padding and are followed by ReLU activations. In addition, Diffused Mix Dropout (DMD) [9] is used with three potential locations inside each block to reduce overfitting.

3.2 Decoder

The decoder aims at predicting and aligning the probabilities of the characters and the CTC blank label for each 2D position of the features f. The decoder is made up of a single convolutional layer with kernel 5×5, stride 1×1 and padding 2×2. It outputs $N + 1$ channels, N being the size of the charset. Finally, the $\frac{H}{32}$ rows are concatenated to obtain the one-dimensional prediction sequence $p \in \mathbb{R}^{(\frac{H}{32} \cdot \frac{W}{8}) \times (N+1)}$ as depicted in Fig. 2. The CTC loss is then computed between this one-dimensional prediction sequence and the paragraph transcription ground truth, without line breaks.

We can highlight some important aspects about the decoder:

- In this work, the CTC blank label has a new function. Indeed, in standard OCR applied to text lines, the CTC blank label enables to recognize two identical successive characters and to predict "nothing", acting like a joker. Here, it is also used to separate lines in a two-dimensional context, as it allows to label line spacing in the input image.
- One should notice that the prediction occurs before reshaping to 1D, which allows to take advantage of the two-dimensional context in the decision layer. This enables to localize the previous and next lines, and to align the predictions of the same text line on the same row *i.e.*, and to separate it from the other text line predictions.

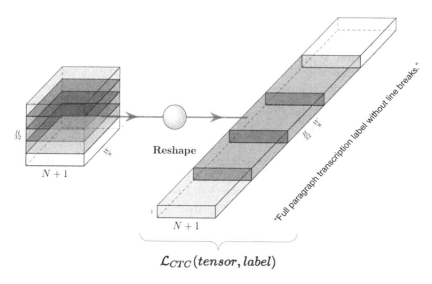

Fig. 2. Reshape operation and loss visualization. No computations are performed in the reshape operation, both left and right tensors represent characters and CTC blank label probabilities. The CTC loss is computed between the one-dimensional probabilities sequence and the paragraph transcription.

– Since the prediction rows are concatenated, they are processed sequentially; nothing prevents the model from predicting the beginning of the text line on one row and the end on the next one as long as there is enough space between this text line and the following one. In Sect. 4.7, we show that this allows us to process inclined lines.

4 Experimental Study

4.1 Datasets

We evaluate our model on three popular datasets at paragraph level: RIMES [11], IAM [13] and READ 2016 [19].

RIMES. We used the RIMES dataset which is made up of French handwritten paragraphs, produced in the context of writing mails scenarios. The images are gray-scaled and have a resolution of 300 dpi. In the official split, 1,500 paragraphs are dedicated to training and 100 paragraphs to evaluation. The last 100 training images are used for validation so as to be comparable with the state of the art.

IAM. The IAM dataset corresponds to handwritten copy of English text passages extracted from the LOB corpus. The images are gray-scaled handwritten paragraph with a resolution of 300 dpi. In this work, we used the unofficial but commonly used split as detailed in Table 1.

READ 2016. The READ 2016 dataset corresponds to Early Modern German handwriting. It has been proposed in the ICFHR 2016 competition on handwritten text recognition. It is a subset of the Ratsprotokolle collection, used in the READ project. Images are in color and we used the paragraph level segmentation. We assume that the images have a resolution of around 300 dpi too.

In Sect. 4, some experiments implies pretraining using the line level images of these three datasets. The corresponding splits are shown in Table 1.

Table 1. Datasets split in training, validation and test sets and associated charset size

Dataset	Level	Training	Validation	Test	Charset size
RIMES	Line	10,532	801	778	100
	Paragraph	1,400	100	100	
IAM	Line	6,482	976	2,915	79
	Paragraph	747	116	336	
READ 2016	Line	8,349	1,040	1,138	89
	Paragraph	1,584	179	197	

Paragraph image examples from these three datasets are depicted in Fig. 3. IAM layout is the more structured and regular. RIMES brings some irregularities in terms of line spacing, text inclination and horizontal text alignment. Finally, the READ 2016 dataset is more complex in terms of noise, text line separation (due to ascents and descents) and size variety.

4.2 Preprocessing

Paragraph images are downscaled by a factor of 2 through a bilinear interpolation leading to a resolution of 150 dpi. Gray-scaled images are converted into RGB images concatenating the same values three times, for transfer learning purposes. They are then normalized (zero mean and unit variance) considering the channels independently.

4.3 Data Augmentation

Data augmentation is applied at training time to reduce over-fitting. The augmentation techniques are used in this order: resolution modification, perspective transformation, elastic distortion and random projective transformation (from [22]), dilation and erosion, brightness, and contrast adjustment and sign flipping. Each transformation has a probability of 0.2 to be applied. Except for perspective transformation, elastic distortion and random projective transformation which are mutually exclusive, each augmentation technique can be combined with the others.

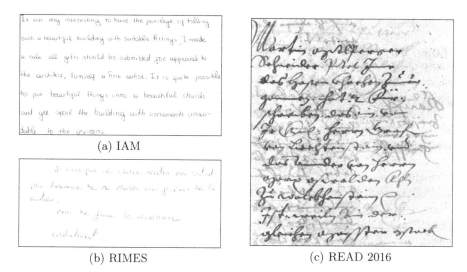

(a) IAM

(b) RIMES

(c) READ 2016

Fig. 3. Paragraph image examples from the RIMES, IAM and READ 2016 datasets.

4.4 Metrics

The Character Error Rate (CER) and the Word Error Rate (WER) are used to evaluate the quality of the text recognition. They are both computed with the Levenshtein distance between the ground truth text and the predicted text at paragraph level, without line breaks. Those edit distances are then normalized by the length of the ground truth. Other metrics are provided in the following experiments such as the number of parameters implied by the models.

4.5 Training Details

We used the Pytorch framework to train and evaluate our models. In all experiments, the networks are trained with the Adam optimizer, with an initial learning rate of 10^{-4}. Trainings are performed on a single GPU Tesla V100 (32Gb), during 2 days, with a mini-batch size of 4 for paragraph images and 16 for text lines images. We used the original paragraph ground truth replacing line breaks by space characters.

4.6 Additional Information

We do not use any post-processing *i.e.* we do not use any language model nor lexicon constraint. Moreover, we only use best path decoding to get the final predictions from the character probabilities lattice. We use exactly the same training configuration from one dataset to another, without model modification, except for the last layer which depends on the charset size.

4.7   Results

Comparison with State of the Art. In this section, we compare our approach to state-of-the-art models on the RIMES, IAM and READ 2016 datasets, at paragraph level and in the same conditions *i.e.* without language model nor lexicon constraint.

Prior to compare the obtained results to the state of the art, it is important to understand the experimental conditions of each of the methods. Table 2 shows model details that should be taken into account to fairly compare the following tables of results. Quantitative metrics are computed for the IAM dataset, without automatic mixed precision (for a fair comparison with respect to the memory usage). From left to right, the columns respectively denote the architecture, the number of trainable parameters, the maximum GPU memory usage during training (for a mini-batch size of 1, data augmentation included), the minimum transcription level required, the minimum segmentation level required, the use of PreTraining (PT) on subimages, the use of specific Curriculum Learning (CL) and finally the Hyperparameter Adaptation (HA) requirements from one dataset to another.

Table 2. Requirements comparison of the SPAN with the state-of-the-art approaches.

Architecture	# Param.	Max memory	Transcription label	Seg. label	PT	CL	HA
[4] RPN+CNN+BLSTM			Word	Word			
[6] RPN+CNN+BLSTM			Word	Word			
[21] RPN+CNN+BLSTM			Line	Line			
[9] FCN+LSTM*	2.7 M	2.2 Gb	Paragraph + line breaks	Paragraph	Line	✗	✗
[2] CNN+MDLSTM**			Paragraph	Paragraph	Line	✓	✗
[1] CNN+MDLSTM*			Paragraph	Paragraph	Line	✓	✗
[22] GFCN	16.4 M	8.8 Gb	Paragraph	Paragraph	✗	✗	✓
[This work - SPAN] FCN	19.2 M	5.1 Gb	Paragraph	Paragraph	Line	✗	✗

* with line-level attention
** with character-level attention

As one can see, models from [4, 6, 21] require transcription and segmentation labels at word or line levels to be trained, which implies more costly annotations. The models from [1, 2, 9] and the SPAN are pretrained on text line images to speed up convergence and to reach better results, thus also using line segmentation and transcription labels even if it is not strictly necessary. While the model from [9] needs line breaks in the transcription annotation, [1, 2] used a specific curriculum learning method for training. In [22], some hyperparameters must be modified from one dataset to another in order to reach optimal performance, namely the fixed input dimension and two intermediate upsampling sizes which are crucial. We do not have such problem since we are working with input images of variable size and we focus on the resolution to be robust to the variety of datasets. Moreover, despite a larger number of parameters (+ 17% compared to [22]), the SPAN requires less GPU memory which is a critical point when training deep neural networks.

Table 3, 4 and 5 show the results of the SPAN compared to the state-of-the-art approaches, for the RIMES, IAM and READ 2016 datasets respectively. One can notice that we reach competitive results on those three datasets, each having its own complexities, without any hyperparameter adaptation. Results here includes model pretraining on line images but the model can be trained without pretraining *i.e.* without using any line-level annotation, while keeping competitive results, as shown in Sect. 4.7.

It has to be noted that the SPAN encoder extracts character representations and aligns them vertically as well, also recognizing interlines. This additional task explains the difference with the results of the other works, notably the VAN [9] which uses a similar encoder. Moreover, the SPAN aligns the prediction with the ground truth at paragraph level whereas the VAN uses the line breaks to make this alignment at line level.

Table 3. Comparison of the SPAN results with the state-of-the-art approaches at the paragraph level on the RIMES dataset.

Architecture	CER (%) validation	WER (%) validation	CER (%) test	WER (%) test
[1]	2.5	12.0	2.9	12.6
[21]			2.1	9.3
[9]	**1.74**	**8.72**	**1.90**	**8.83**
This work - SPAN	3.56	14.29	4.17	15.61

Table 4. Comparison of the SPAN results with the state-of-the-art approaches at the paragraph level on the IAM dataset.

Architecture	CER (%) validation	WER (%) validation	CER (%) test	WER (%) test
[4]*	13.8		15.6	
[6]			8.5	
[21]			6.4	23.2
[9]	**3.04**	**12.69**	**4.32**	**16.24**
[2]			16.2	
[1]	4.9	17.1	7.9	24.6
[22]			4.7	
This work - SPAN	3.57	15.31	5.45	19.83

*Results are given for page level

SPAN Prediction Visualization. Figure 4 presents a visualization of the SPAN prediction for an example of the RIMES test set. Character predictions are shown in red; they seem like rectangle since they are resized to fit the input

Table 5. Comparison of the SPAN results with the state-of-the-art approaches at the paragraph level on the READ 2016 dataset.

Architecture	CER (%) validation	WER (%) validation	CER (%) test	WER (%) test
[9]	**3.75**	**18.61**	**3.63**	**16.75**
This work - SPAN	5.09	23.69	6.20	25.69

image size (the features size is $\frac{H}{32} \times \frac{W}{8}$). Combined with the receptive field effect, this explains the shift that can occur between the prediction and the text. As one can notice, text line predictions are totally aligned, or aligned by blocks; the lines are well separated by blank labels, which act as line spacing labels. As one can see, this block alignment enables to handle downward inclined lines, especially for lines 3 and 4. Moreover, the model does not degrade in the presence of large line spacing.

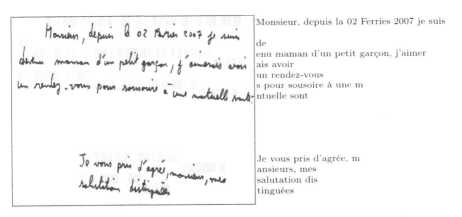

Monsieur, depuis la 02 Ferries 2007 je suis de
venu maman d'un petit garçon, j'aimer
ais avoir
un rendez-vous
s pour sousoire à une m
ntuelle sont

Je vous pris d'agrée, m
ansieurs, mes
salutation dis
tinguées

Monsieur, depuis l**a** 02 Ferries 2007 je suis de*v*enu maman d'un petit garçon, j'aimerais avoir un rendez-vou**s**s pour sousoire à une m**n**tuelle sont*é.* Je vous pris d'agrée, **m**ansieurs, mes salutation**s** distinguées

Fig. 4. SPAN predictions visualization for a RIMES test example. Left: 2D characters predictions are projected on the input image. Red color indicates a character prediction while transparency means blank label prediction. Right: row by row text prediction. Bottom: full text prediction where errors are shown in bold and missing letters are shown in italic.

Impact of Pretraining. In this experiment, we try to highlight the impact of pretraining on the SPAN results. To this end, we compare two pretraining methods at line level: one focusing only on the optical recognition task and the second one focusing on both recognition and prediction alignment. Let's define the following training approaches:

- SPAN-Line-R&A: the SPAN is trained with line-level images. Here, the network has to learn both the recognition and the alignment tasks.
- Pool-Line-R: a new model is trained with line-level images to only focus on the recognition task. This network consists in the previous defined SPAN encoder followed by an Adaptive MaxPooling to collapse the vertical axis; then, a convolutional layer predicts the characters and blank label probabilities. This is the standard way to process text line images, as in [7]. Since the prediction is already in one dimension, the network does not need to care about vertical alignment.
- SPAN-Scratch: the SPAN is trained directly on paragraph images without pretraining.
- SPAN-PT-R: SPAN weights are initialized with Pool-Line-R ones. It is then trained with paragraph images.
- SPAN-PT-R&A: SPAN weights are initialized with SPAN-Line-R&A ones. It is then trained with paragraph images.

One can note that the vertical receptive field is bigger than line image heights. Thus, when the model switches from line images to paragraph images, the decision benefits from more context, which replaces part of the previously used padding.

Results are given in Table 6. Focusing on the line-level section, one can notice that, as expected, we reached better results on text lines when the task is reduced to optical recognition compared to the task of recognition and alignment, whatever the dataset. This leads to a CER improvement of 0.94 point for IAM, 0.79 point for RIMES and 0.38 point for READ 2016. Now, comparing the paragraph level approaches, one can notice that, except for the RIMES CER, pretraining leads to better results, and sometimes by far (-2.93 points of CER for READ 2016); moreover, pretraining on an easier task, *i.e.* only on the optical recognition, is even more efficient.

Table 6. Impact of pretraining the SPAN on line images for the IAM, RIMES and READ 2016 datasets. Results are given on the test sets.

Approach	IAM		RIMES		READ 2016	
	CER (%)	WER (%)	CER (%)	WER (%)	CER (%)	WER (%)
Line-level training						
Pool-Line-R	**4.82**	**18.17**	**3.02**	**10.73**	**4.56**	**21.07**
SPAN-Line-R& A	5.76	21.33	3.81	13.80	4.94	22.19
Paragraph-level training						
SPAN-Scratch	6.46	23.75	**4.15**	16.31	9.13	36.63
SPAN-PT-R	**5.45**	**19.83**	4.74	**15.55**	**6.20**	**25.69**
SPAN-PT-R& A	5.78	21.16	4.17	15.71	6.62	27.38

The RIMES CER value can be explained by the difference between the CTC loss and the Levenshtein distance which are not the same. As a matter of fact, generally, a lower CTC loss implies a lower CER but it is not always true. Indeed, Fig. 5 shows the different loss CTC training curves for the three datasets. This time, we can clearly see that, even for RIMES, pretraining on the recognition task only is more beneficial. This figure also demonstrates the convergence speed up brought by these pretraining approaches.

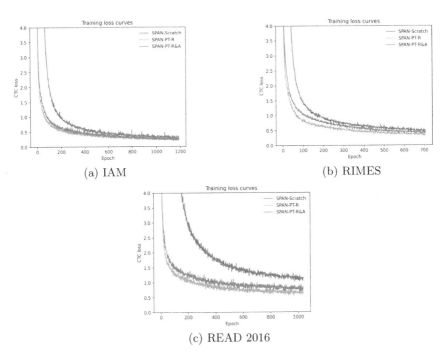

(a) IAM   (b) RIMES

(c) READ 2016

Fig. 5. Training curves comparison between the different pretraining approaches, on the RIMES, IAM and READ 2016 datasets.

Discussion. As we have seen in Fig. 4, the 2D prediction keeps the spatial information. As such, we can assume that the SPAN could be used as a primary stage of a deeper end-to-end network that could handle more complex tasks such as word spotting in handwritten digitized document. Moreover, since we are using the standard CTC loss, one can easily add standard character or word language model to further improve the results. The SPAN could also adapt to full page documents since it is based on the image resolution regardless of its size. However, it has to be noticed that this model is limited to single-column multi-line text images due to the row concatenation operation. Moreover, the SPAN can easily handle downward sloping lines but cannot handle upward ones due to the fixed reshaping order.

5   Conclusion

In this paper, we proposed the Simple Predict & Align Network, an end-to-end recurrence-free segmentation-free FCN model performing OCR at paragraph level. It reaches competitive results on the RIMES, IAM and READ 2016 datasets without any model architecture or training adaptation from one to another. It follows a new training approach bringing several other advantages. First, it only needs transcription label at paragraph level (without line breaks), leveraging the need for handmade annotation, which is a critical point for a deep learning system. Second, training this model is as simple as training a line-level OCR with the CTC loss. Finally, it can handle variable image input sizes, making it robust enough to adapt to multiple datasets; it is also able to deal with downward inclined text lines.

Acknowledgments. The present work was performed using computing resources of CRIANN (Normandy, France) and HPC resources from GENCI-IDRIS (Grant 2020-AD011012155). This work was financially supported by the French Defense Innovation Agency and by the Normandy region.

References

1. Bluche, T.: Joint line segmentation and transcription for end-to-end handwritten paragraph recognition. Adv. Neural. Inf. Process. Syst. **29**, 838–846 (2016)
2. Bluche, T., Louradour, J., Messina, R.O.: Scan, attend and read: end-to-end handwritten paragraph recognition with MDLSTM attention. In: ICDAR, pp. 1050–1055 (2017)
3. Carbonell, M., Fornés, A., Villegas, M., Lladós, J.: A neural model for text localization, transcription and named entity recognition in full pages. Pattern Recognit. Lett. **136**, 219–227 (2020)
4. Carbonell, M., Mas, J., Villegas, M., Fornés, A., Lladós, J.: End-to-end handwritten text detection and transcription in full pages. In: Workshop on Machine Learning, ICDAR, pp. 29–34 (2019)
5. Chollet, F.: Xception: deep learning with depthwise separable convolutions. In: CVPR (2017)
6. Chung, J., Delteil, T.: A computationally efficient pipeline approach to full page offline handwritten text recognition. In: Workshop on Machine Learning, ICDAR, pp. 35–40 (2019)
7. Coquenet, D., Chatelain, C., Paquet, T.: Recurrence-free unconstrained handwritten text recognition using gated fully convolutional network. In: ICFHR, pp. 19–24 (2020)
8. Coquenet, D., Soullard, Y., Chatelain, C., Paquet, T.: Have convolutions already made recurrence obsolete for unconstrained handwritten text recognition ? In: Workshop on Machine Learning, ICDAR, pp. 65–70 (2019)

9. Coquenet, D., Chatelain, C., Paquet, T.: End-to-end handwritten paragraph text recognition using a vertical attention network (2020)
10. Graves, A., Fernández, S., Gomez, F.J., Schmidhuber, J.: Connectionist temporal classification: labelling unsegmented sequence data with recurrent neural networks. In: ICML, vol. 148, pp. 369–376 (2006)
11. Grosicki, E., El Abed, H.: ICDAR 2011-French handwriting recognition competition, pp. 1459–1463 (2011)
12. Grüning, T., Leifert, G., Strauß, T., Michael, J., Labahn, R.: A two-stage method for text line detection in historical documents. Int. J. Doc. Anal. Recognit. **22**(3), 285–302 (2019)
13. Marti, U.V., Bunke, H.: The IAM-database: an English sentence database for offline handwriting recognition. Int. J. Doc. Anal. Recognit. **5**, 39–46 (2002)
14. Michael, J., Labahn, R., Grüning, T., Zöllner, J.: Evaluating sequence-to-sequence models for handwritten text recognition. In: ICDAR, pp. 1286–1293 (2019)
15. Moysset, B., Kermorvant, C., Wolf, C.: Full-page text recognition: learning where to start and when to stop. In: ICDAR, pp. 871–876 (2017)
16. Oliveira, S.A., Seguin, B., Kaplan, F.: dhSegment: a generic deep-learning approach for document segmentation. In: ICFHR, pp. 7–12 (2018)
17. Renton, G., Soullard, Y., Chatelain, C., Adam, S., Kermorvant, C., Paquet, T.: Fully convolutional network with dilated convolutions for handwritten text line segmentation. Int. J. Doc. Anal. Recognit. **21**(3), 177–186 (2018)
18. Schall, M., Schambach, M., Franz, M.O.: Multi-dimensional connectionist classification: reading text in one step. In: 13th International Workshop on Document Analysis Systems, pp. 405–410 (2018)
19. Sánchez, J.A., Romero, V., Toselli, A., Vidal, E.: ICFHR 2016 competition on handwritten text recognition on the read dataset, pp. 630–635 (2016)
20. Tensmeyer, C., Wigington, C.: Training full-page handwritten text recognition models without annotated line breaks. In: ICDAR, pp. 1–8 (2019)
21. Wigington, C., Tensmeyer, C., Davis, B., Barrett, W., Price, B., Cohen, S.: Start, follow, read: end-to-end full-page handwriting recognition. In: Ferrari, V., Hebert, M., Sminchisescu, C., Weiss, Y. (eds.) ECCV 2018. LNCS, vol. 11210, pp. 372–388. Springer, Cham (2018). https://doi.org/10.1007/978-3-030-01231-1_23
22. Yousef, M., Bishop, T.E.: Origaminet: weakly-supervised, segmentation-free, one-step, full page text recognition by learning to unfold. In: CVPR, pp. 14698–14707 (2020)
23. Yousef, M., Hussain, K.F., Mohammed, U.S.: Accurate, data-efficient, unconstrained text recognition with convolutional neural networks. Pattern Recognit. **108**, 107482 (2020)

IHR-NomDB: The Old Degraded Vietnamese Handwritten Script Archive Database

Manh Tu Vu[1]([✉]), Van Linh Le[2]([✉]), and Marie Beurton-Aimar[1]([✉])

[1] LaBRI - Université de Bordeaux, Boredeaux, France
{manh.vu,beurton}@labri.fr
[2] Inria Bordeaux Sud-Ouest, Bordeaux, France
van-linh.le@inria.fr

Abstract. This paper introduces a new handwritten database IHR-NomDB, for an old Vietnamese writing system called ChuNom. Over 260 pages of ChuNom were collected from Vietnamese Nom Preservation Foundation to analyze and annotate the bounding boxes manually to generate more than 5000 patches in which containing the images of handwriting texts, the corresponding digital ChuNom characters and its translation in modern Vietnamese script. Along with this handwriting dataset is a new Synthetic Nom String dataset, which consists of $101,621$ images generated using our collected bank of ChuNom sentences. Totally, $13,254$ characters are presented on the two parts of the database, making this the first and largest publicly available database for researching in this old Vietnamese writing script. For the baseline results, we have performed the testing on the validation set of the handwriting dataset using the Convolution Recurrent Neural Network (CRNN) pretrained on the Synthetic Nom String dataset with CTC Loss and achieved 42.70% accuracy at sentence level and 82.28% accuracy at character level. The database is available to download at https://morphoboid.labri.fr/ihr-nom.html.

Keywords: Handwriting database · Old Vietnamese language · Deep learning · Convolution Recurrent Neural Network

1 Introduction

Handwritten recognition from old degraded documents has been receiving increased attention in the last few decades since it plays a vital role in document analysis and the creation of digital libraries [1]. The searchability, readability and translatability features are extremely important for any kind of documents to be able to reach out to end-users, especially for the old and degraded ones since these documents are representative of a historical period that needs to be preserved. The creating of readily available ground truth for these old and degraded heritage documents, therefore, becoming undoubtedly essential for the development and evaluation of new handwritten recognition technologies. One of these cultured heritages is ChuNom - an old logographic writing system of

J. Lladós et al. (Eds.): ICDAR 2021, LNCS 12823, pp. 85–99, 2021.
https://doi.org/10.1007/978-3-030-86334-0_6

the Vietnamese language that is on the verge of being forgotten. Starting from more than a thousand years ago, ChuNom had been using to record a very long period of Vietnamese's history before being displaced by Vietnamese's modern Latin-based writing system. Thus, there is an enormous number of documents that were written in ChuNom but still have not yet been translated to modern Vietnamese writing script [30]. On top of that, according to Vietnamese Nom Preservation Foundation[1], less than 100 scholars world-wide can read Nom script today, making it even more urgent to preserve this ChuNom heritage.

In this paper, as a step towards the preservation of this cultural heritage, we introduce a new database for this old writing script - the IHR-NomDB[2] database. The database consists of two parts including a handwriting part and a Synthetic Nom String part. Totally, $34,991$ instances of handwriting characters and $704,991$ instances of generated characters are distributed over $106,622$ columns of text characters for the whole database. The underlying lexicon includes 13,254 different ChuNom characters. Along with this database, we also provide some baseline results using the state-of-the-art method Convolution Recurrent Neural Network (CRNN) [26] in handwritten recognition task and our own modified network M-CRNN deep learning network. Results show that by just having some simple modifications, our M-CRNN model can easily reach 82.28% accuracy in character level and 42.70% accuracy in sentence level on the handwritten part of this database. The database described in this paper is freely available for researching purpose.

2   Context and Related Work

In the past, many public handwriting databases have been published and have significantly benefited the research. Among the online datasets that are acquired by using some special writing devices, the UNIPEN [7] and IRONOFF [31] are the first datasets in which including labelled isolated digits, letters, and words in Western scripts. Later on, more and more datasets focus on labelling the whole sentences instead of each isolated characters and words due to the advantaging of recognition technologies, e.g., the IAM-OnDB [16] dataset in which containing a large number of labelled English sentences from the Lancaster-Oslo/Bergen (LOB) corpus, the Japanese online handwriting databases Kuchibue and Nakayosi [18], the Chinese online handwriting databases SCUT-COUCH2009 [11], HIT-OR3C [33] and CASIA [15]. For the more difficult task of offline handwriting recognition where only images of the produced writing are available, CEDAR [8] is one of the first available databases, which contains labelled images of address related words and numbers in Western scripts. Later on, we have the NIST Handprinted Forms and Characters database [6], which contains segmented handprinted digits and alphabetic characters, the IAM database [17] with English sentences from the LOB corpus, the IFN/ENIT database [20] of handwritten Arabic words, the Chinese HIT-MW handwriting database [28]. These modern handwriting document databases

---

[1] http://nomfoundation.org.
[2] IHR = Images Handwritten Recognition.

have been promoting significantly the development of new technologies in hand-written recognition to reach a level that makes commercial applications feasible in some domains, such as postal address or bank check reading [9, 27].

While there are many databases for modern handwriting documents, publicly databases for the old and degraded documents are rather a few due to more difficult and time-consuming in ground truth creation [3]. Among these databases, the work in [22] presents a historical database, which consists of labelled bounding boxes of individual words from 20 pages of George Washington's letters. The IAM-HistDB [3] database contains Latin texts from the 9th, 13th, 18th centuries written in various languages. The ARDIS [13] database contains historical hand-written digits written by priests from Swedish. The READ dataset [23] contains multilingual Latin offline handwriting documents. The Bentham collection [24] consists of a set of images of a collection of works on law and moral philosophy written by the philosopher Jeremy Bentham. These databases play an important role in the task of handwritten recognition for the old and degraded historical documents, which is much more difficult compared to modern handwriting documents [22]. While the languages that have been written in these databases seems to be popular and well known in many countries, it is much more crucial in the task of creating databases for the kind of old, unpopular and nearly lost languages. Since these languages are the evidence as well as the representative of different historical heritages, but having a vulnerable risk of being forgotten. ChuNom is one of these invaluable heritages.

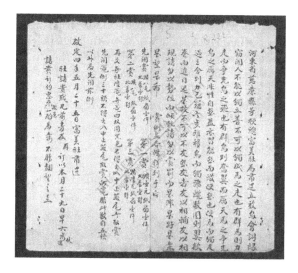

Fig. 1. Example image of Nom document

ChuNom is a logographic writing system formerly used to write the Vietnamese language. It is based on the Chinese writing system but adds a high number of new characters to make it fit with the context of Vietnamese language [12] (example in Fig. 1). Starting from more than a thousand years ago (between the 8th and

10th century [2]), the golden time of ChuNom was between the 15th and 19th centuries. During this period, it was being used by many of Vietnam's cultured elites for popular works. One of the best-known pieces of Vietnamese literature, the Tale-of-Kieu[3], was composed using ChuNom script. After then, during the 1920s, the modern Vietnamese writing system based on Latin alphabet characters displaced ChuNom and became the national and official writing system of Vietnam. However, ChuNom is still an important cultural heritage which was recording the history of Vietnam for more than a thousand years. Therefore, it is extremely necessary to preserve and utilize this cultural heritage.

Up until now, there haven't been many works specifically in ChuNom script recognition. Phan et al. [21,30] tried to segment the characters from documents by training their deep learning classifier network to learn the character's patterns on the combination of ChuNom related languages dataset, which including Chinese, Japanese, Korean, and ChuNom. Although this approach can work on a large number of different categories (characters). However, it gets a lot of noises coming from the other languages and the trained network will certainly be confused if the predicting character is similar to the character from the other languages. Also, their network didn't take into account the relationships between the characters in a sentence. In 2017, Nguyen et al. [19] trained K-means and Convolution Neural Network (CNN) on an artificially generated Nom character dataset to predict about 10,000 Nom characters categories. In 2019, Scius-Bertrand et al. [25] tried to segment the ChuNom columns from the images of historical Vietnamese steles for studying the economical activities of villagers, which is engraved on these steles. The common point of these works is the fact that they don't have a database for either training, evaluating or comparing the performance of their works to the others. Some works using the characters generated from fonts will certainly have their behaviour different from the real handwritten documents. Therefore, a standard publicly available dataset specifically for ChuNom like our database is crucial for the development of the recognition methods of the language's writing system.

3   Data Collection and Annotation

In this part, we describe the content of the IHR-NomDB that we have contructed.

3.1   The Handwriting Part

One of the key requirements of constructing a dataset for training a text recognition system is the ability to identify exactly the location of texts in the image, especially for the kind of document where its characters are densely closed to each other. To ensure this requirement, we have chosen two books collected from Vietnamese Nom Preservation Foundation, which are Luc-Van-Tien[4] and

---

[3] http://nomfoundation.org/nom-project/Tale-of-Kieu/
[4] http://nomfoundation.org/nom-project/Luc-Van-Tien/

Tale-of-Kieu books. These are poem books where the authors are telling stories about different people in the old century. In the case of Luc-Van-Tien, it is a story about two main characters - Luc Van Tien and Kieu Nguyet Nga, who are symbolic of talent, intelligence and human dignity. They are representative of the beautiful dignities of Vietnamese people in general and Southern people in particular. Contrarily with the happy ending of Luc Van Tien is a life of trials and tribulations of Thuy Kieu - a beautiful and talented young woman, who has to sacrifice herself to save her family in Tale-of-Kieu. These poem books were written in six-eight meter verses, column-wise, which is allowing us to extract correctly each sentence in rows and columns, along with its digital characters and translation. An example of these books can be seen in Fig. 2. Totally, 104 pages of Luc-Van-Tien and 163 pages of Tale-of-Kieu books were collected and analyzed. This is including the page images, the digital Nom characters of each of the six-eight meter verses and the translation in Vietnamese modern writing script. These page images were segmented into multiple patches by using the bounding boxes that we had created manually, which can be seen in Fig. 3. Totally, 5001 handwritten patches have been created, making this dataset the first and largest publicly available handwritten database for researching in this old Vietnamese writing script.

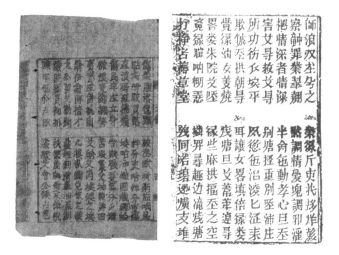

Fig. 2. Example of the page images from the collected books. Page from Luc-Van-Tien's book is on the left and page from Tale Of Kieu is on the right. The words is written from top to bottom and right to left. One column is considered as one sentence.

3.2   The Synthetic Nom String Part

Deep learning in general and text recognition in particular requires to have a very large amount of data to be able to make the prediction correctly. While

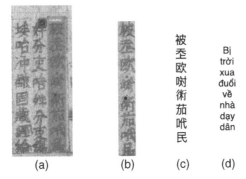

Fig. 3. Example of the patches extracted from the collected data. A bounding box (a) is created manually to indicate the location of the Nom sentence. Then, the patch (b) is extracted along with its ground truth digital characters in ChuNom (c) and its translation in modern writing script (d).

collecting ground truth for this old language's writing system is hard and time consuming due to no longer being used since the early half of the 20th century, generating images from digital texts seems to be the right way to enlarge the dataset. As an attempt to ease the training process for any of the learning algorithms, we have created the Synthetic Nom String dataset by generating ground truth images from our collected bank of sentences. Although we can simply generate sentences by randomly group the Nom characters together, this way there is no link between each of the characters, which is far more different from the texts in the real documents. We collect the sentences from different books and poems to be able to have the sentences with some meaning and have the connection between their characters. This collected bank of sentences is coming from different sources, including:

- The "History of Greater Vietnam" book [5]
- The "Chinh Phu Ngam Khuc" poems[6]
- The "Ho Xuan Huong" poems[7]
- The Corpus documents from chunom.org[8]
- The sample texts coming from 130 different books[9]

The data collected from these sources are the sentences in ChuNom script and their corresponding translation in modern Vietnamese writing script. Totally, it is about 56,820 sentences collected. One can note that the sentences length varies alot, from 1 to 207 characters per sentence. So, we decided to split half

---

[5] http://nomfoundation.org/nom-project/History-of-Greater-Vietnam.

[6] http://nomfoundation.org/nom-project/Chinh-Phu-Ngam-Khuc.

[7] http://nomfoundation.org/nom-project/Ho-Xuan-Huong.

[8] https://chunom.org/shelf/corpus.

[9] http://nomfoundation.org/nom-tools/Tu-Dien-Chu-Nom-Dan_Giai.

of the sentence whenever its length is longer than 10 characters. This limitation is set because we want it to be at the same range as the handwriting part of our database. This resulting 101, 621 sentences with their lengths are ranging from [1–10] characters as showing in Table 4. These Nom sentences are used to generate ground truth dataset by rendering it into some backgrounds. Then, we apply randomly some data augmentation filters to make it as similar to the real handwritten images as possible (see Fig. 4). Particularly, the filters are including Gaussian Blur, Random Dropout Pixels and Grid Distortion operations. Totally, 101, 621 synthetic images have been created for this dataset.

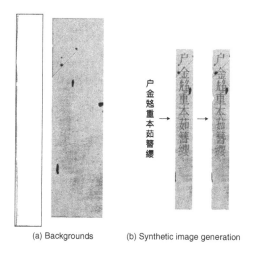

(a) Backgrounds   (b) Synthetic image generation

Fig. 4. Example of the synthetic image generation. A synthetic background is created by cropping randomly from backgrounds (a) with the size of 48×432 pixels. Then, the digital Nom characters are rendered into this synthetic background. Finally, some augmentation filters are applied randomly to generate synthetic images.

4   Statistics on the Database

4.1   The Handwriting Part

As stated, the handwriting part of IHR-NomDB is containing 267 pages from the two books Luc-Van-Tien and Tale-of-Kieu. By annotating the bounding boxes for all of these pages manually, we obtained about 5001 labelled images. Because these data are coming from poem books with the format of six-eight meter verses, which resulting most of the images containing six or eight Nom characters, except some limited exceptional cases when the sentence has 2, 4, 7 or 9 characters (see Table 1).

The total number of different characters in this part is 3441 characters, distributed in different frequency ranges as depicted in Table 2. It can be seen that a

Table 1. Sentence length count of the handwriting dataset

	Sentence length					
	2	4	6	7	8	9
Sentences count	1	1	2498	13	2486	2

Table 2. The distribution in character level of the handwriting part and the Synthetic Nom String part of the IHR-NomDB database

Frequency range	Handwriting part		Synthetic nom string part	
	Distinct count	N appearances	Distinct count	N appearances
1	1119	1119	2847	2847
[2, 5]	1173	3542	4061	12446
[6, 10]	410	3140	1512	11421
[11, 20]	349	5138	1211	17719
[21, 50]	254	7914	1293	42325
[51, 100]	88	6084	843	60023
101+	48	8054	1410	558210

large part of the vocabulary (distinct count) appears only one time in the whole vocabulary of the dataset. This covers about 32% of the dataset's vocabulary while only takes about 3% in frequency appearances. This observation is no surprise since ChuNom is the logographic writing system where each sign represents a meaningful element like a word or a morpheme, thus many Nom characters are created to represent the name of people, places, etc., which are having a small frequency appearance but still take some places in the vocabulary of the dataset. In deep learning's perspective, we need to take into account this characteristic because obviously, the deep learning model can not predict a character that it has never been trained before. Therefore, we need to separate the dataset into two sets (training and validating), in which we have to reduce as minimum as possible the case of having the one-time-appearing samples in the validation set. In order to do so, we have defined a ranking score R for each of the sample s in our dataset D. Lets N be the total number of times each character in s appears in $D \setminus s$. The R score is denoted as:

$$R_s = N_{distinct(s)} \times \max_i^D N_i + N_s \qquad (1)$$

As expressed in Eq. 1, the R score emphasizes firstly, the number of characters in s that appearing in $D \setminus s$, and secondly, the frequency of these appearances. To select k samples for the validation set, we just need to take k samples of the top highest R score. We choose $k = 1000$, which is equivalent to 20% of the dataset. Table 3 shows the samples distribution and the characters intersection of the two sets of the dataset. The characters intersection term refers to the coverage of one set's lexicon to the other. From Table 3 we can see that by applying R score to split the database, we obtained 99.25% of characters intersection of the training set comparing to the validation set, which means the training set consists of almost every characters that appearing in the validation set.

Table 3. The Intersection between training set and validation set in each of the handwriting part and the Synthetic Nom String part of the IHR-NomDB

		N samples	Characters intersection
Handwriting part	Training set	4001	99.25%
	Validation set	1000	42.36%
Synthetic Nom String part	Training set	81297	94.99%
	Validation set	20324	43.54%

4.2 The Synthetic Nom String Part

The Synthetic Nom String dataset consists of $101,621$ images generated from our collected bank of sentences. The sentences length is ranging from 1 to 10 characters, but most of them have six to eight characters (see Table 4). Totally, $13,177$ characters in ChuNom are appearing in this part of the database, distributed in different frequency ranges as showing in Table 2. Similar to the handwriting dataset, this dataset also has a large amount of one-time-appearing characters, about 21.61% of the dataset's vocabulary. Therefore, we applied the same splitting mechanism as we have described previously in the handwriting dataset section, to split this dataset into two sets with its samples distributed as showing in Table 3.

Table 4. Sentence length count of the Synthetic Nom String dataset

	Sentence length									
	1	2	3	4	5	6	7	8	9	10
Sentences count	253	1652	1030	2087	7043	25631	31984	17041	8221	6679

5 Experiments and Baselines

In this work, we conducted several experiments for giving the baseline results for both the two parts of the database. Of course, the database can be used in many different tasks including densely text localization, start-of-column detection, natural language processing, etc. For this initial work, we focus on the handwriting recognition task, which is our main purpose at the database creation. We employed a Convolution Recurrent Neural Network (CRNN) [26] network with Connectionist Temporal Classification (CTC) Loss [4] for our experiments, which is the state of the art in handwritten recognition task [5,26,32].

Along with the original CRNN network, we also created a modification version of this network to make it works with RGB images in higher resolution (48 pixels height) instead of 32 pixels height, gray-scale as the original author [26] proposed. In particular, we set the first Convolution layer to have three input channels instead of just one channel as the original network. Also, we added a sequence of Convolution, Batch Normalization and Rectified Linear Unit (ReLU) layers on top of the CNN part to make sure the output channels have its height equivalent to 1, as this is the requirement to be able to map from CNN outputs to sequence of features [26]. The modified network is called M-CRNN with the network configurations as showing in Table 5. In this table, 'k', 's' and 'p' stand for kernel size, stride and padding size respectively. The modifications are in bold. To estimate the performance of the two networks in this task, we use the Accuracy value, Jaccard similarity [10] and Levenshtein distance [14], which are the usual metrics to estimate the similarity between two sentences. One can note that the Accuracy value is counted when the two candidate sentences (prediction and actual) are exactly match each other.

Table 5. M-CRNN Network configuration summary. The first row is the top layer.

	Type	Configurations
RNN	Transcription	–
	Bidirectional-LSTM	#hidden units: 256
	Bidirectional-LSTM	#hidden units: 256
	Map-to-Sequence	–
CNN	**BatchNorm + ReLU**	–
	Convolution	**#maps:512, k:2, s:1, p:0**
	BatchNorm + ReLU	–
	Convolution	#maps:512, k:2, s:1, p:0
	MaxPooling	k:2, s:2 × 1, p:0 × 1
	Convolution + ReLU	#maps:512, k:3, s:1, p:1
	BatchNorm + ReLU	–
	Convolution	#maps:512, k:3, s:1, p:1
	MaxPooling	k:2, s:2 × 1, p:0 × 1
	Convolution + ReLU	#maps:256, k:3, s:1, p:1
	BatchNorm + ReLU	–
	Convolution	#maps:256, k:3, s:1, p:1
	MaxPooling	k:2, s:2 × 2, p:0
	Convolution + ReLU	#maps:128, k:3, s:1, p:1
	MaxPooling	k:2 × 2, s:2, p:0
	Convolution + ReLU	**#maps:64, k:3, s:1, p:1**
	Input	W × 48 RGB image

Table 6. Experiment results on the Synthetic Nom String dataset. The best results are in bold.

Method	Accuracy	Jaccard similarity	Levenshtein distance
CRNN scratch	97.13%	99.41%	0.0033
M-CRNN scratch	**97.50%**	**99.49%**	**0.0029**

5.1 Experiments on the Synthetic Nom String Dataset

Firstly, we would like to test how the CRNN and M-CRNN networks will work with the logography texts written in vertical mode since the original network is designed for texts written in horizontal mode. In order to align with the requirement of the network, we have rotated all of our images by 90°, counter-clockwise. Although this looks strange for the human observer, we expect that the network will still be able to learn from the rotated characters and make the correct predictions. Table 6 presents the experimental results of two networks on this dataset. From this table, we can see that for both the two networks, we have reached > 97% accuracy at sentence level and > 99% accuracy at character level (Jaccard similarity). The ≈ 3% accuracy missed in the sentence level is probably belonging to the sentences which have one-time-appearing characters that we have mentioned earlier (the training set covers only 94.99% of the whole vocabulary appearing in the validation set, see Table 3). It also can be seen that our modified network M-CRNN performs just slightly better than the original one, this is expected since the characters and drawing in the images of this dataset are clear and easy to recognize.

In the next step, we would like to test these networks with the handwriting dataset when the images are already in a bad shape, to see more clearly the effects of using all three channels RGB in higher resolution compares to gray-scale in lower resolution image.

5.2 Experiments on the Handwriting Dataset

The task of text recognition in this dataset is more challenging compared to the Synthetic Nom String dataset, not only because the number of images is smaller but also because of the degraded in quality of the document and the variation in the way the characters are written. Therefore, this dataset could be a good benchmarking dataset to compare the performance of the different networks. In this section, we will go through step by step of training and evaluating for the two networks on this handwriting dataset.

Image Pre-processing. For the pre-processing step in this dataset, we rotate all of the images in this dataset by 90°, counter-clockwise as we have done with Synthetic Nom String dataset. Since our M-CRNN works with input image resolution of 48×432 pixels, we render these images into a 48×432 pixels background image to eliminate the variation of the image sizes in the dataset. The color of this background will be set with the most dominant color of the image. The main reason for using this operation instead of just rescaling is because these images are already in a bad shape, therefore, rescaling will further degrade the quality of these images.

Training and Evaluating. There are two types of training methods on this dataset, the first is training from scratch, which means we train a completely new model. The second is finetuning from the existing model that has been trained on the Synthetic Nom String dataset (see [29] for more information about finetuning).

Table 7. Experiment results on the handwriting dataset. Best results are in bold.

Method	Accuracy	Jaccard similarity	Levenshtein distance
CRNN scratch	0.1%	13.01%	0.8736
M-CRNN scratch	29.38%	75.83%	0.1591
CRNN finetune	37.00%	79.17%	0.1319
M-CRNN finetune	**42.70%**	**82.28%**	**0.1105**

From Table 7, we can see that our modified network M-CRNN outperforms the original CRNN network in all of our experiments, which proves that using all three channels RGB and higher resolution of the images have a high impact to the performance of the deep learning model for handwriting images. It is also possible that converting from the origin RGB of the old and degraded handwriting image to the gray-scale image ending of losing a lot of information, which lead to bad results when training the CRNN from scratch. Apart from that, it is also interesting to see that, by fine-tuning from the model pretrained with the Synthetic Nom String dataset, it further boosts our results even more. Although 42.7% of accuracy seems to be not very high, it is at sentence level, thus only being counted when predicting perfectly corrected all of its characters. In the case of old and degraded handwriting images, it is easy to understand the reason why the network misses a character or two in a sentence due to faded ink or stained paper. In the character level, the network is still able to reach 82.28% of correctness (Jaccard similarity). The Levenshtein distance reinforces this observation with its distance closes to 0.1. For the most part of the errors, it missed only one character, which can be seen in Fig. 5. A very limited number of samples has more than two errors for one prediction. Figure 6 shows an example of the prediction errors when evaluating a patch image in Tale-of-Kieu's book. It can be seen that in the cases (a) and (b), the prediction of the network is more closed to the shape of the character in the image than the actual ground truth. This probably happens because of losing some sign due to degraded image quality (case a) and the variation in the way the characters are written (case b). In the case of (c), it is interesting to see that the network is able to predict correctly, even when the character (姮) appears just one time in the whole training set.

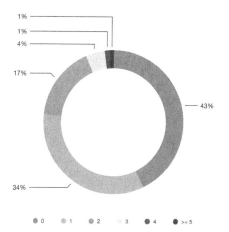

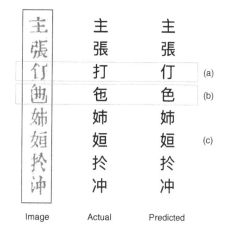

Fig. 5. Character error predictions distribution on the validation set.

Fig. 6. Example of the error predictions of the M-CRNN model.

To conclude these experiments, we could tell that our model is already predicting decent results in this handwriting dataset, and we can be confident with the correctness of the database. However, these is just baseline results, it will be interesting to see how far deep learning can reach on this task of handwriting recognition applying to the old and degraded images. Furthermore, we would like to expand the goal of handwriting recognition from a single patch image to the whole page image in future works.

6   Conclusion

In this paper, we have introduced a new handwriting dataset called IHR-NomDB for the old Vietnamese language. Over 260 pages from two poem books were collected and analyzed to generate more than 5000 patches of old and degraded handwritten images. Along with this dataset is a new Synthetic Nom String dataset, which consists of 101, 621 image generated from the bank of sentences we have collected. The releasing of this database is not only to impulse the development of new technologies in handwritten recognition, but also is a step towards the preservation of a thousand years old heritage, which is on the verge of being forgotten. Besides the releasing of the database, we have also given some results and baselines on the task of handwriting recognition for both two parts of the database. The database can be used for research tasks of handwritten recognition, handwritten document segmentation (vertical mode), start-of-column detection, document retrieval and natural language processing. All the materials are available to download on the website of the IHR-Nom project at https://morphoboid.labri.fr/ihr-nom.html.

Acknowledgements. The authors would like to thank the Vietnamese Nom Preservation Foundation (http://nomfoundation.org) for granting the authorization to access and collect the mentioned data for our analyzing and creating of the database.

References

1. Antonacopoulos, A., Downton, A.: Special issue on the analysis of historical documents. IJDAR **9**, 75–77 (2007)
2. Cam, B.: Nguon Goc Chu Nom, pp. 354–355. Van hoa nguyet san (1960)
3. Fischer, A., Indermühle, E., Bunke, H., Viehhauser, G., Stolz, M.: Ground truth creation for handwriting recognition in historical documents, pp. 3–10, January 2010
4. Graves, A., Fernández, S., Gomez, F., Schmidhuber, J.: Connectionist temporal classification: labelling unsegmented sequence data with recurrent neural networks. In: Proceedings of the 23rd International Conference on Machine Learning, pp. 369–376 (2006)
5. Graves, A., Liwicki, M., Fernández, S., Bertolami, R., Bunke, H., Schmidhuber, J.: A novel connectionist system for unconstrained handwriting recognition. IEEE Trans. Pattern Anal. Mach. Intell. **31**(5), 855–868 (2008)
6. Grother, P.: NIST special database 19 handprinted forms and characters database (1995)
7. Guyon, I., Schomaker, L., Plamondon, R., Liberman, M., Janet, S.: UNIPEN project of on-line data exchange and recognizer benchmarks. In: Proceedings of the 12th IAPR International Conference on Pattern Recognition, Vol. 3 - Conference C: Signal Processing (Cat. No. 94CH3440-5), vol. 2, pp. 29–33 (1994)
8. Hull, J.J.: A database for handwritten text recognition research. IEEE Trans. Pattern Anal. Mach. Intell. **16**(5), 550–554 (1994)
9. Impedovo, S., Wang, P.S.P., Bunke, H.: Automatic Bankcheck Processing, vol. 28. World Scientific, Singapore (1997)
10. Jaccard, P.: Etude de la distribution florale dans une portion des alpes et du jura. Bulletin de la Societe Vaudoise des Sciences Naturelles **37**, 547–579 (1901)
11. Jin, L., Gao, Y., Liu, G., Li, Y., Ding, K.: SCUT-COUCH2009-a comprehensive online unconstrained Chinese handwriting database and benchmark evaluation. IJDAR **14**, 53–64 (2011)
12. Khuê, N.: Chu Nôm: co so va nang cao, pp. 10–15
13. Kusetogullari, H., Yavariabdi, A., Cheddad, A., Grahn, H., Hall, J.: Ardis: a swedish historical handwritten digit dataset. Neural Comput. Appl. **32**, 1–14 (2019)
14. Levenshtein, V.I.: Binary codes capable of correcting deletions, insertions, and reversals. Soviet Physics Dokl **10**, 707–710 (1966)
15. Liu, C.L., Yin, F., Wang, D.H., Wang, Q.F.: CASIA online and offline Chinese handwriting databases, pp. 37–41 (2011)
16. Liwicki, M., Bunke, H.: IAM-OnDB - an on-line English sentence database acquired from handwritten text on a whiteboard. In: Eighth International Conference on Document Analysis and Recognition (ICDAR 2005), vol. 2, pp. 956–961 (2005)
17. Marti, U.V., Bunke, H.: The IAM-database: an English sentence database for offline handwriting recognition. Int. J. Doc. Anal. Recognit. **5**, 39–46 (2002)
18. Matsumoto, K., Fukushima, T., Nakagawa, M.: Collection and analysis of on-line handwritten Japanese character patterns. In: Proceedings of Sixth International Conference on Document Analysis and Recognition, pp. 496–500 (2001)

19. Nguyen, C.K., Nguyen, C.T., Masaki, N.: Tens of thousands of nom character recognition by deep convolution neural networks. In: Proceedings of the 4th International Workshop on Historical Document Imaging and Processing - HIP 2017, pp. 37–41. ACM Press (2017)

20. Pechwitz, M., Maddouri, S.S., Märgner, V., Ellouze, N., Amiri, H., et al.: IFN/ENIT-database of handwritten Arabic words. In: Proceedings of CIFED, vol. 2, pp. 127–136. Citeseer (2002)

21. Phan, T.V., Zhu, B., Nakagawa, M.: Collecting handwritten nom character patterns from historical document pages. In: 2012 10th IAPR International Workshop on Document Analysis Systems, pp. 344–348 (2012)

22. Rath, T., Manmatha, R.: Word spotting for historical documents. Int. J. Doc. Anal. Recognit. (IJDAR) 9(2), 139–152 (2006)

23. Sanchez, J.A., Romero, V., Toselli, A.H., Villegas, M., Vidal, E.: ICDAR 2017 competition on handwritten text recognition on the read dataset. In: 2017 14th IAPR International Conference on Document Analysis and Recognition (ICDAR), vol. 1, pp. 1383–1388. IEEE (2017)

24. Sanchez, J.A., Toselli, A.H., Romero, V., Vidal, E.: ICDAR 2015 competition HTRTS: handwritten text recognition on the transcriptorium dataset. In: 2015 13th International Conference on Document Analysis and Recognition (ICDAR), pp. 1166–1170. IEEE (2015)

25. Scius-Bertrand, A., Voegtlin, L., Alberti, M., Fischer, A., Bui, M.: Layout analysis and text column segmentation for historical Vietnamese steles. In: Proceedings of the 5th International Workshop on Historical Document Imaging and Processing, pp. 84–89 (2019)

26. Shi, B., Bai, X., Yao, C.: An end-to-end trainable neural network for image-based sequence recognition and its application to scene text recognition (2015)

27. Srihari, S.N., Shin, Y.-C., Ramanaprasad, V., Lee, D.-S.: A system to read names and addresses on tax forms. Proc. IEEE 84(7), 1038–1049 (1996)

28. Su, T., Zhang, T., Guan, D.: HIT-MW dataset for offline Chinese handwritten text recognition (2006)

29. Tan, C., Sun, F., Kong, T., Zhang, W., Yang, C., Liu, C.: A survey on deep transfer learning. In: Kůrková, V., Manolopoulos, Y., Hammer, B., Iliadis, L., Maglogiannis, I. (eds.) ICANN 2018. LNCS, vol. 11141, pp. 270–279. Springer, Cham (2018). https://doi.org/10.1007/978-3-030-01424-7_27

30. Van Phan, T., Cong Nguyen, K., Nakagawa, M.: A nom historical document recognition system for digital archiving. Int. J. Doc. Anal. Recognit. (IJDAR) 19(1), 49–64 (2016)

31. Viard-Gaudin, C., Lallican, P.M., Knerr, S., Binter, P.: The IRESTE on/off (IRONOFF) dual handwriting database. In: Proceedings of the Fifth International Conference on Document Analysis and Recognition, ICDAR 1999 (Cat. No. PR00318), pp. 455–458 (1999)

32. Wigington, C., Tensmeyer, C., Davis, B., Barrett, W., Price, B., Cohen, S.: Start, follow, read: end-to-end full-page handwriting recognition. In: Ferrari, V., Hebert, M., Sminchisescu, C., Weiss, Y. (eds.) ECCV 2018. LNCS, vol. 11210, pp. 372–388. Springer, Cham (2018). https://doi.org/10.1007/978-3-030-01231-1_23

33. Zhou, S., Chen, Q., Wang, X.: HIT-OR3C: an opening recognition corpus for Chinese characters, pp. 223–230 (2010)

Sequence Learning Model for Syllables Recognition Arranged in Two Dimensions

Valerii Dziubliuk[(✉)] [iD], Mykhailo Zlotnyk [iD], and Oleksandr Viatchaninov [iD]

Samsung R&D Institute Ukraine (SRK), 57, L'va Tolstogo Str., Kyiv 01032, Ukraine
{v.dzyublyuk,m.zlotnyk,o.viatchanin}@samsung.com

Abstract. The handwritten text recognition from images is a challenging task due to the unique features of human handwriting styles, numerous overlaps and interrupting characters. It is especially difficult for languages where syllables are written by alphabetic letters arranged in two dimensions. Sequence learning architectures have a lot of potential to be applied for solving this type of tasks because they can access global and local contextual information. In this paper, we propose a multi-dimensional sequence learning model for handwriting recognition with residual connections between and inside Separable Blocks and self-attention along horizontal and vertical image directions instead of recurrence. The performance of recurrent and attention-based Directional Blocks on synthetic multi-line MNIST-based datasets is explored. We generated such data to force the model to learn local and global context during recognition. It is shown that a pre-trained model on MNIST-based datasets along with a syllable decomposition can successfully tackle Hangul handwriting recognition. To the best of our knowledge, our approach surpasses state-of-the-art results by achieving accuracy of 97.90% on PE92 and 97.76% on SERI95 for the syllable recognition tasks.

Keywords: Transformer · Self-attention · Recurrent neural network · Synthetic dataset · Offline handwriting recognition

1 Introduction

The transcription of handwritten text images, namely offline handwriting recognition, of short notes (checklists, greeting cards), of notebook writings (ordinary user handwriting) is a desirable feature for many smartphone users [23]. Nevertheless, this task is still not fully solved due to various character deformations, skews, overlaps and interruptions.

Convolutional Neural Network (CNN) architectures have shown a good performance for the object recognition and detection tasks. However, in such architectures sharing of a contextual information is restricted due to fixed-sized receptive fields of convolutions. To tackle such a strict nature of convolutions, it was proposed to use a grid of receptive fields (from different layers/depths) or a pyramid pooling technique [13,16,17]. Another issue for CNN models is the

J. Lladós et al. (Eds.): ICDAR 2021, LNCS 12823, pp. 100–111, 2021.
https://doi.org/10.1007/978-3-030-86334-0_7

data invariance problem. A deformed data can be additionally rotated, translated and scaled. This is important for tasks in which objects can be overlapped or intercepted. In particular, it is crucial for handwriting recognition.

The aforementioned difficulties can be overcome by sequence learning models because of their ability to access a contextual information. Multi-Dimensional Long Short-Term Memory (MDLSTM) model for offline handwriting recognition represents one of them [4]. However, this model has two main drawbacks: poor parallel computations (a computation of each MDLSTM cell depends on output hidden states of top and left neighbors) and tough gradient propagation. A demand for faster training and inference gave MDLSTM a push to a series of breakthroughs in the optical text recognition field. On the one hand, there is an alternative to MDLSTM model which is based on a convolutional encoder and a bidirectional LSTM decoder predicting character sequences [2, 26]. On the other hand, there are several works which speed up MDLSTM [19, 25].

An attention mechanism [7, 20] for images allows to learn long-range dependencies for the semantic image segmentation and super-resolution tasks. It is also applicable to the offline handwriting recognition domain due to its ability to summarize, preserve and inter-connect different parts of script. A recent work [1] with an attention block on top of MDLSTM shows that it is possible to recognize whole text paragraphs without explicit text line segmentation. Such advantage of the attention mechanism for understanding multi-level structures seems promising for languages where syllables are written by letters arranged in two dimensions. Handwriting styles and diverse syllable composition rules of these languages often lead to ambiguities in recognition. Chinese, Japanese, Korean and Bengali languages (CJKB) are the hardest ones for understanding due to a large number of characters (letters) and their possible combinations.

Taking this into account, Korean and Bengali languages are chosen for handwriting recognition as benchmarks for the proposed sequence learning model, because of their different syllable composition rules. These tasks show how well the model tackles the complex context, given that a single syllable can drastically change its meaning by adding/removing/replacing only one stroke [9, 11]. A common approach for CJKB handwriting recognition is primarily based on the segmentation strategy with the following classification of syllables with DNN [9, 11, 14, 25]. In the last work [25] the separable MDLSTM model is applied to Chinese handwriting, however, authors don't inform how to apply their findings in case of new additional syllables. CJKB languages are extremely difficult for recognition mainly due to the large amount of syllables (order of thousands) and their complex composition rules. We handle these problems by decomposing each word on its underlying sequence of characters according to language-specific grammar rules. Then we train images against their decomposed characters' labeling. Such method is used for offline mathematical formulas recognition (e.g., the stochastic context-free grammar in [27].

Based on the architecture studies done in the works [24, 25], and taking into account the aforementioned advantages of the attention mechanism, in the current work we embed self-attention inside horizontal and vertical image passing, thus replacing recurrent connections. Thereby, the contribution of our paper is three-fold:

- We provided a series of experiments for dimensional and separable blocks in proposed architectures with and without recurrent connections to determine their ability to recognize deformed elements in the complex context.
- We showed that our model can be pre-trained on multi-level MNIST-based digit sequences in order to learn the language-specific composition rules.
- We proposed an extendable approach for the offline handwriting recognition task using syllable decomposition rules, which represent a 2D syllable in a 1D sequence of components in an unambiguous way, which greatly reduces amount of output classes.

2   Model Architecture

Designing the model, we assume that it is possible to represent a handwritten two-dimensional syllable as a one-dimensional sequence of components corresponding to the sounds which appear during the pronunciation of a given syllable. To achieve that, we utilized layers with the ability to aggregate global context in both horizontal and vertical dimensions. The overview of the proposed Separable Multi-Dimensional Sequence Learning Model is shown in Fig. 1.

Conv2D Block consists of two layers of convolutions (3×3 kernel size, 16 filters) with ReLU activation and max polling (2×2 pool size). They reduce dimensions of an input data and at the same time extract raw features. The goal of Separable Block is to capture 2D long-range dependencies by transforming multi-dimensional data into a flattened sequence of features.

Separable Block is built with horizontal and vertical Directional Blocks (denoted as Directional Block$_H$ and Directional Block$_V$, respectively). Based on the previous works [7,20,21,24,25] for sequence learning optimization Separable Block reduces computational cost and improves gradient propagation during training. Directional Blocks contain reshape operations (RO) and inner sequence processing. The purpose of reshaping is to provide an inner model the ability to process a horizontal/vertical sequence of vectors (image slices). At the same time, RO after Sequence Layer reverts tensor to its original view (shape). RO remain the same for all experiments. Inner Dense layers make dimension alignment applicable for the next operation. The proposed scheme allows the whole Separable Block to calculate features faster and takes into account a context information along selected dimensions. It is sufficient to apply sequentially two Directional Blocks with horizontal and vertical dimensions for routing out the image context.

Considering the large diversity of sequence models architectures (LSTM [6], GRU [3], Transformer [21], Reformer [12], and others) which perform well on different tasks, Sequence Layer inside Separable Block serves as a wrapper for any of them.

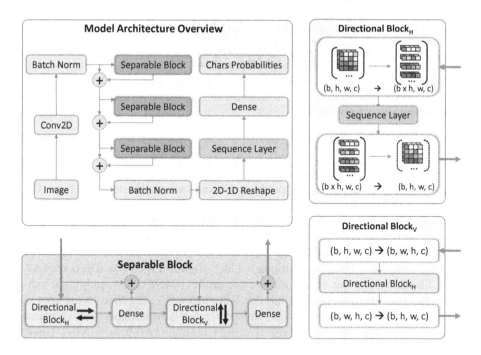

Fig. 1. Schematic overview of the proposed model.

In accordance with our assumption about syllable 2D-1D decomposition inputs should be transformed into 1D sequence at the end of Separable Blocks chain. Therefore, we transform a 2D feature map using 2D-1D Reshape layer into a new shape $(b, w, h \times c)$ and feed output to Sequence Layer (which is the same for Directional Blocks). The last one processes obtained features horizontally along w dimension. The Dense layer with Softmax is at the top of the model. The number of nodes for this last classification layer equals to the number of unique symbols according to a task. Considering advantages of residual connections for CNN models [22], we add inner (between Directional Blocks) and outer (between Separable Blocks) residual connections (Fig. 1). In the Experiments section it will be shown that such modification improves the model generalization and convergence.

3   Datasets

3.1   Synthetic MNIST-based Dataset

To investigate the model's ability to recognize complex multi-line structures we prepared synthetic data based on MNIST dataset[1]. The goal here is to check how well designed models take into account a digit's neighboring context, given

---

[1] http://yann.lecun.com/exdb/mnist/.

Fig. 2. MNIST samples processing . Rows illustrate original MNIST digits, transformed digits and cut digits respectively.

that each digit can be classified with a high accuracy. Two types of images were generated: three-lined sequences with arbitrary lengths and 3×3 rotated blocks composed of digit characters. A performance on such synthetic data shows the model's capability to learn block decomposition rules in the case of a deformed neighborhood. Moreover, such synthetic generation procedure (left-right and top-bottom) partially corresponds to CJKB syllable composition rules. Train and test datasets were generated from the default MNIST train and test split (60 000 train samples, 10 000 test samples). The generation took place in consistent steps. The first step was to preprocess the original MNIST by applying a random perspective transformation and cut boundaries (Fig. 2). Next, sequences of random digits of a given length were shifted randomly in horizontal and vertical directions in a range of $[-5, 5]$ pixels. During training, three-lined MNIST-based ordinary and rotated sequences below are obtained from digit images with cut boundaries and random perspective transformations applied. During the test phase, we used a test part of original MNIST database for generation.

We could use a well-known handwritten IAM dataset [8] for synthetic pre-training, however, we choose MNIST because:

- It provides us independent control over each character in terms of width, height, inclination angle, overlapping;
- Training on MNIST is faster than on IAM (10 vs 26 classes).

Three-Lined MNIST-based Sequences. Three-lined random sequences were composed of lines generated as above with possible vertical intersection up to 4 pixels. The arbitrary length of each line was changed from 1 to 9 digits (Fig. 3 a). Generated images were placed onto a black background of size 84×504 pixels.

Rotated Three-Lined MNIST-based Sequences. We generated three-lined sequences with 3 digits in each line. Then we rotated whole image on a random angle in the range from 0 to $270°$ (Fig. 3 b). Generated images were placed onto a black background of size 112×504 pixels. For both types of synthetic images, labels were formed as a continuous sequence of digits from left to right and from top to bottom. We didn't use any separators inside the labels.

Fig. 3. Examples of generated samples: (a) three-lined MNIST-based digits, (b) rotated block of MNIST digits, (c) single-lined Hangul syllables

3.2   Handwritten Hangul Dataset

Hangul Syllable Dataset. We selected handwritten Hangul datasets PE92 and SERI95[2] to test our approach. PE92 has 2350 classes of syllables, each of which contains about 100 samples. PE92 is split into 199208 train and 23387 test samples. SERI95 has 520 of the most frequently used Hangul syllables, each of which contains about 1000 samples. SERI95 is split into 433575 train and 51494 test samples. Each Hangul syllable from the train split was passed through the same random perspective transformation as MNIST digits. Then it was resized and zero-padded to a 64x64 pixels image. The dataset labels were also changed. Each syllable was decomposed on Hangul characters (i.e., 탈 → ㅌ, ㅏ, ㄹ) in accordance with Unicode syllable specification using *jamotools* library[3]. Since the decomposition/merging rules are unique for each syllable, we were able to drastically reduce the number of output classes (59 unique chars for SERI95 and 67 unique chars for PE92).

Synthetic Single-Line Hangul Dataset. Similar to the synthetic MNIST dataset, we created random sequences of Hangul syllables from concatenated SERI95 and PE92 train splits. The lengths of sequences vary from 1 to 7 syllables or spaces (Fig. 3 c). Generated images were placed onto a 64×504 pixels patch with a black background. *Jamotools* decomposition rules were applied for sequences of syllables. At the end of each syllable, we added a special <END> character.

3.3   Handwritten Bengali Dataset

We used a handwritten Bengali dataset from Kaggle[4]. This language has 49 letters (11 vowels, 38 consonants) and 18 possible diacritics (11 vowel and 7 consonant), or accents. 168 Bengali grapheme roots are composed from 49 letters. Single grapheme consists of 3 components: root, vowel and consonant diacritics. An example of the grapheme composition is ড + ে + ৗ = ড়ৈ . Therefore, there are around 13,000 different grapheme classes. The dataset contains 1295 most frequent Bengali graphemes.

---

[2] https://github.com/callee2006/HangulDB.

[3] https://pypi.org/project/jamotools/.

[4] https://www.kaggle.com/c/bengaliai-cv19/data.

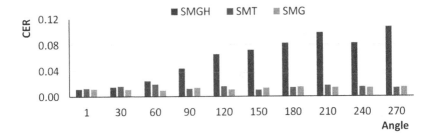

Fig. 4. CER on rotated three-lined MNIST synthetic dataset.

4 Experiments

4.1 Experimental Setup

We equipped the proposed model on Fig. 1 with three different types of Sequence Layers. The first one was based on a bidirectional GRU [3] with 256 hidden units (SMG). The second one was based on the encoder part of a Transformer [21] with the depth 128 and 4 heads (SMT). Finally, the third model was one-dimensional, so that each Separable Block includes only horizontal Directional Blocks with bidirectional GRUs with the same weights as in the SMG model (SMGH). The last model gives an understanding which task type can be solved without applying vertical context coupling. The number of trainable parameters for SMT model is 2,950,734, while SMG and SMGH each has 3,036,803. The logic behind model parameters choosing is based on the restriction of mobile devices inference speed.

We used a cross-validation (5% of train, 5 folds) for model hyperparameters tuning. All models' weights were initialized with He Normal initialization [5]. CTC loss function [4] and Adam optimizer with cycle learning rate [18] (ranging from 3×10^{-4} to 3×10^{-7}) were used for each run. The batch size during training was 64. L2 regularization coefficient was 1×10^{-3} for SMG/SMGH and 1×10^{-5} for SMT. All these hyperparameters are selected the same for all experiments.

As a main quality metric, we adopted Correct Error Rate (CER) [15] for the Hangul handwriting recognition task and synthetic MNIST-based data. Evaluation on Hangul and Bengali datasets additionally employed the accuracy metric used in [9]. The Bengali handwriting recognition task was evaluated with HMAR (a Hierarchical Macro-Averaged Recall)[5] metric in order to compare results with the Kaggle leaderboard.

4.2 Synthetic MNIST-based Recognition

The models' training was performed with a gradually increasing complexity. For three-lined images, we increased the maximum length of arbitrary sequence from

---

[5] https://www.kaggle.com/c/bengaliai-cv19/overview/evaluation.

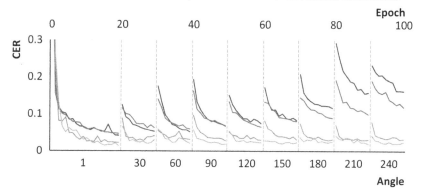

Fig. 5. The chart demonstrates changing of CER on a three-lined MNIST dataset during the training for different configurations of residual connection for SMT model. The training mode gradually increased the complexity by incrementing of maximum possible sample's angle of rotation after some number of epochs passed. The peaks on the chart correspond to hopping of a maximum rotation angle.

3 to 9. For rotated images the maximum rotation angle was incremented from 0 to 270 degrees with 30 degrees step.

Training results on three-lined images are approximately the same for each type of the model (± 0.01 CER). They indicate that in the case of a linear global horizontal context and a moderate vertical one, it is sufficient to use one-dimensional sequence models. Receptive fields of convolutional layers are enough for representation learning. This confirms the results acquired in other papers, where a convolutional encoder and a bidirectional LSTM decoder achieved state-of-the-art results for English handwriting recognition [2].

On the other hand, experiments made with rotated images show that the models with vertical Directional Blocks (SMG, SMT) are more accurate (Fig. 4).

Starting from the angle of 60 degrees, the accuracy of SMGH model decreases. This is especially well observed on the images which include handwriting digits 2, 5, 6, 7, or 9. Such types of sequences can not be properly recognized without an understanding of the neighboring context. This proves the fact that in the case of a diverse global context it is required to use multi-dimensional approaches to achieve better results.

Investigation of the effectiveness of residual connections was done for the SMT model using four types of connections. These modifications were trained on rotated MNIST-based images. The model with inner and outer residual connections outperform others (Fig. 5). The main benefits are faster convergence and better generalization for a larger rotation angle.

4.3 Hangul Recognition

The SMT and SMG architectures were chosen for the Hangul syllable recognition task. We compared both architectures considering different training modes. In the first mode trainable parameters were initialized with the He Normal distribution and model achieved 7% of CER. In the second mode we used pre-trained weights obtained on the rotated MNIST-based dataset, and only the last Dense layer was reinitialized.

Evaluation in Table 1 shows that the SMT model outperforms results obtained in [10] on both Hangul test splits. For synthetic single-lined Hangul images, we got 0.04% CER for SMT and 0.06% CER for SMG. The average CER on validation sets for Hangul models is −0.36% (better) comparing with test splits. However, as it was shown in [23] the accuracy of line recognition can be improved by applying a language modeling. During both experiments on Hangul datasets the SMG model has demonstrated worse results compared to SMT. This could be an evidence that recurrent connections have poorer ability for understanding the context than attention-based architectures. Surely, this phenomenon needs to be investigated more thoroughly. The similar accuracy on the MNIST-based datasets and outperforming results of the SMT model on Hangul dataset show that the model with self-attention better understands decomposition rules of multi-level, more complex character blocks. Despite that, the SMT model's accuracy drops down in the case when an input image length doesn't coincide with the training one. The SMG model works well for different input widths.

Table 1. Recognition results for Hangul syllables datasets

	PE92		SERI95	
Model	Accuracy	CER	Accuracy	CER
SMG	97.70%	0.91%	97.18%	1.20%
SMT	97.90%	0.82 %	97.76%	0.98%
SOTA [10]	96.34%	-	97.67%	-

4.4 Bengali Recognition

Similar to the Hangul evaluation above, the SMT model was chosen for Bengali graphemes recognition. Each grapheme is represented as Unicode characters sequence of length from 2 to 8. In such way we represent each 2D Bengali grapheme as a 1D sequence. As a result, we got only 62 unique classes. During the training from scratch with He Normal initialization we got 81.12% accuracy, while pre-trained on rotated MNIST-based images model gave us 84.67% of accuracy. These results correspond to 0.9 and 0.92 HMAR metric respectively, while a state-of-the-art method achieved 0.9762 on Kaggle competition.

5   Conclusions

In the presented paper it is shown that recurrent connections in a separable multi-dimensional sequence learning model can be replaced by the self-attention mechanism without losing accuracy for offline handwriting recognition, where syllables are written by characters arranged in two dimensions. It is shown that adding inner and outer residual connections can speed up convergence during training. Wherein vertical context coupling increases the accuracy for tasks where it is necessary to understand a diverse directional scope.

We show that it is possible to pre-train the model on synthetics so that it can capture the syllabic structure (form). A model pre-trained on synthetic multi-line MNIST-based data boosts the accuracy for the Hangul handwriting recognition task from 81.95% to 97.9% comparing to training from scratch. However, Bengali syllable (grapheme) composition rules are different from the proposed ones during a synthetic generation. Hence, increase in accuracy is not so significant (84.67% vs 81.12%). Such results point out that each language requires the most appropriate composition rules for a pre-training stage. This statement can be verified in future research.

We propose to consider language-specific composition (writing) rules during the handwriting recognition model design in order to reduce the number of output classes (67 for Hangul and 62 for Bengali) at the top layer of the network. This especially concerns CJKB group of languages, for which common approaches have thousands of output classes.

Without any optimization (i.e., quantization, pruning) during freezing routines, the SMT model trained on synthetic Hangul lines performs with a speed of 500 ms per entire line of syllables on the current flagship smartphone. We are going to extend our results on mathematical expression recognition domain in the nearest future since math formulas allow labeling with LaTeX decomposition into math symbols.

References

1. Bluche, T., Louradour, J., Messina, R.: Scan, attend and read: end-to-end handwritten paragraph recognition with mdlstm attention. In: 2017 14th IAPR International Conference on Document Analysis and Recognition (ICDAR), vol. 01, pp. 1050–1055 (2017). https://doi.org/10.1109/ICDAR.2017.174
2. Bluche, T., Messina, R.: Gated convolutional recurrent neural networks for multilingual handwriting recognition. In: 2017 14th IAPR International Conference on Document Analysis and Recognition (ICDAR), vol. 01, pp. 646–651 (2017)
3. Chung, J., Gulcehre, C., Cho, K., Bengio, Y.: Empirical evaluation of gated recurrent neural networks on sequence modeling. In: NIPS 2014 Workshop on Deep Learning, December 2014 (2014)
4. Graves, A., Schmidhuber, J.: Offline handwriting recognition with multidimensional recurrent neural networks. In: Proceedings of the 21st International Conference on Neural Information Processing Systems, NIPS 2008, pp. 545–552. Curran Associates Inc., Red Hook (2008)

5. He, K., Zhang, X., Ren, S., Sun, J.: Delving deep into rectifiers: surpassing human-level performance on imagenet classification. In: Proceedings of the 2015 IEEE International Conference on Computer Vision (ICCV), ICCV 2015, pp. 1026–1034. IEEE Computer Society (2015). https://doi.org/10.1109/ICCV.2015.123

6. Hochreiter, S., Schmidhuber, J.: Long short-term memory. Neural Comput. **9**(8), 1735–1780 (1997). https://doi.org/10.1162/neco.1997.9.8.1735

7. Huang, Z., Wang, X., Huang, L., Huang, C., Wei, Y., Liu, W.: Ccnet: criss-cross attention for semantic segmentation. In: 2019 IEEE/CVF International Conference on Computer Vision (ICCV), pp. 603–612 (2019)

8. Indermühle, E., Liwicki, M., Bunke, H.: Iamondo-database: an online handwritten document database with non-uniform contents. In: Doermann, D.S., Govindaraju, V., Lopresti, D.P., Natarajan, P. (eds.) The Ninth IAPR International Workshop on Document Analysis Systems, DAS 2010, 9–11 June 2010, Boston, Massachusetts, USA, pp. 97–104. ACM International Conference Proceeding Series, ACM (2010). https://doi.org/10.1145/1815330.1815343

9. Kang, W., Park, K., Zhang, B.: Extremely sparse deep learning using inception modules with dropfilters. In: 2017 14th IAPR International Conference on Document Analysis and Recognition (ICDAR), vol. 01, pp. 448–453 (2017)

10. Kim, I.J., Choi, C., Lee, S.H.: Improving discrimination ability of convolutional neural networks by hybrid learning. Int. J. Doc. Anal. Recogn. (IJDAR) **19**(1), 1–9 (2016). https://doi.org/10.1007/s10032-015-0256-9

11. Kim, I.J., Xie, X.: Handwritten hangul recognition using deep convolutional neural networks. Int. J. Doc. Anal. Recogn. (IJDAR) **18**(1), 1–13 (2015). https://doi.org/10.1007/s10032-014-0229-4

12. Kitaev, N., Kaiser, L., Levskaya, A.: Reformer: the efficient transformer. In: International Conference on Learning Representations (2020). https://openreview.net/forum?id=rkgNKkHtvB

13. Liu, W., et al.: SSD: single shot multibox detector. In: Leibe, Bastian, Matas, Jiri, Sebe, Nicu, Welling, Max (eds.) ECCV 2016. LNCS, vol. 9905, pp. 21–37. Springer, Cham (2016). https://doi.org/10.1007/978-3-319-46448-0_2

14. Purnamawati, S., Rachmawati, D., Lumanauw, G., Rahmat, R.F., Taqyuddin, R.: Korean letter handwritten recognition using deep convolutional neural network on android platform. In: Journal of Physics: Conference Series, vol. 978, pp. 012112 (2018). https://doi.org/10.1088/1742-6596/978/1/012112

15. Read, J., Mazzone, E., Horton, M.: Recognition errors and recognizing errors - children writing on the tablet pc. In: Costabile, M.F., Paternò, F. (eds.) Human-Computer Interaction - INTERACT 2005, pp. 1096–1099. Springer, Heidelberg (2005)

16. Ren, S., He, K., Girshick, R., Sun, J.: Faster R-CNN: towards real-time object detection with region proposal networks. In: Cortes, C., Lawrence, N.D., Lee, D.D., Sugiyama, M., Garnett, R. (eds.) Advances in Neural Information Processing Systems, vol. 28, pp. 91–99. Curran Associates, Inc. (2015). http://papers.nips.cc/paper/5638-faster-r-cnn-towards-real-time-object-detection-with-region-proposal-networks.pdf

17. Ronneberger, O., Fischer, P., Brox, T.: U-Net: convolutional networks for biomedical image segmentation. In: Navab, N., Hornegger, J., Wells, W.M., Frangi, A.F. (eds.) MICCAI 2015. LNCS, vol. 9351, pp. 234–241. Springer, Cham (2015). https://doi.org/10.1007/978-3-319-24574-4_28

18. Smith, L.N.: No more pesky learning rate guessing games. CoRR abs/1506.01186 (2015). http://arxiv.org/abs/1506.01186

19. Van Den Oord, A., Kalchbrenner, N., Kavukcuoglu, K.: Pixel recurrent neural networks. In: Proceedings of the 33rd International Conference on International Conference on Machine Learning ICML 2016, vol. 48, pp. 1747–1756. JMLR.org (2016)
20. Vaswani, A., Parmar, N., Uszkoreit, J., Shazeer, N., Kaiser, L.: Image transformer (2018). https://openreview.net/forum?id=r16Vyf-0-
21. Vaswani, A., et al.: Attention is all you need. In: Guyon, I., (eds.) et al. Advances in Neural Information Processing Systems, vol. 30, pp. 5998–6008. Curran Associates, Inc. (2017). http://papers.nips.cc/paper/7181-attention-is-all-you-need.pdf
22. Veit, A., Wilber, M., Belongie, S.: Residual networks behave like ensembles of relatively shallow networks. In: Proceedings of the 30th International Conference on Neural Information Processing Systems, NIPS 2016, pp. 550–558. Curran Associates Inc., Red Hook (2016)
23. Viatchaninov, O., Dziubliuk, V., Radyvonenko, O., Yakishyn, Y., Zlotnyk, M.: Calliscan: on-device privacy-preserving image-based handwritten text recognition with visual hints. In: The Adjunct Publication of the 32nd Annual ACM Symposium on User Interface Software and Technology, UIST 2019, pp. 72–74. Association for Computing Machinery, New York (2019). https://doi.org/10.1145/3332167.3357119
24. Visin, F., Kastner, K., Cho, K., Matteucci, M., Courville, A.C., Bengio, Y.: Renet: a recurrent neural network based alternative to convolutional networks. CoRR abs/1505.00393 (2015). http://dblp.uni-trier.de/db/journals/corr/corr1505.html
25. Wu, Y., Yin, F., Chen, Z., Liu, C.: Handwritten chinese text recognition using separable multi-dimensional recurrent neural network. In: 2017 14th IAPR International Conference on Document Analysis and Recognition (ICDAR), vol. 01, pp. 79–84 (2017). https://doi.org/10.1109/ICDAR.2017.22
26. Xiao, S., Peng, L., Yan, R., Wang, S.: Deep network with pixel-level rectification and robust training for handwriting recognition. SN Comput. Sci $\mathbf{1}$(3), 145 (2020). https://doi.org/10.1007/s42979-020-00133-y
27. Zhang, J., Du, J., Zhang, S., Liu, D., Hu, Y., Hu, J., Wei, S., Dai, L.: Watch, attend and parse: an end-to-end neural network based approach to handwritten mathematical expression recognition. Pattern Recogn. $\mathbf{71}$, 196–206 (2017). https://doi.org/10.1016/j.patcog.2017.06.017

Transformer for Handwritten Text Recognition Using Bidirectional Post-decoding

Christoph Wick[1(✉)] [ID], Jochen Zöllner[1,2] [ID], and Tobias Grüning[1] [ID]

[1] Planet AI GmbH, Warnowufer 60, 18057 Rostock, Germany
{christoph.wick,jochen.zollner,tobias.gruning}@planet-ai.de
[2] Computational Intelligence Technology Lab, Department of Mathematics,
University of Rostock, 18051 Rostock, Germany
jochen.zollner@uni-rostock.de

Abstract. Most recently, Transformers – which are recurrent-free neural network architectures – achieved tremendous performances on various Natural Language Processing (NLP) tasks. Since Transformers represent a traditional Sequence-To-Sequence (S2S)-approach they can be used for several different tasks such as Handwritten Text Recognition (HTR). In this paper, we propose a bidirectional Transformer architecture for line-based HTR that is composed of a Convolutional Neural Network (CNN) for feature extraction and a Transformer-based encoder/decoder, whereby the decoding is performed in reading-order direction and reversed. A voter combines the two predicted sequences to obtain a single result. Our network performed worse compared to a traditional Connectionist Temporal Classification (CTC) approach on the IAM-dataset but reduced the state-of-the-art of Transformers-based approaches by about 25% without using additional data. On a significantly larger dataset, the proposed Transformer significantly outperformed our reference model by about 26%. In an error analysis, we show that the Transformer is able to learn a strong language model which explains why a larger training dataset is required to outperform traditional approaches and discuss why Transformers should be used with caution for HTR due to several shortcomings such as repetitions in the text.

Keywords: Handwritten Text Recognition · Transformer · Bidirectional

1 Introduction

Even though there was great progress, Handwritten Text Recognition (HTR) is still challenging in general, mainly due to the inherent differences in the writing of individuals and the vague nature of handwritten characters. As a result, single strokes can be ambiguous which is why it is difficult to match strokes to individual characters within a word. Language models help to decipher the writing and consistently yield better recognition rates. Hence, the currently best HTR

J. Lladós et al. (Eds.): ICDAR 2021, LNCS 12823, pp. 112–126, 2021.
https://doi.org/10.1007/978-3-030-86334-0_8

systems combine different approaches from Computer Vision (CV) and Natural Language Processing (NLP).

The current standard in the field of CV is the application of Convolutional Neural Networks (CNNs) and Recurrent Neural Networks (RNNs) which are trained using the Connectionist Temporal Classification (CTC)-loss as objective function. The CTC-algorithm is subject to inherit limitations: strict monotonic input-output alignments and an output sequence length that is bound by the possibly subsampled input length. These limitations, however, hold for HTR which is why this approach is the most popular one. An alternative is the application of Sequence-To-Sequence (S2S) models which are not subject to these limitation by using an encoder-decoder framework [21]. These networks are often combined with attention mechanisms [3] to focus on the most relevant features of the input line. Moreover, this approach intrinsically enables to learn a language model (see, e.g., [7]), but are in most cases still improved by applying an additional statistical language model in a postprocessing step [22].

A most recent approach of Kang et al. [11] applies the Transformer network architecture [23] to HTR and yielded state-of-the-art results. Similar to S2S-approaches, Transformers are based on an encoder-decoder framework, but any traditional RNN is replaced with attention modules. Note that the decoder is still comparable to a RNN in general because, at least during inference, the previous output token is required to predict the next one which forms a recurrent structure. In this paper, we extend the approach of Kang et al. [11] by proposing a Transformer using bidirectional post-decoding (bidirectional Transformer): the forward branch is trained to predict the sequence in order, while the backward branch is processing the sequence in reversed order. A voting mechanism combines the decoded sequences of possibly different length. We evaluated our model on the small IAM-database and on a larger dataset composed of a variety of different data on which the proposed method outperformed a reference model by a significant amount of up to 26%. We observed that the backward in contrast to forward decoding performed equally well. An error analysis revealed that the Transformer was indeed able to learn an intrinsic language model which led to the observed improvements but also showed dangerous shortcomings.

2   Related Work

The fundamental challenge of HTR is to find an alignment of two sequences, the line image as input and the character sequence as output, with arbitrary lengths. To solve this problem, currently two main approaches are established: CTC- and S2S-based models. In the following, we briefly list related work of these two approaches, as well as of more recently proposed recurrent-free approaches such as the Transformers. Finally, related work dealing with voting is mentioned.

2.1   CTC-Based

CTC-based approaches are the most popular ones in HTR (see, e.g., [9,17]) but also for Optical Character Recognition (OCR) in general, e.g., on contemporary or historical prints (see, e.g., [6,26]). The advantage of these approaches is that no alignment information is required during training which simplifies the generation of Ground Truth (GT). Instead, the CTC-loss optimises the log-probability of all possible alignments using a sophisticated forward-backward-algorithm. Most researchers applied (bidirectional) Long-Short-Term-Memory-Cell (LSTM) networks to model the recurrent nature of the sequence. But also Multidimensional LSTMs (MDLSTMs) [10] were proposed which exploit the two-dimensional nature of handwritten images (see, e.g., [16,24]). In almost any of those recent networks, CNNs were inserted before the RNN to extract low-level visual features. This paper uses a state-of-the-art CNN/LSTM-network trained with the CTC-loss as reference.

2.2   S2S-Based

The main idea of S2S architectures is to decouple feature extracting and decoding: first the encoder builds a feature representation of the input sequence, then a decoder emits the output tokens. Current approaches rely on attention mechanisms to allow the network to access the full encoder features during decoding. In [15], Michael et al. summarised and compared several different attention mechanisms on three different HTR datasets.

An attempt to include bidirectional encoding and decoding was proposed by Al-Sabahi et al. [2]. To solve the alignment problem during decoding, they proposed a sophisticated bidirectional beam search: first the backward direction was decoded unidirectional, then the decoding of the forward pass included the information of the backward pass to combine both information.

Dotsch et al. [8] presented a bidirectional decoder for HTR. To solve the alignment problem, another network was introduced to predict the length of the decoded sequence which was then used during decoding. This setup can however lead to errors in the alignment if shifts (e.g., by insertions and deletion) occur.

2.3   Recurrent-Free

Transformer networks were initially introduced by Vaswani et al. [23] which were then successfully applied on a variety of different tasks and datasets. Their idea is to replace any RNN with attention blocks, resulting in a recurrent-free network. Kang et al. [11] were the first to apply transformers using an encoder-decoder framework for HTR. They report overwhelming results on the popular IAM dataset [14] by reducing the Character Error Rate (CER) on the test set to about 4.67% compared to 5.80% [17] and 5.70% [9] if open vocabulary is allowed and if synthetic text lines were included. Without artificially increasing the data, they reached a CER of 7.62%. For comparison, we also report our results on the IAM-dataset (see Sect. 5.1).

Another approach to omit RNNs was published by Yousef et al. [28] who proposed a fully-convolution network trained with the CTC-loss. Experiments yielded competitive results on several tasks and datasets.

2.4   Voting

Since the proposed forward and backward Transformers can produce sequences with different lengths, a sophisticated voting mechanism is required to obtain a single result. Voting is a common technique utilised in text recognition to improve the result of several (independent) transcriptions that can be obtained by different engines [1,5], repeated scans and readings of the same material [12], different editions of the same book [25], or different partitions for training and validation of the same material [18]. The first technique for voting computed the Longest Common Substring (LCS) [19] of multiple sequences and applied majority voting to resolve conflicts and heuristics to break ties. [5] applied a progressive alignment which starts with the two most similar sequences and extends the alignment by adding sequences. Then, the character selection is performed by a Naive Bayes classifier. Azawi et al. [1] introduced weighted finite-state transducers based on edit rules to align the output of two different OCR engines, then, LSTM networks trained on the aligned outputs are used to return a best voting. The voting mechanism of Reul et al. [18] relies on the computation of the LCS but incorporates the confidence of the network to resolve conflicts which results in better results than majority voting.

In this paper, we vote similar to [18]: our two sequences are matched, then conflicts are resolved using different heuristics (see Sect. 3.6).

3   Proposed Methodology

In this section, we describe our proposed generic Transformer (see Fig. 1): a preprocessed image is fed into a CNN to extract visual features and then into the encoder consisting of several attention modules which compute a new feature representation. This allows to learn long range dependencies of the visual features of the CNN. The decoder predicts a sequence until an End-of-Sequence ($\langle$eos$\rangle$) token is reached using the encoded features and the previously predicted token.

3.1   Preprocessing

Preprocessing of the text line images includes contrast normalisation without binarisation of the image, skew correction, and normalisation of slant and height. As last step, all images are scaled to a fixed height of 64 pixels while maintaining their aspect ratio. This ensures that all vectors in the feature sequence conform to the same dimension without putting any restrictions on the sequence length. To artificially increase the amount of training data, we augment the existing preprocessed images by applying minor disturbances to the statistics relevant for the normalisation algorithms. To simulate naturally occurring variations in handwritten text line images, we combine dilation, erosion, and grid-like distortions [27]. These methods are applied to the preprocessed images randomly each with an independent probability of 50%.

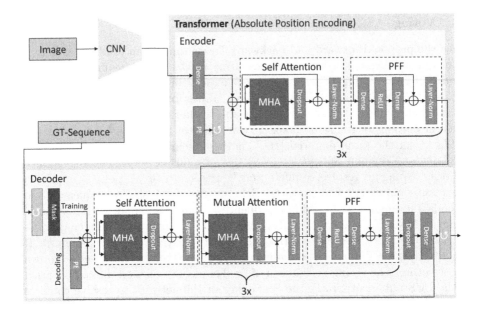

Fig. 1. Architecture of the complete network including the CNN and Transformer using absolute positional encoding. See Fig. 2 for a deeper explanation of the CNN. The transformer consists of an encoder and a decoder. The yellow boxes indicate a temporal reversion applied only in the backwards-transformer. (Color figure online)

3.2   CNN Feature Encoder

The CNN consists of three convolutional and two interleaved max-pooling layers (see Fig. 2) to extract meaningful local visual features. Hereby, the spatial dimensions of the input are reduced while simultaneously increasing the representative depth of the feature vectors. Each convolutional layer uses leaky rectified linear units [13] as activation functions.

3.3   Encoder/Decoder

The schema of the encoder using absolute positional encoding is drawn in the upper half of Fig. 1. It first applies a dense layer on the output of the CNN and optionally adds the positional encodings. The values are fed into three consecutive self-attention and Pointwise Feed-Forward (PFF) blocks. The self-attention block applies a Multi-Head-Attention (MHA) operation followed by a dropout layer, a residual (addition of the previous input), and a layer norm. The PFF comprises two fully connected layers including layer-normalisation, a residual, and a Rectified Linear Unit (ReLU)-activation.

MHA requires three inputs, a query Q, key K, and value V. The output Y is computed as

$$Y_i = \text{Softmax}\left(\frac{Q'_i \cdot K'}{\sqrt{f}}\right) \cdot V' ,$$

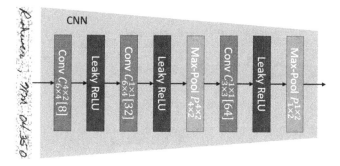

Fig. 2. The CNN architecture used as feature extractor within our proposed Transformer and the reference CNN/LSTM-network. The sub- and superscript in $C/P_{k_y \times k_x}^{s_y \times s_x}$ describe the size of the kernel k and strides s along both dimensions (width x, height y). The depth of the feature map is shown in square brackets, that is, the number of filters for a convolutional layer. The total subsampling factor is $2 \cdot 1 \cdot 2 \cdot 1 \cdot 2 = 8$.

where f is the number of features, i is the i-th entry of the each matrix, and Q', K', V' are obtained by three independent linear transformations using the matrices W_Q, W_K, W_V. Finally, several so-called attention heads using different transformations are computed in parallel and concatenated afterwards. In a self-attention module, all inputs are identical: $Q \equiv K \equiv V$. Due to the MHA block, the encoder is able to see the complete sequence which allows the encoder to learn long range dependencies of the visual features.

The decoder which is shown in the lower half of Fig. 1 comprises two blocks. The self-attention block applies MHA similar to the encoder but on the encoded tokens. This can be understood as learning a language model. The decoder introduces a mutual attention block to combine the language features with the visual ones obtained by the encoder: Q and V come from the encoder, wheres K is the output of the previous module. Three self and mutual attention and PFF blocks are concatenated in total.

Eventually, a dropout and dense layer is appended to obtain a confidence map for each character of the predicted sequence.

3.4   Positional Encoding

Positional encodings are required to inject information about the relative or absolute position of the features or tokens in the sequence. Our Transformer always uses one of the following two approaches: first, absolute positional encodings [23] based on fixed sine and cosine functions of different frequencies are added before the self-attention modules, both in the encoder and decoder (see Fig. 1). In a second approach, we apply relative encodings within the MHA modules as proposed by Shaw et al. [20]. Absolute encodings are easy to use and do not require additional weights in the network, however perform poorly on long sequences especially if not occurring in the training data. By contrast, relative encodings must be learned but extend to sequences with arbitrary lengths.

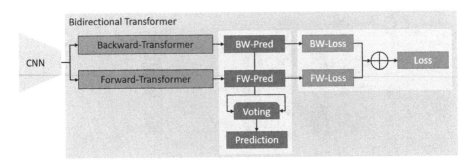

Fig. 3. Overview of our proposed bidirectional Transformer using the *Same CNN*. The backward module equates the one shown in Fig. 1 adding all reverse layers.

3.5 Teacher Forcing

During training, we apply teacher forcing, that is, we feed in the target token instead of the last output of the decoder. This ensures that the decoder sees the correct character of the previous timestep. Nevertheless, we must ensure that the decoder resumes correctly after a wrong token was predicted. Thereto, we added noise to the teacher forcing by replacing a fraction r_{tf} of the tokens with random ones (excluding Start-of-Sequence ($\langle sos \rangle$) and $\langle eos \rangle$). The ratio r_{tf} is initially chosen as 50% which forces the network to focus on visual features. To allow the learning of a language model, r_{tf} is exponentially decayed during training with an exponent of 0.98 resulting in a ratio close to zero at the end of training.

3.6 Bidirectional Transformer

Traditional Transformers and S2S models decode the input in reading order to mimic human behaviour. Machines do not know about our natural reading order which is why processing a line in reversed order is feasible. Since the machine is trained from scratch, i.e., without prior knowledge of the human reading order, reading reverse is as difficult for the machine as reading normal. In this work, we propose bidirectional Transformers which are able to include two-way decoded information by voting the results of a forward and backward pass.

We propose two approaches: first, two individual transformers are trained, whereby one encodes and decodes in reversed order; second, we train both paths simultaneously by sharing their CNN (see Fig. 3). Note, that the CNN is not required to be reversed since all operations performed on the CNN outputs are symmetric. In the following, we call the two approaches *Individual CNN* and *Same CNN*, respectively.

Reversed Encoder/Decoder. The backward transformer reverses its sequences at multiple places (see Fig. 1). First, the absolute positional encodings (if used) of the encoder must be reversed so that the first positional vector is added to the last feature of the CNN output. Thus, the output of the encoder is indirectly reversed by the positional encodings. The reversed decoder prints the reversed sequence: the first token is still a ⟨sos⟩ token, however the following one is the last one. Therefore, during training, the GT sequence presented due to teacher forcing and finally the output must be reversed.

Loss. Training uses a traditional cross-entropy-loss. The *Same CNN* Transformer is optimised by computing a loss for the output of both the forward and backward transformer which is then added to obtain the final loss. Alternatively, due to the correct decoding sequence length caused by teacher forcing, it would also be possible to add both confidence vectors for each time step and compute one combined loss. The first scheme is more reasonable since we not only want the joined Transformer to perform well but also the individual paths.

Voter. During inference, the forward and backward decoder are likely to produce sequences of different lengths which is why a voting mechanism is required to combine these sequences. The proposed voting mechanism is fully deterministic without any trainable parameter. First, we match both predictions using the *Ratcliff and Obershelp algorithm* implementation of pythons difflib-module[1] to obtain their common substrings and their differences. Second, we calculate trust values for each position of the two sequences by multiplying the sequences of character confidences (emitted by the decoders) with a positional factor. In case of relative positional encodings, the positional factor is set to 1.0 for each position, consequently, the trust values equal the character confidences. Since observations show that Transformers using absolute positional encodings are likely to produce errors in long sequences, we fully trust (factor 1.0) the confidences up to the position 30, then linearly decrease the factor down to a limit of 0.1 at position 50. Finally, for each matching or mismatching substring, we apply certain rules based on its type to get the output sequence (an example is shown in Fig. 4):

- Equal substrings do not require voting.
- If there is a replacement with substrings of the same length (middle conflict in Fig. 4), we add the corresponding position-wise confidences and then take the characters with the highest resulting value for prediction.
- If there is a replacement with substrings of different length (last conflict in Fig. 4) or a deletion/insertion (first conflict in Fig. 4), we take the substring with the highest average trust value. Note, the average trust value of the empty substring is defined as the factor corresponding to the position of the empty substring in the full sequence.

---

[1] https://docs.python.org/3/library/difflib.html.

$$\text{Thi} \left\{ \begin{matrix} (1.0 \cdot 1.0) \\ i \ (0.9 \cdot 0.1) \end{matrix} \right\} \text{s} \ ... \ \text{is} \left\{ \begin{matrix} a(0.4), d(0.5), ... \\ a(0.8), d(0.1), ... \end{matrix} \right\} \text{n} \ ... \ \text{exa} \left\{ \begin{matrix} ni \ (0.8 \cdot 0.11 + 0.7 \cdot 0.10)/2 \\ m \ (0.4 \cdot 1.0) \end{matrix} \right\} \text{ple.}$$

Fig. 4. Example of the voting mechanism producing "This ... is an ... example" as result. The brackets show conflicts of the forward/backward transformer (top/bottom). Blue and green numbers correspond to position weight (absolute encoding only) and character confidence, respectively. (Color figure online)

4   Implementation Details

In this section, we summarize the implementation details of our proposed Transformers and the reference network.

4.1   CNN/LSTM-Reference Network

Our reference network comprises the same CNN as our Transformer (see Sect. 3.2 and Fig. 2), three bidirectional LSTM-layers, each with 256 hidden nodes per direction and dropout of 50%, and the final output layer producing the confidence matrix. This network was trained using the CTC-loss, a learning rate of 0.001 with a exponential decay of 0.99 per epoch, and the ADAM optimiser. Each epoch processes 8192 samples with a batch size of 16.

4.2   Bidirectional Transformer

To train the different Transformers, we initialised all weights of the CNN with pretrained weights of the final CNN/LSTM-network. Preliminary experiments showed significantly worse results when training with randomly initialised weights.

All dense layers comprise 1024 hidden nodes (see Fig. 1). The MHA mappings in the encoder and decoder have 512 nodes and use 8 attention heads. Dropout is set to a ratio of 10% within the transformer, whereas the final dropout rate is set to 50%. The concluding dense layer maps the output of the decoder to the alphabet size (the number of characters including ⟨sos⟩ and ⟨eos⟩ tokens).

We train the network using the ADAM optimiser with a batch size of 8 for 500 epochs each processing 8192 samples. The learning rate is warmed up starting from 10^{-6} up to 10^{-4} in the first two epochs. Then, the learning rate is decayed with a factor of 0.5 every 100 epochs. We apply label-smoothing with a factor of 0.4 and gradient clipping of 5.

5   Evaluation

In this section, we introduce our evaluation material and present the results of the performed experiments. For each dataset, we trained the four approaches of the Transformer networks and the reference network. We report the CER on the validation and test sets whereby the network yielding the best obtained validation value was chosen.

Table 1. Reference results on IAM-test whereby usage of external data is forbidden.

Authors	Method	Result
Puigcerver et al. [17]	MDLSTM	5.8%
Michael et al. [15]	S2S	5.24%
Kang et al. [11]	Transformer	7.62%
This paper	CNN/LSTM	5.38%
This paper	Transformer	6.02%
This paper	Bidi transformer	5.67%

Table 2. Results on the IAM and Internal dataset comparing the two stand-alone unidirectional Transformers (FW and BW) and the bidirectional one (Voted).

Dataset	CTC	Pos	Same CNN			Indiv. CNN		
			FW	BW	Voted	FW	BW	Voted
IAM Val	3.45%	ABS	4.92%	4.66%	4.11%	4.54%	4.56%	4.09%
		REL	4.45%	4.50%	3.81%	4.34%	4.08%	3.58%
IAM test	5.38%	ABS	7.09%	6.68%	6.17%	6.57%	6.48%	5.91%
		REL	6.19%	6.22%	5.82%	6.48%	6.02%	5.67%
Internal Val	3.55%	ABS	1.89%	2.12%	1.79%	1.86%	1.78%	1.69%
		REL	2.38%	2.57%	2.22%	2.13%	2.10%	1.94%
Internal test	12.32%	ABS	10.42%	10.22%	9.12%	11.12%	10.78%	9.97%
		REL	11.77%	11.78%	10.85%	11.55%	11.95%	11.34%

5.1   IAM-Dataset

We performed our experiments on the popular IAM-dataset for HTR [14] using Aachen's partition[2] which splits the dataset in 6161, 966, and 2915 lines for training, validation, and test, respectively. The alphabet size is 80, including tokens for $\langle$sos$\rangle$ and $\langle$eos$\rangle$.

Our results are shown in the first half of Table 2 while reference values are listed in Table 1. Both our Transformer and CTC-reference model outperformed the Transformer of Kang et al. [11] if no synthetic data was permitted. They reach however a best value of 4.67% when including additional data. Both our Bidi-Transformer and our reference model outperform Puigcerver et al. [17] whereby our competitive reference model is close to the performance of Michael et al. [15].

The voting mechanism of the forward and backward decoder improves our Transformer by about 5%-10% in every setup. Relative positional encoding and using an *Individual CNN* resulted in a higher performance. Compared to our reference model, our best bidirectional Transformer performs worse (5.67% vs. 5.38% CER). This can be explained by the bigger capacity of the transformer being able to learn language models and long range dependencies. This facilitates overfitting which also manifests in a halved CER on the training data (not

---

[2] https://github.com/jpuigcerver/Laia/tree/master/egs/iam.

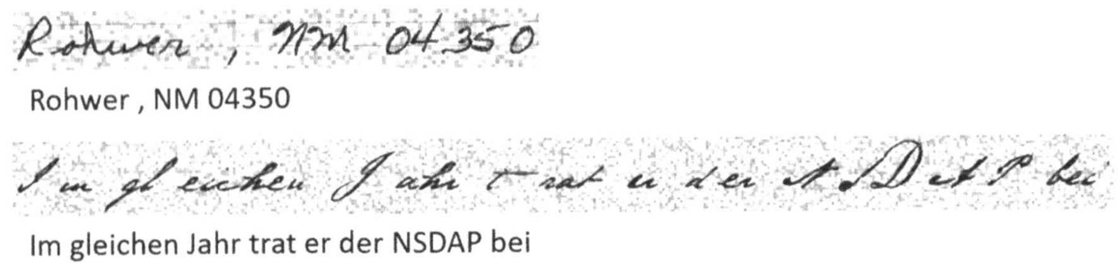

Rohwer , NM 04350

Im gleichen Jahr trat er der NSDAP bei

Fig. 5. Two example lines from the Internal train/val-dataset, the corresponding GT is written below each line. The first line is a real writing with added background noise, the second line was generated synthetically.

shown). In summary, increasing the training data is mandatory to successfully train the proposed Transformer network for HTR which also manifests in the results of Kang et al. [11].

5.2   Internal

Since our experiments on the quite small IAM dataset show that the Transformer model is massively overfitting on the training data even though data augmentation as well as dropout are applied, we collected several datasets and added synthetic lines rendered by computer fonts to create a larger dataset set with the most possible number of real lines in the training dataset. The resulting Internal dataset is considerably larger than IAM comprising 1 Million lines (see Fig. 5) of which 2000 lines are extracted for validation. Unfortunately, the Internal dataset can not be published due to data privacy issues. We chose additional 246 lines (5558 characters) of difficult real-world use-case documents (including bank information, addresses, currencies, dates, letters, check-boxes, etc.) for testing (see lines in Fig. 6). The dataset also comprises lines in different languages (mainly German and English).

The last rows of Table 2 list the results on the Internal dataset. The best CTC model yielded a CER of 3.55% on the validation data which is similar to the results on IAM, whereas only 12.32% were reached on the test set. This big gap is expected because the validation data is extracted from the same data distribution as the training data, but the test data was chosen completely separately with very difficult lines derived from customer problem scenarios.

Any Transformer-model reaches a clearly lower CER on the validation set but also on the test set compared to the CTC model, whereby the *Same CNN*-Transformer-approach using absolute positional encoding yielded with an improvement of improvement of 26% the best results on the test set. In general, there is no clear preference for choosing *Same* and *Indiv.* since their results are similar. *Same* is however faster since the CNN only needs to be computed once. The results also point out that the CNN, as expected, does not store relevant information for the direction of decoding. In contrast to our experiments on IAM, the absolute positional encoding was clearly superior to using relative encodings.

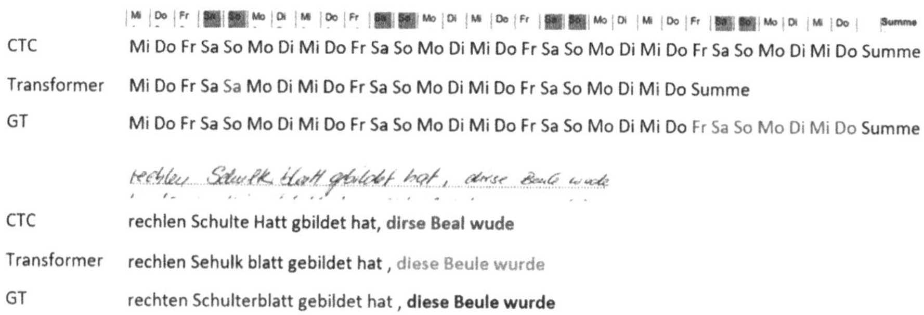

| Mi | Do | Fr | Sa | So | Mo | Di | Mi | Do | Fr | Sa | So | Mo | Di | Mi | Do | Fr | Sa | So | Mo | Di | Mi | Do | Fr | Sa | So | Mo | Di | Mi | Do | Summe |

CTC   Mi Do Fr Sa So Mo Di Mi Do Fr Sa So Mo Di Mi Do Fr Sa So Mo Di Mi Do Fr Sa So Mo Di Mi Do Summe

Transformer Mi Do Fr Sa Sa Mo Di Mi Do Fr Sa So Mo Di Mi Do Fr Sa So Mo Di Mi Do Summe

GT   Mi Do Fr Sa So Mo Di Mi Do Fr Sa So Mo Di Mi Do Fr Sa So Mo Di Mi Do Fr Sa So Mo Di Mi Do Summe

rechlen Schulte Hatt gbildet hat, dirse Beul wude

CTC   rechlen Schulte Hatt gbildet hat, dirse Beal wude

Transformer rechlen Sehulk blatt gebildet hat , diese Beule wurde

GT   rechten Schulterblatt gebildet hat , diese Beule wurde

Fig. 6. Two example showing the input line, the predictions of the CNN/LSTM and the bidirectional Transformer (absolute positional encoding and the same CNN), and the GT. Similar observations were made for any other Transformer approach. (Color figure online)

6 Discussion and Error Analysis

The previous experiments pointed out that Transformers outperform traditional CNN/LSTM approaches if the number of training lines is sufficient. In this section, we analyse which errors our bidirectional Transformer are able to avoid and if this can be explained by its learning of a language model. Moreover, we examine lines where the Transformer performed worse to show its limits.

The first example of Fig. 6 shows the worst line of the Transformer which accounts for about 3% of the total error on the Internal test set. This line is conspicuous because it is a repetition of the abbreviated German weekdays starting at Wednesday ("Mi"). While the CTC-based approach perfectly predicts the printed line, the Transformer omits a complete week (blue). This shows a fundamental problem of Transformers[3]: without positional encoding, the transformer would not know its position during decoding at all, therefore repetitions, especially if they are large-scale, appear the same. Positional encoding solves this problem, however for long lines as in the example, the uncertainty about the position raises. Furthermore, information about the position using neighbours as in relative positional encoding is not helpful here because the (close) environment looks the same at each repetition. To master such an task (plain repetitions), the network must have information about the correct line position during decoding (similar to reading with a finger). This kind of problem poses a great danger for the application of Transformers for HTR since, in contrast to NLP-tasks, repetitions, as in numbers, are common but extremely dangerous if missed.

In contrast, the second example highlights the advantages of the Transformer and is an example why it outperforms the CTC-approach in general. The handwriting in the image is hard to read, even for a human. While the CTC approach just aims to predict letters without a performant language model (red),

---

[3] Note, this behaviour can easily produced, e.g., in Google-translate by translating repeated words (without line breaks). Translating 16 times the German word "Mann" results in 17 repetitions of the English translation "man" (as of 01/11/2021).

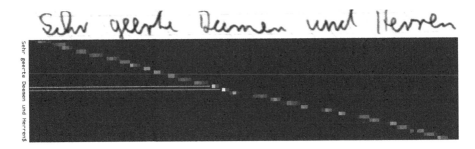

Fig. 7. Example of one head of the mutual attention matrix of the Transformer on the Internal dataset with absolute positional encoding. The decoded characters (including the ⟨eos⟩ token as $) is shown on the left, above the attention matrix is the preprocessed input image (note the visible slant correction). The orange helper lines indicate the location of the error in the prediction "Sehr geerte Deemen und Herren" (GT: "Sehr geerte Damen und Herren"). (Color figure online)

the Transformer recognises the common words (green). Yet, the Transformer is not able to correct all errors, most probably because not all words were presented during training, such as the compound word "Schulterblatt" (shoulder blade), note however, that "blatt" was identified correctly.

Figure 7 shows the values of one head of the mutual attention matrix of the last layer. As expected, the attention when decoding a character (e.g., the capital "S") is at the actual written locations. Figure 7 and observation of the other attention heads also show that the network mainly focuses on the actual character and does not use information from other locations of the image. Therefore, it might be beneficial at least by means of computation time to adapt windowed attention [4] which programmatically restricts the decoding location to a small windows. Note however that this approach can only be applied to the self-attention modules since offsets in the positions are expected for mutual attention (arbitrary space widths, smaller/broader characters, etc.) which manifests in Fig. 7 being not a diagonal matrix.

7 Conclusion

This paper introduced bidirectional Transformers for HTR. While the proposed method performed worse on the rather small IAM dataset, the Transformer significantly outperformed traditional CNN/LSTM approaches when using a larger dataset including data from different sources such as synthetic lines. The error analysis showed a general problem of Transformers that is potentially dangerous for HTR: repetitions of the same text or character as common in digits. The problem of long lines can be solved using relative positional encoding which performs however worse in the overall results. A CTC-based approach does by construction not suffer from these limitations.

A general outcome of our various experiments is that Transformers require a sophisticated selection of hyperparameters and pretrained weights for the CNN which makes training tedious. Further investigations and experiments must be performed to understand why, in contrast to the work of Kang et al. [11], training from scratch, i.e., not using pretrained weights, did not converge.

To better understand if a Transformer is indeed able to learn a language model, we plan to apply a statistical language model in a postprocessing step. We expect that only the CNN/LSTM approach would benefit from it.

A very promising idea would be the combination of the two worlds Transformer and CTC. This could result in a model that does not suffer from repetitions or long sequence-lengths while including a learned language model. However, due to the complete different nature of the two worlds, there is up-to-now no known approach. A promising first attempt might be to use the CTC as teacher forcing like input, even during decoding, however this would only work well to correct substitutions since insertions or deletions can not be mapped by this approach.

Acknowledgments. This work was partially funded by the European Social Fund (ESF) and the Ministry of Education, Science and Culture of Mecklenburg-Western Pomerania (Germany) within the project Neural Extraction of Information, Structure and Symmetry in Images (NEISS) under grant no ESF/14-BM-A55-0006/19.

References

1. Al Azawi, M., Liwicki, M., Breuel, T.M.: Combination of multiple aligned recognition outputs using WFST and LSTM. In: 2015 13th International Conference on Document Analysis and Recognition (ICDAR), pp. 31–35. IEEE (2015)
2. Al-Sabahi, K., Zuping, Z., Kang, Y.: Bidirectional attentional encoder-decoder model and bidirectional beam search for abstractive summarization. arXiv preprint arXiv:1809.06662 (2018)
3. Bahdanau, D., Cho, K., Bengio, Y.: Neural machine translation by jointly learning to align and translate. CoRR abs/1409.0473 (2014)
4. Beltagy, I., Peters, M.E., Cohan, A.: Longformer: the long-document transformer. arXiv preprint arXiv:2004.05150 (2020)
5. Boschetti, F., Romanello, M., Babeu, A., Bamman, D., Crane, G.: Improving OCR accuracy for classical critical editions. In: Agosti, M., Borbinha, J., Kapidakis, S., Papatheodorou, C., Tsakonas, G. (eds.) ECDL 2009. LNCS, vol. 5714, pp. 156–167. Springer, Heidelberg (2009). https://doi.org/10.1007/978-3-642-04346-8_17
6. Breuel, T.: High performance text recognition using a hybrid convolutional-LSTM implementation. In: 14th IAPR International Conference on Document Analysis and Recognition (ICDAR), pp. 11–16. IEEE (2017)
7. Cho, K., Courville, A., Bengio, Y.: Describing multimedia content using attention-based encoder-decoder networks. IEEE Trans. Multimed. **17**, 1875–1886 (2015). https://doi.org/10.1109/TMM.2015.2477044
8. Doetsch, P., Zeyer, A., Ney, H.: Bidirectional decoder networks for attention-based end-to-end offline handwriting recognition. In: 2016 15th International Conference on Frontiers in Handwriting Recognition (ICFHR), pp. 361–366. IEEE (2016)
9. Dutta, K., Krishnan, P., Mathew, M., Jawahar, C.: Improving CNN-RNN hybrid networks for handwriting recognition. In: 2018 16th International Conference on Frontiers in Handwriting Recognition (ICFHR), pp. 80–85. IEEE (2018)

10. Graves, A., Fernández, S., Schmidhuber, J.: Multi-dimensional recurrent neural networks. CoRR abs/0705.2011 (2007)
11. Kang, L., Riba, P., Rusiñol, M., Fornés, A., Villegas, M.: Pay attention to what you read: non-recurrent handwritten text-line recognition. arXiv preprint arXiv:2005.13044 (2020)
12. Lopresti, D., Zhou, J.: Using consensus sequence voting to correct OCR errors. Comput. Vis. Image Underst. **67**(1), 39–47 (1997)
13. Maas, A.L., Hannun, A.Y., Ng, A.Y.: Rectifier nonlinearities improve neural network acoustic models. In: ICML (2013)
14. Marti, U.V., Bunke, H.: The IAM-database: an English sentence database for offline handwriting recognition. Int. J. Doc. Anal. Recognit. **5**(1), 39–46 (2002)
15. Michael, J., Labahn, R., Grüning, T., Zöllner, J.: Evaluating sequence-to-sequence models for handwritten text recognition. In: 2019 International Conference on Document Analysis and Recognition (ICDAR), pp. 1286–1293. IEEE (2019)
16. Pham, V., Kermorvant, C., Louradour, J.: Dropout improves recurrent neural networks for handwriting recognition. CoRR abs/1312.4569 (2013)
17. Puigcerver, J.: Are multidimensional recurrent layers really necessary for handwritten text recognition?. In: 2017 14th IAPR International Conference on Document Analysis and Recognition (ICDAR), vol. 1, pp. 67–72. IEEE (2017)
18. Reul, C., Springmann, U., Wick, C., Puppe, F.: Improving OCR accuracy on early printed books by utilizing cross fold training and voting. In: 2018 13th IAPR International Workshop on Document Analysis Systems (DAS), pp. 423–428 (2018)
19. Rice, S.V., Kanai, J., Nartker, T.A.: An algorithm for matching OCR-generated text strings. Int. J. Pattern Recognit. Artif. Intell. **8**(05), 1259–1268 (1994)
20. Shaw, P., Uszkoreit, J., Vaswani, A.: Self-attention with relative position representations. arXiv preprint arXiv:1803.02155 (2018)
21. Sutskever, I., Vinyals, O., Le, Q.V.: Sequence to sequence learning with neural networks. In: NIPS (September 2014)
22. Tensmeyer, C., Wigington, C., Davis, B., Stewart, S., Martinez, T., Barrett, W.: Language model supervision for handwriting recognition model adaptation. In: 16th International Conference on Frontiers in Handwriting Recognition, pp. 133–138. IEEE (2018)
23. Vaswani, A., et al.: Attention is all you need. In: Advances in Neural Information Processing Systems, pp. 5998–6008 (2017)
24. Voigtlaender, P., Doetsch, P., Ney, H.: Handwriting recognition with large multidimensional long short-term memory recurrent neural networks. In: ICFHR, pp. 228–233 (October 2016)
25. Wemhoener, D., Yalniz, I.Z., Manmatha, R.: Creating an improved version using noisy OCR from multiple editions. In: 2013 12th International Conference on Document Analysis and Recognition (ICDAR), pp. 160–164. IEEE (2013)
26. Wick, C., Reul, C., Puppe, F.: Calamari - a high-performance tensorflow-based deep learning package for optical character recognition. Digit. Humanit. Q. **14**(2) (2020)
27. Wigington, C., Stewart, S., Davis, B., Barrett, B., Price, B., Cohen, S.: Data augmentation for recognition of handwritten words and lines using a CNN-LSTM network. In: ICDAR, pp. 639–645 (November 2017)
28. Yousef, M., Hussain, K.F., Mohammed, U.S.: Accurate, data-efficient, unconstrained text recognition with convolutional neural networks. Pattern Recognit. **108**, 107482 (2020)

Zero-Shot Chinese Text Recognition via Matching Class Embedding

Yuhao Huang[1], Lianwen Jin[1,2(✉)], and Dezhi Peng[1]

[1] South China University of Technology, Guangzhou, China
eelwjin@scut.edu.cn
[2] Guangdong Artificial Intelligence and Digital Economy Laboratory (Pazhou Lab),
Guangzhou, China

Abstract. This paper studies the challenging problem of zero-shot Chinese text recognition, which requires the model to train on text line images containing only the seen characters, and then recognize the unseen characters from new text line images. Most of the previous methods only consider the zero-shot Chinese character recognition problem. They attempt to decompose the Chinese characters into radical representations and then recognize them at the radical level. Some methods developed recently have extended the radical-based recognition model from recognizing characters to recognizing text lines. However, the disadvantages of these methods include the requirement of long training time and a complicated decoding process. In addition, these methods are unsuitable for long text sequences. In this paper, we have proposed a novel zero-shot Chinese text recognition network (ZCTRN) by matching the class embeddings with the visual features. Specifically, our proposed model consists of three components: a text line encoder that extracts the visual features from the text line images, a class embedding module that encodes the character classes into class embeddings, and a bidirectional embedding transfer module that can map the class embeddings into the visual space and preserve the information of the original class embeddings. In addition, we use a distance-based CTC decoder to match the visual features with the class embeddings and output the recognition results. Experimental obtained by applying our proposed network to the MTHv2 dataset and the ICDAR-2013 handwriting competition dataset show that our method not only preserves high accuracy in recognizing text line images containing seen characters, but also outperforms the existing state-of-the-art models in recognizing text line images containing unseen characters.

Keywords: Chinese text recognition · Class embedding · Zero-shot learning

1 Introduction

Recent text recognition methods have achieved great success in recognizing text sequences in many cases, such as, scene text recognition [3,17], offline and online

© Springer Nature Switzerland AG 2021
J. Lladós et al. (Eds.): ICDAR 2021, LNCS 12823, pp. 127–141, 2021.
https://doi.org/10.1007/978-3-030-86334-0_9

handwriting recognition [21], and historical document digitization [14]. Although these methods have proved to be effective, they usually rely heavily on large-scale labeled training data and can hardly be generalized to unseen character classes that are not included in the training data. As for Chinese characters, there are excessive classes in the real world (70224 Chinese characters in GB18030-2005). Thus it is almost impossible to collect a sufficient amount of labeled training data for each Chinese character. However, when teaching a person to learn a Chinese character, he or she only needs to read the character a few times or learn the radicals of the character, which enables him or her to learn that character. The ability of humans to learn a new Chinese character from a few samples or from the radicals of the Chinese character is similar to the few-shot or zero-shot learning. Inspired by this, in this work, we have proposed a zero-shot learning model for Chinese text recognition.

Most of the existing methods have focused on zero-shot Chinese character recognition, whereas only a few have focused on zero-shot Chinese text recognition. Figure 1 illustrates the difference between these two problems.

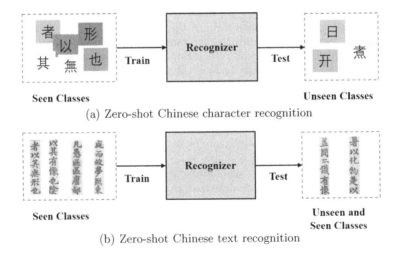

(a) Zero-shot Chinese character recognition

(b) Zero-shot Chinese text recognition

Fig. 1. Schematic showing the (a) zero-shot Chinese character recognition and the (b) zero-shot Chinese text recognition.

In the zero-shot Chinese character recognition problem, a recognizer is trained on the character images containing the seen characters and is then used for recognizing character images containing unseen characters. Previous methods for tackling the zero-shot Chinese character recognition problem, such as those developed by Zhang et al. [35] and Wang et al. [23], first extract the radical sequences of the characters and then transcribe the radical sequences into their corresponding characters. As seen and unseen Chinese characters share common radicals, a model trained with sufficient radical features can be applied to unseen Chinese character recognition. However, the radical-based Chinese character recognition methods work well only when the models have learned all pairs of radicals and the location of these radicals. In order to mitigate this problem, Cao et al. [5] proposed the hierarchical decomposition embedding (HDE) to encode the radicals, structures, and relations of the characters into semantic vectors. Besides, some existing methods [2,12] suggested the use of printed Chinese characters as templates for comparing the unseen characters with their template images and then perform the nearest-neighbor search. Despite their effectiveness, these methods only consider the zero-shot Chinese character recognition problem and cannot be applied simply to text recognition. Therefore, the problem of zero-shot text recognition is worth studying.

In the zero-shot Chinese text recognition problem, we aim to train a recognizer on text line images containing seen characters and then recognize text lines containing unseen characters. The main difference between the zero-shot Chinese character recognition and zero-shot Chinese text recognition is the number of characters in an input image, which implies that the models designed for the latter need to recognize multiple characters in an input image. A method developed recently by Zhang et al. [34] attempted to recognize text line images with radical-based method and proposed the use of a multihead coverage-based attention module for extracting the radical features of the characters. Although it has made some progress, due to the attention mechanism, the disadvantage of this method is the long training time and its unsuitability for long text sequences. And this method requires two-stage decoding: in the first stage, it decodes the radical sequences of the text image, which is typically a long sequence, and, in the second stage, it needs to match the radical representation with its corresponding label using the edit distance.

To overcome these issues, in this work, we have studied the zero-shot Chinese text recognition problem and have proposed a novel zero-shot Chinese text recognition network (ZCTRN) that is simple, effective, and flexible. More specifically, our model contains three components. First, a text line encoder, which is a fully convolutional neural network, is used for extracting the visual features. Second, based on HDE [5], a class embedding module is constructed, which uses the character classes as input and outputs the class embedding of each character. The class embedding considers the entire character and avoids decoding its radical representation directly. Therefore, the time consumed in decoding is reduced and the process of decoding is simplified. Third, to project the class embeddings into a visual space, a bidirectional embedding transfer module, which can map the class embeddings into a visual space and preserve the information of the original class embeddings, has been designed. In addition, a reconstruction loss has been introduced to let the projected class embedding maintain its original information. Furthermore, our model matches the visual features with class embeddings and outputs the recognition results using a distance based Connectionist Temporal Classification (CTC) decoder, which enforces the alignment between the text line features and their labels. As the CTC decoder performs parallel dense prediction on the text images, it is faster and can handle longer text sequences compared to the attention-based methods.

The main contributions of the present work can be summarized as follows:

1. We have proposed a novel ZCTRN that is simple, effective, and flexible for handling the zero-shot text recognition problem.
2. In contrast to most of the previous works on zero-shot Chinese character recognition, we have considered the problem of zero-shot Chinese text recognition, which is more challenging and practical than character recognition.
3. We have conducted experiments on two public datasets, and the results obtained from them demonstrate the ability of our proposed ZCTRN model in effectively recognizing unseen characters in the text line images while preserving high accuracy in recognizing the seen characters in them.

2   Related Works

Text Recognition. The text recognition problem has been studied for many years, and most proposed methods can be applied to Chinese text recognition tasks. The existing methods for text recognition can be roughly divided into two categories: CTC-based [8] methods and attention-based [4] methods. Among the CTC-based methods, Shi et al. [17] proposed a convolutional recurrent neural network (CRNN) to recognize text sequences in images. However, vanilla CTC cannot be applied to two-dimensional (2D) prediction problems. Hence, Wan et al. [20] proposed a 2D-CTC model to address this issue. In addition to the 2D prediction problem, CTC also suffers from the problem of large computation consumption. To solve this problem, Xie et al. [30] proposed an aggregation cross-entropy (ACE) method to replace CTC, and this method is faster and consumes less memory. Unlike the CTC-based methods, the attention-based methods can search for the most relevant parts of the sentences and perform prediction on these parts. Shi et al. [18] proposed a unified network that combines a rectification module and a attention-based recognition module to handle irregular text. However, most attention-based methods suffer from the attention drift problem. To address this problem, Wang et al. [24] proposed to use decoupled attention network (DAN) to decouple the attention decoder into an alignment module and a decoupled text decoder. Our model also includes a CTC-based encoder-decoder network. The main difference between our model and these methods is that our model can recognize unseen characters that are not included in the training data.

Zero-Shot Learning. In zero-shot learning [25], the classes covered by the training samples are referred to as the seen classes and those that only appear in the testing samples, but not in the training samples, are referred to as the unseen classes. Zero-shot learning aims to classify the instances belonging to the unseen classes that are not included in the training samples. In recent years, zero-shot learning has been extensively studied and has achieved tremendous advances. The key idea of the zero-shot learning method is to exploit the semantic relationship between the seen and the unseen classes. Lampert et al. [11] originally proposed to leverage the semantic attributes to detect the unseen object classes, where the seen and the unseen classes share common semantic attributes. However, the attribute-based classifier requires a two-step prediction process that is not straightforward. To mitigate this problem, Akata et al. [1] proposed to view attribute classification as a label embedding problem and introduced a compatible function to evaluate the images and label embedding. However, the methods based on label embedding also suffer from the projection domain shift problem [7] and the hubness problem [19]. Kodirov et al. [10] proposed a semantic autoencoder framework to solve the projection domain shift problem. To address the hubness problem, Zhang et al. [36] proposed to project label embedding into the visual space. Meanwhile, our model performs zero-shot recognition by leveraging the class embeddings of unseen characters, and we design a network to match the visual features with their corresponding class embeddings.

3   Proposed Model

3.1   Problem Definition and Notations

In this paper, we have studied the zero-shot Chinese text recognition problem. Specifically, we have a training set $D_{tr} = \{(x_i, y_i)\}_{i=1}^n$ with $x_i \in X^{tr}$, $y_i \in Y^s$, where X^{tr} is the training set of the text images and Y^s is the set of the corresponding labels containing only the seen character classes. Similarly, we define the testing set as $D_{te} = \{(x_i, y_i)\}_{i=1}^n$ with $x_i \in X^{te}$, $y_i \in Y^s \cup Y^u$, where X^{te} is the testing set of text images and $Y^s \cup Y^u$ is the corresponding labels containing the seen and the unseen character classes. In addition, in order to perform zero-shot recognition, additional information on the seen and unseen characters is provided by the set of the corresponding class embeddings $C = \{c_i\}_{i=1}^N$, where c_i is the corresponding class embedding of class i. Given the training text images X^{tr} containing only the seen character classes and the set of class embeddings $C = \{c_i\}_{i=1}^N$, zero-shot text recognition aims to learn a recognizer $f(\cdot) : x \rightarrow Y^s \cup Y^u$ that can recognize both seen and unseen characters in the testing text images X^{te}.

3.2   Overall Architecture of ZCTRN

The overall architecture of our proposed ZCTRN model is illustrated in Fig. 2. First, a text line encoder is constructed to extract the visual features of the text images, where the height of the visual features is downsampled to 1. Next, based on HDE [5], a class embedding module is constructed to encode the character classes into class embeddings. And then a bidirectional embedding transfer module is proposed to project the class embeddings into the visual space. To preserve the information of the original class embeddings, we apply a condition that the projection must be able to reconstruct the original class embeddings, which are optimized by a reconstruction loss. Furthermore, we propose a distance-based CTC decoder for matching the visual features with the class embeddings and output the recognition results. Specifically, the cosine distance is leveraged for evaluating the similarity between the class embeddings and the visual features in the feature space. And then a cosine similarity map based on the cosine distance is obtained, and the similarity score of each row of the map is maximized to output the predicted classes. Finally, a CTC decoder is proposed for converting the outputs into a label sequence.

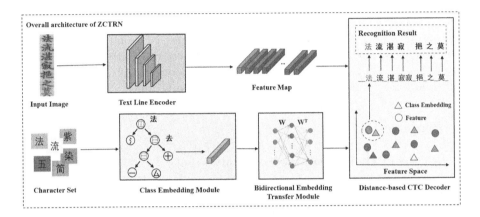

Fig. 2. Overall architecture of the proposed ZCTRN model. Our model consists of three components, including a text line encoder that extracts the visual features of the input image, a class embedding module that encodes the character classes into the class embeddings, and a bidirectional embedding transfer module that projects the class embeddings into the visual space and can preserve the information of the original class embeddings. In addition, we use a distance-based CTC decoder to match the visual features with the class embeddings and decode the final result.

3.3   Text Line Encoder Network

We adopt ResNet18 [9] as the backbone network to extract the visual features from the text images. Specifically, all images are first rescaled to a fixed height while maintaining their aspect ratios. Next, the text line encoder takes these images as input and outputs a sequence of visual features $V = \{v_1, ..., v_W\}$, with $v_i \in \mathbb{R}^C$, where W is the width of the feature map and C is the size of the output channel. In contrast to the original ResNet18, the final fully connected layer is removed in our network and the adaptive average pooling layer is modified to only downsample the height of the feature map to 1, while maintaining its original width. To avoid the problem of overfitting, a dropout rate of 0.3 is applied after the final convolution layer.

3.4   Class Embedding Module

To perform zero-shot Chinese text recognition, we need the radical information of each character because the seen and unseen characters share common
radicals. Therefore, based on HDE [5], we construct a class embedding module
that encodes the Ideographic Description Sequences (IDS) of the seen and the
unseen character classes into class embeddings. Unlike HDE, we have extended
this method to the zero-shot text recognition task. Specifically, given the character classes, the class embedding module first decomposes them into radicals and
structures, and then encodes them into class embeddings. The class embedding
of a character can be defined as follows:

$$\phi(y) = \sum_{n_i \in R} v_{n_i} y_{n_i} + \lambda \sum_{n_j \in S} v_{n_j} y_{n_j}, \tag{1}$$

where n_i is the radical of the set R, n_j is the structure of the set S, y_n is a
one-hot vector of a radical or a structure label, and both the radical vectors and
the structure vectors have same dimensions. In addition, λ is a hyperparameter
that balances the two items, and v_n is the impact value of the radical or the
structure and is calculated by:

$$v = \alpha^l + \sum_{i=1}^{l}(\alpha^i \beta_{p_i}). \tag{2}$$

Here, α and β are hyperparameters, p_i represents the node along the node-
path, and l is the length of the node-path. For λ, α and β, we use the same
values as those used in HDE, since these settings are widely used. Based on the
above embedding strategy, the corresponding class embedding of each character
is obtained. Further, an additional "[blank]" class embedding (for CTC) is also
added using a one-hot vector. As the class embedding module encodes the radical
information of each character class, it can be applied to represent both the seen
and unseen characters.

3.5   Bidirectional Embedding Transfer Module

Although the class embedding module has generated the class embedding of each
character class, the dimensions of the class embeddings and the visual features
are inconsistent. Therefore, our model also has a bidirectional embedding transfer module, which not only projects the class embeddings into the visual space,
but also reconstructs the original class embeddings from the visual space. To
preserve the information of the original class embeddings, we apply the condition that the projection must be able to reconstruct the original class embeddings. Specifically, suppose that the outputs of the class embedding module are
$\Phi = \{\zeta_{c_1}, ..., \zeta_{c_n}\}, \zeta_{c_i} \in \mathbb{R}^D$, where ζ_{c_i} denotes the class embedding vector of
character c_i having the dimension D. Then the class embeddings $\Phi \in \mathbb{R}^{D \times N}$ are
projected into the visual space using a linear mapping function with a projec-
tion matrix $W \in \mathbb{R}^{C \times D}$, resulting in the projected class embeddings $\Phi' \in \mathbb{R}^{C \times N}$.

Note that C denotes the dimension of the visual features and N is the number of class embeddings.

Although our model has learned a projection function to map the class embeddings into the visual space, the projected class embeddings are most likely to lose their original information. To mitigate this problem, an additional constraint is added to the projection, which implies that the projection can also reconstruct the original class embeddings. By adding this constraint, the projected class embeddings Φ' can still preserve the information from the original class embeddings. Formally, the projected class embeddings Φ' are reconstructed back to the original class embedding with a reconstruction matrix $W' \in \mathbb{R}^{D \times C}$. To simplify this process, the transpose of the original projection matrix $W^T \in \mathbb{R}^{D \times C}$ is used, resulting in the reconstruction class embeddings $\hat{\Phi} \in \mathbb{R}^{D \times N}$. In addition, we expect the reconstruction class embeddings $\hat{\Phi}$ to be as similar as possible to the original class embeddings Φ. This objective can be achieved as follows:

$$\mathcal{L}_{re} = \frac{1}{N} \sum_{i=1}^{N} ||\hat{\Phi}_i - \Phi_i||_2^2. \tag{3}$$

where $\mathcal{L}_{re}$ is the reconstruction loss, $\hat{\Phi}_i$ and Φ_i are D dimensional vectors.

3.6   Distance-Based CTC Decoder

Thus far, we have described the procedure for obtaining the visual features from the text line encoder and the projected class embeddings from the bidirectional embedding transfer module. In the next step, our model decodes these visual features by matching similar projected class embeddings. To compare the distance between the visual features and the projected class embeddings, the cosine similarity function is leveraged as follows:

$$d(V, \Phi') = \frac{V^T \Phi'}{||V^T|| \times ||\Phi'||}, V \in \mathbb{R}^{C \times W}, \Phi' \in \mathbb{R}^{C \times N}. \tag{4}$$

where V represents the visual features and Φ' represents the projected class embeddings. In this way, the cosine similarity map $d(V, \Phi') \in \mathbb{R}^{W \times N}$ is obtained, where d_{ij} represents the similarity score between the pixel i of a visual feature and the class embedding j. Further, the similarity score of each row of the cosine similarity map is maximized for outputting the predicted classes. However, this naive prediction is misaligned with the output label. Hence, the distance-based CTC loss is used for aligning the prediction with its ground truth label, which can be defined as:

$$\mathcal{L}_{CTC} = -\sum \log p(l_i | \alpha \cdot d(V, \Phi')). \tag{5}$$

where l_i is the ground truth label, and α is a learnable parameter that can adjust the scale of the cosine similarity. Finally, the decoder result is converted into a label sequence by removing the duplicate labels and the "[blank]" labels.

3.7   Training Objective

During the training process, the reconstruction loss is used for the bidirectional embedding transfer module and the distance-based CTC loss is used for the decoder. Therefore, the overall training objective for our proposed ZCTRN can be expressed as follows:

$$\mathcal{L}_{total} = \mathcal{L}_{CTC} + K\mathcal{L}_{re}. \tag{6}$$

where K is a hyperparameter that balances the reconstruction loss. Note that, we use $K = 1$ for all datasets, because the reconstruction loss is also important, in which the projected class embeddings are used for the further classification task.

4   Experiments

This section describes the experiments conducted on two public datasets under the zero-shot learning setting and the traditional setting, which will be detailed in Sect. 4.1.

4.1   Datasets and Settings

We use two Chinese datasets, namely, MTHv2 [14] and ICDAR-2013 handwriting competition dataset [32], to evaluate our ZCTRN model. In addition, we follow the official training/testing split provided by these datasets. The statistics of these datasets are summarized in Table 1. Note that the ICDAR-2013 handwriting competition dataset (denoted as ICDAR13) only contains a testing set. In this case, we use the CASIA-HWDB2.0-2.2 [13] (denoted as CASIA-HW) as the training set.

In the zero-shot learning setting, during the testing stage, the text lines that contain the unseen characters are used, and the label space is a combination of the seen and the unseen character classes $Y^s \cup Y^u$. To evaluate the accuracy of the unseen characters, we need a metric that can measure the performance of recognizing the unseen characters in the text lines. Therefore, the edit distance is adopted for calculating the accuracy rate of specific characters, and only the substitution and deletion errors are considered. The accuracy rate of the unseen characters is defined as follows:

$$\mathcal{AR}_{unseen} = 1 - \frac{N_s + N_d}{N}. \tag{7}$$

where N_s, N_d, and N denote the number of substitution errors, number of deletion errors, and number of unseen characters, respectively.

For the traditional setting, unlike the zero-shot learning, all text lines in the testing set are used during the testing phase, regardless of the type of text line. Following the work of Yin et al. [32], the accuracy rate (AR) and the correct rate (CR) are used for evaluating the performance of our model.

Table 1. Statistics of the datasets.

Dataset	Radicals	Seen classes	Unseen classes	Text lines (train)	Text lines (test seen/unseen)
MTHv2 [14]	306	6259	466	80317	24698/564
CASIA-HW+ICDAR13 [32]	436	7250	106	52230	2270/1162

4.2   Implementation Details

As described in Sect. 3.3, the ResNet18 is used as the backbone network. For the MTHv2 dataset, the heights of the input images are resized to 96 while maintaining their aspect ratio. For the CASIA-HW and ICDAR13, the data preprocessing strategy described in the work of Peng et al. [16] is adopted, and the heights of the input images are adjusted to 128 while maintaining their aspect ratio. In addition, to increase the diversity of the input images, a data augmentation strategy[1] is applied to the two datasets. For a fair comparison with other methods, the same augmentation strategy is applied for the reproduced methods. Adadelta [33] is chosen as the optimizer and the weight decay is set to 0.0001. For all datasets, our model is trained for 30 epochs with a batch size of 8. The initial learning rate is set to 1.0 and is reduced to 0.1 after 15 epochs.

4.3   Experiments on the MTHv2 Dataset

We compare our model with the recent state-of-the-art method RAN [34] and the conventional text recognition method CRNN [17] using the traditional and the zero-shot settings, and the results are shown in Table 2. As can be seen from Table 2, our method achieves the best result among the traditional and the zero-shot setting, with the result higher by a large margin of 14.18% for the zero-shot setting. On the other hand, CRNN fails to recognize the unseen characters, leading to a value of 0% AR_{unseen}. This shows that our method maintains a good balance between recognizing text images under the zero-shot and the traditional setting, whereas the other methods are unable to perform the trade-off as efficiently in comparison to our model. A few recognition results of images containing unseen characters have been visualized in Fig. 3.

Table 2. Traditional setting and the zero-shot setting results (%) on the MTHv2 dataset. The * symbol denotes the method that we reproduced.

Methods	Traditional setting		Zero-shot setting
	AR	CR	AR_{unseen}
RAN* [34]	91.56	91.79	37.22
CRNN [17]	96.94	97.15	0.00
ZCTRN (ours)	**97.42**	**97.62**	**51.40**

---

[1] https://github.com/Canjie-Luo/Text-Image-Augmentation.

Fig. 3. Examples of images containing unseen characters and recognition results in three methods. Red characters are the wrong results, and green ones are the correct unseen characters. (Color figure online)

4.4 Experiments on the MTHv2 Dataset with Different Ratios

To further explore the effect of zero-shot learning, experiments are conducted on the MTHv2 dataset with different ratios of the training samples. For the same, a certain ratio of the training samples are randomly selected as the training set and the testing samples that contain the unseen character classes are selected as the testing set. Further, we evaluate the performance of the traditional setting with different ratios of training set.

The results of training the model with different ratio of training samples are shown in Table 3. As we can see, values of 33.03% AR_{unseen}, 94.49% AR, and 94.75% CR are obtained when using only 10% of the training set samples. This indicates the benefits of zero-shot learning when the training samples are scarce. In addition, the values of AR_{unseen}, AR, and CR increase to 47.92%, 97.18%, and 97.43% respectively, when the ratio of the training samples is 80%. This is because the number of unseen classes decreases as the ratio of the training samples increases and the model can efficiently learn to match the visual features with the class embeddings.

Table 3. Results on MTHv2 (%) with different ratios (training samples/total samples).

Training samples/total samples	10%	30%	50%	80%
Testing seen/unseen samples	22343/2919	23916/1346	24339/923	24619/643
Testing seen/unseen classes	5051/1674	5776/949	6016/709	6195/530
Traditional setting AR/CR	94.49/94.75	96.35/96.60	96.99/97.25	97.18/97.43
Zero-shot setting AR_{unseen}	33.03	41.71	44.59	47.92

4.5   Experiments on the Handwriting Text Dataset

We also conduct experiments on the handwriting text datasets CASIA-HW and ICDAR13, where the former is used as the training set, whereas the latter is used as the testing set. The ICDAR13 comprises of text collected from different writing styles. For conducting a fair comparison with previous conventional methods, we synthesize 200 thousand additional text line samples using the same strategy as that adopted by Peng et al. [16]. The experimental results are shown in Table 4. As we can see, our method exhibits competitive performance without a language model as compared to previous methods under the traditional setting, while maintaining AR_{unseen} 36.49% in the zero-shot learning setting. We do not report the zero-shot setting result of the method using synthetic data because these methods use samples of all character classes for training.

Table 4. Comparison results (%) of previous methods on ICDAR13. Note that the * symbol denotes the method that does not use synthetic data for training.

Methods	Traditional setting (without LM)		Zero-shot setting
	AR	CR	AR_{unseen}
Messina et al. [15]	83.50	–	–
Wu et al. [28]	86.64	87.43	–
Du et al. [6]	83.89	–	–
Wang et al. [22]	88.79	90.67	–
Wang et al. [27]	89.66	–	–
Wang et al. [26]	91.58	–	–
Peng et al. [16]	89.61	90.52	–
Xiu et al. [31]	88.74	–	–
Xie et al. [29]	91.55	**92.13**	–
ZCTRN* (ours)	89.80	90.16	36.49
ZCTRN (ours)	**91.82**	**92.13**	–

4.6   Ablations

In this section, we describe the tests conducted for estimating whether the bidirectional embedding transfer module is useful or not. For the same, we train our model using two different approaches, namely, unidirectional embedding transfer and bidirectional embedding transfer, on the MTHv2 dataset. In unidirectional embedding transfer, the projecting function does not add an additional constraint. And thus, the projected class embeddings cannot be reconstructed back to the original class embeddings. The experimental results are shown in Table 5. As can be observed from the table, the value of AR_{unseen} increases by a large margin of 9.35% when the bidirectional embedding transfer module is used, which demonstrates the effectiveness of our proposed module.

Table 5. Traditional setting and zero-shot setting results (%) with different embedding transfer approaches on MTHv2.

Methods	Traditional setting		Zero-shot setting
	AR	CR	AR_{unseen}
Unidirectional	97.01	97.25	42.05
Bidirectional	**97.42**	**97.62**	**51.40**

5   Conclusions

In this paper, we have studied the zero-shot Chinese text recognition problem and have proposed the ZCTRN model that contains a text line encoder, class embedding module, and bidirectional embedding transfer module. To guide the training of the ZCTRN model, reconstruction and distance-based CTC losses have been integrated into the framework. Experiments conducted by applying the ZCTRN model on the historical document dataset and handwriting text dataset demonstrate its effectiveness in traditional as well as zero-shot learning settings. To the best of our knowledge, this may be the first zero-shot learning method proposed for long handwritten Chinese text recognition. In the future, it is worth looking for a method to encode characters at the stroke level to achieve zero-shot Chinese text recognition, which contains more fine-grained character information and requires fewer categories for encoding.

Acknowledgement. This research is supported in part by NSFC (Grant No.: 61936003), the National Key Research and Development Program of China (No. 2016YFB1001405), GD-NSF (no. 2017A030312006), Guangdong Intellectual Property Office Project (2018-10-1).

References

1. Akata, Z., Perronnin, F., Harchaoui, Z., Schmid, C.: Label-embedding for attribute-based classification. In: CVPR, pp. 819–826 (2013)
2. Ao, X., Zhang, X., Yang, H., Yin, F., Liu, C.: Cross-modal prototype learning for zero-shot handwriting recognition. In: ICDAR, pp. 589–594 (2019)
3. Baek, J., et al.: What is wrong with scene text recognition model comparisons? Dataset and model analysis. In: ICCV, pp. 4714–4722 (2019)
4. Bahdanau, D., Cho, K., Bengio, Y.: Neural machine translation by jointly learning to align and translate. In: ICLR (2015)
5. Cao, Z., Lu, J., Cui, S., Zhang, C.: Zero-shot handwritten Chinese character recognition with hierarchical decomposition embedding. Pattern Recognit. **107**, 107488 (2020)
6. Du, J., Wang, Z.-R., Zhai, J., Hu, J.: Deep neural network based hidden Markov model for offline handwritten Chinese text recognition. In: ICPR, pp. 3428–3433 (2016)
7. Fu, Y., Hospedales, T.M., Xiang, T., Gong, S.: Transductive multi-view zero-shot learning. IEEE Trans. Pattern Anal. Mach. Intell. **37**(11), 2332–2345 (2015)

8. Graves, A., Fernández, S., Gomez, F., Schmidhuber, J.: Connectionist temporal classification: labelling unsegmented sequence data with recurrent neural networks. In: ICML, pp. 369–376 (2006)

9. He, K., Zhang, X., Ren, S., Sun, J.: Deep residual learning for image recognition. In: CVPR, pp. 770–778 (2016)

10. Kodirov, E., Xiang, T., Gong, S.: Semantic autoencoder for zero-shot learning. In: CVPR, pp. 3174–3183 (2017)

11. Lampert, C.H., Nickisch, H., Harmeling, S.: Learning to detect unseen object classes by between-class attribute transfer. In: CVPR, pp. 951–958. IEEE (2009)

12. Li, Z., Wu, Q., Xiao, Y., Jin, M., Lu, H.: Deep matching network for handwritten Chinese character recognition. Pattern Recognit. **107**, 107471 (2020)

13. Liu, C., Yin, F., Wang, D., Wang, Q.: CASIA online and offline Chinese handwriting databases. In: ICDAR, pp. 37–41 (2011)

14. Ma, W., Zhang, H., Jin, L., Wu, S., Wang, J., Wang, Y.: Joint layout analysis, character detection and recognition for historical document digitization. In: ICFHR, pp. 31–36 (2020)

15. Messina, R., Louradour, J.: Segmentation-free handwritten Chinese text recognition with LSTM-RNN. In: ICDAR, pp. 171–175 (2015)

16. Peng, D., Jin, L., Wu, Y., Wang, Z., Cai, M.: A fast and accurate fully convolutional network for end-to-end handwritten Chinese text segmentation and recognition. In: ICDAR, pp. 25–30. IEEE (2019)

17. Shi, B., Bai, X., Yao, C.: An end-to-end trainable neural network for image-based sequence recognition and its application to scene text recognition. IEEE Trans. Pattern Anal. Mach. Intell. **39**(11), 2298–2304 (2017)

18. Shi, B., Yang, M., Wang, X., Lyu, P., Yao, C., Bai, X.: ASTER: an attentional scene text recognizer with flexible rectification. IEEE Trans. Pattern Anal. Mach. Intell. **41**(9), 2035–2048 (2018)

19. Shigeto, Y., Suzuki, I., Hara, K., Shimbo, M., Matsumoto, Y.: Ridge regression, hubness, and zero-shot learning. In: Appice, A., Rodrigues, P.P., Santos Costa, V., Soares, C., Gama, J., Jorge, A. (eds.) ECML PKDD 2015, Part I. LNCS (LNAI), vol. 9284, pp. 135–151. Springer, Cham (2015). https://doi.org/10.1007/978-3-319-23528-8_9

20. Wan, Z., Xie, F., Liu, Y., Bai, X., Yao, C.: 2D-CTC for scene text recognition. arXiv preprint arXiv:1907.09705 (2019)

21. Wang, Q., Yin, F., Liu, C.: Handwritten Chinese text recognition by integrating multiple contexts. IEEE Trans. Pattern Anal. Mach. Intell. **34**(8), 1469–1481 (2012)

22. Wang, S., Chen, L., Xu, L., Fan, W., Sun, J., Naoi, S.: Deep knowledge training and heterogeneous CNN for handwritten Chinese text recognition. In: ICFHR, pp. 84–89. IEEE (2016)

23. Wang, T., Xie, Z., Li, Z., Jin, L., Chen, X.: Radical aggregation network for few-shot offline handwritten Chinese character recognition. Pattern Recognit. Lett. **125**, 821–827 (2019)

24. Wang, T., et al.: Decoupled attention network for text recognition. In: AAAI, pp. 12216–12224 (2020)

25. Wang, W., Zheng, V.W., Yu, H., Miao, C.: A survey of zero-shot learning: settings, methods, and applications. ACM Trans. Intell. Syst. Technol. **10**(2), 1–37 (2019)

26. Wang, Z.R., Du, J., Wang, J.M.: Writer-aware CNN for parsimonious hmm-based offline handwritten Chinese text recognition. Pattern Recognit. **100**, 107102 (2020)

27. Wang, Z.R., Du, J., Wang, W.C., Zhai, J.F., Hu, J.S.: A comprehensive study of hybrid neural network hidden Markov model for offline handwritten Chinese text recognition. Int. J. Doc. Anal. Recognit. **21**(4), 241–251 (2018)
28. Wu, Y.C., Yin, F., Chen, Z., Liu, C.L.: Handwritten Chinese text recognition using separable multi-dimensional recurrent neural network. In: ICDAR, vol. 1, pp. 79–84. IEEE (2017)
29. Xie, C., Lai, S., Liao, Q., Jin, L.: High performance offline handwritten Chinese text recognition with a new data preprocessing and augmentation pipeline. In: Bai, X., Karatzas, D., Lopresti, D. (eds.) DAS 2020. LNCS, vol. 12116, pp. 45–59. Springer, Cham (2020). https://doi.org/10.1007/978-3-030-57058-3_4
30. Xie, Z., Huang, Y., Zhu, Y., Jin, L., Liu, Y., Xie, L.: Aggregation cross-entropy for sequence recognition. In: CVPR, pp. 6531–6540 (2019)
31. Xiu, Y., Wang, Q., Zhan, H., Lan, M., Lu, Y.: A handwritten Chinese text recognizer applying multi-level multimodal fusion network. In: ICDAR, pp. 1464–1469 (2019)
32. Yin, F., Wang, Q.F., Zhang, X.Y., Liu, C.L.: ICDAR 2013 Chinese handwriting recognition competition. In: ICDAR, pp. 1464–1470. IEEE (2013)
33. Zeiler, M.D.: Adadelta: an adaptive learning rate method. arXiv preprint arXiv:1212.5701 (2012)
34. Zhang, J., Du, J., Dai, L.: Radical analysis network for learning hierarchies of Chinese characters. Pattern Recognit. **103**, 107305 (2020)
35. Zhang, J., Zhu, Y., Du, J., Dai, L.: Radical analysis network for zero-shot learning in printed Chinese character recognition. In: ICME, pp. 1–6. IEEE (2018)
36. Zhang, L., Xiang, T., Gong, S.: Learning a deep embedding model for zero-shot learning. In: CVPR, pp. 2021–2030 (2017)

Text-Conditioned Character Segmentation for CTC-Based Text Recognition

Ryohei Tanaka[✉], Kunio Osada, and Akio Furuhata

Software Systems Research and Development Center, Toshiba Digital Solutions Corporation, 72-34 Horikawa-cho, Saiwai-ku, Kawasaki-shi, Kanagawa 212-8585, Japan
{ryohei3.tanaka,kunio.osada,akio.furuhata}@toshiba.co.jp

Abstract. Segmentation-free text recognition achieves successful performance because it can accurately recognize overlapping and touching characters, which is difficult for over-segmentation approaches. In contrast, character segmentation is helpful for explanations, posterior document layout analysis, and other applications. Some methods have been proposed to balance the capability of character segmentation and accurate recognition, but they cannot predict segmentation candidates for characters that can be differently segmented, which bottlenecks segmentation accuracy. In this paper, we propose Text-conditioned Character Segmentation (TCSeg) to improve segmentation accuracy. TCSeg segments characters differently according to each text candidate prediction by segmentation-free text recognition without affecting recognition accuracy. We also propose Overlap and Skip Error Suppression (OSESup) to suppress unintuitive errors using the estimated segmentation. An experiment on text recognition of handwritten Chinese characters shows that TCSeg segments characters more accurately than an existing segmentation method and that OSESup improves the recognition accuracy.

Keywords: Text recognition · Character segmentation · Connectionist temporal classification · Handwritten chinese character recognition

1 Introduction

Segmentation-free text recognition, which uses connectionist temporal classification (CTC) [1–9] or attention mechanisms [10–12], has achieved successful performance. One reason for this is that it can accurately recognize characters even when they are touching or overlapping, which is difficult to realize using over-segmentation approaches [13,14]. However, character positions obtained by character segmentation are helpful for explanations, posterior document layout analysis, and other applications, such as obscuring confidential letters in a text image [15].

Methods balancing the capability for character segmentation and accurate recognition of overlapping characters have been proposed [15,16], but they cannot predict segmentation candidates for characters that can be differently segmented, which bottlenecks segmentation accuracy. For example, "W" can be

© Springer Nature Switzerland AG 2021
J. Lladós et al. (Eds.): ICDAR 2021, LNCS 12823, pp. 142–156, 2021.
https://doi.org/10.1007/978-3-030-86334-0_10

CTC-based Text Recognition

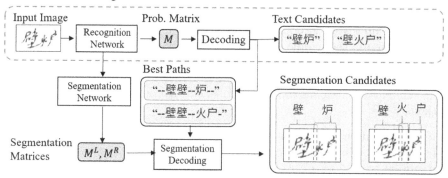

Fig. 1. Overview of TCSeg. The upper flow enclosed in a dashed rectangle is the same as normal CTC-based text recognition. Here, text candidates are two of the most likely recognition results. Segmentation matrices are predicted by a segmentation network and are decoded with each best path to a character segmentation candidate.

visually recognized and segmented as either "W" or "VV," which are almost indistinguishable in handwritten text recognition. Normally, the most likely text should be chosen from among feasible candidates (e.g., "W" and "VV") in a non-visual way, such as context or a language model. Hence, in cases where the image has characters that can be differently recognized and segmented, text candidates and their corresponding character segmentations should be predicted. Such cases are much more common in languages such as Chinese and Japanese, where many characters are composed from multiple radicals.

In this paper, we propose a novel method for accurately segmenting characters in each text candidate predicted from segmentation-free text recognition. We call the proposed method Text-conditioned Character Segmentation (TCSeg) because it segments under the condition that the segmentation is consistent with a given text, such as a text recognition candidate. Because TCSeg can segment a character differently depending on the text candidates, there are more chances to select a correct segmentation using context or language models, enhancing segmentation accuracy. Furthermore, TCSeg can be applied to any CTC-based text recognition method and does not affect recognition accuracy. Figure 1 shows an overview of TCSeg, described below.

We also propose Overlap and Skip Error Suppression (OSESup), which uses character segmentations estimated by TCSeg. Here, an "overlap error" is an error where a particular image area is recognized twice. An example is a "W" image being recognized as "VW," where the left part of the "W" is recognized twice. A "skip error" is an error where a particular image area is not recognized. An example is a "123" image being recognized as "13," where the "2" is skipped. Such non-intuitive errors, which are not seen when over-segmentation approaches are applied, should be particularly avoided in practical use, as users consider them stranger than simple replacement errors. OSESup suppresses overlap errors by penalizing overlapping segmentations and suppresses skip errors by penalizing character-like areas where no characters are assigned. OSESup can be efficiently applied to beam-search decoding algorithms [17,18], because TCSeg can segment characters even in the middle of the beam-search loop.

Finally, we evaluate TCSeg and OSESup on a handwritten Chinese text dataset and demonstrate the effectiveness of TCSeg and OSESup.

In summary, TCSeg has the following benefits as compared with conventional methods:

– its segmentation accuracy is higher because it can segment a character differently depending on text candidates;
– it can be applied to any CTC-based text recognition method and does not affect recognition accuracy; and
– it can be used by OSESup, which suppresses recognition errors.

2   Related Works

Some methods for estimating positions of recognized characters while accurately recognizing overlapping and touching characters have been proposed.

The attention-based approach [10–12] is one of the most accurate segmentation-free text recognition approaches especially for scene-text recognition tasks, where characters in text line images are often distributed two-dimensionally. This approach first estimates an attention map that indicates rough character positions, then estimates character probabilities corresponding to the attention map, repeating this process until the text ends. Attention maps can thus be used to roughly estimate character positions.

Peng et al. [15] proposed a text recognition and character segmentation method that predicts bounding boxes and character-class probabilities for a vertically divided grid of a text image, similar to object detection methods such as YOLO [19]. Evaluating their method on handwritten Chinese character recognition, they demonstrated that its recognition accuracy was competitive with other methods and that segmentation accuracy was not much worse than other object detection methods (YOLO [19] and Faster R-CNN [20]). Qi et al. [16] proposed a similar method that predicts center positions of left and right characters instead of a bounding box. They evaluated their method using their own metrics for scene-text recognition tasks.

However, these methods cannot predict different segmentation candidates for characters that can be segmented differently, such as "W" and "VV," which bottlenecks segmentation accuracy.

TCSeg is an extension of CTC-based text recognition approaches [2–9]. The flow of CTC-based text recognition is as follows. An input image is fed into a recognition network, which is typically composed of convolutional neural networks (CNNs) and recurrent neural networks (RNNs). The recognition network yields a sequence of categorical distributions $M \in \mathbb{R}^{L \times K}$, where L is the sequence

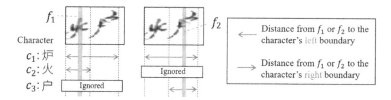

Fig. 2. Concept of TCSeg, which predicts left and right character boundaries from consideration of each frame. Predictions of other characters are omitted in this figure.

length determined by the width of the input image and K represents the number of classes composed of $(K-1)$ character classes and a blank class. The probability of a text y is

$$p_{\mathrm{CTC}}(y) = \sum_{\pi \in \mathcal{B}^{-1}(y)} \prod_{i=1,\ldots,L} M_{i,\pi_i}, \tag{1}$$

where $\mathcal{B}$ is the many-to-one mapping function used in the CTC algorithm [1], which deletes blank and consecutive labels. In the inference stage, a decoder calculates the most likely text $\hat{y}$ as

$$\hat{y} = \arg\max_{y}\{p_{\mathrm{CTC}}(y)p_{\mathrm{LM}}(y)\}, \tag{2}$$

where $p_{\mathrm{LM}}(y)$ is an optional prior probability defined by the language model. The second-most-likely text is obtained by selecting the text with the second-largest score $p_{\mathrm{CTC}}(y)p_{\mathrm{LM}}(y)$. In this way, the CTC-based approach can predict text candidates even when they are segmented differently, which is an essential attribute for TCSeg.

TCSeg is designed not only to segment characters but also to retain the recognition accuracy of CTC-based approaches. CTC-based approaches are known to perform well in regular handwritten text recognition tasks [5,8] and tasks that require complicated language models such as weighted finite state transducers [21].

3   Text-Conditioned Character Segmentation

In this section, we describe the TCSeg algorithm. We consider only vertical segmentation, or left and right character boundaries, because horizontal segmentation is unimportant in text line images. We assume that annotations of character location are available.

3.1   Inference

The main idea of TCSeg is prediction of each character boundary from consideration of each frame. Here, "frame" is a vertically divided area in the input image. Figure 2 shows the desired predictions at two frames f_1 and f_2, where distances from the frames to each character boundary are predicted. For instance, the distance from f_1 to character c_1's right boundary can be larger than that of character c_2's right boundary, depending on the character shape. Other characters, including c_3, do not matter because they do not appear in f_1. This character-wise boundary prediction enables different segmentations depending on a given text.

Figure 1 shows an overview of the TCSeg flow. The upper dashed rectangle denotes the normal CTC-based text recognition flow described in Sect. 2. First, a feature vector extracted from the recognition network is fed into the segmentation network, yielding segmentation matrices $M^L, M^R \in \mathbb{R}^{L \times (K-1)}$, which indicate distances to the left and right boundaries of each character from consideration of each frame. The number of frames is set to the same number as the sequence length L to simplify the calculation. In other words, the input image is vertically divided into L areas, and left and right boundaries of each character are estimated from the viewpoint of the divided areas.

Next, key frames are assigned for each character in the target text. We employ the most likely label sequence (referred as the "best path" in this paper) among the label sequences that can be mapped to the target text. The best path $\hat{\pi}$ corresponding to a text y is calculated as

$$\hat{\pi}(y) = \arg \max_{\pi \in \mathcal{B}^{-1}(y)} \prod_{i=1,\dots,L} M_{i,\pi_i}. \tag{3}$$

The best path can be efficiently calculated by a beam-search algorithm during the text candidate search. Using a feasible path that can be mapped to the target text guarantees that each character in the target text is connected with at least one key frame. Figure 3 (left) shows an example. In this case, the target text is "AB" and the key frames for "A" and "B" are the 4th and the 11–12th frames, respectively.

Finally, the left and right character boundaries are predicted from the viewpoint of corresponding key frames. Formally, from segmentation matrices M^L, M^R and the best path $\hat{\pi}$, left and right character boundaries x_j^L, x_j^R ($j = 1, \dots, |\hat{y}|$) are calculated by a segmentation decoding algorithm, shown as Algorithm 1 and visualized in Fig. 3 (left). Basically, if the i-th character $\hat{\pi}_i$ is not blank, $M_{i,\hat{\pi}_i}^L$ and $M_{i,\hat{\pi}_i}^R$ are used to estimate the left and right boundaries of character $\hat{\pi}_i$. If a character appears consecutively in the best path (i.e., $\hat{\pi}_{i^L} = \hat{\pi}_{i^L+1} = \cdots = \hat{\pi}_{i^R} = c$), like "B" in Fig. 3 (left), then the leftmost and rightmost indices $M_{i^L,c}^L$ and $M_{i^R,c}^R$ are used.

As described above, TCSeg only extracts information from the recognition process and does not interfere with the text prediction process. Because of this independence, TCSeg can be added to any CTC-based text recognition method without affecting its accuracy.

3.2 Training

In this section, we describe segmentation network training. Given a ground truth (GT) text y and left and right boundary positions $x_j^L, x_j^R (j = 1, \dots, |y|)$ for each character, objective matrices $A^L, A^R \in \mathbb{R}^{L \times (K-1)}$ are defined as

Algorithm 1. Segmentation decoding.

Require: segmentation matrices M^L, M^R, best path $\hat{\pi}$, image width W

> $\hat{y} = \mathcal{B}(\hat{\pi})$
> $j = 1$
> **for** $i = 1$ to $|\hat{\pi}|$ **do**
>   **if** $\hat{\pi}_i = \hat{y}_j$ **then**
>   **if** $i > 1$ and $\hat{\pi}_{i-1} \neq \hat{\pi}_i$ **then**
> $x_j^L = (i - 1 - M_{i,\hat{\pi}_i}^L)W/|\hat{\pi}|$
>   **end if**
>   **if** $i = |\hat{\pi}|$ or $\hat{\pi}_i \neq \hat{\pi}_{i+1}$ **then**
> $x_j^R = (i + M_{i,\hat{\pi}_i}^R)W/|\hat{\pi}|$
> $j = j + 1$
>   **end if**
>   **end if**
> **end for**
> **Return** $x_1^L, \ldots, x_{|\hat{y}|}^L, x_1^R, \ldots, x_{|\hat{y}|}^R,$

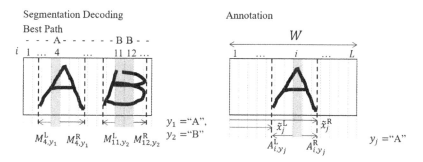

Fig. 3. Visualization of segmentation decoding (left) and annotation (right). "-" denotes a blank label.

$$A_{i,k}^L := \begin{cases} (i - 1) - \tilde{x}_j^L & (\text{if } \exists j, (k = y_j) \cap (\tilde{x}_j^L + 1 \leq i \leq \tilde{x}_j^R)) \\ \text{None} & (\text{otherwise}) \end{cases} \quad (4)$$

$$A_{i,k}^R := \begin{cases} \tilde{x}_j^R - i & (\text{if } \exists j, (k = y_j) \cap (\tilde{x}_j^L + 1 \leq i \leq \tilde{x}_j^R)) \\ \text{None} & (\text{otherwise}), \end{cases} \quad (5)$$

where W is the input image width, $\tilde{x}_j^L = x_j^L L/W$, and $\tilde{x}_j^R = x_j^R L/W$. The formula after "if" is interpreted as "if the i-th frame is in the area of character y_j." A^L and A^R are defined to have the same scale with the frames to simplify the decoding calculation. Figure 3 (right) visualizes the annotation process. When characters overlap, they are annotated as depicted in Fig. 4 (left).

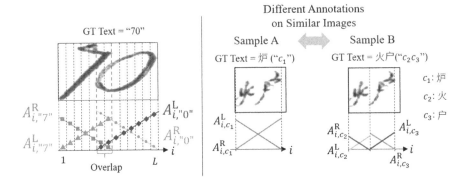

Fig. 4. Annotation on an overlapping character (left) and contradictory samples (right). A^{L} and A^{R} are plotted along the frames. Samples A and B are visually similar, but have different segmentation annotations. An ideal pair of M^{L} and M^{R} makes the loss function $\mathcal{L}$ zero.

For each training sample $\{X, A^{\mathrm{L}}, A^{\mathrm{R}}\}$, where X is an input image and $A^{\mathrm{L}}, A^{\mathrm{R}}$ are annotations for X, the segmentation network is trained to minimize the loss function

$$\mathcal{L}(X, A^{\mathrm{L}}, A^{\mathrm{R}}) := \sum_{\substack{k=1,\ldots,K-1 \\ i=1,\ldots,L}} \left\{ \delta(i,k) \left(|A^{\mathrm{L}}_{i,k} - M^{\mathrm{L}}_{i,k}(X)| + |A^{\mathrm{R}}_{i,k} - M^{\mathrm{R}}_{i,k}(X)| \right) \right\}, \quad (6)$$

where $\delta(i,k) = 0$ when $A^{\mathrm{L}}_{i,k} = $ None and 1 otherwise, and $M^{\mathrm{L}}(X), M^{\mathrm{R}}(X)$ are segmentation network outputs.

This annotation and training strategy allows handling contradictory cases where visually similar training images have different segmentation annotations. Figure 4 (right) shows an example, where samples A and B are visually similar but have different annotations ("c_1" and "c_2c_3"). Even in these contradictory samples, there is an ideal $M^{\mathrm{L}}, M^{\mathrm{R}}$ pair that makes the loss function $\mathcal{L}$ (Eq. 6) zero. In the inference stage, given the ideal $M^{\mathrm{L}}, M^{\mathrm{R}}$ pair, the image can be segmented as both "c_1" and "c_2c_3" as described in Sect. 3.1. This ability of multiple segmentations is attributed to the character-wise boundary prediction approach, the most unique feature of TCSeg.

The segmentation and recognition networks can be trained both separately and simultaneously. When they are trained simultaneously, the recognition network weights should be frozen during back propagation of the segmentation loss.

The segmentation network can be designed to take the image as input rather than the feature vector of the recognition network in cases where calculation times and memory usage are not crucial.

3.3   Limitations

We assume that key frames extracted from the best paths are located within corresponding character areas, but this is not guaranteed. However, we observe that this assumption holds in almost all cases in our experiment, possibly from the CTC learning algorithm. If this assumption does not hold for other tasks, it is helpful to add an extra loss function to increase the character probability at the center frame of the character's area.

4   Overlap and Skip Error Suppression

Next, we consider using the estimated segmentation to suppress overlap and skip errors, or OSESup. We define an overlap penalty $p_{\text{ovlp}}(y)$ and a skip penalty $p_{\text{skip}}(y)$ where y is a given text, and the best text $\hat{y}$ is calculated as

$$\hat{y} = \arg \max_{y}\{p_{\text{CTC}}(y)p_{\text{LM}}(y)p_{\text{ovlp}}(y)p_{\text{skip}}(y)\}, \tag{7}$$

which can be efficiently calculated during beam-search decoding with a small modification.

The overlap penalty is defined as

$$p_{\text{ovlp}}(y) := C_{\text{op}}^{m(y)}, \tag{8}$$

where $m(y)$ $(0 \leq m(y))$ represents the total amount of overlapped area among the estimated character segmentations for text y, and C_{op} $(0 < C_{\text{op}} \leq 1)$ is a parameter for controlling the strength of OSESup. The larger C_{op} is, the weaker OSESup becomes, with OSESup disabled when $C_{\text{op}} = 1$. Note that too small a value for C_{op} results in erroneous suppressions of actually overlapping characters. Figure 5 (left) visualizes the overlap penalty.

The skip penalty is defined as

$$p_{\text{skip}}(y) := \prod_{i=1,\ldots,L} (1 - h_i b_i(y))^{C_{\text{sp}}}, \tag{9}$$

where h_i $(0 \leq h_i \leq 1)$ represents the non-blank probability for the i-th frame, $b_i(y)$ indicates whether a character is segmented in the i-th frame ($b_i(y) = 1$ when no character is segmented and 0 otherwise), and C_{sp} $(0 \leq C_{\text{sp}})$ is a parameter for controlling the strength of OSESup. The smaller C_{sp} is, the weaker OSESup becomes, with OSESup disabled when $C_{\text{sp}} = 0$. Note that too large a value for C_{sp} results in erroneous recognitions of non-character noises such as ink blots or stains. Figure 5 (right) visualizes the skip penalty.

Non-blank probability h_i $(i = 1, \ldots, L)$ represents the likelihood of a character existing in the i-th frame. It can be estimated in many ways, such as analyzing a luminance histogram. In this work, we estimate h_i using a neural network. More precisely, the segmentation network outputs not only $M^{\text{L}}, M^{\text{R}} \in \mathbb{R}^{L \times (K-1)}$, but also $h' \in \mathbb{R}^{L \times 2}$. Then the softmax function is applied to h' to obtain $h \in \mathbb{R}^{L}$.

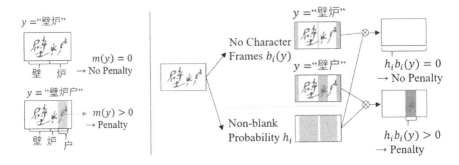

Fig. 5. Visualization of overlap penalty (left) and skip penalty (right). Overlap penalty is applied when segmented areas are overlapping. Skip penalty is applied when no character is segmented in an area with high non-blank probability. h_i and b_i are described in the text.

Prediction of h is regarded as a two-class semantic segmentation task along the frames and is trained by minimizing binary cross-entropy loss alongside minimization of the segmentation loss.

The p_{ovlp} and p_{skip} values can be calculated even in the middle of the beam-search decoding by storing the best path, the latest p_{ovlp} and p_{skip}, and the right boundary of the last character for each candidate during the beam-search loop.

We set the OSESup parameters to empirically chosen values of $C_{\mathrm{op}} = 0.6, C_{\mathrm{sp}} = 0.05$.

5   Experiment on a Handwritten Chinese Text Dataset

5.1   Dataset

Dataset settings follow a previous work [15], except for the minor differences described below.

We use 3,894,998 isolated character images from CASIA-HWDB 1.0–1.2 (HWDB1) [22], 52,230 text line images from CASIA-HWDB 2.0–2.2 (HWDB2) [22], and 3,432 text line images from the ICDAR 2013 competition dataset (ICDAR2013) [23]. We set the total number of character classes to 7,374 after merging all characters in these datasets.

Two training and test dataset settings are used, a standard data setting and a small data setting. In the standard data setting employed in many works [15, 23,24], HWDB1 and HWDB2 are used as training datasets and ICDAR2013 as the test dataset. In the small data setting, which is for fair comparison of segmentation accuracy with a previous work [15], the preset training and test datasets of HWDB2 are used, so the number of training and test images is 41,781 and 10,449, respectively.

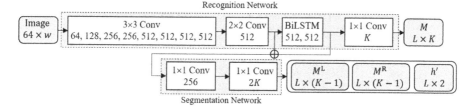

Fig. 6. Architecture of our model. Numbers at the block bottoms denote output channels of the layers or shapes of the tensors. Each convolution layer is followed by batch normalization and ReLU activation. The BiLSTM dropout rate is set to 0.3 during training. $\oplus$ indicates concatenation.

In the standard data setting, the isolated character images from HWDB1 are randomly concatenated to synthesize text line images, as in the previous work [15]. The total number of synthesized images is uncountable, as we dynamically synthesize the training images during training rather than before training. We mix the 52,230 text line images from HWDB2 and the synthetic images at a ratio of 52,230 to 800,000, so synthesized images appear almost 94 times in 100 training images.

As a language model, we employ a trigram trained with a corpus obtained from the Internet.

5.2 Implemetation Detail

The architecture of the recognition network is similar to CRNN [2]. Figure 6 shows the architecture. Input images are resized to a height of 64 px and fed into a recognition network comprising nine convolution layers and two bidirectional LSTM (BiLSTM) layers. The number of classes K is 7,375, including a blank class, and the sequence length L is set to $\lceil \lceil w/2 \rceil /2 \rceil - 1$, where $\lceil \cdot \rceil$ represents the ceiling function. The segmentation network comprises two 1×1 convolutional layers. The features from both convolution layers and BiLSTM layers are used for segmentation because we consider both visual information in the convolutional features and context information in the recurrent features to be important for accurate segmentations.

In training, we first train only recognition networks over 900,000 iterations at a batch size of 10, applying the Adam optimizer and a cosine learning rate decay [25] that gradually declines the learning rate from 10^{-4} to 10^{-5}. Next, we train the segmentation network over 300,000 iterations with the recognition network parameters frozen. At this time, the Adam optimizer with a learning rate of 10^{-4} is applied.

We use the accurate rate (AR) [23] to measure recognition accuracies.

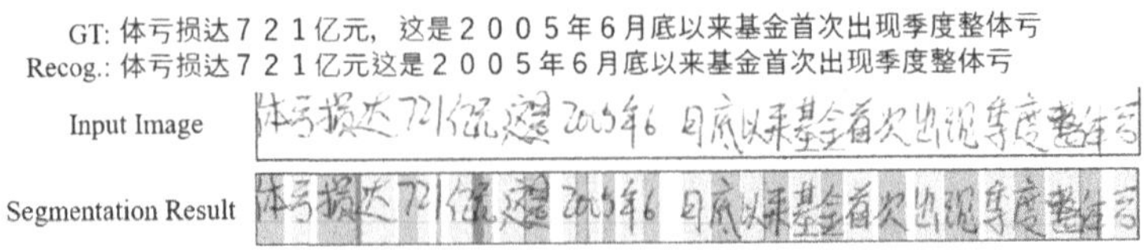

GT: 体亏损达 7 2 1 亿元，这是 2 0 0 5 年 6 月底以来基金首次出现季度整体亏
Recog.: 体亏损达 7 2 1 亿元这是 2 0 0 5 年 6 月底以来基金首次出现季度整体亏

Input Image

Segmentation Result

Fig. 7. Example segmentation result. Segmented areas are alternately colored blue and red. Purple indicates an overlapping area.

Table 1. Character segmentation accuracy. Precision, recall and F-measure are calculated on all characters at an IoU threshold of 0.5.

Method	Small data setting			Standard data setting		
	Precision	Recall	F-measure	Precision	Recall	F-measure
Peng et al. [15]	0.9456	0.9236	0.9344	–	–	–
TCSeg	**0.9628**	**0.9616**	**0.9622**	**0.9509**	**0.9507**	**0.9508**

Application code was implemented in Python 3 with the Tensorflow 2 framework[1], except for beam-search decoding, which is implemented in C++. We use a single Nvidia GeForce GTX 1080Ti GPU for both training and inference.

5.3 Results and Discussion

Segmentation Accuracy. We first evaluate the segmentation accuracy. Figure 7 shows a test sample segmented by TCSeg for qualitative analysis. Although there is one deletion error ("，"), almost all characters, including overlapping or touching characters, are correctly recognized and segmented.

Quantitatively, precision, recall and F-measure at an IoU threshold of 0.5 are calculated to compare segmentation accuracy with the previous work [15]. Because TCSeg cannot estimate upper and lower character boundaries, we simply trim upper and lower white spaces in each segmented character area. We evaluated both small and standard data settings.

Table 1 shows the segmentation accuracy. In the small data setting, TCSeg precision, recall, and F-measure are all higher than under the previous method. In addition, the TCSeg F-measure in the standard data setting is 0.9508, which is sufficiently accurate as compared with the small data setting results, despite the standard data setting being much more difficult than the small data setting when the same texts appear in the training and test images. We consider this substantial progress over the previous work, which reported that synthesized images mislead character segmenting and did not show the results of the standard data setting.

Next, to analyze the segmentation errors, we count the number of segmentation errors for each character in the standard data setting. In this analysis,

---

[1] https://www.tensorflow.org/.

,	。	、	-	"	"	1	0	的	了	Total
2,019	503	416	233	160	160	119	104	58	41	5,443

Fig. 8. Top ten erroneous characters and their counts in the standard data setting. "Total" denotes the total number of errors regardless of characters. Deleted and inserted characters are not included in this table.

Table 2. Recognition accuracy (AR). "Seg." denotes whether the method can segment characters. "LM" denotes the language model.

Method	Seg.	Without LM	With LM
Wang et al. [24]		**91.58**	**96.83**
Peng et al. [15]	✓	89.61	94.88
TCSeg	✓	90.81	94.42
TCSeg + OSESup	✓	91.00	94.63

deleted and inserted characters are ignored for simplicity. Figure 8 shows the top ten erroneous characters and their counts. More than half of the segmentation errors come from small symbols (half- and full-width commas, periods, quotation marks, hyphens, etc.), whose positions are often less important than other characters. TCSeg can thus be considered as having practical segmentation ability exceeding the accuracy shown in Table 1.

Recognition Accuracy. In this section, we evaluate the recognition accuracy. Hereafter, only the standard data setting is used. Table 2 shows the recognition accuracy. The evaluated model is the same as that of the standard data setting in Table 1. As Table 2 shows, the accuracy of TCSeg is competitive with the previous segmentation and recognition method [15] and is not much below state-of-the-art accuracy [24], which cannot segment characters. These results suggest that TCSeg retains competitive recognition accuracy alongside high segmentation accuracy. Note that sources of the language models vary among the methods, but we think this does not affect our conclusions.

In this experiment, we just used a simple architecture for the recognition network. Applying TCSeg to more powerful CTC-based text recognition methods [3–9] is expected to improve recognition accuracy. This flexibility is one of the most important benefits of TCSeg.

OSESup Evaluation. Next, we evaluate OSESup. Figure 9 shows examples of overlap error (left) and skip error (right) corrected by OSESup. In the left example, while the right part of the second character is recognized twice and the segmented areas overlap without OSESup, the correct text and segmentation are ranked as the most likely candidate with OSESup because it penalizes the overlapping segmented areas. In the right example, while the second character is deleted from the recognized text without OSESup, the correct text and

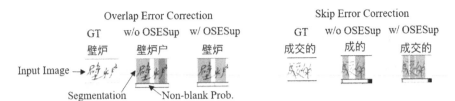

Fig. 9. Overlap error (left) and skip error (right) corrected by OSESup. Estimated character segmentations and non-blank probability (white represents 1.0) are shown below the recognized texts. OSESup successfully corrects the errors.

Table 3. Numbers of deletion, insertion and replacement errors on the result with LM.

Method	Deletion	Insertion	Replacement	AR
TCSeg [a]	598	511	3,996	94.42
TCSeg + OSESup [b]	493	456	3,962	94.63
Improvement [(a − b)/a]	17.6%	10.8%	0.9%	

segmentation are ranked as the most likely candidate with OSESup because it penalizes the area that is not segmented as any character and has high non-blank probability.

Quantitatively, OSESup enhances the recognition from 94.42 to 94.63 with a language model, as Table 2 shows. We further count the number of deletion and insertion errors as an indicator for automatically measuring the number of overlap and skip errors. Table 3 shows the results. Although the improvement in recognition accuracy is not significant, OSESup avoids 17.6% of deletion errors and 10.8% of insertion errors, suggesting OSESup suppresses a substantial number of overlap and skip errors, which as described in Sect. 1 is important for practical use.

Speed. Table 4 shows average speeds. These results suggest that as compared with the baseline, TCSeg has only a minor impact on speed (from 29.7 to 27.7 fps with LM) and OSESup has almost no impact (27.7 vs. 27.6 fps with LM), demonstrating the efficiency of TCSeg and OSESup. Furthermore, deceleration by using a language model is small (from 28.7 to 27.6 fps).

Table 4. Average and standard deviation of speed [fps] among ten trials on the ICDAR2013 dataset, with batch size set to 1. Baseline denotes normal CTC-based text recognition.

Method	Without LM	With LM
Baseline	30.8 ± 0.8	29.7 ± 0.7
TCSeg	28.6 ± 0.6	27.7 ± 0.6
TCSeg + OSESup	28.7 ± 0.8	27.6 ± 0.9

6 Conclusion

We have proposed a novel character segmentation method TCSeg and an error correction method OSESup. Experiments on a handwritten Chinese text dataset show that TCSeg achieves higher segmentation accuracy (0.9344 vs. 0.9622), while retaining competitive recognition accuracy. The results also suggest that OSESup successfully suppresses overlap and skip errors (17.6% deletion and 10.8% insertion error decrease). Moreover, we confirm that the additional computation cost due to TCSeg and OSESup is minor (from 29.7 to 27.6 fps).

References

1. Graves, A., Fernández, S., Gomez, F., Schmidhuber, J.: Connectionist temporal classification: labelling unsegmented sequence data with recurrent neural networks. In: Proceedings of the 23rd International Conference on Machine Learning, pp. 369–376 (2006)
2. Shi, B., Bai, X., Yao, C.: An end-to-end trainable neural network for image-based sequence recognition and its application to scene text recognition. IEEE Trans. Pattern Anal. Mach. Intell. **39**(11), 2298–2304 (2016)
3. Wang, J., Hu, X.: Gated recurrent convolution neural network for OCR. In: Advances in Neural Information Processing Systems, pp. 335–344 (2017)
4. Liu, W., Chen, C., Wong, K.Y.K., Su, Z., Han, J.: Star-net: a spatial attention residue network for scene text recognition. In: BMVC, vol. 2, p. 7 (2016)
5. Strauß, T., Leifert, G., Labahn, R., Hodel, T., Mühlberger, G.: ICFHR 2018 competition on automated text recognition on a read dataset. In: 2018 16th International Conference on Frontiers in Handwriting Recognition (ICFHR), pp. 477–482. IEEE (2018)
6. Hu, W., Cai, X., Hou, J., Yi, S., Lin, Z.: GTC: guided training of CTC towards efficient and accurate scene text recognition. In: AAAI, pp. 11005–11012 (2020)
7. Ingle, R.R., Fujii, Y., Deselaers, T., Baccash, J., Popat, A.C.: A scalable handwritten text recognition system. In: 2019 International Conference on Document Analysis and Recognition (ICDAR), pp. 17–24. IEEE (2019)
8. Xiao, S., Peng, L., Yan, R., Wang, S.: Deep network with pixel-level rectification and robust training for handwriting recognition. SN Comput. Sci. **1**(3), 1–13 (2020)
9. Tanaka, R., Ono, S., Furuhata, A.: Fast distributional smoothing for regularization in CTC applied to text recognition. In: 2019 International Conference on Document Analysis and Recognition (ICDAR), pp. 302–308. IEEE (2019)

10. Cheng, Z., Bai, F., Xu, Y., Zheng, G., Pu, S., Zhou, S.: Focusing attention: towards accurate text recognition in natural images. In: Proceedings of the IEEE international conference on computer vision. pp. 5076–5084 (2017)
11. Shi, B., Yang, M., Wang, X., Lyu, P., Yao, C., Bai, X.: Aster: an attentional scene text recognizer with flexible rectification. IEEE Trans. Pattern Anal. Mach. Intell. **41**(9), 2035–2048 (2018)
12. Baek, J., et al.: What is wrong with scene text recognition model comparisons? Dataset and model analysis. In: Proceedings of the IEEE International Conference on Computer Vision, pp. 4715–4723 (2019)
13. Wu, Y.C., Yin, F., Liu, C.L.: Improving handwritten Chinese text recognition using neural network language models and convolutional neural network shape models. Pattern Recognit. **65**, 251–264 (2017)
14. Wang, Z.X., Wang, Q.F., Yin, F., Liu, C.L.: Weakly supervised learning for over-segmentation based handwritten Chinese text recognition. In: 2020 17th International Conference on Frontiers in Handwriting Recognition (ICFHR), pp. 157–162. IEEE (2020)
15. Peng, D., Jin, L., Wu, Y., Wang, Z., Cai, M.: A fast and accurate fully convolutional network for end-to-end handwritten Chinese text segmentation and recognition. In: 2019 International Conference on Document Analysis and Recognition (ICDAR), pp. 25–30. IEEE (2019)
16. Qi, X., Chen, Y., Xiao, R., Li, C.G., Zou, Q., Cui, S.: A novel joint character categorization and localization approach for character-level scene text recognition. In: 2019 International Conference on Document Analysis and Recognition Workshops (ICDARW), vol. 5, pp. 83–90. IEEE (2019)
17. Graves, A., Liwicki, M., Fernández, S., Bertolami, R., Bunke, H., Schmidhuber, J.: A novel connectionist system for unconstrained handwriting recognition. IEEE Trans. Pattern Anal. Mach. Intell. **31**(5), 855–868 (2008)
18. Scheidl, H., Fiel, S., Sablatnig, R.: Word beam search: a connectionist temporal classification decoding algorithm. In: 2018 16th International Conference on Frontiers in Handwriting Recognition (ICFHR), pp. 253–258. IEEE (2018)
19. Redmon, J., Divvala, S., Girshick, R., Farhadi, A.: You only look once: unified, real-time object detection. In: Proceedings of the IEEE Conference on Computer Vision and Pattern Recognition, pp. 779–788 (2016)
20. Ren, S., He, K., Girshick, R., Sun, J.: Faster R-CNN: towards real-time object detection with region proposal networks. IEEE Trans. Pattern Anal. Mach. Intell. **39**(6), 1137–1149 (2016)
21. Cong, F., Hu, W., Huo, Q., Guo, L.: A comparative study of attention-based encoder-decoder approaches to natural scene text recognition. In: 2019 International Conference on Document Analysis and Recognition (ICDAR), pp. 916–921. IEEE (2019)
22. Liu, C.L., Yin, F., Wang, D.H., Wang, Q.F.: CASIA online and offline Chinese handwriting databases. In: 2011 International Conference on Document Analysis and Recognition, pp. 37–41. IEEE (2011)
23. Yin, F., Wang, Q.F., Zhang, X.Y., Liu, C.L.: ICDAR 2013 Chinese handwriting recognition competition. In: 2013 12th International Conference on Document Analysis and Recognition, pp. 1464–1470. IEEE (2013)
24. Wang, Z.R., Du, J., Wang, J.M.: Writer-aware CNN for parsimonious hmm-based offline handwritten Chinese text recognition. Pattern Recognit. **100**, 107102 (2020)
25. He, T., Zhang, Z., Zhang, H., Zhang, Z., Xie, J., Li, M.: Bag of tricks for image classification with convolutional neural networks. In: Proceedings of the IEEE Conference on Computer Vision and Pattern Recognition, pp. 558–567 (2019)

Towards Fast, Accurate and Compact Online Handwritten Chinese Text Recognition

Dezhi Peng[1], Canyu Xie[1], Hongliang Li[1], Lianwen Jin[1,2(✉)], Zecheng Xie[1], Kai Ding[3], Yichao Huang[3], and Yaqiang Wu[4]

[1] South China University of Technology, Guangzhou, China
{eedzpeng,eecanyuxie,eehugh}@mail.scut.edu.cn, eelwjin@scut.edu.cn
[2] Guangdong Artificial Intelligence and Digital Economy Laboratory (Pazhou Lab), Guangzhou, China
[3] IntSig Information Co., Ltd., Shanghai, China
{danny_ding,charlie_huang}@intsig.net
[4] Lenovo Research, Beijing, China
wuyqe@lenovo.com

Abstract. Although great success has been achieved in online handwritten Chinese text recognition (OLHCTR), most existing methods based on over-segmentation or long short-term memory are inefficient and not parallelizable. Moreover, n-gram language models and beam search algorithm were commonly adopted by many existing systems as a part of post-processing, resulting in extremely low speed and large footprint. To this end, we propose a fast, accurate and compact approach for OLHCTR. The proposed method consists of a global and local relationship network (GLRNet) and a Transformer-based language model (TransLM). A novel feature extraction mechanism, which alternately learns global and local dependencies of input trajectories, is proposed in GLRNet for the recognition of online texts. Based on the output of GLRNet, TransLM captures contextual information through Transformer encoder and further improves the recognition accuracy. The recognition and language modelling are always treated as two separate parts. However, the two components of our methods are jointly optimized, which ensures the optimal performance of the whole model. Furthermore, the non-recurrence design improves the parallelization and efficiency of our method, and the parameterized TransLM avoids the large footprint to store the probabilities of n-grams. The experiments on CASIA-OLHWDB2.0-2.2 and ICDAR2013 competition dataset show that our method achieves state-of-the-art performances with the fastest speed and the smallest footprint. Especially in the situation with language model, our method exhibits 2 times to 130 times acceleration compared with existing methods.

Keywords: Online handwritten Chinese text recognition · Language model · Transformer

© Springer Nature Switzerland AG 2021
J. Lladós et al. (Eds.): ICDAR 2021, LNCS 12823, pp. 157–171, 2021.
https://doi.org/10.1007/978-3-030-86334-0_11

1 Introduction

As the prevalence of pen-based devices, online handwritten Chinese text recognition (OLHCTR) [28,31,32] has been widely applied to the fields including intelligent education, office automation and so on. However, it is still a challenging task to recognize the online handwritten Chinese text because of the diverse writing styles, the large-scale vocabulary and the difficulty of character segmentation. Moreover, plenty of real-world applications that involves OLHCTR also bring very high requirements on accuracy, speed and compactness of models.

During the last decades, lots of solutions have been proposed to address OLHCTR and achieved impressive performances. Existing methods can be roughly divided into three categories, namely, segmentation-based methods, integrated convolutional neural network (CNN) and long short-term memory (LSTM) methods, and LSTM-based methods. Segmentation-based methods [22,24,32,33] derive multiple candidate segmentation-recognition paths from the consecutive segments generated by over-segmentation, and then search for the optimal path. However, the complex design of segmentation-based methods results in inefficiency of the whole system, and the difficulty of character segmentation leads to unsatisfactory recognition accuracy. Approaches combining CNN and LSTM [5,19,28] transforms pen-tip trajectories into image-like representations by path-signature or eight-directional feature maps, which requires domain-specific knowledge and introduces much more redundant computation. LSTM-based methods [10,13,27,31] directly handle the sequential pen-tip trajectories through LSTM or gated recurrent unit (GRU) and achieve impressive performance on accuracy and compactness. But the architecture of recurrent components restricts the model parallelization and reduces the speed during inference.

Language model is another crucial component for OLHCTR yet has not been be well solved. Even for native speakers, the recognition of Chinese texts still heavily relies on linguistic context. Most existing methods [13,28] adopt n-gram language model and beam search algorithm, which will occupy very large memory to store the probabilities of n-grams and consume a very long time to search for the best recognition result. To alleviate the shortcomings above, Cai et al. [3] propose weighted finite-state transducer (WFST) based decoders that have small footprint and fast speed. However, WFST-based decoders are still based on n-grams and not learnable. Besides, Wu et al. [25] use learnable neural network language model to improve handwritten Chinese text recognition, but it is designed based on over-segmentation based methods and may be hard to be applied to other methods. Furthermore, Xie et al. [28] propose an implicit language model, but it is still not fast enough because of the recurrent layers and not accurate enough because of the difficulty of LSTM layers in building long-range context.

To solve the issues mentioned above, we propose a fast, accurate and compact approach for OLHCTR. Recently, Transformer [21] has become a big breakthrough in both natural language processing [6,11,14] and computer vision [4,16]. Partly inspired by Transformer, we design a global and local relationship network (GLRNet) and a Transformer-based language model (TransLM).

GLRNet directly takes pen-tip trajectories as input and is compose of stacked global and local relationship modules (GLRModule). In each GLRModule, 1-dimensional convolution (1D-Conv) layers and Transformer encoder layers are used for modeling local and global dependencies, respectively. By alternately building short-range and long-range relationships, GLRNet attains both effective feature extracting and efficient computing. TransLM receives the embeddings constructed from the predicted probabilities of GLRNet and models linguistic context through Transformer encoder. The whole model, including GLRNet and TransLM, is jointly optimized. The gradients from TransLM can be back-propagated to GLRNet to further improve the performance of GLRNet, and in turn, the improved GLRNet also provide TransLM with a better recognition result. Therefore, the joint optimization guarantees the optimal performance of the whole model. Furthermore, the parameterized TransLM avoids the large footprint to store n-grams and is more flexible than traditional unlearnable n-gram language models. Moreover, although there have already been some works [17,18,23,29] applying Transformer to offline text recognition, they adopt auto-regressive Transformer decoder which outputs characters one by one. Our model runs in parallel, which is more faster than networks containing recurrent layers or auto-regressive decoders that are not parallelizable.

The experiments are conducted using CASIA-OLHWDB2.0-2.2 [12] and ICDAR2013 competition dataset [30]. Compared with previous methods of OLHCTR, our approach achieves state-of-the-art performance on ICDAR2013 competition dataset with accurate rate of 97.36% and correct rate of 97.63%. Moreover, our method exhibits the fastest speed compared with existing methods. Furthermore, the smallest footprint, 73.4 MB, is required to store our whole model including GLRNet and TransLM.

2   Methodology

2.1   Overview

The overall architecture of our method is illustrated in Fig. 1. Our network consists of two parts, namely GLRNet and TransLM. The input pen-tip trajectories first go through preprocessing and feature representation, yielding 6-channel feature sequence L. Then the feature sequence L is fed into GLRNet. The multiple GLRModules of GLRNet alternately build global and local relationships between the features at different time steps. Through a classifier, the prediction of GLR-Net, which contains the classification probabilities of each time step, are obtained based on the features from GLRModules. Next, the prediction of GLRNet is transformed to embeddings through a differentiable prediction embedding module. Based on the embeddings, TransLM uses Transformer encoder to integrate the linguistic context information and then predicts the classification probabilities through the identical classifier of GLRNet.

During training, two connectionist temporal classification (CTC) [8] losses are calculated using the two predictions from GLRNet and TransLM, respectively. The network is optimized to minimize the sum of two losses. The two

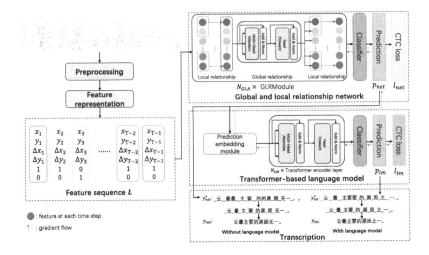

Fig. 1. Overall architecture of our approach. First, through preprocessing and feature representation, feature sequence L is obtained from the input pen-tip trajectory. Then, GLRNet, which contains multiple GLRModules, takes feature sequence L as input and predicts the recognition result without language model. Next, the prediction p_{net} is fed into TransLM to further improve the performance by integrating linguistic context.

parts of our network, GLRNet and TransLM, are jointly optimized since the prediction embedding module is differentiable. The orange and green dashed arrows in Fig. 1 indicate the paths of loss backpropagation. During inference, as shown in the "Transcription" part of Fig. 1, two recognition results are obtained from the predictions of GLRNet and TransLM, for the situations without and with language model, respectively.

2.2   Preprocessing and Feature Representation

A pen-tip trajectory of an online handwritten Chinese text can be represented as:

$$P = \{(x_i, y_i, s_i), t = 1, 2, ..., T\}, \tag{1}$$

where (x_i, y_i) is the coordinate of the i-th point and s_i is the index of the stroke which the i-th point belongs to. Figure 2 shows the pipeline of preprocessing. Given the trajectory P, the three steps of preprocessing and feature representation are described as follows.

Deslope. The texts may not be written horizontally because of the unconstrained situation. Line fitting is first conducted using the coordinates of all the T points. Then, we can estimate the tilt angle of online texts based on the slope of the fitted line. Next, affine transformation is applied to the points to rotate the online texts to be horizontal.

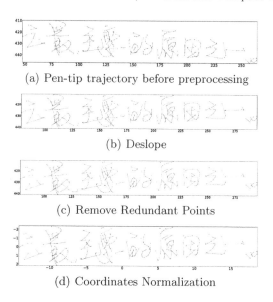

(a) Pen-tip trajectory before preprocessing

(b) Deslope

(c) Remove Redundant Points

(d) Coordinates Normalization

Fig. 2. The pipeline of preprocessing. The horizontal and vertical axes are x-coordinate and y-coordinate of points in the pen-tip trajectory.

Remove Redundant Points. We adopt the same method as [31] to remove redundant points. For point (x_i, y_i, s_i), if $s_i = s_{i-1} = s_{i+1}$ and any one of conditions (2) and (3) is satisfied, this point will be removed.

$$\sqrt{(x_i - x_{i-1})^2 + (y_i - y_{i-1})^2} < T_{dist}, \tag{2}$$

$$\frac{\Delta x_{i-1}\Delta x_i + \Delta y_{i-1}\Delta y_i}{(\Delta x_{i-1}^2 + \Delta y_{i-1}^2)^{0.5}(\Delta x_i^2 + \Delta y_i^2)^{0.5}} > T_{cos}, \tag{3}$$

where $\Delta x_i = x_{i+1} - x_i$ and $\Delta y_i = y_{i+1} - y_i$. Denoting the height of text line as H, two thresholds T_{dist} and T_{cos} are set to $0.01H$ and 0.99, respectively. As shown in Fig. 2(c), the number of points decreases from 401 to 340 through removing redundant points.

Coordinate Normalization. We estimate the mean values u_x and u_y by projecting all lines onto x-axis and y-axis following the coordinate normalization adopted in [31]. But the standard deviation δ_y on y-axis is estimated rather than δ_x used in [31]. Then the coordinates are normalized as:

$$x_{new} = (x - u_x)/\delta_y, \ y_{new} = (y - u_y)/\delta_y. \tag{4}$$

Feature Representation. After preprocessing, the input pen-tip trajectory is represented as a 6-dimensional feature sequence L, where $\mathbb{I}(\cdot) = 1$ when the condition is true and $\mathbb{I}(\cdot) = 0$ otherwise.

$$L = \{(x_i, y_i, \Delta x_i, \Delta y_i, \mathbb{I}(s_i = s_{i+1}), \mathbb{I}(s_i \neq s_{i+1})), i = 1, 2, ..., T - 1\}. \tag{5}$$

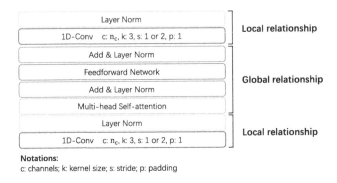

Fig. 3. The architecture of a GLRModule with n_c channels

2.3   Global and Local Relationship Network

Previous methods that directly handle the pen-tip trajectories make use of LSTM or some other recurrent architectures to extract high-level features. However, LSTM is not the best choice for OLHCTR. First, LSTM runs in sequential, which means the output of one time step must be calculated after all the computation before this time step completes. The length of feature sequence L is usually larger than 1,000, resulting in the low speed of model inference and training. Second, the output of one time step is derived only relying on the features of previous time steps rather than reasoning based on the features of all the time steps. This issue can be addressed by replacing LSTM with bidirectional LSTM, but bidirectional LSTM will increase the computational complexity. Third, the distance between different time steps may limit the ability of LSTM in modeling long-range dependencies. To this end, we propose a novel GLRNet which can effectively build local and global relationship between features of different time steps and run in parallel.

As shown in Fig. 1, our proposed GLRNet consists of N_{GLR} GLRModules and one classifier. Given the feature sequence L, GLRNet first extracts high-level features through N_{GLR} GLRModules and then predicts classification probabilities of each time step via a classifier. We will introduce each component in detail in the following sections.

Global and Local Relationship Module. The architecture of GLRModule with n_c channels is depicted in Fig. 3. A GLRModule contains two 1-dimensional convolution layers (1D-Conv) for modeling local relationship and one Transformer encoder layer for modeling global relationship.

When reading the handwritten Chinese texts, people first recognize each character by focusing on local regions. 1D-Conv has been adopted for online handwritten Chinese character recognition [7], but is rarely used for OLHCTR. The output of 1D-Conv at one position only depends on the inputs at several neighboring positions because of the sparse connection characteristic. The usage of 1D-Conv for capturing local contexts also occurs in [26]. Therefore, 1D-Conv is a good choice for building local relationship. Furthermore, although the redundant points have been removed during preprocessing, the features of neighboring time steps still have redundant information, such as neighboring points belonging to one stroke. 1D-Conv can also downsample the size of features by setting the stride to more than one and reduce the computational complexity in the following processes.

Global relationship is built by Transformer [21]. Compared with LSTM, Transformer can capture long-range dependencies regardless of the distance between different time steps and run in parallel. A Transformer encoder layer consists of two sub-layers, namely multi-head self-attention and feedforward network. We denote the output of the first 1D-Conv layer as $f_{in} \in \mathbb{R}^{s \times n_c}$, where s is the number of time steps and n_c is the number of channels. Then the output of h-head self-attention is calculated as:

$$MultiHead(f_{in}) = Concat(head_1, ..., head_h)W^O, \tag{6}$$

$$head_i = Attention(f_{in}W_i^Q, f_{in}W_i^K, f_{in}W_i^V), \tag{7}$$

$$Attention(Q, K, V) = softmax(\frac{QK^T}{\sqrt{n_c/h}})V, \tag{8}$$

where $W_i^Q, W_i^K, W_i^V \in \mathbb{R}^{n_c \times \frac{n_c}{h}}$, $W^O \in \mathbb{R}^{n_c \times n_c}$ and h is set to 8 in our experiments. Another sub-layer, feedforward network, consists of two fully connected layers, which performs further feature extraction based on the output of multi-head self-attention.

Classifier. The final GLRModule outputs features $f_{net} \in \mathbb{R}^{l \times d_{net}}$, where l is the number of time steps and d_{net} is the number of channels. The classifier, which is a fully connected layer, calculates the classification probabilities as:

$$p_{net} = softmax(f_{net}W_{cls} + b_{cls}), \tag{9}$$

where $W_{cls} \in \mathbb{R}^{d_{net} \times (n_{cls}+1)}$ is a weight matrix and b_{cls} is a bias term. n_{cls} is the number of character categories. Then the prediction p_{net} is a $l \times (n_{cls} + 1)$ matrix, which contains classification probabilities of $n_{cls}+1$ categories (including a category for blank) at l time steps.

2.4   Transformer-Based Language Model

TransLM consists of three parts including prediction embedding module, Transformer encoder, and classifier. The prediction p_{net} from GLRNet is transformed to embeddings through prediction embedding module. Then Transformer encoder, which contains N_{LM} Transformer encoder layers, captures the linguistic context based on the embeddings. Finally, we use the identical classifier of GLRNet to predict the classification probabilities of each time step.

Prediction Embedding Module. First, we denote the learnable embeddings of $n_{cls} + 1$ categories as $E \in \mathbb{R}^{(n_{cls}+1) \times d_{net}}$, where d_{net} is the dimension of each embedding which is equal to the number of channels of the last GLRModule in GLRNet. Then the prediction embedding module transforms the prediction p_{net} to embeddings at each time step f_{emb} by multiplying two matrices p_{net} and E:

$$f_{emb} \in \mathbb{R}^{l \times d_{net}} = p_{net}E. \tag{10}$$

The matrix multiplication is a differentiable operation, thus it allows the gradient from TransLM to be backpropagated into GLRNet. This kind of joint optimization can improve the performance of both GLRNet and TransLM.

Transformer Encoder. Transformer encoder consists of N_{LM} Transformer encoder layers described in Sect. 2.3. Since f_{emb} is derived from classification probabilities p_{net}, it only contains semantic information without visual features. Taking embeddings f_{emb} as input, Transformer encoder captures linguistic context between embeddings at different time steps and outputs features $f_{lm} \in \mathbb{R}^{l \times d_{net}}$.

Classifier. The classifier of TransLM is identical to the one of GLRNet, which means these two classifier share the same parameters. Based on the features f_{lm} from Transformer encoder, the classifier predicts p_{lm} in the same manner as p_{net}.

$$p_{lm} = softmax(f_{lm}W_{cls} + b_{cls}). \tag{11}$$

2.5   Transcription and Loss Function

Given predictions p_{net} and p_{lm}, we obtain the sequences y^*_{net} and y^*_{lm}, respectively, by choosing the category with the highest probability per time step. Then the recognition result without language model y_{net} and recognition result with language model y_{lm} are obtained from y^*_{net} and y^*_{lm}, respectively, through the sequence-to-sequence mapping function $\mathcal{B}$ defined in [8]. The function $\mathcal{B}$ first removes repeated categories and then removes the blank category.

CTC loss function [8] is adopted in our approach, since it does not require alignments between time steps and label sequence. Two CTC losses, l_{net} and l_{lm}, are calculated based on p_{net} and p_{lm}, respectively. The network is optimized to minimize the sum of l_{net} and l_{lm}. GLRNet and TransLM are jointly optimized, and the gradients from TransLM can be backpropagated to GLRNet.

3   Experiments

3.1   Datasets

The training data of our experiments includes two parts, namely the training set of CASIA-OLHWDB2.0-2.2 [12] and the samples synthesized using CASIA-OLHWDB1.0-1.2 [12]. The training set of CASIA-OLHWDB2.0-2.2 contains

41,710 text lines from 4,072 pages. The synthetic samples are synthesized using the single character samples from CASIA-OLHWDB1.0-1.2. Furthermore, there are two types of synthetic samples, namely synthetic samples without corpus and synthetic samples with corpus. When synthesizing without corpus, the categories of characters in a text line are randomly selected. When synthesizing with corpus, we synthesize samples according to the randomly selected sentences from corpus. The corpus used in experiments is composed of CLDC [2], SLD [1] and the corpus used in [20]. The synthetic samples without corpus is used when training standalone GLRNet, while the synthetic samples with corpus is used when jointly optimizing GLRNet and TransLM. We evaluate our method on ICDAR 2013 competition dataset [30]. This dataset comprises 3,432 text lines from 300 pages.

3.2 Implementation Details

The network is implemented using PyTorch, and is trained and tested with a 11G RTX 2080ti GPU and a 2.50 GHz Xeon CPU. The number of categories n_{cls} is set to 7,356. During training, there are totally 1 million iterations and the batch size is set to 32. We use AdamW [15] to optimize our network. The initial learning rate is 1e−4 and is multiplied by 0.1 after 0.6 million and 0.95 million iterations. When training the standalone GLRNet, we use 200,000 synthetic samples without corpus. The ratio of real samples and synthetic samples is 0.5:0.5. When jointly training GLRNet and TransLM, the synthetic samples with corpus are synthesized on the fly. The ratio of real samples and synthetic samples is 0.05:0.95. Furthermore, the parameters of GLRNet are initialized by the pretrained standalone GLRNet before joint training of GLRNet and TransLM. During testing, the batch size is set to 1.

3.3 Evaluation Metrics

We evaluate the recognition performance of our approach using accurate rate (AR) and correct rate (CR) [30], which are calculated as:

$$AR = (N_t - D_e - S_e - I_e)/N_t, \qquad (12)$$
$$CR = (N_t - D_e - S_e)/N_t, \qquad (13)$$

where N_t, D_e, S_e and I_e are the total number of characters, deletion errors, substitution errors and insertion errors of all the testing samples, respectively.

We also evaluate our approach in terms of footprint and speed. Footprint is the space required to store the model parameters and language models, while speed is the average time consumption of processing one testing text line of ICDAR2013 competition dataset, considering all the procedures including preprocessing, model inference, transcription and so on.

3.4   Ablation Study on Standalone GLRNet

To verify the effectiveness of the architecture of GLRNet, we present some ablation studies on standalone GLRNet in this section. Note that the experiments in this section adopt the network without TransLM. As mentioned in Sect. 2.3, GLRNet contains N_{GLR} GLRModules. We compare the performance of different N_{GLR} in Table 1. Note that the "Speed" column presents the average time consumption on one testing text line, which considers all the procedures including preprocessing, model inference, transcription and so on. The settings of each GLRModule is described in the format of "$\mathbf{A}c\mathbf{B}s$". $\mathbf{A}$ is the number of channels, namely the n_c in Sect. 2.3, and $\mathbf{B}$ is the stride. There are three different values of $\mathbf{B}$. (1) If $\mathbf{B}$ is 1, the strides of two 1D-Conv layers in the GLRModule are 1. (2) If $\mathbf{B}$ is 2, the stride of the first 1D-Conv layer is 1 while the stride of the second 1D-Conv layer is 2. (3) If $\mathbf{B}$ is 4, the strides of two 1D-Conv layers are 2. As shown in Table 1, the highest accuracy is achieved when N_{GLR} is equal to 2. Unlike most networks in the deep learning era, the performance can not be improved by increasing the depth of a network. This is probably because self-attention can capture the global relationship at different positions, but convolutional network must be deeper to have larger receptive fields. In the remaining experiments, N_{GLR} is set to 2 by default.

Table 1. Comparison of different number of GLRModules in GLRNet

N_{GLR}	Architecture	AR	CR	Footprint	Speed
1	$512c4s$	84.91	86.83	**25.5 MB**	**5.5 ms**
2	$128c4s$-$512c2s$	**91.24**	**91.81**	26.9 MB	7.0 ms
3	$64c2s$-$128c2s$-$512c2s$	90.83	91.26	27.2 MB	9.9 ms
4	$64c2s$-$128c2s$-$256c2s$-$512c1s$	90.52	90.94	31.1 MB	12.2 ms

[a] The "Speed" column presents the average time consumption on one testing text line.

As described in Sect. 2.3, GLRNet contains 1D-Conv layers for building local relationship and Transformer encoder layers for building global relationship. In Table 2, we verify the effectiveness of global and local relationship design. The network only containing global relationship replaces each 1D-Conv layer with a self-attention layer. If the stride of a 1D-Conv layer is 2, an extra maxpooling layer is added after the self-attention layer for downsampling. The network only containing local relationship replaces each self-attention layer with a 1D-Conv layer with the same number of channels. The comparison in Table 2 proves that combining global and local relationships results in a better trade-off of accuracy, speed and compactness.

3.5   Ablation Study on TransLM

As mentioned in Sect. 2.4, the Transformer encoder of TransLM contains N_{LM} Transformer encoder layers. In Table 3, we compare the performances of the

Table 2. Effectiveness of global and local relationship design of GLRNet. (GR: global relationship; LR: local relationship)

GR	LR	AR	CR	Footprint	Speed
✓		3.75	3.76	28.8 MB	11.4 ms
	✓	81.46	84.95	**25.8 MB**	**5.8 ms**
✓	✓	**91.24**	**91.81**	26.9 MB	7.0 ms

whole model, which consists of GLRNet and TransLM, when N_{LM} is set to different values. It can be observed that as N_{LM} increases, the recognition accuracy consistently grows but the model consumes larger footprint to store parameters and more time for inference. When N_{LM} is equal to 4, the footprint is 73.4 MB and the average processing time is 11.9 ms, which achieves a good trade-off between compactness and speed. Compared with the time presented in Table 1, the average time consumption just increases from 7.0 ms to 11.9 ms, which means TransLM only needs 4.9 ms to handle one text line in average. Therefore, in the following experiments, N_{LM} is set to 4 by default.

Table 3. Comparison of different number of layers in TransLM.

N_{LM}	With language model			
	AR	CR	Footprint	Speed
2	97.00	97.30	**57.3 MB**	**9.3 ms**
3	97.14	97.46	65.3 MB	11.1 ms
4	97.24	97.53	73.4 MB	11.9 ms
5	97.28	97.56	81.4 MB	13.7 ms
6	**97.33**	**97.61**	89.4 MB	14.9 ms

Another advantage of the proposed TransLM is the differentiable prediction embedding module, which enables the whole model to be jointly optimized. We conduct ablation studies on TransLM in Table 4. The baseline is the network which replaces the prediction embedding module with a simple mechanism. The mechanism is that for each time step, we select the category with highest probability in p_{net} and assign the embedding of this category as the embedding of this time step. The experiment adding prediction embedding module to baseline has the same architecture of our proposed network, but the gradient backpropagation from TransLM to GLRNet is cut off. The final experiment with joint optimization, which is our proposed approach, allows the gradient backpropagation from TransLM to GLRNet. As shown in Table 4, the proposed approach achieves the best performance. More importantly, the performance of GLRNet, i.e., the performance without language model, is also improved because of joint optimization.

Table 4. Ablation study on TransLM

Description	Without LM		With LM	
	AR	CR	AR	CR
Baseline	91.24	91.81	94.41	94.87
+Prediction embedding module	91.24	91.81	95.73	96.31
+Joint optimization	**94.84**	**95.30**	**97.24**	**97.53**

3.6 Further Improvement

Although impressive performances have been achieved in the sections above, the performance of our approach can be further improved by increasing training iterations or combining with a trigram language model. The trigram language model is derived from the same corpus as described in Sect. 3.1. A beam search algorithm [9] is adopted for the transcription with trigram language model. As shown in Table 5, when increasing the number of training iterations to twice, the accurate rate increase from 94.84% to 95.05% without language model and from 97.24% to 97.36% with language model. Furthermore, the trigram language model can only improve the accurate rate with language model a little, from 97.36% to 97.55%, which verifies that our TransLM has already captured the linguistic context very well. But the use of trigram language model hugely decreases the performance on compactness and speed. The footprint and time consumption increase from 73.4 MB to 792.4 MB and from 7.0 ms to 767.8 ms, respectively.

Table 5. Further Improvement.

Description	Iterations	Without LM		With LM	
		AR	CR	AR	CR
GLRNet+TransLM	1×	94.84	95.30	97.24	97.53
GLRNet+TransLM	2×	95.05	95.46	97.36	97.63
GLRNet+TransLM+Trigram	2×	**95.05**	**95.46**	**97.55**	**97.99**

3.7 Comparison with Existing Methods

We compare our approach with existing methods in Table 6. Most previous methods do not report their speed performance or the speed is tested on different devices. We reimplement these methods and test their speed under the same setting as our method. As mentioned in Sect. 3.2, we evaluate the speed performance, i.e., the average time consumption on one sample of testing dataset, using a device with a RTX 2080ti GPU and a 2.50 GHz CPU.

In Table 6, method "GLRNet" is the standalone GLRNet without TransLM, while the method "GLRNet + TransLM" means the joint optimized GLRNet and TransLM. "impLM" denotes the implicit language model proposed by [28]

Table 6. Comparison with existing methods on ICDAR2013 competition dataset.

Method	Without language model				With language model			
	AR	CR	Footprint	Speed	AR	CR	Footprint	Speed
Zhou et al. [32]	–	–	–	–	94.06	94.76	–	–
Zhou et al. [33]	–	–	–	–	94.22	94.76	–	–
Sun et al. [20]†	89.12	90.18	–	–	93.40	94.43	–	–
2C-FCRN+impLM [28]	88.88	90.17	109.5 MB	15.2 ms	95.46	96.01	174.9 MB	22.1 ms
2C-FCRN+imp & staLM [28]	88.88	90.17	109.5 MB	15.2 ms	96.06	96.58	751.2 MB	856.9 ms
VGG-DBLSTM [5]	87.49	87.98	39.5 MB	16.6 ms	97.03	97.29	153.9 MB	–
CharNet-DBLSTM [5]	87.10	87.71	**19.1 MB**	13.7 ms	96.87	97.15	133.5 MB	–
Liu et al. [13]	91.36	92.37	24.5 MB	18.4 ms	94.89	95.70	600.8 MB	1589.5 ms
GLRNet (ours)	91.24	91.81	26.9 MB	7.0 ms	–	–	–	–
GLRNet+TransLM (ours)	**95.05**	**95.46**	26.9 MB	**7.0 ms**	**97.36**	**97.63**	**73.4 MB**	11.9 ms

and "imp & staLM" denotes implicit language model & statistical language model in [28]. The first two methods [32,33] are based on over-segmentation and use traditional architectures rather than deep learning networks. Besides, the number of categories in the method [20] marked by † is 2,765, which is different from the commonly adopted 7,356 classes. Therefore, we do not compare the speed with these three methods. VGG-DBLSTM [5] and CharNet-DBLSTM [5] adopt WFST-based decoder proposed in [3]. However, Cai et al. [3] test the speed of WFST-based decoder on IAM test set rather than ICDAR2013 competition dataset. Because the number of words in a text line of IAM is much smaller than the number of characters in a text line of ICDAR2013 competition dataset, it is unfair to use the speed reported in [3]. Therefore, we do not list the speed of VGG-DBLSTM and CharNet-DBLSTM with language model. However, our method with language model has already achieved faster speed than these two methods without language model.

As shown in Table 6, in the situation without language model, our approach denoted as "GLRNet + TransLM" reaches the highest recognition accuracy and the fastest speed with footprint comparable to previous methods. However, in the situation with language model, the best performance in accuracy, speed and compactness is achieved by our approach. More importantly, "GLRNet + TransLM" exhibits an average speed of 11.9 ms per line with language model, which is even faster than all the existing methods without language model. Moreover, compared with the standalone GLRNet, the joint optimization of GLRNet and TransLM also brings huge improvement to the performance of GLRNet, increasing AR from 91.24% to 95.05%.

4   Conclusion

In this paper, we propose a novel fast, accurate and compact approach for online handwritten Chinese text recognition. Different from previous methods which mainly rely on recurrent neural network, our approach explores Transformer for better context modelling and faster speed. The proposed method consists of GLRNet and TransLM. GLRNet first extracts features by alternately capturing global and local relationship and then recognizes the text line based on the extracted feature. TransLM integrates linguistic context information based on the prediction from GLRNet to further improve the performance. Thanks to the differentiable prediction embedding module, GLRNet and TransLM can be jointly optimized for the optimal performance of the whole model. Compared with existing methods of OLHCTR, the experiments verify the superiority of our approach in accuracy, speed and compactness. In the future, our approach can be easily combined with plenty of variants of Transformers to further improve the performance.

Acknowledgment. This research is supported in part by NSFC (Grant No.: 6193 6003, 61771199), GD-NSF (no. 2017A030312006).

References

1. Sogou lab data. http://www.sogou.com/labs/resource/cs.php. R&D Center of SOHU
2. Chinese linguistic data consortium (2009). http://www.chineseldc.org. The Contemporary Corpus developed by State Language Commission P.R. China, Institute of Applied Linguistics
3. Cai, M., Huo, Q.: Compact and efficient WFST-based decoders for handwriting recognition. In: ICDAR, pp. 143–148 (2017)
4. Carion, N., Massa, F., Synnaeve, G., Usunier, N., Kirillov, A., Zagoruyko, S.: End-to-end object detection with transformers. In: Vedaldi, A., Bischof, H., Brox, T., Frahm, J.-M. (eds.) ECCV 2020, Part I. LNCS, vol. 12346, pp. 213–229. Springer, Cham (2020). https://doi.org/10.1007/978-3-030-58452-8_13
5. Chen, K., et al.: A compact CNN-DBLSTM based character model for online handwritten Chinese text recognition. In: ICDAR, pp. 1068–1073 (2017)
6. Devlin, J., Chang, M.W., Lee, K., Toutanova, K.: BERT: pre-training of deep bidirectional transformers for language understanding. In: NAACL-HLT, pp. 4171–4186 (2019)
7. Gan, J., Wang, W., Lu, K.: A new perspective: recognizing online handwritten Chinese characters via 1-dimensional CNN. Inf. Sci. **478**, 375–390 (2019)
8. Graves, A., Fernández, S., Gomez, F., Schmidhuber, J.: Connectionist temporal classification: labelling unsegmented sequence data with recurrent neural networks. In: ICML, pp. 369–376 (2006)
9. Graves, A., Jaitly, N.: Towards end-to-end speech recognition with recurrent neural networks. In: ICML, pp. 1764–1772 (2014)
10. Graves, A., Liwicki, M., Fernández, S., Bertolami, R., Bunke, H., Schmidhuber, J.: A novel connectionist system for unconstrained handwriting recognition. IEEE Trans. Pattern Anal. Mach. Intell. **31**(5), 855–868 (2008)
11. Lan, Z., Chen, M., Goodman, S., Gimpel, K., Sharma, P., Soricut, R.: ALBERT: a lite BERT for self-supervised learning of language representations. In: ICLR (2019)
12. Liu, C.L., Yin, F., Wang, D.H., Wang, Q.F.: CASIA online and offline Chinese handwriting databases. In: ICDAR, pp. 37–41 (2011)
13. Liu, M., Xie, Z., Huang, Y., Jin, L., Zhou, W.: Distilling GRU with data augmentation for unconstrained handwritten text recognition. In: ICFHR, pp. 56–61 (2018)

14. Liu, Y., et al.: RoBERTa: a robustly optimized BERT pretraining approach. arXiv preprint arXiv:1907.11692 (2019)
15. Loshchilov, I., Hutter, F.: Decoupled weight decay regularization. In: ICLR (2018)
16. Lu, J., Batra, D., Parikh, D., Lee, S.: Vilbert: pretraining task-agnostic visiolinguistic representations for vision-and-language tasks. arXiv preprint arXiv:1908.02265 (2019)
17. Lu, N., Yu, W., Qi, X., Chen, Y., Gong, P., Xiao, R.: MASTER: multi-aspect non-local network for scene text recognition. Pattern Recognit. **117**, 107980 (2021)
18. Sheng, F., Chen, Z., Xu, B.: NRTR: a no-recurrence sequence-to-sequence model for scene text recognition. In: ICDAR, pp. 781–786 (2019)
19. Shi, B., Bai, X., Yao, C.: An end-to-end trainable neural network for image-based sequence recognition and its application to scene text recognition. IEEE Trans. Pattern Anal. Mach. Intell. **39**(11), 2298–2304 (2016)
20. Sun, L., Su, T., Liu, C., Wang, R.: Deep LSTM networks for online Chinese handwriting recognition. In: ICFHR, pp. 271–276 (2016)
21. Vaswani, A., et al.: Attention is all you need. In: NeuIPS (2017)
22. Wang, D.H., Liu, C.L., Zhou, X.D.: An approach for real-time recognition of online Chinese handwritten sentences. Pattern Recognit. **45**(10), 3661–3675 (2012)
23. Wang, P., Yang, L., Li, H., Deng, Y., Shen, C., Zhang, Y.: A simple and robust convolutional-attention network for irregular text recognition. arXiv preprint arXiv:1904.01375 6 (2019)
24. Wang, Q.F., Yin, F., Liu, C.L.: Handwritten Chinese text recognition by integrating multiple contexts. IEEE Trans. Pattern Anal. Mach. Intell. **34**(8), 1469–1481 (2011)
25. Wu, Y.C., Yin, F., Liu, C.L.: Improving handwritten Chinese text recognition using neural network language models and convolutional neural network shape models. Pattern Recognit. **65**, 251–264 (2017)
26. Wu, Z., Liu, Z., Lin, J., Lin, Y., Han, S.: Lite transformer with long-short range attention. In: ICLR (2019)
27. Xie, Z., Sun, Z., Jin, L., Feng, Z., Zhang, S.: Fully convolutional recurrent network for handwritten Chinese text recognition. In: ICPR, pp. 4011–4016 (2016)
28. Xie, Z., Sun, Z., Jin, L., Ni, H., Lyons, T.: Learning spatial-semantic context with fully convolutional recurrent network for online handwritten Chinese text recognition. IEEE Trans. Pattern Anal. Mach. Intell. **40**(8), 1903–1917 (2018)
29. Yang, L., Wang, P., Li, H., Li, Z., Zhang, Y.: A holistic representation guided attention network for scene text recognition. Neurocomputing **414**, 67–75 (2020)
30. Yin, F., Wang, Q.F., Zhang, X.Y., Liu, C.L.: ICDAR 2013 Chinese handwriting recognition competition. In: ICDAR, pp. 1464–1470 (2013)
31. Zhang, X.Y., Yin, F., Zhang, Y.M., Liu, C.L., Bengio, Y.: Drawing and recognizing Chinese characters with recurrent neural network. IEEE Trans. Pattern Anal. Mach. Intell. **40**(4), 849–862 (2017)
32. Zhou, X.D., Wang, D.H., Tian, F., Liu, C.L., Nakagawa, M.: Handwritten Chinese/Japanese text recognition using semi-Markov conditional random fields. IEEE Trans. Pattern Anal. Mach. Intell. **35**(10), 2413–2426 (2013)
33. Zhou, X.D., Zhang, Y.M., Tian, F., Wang, H.A., Liu, C.L.: Minimum-risk training for semi-Markov conditional random fields with application to handwritten Chinese/Japanese text recognition. Pattern Recognit. **47**(5), 1904–1916 (2014)

HCADecoder: A Hybrid CTC-Attention Decoder for Chinese Text Recognition

Siqi Cai⍟, Wenyuan Xue⍟, Qingyong Li$^{(\boxtimes)}$⍟, and Peng Zhao⍟

Beijing Key Lab of Traffic Data Analysis and Mining, Beijing Jiaotong University,
Beijing, China
{caisiqi,wyxue17,liqy,zhaopeng18}@bjtu.edu.cn

Abstract. Text recognition has attracted much attention and achieved exciting results on several commonly used public English datasets in recent years. However, most of these well-established methods, such as connectionist temporal classification (CTC)-based methods and attention-based methods, pay less attention to challenges on the Chinese scene, especially for long text sequences. In this paper, we exploit the characteristic of Chinese word frequency distribution and propose a hybrid CTC-Attention decoder (HCADecoder) supervised with bigram mixture labels for Chinese text recognition. Specifically, we first add high-frequency bigram subwords into the original unigram labels to construct the mixture bigram label, which can shorten the decoding length. Then, in the decoding stage, the CTC module outputs a preliminary result, in which confused predictions are replaced with bigram subwords. The attention module utilizes the preliminary result and outputs the final result. Experimental results on four Chinese datasets demonstrate the effectiveness of the proposed method for Chinese text recognition, especially for long texts. Code will be made publicly available(https://github.com/lukecsq/hybrid-CTC-Attention).

Keywords: Chinese text recognition · CTC-Attention · Subword

1 Introduction

Benefiting from deep learning approaches and large datasets, researchers have made impressive breakthroughs for the text recognition task in recent years [1–4]. The relevant technologies have been applied in multimedia retrieval [5], indoor positioning [6], and industrial automation [1].

Most existing text recognition methods apply an encoder-decoder framework [1,2,7,8], in which an input text image is encoded as deep features and a decoder transcribes the features into the recognition results. The connectionist temporal classification (CTC) [9] and the attention mechanism [10] are two commonly used methods in the decoding stage. CTC-based methods directly optimize the prediction by mapping the encoded features to the probability space. However, CTC assumes that outputs are independent and neglects the context within a

© Springer Nature Switzerland AG 2021
J. Lladós et al. (Eds.): ICDAR 2021, LNCS 12823, pp. 172–187, 2021.
https://doi.org/10.1007/978-3-030-86334-0_12

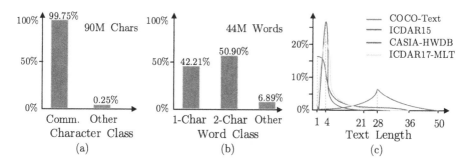

Fig. 1. The word frequency statistics on *The People's Daily* from 1993 to 1996 (subfigure (a) and (b)) and the text length statistics on four text recognition datasets (subfigure (c)). Subfigure (a) shows that 99.75% of 90 million single characters are common Chinese characters (Comm.). From subfigure (b), it can be seen that one-character (1-Char) words and two-character (2-Char) words together account for 93.11% of 44 million words. Subfigure (c) shows that the lengths of most English texts (COCO-Text [3], ICDAR15 [4]) are shorter than 4. However, Chinese texts (CASIA-HWDB [11] and ICDAR17-MLT [12]) usually have longer text lengths.

sequence. Attention-based methods utilize the attention mechanism to capture the semantic dependency between different characters. Compared with CTC-based methods, the attention mechanism helps the model learn a better alignment mode, but falls into the attention drift problem [2], i.e., the predicted attention regions deviate from the proper regions of target characters.

Most above research focus on English datasets [3,4,13]. When transferring to Chinese scenes, they will face some different challenges:

– Chinese characters usually closely arrange in a sentence without significant space between words, which usually results in a longer text length than the English text. As shown in Fig. 1 (subfigure (c)), most samples in Chinese datasets have longer text lengths than texts in another two English datasets.
– Chinese has a huge number of characters but only less than half of them are commonly used. In the corpus of *The People's Daily* from 1993 to 1996 [14], as shown in Fig. 1 (subfigure (a) and (b)), there are about 90 million characters and 44 million words in total, of which 99.75% are common Chinese characters and 93.11% are common Chinese words.
– Chinese characters have complex glyphs, which makes the model easily confuse adjacent characters, especially for a character that can be treated as the combination of others, as shown in Fig. 2.

When English recognition methods are directly applied to the Chinese scene without considering these challenges, some typical failure cases may appear. As shown in Fig. 2, the CTC-based method can result in false predictions when a Chinese character is composed of another two characters. For the attention-based method, the problem of attention drift is obvious, which results in the decrease of recognition accuracy. To solve the above problems, researchers have made great

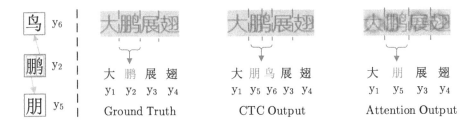

Fig. 2. Typical bad cases for Chinese recognition. On the left of the figure, we present a Chinese character that is the combination of the other two characters. For such a character, the CTC-based method can recognize it into two parts as blue lines separate. For the attention-based method, the recognition result will be wrong when the attention drift happens. (Color figure online)

efforts. The relevant solutions either use more advanced backbone networks to extract robust features [15], or adopt the multi-scale architecture to deal with the complex Chinese glyph [16,17]. Different from these methods, we address the Chinese text recognition problem by utilizing subwords and combining CTC and the attention mechanism.

In this paper, we propose a hybrid CTC-Attention decoder (HCADecoder) that cooperates with bigram mixture labels for Chinese text recognition. Specifically, we add high-frequency bigram subwords into the original unigram characters to construct the mixture label, which can shorten the decoding length. Then, HCADecoder utilizes the results from the CTC module to narrow down the attention scope, which can help the attention module focus on local discriminated features and alleviate the attention drift problem. We evaluate the proposed method on four Chinese datasets and experimental results demonstrate the effectiveness of the proposed method, especially for long texts. The contributions of this work are summarized as follows:

- We propose the bigram mixture label for Chinese text recognition by adding high-frequency bigram subwords into the original unigram characters, which can shorten the label sequence and is conductive to decoding.
- We propose a hybrid CTC-Attention decoder that cooperates with the bigram mixture label for Chinese text recognition.

2 Related Work

2.1 CTC-Based Methods

CTC [9] was first proposed in the speech recognition task. CTC does not need to know whether phonemes in the input audio are aligned with the corresponding labels one by one. Therefore, the network can be trained with only sequence-level labels. Inspired by speech recognition, Shi et al. [1] proposed a classic CTC model, i.e., CRNN, which combines convolutional neural networks (CNN), recursive neural networks (RNN), and CTC. This method has been proved effective when dealing with variable length label sequences and widely used in the industry. Gao et al. [18] introduced a stacked convolutional layer to replace RNN, which can effectively capture the context correlation of input sequences and reduce the computational complexity. All of the above methods treat a text image as a one-dimensional signal, just like speeches. However, a text image is distributed in a two-dimensional space. Therefore, Wang et al. [7] proposed 2D-CTC that gets the position of characters for each frame in two-dimensional feature maps and then decodes with the general CTC. Despite the great success of CTC in sequence decoding, it has an obvious drawback, i.e., CTC assumes that each label is independent. However, in many cases, context labels are associated. Therefore, adding a language model is an effective improvement method [19,20].

2.2   Attention-Based Methods

The attention mechanism [10] was first proposed to improve the modeling performance of machine translation systems. It has been successfully applied in scene text recognition [2,8,21,22]. Lee et al. [21] proposed a recursive neural network with an attention module for text recognition, in which the attention mechanism performs soft feature selection to better utilize image features. Cheng et al. [2] noticed the problem of attention drift in existing attention-based methods and proposed a focusing attention network to solve this problem. Bai et al. [23] proposed an edit probability metric to alleviate the inconsistency between the ground truth and the prediction probability distribution. Though the attention mechanism shows strong modeling ability, the problem of attention drift is still a difficulty, especially for long Chinese texts.

2.3   The Combination of CTC and Attention

Considering the advantages and disadvantages of CTC and the attention mechanism, some researchers have tried to combine them recently. The combined CTC-Attention method was first proposed for speech recognition [24], in which the CTC module is used to assist the attention model training. Based on this work, Hu et al. [25] applied the idea to the field of scene text recognition with some modifications to facilitate CTC decoding. Different from these methods, the proposed hybrid CTC-Attention decoder is designed based on the Chinese word frequency distribution.

2.4   Subword-Based Methods

Subword units are suitable for characters with complex glyphs in the field of text recognition, such as Mongolian and Indian. Fang et al. [26] proposed a Mongolian handwriting recognition system, in which a subword-based language model is used to solve the high out-of-vocabulary rate problem. Saluja et al. [27] proposed a subword-based method as post-processing to correct text recognition results for Indic texts. However, none of these methods tries to apply subwords into the combined CTC-Attention model.

3   Method

In this section, we first introduce how to construct the bigram mixture label. Then, we describe details of the proposed hybrid CTC-Attention decoder and how that cooperates with the mixture label for training and inference.

3.1   Bigram Mixture Label

According to the statistics of the Chinese corpus in Fig. 1, the total usage of one- and two-character words[1] is approximately 93%. Words with three and more characters do not have the statistical significance. Therefore, we use unigram characters and bigram subwords to build the mixture label dictionary. Specifically, let $D = \{d\}^M$ denote the original unigram dictionary in the text recognition task, where d is a non-repetitive character category and M is the total number of character classes. Next, we perform the bigram division on all text labels in the dataset and take the top N with the highest word frequency as the bigram dictionary $S = \{s\}_k^N$, where $s_k = \{\langle d_p, d_q \rangle \mid d_p, d_q \in D\}$. Finally, $D' = D \cup S$ serves as the mixture label dictionary.

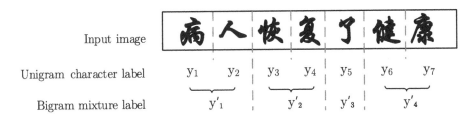

Fig. 3. An example of the bigram mixture label. From the unigram character label $\{(y_1, y_2, y_3, y_4, y_5, y_6, y_7) \mid y_i \in D, i \in [1, 7]\}$, we construct the bigram mixture label $\{(y_1', y_2', y_3', y_4') \mid y_j' \in D', j \in [1, 4]\}$, where $y_1' = \langle y_1, y_2 \rangle$, $y_2' = \langle y_3, y_4 \rangle$, $y_3' = y_5$, $y_4' = \langle y_6, y_7 \rangle$. The label length reduces from 7 to 4.

Based on the mixture label dictionary, we then reconstruct the label sequence for each sample. Let $\{I, Y\}$ represent an input sample in the training set, where I is the text image and $Y = \{(y_1, y_2, \cdots y_l) \mid y_i \in D, i \in [1, l]\}$ is the original unigram label sequence. The mixture label is constructed by replacing all adjacent unigram labels matching $\langle y_i, y_{i+1} \rangle \in S$ with a bigram label. If a unigram label fails to find an appropriate matching, we keep it in the label sequence without any change. Figure 3 presents an example to illustrate this process. The obtained mixture label is denoted as $Y' = \{(y_1', y_2', \cdots, y_{l'}') \mid y_j' \in D', j \in [1, l']\}$, which replaces the original unigram label Y and cooperates with the proposed hybrid CTC-Attention decoder for Chinese text recognition. The mixture label

---

[1] One- and two-character words are defined by the Chinese semantics and have the same length as unigram and bigram words, respectively.

uses high-frequency bigram labels to replace original unigram labels, which can shorten the decoding length and alleviate the long-term dependence problem. This idea can be extended to n-gram mixture labels according to the statistical significance of the word frequency distribution.

3.2   Hybrid CTC-Attention Decoder

The main framework of the proposed HCADecoder is shown in Fig. 4. Given an input text image, we first employ an encoder, e.g., ResNet32 [2] with two BiLSTM [1] layers, to extract the sequence features $H \in \mathbb{R}^{B \times T \times 256}$, i.e., Encoder$(I) = H$. Then, the proposed HCADecoder takes H as input and combines the CTC module with the attention module for decoding. Next, we present details of these two modules.

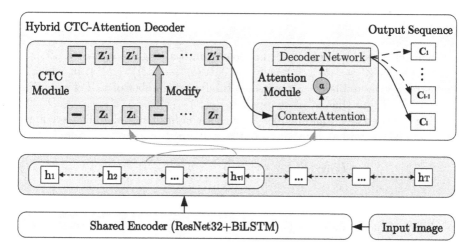

Fig. 4. The framework of the proposed hybrid CTC-Attention decoder or HCADecoder. HCADecoder takes the sequence features as input, which are extracted by the shared encoder. During the decoding stage, the CTC module outputs a preliminary result, in which confused predictions are replaced with bigram subwords and folded into sub-sequences. The attention module works on each sub-sequence and outputs the final prediction.

CTC Module. In the decoding stage, the CTC module first decodes H into a CTC sequence $Z = (z_1, z_2, \cdots, z_T)$, where $z_t \in D' \cup \langle b \rangle$ can be a unigram character or a bigram subword. '$\langle b \rangle$' represents '$blank$'. Then Z is modified to get Z' according to the mapping relations between bigram subwords and unigram characters. As an example shown in Fig. 5, $t_i \in [1, T]$ represents the i-th position in the sequence. We assume that the prediction of z_3 is s_1 at position t_3, where $s_1 = \{\langle d_1, d_2 \rangle \mid d_1, d_2 \in D, \langle d_1, d_2 \rangle \in S\}$. We search forward from t_3 to find t_m

that satisfies $z_m = d_1 \wedge z_{m-1} \neq d_1$ and search backward to find t_n that satisfies $z_n = d_2 \wedge z_{n+1} \neq d_2$. Each element between $z_m \sim z_n$ belongs to $\{d_1, d_2, s_1, \langle b \rangle\}$. After that, the prediction results between $z_m \sim z_n$ are all modified to s_1. In the example of Fig. 5, $t_m = t_2$, $t_n = t_5$. In this way, we have the preliminary prediction Z', in which confused results are changed into bigram-level results.

Position	**T**	t_1		t_2		t_3		t_4		t_5		t_6
Transcription	**Z**	\<b\>	|	d_1	|	s_1	|	\<b\>	|	d_2	|	\<b\>
Modification	**Z'**	\<b\>	|	s_1		s_1		s_1		s_1	|	\<b\>

Fig. 5. An example of modifying Z to Z'. In the CTC sequence Z, we replace confused results at $t_2 \sim t_5$ with bigram subwords to get Z', where $s_1 = \langle d_1, d_2 \rangle$ is a bigram subword.

When repeated elements in Z' are folded into a single occurrence, Z' can be rewritten as $Z' = \left(\langle b \rangle^*, (z')_1^{\varepsilon_1}, \langle b \rangle^*, (z')_2^{\varepsilon_2}, \langle b \rangle^*, \cdots (z')_L^{\varepsilon_L}, \langle b \rangle^* \right)$, in which $*$ denotes zero or more blanks, ε represents the repeat number of z'. Let mapping function $\mathcal{B}$ indicate that Z' is mapped to $\left(((z')_1, (z')_2, \cdots (z')_L), z' \in D' \right)$ by removing repeated labels and blanks. Then, the conditional probability is defined as the sum of the probabilities of all paths π mapped by $\mathcal{B}$ to the ground truth label Y':

$$P_{\text{CTC}}\left(Y' \mid Z'\right) = \sum_{\pi:B(\pi)=Y'} p\left(\pi \mid Z'\right), \tag{1}$$

where the probability of π is defined as $p\left(\pi \mid Z'\right) = \prod_{t=1}^{T} (z')_{\pi_t}^t$, $(z')_{\pi_t}^t$ is the probability of having label π_t at position t. Equation 1 can be efficiently computed using the forward-backward algorithm described in [9].

In theory [9], multiple CTC sequences can be obtained over which we have to marginalize. However, when applying the forward-backward algorithm, i.e. Baum-Welch learning, it requires high computational resources to estimate parameters of the attention module. Besides, it is difficult to handle the backward computation properly, since the attention module needs previous hidden states as input when decoding. Hence, we simplify the problem by using Viterbi learning instead and compute the overall probability of the CTC sequence by forced alignment. Only the CTC sequence with the highest probability is considered. The corresponding modified sequence is denoted as Z'^*.

Attention Module. In the CTC module, confused predictions are changed into bigram-level results and folded together. Based on such preliminary results, the attention module performs decoding on each folded sub-sequence, which is equivalent to narrow down the attention scope to distinguish confused predictions. Specifically, the attention module applies the chain rule to calculate the likelihood of the label sequence based on the conditional probability of y'_l at a given input feature H and the previous label $y'_1, \ldots, y'_{l-1}$:

$$P_{\text{Atten}}\left(Y' \mid Z'^{*}, H\right) = \prod_{l=1}^{L} p\left(y'_l \mid y'_1, \ldots, y'_{l-1}, Z'^{*}, H\right). \tag{2}$$

The alignment information provided by the modified sequence Z'^{*} is only used to help the attention module divide H into sub-sequences:

$$P_{\text{Atten}}\left(Y' \mid Z'^{*}, H\right) = \prod_{l=1}^{L} p\left(y'_l \mid y'_1, \ldots, y'_{l-1}, h_1, \ldots, h_{\tau_l}\right), \tag{3}$$

where $\tau_l = t'_l + \varepsilon_l$, t'_l and ε_l are the position indexes and times in Z'^{*}. We get the conditional probability $p\left(y'_l \mid y'_1, \ldots, y'_{l-1}, h_1, \ldots, h_{\tau_l}\right)$ through a typical attention mechanism, which can be written as follows:

$$
\begin{aligned}
a_{l,t} &= \text{ContextAttention}\left(q_{l-1}, h_t\right), \\
r_l &= \textstyle\sum_{t=1}^{\tau_l} a_{l,t} h_t, \\
p\left(y'_l \mid y'_1, \ldots, y'_{l-1}, h_1, \ldots, h_{\tau_l}\right) &= \text{Decoder}\left(r_l, q_{l-1}, y'_{l-1}\right),
\end{aligned} \tag{4}
$$

where q_{l-1} represents the hidden state of the decoder at the previous position. At each step l, the decoder generates a context vector based on the first τ_l features h and the attention weights a.

Training and Inference. The hybrid CTC-Attention decoder is trained with the multi-task loss that is defined as the weighted sum of the CTC loss and the attention loss:

$$
\begin{aligned}
\mathcal{L}_{Total} &= \lambda \mathcal{L}_{CTC} + (1 - \lambda)\mathcal{L}_{\text{Attention}}, \\
\mathcal{L}_{CTC} &= - \sum_{(I, Y') \in \mathcal{X}} \log P_{\text{CTC}}\left(Y' \mid Z'\right), \\
\mathcal{L}_{\text{Attention}} &= - \sum_{(I, Y') \in \mathcal{X}} \log P_{\text{Atten}}\left(Y' \mid Z'^{*}, H\right)
\end{aligned} \tag{5}
$$

where the adjustment parameter λ controls the weight between the two objective functions. Both the CTC module and the attention module share the same mixture label dictionary. In the HCADecoder, the outputs of the CTC module are just preliminary results that guide the training of the attention module. Therefore, during inference, we take the outputs of the attention module as the final results.

4 Experiments

4.1 Datasets and Evaluation Metrics

Synthetic data are widely used for text recognition because of the requirement of a large number of data for deep learning training. MJSynth [28] and Synth-Text [13] are two popular synthetic English text datasets. We follow the image generation principle of MJSynth with some changes to build training data. The generation process is described as follows:

- Rendering foreground images with more than 20 Chinese fonts.
- Rendering background images by filling with random colors, pure white, and shadows.
- Combining the rendered foreground images with background images.
- Applying perspective and affine transformation.
- Adding noise to the image, such as Gaussian noise blur and image compression randomly.

We generated two Chinese datasets with variable length sequences for experiments:

- **Synthetic Chinese Texts with Variable Length (Synth-CTVL):** The dataset contains five million images that are divided into 8:1:1 for training, validation, and test, respectively. The text lengths range from 5 to 30. The Chinese strings come from two public corpora: *the National Language Commission Modern Chinese Balance Corpus*[2] and *the Wiki Chinese text corpus*[3].
- **Synthetic Real Property Certificates (Synth-RPC):** The dataset contains five million synthetic images that are also divided into 8:1:1 for training, validation, and test, respectively. The text lengths range from 2 to 35. The Chinese strings are selected from a private corpus of real property certificates.

Furthermore, we also evaluate the proposed method on two public datasets. Unlike the Chinese text dataset captured in the wild, these two datasets contain more continuous long texts, especially CASIA-HWDB. This is an important scenario for Chinese text recognition and also the issue that this paper addresses.

- **CASIA-HWDB 2.0-2.2** [11]: This is a Chinese handwritten dataset written by 1,019 people, including 5,091 pages and 52,230 lines with variable text lengths ranging from 1 to 50.
- **ICDAR17 MLT** [12]: This dataset contains text images in multiple languages, from which 2,502 horizontal Chinese text images are selected for evaluation. The text lengths range from 1 to 36. The model is trained with Synth-CTVL.

The prediction accuracy is used as the main evaluation metric. Given a test image, the result is correct if and only if all characters in the prediction are the same as the ground truth without repetition and omission. In CASIA-HWDB, the correct rate (CR) and accurate rate (AR) are also adopted as the metrics for comparison following the practice in [29].

We implement the proposed method with the PyTorch framework. The batch size is set to 32 on each GPU. For each dataset, the height of an image is scaled to 32 pixels. The image width is proportionally scaled with the height and then padded to the maximum image width in the dataset.

---

[2] http://www.cncorpus.org/.
[3] https://dumps.wikimedia.org/zhwiki/latest/.

4.2   Comparison Methods

In Sect. 4.5, we compare the proposed method with two classic text recognition methods, i.e., CRNN [1] and ASTER [30]. Besides, the Joint CTC-Attention [24] is also taken into comparison, which outputs two parallel results and the best one is used for evaluation. For the CASIA-HWDB dataset, we compare the proposed method with recent state-of-the-art methods [32,33] that are designed for handwritten text recognition.

4.3   Hyper-parameters Selection

The number of bigram subwords and the adjustment factor λ are two key parameters in our method. Therefore, we present experiments for hyper-parameters selection in this section.

As shown in Table 1, we first keep $\lambda = 0.2$ and change the number of subwords from 50 to 400. The recognition accuracy increases with the number of subwords. When using 300 subwords, our method achieves the best results. Then we continue to increase the number of subwords to 400, the performance shows degradation. One possible reason is that some low-frequency subwords are taken into the mixture label dictionary, which sparsely distribute in the dataset and make the optimization difficult. After that, we use the top 300 subwords and change λ from 0.1 to 0.5. HCADecoder achieves the best performance when λ is 0.2 or 0.3 and the results show that HCADecoder is insensitive to λ.

Table 1. Experimental results for hyper-parameters selection.

# subwords	λ	Synth-CTVL	Synth-RPC	CASIA-HWDB	ICDAR17 MLT
50	0.2	0.901	0.877	0.737	0.826
100	0.2	0.909	0.886	0.743	0.828
200	0.2	0.918	0.894	0.744	0.832
300	0.2	**0.921**	0.895	**0.746**	**0.836**
400	0.2	0.919	0.894	0.745	0.835
300	0.1	0.913	0.890	0.740	0.830
300	0.3	0.918	**0.905**	0.742	0.834
300	0.4	0.916	0.902	0.740	0.832
300	0.5	0.915	0.900	0.738	0.828

4.4   Ablation Studies

The proposed HCADecoder is composed of a CTC module and an attention module. Based on the result of the CTC module, the attention module performs decoding on each sub-sequence and outputs the eventual prediction. In order to evaluate the effectiveness of such a hybrid mechanism, we conduct ablation studies in this section. Specifically, we degrade the HCADecoder into a vanilla CTC decoder and a typical attention decoder, respectively. The mixture label strategy is kept unchanged. As shown in Table 2, the proposed HCADecoder is superior to any model that only contains a CTC decoder or an attention decoder.

Table 2. Ablation studies for the main components of HCADecoder.

Method	Synth-CTVL	Synth-RPC	CASIA-HWDB	ICDAR17 MLT
Ours (only CTC)	0.883	0.845	0.730	0.810
Ours (only attention)	0.880	0.847	0.732	0.812
Ours (HCADecoder)	**0.921**	**0.905**	**0.746**	**0.836**

4.5   Comparison with Classic Methods

In this section, we first compare the proposed HCADecoder with three classic methods, i.e., CRNN, ASTER, and Joint CTC-Attention. As shown in Table 3, applying the proposed mixture label, the performance for both CRNN and ASTER improves on four datasets. Different from the Joint CTC-Attention that combines two kinds of methods through a multi-task manner without any other optimization, the proposed HCADecoder cooperates with the mixture label and achieves better results than other methods.

To further evaluate the proposed method for handwritten Chinese text recognition, we compare HCADecoder with recent state-of-the-art methods [31–33] on CASIA-HWDB. As shown in Table 4, HCADecoder achieves comparable results with less training data and without the language model.

Table 3. Comparison with classic methods.

Method	Label Type	Synth-CTVL	Synth-RPC	CASIA-HWDB	ICDAR17 MLT
CRNN[1]	Unigram	0.844	0.824	0.712	0.802
	Mixture	0.852	0.834	0.723	0.802
ASTER[30]	Unigram	0.914	0.892	0.732	0.830
	Mixture	0.918	0.897	0.736	0.832
Joint CTC-Attention [24]	Unigram	0.889	0.851	0.732	0.826
HCADecoder	Mixture	**0.921**	**0.905**	**0.746**	**0.836**

Table 4. Comparison with state-of-the-art methods on CASIA-HWDB.

Method	Training data	Language model	CR	AR
Xie et al. [31]	CASIA-HWDB 1.0-1.2 & 2.0-2.2	✓	**0.973**	**0.970**
		✗	0.954	0.949
Wu et al. [32]	CASIA-HWDB 1.0-1.2 & 2.0-2.2	✓	0.959	0.958
Wang et al. [33]	CASIA-HWDB 1.0-1.2 & 2.0-2.2	✓	0.956	0.951
		✗	0.922	0.908
HCADecoder	CASIA-HWDB 2.0-2.2	✗	0.955	0.951

4.6 Comparison for Different Text Lengths

To evaluate the performance of HCADecoder on long text sequences. We present experimental results for different text lengths on the four datasets. For each dataset, the test data are divided into three groups, in which the text lengths are <10, 10–20, and >20, respectively. As shown in Table 5, HCADecoder has a comparable performance with CRNN and ASTER when the text length is shorter than 10. When the text length is longer than 10, it is noticed that both CRNN and ASTER show performance degradation for most datasets. However, HCADecoder can keep a stable performance on long text sequences.

Table 5. Experimental results for different text lengths. The test data are divided into three groups, in which the text lengths are <10, 10–20 and >20, respectively.

Method	Synth-CTVL			Synth-RPC		
	<10	10−20	>20	<10	10−20	>20
CRNN[1]	0.836	0.848	0.846	0.819	0.824	0.828
ASTER[30]	0.917	0.916	0.909	0.900	0.899	0.880
HCADecoder	**0.917**	**0.923**	**0.923**	**0.901**	**0.908**	**0.908**
Method	CASIA-HWDB			ICDAR17 MLT		
	<10	10−20	>20	<10	10−20	>20
CRNN[1]	0.720	0.710	0.712	0.803	0.798	0.782
ASTER[30]	**0.739**	0.731	0.728	0.830	0.837	0.820
HCADecoder	0.738	**0.749**	**0.746**	**0.835**	**0.841**	**0.875**

4.7 Examples of Failure Cases

In this section, we present some failure examples from the proposed method and other comparison methods. In Fig. 6, HCADecoder achieves better results than CRNN and ASTER even on short texts, which is possible because the CTC module helps to narrow down the attention scope and the attention module can focus on local discriminated features. The third example in Fig. 6 demonstrates that HCADecoder can effectively recognize a character that is the combination of other characters. Moreover, we also present some bad cases from HCADecoder. As shown in Fig. 7, most false predictions result from texts that are warped or not clearly written.

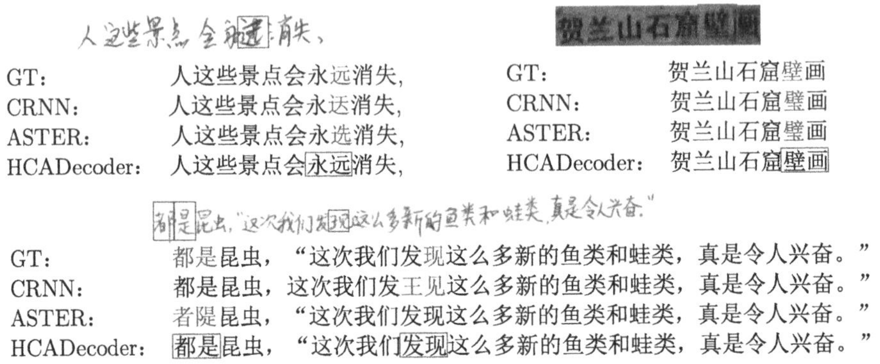

GT: 人这些景点会永远消失，

CRNN: 人这些景点会永远消失，

ASTER: 人这些景点会永选消失，

HCADecoder: 人这些景点会永远消失，

GT: 贺兰山石窟壁画

CRNN: 贺兰山石窟壁画

ASTER: 贺兰山石窟壁画

HCADecoder: 贺兰山石窟壁画

GT: 都是昆虫，"这次我们发现这么多新的鱼类和蛙类，真是令人兴奋。"

CRNN: 都是昆虫，这次我们发王见这么多新的鱼类和蛙类，真是令人兴奋。"

ASTER: 者隉昆虫，"这次我们发现这么多新的鱼类和蛙类，真是令人兴奋。"

HCADecoder: 都是昆虫，"这次我们发现这么多新的鱼类和蛙类，真是令人兴奋。"

Fig. 6. Examples that show HCADecoder is superior to other methods. Red characters and green characters represent false predictions and the corresponding ground truth, respectively. These characters are bounded with red boxes in test images. Bigram subwords are bounded with green boxes. (Color figure online)

GT: 真美香辣虾

CRNN: 姜香辣虾

ASTER: 美香辣虾

HCADecoder: 美香辣虾

GT: 世界人民大团结万岁

CTC: 世界人民大团铺万岁

ASTER: 世界人民大团错万岁

HCADecoder: 世界人民大团错万岁

GT: 而是要告诉人们它们已非往日,旅游胜地亟待

CRNN: 而悬要告诉人们它们已非往日,旅游胜地函待

ASTER: 而是要告诉人们它们已非往日,旅游   地亟待

HCADecoder: 而是要告诉人们它们已非往日,旅游胜地亟待

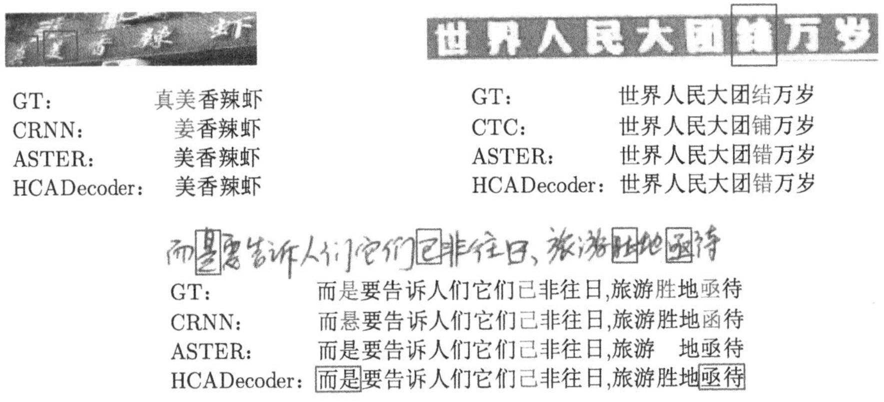

Fig. 7. Examples of failure cases. Red characters and green characters represent false predictions and the corresponding ground truth, respectively. These characters are bounded with red boxes in test images. Bigram subwords are bounded with green boxes. (Color figure online)

5   Conclusion

In this paper, we propose a hybrid CTC-Attention decoder, i.e., HCADecoder, for Chinese text recognition. Specifically, we add high-frequency bigram subwords into the original unigram characters to construct the mixture label for each sample. In HCADecoder, the CTC module first outputs a preliminary prediction, in which confused results are changed into bigram subwords and folded into sub-sequences. Then, the attention module works on each folded sub-sequence and outputs the final prediction. Experiments on four Chinese datasets demonstrate the effectiveness of HCADecoder, especially for long text sequences.

Acknowledgments. This work was supported in part by the National Natural Science Foundation of China under Grant U2034211, 62006017, in part by the Fundamental Research Funds for the Central Universities under Grant 2020JBZD010 and in part by the Beijing Natural Science Foundation under Grant L191016.

References

1. Shi, B., Bai, X., Yao, C.: An end-to-end trainable neural network for image-based sequence recognition and its application to scene text recognition. IEEE Trans. Pattern Anal. Mach. Intell. **39**(11), 2298–2304 (2016)
2. Cheng, Z., Bai, F., Xu, Y., Zheng, G., Pu, S., Zhou, S.: Focusing attention: towards accurate text recognition in natural images. In: Proceedings of the IEEE International Conference on Computer Vision (ICCV), pp. 5076–5084 (2017)
3. Veit, A., Matera, T., Neumann, L., Matas, J., Belongie, S.: Coco-text: dataset and benchmark for text detection and recognition in natural images. arXiv preprint arXiv:1601.07140 (2016)
4. Karatzas, D., et al.: ICDAR 2015 competition on robust reading. In: Proceedings of the International Conference on Document Analysis and Recognition (ICDAR), pp. 1156–1160 (2015)
5. Naphade, M.R., Huang, T.S.: Extracting semantics from audio-visual content: the final frontier in multimedia retrieval. IEEE Trans. Neural Netw. **13**(4), 793–810 (2002)
6. Sadeghi, H., Valaee, S., Shirani, S.: Ocrapose: an indoor positioning system using smartphone/tablet cameras and OCR-aided stereo feature matching. In: Proceedings of the IEEE International Conference on Acoustics, Speech and Signal Processing (ICASSP), pp. 1473–1477 (2015)
7. Wan, Z., Xie, F., Liu, Y., Bai, X., Yao, C.: 2D-CTC for scene text recognition. arXiv preprint arXiv:1907.09705 (2019)
8. Cheng, Z., Xu, Y., Bai, F., Niu, Y., Pu, S., Zhou, S.: AON: towards arbitrarily-oriented text recognition. In: Proceedings of the IEEE Conference on Computer Vision and Pattern Recognition (CVPR), pp. 5571–5579 (2018)
9. Graves, A., Fernández, S., Gomez, F., Schmidhuber, J.: Connectionist temporal classification: labelling unsegmented sequence data with recurrent neural networks. In: Proceedings of the 23rd International Conference on Machine Learning (ICML), pp. 369–376 (2006)
10. Bahdanau, D., Cho, K., Bengio, Y.: Neural machine translation by jointly learning to align and translate. arXiv preprint arXiv:1409.0473 (2014)
11. Liu, C.L., Yin, F., Wang, D.H., Wang, Q.F.: CASIA online and offline Chinese handwriting databases. In: Proceedings of the International Conference on Document Analysis and Recognition (ICDAR), pp. 37–41 (2011)
12. Nayef, N., et al.: ICDAR2017 robust reading challenge on multi-lingual scene text detection and script identification-RRC-MLT. In: Proceedings of the International Conference on Document Analysis and Recognition (ICDAR), vol. 1, pp. 1454–1459 (2017)

13. Gupta, A., Vedaldi, A., Zisserman, A.: Synthetic data for text localisation in natural images. In: Proceedings of the IEEE Conference on Computer Vision and Pattern Recognition (CVPR), pp. 2315–2324 (2016)

14. Ding, X., Wang, Y.: Character Recognition: Principles, Methods and Practice (2017)

15. Wang, Z.R., Du, J., Wang, J.M.: Writer-aware CNN for parsimonious HMM-based offline handwritten Chinese text recognition. Pattern Recognit. **100**, 107102 (2020)

16. Tong, G., Li, Y., Gao, H., Chen, H., Wang, H., Yang, X.: MA-CRNN: a multi-scale attention CRNN for Chinese text line recognition in natural scenes. Int. J. Doc. Anal. Recognit. (IJDAR) **23**, 103–114 (2019)

17. Zhao, Y., Xue, W., Li, Q.: A multi-scale CRNN model for Chinese papery medical document recognition. In: Proceedings of the IEEE Fourth International Conference on Multimedia Big Data (BigMM), pp. 1–5 (2018)

18. Gao, Y., Chen, Y., Wang, J., Lu, H.: Reading scene text with attention convolutional sequence modeling. arXiv preprint arXiv:1709.04303 (2017)

19. Shigeki, K., Soplin, N., Watanabe, S., Delcroix, D., Ogawa, A., Nakatani, T.: Improving transformer-based end-to-end speech recognition with connectionist temporal classification and language model integration. In: Proceedings of INTERSPEECH, vol. 9, pp. 1408–1412 (2019)

20. Chorowski, J., Jaitly, N.: Towards better decoding and language model integration in sequence to sequence models. arXiv preprint arXiv:1612.02695 (2016)

21. Lee, C.Y., Osindero, S.: Recursive recurrent nets with attention modeling for OCR in the wild. In: Proceedings of the IEEE Conference on Computer Vision and Pattern Recognition (CVPR), pp. 2231–2239 (2016)

22. Liu, Z., Li, Y., Ren, F., Goh, W.L., Yu, H.: Squeezedtext: a real-time scene text recognition by binary convolutional encoder-decoder network. In: Proceedings of the AAAI Conference on Artificial Intelligence, vol. 32 (2018)

23. Bai, F., Cheng, Z., Niu, Y., Pu, S., Zhou, S.: Edit probability for scene text recognition. In: Proceedings of the IEEE Conference on Computer Vision and Pattern Recognition (CVPR), pp. 1508–1516 (2018)

24. Kim, S., Hori, T., Watanabe, S.: Joint CTC-attention based end-to-end speech recognition using multi-task learning. In: Proceedings of the IEEE International Conference on Acoustics, Speech and Signal Processing (ICASSP), pp. 4835–4839 (2017)

25. Hu, W., Cai, X., Hou, J., Yi, S., Lin, Z.: GTC: guided training of CTC towards efficient and accurate scene text recognition. In: Proceedings of the AAAI Conference on Artificial Intelligence, vol. 34, pp. 11005–11012 (2020)

26. Fan, D., Gao, G., Wu, H.: Sub-word based Mongolian offline handwriting recognition. In: Proceedings of the International Conference on Document Analysis and Recognition (ICDAR), pp. 246–253 (2019)

27. Saluja, R., Punjabi, M., Carman, M., Ramakrishnan, G., Chaudhuri, P.: Sub-word embeddings for OCR corrections in highly fusional Indic languages. In: Proceedings of the International Conference on Document Analysis and Recognition (ICDAR), pp. 160–165 (2019)

28. Jaderberg, M., Simonyan, K., Vedaldi, A., Zisserman, A.: Synthetic data and artificial neural networks for natural scene text recognition. arXiv preprint arXiv:1406.2227 (2014)

29. Wang, S., Chen, L., Xu, L., Fan, W., Sun, J., Naoi, S.: Deep knowledge training and heterogeneous CNN for handwritten Chinese text recognition. In: Proceedings of the 15th IEEE International Conference on Frontiers in Handwriting Recognition (ICFHR), pp. 84–89 (2016)

30. Shi, B., Yang, M., Wang, X., Lyu, P., Yao, C., Bai, X.: ASTER: an attentional scene text recognizer with flexible rectification. IEEE Trans. Pattern Anal. Mach. Intell. **41**(9), 2035–2048 (2018)
31. Xie, C., Lai, S., Liao, Q., Jin, L.: High performance offline handwritten Chinese text recognition with a new data preprocessing and augmentation pipeline. In: Bai, X., Karatzas, D., Lopresti, D. (eds.) DAS 2020. LNCS, vol. 12116, pp. 45–59. Springer, Cham (2020). https://doi.org/10.1007/978-3-030-57058-3_4
32. Wu, Y.C., Yin, F., Liu, C.L.: Improving handwritten Chinese text recognition using neural network language models and convolutional neural network shape models. Pattern Recognit. **65**, 251–264 (2017)
33. Wang, Z.X., Wang, Q.F., Yin, F., Liu, C.L.: Weakly supervised learning for over-segmentation based handwritten Chinese text recognition. In: Proceedings of the 17th IEEE International Conference on Frontiers in Handwriting Recognition (ICFHR), pp. 157–162 (2020)

Meta-learning of Pooling Layers
for Character Recognition

Takato Otsuzuki, Heon Song, Seiichi Uchida⬥, and Hideaki Hayashi(✉)⬥

Kyushu University, Fukuoka, Japan
{takato.otsuzuki,heon.song}@human.ait.kyushu-u.ac.jp,
{uchida,hayashi}@ait.kyushu-u.ac.jp

Abstract. In convolutional neural network-based character recognition, pooling layers play an important role in dimensionality reduction and deformation compensation. However, their kernel shapes and pooling operations are empirically predetermined; typically, a fixed-size square kernel shape and max pooling operation are used. In this paper, we propose a meta-learning framework for pooling layers. As part of our framework, a parameterized pooling layer is proposed in which the kernel shape and pooling operation are trainable using two parameters, thereby allowing flexible pooling of the input data. We also propose a meta-learning algorithm for the parameterized pooling layer, which allows us to acquire a suitable pooling layer across multiple tasks. In the experiment, we applied the proposed meta-learning framework to character recognition tasks. The results demonstrate that a pooling layer that is suitable across character recognition tasks was obtained via meta-learning, and the obtained pooling layer improved the performance of the model in both few-shot character recognition and noisy image recognition tasks.

Keywords: Convolutional neural network · Pooling layer · Meta-learning · Character recognition · Few-shot learning

1 Introduction

In convolutional neural network (CNN)-based character recognition, pooling layers play an important role in dimensionality reduction and deformation compensation. In particular, the max and average pooling layers, as illustrated in Figs. 1(a) and (b), respectively, are widely used in CNNs. These pooling operations are effective in absorbing the deformations that occur in character images. Even if the convolutional features undergo local changes owing to the deformation of the character image, the dimensionally-reduced feature map is invariant to such changes. As a result, a CNN with pooling layers is robust to character image deformation.

Nevertheless, typical pooling layers have a limitation in that the kernel shape and pooling operation should be determined empirically and manually. In the

---

We provide our implementation at https://github.com/Otsuzuki/Meta-learning-of-Pooling-Layers-for-Character-Recognition.

J. Lladós et al. (Eds.): ICDAR 2021, LNCS 12823, pp. 188–203, 2021.
https://doi.org/10.1007/978-3-030-86334-0_13

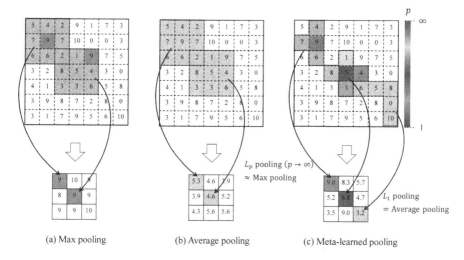

Fig. 1. Meta-learning of pooling and comparison with other pooling methods: (a) max pooling, (b) average pooling, and (c) meta-learned pooling (ours). The matrices in the top and bottom rows represent feature maps before and after pooling processing, respectively, where the number in each cell corresponds to the feature value in each pixel. The colored squares represent kernels, and their colors correspond to the type of operations. In (a) and (b), the maximum and average values are calculated for each kernel. Moreover, all kernels share the same shape and operation. In (c), the proposed meta-learned pooling employs different shapes and operations in individual kernels.

pooling layers, pooling operations are performed by sliding a fixed-size kernel, and the kernel size is predetermined as a hyperparameter. In two-dimensional pooling, the shape of the kernel is fixed to a square generally. Although various pooling operations have been proposed [12,20,23,25,32], the type of pooling operation is chosen manually, and the same operation is performed in the layer.

The purpose of this study is to determine a pooling layer suitable for character recognition tasks in a data-driven approach. Although max and average pooling layers with fixed-size square kernels are generally used in tasks using character images, it is unknown whether this is in fact appropriate. If we can obtain a suitable pooling layer across multiple tasks using character images, we can apply it to subsequent new tasks using character images, thereby leading to improvements in recognition accuracy.

For this purpose, we utilize meta-learning. Meta-learning aims to improve the learning performance of a machine learning model by providing datasets of multiple tasks and then acquiring knowledge shared among the tasks. Specifically, hyperparameters, such as a learning rate and initial weights, that are suitable across all the tasks are obtained via meta-learning. As a result, the meta-learned model can improve its learning efficiency when a new task is given.

In this paper, we propose a meta-learning framework for pooling layers. As part of the framework, a parameterized pooling layer is proposed in which the kernel shape and pooling operation can be trained using two parameters, thereby allowing flexible pooling of the data. We also propose a meta-learning algorithm for the parameterized pooling layer, which allows us to acquire a suitable pooling layer across multiple tasks.

Figure 1 illustrates an example of a pooling layer obtained under our framework, alongside its comparison with max and average pooling layers. In the traditional max and average pooling layers shown in Figs. 1(a) and (b), respectively, fixed-size square kernels are used, and the same operation is performed in each kernel. In contrast, in the pooling layer meta-learned in our framework, flexible kernel shapes are used. Furthermore, the operation in each kernel is defined based on L_p pooling [27], which means various types of operations, including max and average pooling, can be realized depending on the value of p.

In the experiments, we reveal *what pooling layer is suitable for character recognition tasks* in a data-driven approach and demonstrate that the obtained pooling layer improves the learning performance of a CNN for a subsequent new task. Figure 2 shows the flow of the proposed method applied to the meta-learning of character image recognition. First, we prepare a meta-dataset containing a large number of character recognition tasks and apply the proposed meta-learning framework to a pooling layer in a CNN, as shown Fig. 2(a). In this step, a pooling layer suitable for multiple character recognition tasks can be obtained. Then, the CNN with the meta-learned pooling layer is adapted to a new task (Fig. 2(b)).

The main contributions of this study can be summarized as follows:

- We propose a meta-learning framework for pooling layers. As part of the proposed framework, a parameterized pooling layer is proposed to make the kernel shape and pooling operation trainable. A meta-learning algorithm for the parameterized pooling layer is also presented.
- We reveal a pooling layer suitable for character recognition tasks in a data-driven manner. We prepare a large number of character image recognition tasks using the Omniglot dataset [17] and then apply the proposed meta-learning framework to these tasks. After meta-learning, we analyze the obtained kernel shapes and pooling operations by visualizing the meta-parameters of the parameterized pooling layer. The results demonstrate that the meta-learned pooling layer applies max pooling around the character region in an annular shape.
- We demonstrate the effectiveness of the meta-learned pooling layer for few-shot character image recognition and noisy character image recognition tasks. The meta-learned pooling layer is effective in improving accuracy for one-shot character image recognition compared to max and average pooling layers. The robustness to noisy images is also remarkably improved by the proposed method.

2   Related Work

2.1   Pooling Layers

In recent years, many pooling methods have been proposed and have demonstrated remarkable performance in specific tasks. Gao *et al.* [10] proposed a

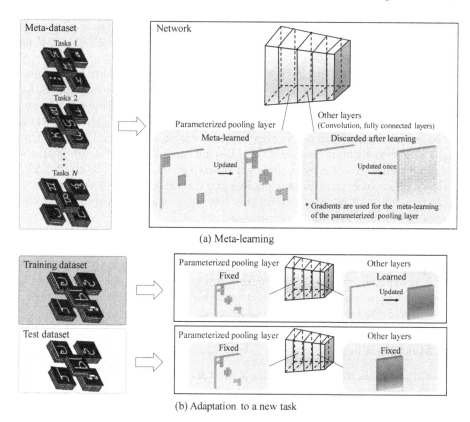

(a) Meta-learning

(b) Adaptation to a new task

Fig. 2. Meta-learning flow. The flow of the proposed framework is divided into two steps: (a) First, the kernel shapes and pooling operations of the parameterized pooling layer are trained via meta-learning. By learning the pooling layer from a large number of tasks, we can acquire knowledge across all tasks. At this time, the parameters of the other layers are updated once for each task batch, and their gradients are used in the meta-learning of the parameterized pooling layer, whereas the updates of the other layers themselves are not used in the subsequent step. (b) Then, the meta-learned pooling layer is adapted to a new task. In this step, with the meta-learned pooling fixed, the parameters of the remaining layers are trained using the training dataset of the new task. Next, with all parameters fixed, the trained model is adapted to the new task by classifying the test data.

global second-order pooling (GSoP) block which can be used to calculate the second-order information in the middle layer of a CNN. Although methods that utilize the mean and variance of feature maps in the last layer have been proposed [4,9,18,19,31], the GSoP block is the first method that can be plugged in at any location of a CNN. Van *et al.* [22] proposed a novel scene text proposal technique, which iteratively applies a pooling operation to the edge feature extracted from an image, thereby improving the accuracy of scene text box detection. Hou *et al.* [13] proposed a strip pooling module, which considers a long but narrow kernel. They demonstrated that short-range and long-range dependencies among different locations can be captured simultaneously by combining a strip pooling module with a pyramid pooling module (PPM) [34] in semantic segmentation tasks.

Instead of utilizing a specific pooling operation, some trials have been aimed to generalize pooling operations for global use. A representative example is L_p pooling [7,27], which calculates an L_p-norm over the input elements. Consequently, L_p pooling involves average pooling when $p = 1$ and max pooling when $p \to \infty$, and it can perform pooling operations between average and max pooling using $1 < p < \infty$ without completely eliminating the influence of input elements with small values. Another example is generalized mean pooling (GeM) [2,5,29], which is a generalization of average pooling and consequently has a similar operation to L_p pooling. A pooling method that can automatically enhance important features by learning has also been proposed [11,33].

The proposed method is inspired by the attempts to generalize pooling operations. In particular, the formulation of the proposed method is an expansion of L_p pooling. The proposed method differs from L_p pooling in that the kernel shape is trainable. Another difference from conventional trainable pooling methods is that the proposed method learns suitable pooling across multiple tasks, instead of a single task, based on meta-learning.

2.2   Meta-learning

Meta-learning aims to learn knowledge shared across multiple tasks to adapt the model to another subsequent task given with a small amount of training data or few learning steps. According to Baik *et al.* [1], meta-learning algorithms can be divided into three categories: network-based, metric-based, and optimization-based. Network-based methods learn a fast adaptation strategy using auxiliary networks [21,26]. Metric-based methods learn the relationships between inputs in the task spaces and acquire object metrics or distance functions to facilitate problem solving [3,28,30]. Optimization-based methods adjust the optimization algorithm so that the model can be adapted to several tasks [1,8,24,35].

Among existing optimization-based meta-learning algorithms, model-agnostic meta-learning (MAML) [8] has attracted much attention because of its simplicity and generality. MAML encodes prior knowledge so that the model can perform well across tasks and learn new tasks rapidly. Some studies have aimed to improve the performance of MAML while maintaining simplicity and generality. Rusu *et al.* [24] proposed an approach that learns an embedding of the model parameters into a low-dimensional space and conducts meta-learning in that space, achieving more efficient adaptation. Baik *et al.* [1] improved the MAML-based framework by attenuating the conflicts among tasks and layers, thereby improving its performance in few-shot learning. Some researchers have attempted to extend MAML. Zhou *et al.* [35] introduced a method called meta-learning symmetries by reparameterization (MSR) for meta-learning neural network architectures. They showed that MSR can learn symmetries shared among tasks from data based on meta-learning and reparameterization of network weights.

Although the algorithm of the proposed method is inspired by MAML and MSR, the proposed algorithm differs from the above optimization-based meta-learning algorithms in that meta-learning is applied to the pooling layer while considering the gradients of other layers. To the best of our knowledge, this is the first attempt in adopting meta-learning in the pooling layer. Another unique feature of the proposed algorithm is that the kernel shape matrix is learned with a constraint to be binary.

3   Meta-learning of Pooling Layers

In this section, we present our meta-learning framework for pooling layers. To make pooling layers trainable, we first introduce a parameterized pooling layer, wherein the shape and operation of each kernel are trained individually. We then describe the meta-learning algorithm of the parameterized pooling layer.

3.1   Parameterized Pooling Layer

The purpose of the proposed method is to acquire the knowledge shared across multiple tasks, thereby developing a pooling layer that can be efficiently applied to new tasks. By parameterizing the pooling layer and making it trainable, flexible determination of kernel shapes and pooling operations can be achieved.

The parameterized pooling layer has two (meta-)parameters that are trained via meta-learning. One is a kernel shape matrix, W, with binary elements that determines the area in which the pooling operation is applied. By multiplying W with the input vector/image, various types of kernel locations and shapes can be realized, including the sliding square kernels of ordinary pooling. The other is an operation parameter, p. The parameterized pooling layer is based on L_p pooling, where the pooling operation in each kernel is defined by the L_p norm. Incidentally, L_p pooling is a generalization of pooling operations that includes max pooling (when $p \to \infty$) and average pooling (when $p = 1$). Whereas the value of p in L_p pooling is determined empirically, the value of p in the proposed method is automatically determined via meta-learning.

Given an input vector $x \in \mathbb{R}_+^J$, the parameterized pooling layer $f : \mathbb{R}_+^J \mapsto \mathbb{R}^I$ ($J \geq I$) is defined as follows:

$$f_i(x) = \left(\frac{1}{J} \sum_{j=1}^{J} W_{ij} x_j^{p_i} \right)^{\frac{1}{p_i}}, \quad i = 1, \ldots I, \tag{1}$$

where $W \in \{0,1\}^{I \times J}$ is the kernel shape matrix, $p \in \mathbb{R}_+^I$ is the operation parameter, and J and I are the input and output dimensions, respectively. Although the assumption that the input elements are positive is relatively strong, in practice, it can be satisfied by applying a rectified linear unit (ReLU) activation function to the previous layer.

The parameterized pooling layer can also be applied to an image input, as with conventional pooling layers. One simple method is to vectorize the image input by raster scanning before inputting the data to the parameterized pooling layer. Another method is to combine Eq. (1) and sliding windows. Taking a small

---

Algorithm 1. Meta-learning of the parameterized pooling layer

---

Inputs: $\{\tau_j\}_{j=1}^N \sim \rho(\tau)$: Meta-learning tasks
Inputs: $\{W, p\}$: Randomly or arbitrarily initialized meta-parameters
Inputs: θ: Randomly initialized weights of other layers
Inputs: α, η: Learning rates for the inner and outer loops
while *not done* **do**
    Sample minibatch $\{\tau_i'\}_{i=1}^n \sim \{\tau_j\}_{j=1}^N$
    for $\tau_i' \in \{\tau_i'\}_{i=1}^n$ **do**
        $\{D_i^{\text{tr}}, D_i^{\text{val}}\} \leftarrow \tau_i'$; // task data
        $\delta_i \leftarrow \nabla_\theta \mathcal{L}(W, p, \theta, D_i^{\text{tr}})$;
        $\theta' \leftarrow \theta - \alpha\delta_i$; // inner step
        /* outer gradient */
        $G_i^{\text{w}} \leftarrow \frac{\mathrm{d}}{\mathrm{d}W}\mathcal{L}(W, p, \theta', D_i^{\text{val}})$, $G_i^{\text{p}} \leftarrow \nabla_p \mathcal{L}(W, p, \theta', D_i^{\text{val}})$;
    end for
    $W \leftarrow W - \eta \sum_i G_i^{\text{w}}$, $p \leftarrow p - \eta \sum_i G_i^{\text{p}}$;
end while

---

window in the input image, we vectorize the pixels in the window and prepare independent W and p for each window. Then, we apply Eq. (1) to the entire image while shifting the window across the image. In practice, the latter method is computationally advantageous; therefore, we employed it in our experiments using image inputs, as reported in Sects. 4 and 5.

To maintain the binariness of W and positiveness of p, we employ the following variable transformation:

$$W_{ij} = \begin{cases} \text{Sigmoid}((\tilde{W}_{ij} - 0.5)/T) & \text{during meta-learning} \\ \text{Step}(\tilde{W}_{ij} - 0.5) & \text{otherwise} \end{cases}, \tag{2}$$

$$p_i = \exp(\tilde{p}_i), \tag{3}$$

where $\text{Sigmoid}(\cdot)$ is the sigmoid function, T is a temperature parameter, $\text{Step}(\cdot)$ is the unit step function, which returns 0 for a negative input and 1 otherwise, and $\tilde{W}_{ij} \in \mathbb{R}$ and $\tilde{p}_i \in \mathbb{R}$ are auxiliary variables. The step function is used to make W binary, but it is not differentiable; therefore, the sigmoid function is used as an approximation during meta-learning to enable gradient calculation. For p, the exponential function is used to satisfy $p_i > 0$. In the meta-learning algorithm described below, the gradients are calculated with respect to the auxiliary variables $\tilde{W}_{ij}$ and $\tilde{p}_i$. However, in the following sections, we will explain as if we deal directly with W_{ij} and p_i instead of $\tilde{W}_{ij}$ and $\tilde{p}_i$ for simplicity.

3.2   Meta-learning Algorithm

The aim of the meta-learning of pooling is to learn and exploit a pooling layer that is suitable for multiple tasks. Given a task distribution $\rho(\tau)$, we assume that there is an optimal pooling layer that is shared across $\rho(\tau)$, and we estimate it by learning meta-parameters W and p through meta-learning.

Algorithm 1 shows the meta-learning algorithm of the parameterized pooling layer. The algorithm is a gradient-based meta-learning algorithm similar to MSR [35] and MAML [8] algorithms. Assuming that we have a neural network with a parameterized pooling layer as well as other layers with weights $\boldsymbol{\theta}$, the meta-parameters $\boldsymbol{W}$ and $\boldsymbol{p}$ are updated during the meta-learning algorithm. Note that even if there are multiple parameterized pooling layers in a neural network, we can independently apply this algorithm to each one.

This algorithm consists of inner and outer loops. The inner loop calculates gradients and is nested in the outer loop, which updates the meta-parameters. In the outer loop, a task minibatch $\{\tau_i'\}_{i=1}^n$ is first sampled from the task set $\{\tau_j\}_{j=1}^N$, where n and N are the minibatch size and number of tasks, respectively. The task minibatch is then split into training and validation data $\{D_i^{\mathrm{tr}}, D_i^{\mathrm{val}}\}$ for any τ_i'. Proceeding to the inner loop, the parameters of other layers $\boldsymbol{\theta}$ are updated as $\boldsymbol{\theta}' \leftarrow \boldsymbol{\theta} - \alpha \nabla_{\boldsymbol{\theta}} \mathcal{L}(\boldsymbol{W}, \boldsymbol{p}, \boldsymbol{\theta}, D_i^{\mathrm{tr}})$, where $\mathcal{L}(\cdot)$ is the loss function and α is the learning rate for the inner loop. After calculating the gradients with respect to the meta-parameters using the updated $\boldsymbol{\theta}'$ and validation data, the meta-parameters are updated in the outer loop as follows:

$$\boldsymbol{W} \leftarrow \boldsymbol{W} - \eta \sum_i \frac{\mathrm{d}}{\mathrm{d}\boldsymbol{W}} \mathcal{L}(\boldsymbol{W}, \boldsymbol{p}, \boldsymbol{\theta}', D_i^{\mathrm{val}}), \tag{4}$$

$$\boldsymbol{p} \leftarrow \boldsymbol{p} - \eta \sum_i \nabla_{\boldsymbol{p}} \mathcal{L}(\boldsymbol{W}, \boldsymbol{p}, \boldsymbol{\theta}', D_i^{\mathrm{val}}), \tag{5}$$

where η is the learning rate for the outer loop.

4   Experiment on Artificial Data

To verify the validity of the proposed method, we conducted an experiment using artificially generated data. The goal of this experiment is to verify whether the proposed method can meta-learn the kernel shapes and pooling operations from a task set that is generated by passing random vectors or images through a certain pooling layer.

Figure 3 outlines the data generation process for this experiment. We prepared datasets for one-dimensional and two-dimensional cases. For the one-dimensional case (Fig. 3(a)), we generated sets of inputs and outputs by passing random input vectors, whose elements were generated from a uniform distribution over $[0, 1]$, through a 1D pooling layer with a filter size of 2 and a stride of 2. For the two-dimensional case (Fig. 3(b)), we generated sets of inputs and outputs by passing a random image, in which each pixel value was generated from a uniform distribution over $[0, 1]$, through a 2D pooling layer with a filter size of 2×1 (vertical rectangle) and a stride of 2. Here, the pooling layer consisted of max pooling for the first half and average pooling for the second half. Each task contains 20 sets of input and output vectors/images, and we generated 8,000 training tasks. We set the dimensions of the input and output data to 60 and 30 in the one-dimensional case, and 28×28 and 14×14 in the two-dimensional case, respectively.

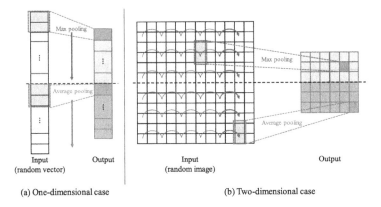

Input   Output   Input   Output
(random vector)   (random image)

(a) One-dimensional case   (b) Two-dimensional case

Fig. 3. Data generation process for the experiment using artificial data. (a) In the one-dimensional case, a set of input and output data was generated by passing a random vector through a 1D pooling layer with a kernel size of 2 and stride of 2. (b) In the two-dimensional case, a random image was passed through a 2D pooling layer with a kernel size of 2×1 and stride of 2, i.e., pooling with a vertical rectangular kernel was performed for every other column. In both cases, max pooling was used in the first half, and average pooling was used in the second half.

In the meta-learning, we used a neural network with only a single parameterized pooling layer. We set the task batch size to 32 and the number of epochs in the outer loop to 10,000, which was sufficient to converge the training. The data in each task were split into 1 training sample and 19 validation samples. In the outer loop, we used the Adam [16] optimizer with a learning rate of 0.001. The temperature parameter T was set to 0.2. In the inner loop, W and p themselves were updated only once instead of θ because the network used in this experiment did not have any layer other than pooling, and the stochastic gradient descent (SGD) optimizer with a learning rate of 0.1 was used. We used the mean squared error as the loss function. After meta-learning, we observed the learned meta-parameters W and p.

Figures 4 and 5 show the learned meta-parameters W and p for the one-dimensional and two-dimensional artificial data, respectively. In Fig. 4(a), higher values are concentrated on the diagonal parts of W, demonstrating that the parameterized pooling layer successfully learned the kernel shape with a size of 2 and stride of 2. It can also be confirmed that p takes large values in the first half and approximates 1 in the second half, meaning that the max and average pooling operations are also learned approximately. In Fig. 5(a), the pooling pattern with a kernel size of 2×1 and stride of 2 was reproduced. The pooling operations with max pooling in the first half and average pooling in the second half were also correctly estimated as shown in Fig. 5(b).

These results demonstrate that the proposed method can learn the kernel shapes and pooling operations in a data-driven manner. In this experiment, we used a network with only a parameterized pooling layer for simplicity. We have also confirmed that the parameterized pooling layer can be meta-learned in a

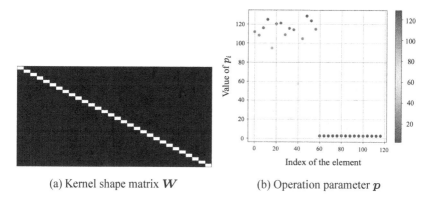

(a) Kernel shape matrix W (b) Operation parameter p

Fig. 4. Learned meta-parameters W and p in the simulation experiment on one-dimensional artificial data. In (a), the black and white pixels represent 0 and 1, respectively. The numbers of columns and rows correspond to the input and output dimensions, respectively.

network that includes other layers, such as convolutional and fully connected layers, but these results are omitted due to page limitations. In the next section, we describe experiments using character images with a network that includes other layers besides a parameterized pooling layer.

5 Experiment on Character Datasets

To evaluate the effectiveness of the proposed method on real-world datasets, we conducted a character image recognition experiment. The aim of this experiment is twofold: One is to determine a pooling layer suitable for character recognition tasks in a data-driven manner, and the other is to evaluate the generalization capability of the meta-learned pooling for new tasks. For the former, we first prepared a CNN that includes a parameterized pooling layer. Then, we meta-learned the parameterized pooling layer using multiple character recognition tasks. After that, we analyzed the learned meta-parameters of the pooling layer, which can be regarded as the kernel shapes and pooling operations that are suitable across multiple tasks. For the latter, we evaluated the model using tasks that were not given during the meta-training. Based on the assumption that the proposed meta-learning of pooling can obtain the knowledge shared across multiple datasets, meta-learned pooling will improve the generalization capability of the model for subsequent new tasks. Therefore, we evaluated the model with the meta-learned pooling layer based on few-shot image recognition and noisy image recognition tasks. Fixing the meta-parameters, we trained the weights of the remaining layers and evaluated the entire model.

5.1 Dataset

We used the Omniglot dataset [17], which is a standard benchmark dataset of handwritten character images. Examples from the Omniglot dataset are shown in

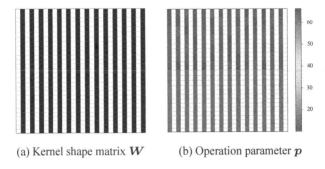

(a) Kernel shape matrix W (b) Operation parameter p

Fig. 5. Learned meta-parameters W and p in the simulation experiment on two-dimensional artificial data. In (a), the black and white pixels represent 0 and 1, respectively.

Fig. 6. Example images from the Omniglot dataset. Pixel colors were inverted from the original images.

Fig. 6. In particular, this dataset is frequently used in few-shot learning tasks [6, 15,30]. In this study, we used this dataset to meta-learn a suitable pooling layer for multiple character recognition tasks and evaluate the meta-learned pooling in few-shot and noisy image recognition tasks. The Omniglot dataset consists of 20 instances of 1,623 characters from 50 different alphabets, and each instance was depicted by a different person. In this experiment, we randomly selected 1,200 characters for meta-learning regardless of the alphabet, and used the remaining 423 characters for few-shot and noisy image recognition tasks. We used the images after inverting pixel colors. We applied data augmentation by rotating the images by 90, 180, or 270°.

5.2   Experimental Setups

Through the experiment, we used the same network architecture. The network consisted of a 3×3 convolutional layer with 64 filters, batch normalization [14], ReLU activation functions, a 2×2 parameterized pooling layer, and a fully connected layer with a softmax activation function. In the parameterized pooling layer, sliding windows with a size of 2×2 and a stride of 2, i.e., no overlap between windows, were used. The temperature parameter T was set to 0.2.

After meta-learning, we conducted few-shot recognition and noisy image recognition tasks. In the few-shot recognition task, the model classified images into one of five classes (five-way recognition) given either one or five images as the training data (one-shot or five-shot, respectively). Among the 423 character classes that were not used in the meta-learning, five classes were randomly sampled 100 times with duplication to generate 100 five-way recognition tasks. In each task, one or five images were used to train the weights of the convolutional and fully connected layers, and the remaining images were used as test data to evaluate the recognition accuracy. In the noisy image recognition task, we reused the network weights obtained in the few-shot recognition tasks and evaluated the recognition accuracy by adding salt-and-pepper noise to the test data. We varied the noise ratio in the interval of 10–60%.

In the meta-learning, the number of task batches was set to 32, and the images in each batch were split into 1 for training and 19 for validation in the one-shot setting and 5 for training and 15 for validation in the five-shot setting. We used the Adam optimizer with a learning rate of 0.001 for the outer loop and the SGD optimizer with a learning rate of 0.1 for the inner loop. The numbers of epochs for inner and outer loops were set to 1 and 10,000, respectively. The initial weights of the convolutional and fully connected layers were determined based on the MAML algorithm [8] using the meta-learning dataset.

For comparison, we prepared two networks with the parameterized pooling layer replaced with either a max or average pooling layer. The results of the few-shot recognition and noisy image recognition tasks were compared to those of the prepared networks. The initial weights of these networks were also determined based on the MAML algorithm.

5.3   Results

The learned meta-parameters W and p are shown in Fig. 7. According to the kernel shape matrix W, the proposed pooling layer learned to extract the annular-shaped area around the region where the character is written. Surprisingly, it did not focus on the center of the image. From the result of p, max pooling was applied around the character region, whereas average pooling was used in other regions. The annular shape observed in both W and p occurred because the character was not always written in the exact center of the image in the Omniglot dataset, and the essential information existed in somewhat peripheral regions. Furthermore, the white annulus in W was thicker than the red annulus in p, indicating that max pooling was learned at the outer edge of the annulus to distinguish the character from the background, and average pooling was used in the area near the character to mitigate the effect of noise in the same manner as a linear filter.

Table 1 lists the recognition accuracies for each pooling method in the one-shot and five-shot settings. As can be seen, the proposed meta-learned pooling showed the best performance in the one-shot setting, whereas max pooling outperformed ours in the five-shot setting. These results indicate that the proposed method is particularly effective when the training data are extremely limited.

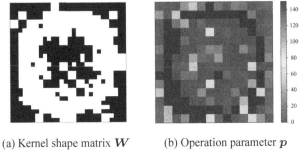

(a) Kernel shape matrix W (b) Operation parameter p

Fig. 7. Learned meta-parameters on the Omniglot dataset. (a) Kernel shape matrix W and (b) operation parameter p. In (a), the black and white pixels represent 0 and 1, respectively. These meta-parameters were learned using non-overlapping sliding windows with a size of 2×2 and a stride of 2. These figures were depicted by placing the learned parameters at the corresponding positions.

Table 1. Accuracies in the few-shot character recognition task.

Pooling operation	Accuracy (%)	
	One-shot	Five-shot
Max pooling	90.52 ± 0.68	$\mathbf{97.3 \pm 0.92}$
Average pooling	90.25 ± 0.81	97.0 ± 0.93
Meta-pooling (proposed)	$\mathbf{93.16 \pm 0.83}$	96.5 ± 0.69

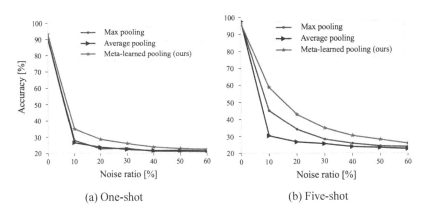

(a) One-shot   (b) Five-shot

Fig. 8. Accuracy in the noisy image recognition task.

The results of noisy image recognition are presented in Fig. 8. The figures show the accuracy of each method with varying noise ratio. The proposed method outperformed max and average pooling when the noise ratio was large. Figure 9 exhibits the pooling features for each pooling operation when the noise ratio was 10%. The figures display a randomly selected channel out of 64 channels for five

Fig. 9. Pooling features for images with noise (noise ratio: 10%): (a) input image with noise, (b) max pooling, (c) average pooling, and (d) meta-learned pooling.

examples of test images. It was confirmed that the max pooling features excessively emphasized noise, and the average pooling produced blurred features due to noise. In contrast, the proposed method could acquire relatively clean pooling features that preserved the strokes of the characters. These results suggest that the proposed method can learn a pooling layer with a high generalization capability, thus resulting in robustness to noise.

6   Conclusion

In this paper, we proposed a meta-learning method for pooling layers. As part of our meta-learning framework, we proposed a parameterized pooling layer in which the kernel shape and pooling operation are trainable using two parameters, thereby allowing flexible pooling of the input data. We also proposed a meta-learning algorithm for the parameterized pooling layer, which allows us to acquire an appropriate pooling layer across the distribution of tasks. In the experiment, we applied the proposed meta-learning framework to character recognition tasks. The results demonstrated that a pooling layer that is suitable across multiple character recognition tasks was obtained via meta-learning, and the obtained pooling layer improved the performance of the model for new tasks of few-shot character image recognition and noisy character image recognition.

In future work, we will conduct a further comparison with other pooling methods since we only compared the proposed method with max pooling and average pooling. A detailed analysis of the computational complexity will also be performed.

References

1. Baik, S., Hong, S., Lee, K.M.: Learning to forget for meta-learning. In: IEEE Conference on Computer Vision and Pattern Recognition, pp. 2379–2387 (2020)

2. Berman, M., Jégou, H., Vedaldi, A., Kokkinos, I., Douze, M.: Multigrain: a unified image embedding for classes and instances. arXiv preprint arXiv:1902.05509 (2019)
3. Chen, J., Zhan, L.M., Wu, X.M., Chung, F.l.: Variational metric scaling for metric-based meta-learning. In: AAAI Conference on Artificial Intelligence, vol. 34, pp. 3478–3485 (2020)
4. Cui, Y., Zhou, F., Wang, J., Liu, X., Lin, Y., Belongie, S.: Kernel pooling for convolutional neural networks. In: IEEE Conference on Computer Vision and Pattern Recognition, pp. 2921–2930 (2017)
5. Dollar, P., Tu, Z., Perona, P., Belongie, S.: Integral channel features. In: British Machine Vision Conference (2009)
6. Elsken, T., Staffler, B., Metzen, J.H., Hutter, F.: Meta-learning of neural architectures for few-shot learning. In: IEEE Conference on Computer Vision and Pattern Recognition, pp. 12365–12375 (2020)
7. Feng, J., Ni, B., Tian, Q., Yan, S.: Geometric l_p-norm feature pooling for image classification. In: IEEE Conference on Computer Vision and Pattern Recognition, pp. 2609–2704 (2011)
8. Finn, C., Abbeel, P., Levine, S.: Model-agnostic meta-learning for fast adaptation of deep networks. In: International Conference on Machine Learning, pp. 1126–1135 (2017)
9. Gao, Y., Beijbom, O., Zhang, N., Darrell, T.: Compact bilinear pooling. In: IEEE Conference on Computer Vision and Pattern Recognition, pp. 317–326 (2016)
10. Gao, Z., Xie, J., Wang, Q., Li, P.: Global second-order pooling convolutional networks. In: IEEE Conference on Computer Vision and Pattern Recognition, pp. 3024–3033 (2019)
11. Gao, Z., Wang, L., Wu, G.: LIP: local importance-based pooling. In: International Conference on Computer Vision, pp. 3355–3364 (2019)
12. Graham, B.: Fractional max-pooling. arXiv preprint arXiv:1412.6071 (2014)
13. Hou, Q., Zhang, L., Cheng, M.M., Feng, J.: Strip pooling: rethinking spatial pooling for scene parsing. In: IEEE Conference on Computer Vision and Pattern Recognition, pp. 4003–4012 (2020)
14. Ioffe, S., Szegedy, C.: Batch normalization: accelerating deep network training by reducing internal covariate shift. In: International Conference on Machine Learning, pp. 448–456 (2015)
15. Khodadadeh, S., Bölöni, L., Shah, M.: Unsupervised meta-learning for few-shot image classification. In: Advances in Neural Information Processing Systems, vol. 32 (2019)
16. Kingma, D.P., Ba, J.: Adam: a method for stochastic optimization. In: International Conference on Learning Representations (2015)
17. Lake, B.M., Salakhutdinov, R., Tenenbaum, J.B.: The Omniglot challenge: a 3-year progress report. Curr. Opin. Behav. Sci. **29**, 97–104 (2019)
18. Li, P., Xie, J., Wang, Q., Gao, Z.: Towards faster training of global covariance pooling networks by iterative matrix square root normalization. In: IEEE Conference on Computer Vision and Pattern Recognition, pp. 947–955 (2018)
19. Lin, T.Y., RoyChowdhury, A., Maji, S.: Bilinear CNN models for fine-grained visual recognition. In: IEEE International Conference on Computer Vision, pp. 1449–1457 (2015)
20. Malinowski, M., Fritz, M.: Learning smooth pooling regions for visual recognition. In: British Machine Vision Conference (2013)
21. Munkhdalai, T., Yu, H.: Meta networks. In: International Conference on Machine Learning, pp. 2554–2563 (2017)

22. NguyenVan, D., Lu, S., Tian, S., Ouarti, N., Mokhtari, M.: A pooling based scene text proposal technique for scene text reading in the wild. Pattern Recogn. **87**, 118–129 (2019)
23. Otsuzuki, T., Hayashi, H., Zheng, Y., Uchida, S.: Regularized pooling. In: Farkaš, I., Masulli, P., Wermter, S. (eds.) ICANN 2020. LNCS, vol. 12397, pp. 241–254. Springer, Cham (2020). https://doi.org/10.1007/978-3-030-61616-8_20
24. Rusu, A.A., et al.: Meta-learning with latent embedding optimization. In: International Conference on Learning Representations (2019)
25. Saeedan, F., Weber, N., Goesele, M., Roth, S.: Detail-preserving pooling in deep networks. In: IEEE Conference on Computer Vision and Pattern Recognition, pp. 9108–9116 (2018)
26. Santoro, A., Bartunov, S., Botvinick, M., Wierstra, D., Lillicrap, T.: Meta-learning with memory-augmented neural networks. In: International Conference on Machine Learning, vol. 48, pp. 1842–1850 (2016)
27. Sermanet, P., Chintala, S., LeCun, Y.: Convolutional neural networks applied to house numbers digit classification. In: International Conference on Pattern Recognition, pp. 3288–3291 (2012)
28. Sung, F., Yang, Y., Zhang, L., Xiang, T., Torr, P.H., Hospedales, T.M.: Learning to compare: relation network for few-shot learning. In: IEEE Conference on Computer Vision and Pattern Recognition, pp. 1199–1208 (2018)
29. Tolias, G., Sicre, R., Jégou, H.: Particular object retrieval with integral max-pooling of CNN activations. In: International Conference on Learning Representations (2016)
30. Vinyals, O., Blundell, C., Lillicrap, T., Kavukcuoglu, K., Wierstra, D.: Matching networks for one shot learning. In: Advances in Neural Information Processing Systems, vol. 29, pp. 3630–3638 (2016)
31. Wang, H., Wang, Q., Gao, M., Li, P., Zuo, W.: Multi-scale location-aware kernel representation for object detection. In: IEEE Conference on Computer Vision and Pattern Recognition, pp. 1248–1257 (2018)
32. Wei, Z., et al.: Building detail-sensitive semantic segmentation networks with polynomial pooling. In: IEEE Conference on Computer Vision and Pattern Recognition, pp. 7115–7123 (2019)
33. Yu, D., Wang, H., Chen, P., Wei, Z.: Mixed pooling for convolutional neural networks. In: Miao, D., Pedrycz, W., Ślęzak, D., Peters, G., Hu, Q., Wang, R. (eds.) RSKT 2014. LNCS (LNAI), vol. 8818, pp. 364–375. Springer, Cham (2014). https://doi.org/10.1007/978-3-319-11740-9_34
34. Zhao, H., Shi, J., Qi, X., Wang, X., Jia, J.: Pyramid scene parsing network. In: IEEE Conference on Computer Vision and Pattern Recognition, pp. 2881–2890 (2017)
35. Zhou, A., Knowles, T., Finn, C.: Meta-learning symmetries by reparameterization. In: International Conference on Learning Representations (2021)

Document Analysis Systems

Text-line-up: Don't Worry About the Caret

Chandranath Adak[1,2(✉)], Bidyut B. Chaudhuri[3,4], Chin-Teng Lin[2], and Michael Blumenstein[2]

[1] JIS Institute of Advanced Studies and Research, JIS University, 700091 Kolkata, India
chandra@jisiasr.org
[2] Australian AI Institute, University of Technology Sydney, Ultimo 2007, Australia
[3] Techno India University, 700091 Kolkata, India
[4] CVPR Unit, Indian Statistical Institute, 700108 Kolkata, India

Abstract. In a freestyle handwritten text-line, sometimes words are inserted using a caret symbol ($^\wedge$) for corrections/annotations. Such insertions create fluctuations in the reading sequence of words. In this paper, we aim to line-up the words of a text-line, so that it can assist the OCR engine. Previous text-line segmentation techniques in the literature have scarcely addressed this issue. Here, the task undertaken is formulated as a path planning problem, and a novel multi-agent hierarchical reinforcement learning-based architecture solution is proposed. As a matter of fact, no linguistic knowledge is used here. Experimentation of the proposed solution architecture has been conducted on English and Bengali offline handwriting, which yielded some interesting results.

Keywords: Handwriting · Hierarchical reinforcement learning · Multi-agent reinforcement learning · Proof-reading.

1 Introduction

Extracting the text-line information from a handwritten page image is a classical problem of document image analysis [1–4], and it is still prominent in this deep learning era [1]. Most of the past works in this direction have focused on text-line segmentation [2], where primarily the text-lines are either separated through a continuous fictitious line [3], or labeled by clusters [4], or marked by baselines [1]. A text-line may contain some words that are written later using the caret symbol ($^\wedge$) for correcting/annotating the manuscript. Here, a text-line segmentation approach may not work well to ascertain whether the inserted words belong to a certain text-line (refer to Fig. 1). As a consequence, it may impede the understanding of the reading sequence of words of a text-line while OCRing.

In this paper, we undertake the task to comprehend the sequence of words of a text-line, so that an OCR engine/automated manuscript-transcriptor can have some prior knowledge of the reading sequence of words. To handle this task, some character recognition followed by natural language processing (NLP) can be

© Springer Nature Switzerland AG 2021
J. Lladós et al. (Eds.): ICDAR 2021, LNCS 12823, pp. 207–222, 2021.
https://doi.org/10.1007/978-3-030-86334-0_14

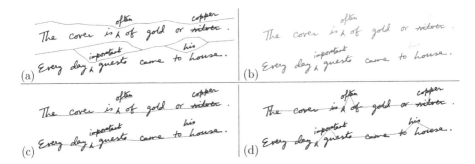

Fig. 1. Text-line representation: (a) separated by a fictitious line [3], (b) clusters [4], (c) marked by a baseline [1], (d) location coordinates and detected path (arrowed-line) of the reading sequence [ours]. (Color figure online)

performed, but that would be a costly procedure. Therefore, we neither use here character recognition nor linguistic knowledge. Another way involves detecting the caret symbols and then attempting to analyze the surrounding areas, but this approach is not fruitful due to false-negative and false-positive cases, e.g., small-sized carets close/overlapping with texts, tiny character graphemes-like carets, word inserted without using a caret, no word insertion after scribbling a caret, etc.

We formulate the above task as a path planning problem, where the system agent attempts to visit every word component exactly once to detect the reading sequence of a text-line. Here, we propose a multi-agent hierarchical reinforcement learning model [5] to impart the machine with a human-like perception of reading a text-line that may be obstructed due to some inserted texts. The motivation to tackle this problem using reinforcement learning is its working strategy of exploring the unknown terrain while exploiting the current knowledge [6]. From the application point of view, this work is significant for analyzing the reading-word sequence of a text-line, especially for freehand writing, where the past methods did not perform well. To the best of our knowledge, our work is the earliest attempt of its kind. From the theoretical perspective, we propose a novel multi-agent hierarchical reinforcement learning architecture, where we define the relationship among agents, their interactions with the environment, and shape the global and internal reward.

We performed the experiments on English and Bengali offline handwriting, which are left-to-right writing systems. The challenges of our employed datasets concerning the undertaken task are discussed in Sect. 2. In Sect. 3, we formulate the problem and propose a solution architecture. Section 4 presents and analyzes the experimental results. Finally, Sect. 5 concludes this paper.

2   Challenges and Dataset Details

In this section, we discuss the offline handwriting dataset employed for our research and the challenges related to the data. The primary aim of this research is to analyze the reading sequence of words from a handwritten text-line without any assistance from OCR and NLP engines.

The freestyle handwritten text-lines are mostly curvilinear rather than linear [4] and pose several challenges for our research. The classical challenges include text-line fluctuation and orientation, e.g., skewness, waviness, curviness. The variation in the inter text-line gap draws significant attention, where the neighboring text-lines may be close, or touching, even overlapping [2]. Similar kinds of issues can be found for the intra text-line gap, i.e., gap between words/inter-word gap.

In daily handwriting and manuscript drafting/corrections, several forms of annotations can be found; therefore, detection of the word-reading sequence is not a straight-forward problem. A common form is inserting a word between two successive words by using a caret symbol ($\wedge$), and writing the word in the inter text-line gap between the current and the previous text-line. Here, the insertion of multiple words can also be noted. The habitual absence of caret symbols is also possible, which leads to more challenging scenarios. The writer often strikes-through formerly written word(s) and inserts some substituting word(s). Such insertions (with/without a caret), deletion (strike-through), update (deletion + insertion) are noted frequently in real-time writing (refer to Fig. 1). The font-size of the inserted word(s) may be smaller to fit into the inter-text-line gap. If this gap is not sufficient to insert the word(s), then the writer may write it in the available marginal-space of the page. Such word insertions depend on the writing space availability. Sometimes, the forceful insertion of words almost removes the usual inter-word and inter-line gaps, which makes it harder to read. False-positive cases may also arise, where some words may be incorrectly thought of as inserted ones, due to the absence of the caret, lower inter text-line gaps, artistic/poetic writing structure, superscript-text, etc. Moreover, the presence of some unconventional annotations performed by the human-writer is quite natural, due to handwriting variations, improper/no formal training of text-annotation, idiosyncratic writing styles, etc.

For our research, we require a database, which contains some handwritten pages having multiple words inserted during text-annotation; so that we can perform our experiments on detecting the reading sequence of words. The hand-written pages of publicly available databases hardly contain text-annotations as per our requirement. Therefore, we manually annotated some handwritten page images of publicly available databases using the GIMP image editor [7]. On a page image, we carefully inserted some words after extracting/cropping those words from the same page. Here, we used 100 English pages from the IAM database [8] and 100 Bengali pages from the PBOK database [9], and called these as DB_{IAM} and DB_{PBOK}, respectively. Here, 50% of the pages of both DB_{IAM} and DB_{PBOK} were annotated/inserted by linguistically meaningful words using carets. For the remaining 50% of pages, we inserted linguistically non-meaningful words with/without a caret, which did not impede our objective to learn the structural pattern and position of inserted handwritten word(s). Moreover, to address the natural flow of annotations during the free-style writing, i.e., insertion of the word(s) by the same writer, we procured 50 pages of English (say,

Table 1. Employed dataset details

Dataset	Script	# page	r_{OTL} ($avg \pm sd$)
DB$_{\text{IAM}}$	English	100	0.5203 ± 0.1841
DB$_{\text{PBOK}}$	Bengali	100	0.4764 ± 0.1160
DB$_{\text{E}}$	English	50	0.3725 ± 0.2748
DB$_{\text{B}}$	Bengali	50	0.3551 ± 0.2377

DB$_{\text{E}}$) and 50 pages of Bengali (say, DB$_{\text{B}}$) offline handwriting by an in-house setup. In this paper, we consider English and Bengali scripts, which are usually written horizontally from left-to-right. In Table 1, we summarize some aspects of our employed datasets. Inserting some word(s) to a text-line obstructs the straight-forward reading path/sequence. On a page, we count such obstructed text-lines (n_{OTL}) and divide it with the total number of text-lines (n_{TL}) of that page, to get a ratio, say, $r_{OTL} = n_{OTL}/n_{TL}$. The average $\pm$ standard deviation ($avg \pm sd$) of r_{OTL} over a dataset is mentioned in Table 1. For the available datasets in the literature, r_{OTL} (avg) ≈ 0, due to their different objectives from ours.

In this paper, we attempt to imitate the human perception of detecting the word sequence while reading a text-line. Therefore, to prepare the ground-truth, we engaged human volunteers having at least professional working proficiency in the English/Bengali language. A volunteer was requested to perform computer-mouse clicks on the words in the same sequence of his/her reading from the digital image of a handwritten page projected on a computer screen, so that we can record the (x_i, y_i) coordinates of the word sequence through mouse-clicks. A very few noisy mouse clicks were manually discarded with opinions from some linguistic experts. We also extracted the CG (x_c, y_c) of a word component semi-automatically. We did not fully rely on the CG due to our objective of mimicking the eye movement during human gazing/reading. At least n_r (> 1) number of readers were engaged for a page. Subsequently, the corresponding ground-truth coordinate (x_{GT}, y_{GT}) of a word is computed as follows. (x_{GT}, y_{GT}) = $(\frac{\alpha_r}{n_r} \sum_{i=1}^{n_r} x_i + \alpha_c x_c, \frac{\alpha_r}{n_r} \sum_{i=1}^{n_r} y_i + \alpha_c y_c)$; where, $\alpha_r + \alpha_c = 1$. For our experimental dataset generation, we chose $n_r = 10, \alpha_r = 0.5$, and $\alpha_c = 0.5$. In our dataset, for a handwritten page, the word ground-truth coordinates are provided sequentially with the demarcation of text-lines.

3   Proposed Method

In this section, we first formulate the undertaken problem, then propose our solution architecture.

3.1   Problem Formulation

In this research work, we are given an image ($\mathcal{I}$) obtained by scanning a hand-written page, which is fed to our system as an input. Our task is to detect the reading-sequence of words in a text-line only by the structural pattern of hand-writing. No linguistic knowledge has been used here. As we mentioned earlier, the sequence of words in a text-line does not always follow a straight path due to the insertion of some words during proof-reading/corrections. The task is formulated here as a path planning problem, where the system-agent visits every word component exactly once. The source, target, and in-between hopping locations of the path are obtained by the agent itself through an exploration-exploitation strategy. The system outputs the sequence of these location coordinates over $\mathcal{I}$, which eventually infers the path.

3.2   Solution Architecture

As we stated previously, we refrain from obtaining assistance from any OCR engine followed by an NLP architecture to comprehend the reading sequence of words in the text-line due to the high cost. Instead, we think of composing our problem as a decision-making task, where an agent interacts with the handwriting image (environment) to find the sequence of word-locations. For such cases, a reinforcement learning (RL)-based agent is a good option, since it learns a policy for maximizing the reward by taking action on the environment [6]. However, a full handwritten page image is a challenging environment for an agent to interact, owing to various complex writing patterns. To alleviate the learning complexity, here, a hierarchical reinforcement learning (HRL)-based architecture can be proposed, which breaks down the task into sub-tasks hierarchically [5,10]. Moreover, at a certain level of the hierarchy, multiple agents can work together for better learning and communication among themselves [5]. Altogether, in this paper, we propose a multi-agent hierarchical reinforcement learning (MAHRL) model to achieve our goal.

Architecture Overview. For the undertaken task, our model contains two levels of hierarchy. In the higher level, one RL agent (manager) is present, while the lower level includes n_w number of RL agents (worker-1, worker-2, ..., worker-n_w) that depends on the count of text-lines on a page. In Fig. 2, we represent our MAHRL model diagrammatically. Formally, the RL problem can be built as an MDP (Markov Decision Process), which consists of a set of agent *states* of the environment (s), a set of *actions* (a) to attain the goal, and a *reward* function (r) to optimize the decision strategy [6]. The manager observes the entire environment ($\mathcal{E}$)/handwritten page to encode its state space (s^m); manager's action (a^m) refers to assigning a sub-task to a worker. The manager receives a *global reward* (r^m) from the environment after completion of the entire work. For the worker RL agent, we formulate a partially observable MDP [11], where the agent stochastically makes a decision in discrete-time without observing the entire environment [6]. The worker-i ($\forall\ i = 1, 2, \ldots, n_w$) observes the partial environment $\mathcal{E}^{w_i}$ from a handwritten page to embed its state (s^{w_i}) and acts (a^{w_i}) to find the sequence of word-locations from a text-line. The worker-i receives an *internal reward* (r^{w_i}) from the manager. The workers share weights among themselves.

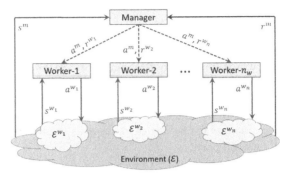

Fig. 2. Proposed multi-agent hierarchical reinforcement learning (MAHRL) model.

Manager. The state of manager involves the raw pixel values of the handwritten page image. The input image $\mathcal{I}$ is resized into size $n_z \times n_z$ by keeping the trace of the aspect ratio and produces the resized image $\mathcal{I}_z$. To maintain the aspect ratio, some columns or rows are filled with zeros. For our task, empirically, we choose $n_z = 1024$. The handwritten page images of our dataset are significantly large in size and are in grayscale. Such resizing assists in reducing the state space, but does not impede our objective. Moreover, the manager should have some perception about the text-line zones, since we aim to inspect the reading sequence at a text-line level. For text-line segmentation, we use an off-the-shelf semantic segmentation network [12], which is basically an encoder-decoder architecture, followed by a softmax layer to capture the pixel-level classification [13]. This network is pre-trained on small handwritten character graphemes to classify into ink-stroke region and background in order to fulfill the objective of text-line segmentation. Actually, the segmented mask ($\mathcal{I}_s$) is of size 1024×1024. At time step t, the manager also observes the workers' location coordinates on image $\mathcal{I}_z$. The workers' location matrix ($\mathcal{I}_l$) is also of size 1024×1024. Now, $\mathcal{I}_z$, $\mathcal{I}_s$, and $\mathcal{I}_l$ are composed as a single image ($\mathcal{I}_{zsl}$) comprising three channels, which is of size $1024 \times 1024@3$. This $\mathcal{I}_{zsl}$ is fed to a convolutional neural network (CNN) f_m to summarize the manager's state-space s^m. We obtain a feature vector v_m with dimension 256 from f_m. The f_m contains some sequentially added convolutional and pooling layers, as follows.

$\mathcal{I}_{zsl}$ $(1024 \times 1024@3)$ $\Rightarrow$ C_1 $(512 \times 512@16)$ $\Rightarrow$ C_2 $(256 \times 256@32)$ $\Rightarrow$ MP $(128 \times 128@32)$ $\Rightarrow$ C_3 $(64 \times 64@64)$ $\Rightarrow$ C_4 $(32 \times 32@128)$ $\Rightarrow$ MP $(16 \times 16@128)$ $\Rightarrow$ C_5 $(8 \times 8@256)$ $\Rightarrow$ GAP $(1 \times 1@256)$ $\Rightarrow$ v_m;

where, "C_i" ($\forall i = 1, 2, \ldots, 5$) denotes the ith convolutional layer, "MP" and "GAP" represent max-pooling and global average pooling layers, respectively [13]. The numeric values in the format of $(n_m \times n_m@n_c)$ symbolize the feature map size $n_m \times n_m$ and the number of channels n_c for a layer. For C_1 and C_2, we use 5×5 sized kernels, while for the rest of convolutions, 3×3 sized kernels are engaged. For all the convolutional layers, the stride size is 2. For max-pooling and global average pooling, the kernel sizes are 2×2 and 8×8, respectively.

Here, each convolution is followed by a batch normalization [14] and a Mish [15] activation function. The batch normalization is used to avoid overfitting. Mish has worked better than major state-of-the-art activation functions, e.g., ReLU, leaky ReLU, GELU, Swish [15], and it has also performed well for our task.

The manager decides which worker when to "*move*" or "*stop*", and commands the workers accordingly. Therefore, the action space (a^m) of the manager is discrete containing two actions (*move* and *stop*) per worker, i.e., a total $2n_w$ number of actions.

The manager focuses on the higher-level goal, and receives a reward $r^m = 1 - r_{LD}$, when a worker reaches at the end of a text-line; otherwise gets zero reward. Here, $r_{LD} = LD(\hat{l}_s, l_s)/max(|\hat{l}_s|, |l_s|)$ is a penalty term within the interval $[0, 1]$. $LD(.,.)$ is Levenshtein distance [16] that measures the distance between actual ($\hat{l}_s$) and predicted (l_s) word sequences.

At time step t, the manager observes the state s_t^m, and selects an action $a_t^m \in \mathscr{A}^m = \{1, 2, \ldots, |\mathscr{A}^m| = 2n_w\}$ to get a reward r_t^m. The fundamentals of reinforcement learning (RL) can be found in [6]. The manager learns a policy π_m to maximize the expected discounted return, which can be defined as the cumulative discounted reward, i.e., $\sum_{t>0} \gamma_m^t r_t^m$. Here, γ_m is a discount factor. To know how good the manager is learning over the policy π_m, values of the state (s^m) and the state-action pair (s^m, a^m) can be defined, which are called the *value* function (V^{π_m}) and the *Q-value* function (Q^{π_m}), respectively [6]. $V^{\pi_m}(s_t^m) = \mathbb{E}[\sum_{t>0} \gamma_m^t r_t^m | s_t^m, \pi_m]$; $Q^{\pi_m}(s_t^m, a_t^m) = \mathbb{E}[\sum_{t>0} \gamma_m^t r_t^m | s_t^m, a_t^m, \pi_m]$. The optimal Q-value function (Q^{*m}) can be iteratively learned via deep Q-learning [5,17], and the optimal action (a^{*m}) is computed as follows. $a^{*m} = \arg\max_{a{_\prime}^m \in \mathscr{A}^m} Q^{*m}(s^m, a{_\prime}^m)$. Here, $Q^{*m}(s^m, a^m) = \max_\pi Q^{\pi_m}(s^m, a^m)$. A deep Q-network $Q^m(s^m, a^m; \theta^m)$ with parameters θ^m can be employed to approximate the value functions [17]. In this paper, the manager adopts the concept of *dueling deep Q-network* due to its better performance than experience replay and prioritized replay-based architectures [18]. Here, a notion of *advantage function* (A^{π_m}) exists, which signifies how much an action is better than the expected. $A^{\pi_m}(s^m, a^m) = Q^{\pi_m}(s^m, a^m) - V^{\pi_m}(s^m)$.

The dueling deep Q-network architecture contains two parallel streams (*value*: f_{mv} and *advantage*: f_{ma}) of fully connected layers. The 256-dimensional feature vector v_m obtained from f_m is now fed to f_{mv} and f_{ma} streams in parallel to produce the separate estimates of the value (V^{π_m}) and advantage (A^{π_m}) functions. The f_{mv} comprises a fully connected layer with 128 nodes followed by ReLU activation, and a sequentially added fully connected single node to produce the output from the value stream. Similarly, f_{ma} contains a fully connected layer with 128 nodes trailed by ReLU activation, and another successive fully connected layer with $|\mathscr{A}^m|$ nodes to obtain the output from the advantage stream. Finally, two streams are combined to produce the Q-value function, i.e., $Q^{\pi_m}(s^m, a^m; \theta^m, \theta^A, \theta^V) = V^{\pi_m}(s^m; \theta^m, \theta^V) + A^{\pi_m}(s^m, a^m; \theta^m, \theta^A)$; where, θ^A, θ^V are the parameters of two sequences f_{mv} and f_{ma}, respectively. To tackle the identifiability issue [18] and to increase the optimization stability, this equation is modified as follows. $Q^{\pi_m}(s^m, a^m; \theta^m, \theta^A, \theta^V) = V^{\pi_m}(s^m; \theta^m, \theta^V) +$

$(A^{\pi_m}(s^m, a^m; \theta^m, \theta^A) - \frac{1}{|\mathscr{A}^m|} \sum_{a_{\prime}^m} A^{\pi_m}(s^m, a_{\prime}^m; \theta^m, \theta^A))$. The parameters $\{\theta^m,$ $\theta^A, \theta^V\}$ are learned by standard policy-based RL strategy [19].

Worker. As mentioned before, while the manager focuses on the higher-level goal, the workers emphasize lower-level fine control to achieve sub-goals. A worker (worker-i) can observe the environment/handwritten page partially $(\mathcal{E}^{w_i})$, which formulates its state space (s^{w_i}). To express this partial observation, we perform a foveal transformation, as follows.

Foveal Transformation (ϑ). A worker agent partially concentrates on $\mathcal{I}_z$ and extracts fragmental information around a location l. We here utilize the idea of foveated imaging [20,21], where the agent focuses at l with foveal vision, corresponding to the highest resolution; and with peripheral vision, as it gradually moves away from l by a lower resolution.

To encode the fragment region, we execute a foveal transformation $\vartheta(\mathcal{I}_z, l)$, which extracts the k (> 1) number of neighboring patches of different resolutions around location l. The foveal transformation is presented in Fig. 3(a). The 1st patch is of size $w_p \times w_p$, 2nd patch is of size $(w_p + d_w) \times (w_p + d_w)$, and so on, the kth patch is of size $(w_p + (k-1).d_w) \times (w_p + (k-1).d_w)$, where d_w is the additional width of the successive patches. All the k patches are resized to $w_p \times w_p$, and thus produce k channels each having $w_p \times w_p$ sized patch-image. As a matter of fact, if l is near the boundary of $\mathcal{I}_z$, then peripheral patches are duly filled with zeros. Resizing a larger size patch to a smaller one lowers the resolution. In our task, empirically, we fix $k = 6$, $w_p = 64$, $d_w = 32$. Therefore, at time step t, by gazing at location l_{t-1} of image $\mathcal{I}_z$ and performing a foveal transformation $\vartheta(\mathcal{I}_z, l_{t-1})$, the agent produces a $64 \times 64 \times 6$ sized fragment p_t. The location l is encoded with a real-valued coordinate (x, y), where $0 \leq x, y \leq 1$. Here, the top-left and bottom-right coordinates of $\mathcal{I}_z$ are $(0, 0)$ and $(1, 1)$, respectively. In our task, the manager initializes the location l of a worker agent on the leftmost ink-stroke pixel of a text-line mask (with prior knowledge from $\mathcal{I}_s$) corresponding to the page image $\mathcal{I}_z$. A worker comprises fragment network (f_g), core network (f_h), and location network (f_l). The inside view of a worker is shown in Fig. 3(b), and its workflow is discussed as follows.

Fragment Network (f_g). At time step t, a $64 \times 64 \times 6$ sized fragment p_t is inputted to a deep neural architecture f_g to extract features g_t. Our f_g architecture adopts ResNet-34 [22] with some minor amendments. The details of f_g are shown in Table 2, where the building residual blocks [22] are shown in brackets with the number of stacked blocks. For example, in *conv*1 (first convolutional layer), the input p_t of size $64 \times 64@6$ is convoluted with 32 filters of size $3 \times 3@6$ to produce a $64 \times 64@32$ sized feature map. Here, "@n_c" represents n_c number of channels. We use $stride = 1$ in *conv*1, whereas $stride = 2$ in *conv*2_x, *conv*3_x, *conv*4_x, *conv*5_x to perform down-sampling. Here also, each convolution is trailed by batch normalization [14] and a Mish [15] activation function. Intending to use f_g as a feature extractor, we discard the last fully connected layer of ResNet-34, which turns f_g into a 33-layered architecture. From the avg_pool (global average pooling) layer of f_g, we obtain a 512-dimensional feature vector g_t by flattening, at time step t.

Table 2. Architecture of f_g

Layer name	Output size	33-layer
input	$64 \times 64@6$	
*conv*1	$64 \times 64@32$	$3 \times 3, 32$
*conv*2_x	$32 \times 32@64$	$\begin{bmatrix} 3 \times 3, 64 \\ 3 \times 3, 64 \end{bmatrix} \times 3$
*conv*3_x	$16 \times 16@128$	$\begin{bmatrix} 3 \times 3, 128 \\ 3 \times 3, 128 \end{bmatrix} \times 4$
*conv*4_x	$8 \times 8@256$	$\begin{bmatrix} 3 \times 3, 256 \\ 3 \times 3, 256 \end{bmatrix} \times 6$
*conv*5_x	$4 \times 4@512$	$\begin{bmatrix} 3 \times 3, 512 \\ 3 \times 3, 512 \end{bmatrix} \times 3$
avg_pool	$1 \times 1@512$	

Core Network (f_h). At this point, g_t is fed to the core network f_h. The worker agent needs to memorize the past explored fragment information, therefore, a recurrent neural network (RNN) is employed as f_h. The GRU (Gated Recurrent Unit) [23] is used here as an RNN unit, due to its similar performance compared to LSTM (Long Short-Term Memory) for our task, while having fewer learning parameters [24]. The f_h contains 256 GRU units. Here, the agent maintains an internal state to encode information about *where* it gazed as well as *what* was observed. This information is crucial in deciding the action and finding the next location. The internal state h_t at time step t is updated over time by f_h. The present internal state h_t is a function of the previous state h_{t-1}

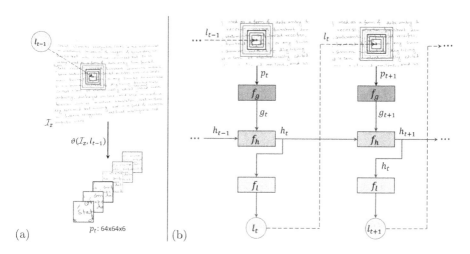

Fig. 3. (a) Foveal transformation $\vartheta(\mathcal{I}_z, l)$, (b) Internal workflow of a worker.

and the external input g_t, which is formulated by GRU gates, as follows. $h_t = f_h(h_{t-1}, g_t) = \Gamma_u * \tilde{h}_t + (1 - \Gamma_u) * h_{t-1}$; where, $\Gamma_u = \sigma(linear(h_{t-1}, g_t))$, $\tilde{h}_t = tanh(linear(\Gamma_r * h_{t-1}, g_t))$, $\Gamma_r = \sigma(linear(h_{t-1}, g_t))$. Γ_u and Γ_r denote *update* and *relevant* gates of the GRU, respectively [23]. *Sigmoid* (σ) and *tanh* non-linear activation functions are used here [13]. $linear(\bar{v})$ represents the linear transformation of a vector $\bar{v}$.

Location Network (f_l). The h_t is embedded into f_l to find the next location l_t. The location policy is defined by a 2-component Gaussian with a fixed variance [25]. At time step t, we obtain the mean of the location policy from f_l, which is defined as $f_l(h_t) = linear(h_t)$. Here, the fully connected layer is trailed by a sigmoid (σ) activation to clamp the location coordinates in $0 \leq x, y \leq 1$. The f_l is trained using RL [6] to find l_t for emphasizing to the next fragment p_{t+1}.

In RL, the worker agent interacts with the state s^w of the environment and takes action a^w to get the reward r^w. At time step t, the state s_t^w engages p_t around l_{t-1} and summarized into h_t. The action a_t^w at t is actually the location-action l_t chosen stochastically from a distribution θ_l-parameterized by $f_l(h_t)$. We shape the reward r_t^w at t internally, as follows.

$$r_t^w = \begin{cases} 1 - \alpha_d D^2(l_t, \hat{l}_t) - \alpha_y(y_t - y_{t-t_d}) & ; \text{if } t > t_d \\ 1 - \alpha_d D^2(l_t, \hat{l}_t) & ; \text{otherwise} \end{cases} \tag{1}$$

where, $D(.,.)$ is the Euclidean distance. The term $\alpha_d D^2(l_t, \hat{l}_t)$ signifies the loss due to the difference between actual ($\hat{l}_t := (\hat{x}_t, \hat{y}_t)$) and predicted ($l_t := (x_t, y_t)$) locations. The ground-truth was required here for the reward calculation. If the worker strays away from the designated horizontal text-line, then the manager penalizes the worker slightly with an amount of $\alpha_y(y_t - y_{t-t_d})$. Here, the hyper-parameters, i.e., $\alpha_d = 0.5$, $\alpha_y = 0.2$, and $t_d = 2$, are set empirically.

The RL-based agent learns a stochastic policy $\pi_\theta(l_t | s_{1:t}^w)$ at every t, which maps the past trajectory of the environmental interactions $s_{1:t}^w$ to the location-action distribution l_t. For our task, the policy π_θ is defined by early mentioned RNN, and s_t^w is summarized into h_t. The parameter $\theta = \{\theta_g, \theta_h\}$ is acquired from the parameters θ_g and θ_h of f_g and f_h, respectively. The agent learns θ to find an optimal policy π^* that maximizes the expected sum of discounted rewards. The cost function J_l is defined as follows. $J_l(\theta) = \mathbb{E}_{\rho(s_{1:T}^w; \theta)}[\sum_{t=1}^{T} \gamma^t r_t^w] = \mathbb{E}_{\rho(s_{1:T}^w; \theta)}[R]$; where, ρ is the transition probability from one state to another, which depends on π_θ [6]. T is the episodic time step and γ is a discount factor.

The optimal parameter is decided by $\theta^* = \arg\max_\theta J_l(\theta)$, where we employ gradient ascent using the tactics from RL literature [19], as follows. $\nabla_\theta J_l(\theta) = \sum_{t=1}^{T} \mathbb{E}_{\rho(s_{1:T}^w; \theta)}[R \nabla_\theta \log \pi_\theta(l_t | s_t^w)] \approx \frac{1}{N} \sum_{n=1}^{N} \sum_{t=1}^{T} R^{(n)} \nabla_\theta \log \pi_\theta(l_t^{(n)} | s_t^{w(n)})$; where, trajectories $s^{w(n)}$'s are generated by executing the agent on policy π_θ for $n = 1, 2, \ldots, N$ episodes. The $\nabla_\theta log \pi_\theta(l_t | s_t^w)$ portion is calculated from the gradient of RNN with standard backpropagation [11].

To avoid the high variance problem of the gradient estimator, variance reduction is performed here [26]. We employ variance reduction with the baseline ($\mathcal{B}$) that comprehends whether a reward is better than the expected one, as follows. $\nabla_\theta J_l(\theta) \approx \frac{1}{N} \sum_{n=1}^{N} \sum_{t=1}^{T} (R_t^{(n)} - \mathcal{B}_t) \nabla_\theta \log \pi_\theta(l_t^{(n)} | s_t^{w^{(n)}})$; where, $R_t = Q^{\pi_\theta}(s_t^w, l_t) = \mathbb{E}[\sum_{t \geq 1} \gamma^t r_t^w | s_t^w, l_t, \pi_\theta]$ is Q-value function and $\mathcal{B}_t = V^{\pi_\theta}(s_t^w) = \mathbb{E}[\sum_{t \geq 1} \gamma^t r_t^w | s_t^w, \pi_\theta]$ is value function [6, 26]. The learning of the baseline is performed by reducing the squared error between Q^{π_θ} and V^{π_θ}.

Finally, the detected series of the location (l_s) signifying the reading-sequence of words in a text-line, is reverted to the original input handwritten page image $\mathcal{I}$ by obtaining assistance from the previously-traced aspect ratio.

4   Experiments and Discussion

In this section, we present the dataset employed, followed by experimental results with discussions.

4.1   Database Employed

As we discussed earlier in Sect. 2, for the experimental analysis, we have procured 300 handwritten pages with the ground-truth information of the reading sequence of text-lines. Our database comprises 4 sets of data, i.e., DB_{IAM}, DB_{PBOK}, DB_E, DB_B containing 100 English, 100 Bengali, 50 English, and 50 Bengali pages, respectively.

Each dataset DB_i (for $i \in \{IAM, PBOK, E, B\}$) is split into a training (DB_i^{tr}), validation (DB_i^v) and testing (DB_i^t) set with a ratio of 5:2:3. To reduce overfitting during training, we augment our training data DB_i^{tr}. For data augmentation, we randomly drop some word components from a handwritten page to dilute [27] the reading sequences. From a page, we generated 10 augmented pages.

4.2   Results and Evaluation

In this subsection, we present the experimental results to analyze our model performance. All results presented here were executed on the testing set DB_i^t (for $i \in \{IAM, PBOK, E, B\}$). The hyperparameters of our model were tuned and fixed during system training based on the validation/development set DB_i^v. Empirically, we set $initial\_learning\_rate = 10^{-3}$, $discount\_factor = 0.95$, mini-batch_size = 32, episodic_time_step = 64, and episode = 512. We analyzed our model performance based on finding the location coordinates and subsequent detection of the reading sequence.

Location Finding. The effectiveness of our model depends on finding the precise location to gaze at while reading. Here, we used the $RMSE$ (Root Mean Squared Error) [28] as a performance measure due to its efficacy in addressing the deviation of location-coordinate from the ground-truth (x_{GT}, y_{GT}). On a page, we computed the $RMSE$ over all the location coordinates. In Table 3, we present our model performance on location finding over DB_i^t (for $i \in$ IAM, PBOK, E, B) in terms of $RMSE$ ($avg \pm sd$).

From Table 3, we can observe that the overall performance of location finding was the best for DB_B and the worst for DB_{IAM}. It is evident from this table that

Table 3. Performance on location finding

Dataset	$RMSE$ $(avg \pm sd)$			
	DB_{IAM}	DB_{PBOK}	DB_E	DB_B
OTL	0.02359 ± 0.00347	0.01832 ± 0.00127	0.01813 ± 0.00753	0.01621 ± 0.00283
non-OTL	0.01956 ± 0.00147	0.01682 ± 0.00358	0.01470 ± 0.00098	0.01262 ± 0.00489
Overall	0.02162 ± 0.00433	0.01741 ± 0.00445	0.01684 ± 0.00564	0.01437 ± 0.00389

Table 4. Performance on reading sequence detection

Dataset	MLD $(avg \pm sd)$			
	DB_{IAM}	DB_{PBOK}	DB_E	DB_B
OTL	1.4331 ± 0.4745	1.1295 ± 0.2387	1.1581 ± 0.4462	0.8573 ± 0.2740
non-OTL	1.0936 ± 0.4033	0.9454 ± 0.3631	0.8096 ± 0.3400	$0.7695 \pm 0. 3173$
Overall	1.2673 ± 0.8485	1.0330 ± 0.6778	0.9870 ± 0.8748	0.8156 ± 0.5030

our system found the location better on unobstructed text-lines (non-OTL) than the obstructed text-lines (OTL).

Reading Sequence Detection. From the series of ground-truth coordinates (x_{GT}, y_{GT}), we can obtain the actual reading sequence $(\hat{l}_s)$ of a text-line. Similarly, from the predicted series of coordinates, we obtained the predicted reading sequence (l_s) of a text-line. Now, we measured the distance between the actual and predicted sequences with a small data-driven relaxation. Here, we used the Levenshtein distance (LD) [16] due to its efficiency in measuring the difference between the ground-truth and predicted sequences undertaking the pairwise sequence alignment. On a page, we computed the LDs for all the sequences over text-lines and took its page-level arithmetic mean, say, MLD. In Table 4, we present our model performance on reading sequence detection over DB_i^t (for $i \in$ IAM, PBOK, E, B) in terms of MLD $(avg \pm sd)$.

From Table 4, we can see that our system performed better for unobstructed text-lines (non-OTL) than the obstructed ones (OTL) concerning the task of reading sequence detection. Here, the overall result was the poorest for DB_{IAM} as it contains the highest number of OTLs among the datasets employed.

In Fig. 4, we present some qualitative results of our system. The word insertions in OTLs of Fig. 4(b), (d) are done synthetically. Figure 4(a) shows an example, where a single word is inserted by a caret after striking-out the mistaken word. In Fig. 4(c), two successive words are inserted by placing a single caret. Without any caret, a word is inserted in Fig. 4(b). In Fig. 4(d), two words are inserted separately by two respective carets, where our system fails to detect an insertion since the inserted word seems part of the prior text-line. Our system succeeds in Fig. 4(a), (b), (c).

4.3   Comparison

As we mentioned in Sect. 1, the approach undertaken in this paper is the earliest attempt of its kind, and we did not find any direct work for comparison purposes.

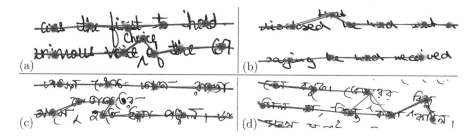

Fig. 4. Qualitative results of our system on samples of (a) DB$_E$, (b) DB$_{IAM}$, (c) DB$_B$, and (d) DB$_{PBOK}$. Detected locations and reading sequence paths are shown by red dots and blue arrowed-lines, respectively. Ground-truth locations and paths are also shown. (Softcopy exhibits better display.) (Color figure online)

However, for comparison analysis, we adapted some indirect methods of the literature to fit into our problem, which is briefly discussed as follows.

Method-SN: We used SegNet [12] for semantic segmentation of both text-lines and words. Prior training was performed using small handwritten character graphemes. We obtained the CGs of the segmented words to treat as the location coordinates. The sequence of these locations was detected with a data-driven threshold.

Method-Ga: This method follows a similar approach to Method-SN in finding the location coordinates and subsequent sequence detection. The only difference is in the line and word segmentation module, for which we employed the 2D Gaussian filtering-based technique of GOLESTAN-a [29].

We also engaged some human knowledge to compare with our system.

Human-E: Here, 5 healthy persons without any known reading/writing disorders, were appointed to manually record the coordinates on the test data, similar to the ground-truth generation scheme (refer to Sect. 2). We took an average of location coordinates provided by 5 persons, which subsequently contributed to finding the reading sequence. All 5 persons appointed here had at least professional proficiency in the English language and they operated only on the English datasets, i.e., DB$_{IAM}$, DB$_E$. In this paper, we present the average result obtained from these 5 individuals.

Human-B: This setup is similar to the Human-E with the only difference is that all 5 persons appointed here had native proficiency in the Bengali language. The appointed persons here acted on Bengali datasets only, i.e., DB$_{PBOK}$, DB$_B$.

Human-nonB: This setup is similar to the Human-E and Human-B setups. The only difference is that all 5 adults appointed here had no proficiency in Bengali language. Still, they operated on Bengali datasets, i.e., DB$_{PBOK}$, DB$_B$. Here, the appointed individuals were provided some basic information of Bengali writing, such as Bengali is a left-to-right writing system similar to English, here too a word can be inserted with/without caret symbol, it maintains inter-text-line and inter-word gaps, etc.

Table 5. Comparative analysis

Method	Location finding				Reading sequence detection			
	$RMSE$ (avg)				MLD (avg)			
	DB_{IAM}	DB_{PBOK}	DB_E	DB_B	DB_{IAM}	DB_{PBOK}	DB_E	DB_B
Method-SN	0.04072	0.03908	0.03676	0.03530	2.3574	2.1206	1.8938	1.8370
Method-Ga	0.07427	0.07106	0.06861	0.06365	3.2776	3.0461	2.8664	2.6894
Human-E	0.00926	-	0.01038	-	0.0126	-	0.0065	-
Human-B	-	0.01161	-	0.01424	-	0.0232	-	0.0190
Human-nonB	-	0.01057	-	0.01591	-	0.8779	-	0.5603
Ours	0.02162	0.01741	0.01684	0.01437	1.2673	1.0330	0.9870	0.8156

As a matter of fact, the linguistic/readability knowledge of human-beings was used in the Human-E and Human-B setups, whereas such knowledge was missing in the Human-nonB setup. Owing to our limited opportunity, we were unable to make a rational choice of the experiment, i.e., Human-nonE, where the appointed individuals do not have any expertise in English/Latin script.

In Table 5, we compare our overall results with the performances of some baseline methods and humans, with respect to the location coordinate finding and reading sequence detection, while keeping the same experimental setup as before. For both cases, our system performed better than Method-SN and Method-Ga. From Table 5, it is interesting to observe that for a human (Human-E/Human-B) having linguistic knowledge, MLD (avg) ≈ 0; but without linguistic knowledge, even human (Human-nonB) performance is not sufficiently reliable in finding the reading sequence. However, our system attempts to learn the reading sequence without any linguistic knowledge through the exploration-exploitation mechanism of RL.

4.4 Limitation

In this current research, we considered word insertion in principal text-lines [2], and did not tackle the (oriented) insertions in the margins. However, our work can be extended to address this issue. We here worked with obstructed text-lines at the word-level. Sometimes, in a word, character-level insertion/annotation can be noted, which we did not handle here.

We did not pay additional attention to the words with strike-through. However, a special module [30] for this can be added. Experiments were performed on handwriting samples written on white pages. However, if a sample is written on a rule-lined page, a preprocessing module [31] can be added.

We experimented on left-to-right writing systems (English and Bengali); however, our architecture can be extended to right-to-left writing systems [31] with some minor changes. Some unconventional annotations, due to writing idiosyncrasies, such as inserting a word below a certain text-line using a down-caret symbol ($_\vee$), are out of the scope of this paper.

5   Conclusion

In this paper, we studied the detection of the reading sequence of words in a handwritten text-line, including those that are obstructed due to inserting some words during amendments. We proposed a multi-agent hierarchical reinforcement learning-based architecture for our work. For experimental analysis, we took English and Bengali offline handwriting. We compared our system performance with some baseline approaches and obtained encouraging outcomes. However, still, there is a room for improving our system performance, which we will endeavor to address in the future.

Acknowledgment. All the people who contributed to generating the database are gratefully acknowledged. The authors also heartily thank all the consulted linguistic and handwriting experts.

References

1. Grüning, T., et al.: A two-stage method for text line detection in historical documents. IJDAR **22**, 285–302 (2019)
2. Survey, A., Sulem, L.L., Zahour, A., Taconet, B.: Text line segmentation of historical documents. IJDAR **9**, 123–138 (2007)
3. Surinta, O., et al.: A* path planning for line segmentation of handwritten documents. In: ICFHR, pp. 175–180 (2014)
4. Li, X.Y., et al.: Script-independent text line segmentation in freestyle handwritten documents. IEEE TPAMI **30**(8), 1313–1329 (2008)
5. Arulkumaran, K., et al.: Deep reinforcement learning: a brief survey. IEEE Sig. Process. Mag. **34**(6), 26–38 (2017)
6. Sutton, R.S., Barto, A.G.: Reinforcement Learning: An Introduction, 2nd edn. MIT Press, Cambridge (2018). ISBN: 9780262039246
7. Wilber: GIMP 2.10.22 Released (2020). Online: gimp.org. Accessed 3 May 2021
8. Marti, U., Bunke, H.: The IAM-database: an English sentence database for off-line handwriting recognition. IJDAR **5**, 39–46 (2002)
9. Alaei, A., Pal, U., Nagabhushan, P.: Dataset and ground truth for handwritten text in four different scripts. IJPRAI **26**(4), 1253001 (2012)
10. Berliac, Y. F.: The Promise of Hierarchical Reinforcement Learning. The Gradient (2019)
11. Wierstra, D., Foerster, A., Peters, J., Schmidhuber, J.: Solving deep memory POMDPs with recurrent policy gradients. In: ICANN, pp. 697–706 (2007)
12. Badrinarayanan, V., et al.: SegNet: a deep convolutional encoder-decoder architecture for image segmentation. IEEE TPAMI **39**(12), 2481–2495 (2017)
13. Zhang, A., et al.: Dive into Deep Learning (2020). Online: d2l.ai. Accessed 3 May 2021
14. Ioffe, S., Szegedy, C.: Batch normalization: accelerating deep network training by reducing internal covariate shift. ICML **37**, 448–456 (2015)
15. Misra, D.: Mish: a self regularized non-monotonic activation function. In: Paper # 928, BMVC 2020 (2020)
16. Levenshtein, V.I.: Binary codes capable of correcting deletions, insertions, and reversals. Doklady Akademii Nauk SSSR **163**(4), 845–848 (1965)

17. Mnih, V., et al.: Human-level control through deep reinforcement learning. Nature **518**, 529–533 (2015)
18. Wang, Z., et al.: Dueling network architectures for deep reinforcement learning. ICML **48**, 1995–2003 (2016)
19. Williams, R.J.: Simple statistical gradient-following algorithms for connectionist reinforcement learning. Mach. Learn. **8**(3–4), 229–256 (1992)
20. Wandell, B.A.: Foundations of Vision. Sinauer Asso. Inc. (1995). ISBN: 9780878938537
21. Larochelle, H., Hinton, G.E.: Learning to combine foveal glimpses with a third-order Boltzmann machine. In: NIPS, pp. 1243–1251 (2010)
22. He, K., et al.: Deep residual learning for image recognition. In: CVPR, pp. 770–778 (2016)
23. Cho, K., et al.: Learning phrase representations using RNN encoder-decoder for statistical machine translation. In: EMNLP, pp. 1724–1734 (2014)
24. Chung, J., et al.: Empirical evaluation of gated recurrent neural networks on sequence modeling. In: NIPS Workshop on Deep Learning (2014)
25. Mnih, V., et al.: Recurrent models of visual attention. In: NIPS, pp. 2204–2212 (2014)
26. Sutton, R.S., et al.: Policy gradient methods for reinforcement learning with function approximation. In: NIPS, pp. 1057–1063 (1999)
27. Hertz, J., Krogh, A., Palmer, R.G.: Introduction to the Theory of Neural Computation. CRC Press, Boca Raton (1991). https://doi.org/10.1201/9780429499661
28. Botchkarev, A.: Performance metrics (error measures) in machine learning regression, forecasting and prognostics: properties and typology arXiv:1809.03006 (2018)
29. Stamatopoulos, N., et al.: ICDAR 2013 handwriting segmentation contest. In: ICDAR, pp. 1402–1406 (2013)
30. Chaudhuri, B.B., Adak, C.: An approach for detecting and cleaning of struck-out hand-written text. Pattern Recogn. **61**, 282–294 (2017)
31. Almageed, W.A., et al.: Page rule-line removal using linear subspaces in monochromatic handwritten Arabic documents. In: ICDAR, pp. 768–772 (2009)

Multimodal Attention-Based Learning for Imbalanced Corporate Documents Classification

Ibrahim Souleiman Mahamoud[1,2]($\boxtimes$), Joris Voerman[1,2], Mickaël Coustaty[1],
Aurélie Joseph[2], Vincent Poulain d'Andecy[2], and Jean-Marc Ogier[1]

[1] La Rochelle Université, L3i Avenue Michel Crépeau, 17042 La Rochelle, France
`{joris.voerman,mickael.coustaty,jean-marc.ogier}@univ-lr.fr`
[2] Yooz 1 Rue Fleming, 17000 La Rochelle, France
`{ibrahim.souleimanmahamoud,aurelie.joseph,`
`vincent.poulaindandecy}@getyooz.com`

Abstract. The corporate document classification process may rely on the use of textual approach considered separately of image features. On the opposite, some methods only use the visual content of documents while ignoring the semantic information. This semantic corresponds to an important part of corporate documents which make some classes of document impossible to distinguish effectively. The recent state-of-the-art deep learning methods propose to combine the textual content and the visual features within a multi-modal approach. In addition, corporate document classification processes offer a particular challenge for deep learning-based systems with an imbalanced corpus. Indeed the neural network performances strongly depend on the corpus used to train the network, and an imbalanced set generally entails bad final system performances. This paper proposes a multi-modal deep convolutional network with an attention model designed to classify a large variety of imbalanced corporate documents. Our proposed approach is compared to several state-of-the-art methods designed for document classification task using the textual content, the visual content and some multi-modal approaches. We obtained higher performances on our two testing datasets with an improvement of 2% on our private dataset and a 3% on the public RVL-CDIP dataset.

Keywords: Document classification · Imbalanced classifcation · Multimodal classification · Attention mechanism

1 Introduction

Companies need to manage each day a large number of documents. Those documents represent the life of the company and can be of a large variety of types, classes and origins. They are generally linked to the administrative part (like

© Springer Nature Switzerland AG 2021
J. Lladós et al. (Eds.): ICDAR 2021, LNCS 12823, pp. 223–237, 2021.
https://doi.org/10.1007/978-3-030-86334-0_15

invoices, letters, receipts) or to the core activity of the company. Such documents are of primary importance as they generally validate an action or a decision inside and/or outside the company. The management of those documents is a challenge between speed and precision. Indeed, an error could have an heavy cost by causing a wrong action or decision. In this context, precision is then preferred to recall.

Many companies use Digital Mailroom system [1] to automatize document processing and so reduce workload and time required. Those systems entries can be modeled as document streams which define many constraints. One of these constraints is a strong imbalanced representation between classes that could affect any learning methods based on a training set. Even if deep learning models recently offer impressive performance in multiple difficult situations, the document stream classification process remains a challenge. This challenge is linked to the training set used to define the model: a strong imbalance inside the training set reduce neural network performances for low represented classes and they become like noises for greater classes. Moreover, all the classes need to be known when training the system or they will be ignored and classified as a wrong known class.

Document classification processes are traditionally divided into two categories: The image processing and the textual analysis categories. Language processing methods offer good performances for classification of corporate documents where textual content is the main resource (Fig. 1a). But their performances depend on the quality of the text extraction process (generally done by Optical Characters Recognition—OCR). OCR is not perfect and an error on most important words could make the classification impossible. OCR quality has recently greatly increased but remains affected by noises and can not recognize handwritten texts (Fig. 1b). In addition, some documents classes contain few text, like advertisement (Fig. 1c). It is then hard to classify a document with this type of method. This is where the image-based approaches are more suitable. In the same way, image processing classify with efficiency documents classes with a template or with a common structure but many classes are visually very close and could only be distinguished by their semantic or few keywords.

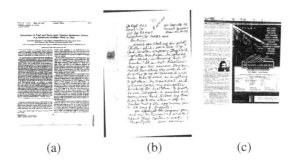

(a)   (b)   (c)

Fig. 1. Three examples of documents that can only be classified based on their textual content (a), graphical content (b) or a combination of them (c)

With the spread of multimodal approaches in the deep-learning era [2], some methods have been proposed to benefit from both sides. Multimodal or Cross-modal systems propose a solution by making a decision based on a combination of graphical and textual information. The way to combine those two modalities then becomes the new main problem in order to keep the best of the two previous approaches. Recently, cross-modal methods demonstrate that they could be considered as the best ones compared to the state-of-the-art image processing system on RVL-CDIP dataset [3] and offering impressive performances.

In our paper, we want to deepen the evaluation of these new approaches in the field of imbalanced classification of document flows by comparing them with more classical methods of the state-of-the-art. In addition, we propose our multimodal system integrating an attention model per modality designed to force the system to learn the most relevant features even with the least represented classes. The second advantage is its ability to visualize features used by the deep neural network in the decision process, thus reducing the black box effect of our architectures.

2 Related Work

Few works have been proposed on multimodal analysis and classification of documents, and even fewer on the impact analysis of imbalanced number of document per class during the learning process. We will first provide an overview of articles related to one modality (based on visual or textual features), followed by the multi-modal/cross-modal approaches. The last part will focus on attention-based architectures.

Regarding the visual-based approaches, many image classification methods have been efficiently applied on the document classification task since the RVL-CDIP dataset [4] have been proposed. They are mainly based on deep convolution architecture pre-trained on ImageNet dataset [5]. Multiple architectures were proposed for this task like the InceptionResNet [6], NasNet [7] and VGG [8] for naming the most used ones. In first hand, a majority of those methods isn't especially adapted for document. In another hand, some methods propose to take into account structural aspect of documents and divided them in multiple sections like in [4].

Regarding text content based neural network, they need firstly a word representation system. Most used systems are currently word embedding like Fast-Text [9], BERT [10] has became probably the most interesting word embedding technique and seems to be more efficient that previous methods. Those methods are based on various elder statistical strategy like bag-of-words and co-occurrences matrix. They reinforce them with deep learning method trained on huge corpus of text, with more than one million words, like Wikipedia's articles. For the network himself, current research are concentrated on recurrent strategy to keep sentence sequential information with Recurrent Neural Network (RNN) [11] and bidirectional-RNN [12].

Some recent works proposed to combine textual and visual features to take the whole document's content into account. [13] uses a MobileNetV2 [14] architecture combined with a CNN model inspired of [15] by concatenating the network outputs. [3] introduced another architecture with a NasNet [7] network for the image part and a bidirectional BERT architecture for the text. The paper tested several combinations between the networks. Even if the performances improved recently, the neural network results interpretation remains difficult. This is moreover becoming more important with the development of multimodal classification process and the need to a better understanding of limitations (which modality is penalising the other one).

Attention is one of the most influential ideas in the Deep Learning community. Even though this mechanism is now used in various problems like image captioning and others, it was initially designed in the context of Neural Machine Translation using Seq2Seq Models [16]. One important advantage of those attention models is their ability to visualize the features used by the network to make a decision, and then offer a possible interpretation of errors. This system could also be used to reinforce network performance as in [17] where the integration of an attention model into a recurrent LSTM offers better performances for image classification. Another work [18] proposed an attention model architecture combined with the VGG16 network with several layers at different resolutions to generate more global features. The attention model could force the networks to learn relevant patterns. [19] propose to use attention model to identify salient image regions at different stages of the CNN. During decision, the system reinforces their importance and therefore suppress irrelevant or confusing information.

Our work is particularly interested in the classification of an imbalanced corpus. This subject is close to two image classification challenges known as zero-shot and one-shot learning. The first introduces a context where a part of classes is unknown during training. The second is close but classes are represented by at least one or few documents. Zero-shot solutions [20] mainly use a transfer learning strategy based on semantic description of image. Our context does not offer access to such descriptions and we do not have any prior knowledge about unknown classes. So those solutions can not be used in our situation. One-shot learning methods offer more interesting option with Bayesian-based techniques like in [21] or adapted deep learning methods like in [22] and [23] but those methods are not directly compatible with our multi-modal approach. In addition, some methods are designed for imbalanced classification out of those challenge like in [24]. This propose to apply reinforcement learning with higher rewards and penalties on low represented classes.

The originality of our approach lies therefore in the proposal of a model based on attention models that allow us to have a better performance and a better interpretation of the results. In the next section, we will describe our own method for imbalanced multi-modal documents classification.

3   Problem Definition

The main objective of our proposal is to provide solutions that improve the predictive accuracy in forecasting tasks involving imbalanced dataset. The task of prediction of document classes in an industrial environment involves input flows $s_1, s_2, s_3, ..., s_n \in S$, where s_n is the number of samples from classes n in the training set. The number of samples per class is by definition imbalanced due to the industrial context. Some document classes are very recurrent (i.e. there will be thousands of invoices) while others are very rare with a very few number of samples (ten or even 1 document) as displayed in Fig. 2.

The objective of this predictive task is to predict the class of samples S. The overall assumption is that an unknown function correlates the samples and ground truth classes of S, i.e. $Cn = f(sn)$. The goal of the learning process is to provide an approximation of this unknown function whatever the quantity of samples available by class. To better approximate this function f, we must know the are significant intra-class variation and inter-class similarity caused by different structure documents for each client brings a great deal of difficulties to classify (see the Fig. 3).

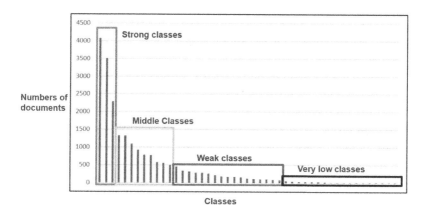

Fig. 2. Distribution of samples by classes for the private data YOOZ

The hypotheses tested in our experimental evaluation are:

1. A higher inter-class variance in the representation space will reduce the confusion between classes and also better predict weakly represented cases
2. The use of an attention mechanic for textual and visual parts will allow to better focus on the important part of the document content and thus reduce intra-class variation errors

4   Proposed Approach

To see the relevance of our hypotheses, we inspired by state-of-the-art models to propose a classification architecture with three attention models. The combination of these three attention models has not been studied in the state-of-the-art and moreover on imbalanced data. In a first step we will present the architecture without the attention mechanism and then we discuss each of the attention mechanisms.

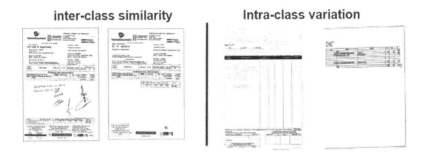

Fig. 3. The two documents on the left come from two different classes even though their structure and contents are similar. The documents on the right come from the same class and we can see that they have a different structure.

4.1   Model Without Attention

The initial system is composed of two classifiers, one for the visual part and the other one for the text extracted from the image. We use a bidirectional recurrent LSTM network with bert embedding for the text classification and a VGG16 network pre-trained on ImageNet for the image part. The choice of this architecture is inspired by the methods from the state of art for document and image classification. Here, the idea is to use the best systems from the literature as a baseline.

Each of these models has different input feature (denoted X and Z in the Eq. 1). The features of each sample i are decomposed in $x_i \in R^{d_1}$ for the textual branch and $z_i \in R^{d_2}$ for the visual branch. d_1 and d_2 are the features dimension respectively for textual and visual parts. N is the size of the corpus (i.e. $1 \leq i \leq N$).

The one-hot vector corresponding the label as $y_i \in R^L$ and L corresponds the total number of classes. The one-hot vector values are defined as 1 for the corresponding class in the groundtruth, and 0 for the others.

For the textual content, we uses the first 150 words and extracted the Camembert features [25], where each word is represented by 768 values. We then have d_1 of size (150,768). The input images are color images of size $d_2 = (224,224,3)$ as an input for the VGG16 network.

$$X = [x_1, x_2, ..., x_i]^T \in \mathbf{R}^{N \times d_1}$$
$$Z = [z_1, z_2, ..., z_i]^T \in \mathbf{R}^{N \times d_2} \qquad (1)$$
$$Y = [y_1, y_2, .., y_i]^T \in \mathbf{R}^{N \times L}$$

4.2 Attention Model

The use of Attention Mechanism aims at focusing the network to the most relevant features related to our task (*i.e.* document classification). This corresponds to learning a matrix (or a mask) which weights features for each class. We propose to adapt this to each branch of our architecture.

Self-attention Mechanism for Label Fitting. The self-attention mechanism used in this article is inspired by [26]. They propose a self-attention mechanism on the input image to consider the inherent correlation (attention) between the input features themselves, and then use a graph neural network for the classification task. We use this self-attention on both the image and the text to focus our network on common features from the input. This will allow us to exploit the interclass and intersample correlation at the initial stage. Contrary to our baseline, we use the label Y as input and not only to help error propagation. We calculate the correlation between the input data and the label to predict. To do this, we transform X, Z to X', Z' which will be the input of our two classifiers. To achieve that, we follow several steps, the first step is to calculate the sample and label correlation matrices as:

$$C^x = softmax(XX^T)$$
$$C^z = softmax(ZZ^T) \qquad (2)$$
$$C^y = softmax(YY^T)$$

Here softmax($\cdot$) denotes a softmax operator. The inputs of this function have the same dimension where $XX^T, ZZ^T, YY^T \in R^{N \times N}$

The self-attention module exploits C^x, C^z, C^y. Thus, the next step is to fuse Cx and Cy using trainable 1×1 kernels as:

$$C^{xy} = fusion([Cx, Cy]) \in R^{N \times N}$$
$$C^{zy} = fusion([Cz, Cy]) \in R^{N \times N} \qquad (3)$$

where e.g. [Cx, Cy] denotes the attention map concatenation, This fusion function $fusion$ is equivalent e.g. $C_x y = w_1 Cx + w_2 Cy$, where the weighted parameters w_1, w_2 are learned adaptively. We will use C^{xy}, C^{zy} to update both the visual feature X and the textual feature Z.

$$X' = XC^{xy}, Z' = ZC^{zy} \qquad (4)$$

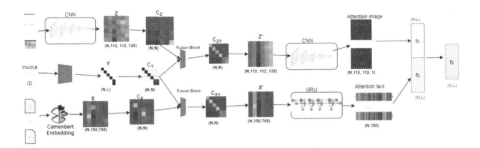

Fig. 4. The proposed multimodal model uses the three attention mechanisms described above. We used a small colored matrix to illustrate how the attention mechanism is used (Figure description see part 4)

Attention for the Textual Content. The Text Attention Model allows us to see the relevancies of the 150 words, to determine which ones have which has more impact in the classification.

$$A_t = f(softmax(h_f)) \odot h \tag{5}$$

Here h is the hidden LSTM and h_f is after having flatten h, $\odot$ denotes a point-wise product operator, and softmax($\cdot$) denotes a row-wise softmax operator for the sample. The f function is a dropout which is intended to prevent overfitting on the data by activating neurons with probability p or kept with probability $1 - p$ (p fixed to 0.3 in our experiments). We analyze the output of the softmax to determine the words the models of which gave more weight.

Attention for the Visual Content. This mechanism is inspired by the work done by [18]. Their original goal was to predict the age of people from fixed flexion knee X-ray images. We use this image attention model as it brings two advantages: it helps understanding which part of the image is the most relevant (like the titles of document for instance); it allows understanding which part the attention model focused on, and then to understand its mistakes (for instance, we can determine the area of zones of interest it concentrates whereas it should not. The attention model embedded in the image classification network uses the output of the 5th layer of the VGG16 model D. This layer is composed of 128 filters and has a dimension of (112, 112, 128). The last layer is our attention layer A and is of dimension (112, 112, 1). The matrix D is multiplied by D to obtain the matrix D'. Finally, the dimension is reduced by applying an average pooling layer on D' to generate a vector of size (128). The last step consists in normalizing this attention vector by a pooling layer on A. This normalization is the function $g(x1, x2) = x1/x2$ and the input of the visual classifier will be the output of this normalization function.

4.3   Combination of Modalities

As stated at the beginning of this paper, combining the two previous modalities should allow to take advantage of each other to increase the overall performance of the model. The proposed multimodal architecture is illustrated in Fig. 4.

First we apply self-attention mechanism on the inputs for a better separation of features in their representation space. Then we use the updated inputs by the self-attention mechanism to classify textual and visual parts, and each part is reinforced with its own specific attention mechanism. Our final classifier is based on the prediction output of the two classifiers. The following equation explain how we combined the two models:

$$[Y_x, Y_z] \in R^{N^2 \times C} \tag{6}$$

Where [Yx, Yz] is the concatenation of the outputs of the text Y_x and image Y_z classifier. We noticed that the model had some difficulties to classify the weakly represented classes. We then decided to use a weight on the Cross Entropy loss function presented in Eq. 7. The Cross Entropy [27] tends to counterbalance the weight of the under-represented classes towards the over-represented ones (the minority classes are often ignored as they represent only a small part of the total loss). The weight assigned to each class corresponds to the inverse percentage of the examples present in the training set. The smaller the presence of a class, the greater its weight on the loss will be.

$$Loss_{CE}(i) = -Wt_i log(P(i)) \tag{7}$$

5   Experiments

5.1   Dataset

To evaluate our multimodal attention-based model, we used two different datasets. The first one named "YOOZ dataset" is a private and already imbalanced set issuing from our customers. In order to assess the relevance of our method, and to have a comparison with the best approaches from the literature, we also use the public RVL-CDIP [4] dataset. This large public dataset composed of a large variety of document classes equally balanced between them. In order to evaluate our contribution, we propose a protocol to unbalance the RVL-CDIP dataset.

The YOOZ dataset is composed of 15 thousands training documents, 2 thousand for the validation set and 5 thousand for testing. This dataset is composed of 47 classes (Invoice, Quotation, general sales condition, check, etc.). Some classes are very similar from the layout and the content point of view while other classes are easily distinguishable. The Fig. 2 proposes an illustration of the classes distribution of our imbalanced dataset. One can observe that four groups appear (from the most frequent on the left to the least frequent classes on the right).

The second dataset, The RVL-CDIP, is balanced by definition with 20 000 images per class in the training set and 2500 images per class in the validation and test sets. In order to assess the relevance of our proposed approach (a multi-modal attention-based classification model for imbalanced dataset), we reduced the number of document for each class. This reduction mimic the configuration of real-life conditions (some very frequent classes opposed to classes with very few samples).

To reproduce this unbalanced dataset, in the first we took the first 90 first thousand dataset(image and text) of RVL CDIP according to the order present in the train.txt in the folder labels. The exact document partition ordered by classes (0 to 15) is the following, for each class we recover a proportion to have an unbalanced distribution [5673*1, 5662*1, 5630*1, 5735*1, 5613*0.5, 5584*0.5, 5559*0.5, 5622*0.5, 5658*0.1, 5649*0.1, 5593*0.1, 5642*0.1, 5553*0.05, 5639*0.05, 5619*0.05, 5569*0.05]. So we get a total of 33 thousand documents for training. we did the same procedure to unbalance the data for the test and validation data, for each we had 16 thousand documents.

The textual content of both datasets were extracted from documents with an OCR. For our private YOOZ dataset, we used ABBYY FineReader OCR, while the OCRed version of the RVL-CDIP documents is provided by their authors (more details are available in [4]).

5.2 Implementation Details

For all our experiments, we used the ADAM optimizer [28] with $\beta_1 = 0.9$, $\beta_2 = 0.999$ and a batch size of 64, an initial learning rate of 10^{-3} scaled from 0.1 every 3 epochs without improvement in validation loss and an early stopping after 5 epochs without improvement.

5.3 Performance Evaluation and Discussions

The first part of our evaluation protocol consists in comparing the proposed approach with the best approaches from the state-of-the-art. Table 1 presents the precision and recall values of all these approaches on the RVL-CDIP original dataset, its imbalanced proposed version and our internal dataset.

When looking at the textual part of the documents, we can see that even if the work done by [3] used a similar approach than ours, we can observe that we have better performances. This can be explained by the fact that the authors of [3] used a LSTM network with BERT embedding. We can here observe that the proposed attention model allows us to increase the precision by 3 points on the whole dataset.

When focusing on the visual part of the architecture, we compared our proposed attention-based pattern mechanism compared to the state-of-the-art model [29]. This model uses both holistic image and image split by region(right, left, top and bottom). We can here again observe that or complex model have better performance (+1%) with less parameters.

Finally, we present the performances of our multimodal architecture on imbalanced dataset to highlight the complementary effects of the textual and visual modalities. We can observe that the increase is much important on the imbalanced RVL-CDIP dataset than on our private one. This can be explained by the fact that classes are mostly distinguishable by their structure in the RVL-CDIP dataset, so the model relies on the image when it does not reach class with the text.

Table 1. The results of the models, the models with Att are the models with attentions either on the text or the image while S_{Att} is Att added the self attention. BiRNN is the bi-directional RNN model and VGG16 is the image classification model.

Modality	Data	RVL-CDIP		RVL-CDIP unbalanced		YOOZ	
Model		Precision	Recall	Precision	Recall	Precision	Recall
Text	*Bert (Souhail_CVPR2020)*	86%	86%	—	—	—	—
	biRNN	88.58%	87.95%	79.7%	68.5%	95.7%	94.3%
	$biRNN_{Att}$	89.5%	88.1%	80.3%	69.8	96.8%	95.2%%
	$biRNN_{S_{Att}}$	—	—	80.7%	70.2	97.5%	97.3%%
Image	*VGG16_pretrained (das2018document)*	91.1%	—	—	—	—	—
	VGG16	91.8%	90.4%	87.3%	86.8%	88.1%	83.5%
	$VGG16_{Att}$	**92.2%**	**91.3%**	88.5%	87.4%	88.7%	83.2%
	$VGG16_{S_{Att}}$	—	—	88.8%	87.6%	89.2%	83.4%
Multimodality	*Multimodality (aude-bert19multimodal)*	90.6%	—	—	—	—	—
	Multimodality	—	—	90.8%	88.7%	96.6%	96.2%
	$Multimodality_{Att}$	—	—	92.03%	90.06%	97.3%	93.18%
	$Multimodality_{S_{Att}}$	—	—	**92.90%**	**91.40%**	**98.2%**	**97.5%**

The $Multimodality_{S_{Att}}$ propose allows us to have real gains on both RVL and YOOZ data. As previously mentioned in the industrial context prefers a model with little bad prediction, for this we took a threshold of 0.99 to filter only the cases where the model has more confidence on its prediction.

The multimodal model without model attention and with attention we almost better classify about 64 documents in dataset YOOZ as illustrated in Fig. 5(a) and (b) with threshold 0.9. When we further analyze the results class by class, we find out that the true contribution of this model lies in the reduction of intra-class errors. This contribution on almost all the classes highlight a better separation in the feature representation space.

Following this idea, we propose in our evaluation to illustrate the impact of our attention mechanism on the inter-class correlation. The proposed multimodal module exploits three attention mechanisms and this combination of attention mechanisms effectively guide the feature representation for each document and was designed to better separate classes in the feature space (thus reducing the confusion in the decision process). We then first propose to visualize the way documents are distributed in the feature space using a t-SNE visualization tool. One can observe in Fig. 5 (second line) that the classes are better seperated after applying our self-attention mechanism, demonstrating the effect of our architecture. In order to illustrate the impact of a better distributed and separated classes in the feature space, we also display the total number of error per class before and after applying our attention mechanisms (first line of Fig. 5). We can notice that this better separation entails a much lower number of errors for each class and then validate our initial assumptions.

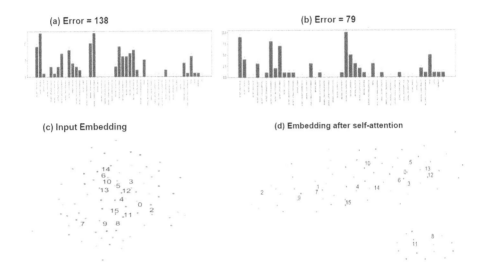

Fig. 5. In this figure we compare the errors between (a) the multimodal model with no attention and (b) the model with attentions (YOOZ dataset). These bars represent the error rate after a 99% threshold has been applied. We note that the multimodal model proposes to increase these cases of errors, which is highly appreciated in the industrial context. T-SNE visualization: (c): The embeddings without correlation, (d): The embedding created after correlation with labels (RVL-CDIP dataset). The numbers represent the class in the figure it's placed in the center of each class.

Finally, the past part of our evaluation process tends to illustrate the impact of this better separation of classes class by class. Table 2 presents the number of documents misclassified despite a high threshold of 0.99. The multimodality reduces the misclassified with threshold compared to the separate modalities for almost all classes because it manages to exploit the best of each modality. For some classes the contribution of multimodality is very low, which can be explained because one of the modalities is much more confident than the other so multimodality will use the prediction of only one modality (e.g. image) and so they will have almost the same misclassified.

Discussion. The challenge we hope to meet is to combine several modalities so that each one brings its own contribution in order to have better performance for both frequent and non-frequent classes.

Our future research will be much more focused on the training technique of our methods in order to better take into account the imbalanced data. We

Table 2. The number document of misclassified with a threshold 99 test data. hw = handwritten, adv = advertisement, s.r = scientific report, s.p=scientific publication spec = specification, fileF = file folder, newAr = news article pre = presentation, ques = questionnaire

Method	Letter	Form	email	hw	adv	s.r	s.p	spec
Data	1217	1261	1289	1167	620	641	640	625
Text	380	441	220	352	467	380	322	117
Image	320	412	219	322	167	219	208	78
Multimodality	**310**	**384**	**201**	**317**	**164**	**210**	**204**	**71**
Method	fileF	newAr	Budget	Invoice	pre	ques	Resume	Memo
Data	123	129	123	130	61	64	58	61
Text	28	48	36	21	30	34	26	21
Image	17	24	29	21	22	28	15	12
Multimodality	**15**	**21**	**26**	**16**	**21**	**26**	**13**	**12**

will test several types of specific loss for imbalanced data and will also test reinforcement learning to better train our model to force the network to be performant on the weakly represented classes.

6   Conclusion

In this article, we have proposed methods using multimodality and attention patterns on images and text. The use of multimodality is necessary in order to have better performance and take advantage of both the features extracted from the text and the image. The combination of three attention mechanisms allows focusing on the most relevant visual or semantic features to classify documents, and ease the understanding of results and mistakes. Our best proposed multi-modal weighted system is able to increase by 2% the global precision compared to state-of-the-art architecture. We can also drastically reduce the number of errors, mainly by reducing the confusion between classes.

Even if we obtained good performances with our proposed multimodality approach on both the YOOZ and the RVL-CDIP dataset, many perspectives appear. The use of weighted categorical cross-entropy has certainly slightly improved the results on some classes with a low presence in the training dataset, but nevertheless we hope to get higher performance with the use of reinforcement learning to strengthen the answer when the models correctly predicts the less frequent classes. This could be done using the work proposed in [30] which describes the use of reinforcement learning in the case where the data is out of balance. We then expect improving our performances.

Acknowledgment. This research has been funded by the LabCom IDEAS under the grand number ANR-18-LCV3-0008, by the French ANRT agency (CIFRE program) and by the YOOZ company.

References

1. Schuster, D., et al.: Intellix-end-user trained information extraction for document archiving. In: 2013 12th International Conference on Document Analysis and Recognition. IEEE, pp. 101–105 (2013)
2. Srivastava, N., Salakhutdinov, R.R.: Multimodal learning with deep Boltzmann machines. In: Pereira, F., Burges, C.J.C., Bottou, L., Weinberger, K.Q. (eds.) Advances in Neural Information Processing Systems, vol. 25, pp. 2222–2230. Curran Associates Inc. (2012). http://papers.nips.cc/paper/4683-multimodal-learning-with-deep-boltzmann-machines.pdf
3. Bakkali, S., Ming, Z., Coustaty, M., Rusinol, M.: Visual and textual deep feature fusion for document image classification. In: Proceedings of the IEEE/CVF Conference on Computer Vision and Pattern Recognition Workshops, pp. 562–563 (2020)
4. Harley, A.W., Ufkes, A., Derpanis, K.G.: Evaluation of deep convolutional nets for document image classification and retrieval. In: 2015 13th International Conference on Document Analysis and Recognition (ICDAR), pp. 991–995. IEEE (2015)
5. Russakovsky, 0, et al.: ImageNet large scale visual recognition challenge. Int. J. Comput. Vis. **115**(3), 211–252 (2015)
6. Szegedy, C., Ioffe, S., Vanhoucke, V., Alemi, A.A.: Inception-v4, inception-resnet and the impact of residual connections on learning. In: Thirty-First AAAI Conference on Artificial Intelligence (2017)
7. Zoph, B., Vasudevan, V., Shlens, J., Le, Q.V.: Learning transferable architectures for scalable image recognition. In: Proceedings of the IEEE Conference on Computer Vision and Pattern Recognition, pp. 8697–8710 (2018)
8. Simonyan, K., Zisserman, A.: Very deep convolutional networks for large-scale image recognition. arXiv preprint arXiv:1409.1556 (2014)
9. Joulin, A., Grave, E., Bojanowski, P., Douze, M., Jégou, H., Mikolov, T.: Fasttext.zip: compressing text classification models, arXiv preprint arXiv:1612.03651 (2016)
10. Devlin, J., Chang, M.-W., Lee, K., Toutanova, K.: Bert: pre-training of deep bidirectional transformers for language understanding, arXiv preprint arXiv:1810.04805 (2018)
11. Graves, A., Mohamed, A.-R., Hinton, G.: Speech recognition with deep recurrent neural networks. In: IEEE International Conference on Acoustics, Speech and Signal Processing, pp. 6645–6649. IEEE (2013)
12. Zhou, P., Qi, Z., Zheng, S., Xu, J., Bao, H., Xu, B.: Text classification improved by integrating bidirectional LSTM with two-dimensional max pooling, arXiv preprint arXiv:1611.06639 (2016)
13. Audebert, N., Herold, C., Slimani, K., Vidal, C.: Multimodal deep networks for text and image-based document classification. In: Cellier, P., Driessens, K. (eds.) ECML PKDD 2019. CCIS, vol. 1167, pp. 427–443. Springer, Cham (2020). https://doi.org/10.1007/978-3-030-43823-4_35
14. Sandler, M., Howard, A., Zhu, M., Zhmoginov, A., Chen, L.-C.: MobileNetV2: inverted residuals and linear bottlenecks. In: Proceedings of the IEEE Conference on Computer Vision and Pattern Recognition, pp. 4510–4520 (2018)
15. Kim, Y.: Convolutional neural networks for sentence classification, arXiv preprint arXiv:1408.5882 (2014)
16. Luong, M.-T., Manning, C.D.: Stanford neural machine translation systems for spoken language domains. In: Proceedings of the International Workshop on Spoken Language Translation, pp. 76–79 (2015)

17. Jain, R., Wigington, C.: Multimodal document image classification. In: 2019 International Conference on Document Analysis and Recognition (ICDAR), pp. 71–77. IEEE (2019)
18. Górriz, M., Antony, J., McGuinness, K., Giró-i Nieto, X., O'Connor, N.E.: Assessing knee OA severity with CNN attention-based end-to-end architectures, arXiv preprint arXiv:1908.08856 (2019)
19. Jetley, S., Lord, N.A., Lee, N., Torr, P.H.: Learn to pay attention, arXiv preprint arXiv:1804.02391 (2018)
20. Xian, Y., Schiele, B., Akata, Z.: Zero-shot learning-the good, the bad and the ugly. In: Proceedings of the IEEE Conference on Computer Vision and Pattern Recognition, pp. 4582–4591 (2017)
21. Lake, B.M., Salakhutdinov, R., Tenenbaum, J.B.: Human-level concept learning through probabilistic program induction. Science **350**(6266), 1332–1338 (2015)
22. Santoro, A., Bartunov, S., Botvinick, M., Wierstra, D., Lillicrap, T.: One-shot learning with memory-augmented neural networks, arXiv preprint arXiv:1605.06065 (2016)
23. Koch, G., Zemel, R., Salakhutdinov, R.: Siamese neural networks for one-shot image recognition. In: ICML Deep Learning Workshop, vol. 2. Lille (2015)
24. Lin, E., Chen, Q., Qi, X.: Deep reinforcement learning for imbalanced classification. Appl. Intell. **50**(8), 2488–2502 (2020). https://doi.org/10.1007/s10489-020-01637-z
25. Martin, L.: CamemBERT: a tasty French language model. In: Proceedings of the 58th Annual Meeting of the Association for Computational Linguistics (2020)
26. Cheng, H., Zhou, J.T., Tay, W.P., Wen, B.: Attentive graph neural networks for few-shot learning (2020)
27. Nasr, G.E., Badr, E.A., Joun, C.: Cross entropy error function in neural networks: forecasting gasoline demand. In: Applied Intelligence, pp. 1–15 (2002)
28. Kingma, D.P., Ba, J.: Adam: A method for stochastic optimization (2017)
29. Das, A., Roy, S., Bhattacharya, U., Parui, S.K.: Document image classification with intra-domain transfer learning and stacked generalization of deep convolutional neural networks (2018)
30. Lin, E., Chen, Q., Qi, X.: Deep reinforcement learning for imbalanced classification (2019)

Light-Weight Document Image Cleanup Using Perceptual Loss

Soumyadeep Dey$^{(\boxtimes)}$ and Pratik Jawanpuria

Microsoft India Development Center, Hyderabad, India
{soumyadeep.dey,pratik.jawanpuria}@microsoft.com

Abstract. Smartphones have enabled effortless capturing and sharing of documents in digital form. The documents, however, often undergo various types of degradation due to aging, stains, or shortcoming of capturing environment such as shadow, non-uniform lighting, etc., which reduces the comprehensibility of the document images. In this work, we consider the problem of document image cleanup on embedded applications such as smartphone apps, which usually have memory, energy, and latency limitations due to the device and/or for best human user experience. We propose a light-weight encoder decoder based convolutional neural network architecture for removing the noisy elements from document images. To compensate for generalization performance with a low network capacity, we incorporate the perceptual loss for knowledge transfer from pre-trained deep CNN network in our loss function. In terms of the number of parameters and product-sum operations, our models are 65–1030 and 3–27 times, respectively, smaller than existing state-of-the-art document enhancement models. Overall, the proposed models offer a favorable resource versus accuracy trade-off and we empirically illustrate the efficacy of our approach on several real-world benchmark datasets.

Keywords: Document cleanup · Perceptual loss · Document binarization · Light-weight model

1 Introduction

The smartphone camera have simplified the capture of various physical documents in digital form. The ease of share of digital documents (e.g., via messaging/networking apps) have made them a popular source of information dissemination. However, readability of such digitized documents is hampered when the (original) physical document is degraded. For instance, the physical document may contain extraneous elements like stains, wrinkles, ink spills, or can undergo degradation over time. As a result, while scanning such documents (e.g., via a flat-bed scanner), these elements also get incorporated into the document image. In case of capturing document images via mobile cameras, the images are prone to being impacted by shadow, non-uniform lighting, light from multiple sources, light source occlusion, etc. Such *noisy* elements not only effects the comprehensibility of the corresponding digitized document to the human readers, it may also

© Springer Nature Switzerland AG 2021
J. Lladós et al. (Eds.): ICDAR 2021, LNCS 12823, pp. 238–253, 2021.
https://doi.org/10.1007/978-3-030-86334-0_16

Fig. 1. Typical examples of noisy document images; (a) real world degraded images from various datasets [5, 35, 51], (b) noisy real world images captured via mobile devices.

break down the automatic (document-image) processing/understanding pipeline in various applications (e.g., OCR, bar code reading, form detection, table detection, etc.). Few instances of *noisy* document images are shown in Fig. 1.

Given an input noisy document image, the aim of document image cleanup is to improve its readability and visibility by removing the noisy elements. While general (natural scene) image restoration has been traditionally explored by the computer vision community, recent works have also focused on developing cleanup techniques for document images depending on the type of noise and document-class. These include foreground background separation [23, 24, 40], differential fading problem [22, 44], removal of shadow/smear/strain [21, 22, 39, 46, 47, 49, 50], and handling ink bleed [41, 46], etc.

Recent works [6, 20, 52] view document cleanup as an image to image translation problem, modeled using deep networks. A general direction of research has been to explore deeper and more complicated networks in order to achieve better accuracy [43, 52]. However, such deep networks often require high computational resources, which is beyond many mobile and embedded applications on a computationally limited platform. Deeper networks also usually entail a higher inference time (latency), which document image processing mobile apps such as Adobe Lens, CamScanner, Microsoft Office Lens, etc., aim to minimize for best human user experience.

In this work, we propose a light-weight encoder-decoder based convolutional neural network (CNN) with skip-connections for cleaning up document images. Focusing on memory constrained mobile and embedded devices, we design a light-weight deep network architecture. It should be noted that light-weight deep network architecture usually costs generalization performance when compared with deeper networks. Hence, in order to obtain a healthy interplay between resource/latency and accuracy, we propose to employ perceptual loss function (instead of the more popular per-pixel loss function) for document image cleanup. The perceptual loss function [11] enables transfer learning by comparing high-level representation of images, obtained from a pre-trained CNNs (e.g., trained on image classification tasks). We empirically show the effectiveness of the proposed network on several real-world benchmark datasets.

The outline of the paper is as follows. We discuss the existing literature in Sect. 2. In Sect. 3, we detail our methodology. The empirical results are presented in Sect. 4, while Sect. 5 concludes the paper.

2   Related Work

In this section, we briefly discuss existing approaches that aim to recover/enhance images of degraded documents via techniques involving binarization, and illumination/shadow correction, and deblurring, among others.

Document Image Binarization: A popular framework for document image cleanup is background foreground separation [14], where the foreground pixels are preserved and enhanced and the background is made uniform. Binarization is a technique to segment foreground from the background pixels. Analytical techniques for document image binarization involve segmenting the foreground pixels and background pixels based on some thresholding. Traditional image binarization technique such as [26] compute a global threshold assuming that the pixel intensity distribution follows a bi-modal histogram. As estimating such thresholds may difficult for degraded document images, Moghaddam and Cheriet, in [23], proposed an adaptive generalization of the Otsu's method [26] for document image binarization. In [40], Sauvola and Pietikäinen proposed a local adaptive thresholding method for the image binarization task. To improve Sauvola's algorithm's performance in low contrast setting, Lazzara and Geraud developed its multi-scale generalization in [19]. Recent works on document image binarization have also explored techniques based on conditional random fields [27], fuzzy C-means clustering [24], robust regression [48], and maximum entropy classification [22].

Deep convolutional neural networks (CNNs) have become all-pervasive in computer vision ever since AlexNet [18] won the ILSVRC 2012 ImageNet Challenge [38]. Tensmeyer and Martinez [46] posed document image binarization as a pixel classification problem and developed a fully connected convolution network for it. An encoder-decoder network was proposed in [9] to estimate the background of a document image. Then, Otsu's global thresholding technique [26] is used to obtain a binarized image with uniform background. Afzal *et al.* [1] employed a long short-term memory (LSTM) network to classify each pixel as background and foreground by considering images to be a two-dimensional sequence of pixels. In [28], Peng *et al.* proposed a multi-resolutional attention model to learn the relationship between the text regions and background through convolutional conditional random field [16,45]. To bypass the need of large training datasets with ground truths, Kang *et al.* [12] employed modular U-Nets [37] pre-trained for specific tasks such as dilation, erosion, histogram equalization, etc. These U-Nets are cascaded using inter-module skip connections and the final network is fine-tuned for the document image binarization task.

Document Image Enhancement: In addition to working within the background foreground separation framework, existing works have developed noise-specific document image cleanup methods such as shadow removal. Bako *et al.* [2]

assumes a constant background color generates a shadow map that matches local background colors to a global reference. Similar to Bako's method, local and background colors are estimated to remove shadow from document images in [49,50]. Inspired by the topological surface filled by water, Jung *et al.* proposed an illumination correction algorithm for document images in [39]. A document image enhancement approach have been proposed by Krigler *et al.* by representing the input image as 3D point cloud and adopting the visibility detection technique to detect the pixels to enhance [15]. Recently, Lin *et al.* [21] proposed a deep architecture to estimate (i) the global background color of the document, and (ii) an attention map which computes the probability of a pixel belonging to the shadow-free background. An illumination correction and document rectification technique using patch based encoder-decoder network is proposed in [20].

Existing works have also explored deep networks for overall document enhancement rather than focusing on correcting specific document degradations. A skip-connected based deep convolutional auto-encoder is proposed in [52]. Instead of learning the transformation function from input to output, this network learns the residual between input and output. This residual when subtracted from the input image results in a noise free enhanced image. An end to end document enhancement framework using conditional Generative Adversarial Networks (cGAN) is proposed in [43], where an U-Net based encoder-decoder architecture is used for the generator network.

Document Image Cleanup for Mobile and Embedded Applications: Low-resource consuming models are desirable for mobile document image processing, e.g., in apps like Adobe Lens, CamScanner, Microsoft Office Lens, etc. However, existing CNN based methods [43,52], discussed above, propose deep architectures with huge number of parameters, making them unsuitable for memory and energy constrained devices. In this work, we propose a comparatively light-weight deep encoder-decoder based network for document image enhancement task. We employ the perceptual loss based transfer learning technique to compensate for generalization performance with a low network capacity.

3 Proposed Approach

As discussed, we propose a light-weight deep network, suitable for mobile document image cleanup applications.

3.1 Network Architecture

We design an encoder decoder based image to image translation network. The encoder part of the model consists of three convolution layers followed by five residual blocks. The residual blocks were first introduced in [8] for generic image processing tasks. We modify the residual blocks from the original design [8] to suite our network design. The decoder part of the model consists of five convolution layer along with skip connections from the encoder layers. This type of

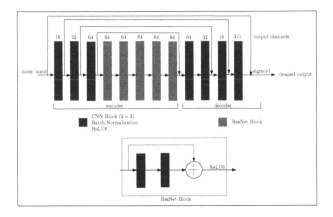

Fig. 2. Proposed light-weight CNN architecture used for document image cleanup

skip connection helps mitigate the vanishing gradient and the exploding gradient issues [8,43,52]. Hence, the skip connections help to simplify the overall learning of the network. Each convolution layer is followed by batch normalization layer and ReLU6 [17] activation layer.

The kernel size of the convolution layer is 3×3 and strides for all the layers is set to 1. The padding at each layer is set as *"same"*, which helps to pad the input such that it is fully covered by the filter. Padding *"same"* with stride 1 helps to keep the spatial dimension of the convolution layer output same as its input. At the end of the network a sigmoid activation function is used to obtain a normalized output between 0 and 1. The output dimension of the last layer of the decoder is either one or three depending on the end task of the network. If the network is trained for the task of binarization or gray scale cleanup then the output dimension of the last layer is set to one. For color cleanup task the output dimension of the last layer is set to three.

We term our models as M-x, where x represents the value of the maximum width of the network. In our experiments, we have considered x to be 16, 32 and 64. The M-64 model is shown in Fig. 2. In M-32 model's architecture, the output dimensions of the residual blocks are 32 and the CNN blocks with output dimensions 64 are removed. Similarly, in the case of M-16, the output dimensions of the residual blocks are set to 16 and the CNN blocks with output dimensions 32 and 64 are removed.

3.2 Loss Function

The network is optimized by minimizing the loss function L computed using Eq. 1:

$$L(I_t, I_g) = \lambda_1 \ell_1(I_t, I_g) + \lambda_2 \ell_2(I_t, I_g) + \lambda_3 \ell_3(I_t, I_g). \tag{1}$$

Here, $\ell_1(I_t, I_g) = \|I_t - I_g\|_1$ is the 1-norm loss between the translated image I_t and the ground truth image I_g in YC_bC_r color space for color image cleanup.

For gray scale cleanup, $\ell_1(I_t, I_g) = \|I_t - I_g\|_1$, refer to the 1-norm loss between the two images in gray scale. In addition to pixel-level loss function $\ell_1(I_t, I_g)$, we also employ the perceptual loss functions $\ell_2(I_t, I_g)$ and $\ell_3(I_t, I_g)$ in Eq. 1.

Perceptual loss functions [11,36,53] compute the difference between images I_t and I_g at high level feature representations extracted from a pre-trained CNNs such as those trained on ImageNet image classification task. They are more robust in computing distance between images than pixel-level loss functions. In the context of developing light-weight document image cleanup models, perceptual loss functions serve an additional role of enabling transfer learning. The perceptual loss functions in Eq. 1 helps to transfer the semantic knowledge already learned by the pre-trained CNN network to our smaller network.

The perceptual loss has two components [11]: feature reconstruction loss $\ell_2(I_t, I_g)$ and style loss $\ell_3(I_t, I_g)$. Feature reconstruction loss encourages the transformed image to be similar to ground truth image at high level feature representation as computed by a pre-trained network $\wp$. Let $\wp_j(I)$ be the activations of the jth layer of the pre-trained network $\wp$. Then, Eq. 2 represents the feature reconstruction loss:

$$\ell_2(I_t, I_g) = \frac{1}{H_j W_j C_j} \|\wp_j(I_t) - \wp_j(I_g)\|_1, \tag{2}$$

where the shape of $\wp_j(I)$ is $H_j \times W_j \times C_j$. The feature reconstruction loss penalizes the transformed image when it deviate from the content of the ground truth image. Additionally, we should also penalize the transformed image if it deviate from the ground truth image in terms of common feature, texture, etc. To achieve this style loss is incorporated as proposed in [11]. The style loss is represented in Eq. 3 as follows:

$$\ell_3(I_t, I_g) = \sum_{\forall j \in J} \|G_j^{\wp}(I_t) - G_j^{\wp}(I_t)\|, \tag{3}$$

where $\wp$ represent pre-trained CNN network, J represent set of layers of $\wp$ used to compute style loss, and $G_j^{\wp}(I)$ represent a Gram matrix containing second-order feature covariances. Let $\wp_j(I)$ be the activation of the jth layer of the pre-trained network $\wp$, where the shape of $\wp_j(I)$ is $H_j \times W_j \times C_j$. Then, the shape of the Gram matrix $G_j^{\wp}(I)$ is $C_j \times C_j$ and each element of $G_j^{\wp}(I)$ is computed according to Eq. 4 as follows:

$$G_j^{\wp}(I)_{c,c'} = \frac{1}{H_j W_j C_j} \sum_{h=1}^{H_j} \sum_{w=1}^{W_j} \wp_j(I)_{h,w,c} \wp_j(I)_{h,w,c'}. \tag{4}$$

In our work, we use the $VGG19$ network [42] trained on the ImageNet classification task [4] as our pre-trained network $\wp$ for the perceptual loss. Here, feature reconstruction loss is computed at layer conv1-2 and style reconstruction loss is computed at layers conv1-1, conv2-1, conv3-1, conv4-1, and conv5-1.

Table 1. No. of parameters, product-sum operations, and other statistics of various models. Average per-patch inference time is reported. As an example, a 2560 × 2560 image has 100 patches.

Model	Mult-Adds (in billions)	Parameters (in millions)	Size (in KB)	Inference time (in seconds)	Load time (in seconds)
DE-GAN [43]	46.1	31.00	121215	0.36	9.96
SkipNetModel [52]	106.8	1.64	6380	0.34	3.62
M-64 (proposed)	15.1	0.46	1779	0.24	3.25
M-32 (proposed)	6.7	0.11	445	0.07	2.89
M-16 (proposed)	1.7	0.03	111	0.02	2.68

4   Experimental Results and Discussion

We evaluate the generalization performance of the proposed models on binarization, gray scale, and color cleanup tasks.

Experimental Setup: In our experiment, the input of the network is set as 256 × 256. The input to the network is a 3 channel RGB image, whereas the output dimension of the network is set as 1 or 3 depending on the downstream task. If the downstream task is to obtain an image in gray scale or a binary image, then the output dimension is set as 1. The output dimension is set as 3 for color cleanup task.

To handle different type of noise at various resolution, the training images are scaled at scale 0.7, 1.0, and 1.4. Further at each scale, the training images are divided into overlapping blocks of 256 × 256. During training, a few random patches from the training images are also used for data augmentation using random brightness-contrast, jpeg noise, ISO noise, and various types of blur [3]. Randomly selected 80% of the training patches is used to train the model while the remaining 20% is kept for validation. The model with best validation performance is saved as the final model. The network is optimized using Adam algorithm [13] with default parameter settings. The parameters λ_1, λ_2, and λ_3 of Eq. 1 is set to $1e1$, $1e-1$, and $1e1$ respectively. During inference, an input image is divided into overlapping blocks of 256 × 256. Each patch is inferred using the trained model. Finally, all the patches are merged to obtain the final result. We use simple averaging for the overlapping pixels of the patches.

Compared Algorithms: We compare the proposed models with recently proposed deep CNN based document image cleanup models: SkipNetModel [52] and DE-GAN [43]. Table 1 presents a comparative analysis of our proposed models with SkipNetModel and DE-GAN in terms of: (i) number of multiplication and addition operations (Mult-Adds) associated with the model [10], (ii) number of parameters, (iii) actual size on device, (iv) model load time, and (v) model inference time. A comparison with respect to these parameters is essential if the applicability of any model for memory and energy constrained devices is to

Table 2. Results on DIBCO13 [32]

Model	F-measure	F_{ps}	PSNR	DRD
Otsu [26]	83.9	86.5	16.6	11.0
Sauvola *et al.* [40]	85.0	89.8	16.9	7.6
Tensmeyer *et al.* [46]	93.1	96.8	20.7	2.2
Vo *et al.* [48]	94.4	96.0	21.4	1.8
DE-GAN [43]	99.5	99.7	24.9	1.1
SkipNetModel [52]	95.3	96.6	22.8	1.5
M-64 (proposed)	94.1	95.7	21.7	2.1
M-32 (proposed)	92.3	93.3	20.4	2.5
M-16 (proposed)	90.4	91.6	19.9	3.1

determined. We implemented the models using TensorFlow Lite (https://www.tensorflow.org/lite) on an Android device with Qualcomm SM8150 Snapdragon 855 chipset and 6 GB RAM size. On the device, we observe that our models are 65–1090 and 3–55 times lighter in size than DE-GAN and SkipNetModel, respectively. Similarly, our models has lesser product-sum operations and prediction time during the inference stage, making them suitable to mobile and embedded applications.

4.1 Binarization

We begin by discussing our results on document image binarization task. For our experiment, we have considered the publicly available binarization dataset DIBCO13 [32] and DIBCO17 [35] as test sets. The proposed models and Skip-NetModel are trained on the datasets [7,25,29–31,33,34]. While training the models for the test set DIBCO13 [32], we also include the dataset DIBCO17 [35] into our training data. The models for this task are trained using the augmentation strategy described in Sect. 4. The same training strategy is also followed while training the models for the task DIBCO17 [35]. The models are compared using the DIBCO13 [32] evaluation criteria: F-measure, pseudo F-measure (F_{ps}),

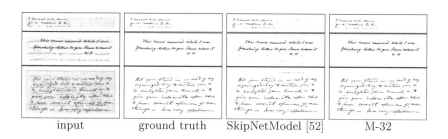

| input | ground truth | SkipNetModel [52] | M-32 |

Fig. 3. Typical examples of DIBCO13 [32].

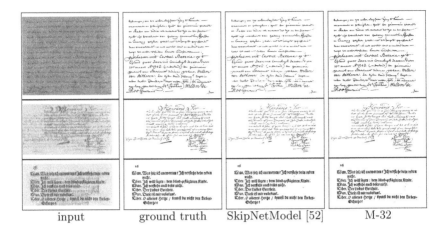

| input | ground truth | SkipNetModel [52] | M-32 |

Fig. 4. Examples from DIBCO17 [35].

peak signal to noise ratio (PSNR), and distance reciprocal distortion (DRD). For the metrics F-measure, F_{ps}, and PSNR, higher values correspond to better performance whereas, in case of the metric DRD lower is better. While evaluating our methods on DIBCO13 dataset, we have compared our methods with traditional binarization algorithms [26,40], state of the art binarization techniques [46,48], DE-GAN [43] and SkipNetModel [52]. Overall performance of these methods are reported in Table 2. In this table, performance of the methods [26,40,43,46,48] are reported as they are reported in [43]. From this table, it is evident that DE-GAN outperforms all other methods in terms of all metrics. However, the proposed method performs better than the traditional binarization algorithms [26,40] and they perform more or less similar to other state of the art techniques. Moreover, from Tables 1 and 2, we can observe that though the proposed method can not outperform the state of the art techniques but they perform similar to most of the state of the art techniques with much lesser computational and memory cost. We have also reported the performance of the proposed methods with the top 5 methods of DIBCO17 competition [35], SkipNetModel and DE-GAN in Table 3. A similar performance of the proposed methods is also observed from this table in comparison to the state of the art techniques. Typical examples from the datasets DIBCO13 and DIBCO17 are shown in Figs. 3 and 4.

4.2   Gray Scale and Color Cleanup

For the purpose of document cleanup, we first show the effectiveness of the proposed methods in gray scale. **The gray scale cleanup** part of our experiment is conducted on the publicly available dataset _N_oisyOffice [5,51]. This dataset consists of two parts, first one is real noisy images consisting of 72 files, and a synthetic dataset consisting of 216 files. There were no groundtruth images

Table 3. Results on DIBCO17 [35]. Here 10, 17a, 12, 1b, and 1a are the top 5 methods from DIBCO 2017 competition [35]

Model	F-measure	F_{ps}	PSNR	DRD
10 [35]	91.04	92.86	18.28	3.40
17a [35]	89.67	91.03	17.58	4.35
12 [35]	89.42	91.52	17.61	3.56
1b [35]	86.05	90.25	17.53	4.52
1a [35]	83.76	90.35	17.07	4.33
DE-GAN [43]	97.91	98.23	18.74	3.01
SkipNetModel [52]	91.13	92.91	18.01	3.22
M-64 (proposed)	90.80	91.73	17.84	3.32
M-32 (proposed)	89.93	90.61	17.32	3.74
M-16 (proposed)	87.81	89.40	16.91	4.15

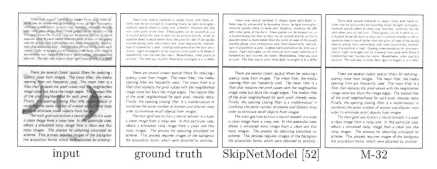

input   ground truth   SkipNetModel [52]   M-32

Fig. 5. Typical examples of noisy images from our test set of synthetic data from [51].

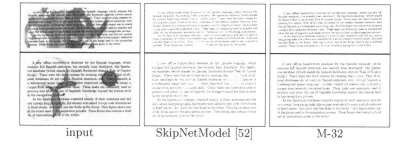

input   SkipNetModel [52]   M-32

Fig. 6. Typical examples of noisy images of real data from [51].

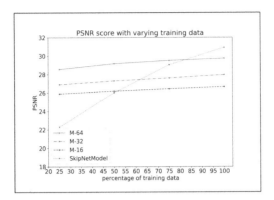

Fig. 7. PSNR scores of the models on NoisyOffice dataset [51] with varying the training data from 25% to 100%

available for the real data, therefore, we are not able to include the real dataset for quantitative analysis of our experiment. The model is trained and evaluated on the synthetic data. We divide the synthetic data into two parts - 172 images for training and 44 images for testing. The authors of [43] did not share the saved model for gray scale cleanup. Therefore, for this experiment, only the method proposed in [52] is used for comparison. To measure the capability of the proposed models with respect to removing noise, we adopt peak signal to noise ratio (PSNR) as the quality metric. In order to determine the dependence of the model performance on the amount of available training data, we have trained the SkipNetModel and the proposed models M-64, M-32, and M-16 by varying the amount of training data from 25% to 100%. The performance of the models with respect to PSNR score is shown in Fig. 7. It can be observed from this figure that in presence of 100% training data, SkipNetModel [52] performs better than the proposed models. However, it can also be observed from this figure that the performance of the proposed models is more or less remains the same. It can also be observed from this figure that the performance of SkipNetModel varies a lot with the variation in the amount of training data. Typical examples of inputs, groundtruths from the test set along with the outputs of the models trained on 100% training data are shown in Fig. 5. We have also shown a few examples of inputs and outputs of the trained models on real data in Fig. 6. It can be seen

Table 4. Color cleanup performance of the proposed model: SSIM and PSNR score of the noisy input images with respect to the groundtruth are 0.87 and 16.3 respectively.

Model	SSIM	PSNR
M-64 (proposed)	0.967	22.8
M-32 (proposed)	0.950	21.4
M-16 (proposed)	0.923	19.8

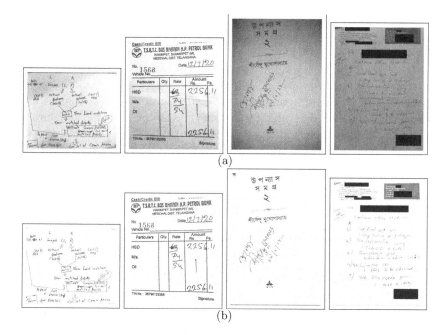

Fig. 8. Typical examples of color clean up. (a) random inputs images; (b) cleaned outputs using M-32 based model on a mobile device

from this figure that the model M-32 performs better than SkipNetModel in few of the examples. From Figs. 6 and 7, we can conclude that the proposed model is more generalized and performs more robustly with respect to SkipNetModel.

Finally, we present our experimental results with respect to document **color cleanup** task. One challenging aspect of color cleanup is the preservation of color of the foreground pixels. For this experiment, we used a color dataset consisting of 250 mobile captured images. Each of the images are manually cleaned. We followed the same training strategy for training our model as described in Sect. 4. Random real life images (not belonging to the train/test set) and their corresponding outputs are shown in Fig. 8. From this figure, we can observe a decent performance of the proposed model in performing color cleanup of document images. However, to provide a quantitative measure of our method, we compute PSNR, and structural similarity index (SSIM) score on the test set of the data in Table 4.

5   Conclusion

We have proposed an encoder-decoder based document cleanup model for resource constrained environments. To this end, we design a light-weight deep network with only a few residual blocks and skip connections. Our loss function incorporates the perceptual loss, which enables transfer learning from pre-trained deep CNN networks.

We develop three models based on our network design, with varying network width. In terms of the number of parameters and product-sum operations, our models are 65–1030 and 3–27 times, respectively, smaller than a recently proposed GAN based document enhancement model [43]. In spite of our relatively low network capacity, the generalization performance of our models on various benchmarks are encouraging and comparable with several document image cleanup techniques with deep architectures such as [46,52]. In addition, our models are more robust to low training data regime than [52]. Hence, the proposed models offer a favorable trade-off between memory/latency and accuracy, making them suitable for mobile document image cleanup applications.

References

1. Afzal, M.Z., Pastor-Pellicer, J., Shafait, F., Breuel, T.M., Dengel, A., Liwicki, M.: Document image binarization using LSTM: a sequence learning approach. In: Proceedings of the 3rd International Workshop on Historical Document Imaging and Processing, pp. 79–84, HIP 2015. Association for Computing Machinery, New York (2015)
2. Bako, S., Darabi, S., Shechtman, E., Wang, J., Sunkavalli, K., Sen, P.: Removing shadows from images of documents. In: Lai, S.-H., Lepetit, V., Nishino, K., Sato, Y. (eds.) ACCV 2016. LNCS, vol. 10113, pp. 173–183. Springer, Cham (2017). https://doi.org/10.1007/978-3-319-54187-7_12
3. Buslaev, A., Iglovikov, V.I., Khvedchenya, E., Parinov, A., Druzhinin, M., Kalinin, A.A.: Albumentations: fast and flexible image augmentations. Information **11**(2), 125 (2020)
4. Deng, J., Dong, W., Socher, R., Li, L.J., Li, K., Fei-Fei, L.: ImageNet: a large-scale hierarchical image database. In: CVPR 2009 (2009)
5. Dua, D., Graff, C.: UCI machine learning repository (2017)
6. Gangeh, M.J., Tiyyagura, S.R., Dasaratha, S.V., Motahari, H., Duffy, N.P.: Document enhancement system using auto-encoders. In: Workshop on Document Intelligence at NeurIPS 2019 (2019)
7. Gatos, B., Ntirogiannis, K., Pratikakis, I.: ICDAR 2009 document image binarization contest (DIBCO 2009). In: 10th International Conference on Document Analysis and Recognition, pp. 1375–1382 (2009)
8. He, K., Zhang, X., Ren, S., Sun, J.: Deep residual learning for image recognition. In: 2016 IEEE Conference on Computer Vision and Pattern Recognition (CVPR), pp. 770–778 (2016)
9. He, S., Schomaker, L.: DeepOtsu: document enhancement and binarization using iterative deep learning. Pattern Recogn. **91**, 379–390 (2019)
10. Howard, A.G., et al.: MobileNets: efficient convolutional neural networks for mobile vision applications (2017)
11. Johnson, J., Alahi, A., Fei-Fei, L.: Perceptual losses for real-time style transfer and super-resolution. In: Leibe, B., Matas, J., Sebe, N., Welling, M. (eds.) ECCV 2016. LNCS, vol. 9906, pp. 694–711. Springer, Cham (2016). https://doi.org/10.1007/978-3-319-46475-6_43
12. Kang, S., Iwana, B.K., Uchida, S.: Cascading modular U-Nets for document image binarization. In: 2019 International Conference on Document Analysis and Recognition (ICDAR), pp. 675–680 (2019)

13. Kingma, P.D., Ba, L.J.: Adam: a method for stochastic optimization. In: International Conference on Learning Representations (2015)
14. Kise, K.: Page segmentation techniques in document analysis. In: Doermann, D., Tombre, K. (eds.) Handbook of Document Image Processing and Recognition, pp. 135–175. Springer, London (2014)
15. Kligler, N., Katz, S., Tal, A.: Document enhancement using visibility detection. In: 2018 IEEE/CVF Conference on Computer Vision and Pattern Recognition, pp. 2374–2382 (2018)
16. Krähenbühl, P., Koltun, V.: Efficient inference in fully connected CRFs with Gaussian edge potentials. In: Advances in Neural Information Processing Systems (2011)
17. Krizhevsky, A.: Convolutional deep belief networks on CIFAR-10 (2010)
18. Krizhevsky, A., Sutskever, I., Hinton, G.E.: ImageNet classification with deep convolutional neural networks. In: Advances in Neural Information Processing Systems (2012)
19. Lazzara, G., Géraud, T.: Efficient multiscale Sauvola's binarization. Int. J. Doc. Anal. Recogn. **17**(2), 105–123 (2014)
20. Li, X., Zhang, B., Liao, J., Sander, P.V.: Document rectification and illumination correction using a patch-based CNN. ACM Trans. Graph. **38**(6), 1–11 (2019)
21. Lin, Y.H., Chen, W.C., Chuang, Y.Y.: BEDSR-Net: a deep shadow removal network from a single document image. In: 2020 IEEE/CVF Conference on Computer Vision and Pattern Recognition (CVPR), pp. 12902–12911 (2020)
22. Liu, N., et al.: An iterative refinement framework for image document binarization with bhattacharyya similarity measure. In: 14th International Conference on Document Analysis and Recognition, pp. 93–98. ICDAR 2017, IEEE Computer Society (2017)
23. Moghaddam, R.F., Cheriet, M.: AdOtsu: an adaptive and parameterless generalization of Otsu's method for document image binarization. Pattern Recogn. **45**(6), 2419–2431 (2012)
24. Mondal, T., Coustaty, M., Gomez-Krämer, P., Ogier, J.: Learning free document image binarization based on fast fuzzy c-means clustering. In: 2019 International Conference on Document Analysis and Recognition (ICDAR), pp. 1384–1389 (2019)
25. Ntirogiannis, K., Gatos, B., Pratikakis, I.: ICFHR 2014 competition on handwritten document image binarization (H-DIBCO 2014). In: 2014 14th International Conference on Frontiers in Handwriting Recognition, pp. 809–813 (2014)
26. Otsu, N.: A threshold selection method from gray-level histograms. IEEE Trans. Syst. Man Cybern. **9**(1), 62–66 (1979)
27. Peng, X., Cao, H., Subramanian, K., Prasad, R., Natarajan, P.: Exploiting stroke orientation for CRF based binarization of historical documents. In: 2013 12th International Conference on Document Analysis and Recognition, pp. 1034–1038 (2013)
28. Peng, X., Wang, C., Cao, H.: Document binarization via multi-resolutional attention model with DRD loss. In: 2019 International Conference on Document Analysis and Recognition (ICDAR), pp. 45–50 (2019)
29. Pratikakis, I., Gatos, B., Ntirogiannis, K.: H-DIBCO 2010 - handwritten document image binarization competition. In: 2010 12th International Conference on Frontiers in Handwriting Recognition, pp. 727–732 (2010)
30. Pratikakis, I., Gatos, B., Ntirogiannis, K.: ICDAR 2011 document image binarization contest (dibco 2011). In: 2011 International Conference on Document Analysis and Recognition, pp. 1506–1510 (2011)

31. Pratikakis, I., Gatos, B., Ntirogiannis, K.: ICFHR 2012 competition on handwritten document image binarization (H-DIBCO 2012). In: 2012 International Conference on Frontiers in Handwriting Recognition, pp. 817–822 (2012)
32. Pratikakis, I., Gatos, B., Ntirogiannis, K.: ICDAR 2013 document image binarization contest (DIBCO 2013). In: 2013 12th International Conference on Document Analysis and Recognition, pp. 1471–1476 (2013)
33. Pratikakis, I., Zagori, K., Kaddas, P., Gatos, B.: ICFHR 2018 competition on handwritten document image binarization (H-DIBCO 2018). In: 2018 16th International Conference on Frontiers in Handwriting Recognition (ICFHR), pp. 489–493 (2018)
34. Pratikakis, I., Zagoris, K., Barlas, G., Gatos, B.: ICFHR 2016 handwritten document image binarization contest (H-DIBCO 2016). In: 2016 15th International Conference on Frontiers in Handwriting Recognition (ICFHR), pp. 619–623 (2016)
35. Pratikakis, I., Zagoris, K., Barlas, G., Gatos, B.: ICDAR 2017 competition on document image binarization (DIBCO 2017). In: 2017 14th IAPR International Conference on Document Analysis and Recognition (ICDAR), vol. 1, pp. 1395–1403 (2017)
36. Rad, M.S., Bozorgtabar, B., Marti, U., Basler, M., Ekenel, H.K., Thiran, J.: SROBB: targeted perceptual loss for single image super-resolution. In: International Conference on Computer Vision (2019)
37. Ronneberger, O., Fischer, P., Brox, T.: U-net: Convolutional networks for biomedical image segmentation. In: Medical Image Computing and Computer-Assisted Intervention (2015)
38. Russakovsky, O., et al.: ImageNet large scale visual recognition challenge. Int. J. Comput. Vis. **115**(3), 211–252 (2015)
39. Jung, S., Hasan, M.A., Kim, C.: Water-filling: an efficient algorithm for digitized document shadow removal. In: Jawahar, C.V., Li, H., Mori, G., Schindler, K. (eds.) ACCV 2018. LNCS, vol. 11361, pp. 398–414. Springer, Cham (2019). https://doi.org/10.1007/978-3-030-20887-5_25
40. Sauvola, J., Pietikäinen, M.: Adaptive document image binarization. Pattern Recogn. **33**, 225–236 (2000)
41. Silva, J.M.M.D., Lins, R.D., Martins, F.M.J., Wachenchauzer, R.: A new and efficient algorithm to binarize document images removing back-to-front interference. J. Univ. Comput. Sci. **14**(2), 299–313 (2008)
42. Simonyan, K., Zisserman, A.: Very deep convolutional networks for large-scale image recognition (2014)
43. Souibgui, M.A., Kessentini, Y.: DE-GAN: a conditional generative adversarial network for document enhancement. IEEE Trans. Pattern Anal. Mach. Intell. 1–12 (2020). Early access
44. Tabatabaei, S.A., Bohlool, M.: A novel method for binarization of badly illuminated document images. 17th IEEE International Conference on Image Processing, pp. 3573–3576 (2010)
45. Teichmann, M., Cipolla, R.: Convolutional CRFs for semantic segmentation. In: British Machine Vision Conference (2019)
46. Tensmeyer, C., Martinez, T.: Document image binarization with fully convolutional neural networks. In: 14th International Conference on Document Analysis and Recognition, pp. 99–104, ICDAR 2017. IEEE Computer Society (2017)
47. Valizadeh, M., Kabir, E.: An adaptive water flow model for binarization of degraded document images. Int. J. Doc. Anal. Recogn. **16**, 1–12 (2013)
48. Vo, G.D., Park, C.: Robust regression for image binarization under heavy noise and nonuniform background. Pattern Recogn. **81**, 224–239 (2018)

49. Wang, B., Chen, C.L.P.: An effective background estimation method for shadows removal of document images. In: 2019 IEEE International Conference on Image Processing (ICIP), pp. 3611–3615 (2019)

50. Wang, J., Chuang, Y.: Shadow removal of text document images by estimating local and global background colors. In: ICASSP 2020–2020 IEEE International Conference on Acoustics, Speech and Signal Processing (ICASSP), pp. 1534–1538 (2020)

51. Zamora-Martínez, F., España-Boquera, S., Castro-Bleda, M.J.: Behaviour-based clustering of neural networks applied to document enhancement. In: Sandoval, F., Prieto, A., Cabestany, J., Graña, M. (eds.) Computational and Ambient Intelligence, pp. 144–151. Springer, Heidelberg (2007)

52. Zhao, G., Liu, J., Jiang, J., Guan, H., Wen, J.: Skip-connected deep convolutional autoencoder for restoration of document images. In: 2018 24th International Conference on Pattern Recognition (ICPR), pp. 2935–2940 (2018)

53. Zhao, H., Gallo, O., Frosio, I., Kautz, J.: Loss functions for image restoration with neural networks. IEEE Trans. Comput. Imaging **3**(1), 47–57 (2017)

Office Automation

A New Semi-automatic Annotation Model via Semantic Boundary Estimation for Scene Text Detection

Zhenzhou Zhuang[1], Zonghao Liu[1], Kin-Man Lam[2], Shuangping Huang[1,3(✉)], and Gang Dai[1]

[1] School of Electronic and Information Engineering,
South China University of Technology, Guangzhou, China
{zhenzhouzhuang,zonghaoliu_work}@foxmail.com, eehsp@scut.edu.cn,
eedaigang@mail.scut.edu.cn
[2] Department of Electronic and Information Engineering,
The Hong Kong Polytechnic University, Hong Kong, China
enkmlam@polyu.edu.hk
[3] Pazhou Lab, Guangzhou, China

Abstract. Manually annotating a data set for scene text detection is extremely time-consuming. In this paper, we propose a new semi-automatic annotation model to produce tight polygonal annotations for text instances in scene images, based on the input of manually annotated text center lines. Our approach first generates multiple candidate boundaries, which share the same input center line. Then, by training a fastidious content recognizer, optimal boundary selection is performed. The bounded text region, which achieves the smallest recognition loss, is selected as the tightest of the text. As this optimal boundary estimation is guided by semantic recognition, our method is called Semantic Boundary Estimation. Experiment results show that only half clicks compared to manually annotated polygon, are input to annotate center line, and precise polygon text region annotation is automatically produced. A high recall of more than 95% at IoU > 0.5 and 80% at IoU > 0.7 is achieved, demonstrating the high agreement with the original ground truth. In addition, using the generated annotations on benchmarks, such as Total-Text, CTW1500 and ICDAR2015, to train state-of-the-art detectors can achieve similar performance to those trained with manual annotations. This further verifies the good annotation performance. A annotation toolkit based on the proposed model is available at CenterlineAnnotation.

Keywords: Semi-automatic annotation algorithm · Scene text detection · Semantic boundary estimation

1 Introduction

Detecting texts in natural scene images plays a critical role in a wide range of applications, such as traffic monitoring, multimedia retrieval, semantic natural scene understanding [24,27], etc. Recently, deep learning methods have made

© Springer Nature Switzerland AG 2021
J. Lladós et al. (Eds.): ICDAR 2021, LNCS 12823, pp. 257–273, 2021.
https://doi.org/10.1007/978-3-030-86334-0_17

remarkable achievements in scene text detection [9,19,20,25,28,36–38,41,43,46]. However, these methods require a large amount of data with high-quality annotations for training.

So far, well-labeled data for the task of scene text detection is obtained basically by manual annotation, which is a laborious process. For example, to annotate a multi-oriented text region, 4 points are often required to draw the polygonal bounding box. For a more complicated curved text, more than 14 points are required to represent the bounding box. According to [45], manually annotating a text instance usually needs $13 - 40$ s to click all the vertices of the polygonal outline.

To reduce the annotation time, two main approaches are available in the field of computer vision. One approach is to use existing text detectors, such as MSER [7] and EAST [46], to give some annotation suggestions [1,8], followed by manual adjustments and corrections to obtain tight boundaries. Due to the limited performance of the detectors, users still need to carefully check all the annotations and re-annotate the missing or incorrect bounding boxes. Another approach attempts to reduce the annotation time in a semi-automatic way. For example, Interactive Instance Segmentation (IIS) models [31] are the widely used semi-automatic method for the segmentation task [47]. This approach can provide accurate annotation with little manual interaction, such as requiring a few points, or an easy-drawn region. The IIS-based algorithms [5,32,34] work normally under the assumption of intensity homogeneity and continuity within the region to be annotated, and hence are usually used for the object-segmentation tasks. However, the intensity inside a text region is inhomogeneous and discontinuous, which means that IIS is not suitable for text annotation.

In this paper, we develop a new semi-automatic annotation model for scene text detection, i.e., Semantic Boundary Estimation (SBE). Simply providing a manually annotated text center line as input (as shown in Fig. 5), this model can automatically produce tight polygonal annotations for the text instances. Different from the existing manual tools that click all the vertices of the polygonal bounding boxes, our method can reduce the number of human clicks by half [40,45], while achieving high agreement with the original ground truths. The contributions of this paper are as follows:

- We propose a new semi-automatic annotation model for scene text detection, which can reduce manual annotation costs by half, while achieving high agreement with original ground truths.
- We introduce a new concept, namely Semantic Boundary, which is the optimum region boundary for a text recognizer to understand its content. In particular, a pre-trained fastidious recognizer is used to find the semantic boundary from multiple candidate boundaries.
- Intersection-over-Union (IoU) between the predicted and ground-truth polygons are computed over several text detection benchmarks, demonstrating the effectiveness of the proposed model. Detectors trained with the generated annotations can achieve similar performance to those trained with manual annotations.

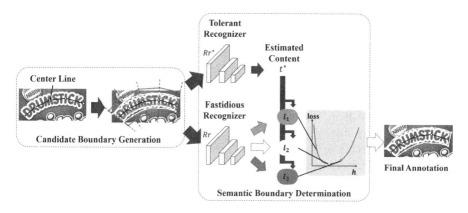

Fig. 1. The pipeline of our proposed Semantic Boundary Estimation. The black curve is the text center line. For demonstration, the generated multiple candidate boundaries are highlighted in different colors (in red, yellow and blue in this example). All the polygons, surrounded by the candidate boundaries, are fed into both the specifically trained Tolerant Recognizer and Fastidious Recognizer. The former recognizer produces an estimated text content, while the latter recognizer computes the recognition loss of all the candidate regions, with respect to the estimated text content t^*. The candidate with the smallest recognition loss (the yellow one in this example) will be determined as the final annotation. (Color figure online)

2   Related Works

Annotation Tools for Text Detection. Annotation tools for text detection are relatively rare. The early annotated scene text dataset used the tools designed for object detection[16,35], which only offered a platform to manually annotate axis-aligned or multi-oriented text regions. Recently, the manual annotation tools provided by [26] can support three annotation modes: axis-aligned, arbitrary quadrilateral, and curved quadrilateral. For axis-aligned text, the tool only needs to drag a box from the top-left corner to the bottom-right corner of a text region. For quadrilateral text, the tool needs to mark the four vertices of a quadrilateral to represent the text region. For curved text, the tool needs manually annotating 14 vertices of a text polygon. To reduce the workload of annotation, the tools in [1,8] introduced some text detectors [7,9,19,20,46] to provide some annotation suggestions [1,7] for images with scene text. However, due to the imperfect detection performance, users still need to carefully check all the annotations and re-annotate the missing or incorrect bounding boxes. Our model can reduce the annotation costs by just manually drawing the center line of scene text. And the model still obtains a tight text region with higher labeling efficiency.

Interactive Instance Segmentation. The irregular-shape annotations for instance segmentation tasks [47] are difficult to obtain. Some researchers attempted to solve this annotation difficulty in a human-in-loop way, i.e., Interactive Instance Segmentation (IIS) [31]. Interactive algorithms were developed, which can reduce annotation costs by using manual interaction, such as clicking some points, or drawing a simple region. Specifically, previous algorithms [5,32] rely on intensity homogeneity and continuity within the region to be annotated. Graph-Cut [5] models were used to extract homogeneous regions for annotation. Recently, deep learning methods have been used in IIS. For example, the seed-based methods [13,15,18,21,22] usually interactively inject user clicked points into Gaussian heat-map. Together with the image to be annotated, these maps are fed into a segmentation model for object segmentation. The ROI-based methods [2,6,23,44] used the bounding-box region provided by users as input, and predict the segmentation regions in the form of polygon's vertices or an irregular mask. Furthermore, [2,6,23] use sequential model or GCN model to perform semi-automatic adjustments, and speed-up the refining process when segmentation results are not accurate. Our work is inspired by the idea of combining intelligent algorithms with manual interaction for annotation. However, we can not directly use the IIS algorithm, due to the lack of homogeneity and continuity within the text region. In this paper, we propose to use semantic information to guide the automatic generation of a polygon-shaped text region, with only one center line annotated manually.

3 Methodology

The overall process of SBE is shown in Fig. 1. There are two main processes in SBE, which are "Candidate Boundary Generation" and "Semantic Boundary Determination" (SBD). In the first process, by utilizing the input text center line, multiple candidate boundaries are generated. Then, in the process of "Semantic Boundary Determination", the regions bounded by the above candidates are fed into the specifically trained Tolerant Recognizer and Fastidious Recognizer simultaneously. The former recognizer gives an estimated text content, while the latter recognizer computes the recognition loss of all the candidate regions, with respect to the estimated text content. Finally, if the text region surrounded by a boundary has the minimum recognition loss, it will be selected as the final annotation. As the text boundary is selected with the help of the text semantic content, we call it Semantic Boundary.

3.1 Candidate Boundary Generation

The process of Candidate Boundary Generation is illustrated in Fig. 2. Let $\{c_i | i = 1, 2, ..., K\}$ denote K points on the input text center line. K normal lines, passing through these K points, are constructed, and are denoted as $\{n_i | i = 1, 2, \ldots, K\}$. Then, a candidate boundary can be represented by a sequence of the normal line segments(as highlighted in blue in Fig. 2). The length of the segment n_i is denoted as h_i, and the point c_i bisects the segment. By connecting all the end points of the segments in turn, a candidate boundary is generated. In other words, a polygonal boundary can be depicted as a series of normal line segments, which has variable lengths and orientations. The orientation of the normal line n_i is perpendicular to the tangent of the center line at

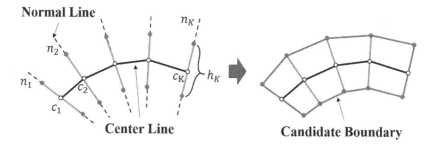

Fig. 2. The process of Candidate Boundary Generation. The black dash lines are normal to the center line. In each of the normal directions, a line segment (highlighted in blue), accompanied with a pair of end points(highlighted as red dot), is bisected by the center line (the black line). Then, by connecting all the adjacent points in turn, the candidate boundary is generated, as illustrated by the red polygon. (Color figure online)

c_i, so it is determined by the input. The length of n_i, i.e., h_i is estimated by the Semantic Boundary Determination process (as described in Sect. 3.2).

Therefore, each candidate boundary can be parameterized by a vector $\mathbf{H} = (h_1, h_2, \ldots, h_K)$. In this section, we will first discuss the generation of a candidate boundary, with a simple setting, i.e., a uniform $\mathbf{H}$, with $h_1 = h_2 = \ldots = h_K = h$. Then, in Sect. 4, each h_i will be adjusted under the guidance of the recognition loss to obtain a tighter boundary, as shown in Fig. 4(a).

For a uniform vector, the Candidate Boundary Generation is controlled by h. To generate candidates with different scales, we need to set a large range for h, which will lead to more computation resources. Our solution is to search the ratio λ of h to L, i.e. $\lambda = h/L$, where L denotes the length of the given center line. Most text regions have a shorter h than their center line, so $\lambda \leq 1$. In considering some exception cases, e.g., for the single text 'I', we sample λ in the range $(0, 1.5]$.

3.2 Semantic Boundary Determination

Semantic Boundary and Text Annotation. Our goal is to find a tight boundary surrounding the target text region. Enlarging this tight boundary will include more background noise, while shrinking it will exclude some parts of the text. Both these operations will increase the recognition loss. As the tight boundary will result in minimum recognition loss, with respect to the text semantic content, the tight boundary is called as the "semantic boundary".

SBD takes the regions surrounded by each candidate boundary as its input. The Fastidious Recognizer, denoted as Rr, will read out text content inside the candidate boundaries, and compute the corresponding recognition loss. The candidate boundary with the smallest recognition loss is selected, and the corresponding text region is output as the text annotation.

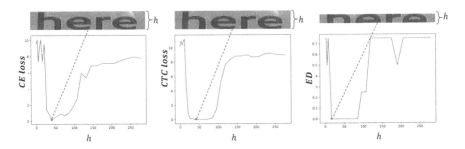

Fig. 3. Effect of different recognition loss functions: CE loss, CTC loss, and ED. The red point on the curve is the minimum loss point. The text region linked by the dash line is the selected region corresponding to the point of minimum loss. The quantization effect of ED cannot tell the difference between the tight region and the over-cut region. (Color figure online)

Fastidious Recognizer. One of the main factors that affects the performance of SBD for annotation is the Fastidious Recognizer. The recognizer is specially trained to be sensitive to interference, such as background noise, partial occlusion of text, etc. This idea is completely different from the design of those usual text recognizers [3,4,29,33,42], which strive to be robust to all kinds of noisy text. Specifically, we design the recognizer from the following two aspects:

1. The architecture of the recognizer. We follow [3] to select a normal encoder-decoder framework as the architecture of the Fastidious Recognizer. It consists of an encoder, a sequence model and a decoder [3], with the rectification subnetwork, such as TPS [39], being removed. This is because introducing rectification to improve the robustness of a text recognizer to irregular shapes will lead to insensitiveness to the tightness of the text boundary, and thus, yield loose annotation in our situation.
2. The training of the recognizer. We train the Fastidious Recognizer with tightly annotated text regions, which can be easily implemented, as we use the synthetic data set, SynthText [11], in the training process. SynthText provides character location for every word. For each word, we crop a minimum-area quadrilateral surrounding all its characters, and construct a training set with tight regions for the recognizer. This training strategy is different from that adopted by most scene text recognizers [3,4,29,33,42], whose recognition models are trained with axis-aligned annotated regions, thus inevitably introducing background noise and becoming robust to the interference. However, this robustness does not benefit our Fastidious Recognizer in becoming tightness-sensitive. Detailed analysis about tightness will be discussed in Sect. 5.3.

Loss Function. Another factor that affects the performance of SBE annotation is the recognition loss function being used. There are three kinds of recognition loss functions being adopted by existing text recognizers: Cross Entropy Loss (CE), Connectionist Temporal Classification Loss (CTC) [10] and Edit Distance

(ED). CE and CTC are both continuous loss, while ED is discrete. The discrete loss turns the softmax output of the recognizer into a hard code, such as the one-hot code in the calculation of loss. Such quantization effect will distort the measurement of the recognition loss. This will cause a loose or over-cut annotation (as shown in Fig. 3). Therefore, we choose the continuous loss, i.e., CE and CTC, to be our loss function.

Estimated Content. The Fastidious Recognizer requires a text content to compute the recognition loss. Manually typing the text content is a good way to provide an accurate recognition supervision, but this process is time-consuming and laborious.

To this end, we use a Tolerant Recognizer, denoted as Rr^* in Fig. 1, to estimate the text content. In this solution, the multiple candidate boundaries generated in Sect. 2 are taken as input. By feeding these candidates into our Tolerant Recognizer, multiple predicted content, denoted as $\{s_1, s_2, ..., s_N\}$, are obtained. Then, one of the predicted contents is selected as the estimated content.

Our method is to choose the prediction s_i that has the minimum difference with the adjacent prediction s_{i+1}. The difference d_i is computed as follows:

$$d_i = |l(s_i, s_{i+1})| \tag{1}$$

where $l(\cdot)$ is the recognition loss function, such as CE loss and CTC loss. We regard s_{i+1} as the pseudo ground-truth of s_i, in order to compute difference d_i. The recognition result s_{i*} of the minimum difference, where $i^* = arg\min_i d_i$, will be used as the estimated text content.

Tolerant Recognizer. To make the estimated text content as accurate as possible, the Tolerant Recognizer should be as robust as possible to all kinds of interference. For example, a bigger text region will result in introducing more background. To this end, the rectification subnetwork is included in the Tolerant Recognizer. This is the difference from the Fastidious Recognizer. In addition, we train this Tolerant Recognizer with rectangular axis-aligned text regions that are cropped from SynthText [14] and Synth90k [14].

4   Tighter Boundary

As stated in Sect. 3, we produce a semantic boundary by SBE, using a uniform vector **H**. However, it cannot tightly surround an irregular text with a more complicated shape, as an example shown in Fig. 4(a). To this end, we propose a refining process, called Tighter Boundary Estimation (TBE), to obtain a tighter boundary. **H** is still updated under the guidance of the recognition loss as adopted in SBE, yet a step-by-step strategy is used here. In every updating step, the updated **H** will reduce the loss of the region. Therefore, the recognition loss, based on our Fastidious Recognizer, is taken as the objective function, and

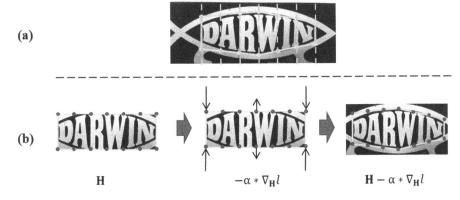

Fig. 4. (a) An irregular text with non-uniform **H**. (b) The process for optimizing **H**. We first obtain $-\alpha * \nabla_{\mathbf{H}} l$ to represent a non-uniform updating vector, which is the black arrow. Then, we update **H** to generate a tighter region.

H is optimized with Gradient Descent. Each updating step can be summarized as follows:

$$\mathbf{H} = \mathbf{H} - \alpha * \nabla_{\mathbf{H}} l \qquad (2)$$

where l is the recognition loss with respect to the estimated content, $\nabla_{\mathbf{H}} l$ is the gradient of the loss with respect to the vector **H**, and α is the updating stride. The minus gradient $-\nabla_{\mathbf{H}} l$ gives **H** a non-uniform updating vector, to make the updated boundary tighter(as the black arrows in Fig. 4(b)). Since the loss l is not differentiable to **H**, we propose an approximate calculation of $\nabla_{\mathbf{H}} l$ as follow:

1. For each h_i in $\mathbf{H} = (h_i | i = 1, 2, ..., K)$, we first add a small $\Delta h = 0.15 h_i$ to h_i, then generate a new candidate with the new **H**.
2. Then we compute the recognition-loss difference Δl_i between the new candidate and the original one.
3. Finally, the gradient $\nabla_{\mathbf{H}} l$ can be calculated as follows:

$$\nabla_{\mathbf{H}} l = (\frac{\Delta l_1}{\Delta h}, \frac{\Delta l_2}{\Delta h}, ..., \frac{\Delta l_K}{\Delta h}) \qquad (3)$$

The calculation of $\frac{\Delta l_i}{\Delta h}$ requires double times of the Fastidious Recognizer's forward to compute the loss of the new candidate and the original one. In order to save time, we only calculate $\frac{\Delta l_1}{\Delta h}$, $\frac{\Delta l_{\lfloor K/2 \rfloor}}{\Delta h}$ and $\frac{\Delta l_K}{\Delta h}$, and the other components in $\nabla_{\mathbf{H}} l$ are obtained by using linear interpolation. In addition, we only update the **H** 10 times in our TBE.

5 Experiment

5.1 Experiment Setup

Table 1. The annotation quality of SBE in terms of IoU between the generated and ground-truth annotations and the detection performance. **mIoU, IoU > 0.7** and **IoU > 0.5** represent the mean IoU(%), and the recall(%) at IoU > 0.5 and IoU > 0.7. **P, R** and **F** represent *Precision*(%), *Recall*(%) and *F1 measure*(%), respectively, evaluating the detector's performance. **ACC** represents the recognition accuracy.

Datasets	Annotation approach	Annotation quality			DB [20]			ABCNet [25]			TRBA [3]	TRBC [3]
		mIoU	IoU > 0.7	IoU > 0.5	P	R	F	P	R	F	ACC	ACC
ICDAR2015	Center Line	80.1	82.2	95.9	90.9	79.3	84.7	88.0	81.6	84.7	70.8	67.5
	GT polygon	–	–	–	91.8	83.2	87.3	89.1	80.2	84.6	70.2	67.0
Total-text	Center line	79.5	81.9	96.8	87.2	81.0	84.0	87.6	84.5	86.0	76.9	74.1
	GT polygon	–	–	–	87.1	82.5	84.7	87.2	85.9	86.5	77.0	74.1
CTW1500	Center line	79.9	82.2	96.1	84.0	80.4	82.2	87.0	81.4	84.1	48.0	47.6
	GT polygon	–	–	–	86.9	80.2	83.5	87.9	81.5	84.6	48.8	48.2

Dataset. The performance of our model is evaluated on scene text datasets: Total-Text [8], CTW1500 [26] and ICDAR2015 [17], which include multi-oriented and curve texts. All of them have about 1,000 training images, while both CTW1500 and ICDAR2015 contain 500 testing images, and Total-Text contains 300 testing images. Since these datasets only offer polygonal annotations, we generate text center-line annotations based on the origin polygonal annotations for all the training sets, and treat these center lines as the input of SBE.

Evaluation Measure. To evaluate the quality of our annotated polygon boxes, we follow [30] to compute intersection-over-union (IoU) of the generated annotations with respect to the original ground-truth. On the one hand, we calculate the average IoU for all instances and average them across the whole dataset to obtain the mean IoU (mIoU). On the other hand, we measure the recall of annotated text polygons with IoU greater than 0.5 and 0.7(IoU > 0.5, IoU > 0.7), respectively. For further evaluation of the annotation quality, we train the text detectors DB [20] and ABCNet [25] with generated annotations for measuring their detection performance, in terms of *Precision*, *Recall* and *F1 measure*. We also evaluate the quality of the generated annotations through recognition experiments. Two pretrained recognition models proposed by [3], TPS-ResNet-BiLSTM-Attn(TRBA) and TPS-ResNet-BiLSTM-CTC(TRBC), are used to recognize the contents of the generated regions and ground-truth regions. We compare the recognition accuracies between the generated annotations and the ground-truth.

Implementation Details. The text recognizer proposed in [3] with a ResNet [12] encoder and a CTC [10] decoder is used to form the main architecture of the Fastidious Recognizer and the Tolerant Recognizer. We optimize the recognizers with the ADAM algorithm, and set the initial learning rate as 1, momentum as 0.9, minibatch size as 64 and maximum iterations of 3 million. We traverse λ in a value set $\{\lambda | \lambda = x^2\}$ to generate multiple candidate boundaries, where x increases uniformly from 0 to $\sqrt{1.5}$ in 30 steps. For training the detectors, we use the same training setting, including the learning rate, momentum, maximum iterations and other training details as [20] and [25]. We implement all the models using PyTorch and train them using Intel i7-7700K CPU, 64G RAM and 4 NVIDIA GTX1080Ti GPUs.

Center Line

Annotation Result

Fig. 5. Qualitative results of the proposed semi-automatic SBE annotation model. Top row: input center line highlighted in green. Bottom row: generated polygons highlighted in blue, and the corresponding ground-truth polygons in green. (Color figure online)

5.2   Annotation Results

Figure 5 shows some example predictions obtained by the proposed SBE model. We can see that a high annotation agreement with human is achieved. We can obtain visually satisfactory performance just with the use of only a half number of clicks to annotate the center line. Furthermore, the quantitative results, in terms of IoU, are summarized in Table 1. The generated annotations have more or less 80% mean IoU with respect to the ground-truth of all three benchmarks. Moreover, the recall for IoU greater than 0.5 can reach more than 95%. The recall for IoU greater than 0.7 can reach about 82%. Both DB [20] and ABCNet [25] achieve a similar, or even better *Precision*, *Recall* and *F1 measure*, when they are trained with generated annotations, compared to the detectors trained with fully manual annotations. And the recognition accuracy of generated text regions is close to that of ground truth ones. To sum up, both the qualitative or quantitative evaluations demonstrate that the generated annotations are highly consistent with the ground truth given manually.

5.3   Ablation Study

Fastidious Recognizer. We train several versions of the Fastidious Recognizer with different degrees of tightness of the text regions. As shown in Fig. 6, assume that we have a tight quadrilateral region, with a vertex denoted as p_i. To simulate different degrees of tightness, we sample a new vertex q_i to replace p_i, resulting in a polygon bigger than the text region. Specifically, $q_i = p_i + (1 - \delta) \times t_i$, where t_i is a random vector. It starts from the point p_i and ends at a random point distributed in the region A_i, which is a region with the same shape as the tight region, with one corner at p_i, as the blue shaded region in Fig. 6. Then, $\delta \in [0, 1]$ is to control the tightness of the text region. When δ is larger, we obtain a tighter instance. These text regions, with different degrees of tightness, are used to train different versions of the Fastidious Recognizer, and we can obtain different generated annotations with SBE. Recalls at IoU > 0.5 and IoU > 0.7 are measured and plotted in Fig. 6. We can see that using looser text regions used to train the Fastidious Recognizer, will make the annotation quality worse.

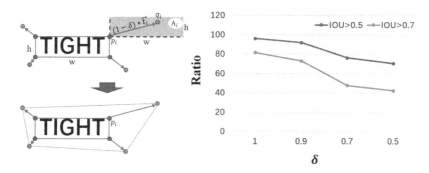

Fig. 6. Left: Illustrating the process of generating loose regions. The red arrows are the random vectors t_i directing the point p_i to a new position in the region A_i. The tightness parameter $\delta \in [0, 1]$ is used to control the length of the arrows. The region bounded by the green lines is a loose region, and the blue lines form a tight region. Right: The curve of recall at IoU > 0.5 and IoU > 0.7 versus δ. The evaluation was conducted on ICDAR2015. (Color figure online)

Without Tighter Boundary Estimation. Figure 7 shows example annotations obtained by the proposed SBE model with and without TBE for further refining. Furthermore, the quantitative comparisons in terms of IoU and the detectors' performances, are summarized in Table 2. As shown in Fig. 7, the annotations generated by TBE(polygons in blue) have more agreement with the ground-truth(polygons in green) than the annotations generated without TBE(polygons in red). As shown in Table 2, on CTW1500, the generated annotations with TBE outperform those without TBE by 2.3% and 0.7% on recall at IoU > 0.7 and IoU > 0.5, respectively. Detector DB [20] trained with TBE-generated annotations outperforms the one trained with only SBE-generated annotations by 1.6% and 1.1% on precision and F-measure, respectively.

Table 2. The annotation quality with and without TBE. Experiments were conducted on CTW1500.

Method	Annotation quality			DB[20]		
	mIoU	IoU > 0.7	IoU > 0.5	P	R	F
SBE without TBE	79.7	79.9	95.4	82.5	79.7	81.1
SBE with TBE	**79.9**	**82.2**	**96.1**	**84.1**	**80.3**	**82.2**

5.4   Discussion

Fig. 7. Qualitative annotation results with and without TBE. The generated polygons with TBE are highlighted in blue, the generated polygons without TBE are highlighted in red, and the corresponding ground-truth polygons are in green. (Color figure online)

Center Line Deviation. To evaluate the robustness of our SBE to the deviation of the center line, we simulate the center-line annotation deviation by a K-dimensional random vector sampled from a Gaussian distribution $N(0, \sigma \times h_m \times \mathbf{I})$, where $\mathbf{I}$ is the $K \times K$ identity matrix, and h_m is the maximum normal line-segments length of each text instance, and K is the number of points used to annotate a center line. The random vector represents the jitter amplitude of the points on the center line in its normal direction. σ is used to control the degree of deviation. The larger the σ is, the more deviation the center line will have. The center lines and the corresponding annotation quality at different noise levels are shown in the upper and the lower parts of Fig. 8. It can be seen that the proposed SBE has only a slight performance degradation, when the degree of noise changes from 0 to 0.2. In other words, our method is robust to the deviation of center line.

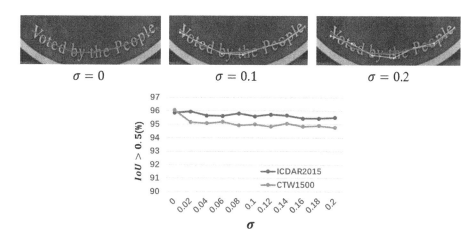

Fig. 8. Top row: Center line under different degrees of noise. The red line is the center line generated from the ground-truth. The lines in other colors have deviated with a different σ. Visually, when $\sigma = 0.2$, some points on the center lines are very close to the boundary of the text region, which is generally unlikely happen in the manual annotation process. Bottom row: The annotation quality at different degrees of center-line deviation. (Color figure online)

Table 3. The annotation quality with/without human-drawn center lines.

Dataset	Human-drawn center line	mIoU	IoU > 0.7	IoU > 0.5
Subset of IC15	Y	70.4	78.2	93.9
	N	74.1	79.6	94.5
Subset of total-text	Y	71.6	77.1	91.3
	N	72.9	77.7	90.9
Subset of CTW1500	Y	79.9	87.2	98.3
	N	78.7	85.6	98.3

User Study. We further set up a user study to evaluate the performance of SBE when using human-drawn center lines. For each experiment dataset, we randomly select 50 images and annotate the text center lines manually. Then, SBE takes these center lines as input and generates polygon annotations. For the same batch of images, annotation results based on center lines that generated from ground-truth are used for comparison. As shown in Table 3, no matter whether the manually marked center line is adopted or not, SBE achieves more than 90% recall at IoU > 0.5, and about 80% recall at IoU > 0.7. Compared to results in Table 1, SBE has a performance degradation in subset of IC15 and Total-Text due to the sampling bias between the subset and the trainset. In a word, the proposal method is still effective when using human-drawn center lines.

6 Conclusion

In this paper, we propose a new semi-automatic annotation algorithm, i.e., Semantic Boundary Estimation, for text detection. Our method only needs to take manually annotated center line as input, then can automatically generate high-quality text region annotation. In particular, the annotation is represented by a series of normal line segments, bisected by the input center line, to depict a polygonal boundary, which can be of variable orientations and lengths. The orientation is perpendicular to the tangent line of the center line, which is completely determined by the input center line. While the length is estimated by the guidance of text content recognition. To this end, a Fastidious Recognizer is specifically trained to be sensitive to boundary tightness, and is used to select a semantic boundary that results in the smallest recognition loss. This can guarantee the polygon text region to be the tightest as final annotation. Experiment results have demonstrated that the generated annotations by SBE have a high agreement with manual annotations.

Acknowledgements. This research is supported in part by Guangdong Basic and Applied Basic Research Foundation (No. 2021A1515012282) and the Alibaba Innovative Research (AIR) Program.

References

1. https://github.com/PaddlePaddle/PaddleOCR#
2. Acuna, D., Ling, H., Kar, A., Fidler, S.: Efficient interactive annotation of segmentation datasets with polygon-RNN++. In: Proceedings of the IEEE conference on Computer Vision and Pattern Recognition, pp. 859–868 (2018)
3. Baek, J., et al.: What is wrong with scene text recognition model comparisons? Dataset and model analysis. In: Proceedings of the IEEE International Conference on Computer Vision, pp. 4715–4723 (2019)
4. Bartz, C., Bethge, J., Yang, H., Meinel, C.: Kiss: keeping it simple for scene text recognition. arXiv preprint arXiv:1911.08400 (2019)
5. Boykov, Y.Y., Jolly, M.P.: Interactive graph cuts for optimal boundary & region segmentation of objects in nd images. In: Proceedings Eighth IEEE International Conference on Computer Vision, ICCV 2001, vol. 1, pp. 105–112. IEEE (2001)
6. Castrejon, L., Kundu, K., Urtasun, R., Fidler, S.: Annotating object instances with a polygon-RNN. In: Proceedings of the IEEE Conference on Computer Vision and Pattern Recognition, pp. 5230–5238 (2017)
7. Chen, H., Tsai, S.S., Schroth, G., Chen, D.M., Grzeszczuk, R., Girod, B.: Robust text detection in natural images with edge-enhanced maximally stable extremal regions. In: 2011 18th IEEE International Conference on Image Processing, pp. 2609–2612. IEEE (2011)
8. Ch'ng, C.K., Chan, C.S.: Total-text: toward orientation robustiness in scene text detection. Int. J. Doc. Anal. Recogn. (IJDAR) **23**, 31–52 (2019)

9. Deng, D., Liu, H., Li, X., Cai, D.: Pixellink: Detecting scene text via instance segmentation. In: Proceedings of the AAAI Conference on Artificial Intelligence, vol. 32 (2018)
10. Graves, A., Fernández, S., Gomez, F., Schmidhuber, J.: Connectionist temporal classification: labelling unsegmented sequence data with recurrent neural networks. In: Proceedings of the 23rd International Conference on Machine Learning, pp. 369–376 (2006)
11. Gupta, A., Vedaldi, A., Zisserman, A.: Synthetic data for text localisation in natural images. In: Proceedings of the IEEE Conference on Computer Vision and Pattern Recognition, pp. 2315–2324 (2016)
12. He, K., Zhang, X., Ren, S., Sun, J.: Deep residual learning for image recognition. In: Proceedings of the IEEE Conference on Computer Vision and Pattern Recognition, pp. 770–778 (2016)
13. Hu, Y., Soltoggio, A., Lock, R., Carter, S.: A fully convolutional two-stream fusion network for interactive image segmentation. Neural Netw. **109**, 31–42 (2019)
14. Jaderberg, M., Simonyan, K., Vedaldi, A., Zisserman, A.: Synthetic data and artificial neural networks for natural scene text recognition. arXiv preprint arXiv:1406.2227 (2014)
15. Jang, W.D., Kim, C.S.: Interactive image segmentation via backpropagating refinement scheme. In: Proceedings of the IEEE Conference on Computer Vision and Pattern Recognition, pp. 5297–5306 (2019)
16. Karatzas, D., Gómez, L., Nicolaou, A., Rusinol, M.: The robust reading competition annotation and evaluation platform. In: 2018 13th IAPR International Workshop on Document Analysis Systems (DAS), pp. 61–66. IEEE (2018)
17. Karatzas, D., et al.: ICDAR 2015 competition on robust reading. In: 2015 13th International Conference on Document Analysis and Recognition (ICDAR), pp. 1156–1160. IEEE (2015)
18. Li, Z., Chen, Q., Koltun, V.: Interactive image segmentation with latent diversity. In: Proceedings of the IEEE Conference on Computer Vision and Pattern Recognition, pp. 577–585 (2018)
19. Liao, M., Shi, B., Bai, X.: Textboxes++: a single-shot oriented scene text detector. IEEE Trans. Image Process. **27**(8), 3676–3690 (2018)
20. Liao, M., Wan, Z., Yao, C., Chen, K., Bai, X.: Real-time scene text detection with differentiable binarization. In: Proceedings of AAAI (2020)
21. Liew, J., Wei, Y., Xiong, W., Ong, S.H., Feng, J.: Regional interactive image segmentation networks. In: 2017 IEEE International Conference on Computer Vision (ICCV), pp. 2746–2754. IEEE Computer Society (2017)
22. Lin, Z., Zhang, Z., Chen, L.Z., Cheng, M.M., Lu, S.P.: Interactive image segmentation with first click attention. In: Proceedings of the IEEE/CVF Conference on Computer Vision and Pattern Recognition, pp. 13339–13348 (2020)
23. Ling, H., Gao, J., Kar, A., Chen, W., Fidler, S.: Fast interactive object annotation with curve-GCN. In: Proceedings of the IEEE Conference on Computer Vision and Pattern Recognition, pp. 5257–5266 (2019)
24. Liu, X., Meng, G., Pan, C.: Scene text detection and recognition with advances in deep learning: a survey. Int. J. Doc. Anal. Recogn. (IJDAR) **22**(2), 143–162 (2019). https://doi.org/10.1007/s10032-019-00320-5
25. Liu, Y., Chen, H., Shen, C., He, T., Jin, L., Wang, L.: ABCNet: real-time scene text spotting with adaptive Bezier-curve network. In: Proceedings of the IEEE/CVF Conference on Computer Vision and Pattern Recognition, pp. 9809–9818 (2020)

26. Liu, Y., Jin, L., Zhang, S., Luo, C., Zhang, S.: Curved scene text detection via transverse and longitudinal sequence connection. Pattern Recogn. **90**, 337–345 (2019)
27. Long, S., He, X., Yao, C.: Scene text detection and recognition: the deep learning era. Int. J. Comput. Vis. **129**, 161–184 (2020)
28. Long, S., Ruan, J., Zhang, W., He, X., Wu, W., Yao, C.: Textsnake: a flexible representation for detecting text of arbitrary shapes. In: Proceedings of the European Conference on Computer Vision (ECCV), pp. 20–36 (2018)
29. Luo, C., Jin, L., Sun, Z.: Moran: a multi-object rectified attention network for scene text recognition. Pattern Recogn. **90**, 109–118 (2019)
30. Papadopoulos, D.P., Uijlings, J.R., Keller, F., Ferrari, V.: Extreme clicking for efficient object annotation. In: Proceedings of the IEEE International Conference on Computer Vision, pp. 4930–4939 (2017)
31. Ramadan, H., Lachqar, C., Tairi, H.: A survey of recent interactive image segmentation methods. Comput. Vis. Media **6**, 355–384 (2020)
32. Rother, C., Kolmogorov, V., Blake, A.: "GrabCut" interactive foreground extraction using iterated graph cuts. ACM Trans. Graph. (TOG) **23**(3), 309–314 (2004)
33. Shi, B., Yang, M., Wang, X., Lyu, P., Yao, C., Bai, X.: ASTER: an attentional scene text recognizer with flexible rectification. IEEE Trans. Pattern Anal. Mach. Intell. **41**, 2035–2048 (2019)
34. Tang, M., Gorelick, L., Veksler, O., Boykov, Y.: GrabCut in one cut. In: Proceedings of the IEEE International Conference on Computer Vision, pp. 1769–1776 (2013)
35. Tzutalin: Labelimg. https://github.com/tzutalin/labelImg#
36. Wang, W., et al.: Shape robust text detection with progressive scale expansion network. In: Proceedings of the IEEE/CVF Conference on Computer Vision and Pattern Recognition, pp. 9336–9345 (2019)
37. Wang, X., Jiang, Y., Luo, Z., Liu, C.L., Choi, H., Kim, S.: Arbitrary shape scene text detection with adaptive text region representation. In: Proceedings of the IEEE Conference on Computer Vision and Pattern Recognition, pp. 6449–6458 (2019)
38. Wang, Y., Xie, H., Zha, Z.J., Xing, M., Fu, Z., Zhang, Y.: ContourNet: taking a further step toward accurate arbitrary-shaped scene text detection. In: Proceedings of the IEEE/CVF Conference on Computer Vision and Pattern Recognition, pp. 11753–11762 (2020)
39. Wood, S.N.: Thin plate regression splines. J. R. Stat. Soc. Ser. B (Stat. Methodol.) **65**(1), 95–114 (2003)
40. Wu, W., Xing, J., Yang, C., Wang, Y., Zhou, H.: Texts as lines: text detection with weak supervision. Math. Probl. Eng. **2020**, 3871897 (2020)
41. Xu, Y., Wang, Y., Zhou, W., Wang, Y., Yang, Z., Bai, X.: TextField: learning a deep direction field for irregular scene text detection. IEEE Trans. Image Process. **28**(11), 5566–5579 (2019)
42. Zhan, F., Lu, S.: ESIR: End-to-end scene text recognition via iterative image rectification. In: Proceedings of the IEEE/CVF Conference on Computer Vision and Pattern Recognition, pp. 2059–2068 (2019)
43. Zhang, C., et al.: Look more than once: An accurate detector for text of arbitrary shapes. In: Proceedings of the IEEE Conference on Computer Vision and Pattern Recognition, pp. 10552–10561 (2019)
44. Zhang, S., Liew, J.H., Wei, Y., Wei, S., Zhao, Y.: Interactive object segmentation with inside-outside guidance. In: Proceedings of the IEEE/CVF Conference on Computer Vision and Pattern Recognition, pp. 12234–12244 (2020)

45. Zhang, W., Qiu, Y., Liao, M., Zhang, R., Wei, X., Bai, X.: Scene text detection with scribble lines. arXiv preprint arXiv:2012.05030 (2020)
46. Zhou, X., et al.: East: an efficient and accurate scene text detector. In: Proceedings of the IEEE conference on Computer Vision and Pattern Recognition, pp. 5551–5560 (2017)
47. Zhu, H., Meng, F., Cai, J., Lu, S.: Beyond pixels: a comprehensive survey from bottom-up to semantic image segmentation and cosegmentation. J. Vis. Commun. Image Representation **34**, 12–27 (2016)

Searching from the Prediction of Visual and Language Model for Handwritten Chinese Text Recognition

Brian Liu[✉], Weicong Sun, Wenjing Kang, and Xianchao Xu

Intel Flex, Beijing, China
{weicong.sun,wenjing.kang,james.xu}@intel.com

Abstract. In this paper, we build the deep neural networks for offline handwritten Chinese text recognition (HCTR) with only convolutional layers and one of the mainstream learning methods Connectionist Temporal Classification (CTC). Iteratively, different configurations of network architectures with residual and squeeze-and-excitation structures are explored. We ease the serious overfitting issue by applying high dropout rate 0.9 at the input of the last classification layer, and synthesize new text samples with isolated characters by reusing the character-level bounding boxes from the CASIA-HWDB train set. These empirical and intuitive tricks help us achieve the character error rate (CER) at 6.38% on the ICDAR2013 competition set. To further improve the performance, at each step of the CTC decoding, we propose a novel context beam search (CBS) algorithm, which conducts decoding from both the prediction of the basic visual model and another customized transformer-based language model simultaneously. The final CER is reduced to 2.49%. Code will be available online at https://github.com/intel/handwritten-chinese-ocr.

Keywords: Offline handwritten Chinese text recognition · Se-ResNet · CTC · Context Beam Search · Transformer · Language model · Dropout · Data synthesis

1 Introduction

In the current era of deep learning, many works have demonstrated great success in text image recognition tasks by taking full advantage of different kinds of neural network architectures. For instance, RNN or LSTM based networks in [2,3], hybrid networks with CNN and RNN in [4,5], fully convolutional networks in [6,7] or CNN with gated mechanism networks in [8,9], CNN-only networks in [10,11], and recent Transformer-based networks in [12,13].

HCTR is one of the most challenging text recognition task not only because of its thousands of classes of characters but also the complexity of some characters and similarity between many of them. During the recent decade, there are mainly two branches of methods to solve this problem. One is the segmentation-based method, and the other is segmentation-free method.

© Springer Nature Switzerland AG 2021
J. Lladós et al. (Eds.): ICDAR 2021, LNCS 12823, pp. 274–288, 2021.
https://doi.org/10.1007/978-3-030-86334-0_18

The performance of traditional segmentation-based methods is led by the work in [14] which proposed a segmentation and recognition framework. It consists of the steps of over-segmentation of a text line image, construction of the segmentation and recognition candidate lattices, and path search in the lattices with context fusion. Another recent work [7] shared a new perspective and proposed an one-step segmentation and recognition architecture by taking advantage of fully convolutional networks to predict the location of characters and recognize them simultaneously.

Segmentation-free methods are mainly based on the HMM framework, CTC alignment [30] and attention mechanism. The published work in [15] leads the accuracy of HCTR, which had being improved year over year by applying DNN [16], DCNN [17], parsimonious HMM [18], and writer adaptation networks [15]. And the work in [19] was the first to introduce CTC for HCTR. Its aim was to extend and access the capacity of proved MDLSTM-RNN and CTC-based line recognition systems for Chinese language specifically. Later, this CTC-based segmentation-free method was adopted by the work [20] which proposed a hybrid network with CNN and ResLSTM, and a new data pre-processing and augmentation pipeline. For the attention-based HCTR, the only successful work in [21] proposed a novel multi-level multimodual fusion network and properly embedded it into the attention-based LSTM so that both the visual information and the linguistic information can be fully leveraged.

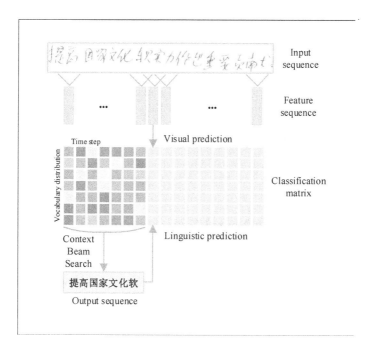

Fig. 1. Recognition and decoding overview.

In this work, we follow the CTC-based segmentation-free method for HCTR considering about its wide acceptance and simplicity. The open-source implementation of hybrid network with CNN and BLSTM for printed text recognition [5] on DAS2018 is our starting point. During the exploration of optimal network architecture for feature extraction, we iteratively introduce the VGG-16 [22] network without its fully-connection layers, the residual [23] and squeeze-and-excitation [24] structures inside each network block, and stack different numbers of blocks under target training limitation. Meanwhile, with the rise of recurrent-free architectures for text recognition tasks and the result of our quick experiment on smaller task, we find the BLSTM is indeed optional for HCTR so that we remove it. We then focus on easing the serious overfitting issue by adopting the best known methods from the practice, especially the dropout [25] strategy. Given no contextual requirement in convolutional-only networks, we synthesize more text samples with randomly selected characters from CASIA-HWDB1.x [1] database by reusing the character-level bounding boxes in CASIA-HWDB2.x [1] train set as the template. These tricks surprisingly lead to big reduction of the character error rate on evaluation set. To further improve the performance, we also introduce language model as other works did but train it with the emerging transformer [26] architecture, and propose a novel CTC decoding algorithm called context beam search (CBS) which takes account of both visual and semantic information at each step. Figure 1 shows the overview of text recognition and decoding process.

Our main contributions to this HCTR are summarized as follows: (1) One of the simplest method with convolutional-only networks and CTC loss for hand-written Chinese text recognition. (2) Two differentiation for performance boost: high dropout rate 0.9 at the input of last classification layer and data synthesis by reusing pre-labelled bounding boxes. (3) A novel context beam search (CBS) algorithm for CTC decoding by using visual and semantic information together. (4) The first work to achieve character error rate (CER) under 2.5% on ICDAR 2013 competition set.

The rest of this paper includes our methodology in Sect. 2, the experiments in Sect. 3, and the conclusion in Sect. 4.

2   Methodology

2.1   Convolutional-Only Neural Networks and CTC

The shallow VGG-based [22] networks have been adopted as fundamental architectures for broad range of computer vision tasks. And in the topologies of many text recognition models, similar patterns are referred to stack layers for feature extraction. Our proposed convolutional-only network topology illustrated in Fig. 2, which reuses all the 13 convolution layers from the official VGG-16 network except those 3 fully-connection layers (one convolution layer with 128 output channels added for unified block structure). Batch normalization [28] and ReLU [29] layer follow each of these convolution layers, and max-pooling with 2×2 and dropout layer follow each group of convolution layers with different output channels (highlighted with different colors in Fig. 2). After these back-bone layers, the 3D (C, H and W) extracted features are flatten to 2D with [C and H] (or [H and W] as alternative) dimensions for being able to connect to the last linear layer (or BLSTM first for experiment only) for prediction per step.

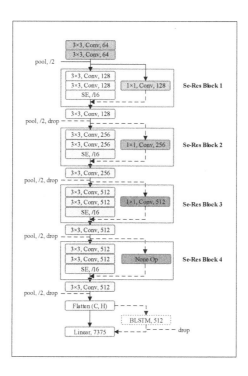

Fig. 2. The architecture of convolutional-only network.

Residual connections [23] can ease the training of networks that are substantially deeper than those used previously. Official ResNet architectures in that paper were mainly designed for ImageNet challenges with coarse-grained objects in 224×224 resolution images. While for the fine-grained characters in text images with much longer width and several thousands of classes in HCTR, those networks take too long to train and do not perform well experimentally. Introducing only the residual connection is more practical. Meanwhile, the Squeeze-and-Excitation [24] block, with simple structure and slight additional computational cost, brings in significant improvement in performance for existing state-of-the-art CNNs by explicitly modelling the interdependencies between channels of convolutional features. So we construct our network with Se-Res block including both SE block and residual connections in order to stack more layers to get better performance. The structure of SE block in those four Se-Res blocks in Fig. 2 are the same as the default SENet [24] but with a reduction equal to 16, and the structure of residual connection also reserves the same as default ResNet implementation, but with different number of output channels and stride size which is 1 instead of the original 2.

CTC [30] learning method eliminates the need to segment the input sequences, opening the way to data driving training. During the training, probabilities distribution over all label tokens and one additional blank token at each step are calculated along the feature sequences extracted from input data at first. CTC loss is then represented as the negative log probability of output for all valid alignments [31], given one input sequence X and target Y as an example,

$$L = -\log P(Y|X) = -\log \sum_{A \in \pi} \prod_{t=1}^{T} P_t(a_t|X) \tag{1}$$

where π is the set of valid alignments.

2.2 Transformer-Based Language Model

The text in the image being recognized usually has meaningful context. Above visual model does not use the underlying semantic information across steps. In some of the state-of-the-art approaches which integrated the language model into their optimal path search [14] or for evaluating character sequences [15] for prior probability, they have successfully gained the performance boost, at about 5%.

Different from adopting hybrid language model in [14,15], and considering the limitation of N-grams, we drop N-grams, skip RNN-based language model, and directly leverage transformer-based language model, because it has proved itself in recent years in generalized language model tasks [32].

The original transformer [26] has both encoder and decoder which was designed for machine translation or other NLP tasks. While training language model can only use part of this transformer to learn from character sequences in target language corpus with multi-head attention and computational efficiency. Initially, we reuse the very basic language model architecture with 6 decoders from the open-source sequence learning framework Fairseq [33], developed by Facebook. Its loss function which is also called negative log likelihood loss is represented as,

$$L(X,Y) = -\sum_{c=1}^{C} \log P_c \tag{2}$$

where X is the input, Y is the target, C is the length of sentences, and P_c is the log probability of X_c. At the inference stage, given a sentence with any length, the trained language model can be deployed as a normal language model to predict the index of next word as,

$$C_{next} = argmax(P_c) \tag{3}$$

or to score the whole sentence by calculating the perplexity (ppl) according to the loss of this sentence.

More powerful transformers with more layers stacked for language model training are not researched in this work.

2.3   Context Beam Search Algorithm

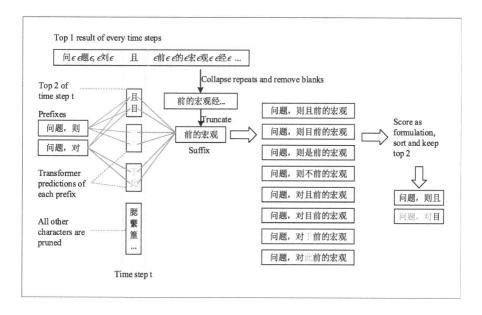

Fig. 3. Illustration of the Context Beam Search Algorithm.

The task of finding the text sequence from the CTC output is referred as decoding. It can be done by greedy search, also called best path decoding, using a max function along the sequence axis. But it is not guaranteed to find the most probable sequence [30]. Given the existence of linguistic information, the combination of beam search and language model is widely accepted, i.e. prefix beam search [34]. Here we propose the context beam search, to fully use language model to improve the recognition result. It leverages language model to do an approximate search for the most probable sequence, and takes account of all the context information from prefix, suffix, and language model prediction of each prefix at each time step during the process.

Due to the large size of Chinese vocabulary, character candidates for sequence search at every time step need to be pruned in order to prevent enormous calculation. Selecting character candidates turns to be the key to balance the performance and accuracy. Prior works ignored the underlying semantic information in historical sequence adjust to current step, so in this work, a transformer-based language model is introduced for predicting more meaningful character candidates. As is shown in Fig. 3, characters with top-k predictions from the visual model at each time step are saved; Besides, each beam is input into the transformer to predict the next top-k characters. All these characters are concatenated as candidates (do not remove duplicated characters), while others in the vocabulary are pruned.

A lot of sentences show that characters at one position does not only have a semantic relationship with the prefix, but they also have good semantic connections with the characters behind it, i.e. the suffix. If a meaningful prefix with a meaningful candidate character is connected with a meaningful suffix, it will have a lower perplexity than not connected. The connection with suffix makes the meaningful text sequence more coherent, then the language model can notice and select the most probable sequence.

Usually the suffix is unknown for beam search. In this proposed algorithm, especially for vision task, the best path decoding result can be firstly calculated for reference, then the context beam search involving suffix can be conducted. During the searching, all these prefixes together with all candidates at current step and the head part of the reference suffixes are input to language model to be scored. Specifically, assume the initial best path decoding result is S, whose length equals to CTC output length T, and at time step t, each beam Y (prefix and current candidate) is scored by,

$$P = P_{net}(Y|X_{1:t})P_{LM}(Y + B(S_{t+1:T})_{1:n})^{\alpha}|Y|^{\beta} \tag{4}$$

where B is the CTC many-to-one mapping (by collapsing repeats and removing blanks), and n is the specified length of suffix. If language model is N-gram based, n should be less than N, while language model is RNN or other sequence model, n is selected through cross validation. α serves as the weight of P_{LM} and $|Y|^{\beta}$ is a bonus for a longer sequence.

Note that the same language model in our CBS algorithm can be used for both predicting the next word of each prefix and scoring each newly constructed beam (and suffix) for selecting the best sequence. The language model prediction part can be disabled in order to reducing computational cost.

Compared to the traditional beam search, the proposed decoding algorithm not only adopts the language model, but also brings in more context information by using language model prediction and suffix. It significantly helps correct lots of recognition errors and achieves a vast improvement of the accuracy.

3   Experiments

3.1   Datasets

The basic training data for visual model is the train set of offline handwritten Chinese text databases [1] CASIA-HWDB2.x which includes 41,781 text samples. With data synthesis method (3.2), additional 153,955 text samples generated with the isolated offline handwritten Chinese character databases CASIA-HWDB1.x. The size of vocabulary our models support is 7,375 which covers 7,373 character classes from the training set, one "blank" token for CTC learning and one "unknown" token reserved for characters in training set but not in target character dictionary.

The training data for transformer-based language model is the train set of open-source news2016zh from NLP Chinese corpus [35], which has 2,430,752 news. We extract the "title" and "content" fields, filter out all characters out of above 7,373 vocabulary and insert one space between every two characters.

The CASIA-HWDB2.x test set which has 10,449 text samples written by the same people as its train set is our validation set. And the ICDAR2013 competition set with 3,432 text samples written by 60 new writers is the evaluation set. Considering that the punctuation, numbers and English alphabet are originally labelled using Chinese input method in the ground truth of ICDAR 2013 competition set, but they do not exist in the vocabulary of training set, we thus replace these characters with the same ones in ground truth with English input method. We believe that this correction of labels is acceptable as we do not change the data itself.

About the metric, for our all experiments, we use the character error rate (CER),

$$CER = (N_i + N_d + N_s)/N \tag{5}$$

where the N is the length of character sequences, and the N_i, N_d and N_s are the number of insertion error, the deletion error and the substitution errors, respectively. It is calculated using standard python library edit distance (insertion error cost $= 1$).

3.2   Input Pre-processing and Data Synthesis

In reality, the captured images can be in any resolution. Resizing images to fixed-length is popular for implementation simplicity or limitation, but images with too short or long width will get much distorted in character level. We thus adopt the fixed-height and fixed-ratio resizing method on gray-scale images, after that the input image will have dynamic length. During the batch mode training and inference, we pad the extra region of those short samples using the rightmost column of their data which is normalized in advance.

With the fact that both the text and character version of the well-known CASIA-HWDB [1] databases are compatible with each other, an intuitive way for data synthesis is to construct new text data from character samples. As the work [10] did, we reuse the pre-labelled bounding box of each character in text samples and reserve the nature writing habits of people, by just replacing all the Chinese characters (non-Chinese characters are relatively small) in text samples from CASIA-HWDB2.x train set with randomly selected characters from CASIA-HWDB1.x database.

3.3   Visual Model Training

Our proposed networks are built with version 1.7.1 of PyTorch framework and are trained on two Nvidia TITAN V GPUs with warp-ctc loss [36]. We choose 128 pixels as the fixed height of input images for model training considering the complexities of Chinese characters, and this number can also be scaled down for performance and accuracy trade-off. The optimizer, momentum and weight decay are set to SGD, 0.9, and 1e-4 respectively by default. We mainly tune the hyperparameters: batch size and learning rate. Since the average length of text samples is about in 1500 pixels in the training set, big batch size will consume almost all the graphic memory. Limited by the target GPUs we set the batch size to 8 and initial learning rate to 1e-4 accordingly. Learning rate will be automatically adjusted by timing 0.1 after 30 epochs. Generally, these models training will converge around 40 epochs.

Mixed Precision. For quicker training, we adopt the Nvidia apex [37] library which enables the mixed precision data format training and reserves the accuracy between original FP32 and FP16. Thanks to the simplicity of our models, the training gets 2-3x speed up without obvious accuracy drop by setting the opt level to 2.

Best Checkpoint Selection. In order to reduce the total training time and save the checkpoint as better as possible, we validate the training in epoch level when the epoch $<= 30$, and then valid it every 1000 iterations (2000 iterations for training with synthesized data) when epoch > 30. We stop the training after the validated CER will not decrease within 5 epochs. The model with lowest CER on validation set is selected for evaluation.

3.4   Results Without LM

Table 1. Recognition results of base models on the ICDAR 2013 competition set without LM.

Models	Parameter size (MB)	Training time (min/ep)	CER (%)
SeRes-1111 + w/o drop	115	16	14.06
SeRes-1111 + w/o drop + BLSTM	127	18	12.18
SeRes-1111 + drop + BLSTM	127	18	10.28
SeRes-1111 + drop	115	16	**9.96**

Dropout. The base model SeRes-1111 (number of every different blocks are equal to 1) in Table 1 suffers from serious overfitting. Similar issue was found and eased in the work [10]. After applying the same dropout strategy at the output of first four max-pooling layers (0.0-0.3-0.3-0.3) and especially at the input of final linear layer with high rate 0.9, in this work, the error rate drops over 4%, which is an amazing trick.

Hybrid Model. At the very beginning of the experiments, without dropout applied, the hybrid model with one more BLSTM layer (512 hidden units) than base convolutional-only model indeed got better result, as shown in Table 1. But for the same task and backbone layers, the hybrid model takes more time to train and has more memory footprints. With the same dropout strategy applied for hybrid model (0.0-0.3-0.3-0.3-0.3 after max-pooling layers and 0.9 at the input of linear layer), the error rate can further drop about 2%. Considering parameter size, computational cost and the accuracy, we choose the convolutional-only architecture as the base model.

Table 2. Recognition results of models with different blocks and synthesized data on the ICDAR 2013 competition set without LM.

Models	Parameter size (MB)	Training time (min/ep)	CER (%)
SeRes-1111 + drop	115	16	9.96
SeRes-1221 + drop	138	20	9.49
SeRes-2451 + drop	203	30	9.04
SeRes-2451 + drop + synth	203	120	**6.38**
SeRes-2451 + drop + BLSTM + synth	214	128	6.77

Se-Res Block Configurations. Under the limitation of training resource, we provide a group of network block configurations for performance and accuracy trade-off which is essential for model deployment. In the Fig. 2, the last Se-Res block which does not have a residual connection has the same structure as its previous layers, so that the different configurations are mainly explored on the first three blocks. This step totally contributes 0.9% reduction to CER, as listed in Table 2.

Training with Synthesized Data. With the best model configuration and without training procedure changed, the training with synthesized data needs 3x more time, but the reduction is significant, another 2.6%+ as shown in Table 2, which proves that this data synthesis method is very successful. Interestingly, model with BLSTM layer is also trained and its result is close to our best. We argue that the recurrent layers used here or in other similar text recognition tasks are not for learning context information from corpus, as 75% of the training data in which all characters are in random order.

3.5 Context Beam Search Results with Transformer-Based LM

Transformer-Based LM Training. We start the LM training from the language model example in Fairseq [33] project with our customized corpus. The architecture of transformer which includes 6 decoders is not changed. We only modify the warmup updates (from 4,000 to 400) and max tokens (from 2,048 to 32,768) accordingly for the two Nvidia TITAN V GPUs. The training is converged at epoch 27, with ppl equal to 29.53 on page level labels of CASIA-HWDB2.x test set.

Hyper-parameters Searching. As described in the Eq. 4, there are two hyper-parameters used for the context beam search. Finding the best combination is done through grid search on 10% of the validation data. The ranges and strides of both α and β are specified manually, and the metric is guided by CER. For faster searching, beam size is set to 5 and language model prediction is disabled.

Effectiveness of Different Context. In Table 3, the traditional prefix beam search, which can be configured with CBS but not using language model prediction and suffix, works as expected and the CER drops over 3.3% in row 2. Then our context beam search with language model prediction enabled contributes another 0.36% decrease in row 4. Suffix can also help but is not that obvious in row 3 and 5. The last row shows that a larger beam size and more predictions provided can further lead to even better result, but the computational cost will increase dramatically. One practical solution would be dynamically skipping the language model prediction at some time steps if the visual model can give high enough confidence.

Table 3. Recognition results of best models with context beam search on the ICDAR 2013 competition set.

Beam size	Visual Top-K pred.	LM Top-K pred.	Suffix length	CER (%)
1	1	0	0	6.38
5	5	0	0	3.07
5	5	0	4	3.03
5	5	5	0	2.71
5	5	5	4	2.66
10	10	10	4	**2.49**

3.6 Comparison with Other State-of-the-Art Works

We make a comparison with recent successful approaches since the ICDAR 2013 competition in Table 4. Our proposed method reduces the previous published lowest CER by at least 20%, from 3.17% to 2.49%, on the competition set of ICDAR 2013, to the best of our knowledge. What's more, the CER on the CASIA-HWDB2.x test set (our validation set) reaches 3.59% without LM, and 1.92% with LM, respectively.

Besides the new CER record, our method is more straightforward and can be easily trained end-to-end or deployed with a popular deep learning framework. While the traditional two-stage method [14] depends greatly on accurate

Table 4. Comparison with recent works on both the CASIA-HWDB test set and the ICDAR 2013 competition set with and without LM in CER (100% - AR).

Methods	Without LM		With LM	
	CASIA-HWDB	ICDAR2013	CASIA-HWDB	ICDAR2013
HIT-2 2013 [38]	-	-	-	13.27
LSTM-RNN-CTC [19]	-	16.50	-	11.60
DNN-HMM [16]	-	16.11	-	6.50
NA-CNN [39]	7.96	11.21	4.79	5.98
Over-segmentation [14]	-	-	4.05	3.68
SMDLSTM [40]	-	13.36	-	7.39
DCNN-HMM [17]	-	10.34	-	3.53
FCN [7]	-	9.48	-	4.49
Attention-based LSTM [21]	-	11.26	-	3.65
WCNN-PHMM [15]	-	8.42	-	3.17
CNN-ResLSTM-CTC [20]	5.10	8.45	3.03	3.28
CNN-CTC [10]	-	6.81	-	-
CNN-CTC-CBS (this)	**3.59**	**6.38**	**1.92**	**2.49**

segmentation result and needs to train and optimize over multiple standalone models. HMM-based methods lead the HCTR task for a long time. The WCNN-PHMM [15] needs to train an additional writer adaptation model, whose iterative decoding manner is unpredictable, and this method potentially suffers from tens of thousands of character states. Another related work [20] which is also based on CTC, mainly takes advantage of data augmentation. Due to the ResLSTM layer in its network, it increases the complexity for reproducing and deployment.

There are many powerful pre-trained language models based on transformer recently, which can be directly used for error correction at the post-processing stage of text recognition [27], but our one-stage method has the advantage of leveraging the visual information in the field and is simple for deployment.

3.7   Recognition Examples

The example at the top of Fig. 4 demonstrates the great advantage and context beam search. But the one at the bottom, which is very difficult to recognize because of the overly skewed text.

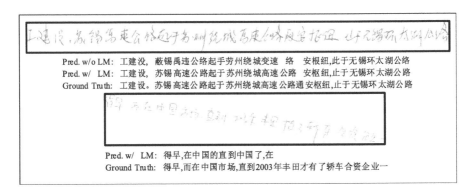

Fig. 4. Good (top) and hard (bottom) examples.

4   Conclusion

In this paper, we presented a simple and effective method for the offline hand-written Chinese text recognition task. Driven by powerful feature extractor, high dropout rate and successful data synthesis, this method with convolutional-only neural networks and CTC criterion demonstrated its great superiority over other state-of-the-art approaches. Furthermore, we improve the performance up to a newly record by searching optimal candidates from the prediction of both visual model and transformer-based language model during the CTC decoding with our proposed context beam search algorithm.

Nevertheless, we are open to the network architecture and dropout rate search. We also notice that there is still much space left to optimize the training, for example, better learning rate scheduler, bigger batch size, practical data augmentation etc.

References

1. Liu, C.-L., et al.: CASIA online and offline Chinese handwriting databases. In: 2011 International Conference on Document Analysis and Recognition. IEEE (2011)
2. Breuel, T.M., et al.: High-performance OCR for printed English and Fraktur using LSTM networks. In: 2013 12th International Conference on Document Analysis and Recognition. IEEE (2013)
3. Voigtlaender, P., Doetsch, P., Ney, H.: Handwriting recognition with large multidimensional long short-term memory recurrent neural networks. In: 2016 15th International Conference on Frontiers in Handwriting Recognition (ICFHR). IEEE (2016)
4. Shi, B., Bai, X., Yao, C.: An end-to-end trainable neural network for image-based sequence recognition and its application to scene text recognition. IEEE Trans. Pattern Anal. Mach. Intell. 39(11), 2298–2304 (2016)
5. Breuel, T.M.: High performance text recognition using a hybrid convolutional-LSTM implementation. In: 2017 14th IAPR International Conference on Document Analysis and Recognition (ICDAR), vol. 1, pp. 11–16. IEEE (2017)

6. Busta, M., Neumann, L., Matas, J.: Deep TextSpotter: an end-to-end trainable scene text localization and recognition framework. In: Proceedings of the IEEE International Conference on Computer Vision (2017)

7. Peng, D., et al.: A fast and accurate fully convolutional network for end-to-end handwritten Chinese text segmentation and recognition. In: 2019 International Conference on Document Analysis and Recognition (ICDAR). IEEE (2019)

8. Yousef, M., Hussain, K.F., Mohammed, U.S.: Accurate, data-efficient, unconstrained text recognition with convolutional neural networks. arXiv preprint arXiv:1812.11894 (2018)

9. Reeve Ingle, R., et al.: A scalable handwritten text recognition system. In: 2019 International Conference on Document Analysis and Recognition (ICDAR). IEEE (2019)

10. Liu, B., Xu, X., Zhang, Y.: Offline handwritten Chinese text recognition with convolutional neural networks. arXiv preprint arXiv:2006.15619 (2020)

11. Ptucha, R., et al.: Intelligent character recognition using fully convolutional neural networks. Pattern Recogn. **88**, 604–613 (2019)

12. Sheng, F., Chen, Z., Xu, B.: NRTR: a no-recurrence sequence-to-sequence model for scene text recognition. In: 2019 International Conference on Document Analysis and Recognition (ICDAR). IEEE (2019)

13. Lu, N., et al.: Master: multi-aspect non-local network for scene text recognition. arXiv preprint arXiv:1910.02562 (2019)

14. Wu, Y.-C., Yin, F., Liu, C.-L.: Improving handwritten Chinese text recognition using neural network language models and convolutional neural network shape models. Pattern Recogn. **65**, 251–264 (2017)

15. Wang, Z.-R., Du, J., Wang, J.-M.: Writer-aware CNN for parsimonious HMM-based offline handwritten Chinese text recognition. Pattern Recogn. **100**, 107102 (2020)

16. Du, J., et al.: Deep neural network based hidden Markov model for offline handwritten Chinese text recognition. In: 2016 23rd International Conference on Pattern Recognition (ICPR). IEEE (2016)

17. Wang, Z.-R., Du, J., Wang, W.-C., Zhai, J.-F., Hu, J.-S.: A comprehensive study of hybrid neural network hidden Markov model for offline handwritten Chinese text recognition. Int. J. Doc. Anal. Recogn. (IJDAR) **21**(4), 241–251 (2018). https://doi.org/10.1007/s10032-018-0307-0

18. Wang, W., Du, J., Wang, Z.-R.: Parsimonious HMMS for offline handwritten Chinese text recognition. In: 2018 16th International Conference on Frontiers in Handwriting Recognition (ICFHR). IEEE (2018)

19. Messina, R., Louradour, J.: Segmentation-free handwritten Chinese text recognition with LSTM-RNN. In: 2015 13th International Conference on Document Analysis and Recognition (ICDAR). IEEE (2015)

20. Xie, C., Lai, S., Liao, Q., Jin, L.: High performance offline handwritten Chinese text recognition with a new data preprocessing and augmentation pipeline. In: Bai, X., Karatzas, D., Lopresti, D. (eds.) DAS 2020. LNCS, vol. 12116, pp. 45–59. Springer, Cham (2020). https://doi.org/10.1007/978-3-030-57058-3_4

21. Xiu, Y., et al.: A handwritten Chinese text recognizer applying multi-level multimodal fusion network. In: 2019 International Conference on Document Analysis and Recognition (ICDAR). IEEE (2019)

22. Simonyan, K., Zisserman, A.: Very deep convolutional networks for large-scale image recognition. arXiv preprint arXiv:1409.1556 (2014)

23. He, K., Zhang, X., Ren, S., Sun, J.: Deep residual learning for image recognition. In: Proceedings of the IEEE Conference on Computer Vision and Pattern Recognition, pp. 770–778 (2016)
24. Hu, J., Shen, L., Sun, G.: Squeeze-and-excitation networks. In: Proceedings of the IEEE Conference on Computer Vision and Pattern Recognition (2018)
25. Srivastava, N., Hinton, G., Krizhevsky, A., Sutskever, I., Salakhutdinov, R.: Dropout: a simple way to prevent neural networks from overfitting. J. Mach. Learn. Res. **15**(1), 1929–1958 (2014)
26. Vaswani, A., et al.: Attention is all you need. arXiv preprint arXiv:1706.03762 (2017)
27. Zhang, S., et al.: Spelling error correction with soft-masked BERT. arXiv preprint arXiv:2005.07421 (2020)
28. Ioffe, S., Szegedy, C.: Batch normalization: Accelerating deep network training by reducing internal covariate shift. arXiv preprint arXiv:1502.03167 (2015)
29. Nair, V., Hinton, G.E.: Rectified linear units improve restricted Boltzmann machines. In: ICML (2010)
30. Graves, A., Fernández, S., Gomez,F., Schmidhuber, J.: Connectionist temporal classification: labelling unsegmented sequence data with recurrent neural networks. In: Proceedings of the 23rd International Conference on Machine Learning, pp. 369–376 (2006)
31. Hannun, A.: Sequence modeling with CTC. Distill **2**(11), e8 (2017)
32. Weng, L.: Generalized Language Models. http://lilianweng.github.io/lil-log/2019/01/31/generalized-language-models.html (2019)
33. Ott, M., et al.: fairseq: a fast, extensible toolkit for sequence modeling. arXiv preprint arXiv:1904.01038 (2019)
34. Hannun, A.Y., et al.: First-pass large vocabulary continuous speech recognition using bi-directional recurrent DNNs. arXiv preprint arXiv:1408.2873 (2014)
35. Xu, L.: NLP Chinese corpus: large scale Chinese corpus for NLP (2019). https://doi.org/10.5281/zenodo.3402033
36. Naren, S.: PyTorch bindings for Warp-CTC. https://github.com/SeanNaren/warp-ctc
37. Micikevicius, P., et al.: Mixed precision training. arXiv preprint arXiv:1710.03740 (2017)
38. Yin, F., et al.: ICDAR 2013 Chinese handwriting recognition competition. In: 2013 12th International Conference on Document Analysis and Recognition. IEEE (2013)
39. Wang, S., et al.: Deep knowledge training and heterogeneous CNN for handwritten Chinese text recognition. In: 2016 15th International Conference on Frontiers in Handwriting Recognition (ICFHR). IEEE (2016)
40. Wu, Y.-C., et al.: Handwritten Chinese text recognition using separable multi-dimensional recurrent neural network. In: 2017 14th IAPR International Conference on Document Analysis and Recognition (ICDAR), vol. 1. IEEE (2017)

Towards an IMU-based Pen Online Handwriting Recognizer

Mohamad Wehbi[1]([✉]), Tim Hamann[2], Jens Barth[2], Peter Kaempf[2],
Dario Zanca[1], and Bjoern Eskofier[1]

[1] Machine Learning and Data Analytics Lab, Friedrich-Alexander-Universität
Erlangen-Nürnberg, Erlangen, Germany
{mohamad.wehbi,dario.zanca,bjoern.eskofier}@fau.de
[2] STABILO International GmbH, Heroldsberg, Germany
{tim.hamann,jens.barth,peter.kaempf}@stabilo.com

Abstract. Most online handwriting recognition systems require the use of specific writing surfaces to extract positional data. In this paper we present a online handwriting recognition system for word recognition which is based on inertial measurement units (IMUs) for digitizing text written on paper. This is obtained by means of a sensor-equipped pen that provides acceleration, angular velocity, and magnetic forces streamed via Bluetooth. Our model combines convolutional and bidirectional LSTM networks, and is trained with the Connectionist Temporal Classification loss that allows the interpretation of raw sensor data into words without the need of sequence segmentation. We use a dataset of words collected using multiple sensor-enhanced pens and evaluate our model on distinct test sets of seen and unseen words achieving a character error rate of 17.97% and 17.08%, respectively, without the use of a dictionary or language model.

Keywords: Online handwriting recognition · Digital pen · Inertial measurement unit · Time-series data

1 Introduction

The field of handwriting recognition has been studied for decades, increasing in popularity with the advancements of technology. This increase in popularity is due to the substantial number of people using handheld digital devices that provide access to such technologies, and the desire of people to save and share digital copies of written documents. The aim of a handwriting recognition system is to allow users to write without constraints, then digitize what was written for a multitude of uses.

Handwriting recognition (HWR) is widely known to be separated into two distinct types, offline and online recognition [37,38]. For offline recognition, a static scanned image of the written text is given as input to the system. Offline recognition, also known as optical character recognition (OCR), is the more common recognition technique used in a wide range of applications for reading

© Springer Nature Switzerland AG 2021
J. Lladós et al. (Eds.): ICDAR 2021, LNCS 12823, pp. 289–303, 2021.
https://doi.org/10.1007/978-3-030-86334-0_19

specific details on documents, such as in healthcare and legal industry [42], banking [36], and postal services [43]. Online handwriting recognition (OHWR), alternatively, requires input data in the form of time series and includes the use of an additional time dimension within the data to be digitized. This results in a dynamic spatio-temporal signal that characterizes the shape and speed of writing [25]. OHWR systems are deployed in applications on tablets and mobile phones for users to digitize text using stylus pens or finger inputs on touch screens. Such systems use precise positions of the writing tip. However, one drawback that can be perceived is the need for a positional tracking system, whether it be a mobile touch screen, or any other pen tracking application. The need for such a tracking system restrains the user from writing on any surface and limits the usability of the system as well as the capability of the writer [13].

Another approach to applying OHWR is the use of inertial measurement units (IMUs) as, or integrated within, a writing tool. These tools provide movement data, such as accelerometer or gyroscope signals, which can be used for classification and recognition tasks. The major disadvantage of IMU sensors is that they are prone to error accumulation over time which, if not corrected, can lead to significant errors in the data recordings. Furthermore, IMU sensors generate noisy output signals, which is further intensified when touching a surface due to surface friction, which successively leads to lower performance at the task required due to deficient input data quality. However, when coupled with the correct models, IMUs produce beneficial data from which precise information can be extracted, such as the specific movements of a pen during handwriting. Moreover, IMUs are sourceless, self-contained and require no additional tools for data collection and extraction, and hence, a major advantage of IMU-based recognition systems is that no specific writing surface is required and systems rely only on the signals collected from the sensors.

In this paper, we discuss further the latter approach and introduce an OHWR system that uses sensor data recordings from pen movements to recognize writing on regular paper. We present an end-to-end system that processes sensor recordings in the form of time series data, and outputs the interpreted digital text on a tablet. Our system surpasses previous sensor-based pen systems in recognition rates, and is the first IMU-based pen recognizer that recognizes complete words and is not restricted to single character or digit recognition on paper. We use, as a digitizer, a regular ballpoint pen integrated with multiple sensors, and designed with a soft grip, that allows the user to write on a plain paper surface without constraints.

The rest of the paper is structured as follows: Sect. 2 summarizes available OHWR systems, distinguishing between positional-based and IMU-based systems. Section 3 presents the digitizer used in our system and explains the data acquisition process. Section 4 describes our end-to-end neural network architecture and describes the model training process, the hyperparameters used, and the data splitting. Section 5 reports the results and discusses the results obtained on distinct test sets. Section 6 outlines the future work to be implemented in our system.

2   Related Work

HWR has been a topic of interest in research for many years. Reviews about recognition systems [25,37] present pre-processing techniques, extracted features, in addition to different recognition models such as segment-and-decode methods and end-to-end recognition systems. In our work, we focus on the difference in the type of data used rather than methods of recognition. We briefly describe some previously developed recognition systems while distinguishing between systems using positional data and ones using IMU data.

2.1   Positional-Data Based Systems

Basic OHWR systems were presented by [6,48] in the late 1990s using Hidden Markov Model (HMMs) and Artificial Neural Networks (ANNs) that model the spatial structure of handwriting. A system developed using a multi-state time delay neural network was presented in [21] using a dictionary of 5000 words with pen position and pen-up/pen-down data. Models in which both an image of the text along with pen tracking data were used to develop a Japanese handwriting recognition system [22]. Different language specific systems were implemented to apply recognition systems for different writing styles, such as Arabic [45] and Chinese [28].

The availability of public datasets considerably increased research in this field. The UNIPEN dataset [18] is a collection of characters with recorded pen trajectory information including coordinate data with pen-up/down features, which was used to implement a character recognition model using time delay [19] and Convolutional [33] neural networks. An Arabic recognition system [4] applied HMMs using the SUSTOLAH dataset [34].

The IAM-ONDB dataset [30] is considered the most popular dataset in the OHWR domain. It includes pen trajectories of sentences written on a smart whiteboard with an infrared device mounted on the corner of the board to track the position of writers' pens, in addition to image data of written text from a collection of 86,272 word instances. [30] also introduced a HMM-based model with segmented data reaching up to about 66% recognition rate. This rate increased to 74% and consecutively to 79% when recurrent neural networks (RNNs) and bi-directional Long Short-Term Memory-Networks (BLSTMs) were implemented with non-segmented data [16,32]. An unconstrained recognition system was introduced in [15] with the integration of External Grammar models, achieving a word error rate (WER) of 35.5% using HMMs and 20.4% using BLSTMs. The combination of diverse classifiers led to a word level accuracy of 86.16% [31].

A multi-language system [24], supporting up to 22 different scripts for touch enabled devices, was based on several components: character model, segmentation model, and feature weights. The model was trained and evaluated on different public and internal datasets, leading with error rates as low as 0.8% to 5.1% on different UNIPEN test sets, also achieving a character error rate (CER) and WER of 4.3% and 10.4%, respectively, on the IAM-ONDB dataset.

The model results improved with the introduction of Bézier Curves, an end-to-end BLSTMs architecture, and language specific models with CER and WER of 2.5% and 6.5%, respectively [8].

The above metioned systems describe different methods for data pre-processing, feature extraction, and classification models, achieving different results for OHWR, while using input of either raw coordinate strokes or features extracted from these strokes. These systems were designed for position-based data that was extracted using specially designed hardware writing surfaces or touch screens and thus still pose a limitation if a system for digitzing paper-writing is required.

2.2 IMU-data Based Systems

The use of IMU data for HWR has been presented in different forms throughout the past years. Accelerometer-based digital pens for handwritten digit and gesture trajectory were developed in [23,46] with an accuracy of 98% & 84.8% over the ten digits, respectively. A 26 uppercase alphabet recognition system was developed using an inertial pen with a KNN classifier with 82% accuracy [41]. Pentelligence [40] combined the use of writing sounds with pen-tip motion from a digital pen equipped with microphones and IMU sensors for digit recognition reaching an accuracy of 98.33% for a single writer.

More recent studies used the Digipen [27] for the recognition of lowercase Latin alphabet characters. A recognition rate of 52% was achieved using LSTMs. The Digipen [47] was also used for the classification of uppercase and lowercase Latin alphabet characters separately, with a different dataset than what was used in [27]. A 1-Dimensional Convolutional Neural Network (1D-CNN) model achieved an accuracy of 86.97%.

At the time of the development of these systems, no public dataset for this task was available for a concrete evaluation of different systems. The On-HW dataset [35] was the first published IMU-based dataset and consisted of recordings of the complete Latin alphabet characters. It was released with baseline methods having an accuracy of 64.13% for the classification of 52 classes. These results were based on the writer-independent scenarios described in the papers (when available), since a writer-dependent model is not a feasible model when developing an OHWR system for general use.

For word level recognition, wearable technologies were implemented as approaches for HWR using IMUs. Airwriting is a tracked motion of continuous sensor stream, in which writing is of a single continuous stroke. It suffers from no surface friction and allows writing in free space. A digital glove equipped with accelerometers and gyroscopes for airwriting was designed in [5], achieving a WER of 11% using an HMM model following the segment-and-decode approach, evaluated on nine users writing 366 words using a language model consisting of 60000 words. A CNN-RNN approach for in-air HWR [12] achieved a word recognition rate (WRR) 97.88% using BLSTMs. Similar work was presented in [11] with a recognition rate of 97.74% using an encoder-decoder model. More recent

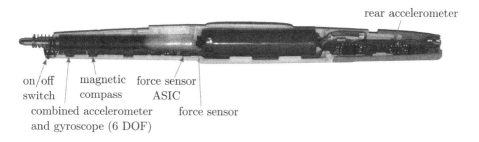

Fig. 1. The Digipen sensor placement

work presented a wearable ring for on-surface HWR [29] which provided acceleration and the angular velocity data from the finger resulting in 1.05% CER and 7.28% WER on a dataset of 643 words collected by a single writer.

Differently from positional-based systems or wearable systems, IMU-based digital pen systems are still limited to single digit or character recognition, with a solution for word recognition not demonstrated yet. This is due to the fact that writing on a paper surface introduces a considerable amount of noise in the data which makes the learning of a recognition model challenging. Furthermore the evaluation of such systems has been conducted on very different setups and on limited data. Here we propose a system that aims at filling the gap between pen-based systems and other approaches. We show that our IMU-based pen recognizer is practical for word recognition, and achieves significant improved results in comparison to previous pen devices.

3   Data Acquisition and Description

In this section, we introduce the pen used as a digitzer and the data collection process, and describe the data that was used to train our model. We base our system on a set of digital pens of the same model to ensure that our work is not biased towards one single instance of the device. The pen model used in our system is the STABILO Digipen which was used in [27,35,47]. The selection of this digitizer was based on the two main factors:

- **Suitability**: The Digipen is a ballpoint pen that can be used to write on paper like any regular pen. It is equipped with five different sensors, a combined accelerometer and gyroscope module at one side of the pen close to the pen tip, another single accelerometer module close to the other end of the pen, a magnetometer module, in addition to a force sensor at the tip of the pen that provides data about when the tip touches a surface, displayed in Fig. 1. This tool also includes a Bluetooth module that allows the transmission of collected sensor data to other devices in real-time. Accordingly, a trained recognition model can be integrated within a mobile app and can be used without the need for any further equipment but a mobile device. The Digipen streams sensor data via Bluetooth Low Energy with a sampling frequency 100 Hz.

Detailed information about the pen dimensions, sensor modules and ranges can be found in [2,35].

- **Availability**: The Digipen was not specifically designed for our work, and is available in a line of products. This implies that the system we developed is not for a single use case study, but can be further extended for different use case scenarios when required. Additionally, the availability of the pen allows the collection of data in a parallel manner which can accelerate a study that uses this tool.

3.1   Data Collection Application

To collect the ground truth labels of the data, the Digipen is provided with a Devkit [1] guide for the development of a mobile application for interaction with the pen. The application provides two files with similar timestamps of the recording session that can be used to extract data samples in the form of training data with the relative labels. A sample is defined as a complete word recording that consists of a series of timesteps.

3.2   Data Recording

Recording sessions were conducted in parallel using 16 different Digipens, with each session taking up to 45 min of recording time. A set of 500 words was used to collect the main set of data used for the system. Single words were displayed on the screen of a tablet, the users were asked to write the word on paper, in their own handwriting style, using the Digipen.

The dataset included recordings from 61 participants who volunteered to contribute to our study, with some participants contributing less than the 500 required words due to time constraints. The number of samples collected was 27961 word samples.

In addition to the main dataset, a separate dataset (unseen words set) was recorded from two other individuals. This recording consisted of random words selected from a set of 98463 words, different from the main set, serving as a second test set, with the purpose of testing the results of the system on unseen words. The final count of this set was 1006 sample recordings. Figure 2 shows histograms of the data count of both datasets with respect to the lengths of the samples and labels separately.

3.3   Data Preparation

Data recording is subject to faults during the process. To ensure that our model was trained on valid data, all hovering data before and at the end of each single sample recording was trimmed out. This was achieved by removing the data associated with force sensor readings below a pre-specified threshold at the beginning and the end of a recording. From the first time this threshold was exceeded within that recording, all data was kept even when the force reading temporarily

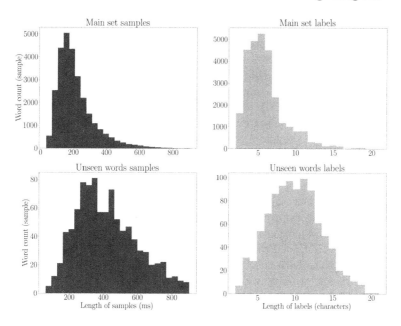

Fig. 2. A histogram displaying the number of samples in relation with (left) length of samples and (right) length of labels, in both (above) the main set and (below) the second test set.

fell below that threshold. The threshold was determined experimentally through monitoring the highest force sensor values while hovering with the pen. Additionally, samples which appeared too short or long to be correct recordings were considered as faulty recordings and removed from the dataset. The data was then normalized per sample using the z-score normalization in order to input data features of a similar scale into the model. No further preprocessing or feature extraction was applied.

4   End-to-End Models

The architectures described in this section were inspired by research aimed at developing end-to-end recognition models in handwriting [8,16] and speech recognition [10]. The use of Recurrent Neural Networks (RNNs), distinctively, Long Short-Term Memory networks (LSTMs), is common in the applications of handwriting and speech recognition due to the ability to transcribe data into sequences of characters or words while preserving sequential information. Bidirectional RNNs (BRNNs) make use of both past and future contextual information at every position of the input sequence in order to calculate the output sequences, and Bidirectional LSTMs (BLSTMs) have shown to achieve the best recognition results in the context of phoneme recognition when compared to other neural networks [17].

The models presented in this paper take input multivariate time series data samples of different lengths, comprised of 13 channels, representing the tri-axial measurements of the three IMU sensors and the magnetometer, in addition to the force sensor. In this section we evaluate different model architectures, describe the data splitting and the training process using raw sensor data.

4.1 Model Architectures

In the context of handwriting recognition using positional data, a model consisting of BLSTM layers proved sufficient to achieve the best recognition rates with the use of extracted feature vectors [16], or resampled raw stroke data [8], while CNN models obtained the best character classification accuracy in systems using raw sensor data with the Digipen [35, 47].

Following recent studies, we included the CNN model in our study. The model included four 1D-Convolutional layers, consisting of $1024, 512, 256, 128$ feature maps, consecutively, with kernel sizes of $5, 3, 3, 3$, respectively, and a fully connected layer of 100 units.

In contrast to positional data, in our case the input sequences are long due to a high sampling rate. Downsampling is not a viable option with IMU data because it leads to the loss of critical information [7]. Therefore, in addition to the CNN model, we implemented a CLDNN model (including Convolutional, LSTMs, and fully connected layers), which is typically used in speech recognition [10], where data samples are of high sampling rates and BLSTM models lead to latency constraints. The Convolutional layers reduce the dimensionality of the input features, which reduces the temporal variations within the LSTMs, which are then fed into the Dense layers where the features are tranformed into a space that makes that output easier to classify [39]. Hence, a CLDNN model allows to avoid latency constraints and slow training and prediction times which occur with BLSTM models. The model consisted of three Convolutional layers, followed by two BLSTM layers and a single fully connected layer. The Convolutional layers comprised of $512, 256, 128$ feature maps with kernel sizes of $5, 3, 3$, respectively. The BLSTM layers were of 64 units each, and the fully connected layer included 100 units. A grid search was implemented to determine the optimal hyperparameters setup.

In both described models, Batch Normalization [20] and Max Pooling (of size 2) were applied after each Convolutional layer. The Relu activation was used in the Convolutional and the fully connected layers, while Tanh was used in the BLSTM layers. Random dropout [44] with a dropout rate of 0.3 was applied after each layer to prevent overfitting and improve robustness of the system.

Similarly to the current developed systems in the field, we relied in our model on the Connectionist Temporal Classification (CTC) loss [14] with a Softmax output layer which provides an implicit segmentation of the data. The CTC is an RNN loss function that enables labeling whole sequences at once. It uses the

Table 1. Number of samples in each set per fold (word samples).

Sets	Folds				
	1	2	3	4	5
Training	18452	18655	17588	17579	17598
Validation	4614	4664	4397	4395	4400
Test	4895	4642	5976	5987	5963

network to provide direct mapping from an input sequence to an output label without the need of segmenting the data. It introduces a 'blank' character that is used to find the best alignment of characters that best interprets the input.

4.2   Model Training

We split the main dataset into five folds, distributed into 49 users in the training set and 12 users in the test set, and train our model on the different folds separately. No writer appears in both sets to consider a writer-independent recognition task. The training data for each fold was divided into an 80/20 (training/validation) split. The unseen words dataset was used to test the effectiveness of the models for unseen word data. Table 1 shows the different training, validation, and seen test sets, per each fold, not including any unseen words data.

For the implementation of our models, we used Keras/Tensorflow(v1) python libraries [3,9], which include standard functions required for our work. The models were trained using a batch size of 64 samples and optimized with the Adam Optimizer [26] with a starting learning rate of 10^{-2}. A learning rate scheduler was implemented to monitor the validation loss and decrease the learning rate with a patience of 10 epochs and a factor of 0.8. We trained the models until the validation loss showed no decrease for 20 iterations after the minimum learning rate of 10^{-4} was reached, and saved the best model determined by the lowest validation loss during training. Finally, the evaluation of our model required the decoding of the CTC output into a word interpretation for which we used the Tensorflow standard CTC decoder function with a greedy search that returns the most likely output token sequence without the use of a dictionary.

5   Evaluation and Discussion

Table 2 presents the average results obtained. In terms of word recognition, the CNN model achieved the higher error rate of 35.9% and 31.65% average CER for seen and unseen words, respectively, which implies that the even though a CNN model achieved good results in character recognition [35,47], it was not sufficient for the CTC to find the best character alignment within a word sample. The higher recognition rates were achieved by the CLDNN model, with an average of 17.97% and 17.10% CER. The models recognized unseen words

Table 2. Average error rates with respect to model properties.

Models	Seen words Avg. % CER	UnSeen Words Avg. % CER	Avg. epochs	Trainable parameters	Seconds per epoch
CNN	35.90 (± 2.01)	31.65 (± 1.07)	153	2,154,957	53
CLDNN	17.97 (± 1.98)	17.10 (± 1.68)	236	743,373	102

without distinction from seen words, since the CTC learns to identify individual characters within the data. Additionally, having users in two distinct test sets different from users in the training sets provided a user-independent recognition model.

Considering the different models in regard with the model complexity and time performance, Table 2 shows the training time with respect to the trainable parameters of each model, in addition to the training iterations required to converge to the best performance. The CNN model consisted of a larger number of training parameters, however required lower training and prediction times. The CLDNN model achieved the better complexity to performance ratio with a significantly better recognition rate yet a longer training time relative to the CNN.

In addition to the average CER, Fig. 3 reports the average Levenshtein Distance per label length for both test sets using the CLDNN model. This shows the minimum number of character edits, including insertions, deletions and substitutions, required to change a predicted word into the ground truth label. This means that the prediction of our model was on average divergent by 0.98 and 1.66 character edits for the average length 5.59 and 9.72 characters for the seen and unseen test sets, respectively. A detailed analysis of the errors showed that an average of 68% of the predicted words were missing characters, which is due to cursive writing. 26% of the prediction were of a substitution nature, which occurs between characters that look similar in both uppercase and lowercase, such as 'P-p', 'K-k', and 'S-s', while 6% only included more characters than the relative ground truth, which occurs with multiple stroke characters.

The model used in our system followed the common used model in HWR systems, both offline and online, which is a stack of Convolutional or Recurrent layers trained with the CTC loss, and achieved an overall recognition rate similar to previous position-based models that did not make use of languages models [16]. However, this result is not directly comparable with previous systems, since these systems were trained on different data types, with sentence data, while our dataset consists of word data. Additionally, the state-of-the-art models in positional-based systems make use of complex language models. Moreover, the public IAM-OnDB dataset includes a higher number of classes in comparison to our dataset. Nonetheless, the presented results suggest that our system is on an adjacent level in terms of recognition rates without the use of a dictionary.

Our system did not show the same level of recognition rates in comparison with the wearable systems described, which were trained on different datasets using distinct hardware. These systems followed the segment-and-decode approach with separate system-specific extracted features from uni-stroke data. Such

Table 3. IMU-based pen recognizers with respective data type and performance, including the results of the presented CLDNN model. Digipen represents the pen used in this work.

	Data type	Recognition type	Accuracy (%)
[23]	Acc	10 Digits	84.8
[46]	Acc	10 Digits	98
[40]	Acc, Gyr, Sound	10 Digit	98.33
[27]	**Digipen**	26 Characters	52
[41]	Acc, Gyr, Mag	26 Characters	82
[47]	**Digipen**	26 Characters	86.97
[35]	**Digipen**	52 Characters	64.13
Ours	**Digipen**	**Words**	**82.92 (CRR)**

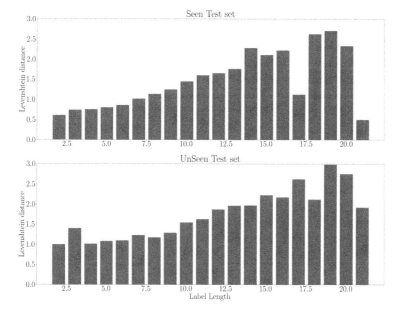

Fig. 3. A bar graph displaying the average Levenshtein Distance per label length for the seen and unseen test sets evaluated using the CLDNN model.

systems provided air-writing capability, which does not fit for our paper-writing recognizer. The wearable ring presented in [29] was designed for on-surface writing, however, the system was developed and evaluated for a single specific writer. Also, writing with finger does not present the same efficiency in comparison with pen writing.

Considering paper-writing recognition using sensor-equipped pens, our system achieved significant results in comparison to previously developed systems. Even though some previous systems used different hardware, our system, to the best knowledge of the authors, is the first IMU-based pen system that enables word recognition. Table 3 shows a summary of the described sensor-equipped pens in Sect. 2. Moreover, our model achieved an improved character recognition rate (CRR) by 18.79% for the 52 Latin alphabet characters relatively to previous systems using the same hardware.

6   Conclusion and Future Work

In this paper, we presented a system that applies OHWR by writing on normal paper using an IMU-enhanced ballpoint pen. We described the data collection tools and process in detail, and provided a complete system setup. We trained CNN and CLDNN end-to-end models that take normalized raw sensor data as input, and output word interpretations using the CTC loss with a greedy search decoder. The models were trained and evaluated using a five-fold cross-validation method, with test users being different from the users in the training set. We also evaluated the models on a separate test set to evaluate the efficiency of our system for unseen words. The presented CLDNN model showed the best performance without distinction between seen and unseen words.

Our system showed significant improvements in comparison with previously presented character recognition systems using digital pens. With the results presented in this work, we showed that sensor-enhanced pens are efficient and yield promising results in the OHWR field in which digitizing writing on paper is required. Accordingly, to further improve the applications of OHWR using digital pens, the dataset used in this work is planned to be published for use in the scientific community. Future work following this will include complete sentence recognition, in addition to including digits and punctuation marks. Finally, the end-to-end model we presented requires minimal preprocessing, and mainly depends on the data, and thus to increase the robustness of a language recognizer, we plan to pair our model with a distinct dictionary or language model specific to the language to be recognized.

Acknowledgments. This work was supported by the Bayerisches Staatsministerium für Wirtschaft, Landesentwicklung und Energie as part of the EINNS project (Entwicklung Intelligenter Neuronaler Netze zur Schrifterkennung) (grant number IUK-1902-0005 // IUK606/002). Bjoern Eskofier gratefully acknowledges the support of the German Research Foundation (DFG) within the framework of the Heisenberg professorship program (grant number ES 434/8-1).

References

1. The digipen devkit. https://stabilodigital.com/devkit-demoapp-introduction/. Accessed 30 Jan 2021
2. The digipen hardware. https://stabilodigital.com/sensors-2021/. Accessed 30 Jan 2021

3. Abadi, M., et al.: Tensorflow: a system for large-scale machine learning. In: 12th {USENIX} Symposium on Operating Systems Design and Implementation ({OSDI} 16), pp. 265–283 (2016)
4. Abd Alshafy, H.A., Mustafa, M.E.: Hmm based approach for online Arabic handwriting recognition. In: 2014 14th International Conference on Intelligent Systems Design and Applications, pp. 211–215. IEEE (2014)
5. Amma, C., Georgi, M., Schultz, T.: Airwriting: hands-free mobile text input by spotting and continuous recognition of 3d-space handwriting with inertial sensors. In: 2012 16th International Symposium on Wearable Computers, pp. 52–59. IEEE (2012)
6. Bengio, Y., LeCun, Y., Nohl, C., Burges, C.: LEREC: a NN/HMM hybrid for on-line handwriting recognition. Neural Comput. **7**(6), 1289–1303 (1995)
7. Bersch, S.D., Azzi, D., Khusainov, R., Achumba, I.E., Ries, J.: Sensor data acquisition and processing parameters for human activity classification. Sensors **14**(3), 4239–4270 (2014)
8. Carbune, V., et al.: Fast multi-language LSTM-based online handwriting recognition. In: International Journal on Document Analysis and Recognition (IJDAR), pp. 1–14 (2020)
9. Chollet, F., et al.: Keras (2015). https://keras.io
10. Feng, Y., Zhang, Y., Xu, X.: End-to-end speech recognition system based on improved CLDNN structure. In: 2019 IEEE 8th Joint International Information Technology and Artificial Intelligence Conference (ITAIC), pp. 538–542. IEEE (2019)
11. Gan, J., Wang, W.: In-air handwritten English word recognition using attention recurrent translator. Neural Comput. Appl. **31**(7), 3155–3172 (2019)
12. Gan, J., Wang, W., Lu, K.: A unified CNN-RNN approach for in-air handwritten English word recognition. In: 2018 IEEE International Conference on Multimedia and Expo (ICME), pp. 1–6. IEEE (2018)
13. Gerth, S., et al.: Is handwriting performance affected by the writing surface? comparing tablet vs. paper. Frontiers in psychology 7 (2016)
14. Graves, A., Fernández, S., Gomez, F., Schmidhuber, J.: Connectionist temporal classification: labelling unsegmented sequence data with recurrent neural networks. In: Proceedings of the 23rd International Conference on Machine Learning, pp. 369–376 (2006)
15. Graves, A., Fernández, S., Liwicki, M., Bunke, H., Schmidhuber, J.: Unconstrained online handwriting recognition with recurrent neural networks. In: Advances in Neural Information Processing Systems 20, NIPS 2008 (2008)
16. Graves, A., Liwicki, M., Fernández, S., Bertolami, R., Bunke, H., Schmidhuber, J.: A novel connectionist system for unconstrained handwriting recognition. IEEE Trans. Pattern Anal. Mach. Intell. **31**(5), 855–868 (2008)
17. Graves, A., Schmidhuber, J.: Framewise phoneme classification with bidirectional LSTM and other neural network architectures. Neural Netw. **18**(5–6), 602–610 (2005)
18. Guyon, I., Schomaker, L., Plamondon, R., Liberman, M., Janet, S.: Unipen project of on-line data exchange and recognizer benchmarks. In: Proceedings of the 12th IAPR International Conference on Pattern Recognition, Vol. 3-Conference C: Signal Processing (Cat. No. 94CH3440-5), vol. 2, pp. 29–33. IEEE (1994)
19. Halder, A., Ramakrishnan, A.: Time delay neural networks for online handwriting recognition, June 2007. https://doi.org/10.13140/RG.2.2.25975.52641

20. Ioffe, S., Szegedy, C.: Batch normalization: accelerating deep network training by reducing internal covariate shift. In: International Conference on Machine Learning, pp. 448–456. PMLR (2015)
21. Jaeger, S., Manke, S., Reichert, J., Waibel, A.: Online handwriting recognition: the NPEN++ recognizer. Int. J. Doc. Anal. Recogn. **3**(3), 169–180 (2001)
22. Jäger, S., Liu, C.L., Nakagawa, M.: The state of the art in Japanese online handwriting recognition compared to techniques in western handwriting recognition. Doc. Anal. Recogn. **6**(2), 75–88 (2003)
23. Jeen-Shing, W., Yu-Liang, H., Cheng-Ling, C.: Online handwriting recognition using an accelerometer-based pen device. In: 2nd International Conference on Advances in Computer Science and Engineering (CSE 2013), pp. 231–234. Atlantis Press (2013)
24. Keysers, D., Deselaers, T., Rowley, H.A., Wang, L.L., Carbune, V.: Multi-language online handwriting recognition. IEEE Trans. Pattern Anal. Mach. Intell. **39**(6), 1180–1194 (2016)
25. Kim, J.H., Sin, B.-K.: Online handwriting recognition. In: Doermann, D., Tombre, K. (eds.) Handbook of Document Image Processing and Recognition, pp. 887–915. Springer, London (2014). https://doi.org/10.1007/978-0-85729-859-1_29
26. Kingma, D.P., Ba, J.: Adam: a method for stochastic optimization. arXiv preprint arXiv:1412.6980 (2014)
27. Koellner, C., Kurz, M., Sonnleitner, E.: What did you mean? an evaluation of online character recognition approaches. In: 2019 International Conference on Wireless and Mobile Computing, Networking and Communications (WiMob), pp. 1–6. IEEE (2019)
28. Liu, C.L., Jaeger, S., Nakagawa, M.: 'online recognition of Chinese characters: the state-of-the-art. IEEE Trans. Pattern Anal. Mach. Intell. **26**(2), 198–213 (2004)
29. Liu, Z.T., Wong, D.P., Chou, P.H.: An IMU-based wearable ring for on-surface handwriting recognition. In: 2020 International Symposium on VLSI Design, Automation and Test (VLSI-DAT), pp. 1–4. IEEE (2020)
30. Liwicki, M., Bunke, H.: IAM-ONDB-an on-line English sentence database acquired from handwritten text on a whiteboard. In: Eighth International Conference on Document Analysis and Recognition (ICDAR 2005), pp. 956–961. IEEE (2005)
31. Liwicki, M., Bunke, H., Pittman, J.A., Knerr, S.: Combining diverse systems for handwritten text line recognition. Mach. Vis. Appl. **22**(1), 39–51 (2011)
32. Liwicki, M., Graves, A., Fernàndez, S., Bunke, H., Schmidhuber, J.: A novel approach to on-line handwriting recognition based on bidirectional long short-term memory networks. In: Proceedings of the 9th International Conference on Document Analysis and Recognition, ICDAR 2007 (2007)
33. Mandal, S., Prasanna, S.M., Sundaram, S.: Exploration of CNN features for online handwriting recognition. In: 2019 International Conference on Document Analysis and Recognition (ICDAR), pp. 831–836. IEEE (2019)
34. Musa, M.E.: Towards building standard datasets for Arabic recognition. Int. J. Eng. Adv. Res. Technol. (IJEART) **2**(2), 16–19 (2016)
35. Ott, F., Wehbi, M., Hamann, T., Barth, J., Eskofier, B., Mutschler, C.: The onhw dataset: Online handwriting recognition from imu-enhanced ballpoint pens with machine learning. In: Proceedings of the ACM on Interactive, Mobile, Wearable and Ubiquitous Technologies, vol. 4, no. 3, pp. 1–20 (2020)
36. Palacios, R., Gupta, A., Wang, P.S.: Handwritten bank check recognition of courtesy amounts. Int. J. Image Graphics **4**(02), 203–222 (2004)
37. Plamondon, R., Srihari, S.N.: Online and off-line handwriting recognition: a comprehensive survey. IEEE Trans. Pattern Anal. Mach. Intell. **22**(1), 63–84 (2000)

38. Priya, A., Mishra, S., Raj, S., Mandal, S., Datta, S.: Online and offline character recognition: a survey. In: 2016 International Conference on Communication and Signal Processing (ICCSP), pp. 0967–0970. IEEE (2016)
39. Sainath, T.N., Vinyals, O., Senior, A., Sak, H.: Convolutional, long short-term memory, fully connected deep neural networks. In: 2015 IEEE International Conference on Acoustics, Speech and Signal Processing (ICASSP), pp. 4580–4584. IEEE (2015)
40. Schrapel, M., Stadler, M.L., Rohs, M.: Pentelligence: Combining pen tip motion and writing sounds for handwritten digit recognition. In: Proceedings of the 2018 CHI Conference on Human Factors in Computing Systems, pp. 1–11 (2018)
41. Shaikh Jahidabegum, K.: Character recognition system for text entry using inertial pen. Int. J. Sci. Eng. Technol. Res. (IJSETR) 4 (2015)
42. Singh, A., Bacchuwar, K., Bhasin, A.: A survey of OCR applications. Int. J. Mach. Learn. Comput. **2**(3), 314 (2012)
43. Srihari, S.N.: Recognition of handwritten and machine-printed text for postal address interpretation. Pattern Recogn. Lett. **14**(4), 291–302 (1993)
44. Srivastava, N., Hinton, G., Krizhevsky, A., Sutskever, I., Salakhutdinov, R.: Dropout: a simple way to prevent neural networks from overfitting. J. Mach. Learn. Res. **15**(1), 1929–1958 (2014)
45. Tlemsani, R., Belbachir, K.: An improved Arabic on-line characters recognition system. In: 2018 International Arab Conference on Information Technology (ACIT), pp. 1–10. IEEE (2018)
46. Wang, J.S., Chuang, F.C.: An accelerometer-based digital pen with a trajectory recognition algorithm for handwritten digit and gesture recognition. IEEE Trans. Industr. Electron. **59**(7), 2998–3007 (2011)
47. Wehbi, M., Hamann, T., Barth, J., Eskofier, B.: Digitizing handwriting with a sensor pen: a writer-independent recognizer. In: 2020 17th International Conference on Frontiers in Handwriting Recognition (ICFHR),pp. 295–300. IEEE (2020)
48. Yaeger, L.S., Webb, B.J., Lyon, R.F.: Combining neural networks and context-driven search for online, printed handwriting recognition in the newton. AI Mag. **19**(1), 73–73 (1998)

Signature Verification

2D vs 3D Online Writer Identification: A Comparative Study

Antonio Parziale[1]($\boxtimes$)(iD), Cristina Carmona-Duarte[2](iD), Miguel Angel Ferrer[2](iD), and Angelo Marcelli[1](iD)

[1] DIEM, University of Salerno, 84084 Fisciano, SA, Italy
{anparziale,amarcelli}@unisa.it
[2] IDETIC, Universidad de Las Palmas de Gran Canaria,
Las Palmas de Gran Canaria, Spain
ccarmona@idetic.eu, miguelangel.ferrer@ulpgc.es

Abstract. Nowadays, different automatic systems for writer identification and verification are available. On-line writer identification through automatic analysis of handwriting acquired with a tablet has been widely studied. Furthermore, the recent development of Commercial Off-The-Shelf (COTS) wearables with integrated inertial measurement units (IMUs) recording limbs movement allows the study of handwriting movements executed on the air. The goal of this paper is to compare the performance of an online writer identification system while processing 2D data acquired by a tablet while writing on-paper and 3D data acquired by a smartwatch while writing on-air. To this end, a database of handwriting samples produced by the same writers while writing the same symbols in the two modalities has been built up. The results of the study show a performance gap smaller than 5% between the 2D and 3D top implementations of the system, confirming that 3D handwriting is a promising alternative for developing wearable user authentication system.

Keywords: Online writer identification · User identification · On-air handwriting

1 Introduction

User authentication is the process of identifying a person from his/her physical, behavioral, and physiological characteristics such as fingerprint, face, iris, gait, handwriting, gestures, etc. [22].

Handwriting is likely one of the first traits that has been used by human beings for proofing their identity. Studies have been led to validate the hypothesis that handwriting is individualistic, and it can be used as biometric trait [36,37].

Thanks to the diffusion of digital tablets, a lot of commercial solutions have been designed to sign any kind of documents and automatically verify the identity of the writer [12,21,30]. Signature is not the unique handwritten pattern used for authenticating a subject. Many studies have been proposed for writer

J. Lladós et al. (Eds.): ICDAR 2021, LNCS 12823, pp. 307–321, 2021.
https://doi.org/10.1007/978-3-030-86334-0_20

identification by analyzing handwritten pin codes [24], characters and handwritten documents in general [18,32,41,45,46].

Studies on human movement learning and execution have shown that when writing movements learned in a 2D space (as a paper) through the dominant hand are repeated using other body parts, the same abstract representation of the movement (named motor program) is retrieved by the brain even if a different group of muscles is activated [8,14,26,31,34,43]. Handwritten patterns executed with groups of muscles with whom the subject has no or little experience (as when writing in a 3D space) show a trajectory similar to the movements executed on the 2D surface with the group of muscles usually activated by the subject. The differences are visible especially in the smoothness of trajectories and fluency of movements.

Several systems adopting either Microsoft Kinect and Leap Motion have been proposed for gesture recognition and writer identification [5,13,15,23,27,28,35, 42]. However, to capture the gestures with this kind of system the movement has to be executed in the acquisition area of the device, making the person movement conditioned by the device.

In the last years smart personal devices, such as smartphones or smartwatches are widely adopted making way to a new generation of applications, such as contactless payment, access to user's accounts, etc. [7]. These devices are equipped with an array of built-in sensors that make them very suitable for biometric authentication.

In particular, smartphones and smartwatches are usually equipped with inertial measurement units (IMUs), which encompass an accelerometer, gyroscope and magnetometer, whose signals can be recorded and processed for reconstructing the movement performed by the user. These new technologies and the previous studies on handwriting as biometric trait have motivated the development of authentication methods based on 3D handwriting-like gestures recorded from IMUs.

Using mobile phones to get the gestures, Bailador et al. proposed a user identification system that evaluates the execution of a signature on-air [4]. Moreover, Sun et al. [38] used the gestures captured by the IMU integrated in the phone to authorize the access of a user. In this way, each subject has a different signature or gesture to get access.

Furthermore, user identification with different smartbands has been proposed. The systems differ for the allowed gestures and the signals analyzed. The authentication system presented in [25] exploits only signals from the accelerometer, the system presented in [20] is based only on signals from the gyroscope while the system in [44] processes signals from both the IMUs of the smartbands. Each system adopts a different set of gestures for user authentication: simple and natural gestures like a single movement (arm up/down) [44], gesture with a complex shape [20] and user-defined gesture [25].

Moreover, recent works [9,16,17] have shown that it is possible to identify a subject that is writing a sentence or a signature on a sheet of paper using the signals collected from the smartwatch he/she is wearing.

In this paper, we compare the performance of a writer identifier dealing with 2D and 3D equivalent symbols of the same writers. In particular, we compare the performance of an authentication system when a digit is written by subjects on-paper over a tablet and on-air with a smartwatch. To the best of our knowledge, that is the first study that evaluates the performance of a user authentication system when the same group of users execute the same gesture collected in 2D and 3D space. The comparison was performed in terms of accuracy of three different classification paradigms.

Experiments performed on a dataset involving 98 subjects show that the 2D systems outperform the corresponding 3D ones, with a difference in accuracy between 1.73% and 4.98%, depending on the classifier. Such a performance gap, however, can be explained by taking into account the much smaller skill of the subjects while writing on-air in comparison with the lifelong one they have developed writing on-paper, resulting in a much bigger intra-writer variability of the former with respect to the latter.

The remaining of the paper is organized as follows: Sect. 2 describes the protocol adopted for data acquisition and how has been designed the authentication system, Sect. 3 reports the experimental results and the statistical analysis, Sect. 4 discusses the results and, eventually, Sect. 5 concludes the paper. The feature sets used in the experimentation are freely available for research purposes, as described in Section "Data availability".

2 Methods and Materials

In this section, we present the devices and the protocol adopted for data collection. We describe how data are processed and features extracted and eventually we introduce the classification paradigms adopted for implementing the six user identification systems that are then compared in the experimental section.

2.1 Devices

Figure 1 shows the two devices used in this paper for collecting data: Wacom Intuos 2 and Apple Watch series 4.

Wacom Intuos 2 is a digitizing tablet with 100 Hz sampling rate. It records the movements of a subject that is writing with an electronic ballpoint pen on sheets of paper placed over the device's surface. The device records the x and y position of the ballpoint pen up to a distance of 3 cm from the surface. The pressure (p) exerted by the pen on the device surface is also recorded. We adopted an ink-and-paper digitizing tablet instead of a stylus-and-screen digitizing tablet to avoid changes in subjects' writing style caused by changes in the writing conditions introduced by the latter class of tablet, as for example a different attrition.

(a) Wacom Intuos 2   (b) Apple Watch

Fig. 1. The two devices used for data collection. (a) Wacom Intuos 2 (b) Apple Watch series 4 with custom application

Apple Watch series 4 is a smartwatch equipped with an accelerometer whose signal is obtained from the Device-Motion Data. The accelerometer signal is acquired at 100 Hz sampling rate and it is elaborated in order to remove the environmental bias, such as the effect of gravity. We developed an application running on the device to easily record data during the experimentation. The application allows to record up to 5 different tasks. Once the subject presses on one of the buttons a sequence of sound is emitted for synchronizing his/her activity with data recording.

2.2 Dataset

We collected data from 98 subjects, 55 males and 43 females, whose age ranges in the interval 14–67 years with a mean value of 32.60 and a standard deviation of 13.94. 6 out of 98 participants were left-handed. Participants volunteered to take part in the experiment and expressed their formal consent. A questionnaire was administered to each subject in order to define the level of education, the health conditions, whether they use routinely drugs or other substances that are known to affect motor control, as well as the questions required to evaluate the Edinburgh Handedness score.

Each subject took part in two sessions of data acquisition, one for 3D data and the other for 2D data, with a recess of 5 min between them. The order of the sessions was randomized across the participants, to avoid possible biases.

During the 3D session, the subject wore the smartwatch on the wrist of the dominant hand and wrote on air the digit "8" ten times. Every time the subject was ready to write, he/she pressed a button in the application we designed for data collection and waited for a sequence of two sounds. Once he/she completed the execution of the movement, he/she kept the arm relaxed and motionless until the emission of a third sound by the smartwatch. We asked the subjects to change their postures (sitting and standing) and poses for each of the ten repetitions.

During the 2D session, the subject wrote the digit "8" ten times on a sheet of paper placed on the tablet. We used MovAlyzeR®v6.1 [39] as computer application for recording the (x(t), y(t), p(t)) signals from tablet. The computer emitted a sound every time the subject could write an instance of the digit. We asked the subjects to move their arm far from the tablet each time they finished writing a digit, in order to stop the sample recording and to wait until the next sound. Sounds were emitted every five seconds for being sure that the subject had the time to write the digit and to rest before writing again.

Overall, 980 samples were collected with the tablet and 977 samples were collected with the smartwatch. Three samples were removed by the data acquired with the smartwatch due to an error in one execution from three subjects.

Figure 2 shows the (x(t), y(t)) trajectories collected with the Wacom and the (x(t), y(t), z(t)) trajectories recorded from the Watch after processing the acceleration signal.

2.3   Feature Set

Data Preprocessing. The linear acceleration recorded by the accelerometer inside the Apple Watch was processed to obtain the movement trajectory along the (x,y,z) axis. First of all, the mean acceleration is computed and subtracted from the acceleration signal. Then, this signal is integrated to work out the velocity. The obtained velocity signal is filtered by a low pass filter (30 Hz) and the mean of the resulted velocity is subtracted to it. Finally, the velocity signal is integrated to recover the wrist trajectory (x(t), y(t), z(t)).

Data recorded with the tablet are sharpened and smoothed using a Fast Fourier Transform (FFT) low-pass filter with a frequency equal 12 Hz. The frequency components of the signal are extracted using FFT [40]. Eventually, pen-ups at the beginning and at the end of each sample are removed.

Feature Extraction. Samples acquired with both the devices are represented by features devoted to highlighting their spatial and temporal characteristics. The feature sets used for representing the 2D and 3D samples share more than half of the features and they also include features that are tailored for the specific writing modality. Table 1 reports the description of the 37 features used for representing the samples acquired with the tablet, and the 42 features used for representing the samples acquired with the smartwatch.

Most of the features are computed over the signals acquired by the devices (the coordinates $x(t)$, $y(t)$, $z(t)$ and the pressure $p(t)$) and signals derived from them (absolute velocity $v(t)$, absolute acceleration $a(t)$, absolute jerk $j(t)$, velocity along x-axis $v_x(t)$, y-axis $v_y(t)$ and z-axis $v_z(t)$, acceleration along x-axis $a_x(t)$ y-axis $a_y(t)$ and z-axis $a_z(t)$, jerk along x-axis $j_x(t)$ y-axis $j_y(t)$ and z-axis $j_z(t)$) [29]. Moreover, coordinates in the Cartesian plane are converted into polar coordinates in order to compute the angular velocity $\omega(t)$, as suggested in [6], and the angular acceleration $\alpha(t)$ by derivation. The direction of velocity vector in the xy plane and in the xz plane are also used as features and are computed as in Eq. 1 and Eq. 2, respectively.

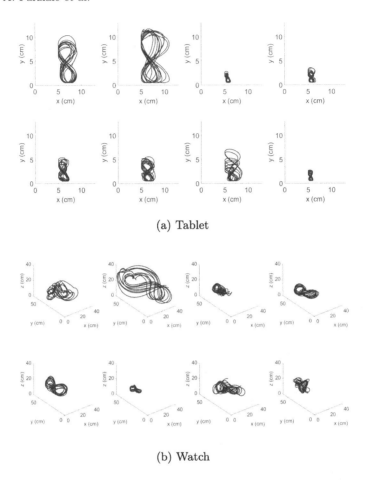

(a) Tablet

(b) Watch

Fig. 2. Data collected from 8 different subjects with (a) the tablet and (b) the smart-watch (axes in cm). Each box contains the superposed samples produced by the same user.

$$\theta_{xy}(t) = \tan^{-1}\left(\frac{v_y(t)}{v_x(t)}\right) \tag{1}$$

$$\theta_{xz}(t) = \tan^{-1}\left(\frac{v_z(t)}{v_x(t)}\right) \tag{2}$$

Another group of features is devoted to representing the spatial properties of each sample. The *displacement* measured the difference between the first and the last point of a sample along the axes. *Width*, *height* and *depth* measured the maximum distance among points along the x-axis, y-axis and z-axis, respectively.

Samples acquired with tablet were segmented in strokes by zero crossings of the vertical velocity after peak velocity [40] in order to compute the following

Table 1. Features computed for samples acquired with tablet and smartwatch.

Device	Number of features	Description
Tablet/Watch	20	Standard deviation and Average of $x(t)$, $y(t)$, $v_x(t)$, $v_y(t)$, $a_x(t)$, $a_y(t)$, $j_x(t)$, $j_y(t)$, $w(t)$, $\alpha(t)$
Tablet/Watch	2	Number of local maxima of: $v(t)$, $a(t)$
Tablet/Watch	2	Number of local minima of: $v(t)$, $a(t)$
Tablet/Watch	2	Displacement along: $x(t)$, $y(t)$
Tablet/Watch	2	Width and Height of symbol
Tablet/Watch	2	Standard deviation and Average of $\theta_{xy}(t)$
Tablet	2	Standard deviation and Average of $p(t)$
Tablet	1	Number of strokes
Tablet	1	Average of absolute stroke size
Tablet	1	Number of pen-ups
Tablet	1	Ratio between on-air time and total duration of symbol
Tablet	1	Average of straightness error
Watch	8	Standard deviation and Average of $z(t)$, $v_z(t)$, $a_z(t)$, $j_z(t)$
Watch	1	Displacement along $z(t)$
Watch	2	Standard deviation and Average of $\theta_{xz}(t)$
Watch	1	Depth of digit

features: number of strokes, average of the absolute size of the sample strokes, average of the straightness error of the strokes [10]. This subset of features is widely proposed in the literature for 2D systems for writer identification but is not used in 3D systems.

Eventually, the number of pen-ups and the ratio between the time spent on-air and the total duration of a symbol are two more features used for representing only 2D samples because they cannot be computed over 3D samples.

Feature Normalization. The value of each feature is normalized as described by Eq. 3, where $\bar{f}_i$ and σ_i are the mean and the standard deviation of the i-th feature f_i computed over the whole dataset.

$$f_i^{norm} = \frac{f_i - \bar{f}_i}{\sigma_i} \qquad (3)$$

Feature Selection. Each set of features was analyzed in order to extract a subset of non-redundant features and improving the performance of the classifier and for speeding up the training and the identification phases. We adopted the Pearson correlation coefficient for selecting the best subset of features. In particular, we computed the absolute value of the Pearson correlation coefficient between each pair of features obtaining a correlation matrix. We scanned the upper triangle of the correlation matrix from left to right and we dropped the features with at least one coefficient greater than 0.9. At the end, each sample was represented by 28 features, both for the data acquired with the tablet and the smartwatch.

2.4 Classifiers

For a thorough comparison, we evaluated three different classification paradigms for implementing our user identification system: Random Forest (RF), Support Vector Machine (SVM), K-nearest-neighbors (KNN).

RF is a meta classifier that aggregates several decision trees on various sub-samples of the dataset. The predictive accuracy is improved, and the over-fitting is mitigated with averaging. The process of finding the root node and splitting the feature nodes is random [19].

SVM has the objective of finding the best hyperplane that distinctly classifies the data points in the space. The best hyperplane is the one with the maximum margin, i.e., the maximum distance among data points of the classes [11].

KNN is a non-parametric method and a type of instance-based learning that classifies a sample by evaluating what is the class most common among its nearest neighbors in the space [1,2].

3   Experimental Results

In this section, we measured and compared the performance of the systems based on the smartwatch and the systems based on the tablet in a non-critical authentication scenario: to recognize the user out of a set of registered users.

Overall, we compared six versions of the system that differs for the device used for data collection (tablet or smartwatch) and for the classification stage (RF, SVM or KNN).

The systems were realized using the classifiers and the feature selection tool implemented in Scikit-learn [33].

3.1   User Identification Experiment

We evaluated the classification accuracy of the six systems with 98 subjects enrolled in each of them.

Both the datasets, the one collected with the smartwatch and the one collected with the tablet, were split in a training and in a test set. In particular, for each subject, 70% of the samples were used as training set and the remaining samples as test set. The partition in training set and test set was repeated 20 times by randomly shuffling the data at each split.

We compared the 2D and 3D systems in terms of classification accuracy measured over the 20 partitions of the dataset. For each data partition, we set the classifiers' parameters to the values that maximized the accuracy measured on the training set. In particular, the parameters of each classifier were fine-tuned through a cross-validated grid-search. We adopted a 3-folds cross-validation over

Table 2. Analyzed range for classifiers' parameters

	Parameters
RF	*n_estimators*: [500,1000] (step:100), *criterion*: entropy, *max_depth*: [80, 90, 100], *min_samples_split*: [2, 4], *min_samples_leaf*: [1, 2] *max_features*: [sqrt, log2], *bootstrap*: [True, False]
SVM	*C*: [0.001,0.01, 0.1, 1, 10, 100, 1000], *gamma*: [0.0001, 0.001, 0.01, 0.035, 1], *kernel*: [rbf, sigmoid], *decision function shape*: [ovo, ovr]
KNN	*weights*: [uniform, distance], *algorithm*: [ball_tree, kd_tree], *metric*: [euclidean, manhattan, chebyshev], *n_neighbors*: [1,6] (step: 1), *leaf_size*: [10,40] (step: 10)

each training set in order to fine-tune the parameters of the classifiers. A 3-folds cross-validation divides the data into 3 disjointed folds preserving the class distribution and then, in turns, it uses 2 folds as training set and 1 as validation set. Table 2 reports the classifiers' parameters, with the corresponding analyzed ranges, which were fine-tuned with the cross-validated grid-search. The grid search returned different parameter sets for the 2D and 3D systems depending on the 20 training set configurations. By comparing the two systems equipped with the RF classifier, we noticed that, in most of the cases, the best accuracies on the 3D training sets were obtained with a number of estimators greater than the one required for obtaining the best accuracies on the 2D training sets. By comparing the two systems based on the SVM classifier, we noticed that the gamma values were greater when 3D feature sets were processed, resulting in a coarser approximation of the decision boundary. Eventually, for some training set configurations, the 3D system equipped with the KNN classifier obtained the best accuracy on the training set with a number of neighbors greater than the one required by the 2D system. Overall, the results of the grid-search confirm that the distribution of the features' values obtained from the 3D points is harder to model than in the case of 2D.

Table 3 reports the mean percentage of correct identifications obtained with the six systems repeating 20 times the partition in training and test sets.

3.2   Statistical Analysis

A one-way analysis of variance (ANOVA) was performed to determine whether there were statistically significant differences between the mean accuracies of the

Table 3. Classification accuracies (mean ± st.dev)

Classifier	Tablet (2D)	Smartwatch (3D)
RF	94.747 ± 1.432	89.766 ± 1.103
SVM	90.735 ± 1.500	89.000 ± 1.547
KNN	90.531 ± 1.431	87.620 ± 1.366

six systems (named treatments in the ANOVA framework). In particular, we tested the hypothesis that the six treatments were drawn from populations with the same mean against the alternative hypothesis that the population means are not all the same. Figure 3 shows the box plot of the accuracies obtained with the six systems repeating 20 times the partition in training and test sets. Before the ANOVA, we performed the Anderson-Darling test in order to verify that the accuracy values of each treatment were drawn from a normal distribution. For all the treatments, it was not possible to reject the hypothesis that samples are from a population with a normal distribution at 5% significance level.

The ANOVA returned a p-value, which is the probability that the test statistic can take a value greater than the computed F-statistic (59.11), equal to 4.54317e-30. This small p-value suggests rejection of the null hypothesis, therefore the differences among the accuracies of the six systems are significant.

We performed a multiple comparison with Bonferroni method at the 5% significance level for understanding which pairs of means are significantly different. The comparison between the systems *Tablet-KNN* and *Tablet-SVM* returns that their mean accuracies are not significantly different, with a p-value equal to 1. The same result is obtained by comparing the systems *Tablet-KNN* and

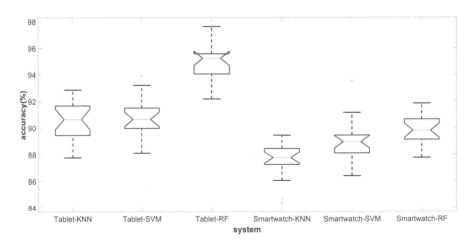

Fig. 3. Box plot of the accuracies obtained with the six systems and 20 configurations of training and test sets.

Smartwatch-RF and the systems *Smartwatch-SVM* and *Smartwatch-RF*. Eventually, the comparison between *Tablet-SVM* and *Smartwatch-RF* returns that the mean accuracies are not significantly different with a p-value equal to 0.4662.

The Anderson-Darling test, the ANOVA and the Bonferroni method were performed using the functions *adtest*, *anova1*, *multiplecompare* implemented in MATLAB R2021a, respectively.

4   Discussion

The results reported in Table 3 show that the on-air handwriting system achieves lower performance than the on-paper one.

We conjecture that such a difference may be due to the fact that the subjects have never performed on-air writing before. As a consequence, they are learning on the fly how to execute the motor program they have developed for traditional writing to the new task. Moreover, the subjects do not have a visible trace of the movement to be exploited by the visual system in order to evaluate the quality of the movement and eventually improve it. Thus, there is no feedback to drive the learning besides the one provided by the proprioceptive system, and therefore we expect to observe a larger intra-subject variability with respect to the 2D case, whose motor program has been fully learnt by the subject. To prove such a conjecture, we evaluated the intra-subjects variability in both cases by computing the SNR between the velocity profile for each pair of samples:

$$SNR = 10log\left(\frac{\int_{t_s}^{t_e} |\boldsymbol{v}_1(t)|\, dt}{\int_{t_s}^{t_e} |\boldsymbol{v}_1(t) - \boldsymbol{v}_2(t)|^2\, dt}\right) \qquad (4)$$

where $\boldsymbol{v}_1(t)$ and $\boldsymbol{v}_2(t)$ are the velocity profiles of the two samples, and t_s and t_e are the starting and the ending time of $\boldsymbol{v}_1(t)$. Note that in case two samples have different duration, the shorter one is elongated by adding zero values until its duration becomes the same as the longer ones. The intra-subject variability is then evaluated by computing, for each device, the SNR values among the 90 pairs of digits produced by the same subject. For each subject, the two SNR samples are compared using the Ansari-Bradley test, which evaluates the null hypothesis that the SNR samples come from the same distribution against the alternative hypothesis that the samples come from distributions with the same median and shape but different variance [3]. The Ansari-Bradley test reports that the null hypothesis has to be rejected at 5% significance level for 36 out of 98 subjects. In particular, 25 out of 36 subjects have the variance of the SNR values greater when the digits are executed on-air than on-paper. Therefore, 25 subjects show an intra-writer variability on-air larger than their intra-writer variability on-paper, while 11 subjects show the opposite. This means that 14 out of 98 subjects, about 14% of the total, show an on-air variability greater than their on-paper variability, thus supporting our conjecture to explain, at least partially, the observed performance gap.

5   Conclusion

Handwriting is widely adopted as biometric trait for identifying a subject and verifying a claimed identity. The diffusion of the new smart devices with in-built IMUs has opened the door to new biometric systems that identify users by the simple or complex gestures they perform. With respect to tablets or pen-based devices, these devices offer the advantage of following the users anywhere, instead of requiring the user to reach a specific location, are of personal use, instead of being shared by many users, and the user model can be stored on them, instead of being stored on a server. In a nutshell, they are simpler to use, safer and potentially more secure than currently used devices.

Motivated by the previous observations, we have designed a user authentication system capable of using both modalities and envisaging different classifiers. Then, we compared the performance in a user identification task by means of an ad-hoc dataset, obtained by 98 subjects, each one executing the same pattern, the trajectory of the digit "8", both on-paper and on-air.

Experiments, envisaging different implementations of the system for each modality, show that 2D systems outperform 3D ones with an improvement of the accuracy ranging between 1.73% and 4.98%, depending on the classifier.

The experimental results have shown that on-air and on-paper modalities reach similar performance, with the top ones achieved when a Random Forest classifier is used. Even more relevant, those performances are obtained by using as handwriting pattern a single symbol, namely the digit "8".

The analysis of the SNR between each pair of samples showed that, because of the simplicity of the task, there were cases when samples produced by a given writer were more similar to samples produced by other writers than to its own ones. This means that the data set contains samples that can be considered as skilled imitations of other writers, thus making the performance reported in Sect. 3 more suitable to provide an overall characterization of the system behavior.

The analysis of the data supported the conjecture that the better performance of the tablet-based system can be explained with the higher variability exhibited by the on-air writing patterns with respect to on-paper writing patterns. In fact, and according to the studies on motor control, the analysis of variability of both on-paper and on-air movements has confirmed that for a certain number of subjects the intra-writer variability is greater when they write on-air. On-air skill execution is much more variable with respect to on-paper one due to the much shorter time the users spent experiencing the on-air modality than the lifelong experience they have acquired on-paper.

Further investigations will consider patterns including many symbols, to evaluate to which extent longer patterns effects both intra-writer and inter-writer variability and eventually performance, with the ultimate goal of establishing whether or not there is an optimal number of symbols to be included to achieve the best performance. Next, we will further develop the approach we have adopted to evaluate the intra-writer and inter-writer variability to estimate the level of skillfulness of the forgeries and to which extent such a measure correlates with the performance of both the systems.

Acknowledgment. This study was funded by the Spanish government's MIMECO PID2019-109099RB-C41 research project and European Union FEDER program/funds. C. Carmona-Duarte was supported by a Juan de la Cierva grant (IJCI-2016-27682), and Viera y Clavijo grant from ULPGC.

Data Availability Statement. The feature sets used in the experiments are freely available for research purposes at https://github.com/Natural-Computation-Lab/2D_vs_3D_handwriting.

References

1. Aha, D.W., Kibler, D., Albert, M.K.: Instance-based learning algorithms. Mach. Learn. **6**(1), 37–66 (1991)
2. Altman, N.S.: An introduction to kernel and nearest-neighbor nonparametric regression. Am. Stat. **46**(3), 175–185 (1992)
3. Ansari, A.R., Bradley, R.A., et al.: Rank-sum tests for dispersions. Ann. Math. Stat. **31**(4), 1174–1189 (1960)
4. Bailador, G., Sanchez-Avila, C., Guerra-Casanova, J., de Santos Sierra, A.: Analysis of pattern recognition techniques for in-air signature biometrics. Pattern Recogn. **44**(10–11), 2468–2478 (2011)
5. Berman, S., Stern, H.: Sensors for gesture recognition systems. IEEE Trans. Syst. Man Cybern. Part C (Appl. Rev.) **42**(3), 277–290 (2011)
6. Bhatia, S., Bhatia, P., Nagpal, D., Nayak, S.: Online signature forgery prevention. Int. J. Comput. Appl. **75**(13), 21–29 (2013)
7. Blasco, J., Chen, T.M., Tapiador, J., Peris-Lopez, P.: A survey of wearable biometric recognition systems. ACM Comput. Surv. (CSUR) **49**(3), 1–35 (2016)
8. Carmona-Duarte, C., Ferrer, M.A., Parziale, A., Marcelli, A.: Temporal evolution in synthetic handwriting. Pattern Recogn. **68**, 233–244 (2017)
9. Ciuffo, F., Weiss, G.M.: Smartwatch-based transcription biometrics. In: 2017 IEEE 8th Annual Ubiquitous Computing, Electronics and Mobile Communication Conference (UEMCON), pp. 145–149. IEEE (2017)
10. Contreras-Vidal, J.L., Teulings, H., Stelmach, G.: Elderly subjects are impaired in spatial coordination in fine motor control. Acta Physiol. **100**(1–2), 25–35 (1998)
11. Cortes, C., Vapnik, V.: Support-vector networks. Mach. Learn. **20**(3), 273–297 (1995)
12. Diaz, M., Ferrer, M.A., Impedovo, D., Malik, M.I., Pirlo, G., Plamondon, R.: A perspective analysis of handwritten signature technology. ACM Comput. Surv. (CSUR) **51**(6), 1–39 (2019)
13. Erol, A., Bebis, G., Nicolescu, M., Boyle, R.D., Twombly, X.: Vision-based hand pose estimation: a review. Comput. Vis. Image Underst. **108**(1–2), 52–73 (2007)
14. Ferrer, M.A., Diaz, M., Carmona-Duarte, C., Morales, A.: A behavioral handwriting model for static and dynamic signature synthesis. IEEE Trans. Pattern Anal. Mach. Intell. **39**(6), 1041–1053 (2016)
15. Gan, J., Wang, W., Lu, K.: In-air handwritten Chinese text recognition with temporal convolutional recurrent network. Pattern Recogn. **97**, 107025 (2020)

16. Griswold-Steiner, I., Matovu, R., Serwadda, A.: Handwriting watcher: a mechanism for smartwatch-driven handwriting authentication. In: 2017 IEEE International Joint Conference on Biometrics (IJCB), pp. 216–224. IEEE (2017)
17. Griswold-Steiner, I., Matovu, R., Serwadda, A.: Wearables-driven freeform handwriting authentication. IEEE Trans. Biom. Behav. Identity Sci. **1**(3), 152–164 (2019)
18. He, S., Schomaker, L.: Writer identification using curvature-free features. Pattern Recogn. **63**, 451–464 (2017)
19. Ho, T.K.: Random decision forests. In: Proceedings of 3rd International Conference on Document Analysis and Recognition, vol. 1, pp. 278–282. IEEE (1995)
20. Huang, C., Yang, Z., Chen, H., Zhang, Q.: Signing in the air w/o constraints: robust gesture-based authentication for wrist wearables. In: GLOBECOM 2017– 2017 IEEE Global Communications Conference, pp. 1–6. IEEE (2017)
21. Impedovo, D., Pirlo, G., Plamondon, R.: Handwritten signature verification: new advancements and open issues. In: 2012 International Conference on Frontiers in Handwriting Recognition, pp. 367–372. IEEE (2012)
22. Jain, A.K., Flynn, P., Ross, A.A.: Handbook of Biometrics. Springer, Heidelberg (2007). https://doi.org/10.1007/978-0-387-71041-9
23. Kamaishi, S., Uda, R.: Biometric authentication by handwriting using leap motion. In: Proceedings of the 10th International Conference on Ubiquitous Information Management and Communication, pp. 1–5 (2016)
24. Kutzner, T., Pazmiño-Zapatier, C.F., Gebhard, M., Bönninger, I., Plath, W.D., Travieso, C.M.: Writer identification using handwritten cursive texts and single character words. Electronics **8**(4), 391 (2019)
25. Liu, J., Zhong, L., Wickramasuriya, J., Vasudevan, V.: User evaluation of lightweight user authentication with a single tri-axis accelerometer. In: Proceedings of the 11th International Conference on Human-Computer Interaction with Mobile Devices and Services, pp. 1–10 (2009)
26. Marcelli, A., Parziale, A., Senatore, R.: Some observations on handwriting from a motor learning perspective. In: AFHA, vol. 1022, pp. 6–10. Citeseer (2013)
27. Mitra, S., Acharya, T.: Gesture recognition: a survey. IEEE Trans. Syst. Man Cybern. Part C (Appl. Rev.) **37**(3), 311–324 (2007)
28. Murata, T., Shin, J.: Hand gesture and character recognition based on kinect sensor. Int. J. Distrib. Sens. Netw. **10**(7), 278460 (2014)
29. Parziale, A., Senatore, R., Della Cioppa, A., Marcelli, A.: Cartesian genetic programming for diagnosis of parkinson disease through handwriting analysis: performance vs. interpretability issues. Artif. Intell. Med. **111**, 101984 (2021)
30. Parziale, A., Diaz, M., Ferrer, M.A., Marcelli, A.: SM-DTW: stability modulated dynamic time warping for signature verification. Pattern Recogn. Lett. **121**, 113–122 (2019)
31. Parziale, A., Parisi, R., Marcelli, A.: Extracting the motor program of handwriting from its lognormal representation. In: The Lognormality Principle and its Applications in e-Security, e-Learning and e-Health, pp. 289–308 (2021). https://doi.org/10.1142/9789811226830_0013
32. Parziale, A., et al.: An interactive tool for forensic handwriting examination. In: 2014 14th International Conference on Frontiers in Handwriting Recognition, pp. 440–445. IEEE (2014)
33. Pedregosa, F., et al.: Scikit-learn: machine learning in Python. J. Mach. Learn. Res. **12**, 2825–2830 (2011)
34. Raibert, M.H.: Motor control and learning by the state space model. Ph.D. thesis, Massachusetts Institute of Technology (1977)

35. Shin, J., Kutsuoka, T., Kim, C.M.: Writer verification based on three-dimensional information using kinect sensor. In: Proceedings of the International Conference on Research in Adaptive and Convergent Systems, pp. 89–90 (2016)
36. Srihari, S.N., Cha, S.H., Arora, H., Lee, S.: Individuality of handwriting. J. Forensic Sci. **47**(4), 1–17 (2002)
37. Srihari, S.N., Tomai, C.I., Zhang, B., Lee, S.: Individuality of numerals. In: ICDAR, vol. 3, pp. 1096–1100. Citeseer (2003)
38. Sun, Z., Wang, Y., Qu, G., Zhou, Z.: A 3-D hand gesture signature based biometric authentication system for smartphones. Secur. Commun. Netw. **9**(11), 1359–1373 (2016)
39. Teulings, H.L.: MovAlyzeR. Version 6.1. Neuroscript LTD (2021). https://www.neuroscript.net
40. Teulings, H.L., Maarse, F.J.: Digital recording and processing of handwriting movements. Hum. Mov. Sci. **3**(1–2), 193–217 (1984)
41. Venugopal, V., Sundaram, S.: Online writer identification system using adaptive sparse representation framework. IET Biom. **9**(3), 126–133 (2020)
42. Wang, X., Tanaka, J.: GesID: 3D gesture authentication based on depth camera and one-class classification. Sensors **18**(10), 3265 (2018)
43. Wing, A.M.: Motor control: mechanisms of motor equivalence in handwriting. Curr. Biol. **10**(6), R245–R248 (2000)
44. Yang, J., Li, Y., Xie, M.: Motionauth: motion-based authentication for wrist worn smart devices. In: 2015 IEEE International Conference on Pervasive Computing and Communication Workshops (PerCom Workshops), pp. 550–555. IEEE (2015)
45. Yang, W., Jin, L., Liu, M.: Deepwriterid: an end-to-end online text-independent writer identification system. IEEE Intell. Syst. **31**(2), 45–53 (2016)
46. Zhang, X.Y., Xie, G.S., Liu, C.L., Bengio, Y.: End-to-end online writer identification with recurrent neural network. IEEE Trans. Hum. Mach. Syst. **47**(2), 285–292 (2016)

A Handwritten Signature Segmentation Approach for Multi-resolution and Complex Documents Acquired by Multiple Sources

Celso A. M. Lopes Junior[1], Murilo C. Stodolni[1],
Byron L. D. Bezerra[1(✉)], and Donato Impedovo[2]

[1] Polytechnic School of Pernambuco, University of Pernambuco, Pernambuco, Brazil
{camlj,mcs2}@ecomp.poli.br, byron.leite@upe.br
[2] Department of Computer Science, University of Bari, Bari, Italy
donato.impedovo@uniba.ite

Abstract. Handwritten Signature is a biometric feature, which enables personal verification. Thus, it constitutes an alternative authentication used in several applications, such as bank checks, contracts, certificates, and forensic science. Signatures may be presented on a complex background with different textures, turning automatic signature segmentation into a difficult task. In this work, we propose an approach to locate and segment only the signature image pixels in documents with complex backgrounds, acquired by smartphone cameras in different environments, without any prior information about the signature location in these documents, the pen used by the signer, among others issues. Our approach is based on the U-net network architecture, combined with a pre-processing stage that allows dealing with images having different resolutions and distortions due to the document acquiring process. To make our model more robust to background and texture variations, we have generated a data set consisting of 20,000 document photos with different sizes, textures, and documents, named DSSigDataset-2. Our experiments show that the proposed method achieved encouraging results, over precision, recall, and F1-score measures, in all evaluated data sets (ours and benchmark ones).

Keywords: Handwritten signatures · Segmentation · FCN · CNN · Forensic science · Image processing

1 Introduction

Handwritten signatures are still widely used to authenticate and validate various documents, whether by private or public institutions. Besides, forensic document analysis specialists examine handwritten signatures to certify the authenticity of the spelling and reveal potential fraud, which in some cases could mean high-value financial losses. However, many of the signature verification approaches found in the literature feature a controlled data set environment for extracting handwritten

© Springer Nature Switzerland AG 2021
J. Lladós et al. (Eds.): ICDAR 2021, LNCS 12823, pp. 322–336, 2021.
https://doi.org/10.1007/978-3-030-86334-0_21

signatures [2,5,6,20]. Some reference data sets, such as GPDS [22] and MCYT [16], are examples of light background images and dark signatures. This characteristic does not present a scenario found in daily-life environments. For instance, document images captured by mobile devices are increasingly common and accepted by institutions as copies of original documents. Most of these images can compose the characteristics of a more complex scenario, as shown in Fig. 1. Therefore, treating these images so to can extract only handwritten signatures is a challenging task.

Fig. 1. An illustrative example of the different documents that show the diversity of images and backgrounds. Subscriptions are added from a public database.

Another condition that can impair the quality of features in handwritten signatures happens when the image presents some image distortion such as perspective, skew, unexpected scale, or resolution, frequently in photo scanning. Thus, a public database that met such characteristics of a real scenario could be of great interest to the research community. In previous works, the signature segmentation methods require as input an image in a predefined dimension. So, if the image is larger than the expected size, these systems scale down the target image, losing some crucial pixels in the signature for forensic purposes. Otherwise, if the image is smaller than the expected size, it must be scaled up. Additional pixels can be introduced in the signature boundary, affecting the signature verification accuracy. Figure 2 show a result of resizing signatures in the image. The font size, inclination, and width of the handwritten signature were affected.

These images in a real scenario tend to different resolutions and different dimensions. Therefore, applying changes to these images at a fixed size may cause a loss in signature characteristics. Therefore, to better cope with daily-life scenarios, the Handwritten Signature Verification (HSV) systems must look to preserve the attributes of handwritten signatures when extracting them from documents. This can contribute to a more accurate analysis of automatic HSV

(a) original   (b) resized

Fig. 2. (a) original handwritten signature, and (b) handwritten signature after resizing. Red circles and lines point to morphological changes. (Color figure online)

since it conserves important information essential to graphotechnical features used by forensic science during signature analysis.

This paper proposes an approach to the segmentation of handwritten signatures in ID documents images of any size and dimension used as input to a convolutional neural network. We used a Fully Convolutional Network (FCN) [12] for segmentation images. One of the most contributions of our proposal is the ability to take as input the document image in any size, yielding, as a consequence, the maintenance of relevant signature information to feature extraction and signature classification.

2   Related Works

Several researchers have explored the power of Deep Learning for activities related to segmentation and verification of signatures. The CNN models have been widely used for this purpose. Approaches were introduced for offline signature recognition using a rough neural network and rough set. For instance, in [3] a proposal achieves good results because the rough sets can discover hidden patterns and regularities in the application, while the neural network tries to find better recognition performance to classify the input offline signature images. However, the presented approach scales the image to a fixed size.

A model based on a convolutional neural network (CNN-based) is used in [8] to detect handwritten texts on whiteboard images. To identify instances of texts at different scales, the authors adopted a feature pyramid network. The feature pyramid network acts as a backbone network to extract three feature maps of different scales in an input image. Then a detection module is attached for each resource map. However, the images used in this work have characteristics of a light background and dark texts.

In work proposed by [11], the authors adopted the architecture of the U-Net convolutional network for the process of segmenting machine texts and manuscripts. The data set used in this work is the CEDAR signature database [9]. The authors present some qualitative results of the output images. However, despite reporting success in the results with the approach, the authors were more concerned with the extraction of machine texts. Therefore, two important points must be noted: i) the metric used to measure the similarity between the original handwritten texts and the segmentation results are not exposed; ii) the authors do not provide quantitative results for segmenting the manuscripts.

Some authors have used post-processing techniques in conjunction with a model based on CNN. In work proposed by [15], the authors address several tasks simultaneously, such as extracting the baseline for handwritten texts, extracting the page, and analyzing the layout. The authors used a CNN model with the ResNet-50 architecture to perform these tasks, followed by some post-processing techniques, such as binarization, morphological operations, analysis of connected components, and shape vectorization. The databases adopted in this work are private and are not available. Despite this, the exposed images show characteristics similar to other databases of signatures and manuscripts, in which the background is light and the texts are dark.

In [21] is presented a feature extraction approach through Deep Convolutional Neural Network combined with the SVM classifier to writer-independent (WI) HSV. The proposed approach outperformed other WI-HSV methods from the literature and outperformed writer-dependent methods from literature in the Brazilian database. Nevertheless, both works, which reach the state-of-the-art in the HSV task, assume the image signature pixels are available in a clean area.

Studies have been carried out to investigate the performance of the Deep Learning algorithms from literature facing signature and logo detection. In [18], the deep learning-based object detectors, namely, Faster R-CNN, ZF, $VGG16$, VGG_M YOLO, among others, where examined for this task. The proposed approach detects Signatures and Logos simultaneously. Mainly, in this study, the authors worked to detect signatures rather than the segmentation of signature traits. Thus, bounding boxes were generated around the detected signatures and logos. The database used was the Tobacco-800 [1], which has a clear background and is composed of scanned documents comprising printed text, signatures, and logos. The model architecture requires a fixed size (416 x 416) of the input images. Many models of deep learning have excelled in image classification and segmentation. A deep network model called Mask-RCNN, which is an extension of the Faster R-CNN model, produced interesting results and currently belongs to the state-of-the-art Deep CNN for image segmentation [7]. This architecture adds a branch to the faster R-CNN model to predict object masking and the existing branch to predict bounding box recognition. The work does not directly address signature segmentation. However, due to the good results that this model presented, we described some experiments with this model for stroke level signature segmentation in this paper.

Other works in the literature have presented methods for the stroke-based extraction of signatures from document images. The proposed approach in [13] is based on an FCN trained to learn, mapping, and extract the handwritten signatures from documents. Although the proposal achieved good results, the network architecture requires a fixed size (512 x 512) of the input images [13].

Our proposal improved this approach, as it presents a segmentation model at the signature level with promising results for our objectives.

In [19], a similar approach is used for signature extraction in identification documents. The authors used an optimized U-net network with less trainable parameters and input nodes than [13]. To increase the model's generalization, the authors applied the data augmentation technique in the database, generating greater image diversity during training. This proposed model achieved higher rates than [13], despite having fewer parameters.

In fact, stroke pixel integrity is of great importance to the offline signature verification process. Maintaining this integrity to the maximum can increase confidence in HSV systems, especially if more technical approaches, such as graphoscopy, are used. In forensic science, graphoscopy analyzes the properties present in signature traits, such as inclination, line behaviors, upward or downward direction, the pressure, between others. Therefore, the fewer changes made to the signature pixels, the more signature characteristics will be preserved.

3   Proposed Method

Our proposal was motivated because when images are resized, various signature properties may be lost, limiting the signature verification systems to apply the methods usually done in the forensic analysis of handwritten signatures. In addition to the possible loss of intrinsic characteristics of signatures, we see a proportionality problem. This issue is related to the relative size of the signature and the total size of the image. We observed that tiny signatures had more significant losses or were not detected in the inference process. Figure 3 presents an example of small signatures relative to the total image size.

We propose a method of image processing and segmentation to cope with this problem. Our approach can reduce changes in signature trait properties. This way, we can preserve the handwritten signature minimizing the information loss.

Our proposed method is divided into two steps, in which the first one (Subsect. 3.1) splits some input images into blocks. In the second step (Subsect. 3.2), the blocks are processed by an FCN and combined to compose the full resolution image of the signature.

3.1   Stage 1 - Building Blocks Process

This stage starts with the verification of dimensions $h \times w$ of the original image f. In cases where the image has exactly the dimensions $n \times n$ where $n = h = w$, it goes directly to the FCN. However, if the image has larger ($n < h$ or $n < w$) or smaller ($n > h$ or $n > w$) dimensions, it goes to the building blocks process.

Then, the input image is divided into blocks of equal dimensions. After some preliminary experiments, the best block size n was chosen as 512, given a block of $n \times n$ dimension. Smaller dimensions increase a lot of processing time, and feature gain was not significantly better. Oppositely, much larger values of n generated a considerable increase in the edges and created unnecessarily large

Fig. 3. Signature detection issue too small. The signature size relative to the image is relatively small. The face and personal data were pixelated.

images. This way, our block is a matrix B, of dimensions $n \times n$. Therefore, since the dimension of our block unit mask B in the model is 512×512, an image with the dimensions 1024×1024 will have a mask with four blocks. To generate the first blocks, we calculate the blocks horizontally and vertically from the original image. We determine the number of blocks in the horizontal direction by the ratio between the number of columns (w) of the original image f and the width n of our block B, so that $WB = w/n$. The same computation is performed to define the number of blocks in the vertical direction, such as $HB = h/n$.

In some cases, it is possible to have image dimensions not divisible by 512, such as in Fig. 4, producing blocks with dimensions smaller than 512 in height or width. To cope with this problem, we pad the image to complete the blocks in the right and bottom of the original image, as shown in Fig. 5, through the Eqs. 1 and 2.

$$mask_h = \begin{cases} mask_h + 1, \text{ if } n \times HB \le h \\ mask_h, \quad\quad \text{otherwise} \end{cases} \tag{1}$$

$$mask_w = \begin{cases} mask_w + 1, \text{ if } n \times WB \le w \\ mask_w, \quad\quad \text{otherwise} \end{cases} \tag{2}$$

We created a sliding window that has the same dimension as our block. Therefore, all blocks must be the same size for the sliding window to cover the entire image, including the edges. For this, the total image size is calculated to receive block masks with the dimensions of the defined pattern.

Thus, the window reads the image region in the block dimensions sent to the input layer of our FCN. With this, the network performs detecting the handwritten signature traces that will be mounted on the block mask. But we still need to solve the boundary problem between the blocks, shown in Fig. 6.

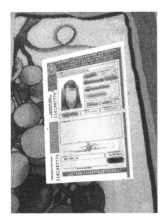
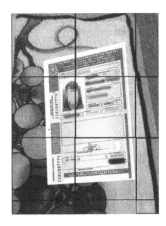

Fig. 4. Splitting the original image (1210 × 1613) in image blocks (512 × 512).

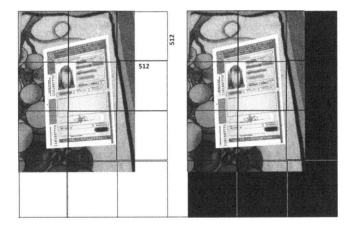

Fig. 5. Adding and padding borders in blocks.

Fig. 6. Example of the boundary problem: (a) The original image; (b) The building block process, where the black vertical line shows the boundary between two blocks; (c) The image result output shown lost pixels in the boundary area.

To avoid losing boundary information between blocks that behave like edges of an image, we move the sliding window in steps smaller than the size of a block. Therefore, the window moves until the full height and width are covered. With this strategy, we assure an overlap in border regions, minimizing the information loss across borders. Formally, the sliding window is represented by W, and the step taken in the displacement is indicated by the constant S. Then, sliding window process is performed on all image I according with the Eqs. 3 and 4.

$$W_{c,r}(x,y) = I(x + c \times n - B_x, y + r \times n - B_y) \qquad (3)$$

$\forall x \in [0, W_w - 1], y \in [0, W_h - 1], c \in [0, max_c], r \in [0, max_r]$, such that:

$$max_c = \lfloor \frac{I_w + 2 \times B_x - W_w}{n} \rfloor, max_r = \lfloor \frac{I_h + 2 \times B_y - W_h}{n} \rfloor \qquad (4)$$

where W are the sliding window that generates the image block I, B_x and B_y are the image border, n is the step of the sliding window, x and y are the relative pixels of the image I, regarding the window position through c and r variables.

3.2   Stage 2 - FCN Processing

To choose a neural network model for image segmentation in the second stage of our proposed pipeline, we trained two state-of-the-art models and did some preliminary experiments. The first model is based on the U-net neural network, and the second is based on the Mask-RCNN.

We tested the U-net and the Mask-RCNN models in a database with static dimensions (512×512), and low background complexity, named the DSSig-Dataset [13]. The U-net based model presented the best qualitative results. U-net was able to segment at the signature pixel level. But the Mask-RCNN-based model, despite finding the signature region, failed to segment at a signature trace level.

Therefore, based on the preliminary results, the U-net based model described in [19] was used in the second stage of our proposed approach, to receive the image blocks generated by the sliding window procedure. Then, the model outputs the corresponding mask with the estimated handwritten pixels in the foreground. To produce the final image mask of the signature pixels preserving the size and aspect ratio of the original image, each output of the U-net is combined into a background in corresponding coordinates of the original block through an OR operation to recover all handwritten pixels during the sliding window process. Figure 7 shows the result around the signature region in one example.

The proposed technique allows the FCN to process data with different resolutions and different perspectives of the identification document images. As described before, the FCN model we selected is based on the U-net architecture proposed in [17]. However, we have fine-tuned the original architecture reducing the number of trainable parameters from $30,299,233$ to $8,873,889$, as proposed by [19]. This reduction provided a significant advantage in training times and image inference. Besides, as a consequence of parameter reduction, we reduce the complexity of the model.

The architecture is shown in Fig. 8. The convolutional operations (in blue) are set to 3×3 of size with ReLU activation function. The max-pooling layers (in red) are set to 2×2 with stride 2. In the expansive path, there are upstream operations (in green) of size 2×2 concatenating with the corresponding characteristics of the path of contraction (gray arrows), followed by two convolutions of size 3×3 followed by ReLU operation.

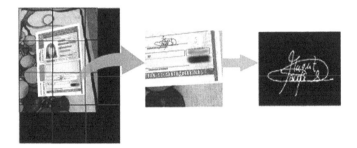

Fig. 7. Magnification of a block of the input and output images with the handwritten signature.

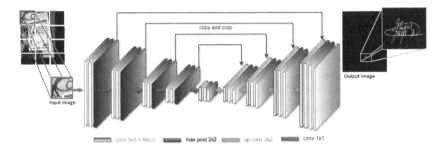

Fig. 8. Fully convolutional network architecture adapted of [17]. In this example, the sliding window scrolls through the image and delivers the block projection to the FCN input. FCN performs segmentation through pixel classification, giving as output an image with the estimated signature pixels as foreground. (Color figure online)

4 Experiments and Results

4.1 Databases

To train and verify the proposed model, images of identification documents captured by smartphone devices were required. These images must be presented in the real world, with different resolutions, distortions, and various qualities. Due to the absence of a public database with these properties, we created a database from images captured by different smartphones.

Therefore, we produced 20,000 images with identification documents on various background types, different resolutions, distortions, and noise. To compose the database, we used 200 background variations and 40 distortion variations in the image. The handwritten signatures blended in the document image were selected from the MCYT database [4] and voluntarily generated signatures. All signatures were stamped in the image document with different strokes, preventing the network from decorating the strokes by color. The images were as similar as possible as they are presented in the real world.

Through this procedure, we built the DSSigDataset-2 composed of 20,000 documents images. For each image, we created the corresponding ground truth, which is used for training and computing the overall metrics of each model evaluated in our experiment. In addition to the DSSigDataset-2, we selected two other databases to evaluate the proposed system. The first one was the DSSigDataset [13], which consists of ID documents with fixed resolutions, no distortion, and low texture variation. The second database was the Tobacco-800 [1], composed of scanned images containing handwritten text and signatures. The Tobacco-800 database has images of different dimensions and light backgrounds.

The brand new database we created, DSSigDataset-2, has real ID document images, contributing to composing a database of handwritten signatures to complement the existing ones. This database is closer to real-world images, where it does not have a controlled environment and has similarities to possible problems that may occur in daily life. With these three databases, we benchmarked the proposed system with approaches in the state-of-the-art. Besides, we used the three databases to evaluate the contribution of the Building Block process.

4.2   Training Procedure

We divided the DSSigDataset-2 for the training and testing of each model. We assigned 80% for training and 20% for testing, which respectively resulted in 16,000 training images and 4,000 images in the test set, taking into account different handwritten signatures in both partitions also, since we had random image transformations applied in the document and the background during the DSSigDataset-2 construction. We employed the transfer learning technique with the weights adjusted by a model. The model used for transfer learning was trained from a set of fixed resolution images with a similar purpose in the handwritten signatures segmentation proposed in [13]. We used the Adam optimizer [10] to minimize the objective function, which was the similarity coefficient Dice [14], shown in Eq. 5:

$$DC = 2\frac{|A \cap B|}{|A| + |B|} \tag{5}$$

where A represents the ground truth image and B represents the segmented image at the network output.

Therefore, the similarity between the output image and the ground truth was evaluated by the Dice coefficient. Both images are represented by binary values, 0 (dark pixels, related to the background) and 255 (bright pixels, the handwritten signature pixels).

For the structure of the layers of the pre-trained model, we deactivate the Dropout and the Batch-Normalization, since they do not present any impact on the overall rate, possibly due to the amount of data available in the DSSigDataset-2. We modified other parameters to select the best setting, such as the learning rate (0.0001) and the value of mini-batches (3). We run the model through 10,000 iterations.

4.3   Experimental Results

We performed several experiments to determine the best possible configuration and to validate the model's ability. We tested different configurations by evaluating the effects on optimization of hyper-parameters. We also tested the effects on the pre-processing of the images for the network's input with different resolutions for the image blocks. However, we report in this paper the best-validated configurations after all the preliminary experiments. According to the results presented in Table 1, our proposed model achieved comparable accuracy, recall, and F1-score rates in both DSSigdatasets (1 and 2). The results show that our model can work dynamically for different types of images, resolutions, textures, and distortions. It is also possible to see the standard deviation is low for both databases. The method allows greater stability to the model in the segmentation of handwritten signatures.

Table 1. The U-Net model performance With/Without the Building Block Process (BBP). The results using Dice Coefficient metric are presented in the following order: mean rate (standard deviation) ... max rate

DSSigDataset-2	With BBP	Without BBP
precision	0.84 ($\pm$0.07) ... 0.93	0.37 ($\pm$0.22) ... 0.82
recall	0.73 ($\pm$0.08) ... 0.87	0.65 ($\pm$0.40) ... 0.89
F1-Score	0.78 ($\pm$0.06) ... 0.89	0.43 ($\pm$0.28) ... 0.77
DSSigDataset[13]	With BBP	Without BBP
precision	0.81 ($\pm$0.06) ... 0.90	0.80 ($\pm$0.06) ... 0.90
recall	0.70 ($\pm$0.09) ... 0.87	0.69 ($\pm$0.09) ... 0.86
F1-Score	0.74 ($\pm$0.06) ... 0.88	0.73 ($\pm$0.06) ... 0.87
Tobacco-800[1]	With BBP	Without BBP
precision	0.64 ($\pm$0.16) ... 0.97	0.28 ($\pm$0.51) ... 0.79
recall	0.69 ($\pm$0.15) ... 0.93	0.38 ($\pm$0.25) ... 0.61
F1-Score	0.64 ($\pm$0.12) ... 0.82	0.42 ($\pm$0.27) ... 0.68

On the other side, we observe a significant drop in the results in DSSigDataset-2 without the Building Block process. Also, it presents high values in the standard deviation and low rates of precision, recall, and F1-score. This result shows that the absence of the proposed method negatively affected the segmentation process of handwritten signatures for the U-Net-based model, despite having the same architecture as the previous test. On the other side, the set of images with fixed resolution and without distortion practically did not change. The data reported from Table 1 show very close results between the three databases. This behavior is positive for the model submitted to the Building Block process method since the model can treat different types of images with the most various interference, as it happens in the real world.

4.4   Qualitative Analysis

We performed tests on different images of real documents to evaluate the qualitative results of our model. Figure 9 presents some of the results applying the Building Block process on the right column and without using it in the center. The output images were up-scaled to visualize the models' outputs better. We see in the images built without the proposed approach, sometimes a high information loss of signature, and in others too much noise in the image. Also, it is easy to note the lower signature quality, with deformations in the signatures. Although, the results produced by the Building Block process are more stable and very similar to the original handwritten signatures, conserving the stroke level signatures properties.

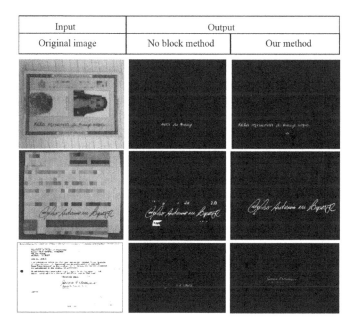

Fig. 9. Sample of qualitative results of segmentation of handwritten signatures. We compare the standard resizing method and our building block process for different types of images.

It is noteworthy that the first two rows of images in Fig. 9 are photos of actual documents. In the third row, we use an image from the Tobacco-800 database, in which the result for our method presents noises close to the signature. This behavior has been observed in all images of the Tobacco-800 database. We do not perform training with this database, and these images come from scanning. However, signature pixels were retrieved in all results from our proposed method. We enlarged the images in Fig. 10 to better visualize the results between the reference model and our proposed model.

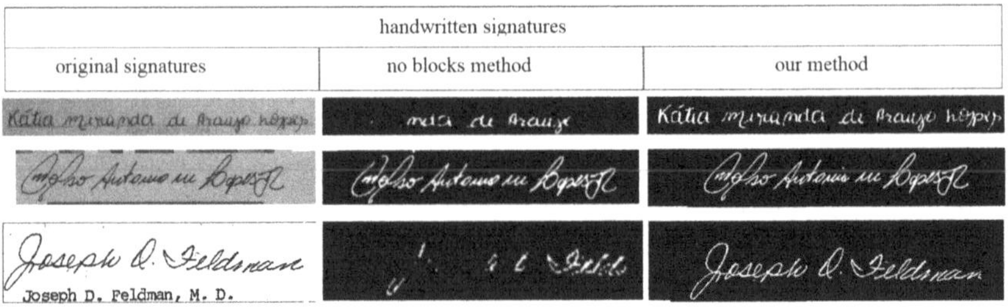

Fig. 10. Amplification of handwritten signatures to view results. On the left, the input images signatures; In the center, the result of the standard method, without applying block building; On the right are the images with the results of our building block process.

5   Conclusions

In this paper, we proposed a method for pre-processing ID document images in a fully convolutional network based on the U-Net architecture. The method allows the network to segment handwritten signatures on images captured by smartphone cameras at different resolutions, textures, and distortions often in daily-life environments. Our approach can mitigate the application of transformations that cause deformations and data loss in signature images. In this way, our method can preserve a large part of the graphoscopic properties and maintain greater fidelity at a stroke line level. These characteristics are the basis for applying the criteria used in forensic science by experts in graphotechnical analysis. To meet the need for a database with the same characteristics and noise problems in photo images, we created the DSSigDataset-2 database, with 20,000 document images captured by smartphone devices. To perform a comparative evaluation and the ability of the proposed model, we evaluated it in two additional databases: the DSSigDataset with fixed characteristics for resolution and without distortion; and the Tobacco-800 with scanned documents, completely different from the previous databases. The results with applying the proposed method, the Building Blocks process, show that the model is dynamic and able to work in different image resolutions and less loss of signatures information. We observed promising results in our experiments for the three databases: DSSigDataset-2, DSSigDataset, and Tobacco-800. The results for the Tobacco-800 database are inferior to those of other databases. However, it is noteworthy that the FCN model was not trained with images of this base and still managed to segment a good part of the signature images. We also emphasize that the results obtained encourage using the proposed method in real-world applications to segmentation signatures with greater preservation of the information and characteristics of the manuscript.

As future work, we will investigate classification models to propose an end-to-end signature verification system that takes as input two different signed documents and returns as output the probability of these documents been signed by the same person.

Acknowledgment. This study was financed in part by: Coordenação de Aperfeiçoamento de Pessoal de Nïvel Superior - Brasil (CAPES) - Finance Code 001, FACEPE, and CNPq - Brazilian research agencies.

References

1. Agam, G., Argamon, S., Frieder, O., Grossman, D., Lewis, D.: The Complex Document Image Processing (CDIP) test collection. Illinois Institute of Technology (2006). http://ir.iit.edu/projects/CDIP.html
2. Diaz, M., Ferrer, M.A., Eskander, G.S., Sabourin, R.: Generation of duplicated offline signature images for verification systems. IEEE Trans. Pattern Anal. Mach. Intell. **39**(5), 951–964 (2016)
3. Elhoseny, M., Nabil, A., Hassanien, A.E., Oliva, D.: Hybrid rough neural network model for signature recognition. In: Hassanien, A.E., Oliva, D.A. (eds.) Advances in Soft Computing and Machine Learning in Image Processing. SCI, vol. 730, pp. 295–318. Springer, Cham (2018). https://doi.org/10.1007/978-3-319-63754-9_14
4. Fierrez-Aguilar, J., Alonso-Hermira, N., Moreno-Marquez, G., Ortega-Garcia, J.: An off-line signature verification system based on fusion of local and global information. In: Maltoni, D., Jain, A.K. (eds.) BioAW 2004. LNCS, vol. 3087, pp. 295–306. Springer, Heidelberg (2004). https://doi.org/10.1007/978-3-540-25976-3_27
5. Guerbai, Y., Chibani, Y., Hadjadji, B.: The effective use of the one-class SVM classifier for handwritten signature verification based on writer-independent parameters. Pattern Recogn. **48**(1), 103–113 (2015)
6. Hafemann, L.G., Sabourin, R., Oliveira, L.S.: Learning features for offline handwritten signature verification using deep convolutional neural networks. Pattern Recogn. **70**, 163–176 (2017)
7. He, K., Gkioxari, G., Dollár, P., Girshick, R.: Mask R-CNN. In: Proceedings of the IEEE International Conference on Computer Vision, pp. 2961–2969 (2017)
8. Jia, W., Zhong, Z., Sun, L., Huo, Q.: A CNN-based approach to detecting text from images of whiteboards and handwritten notes. In: 2018 16th International Conference on Frontiers in Handwriting Recognition (ICFHR), pp. 1–6. IEEE (2018)
9. Kalera, M.K., Srihari, S., Xu, A.: Offline signature verification and identification using distance statistics. Int. J. Pattern Recogn. Artif. Intell. **18**(07), 1339–1360 (2004)
10. Kingma, D.P., Ba, J.: Adam: A method for stochastic optimization. arXiv preprint arXiv:1412.6980 (2014)
11. Kubo, D.A., de Nazare, T.S., Aguirre, P.L., Oliveira, B.D., Duarte, F.S.: The usage of U-Net for pre-processing document images (2018)
12. Long, J., Shelhamer, E., Darrell, T.: Fully convolutional networks for semantic segmentation. In: Proceedings of the IEEE Conference on Computer Vision and Pattern Recognition, pp. 3431–3440 (2015)
13. Melo, V.K.S.L., Bezerra, B.L.D.: A fully convolutional network for signature segmentation from document images. In: 2018 16th International Conference on Frontiers in Handwriting Recognition (ICFHR), pp. 540–545. IEEE (2018)
14. Novikov, A.A., Major, D., Lenis, D., Hladuvka, J., Wimmer, M., Bühler, K.: Fully convolutional architectures for multi-class segmentation in chest radiographs. CoRR abs/1701.08816 (2017)
15. Oliveira, S.A., Seguin, B., Kaplan, F.: dhSegment: a generic deep-learning approach for document segmentation. In: 2018 16th International Conference on Frontiers in Handwriting Recognition (ICFHR), pp. 7–12. IEEE (2018)

16. Ortega-Garcia, J., Fierrez-Aguilar, J., Simon, D., Gonzalez, J., Faundez-Zanuy, M., Espinosa, V., Satue, A., Hernaez, I., Igarza, J.J., Vivaracho, C., et al.: MCYT baseline corpus: a bimodal biometric database. IEE Proc. Vis. Image Sig. Process. **150**(6), 395–401 (2003)
17. Ronneberger, O., Fischer, P., Brox, T.: U-Net: convolutional networks for biomedical image segmentation. In: Navab, N., Hornegger, J., Wells, W.M., Frangi, A.F. (eds.) MICCAI 2015. LNCS, vol. 9351, pp. 234–241. Springer, Cham (2015). https://doi.org/10.1007/978-3-319-24574-4_28
18. Sharma, N., Mandal, R., Sharma, R., Pal, U., Blumenstein, M.: Signature and logo detection using deep CNN for document image retrieval. In: 2018 16th International Conference on Frontiers in Handwriting Recognition (ICFHR), pp. 416–422. IEEE (2018)
19. Silva, P.G.S., Lopes Junior, C.A.M., Lima, E., Bezerra, B.L.D., Zanchettin, C.: Speeding-up the handwritten signature segmentation process through an optimized fully convolutional neural network. In: 2019 15th International Conference on Document Analysis and Recognition (ICDAR), September 2019. https://doi.org/10.1109/ICDAR.2019.00228
20. Soleimani, A., Araabi, B.N., Fouladi, K.: Deep multitask metric learning for offline signature verification. Pattern Recogn. Lett. **80**, 84–90 (2016)
21. Souza, V.L., Oliveira, A.L., Sabourin, R.: A writer-independent approach for offline signature verification using deep convolutional neural networks features. In: 2018 7th Brazilian Conference on Intelligent Systems (BRACIS), pp. 212–217. IEEE (2018)
22. Vargas, F., Ferrer, M., Travieso, C., Alonso, J.: Off-line handwritten signature GPDS-960 corpus. In: Ninth International Conference on Document Analysis and Recognition (ICDAR 2007), vol. 2, pp. 764–768. IEEE (2007)

Attention Based Multiple Siamese Network for Offline Signature Verification

Yu-Jie Xiong[1,2](✉) and Song-Yang Cheng[1]

[1] School of Electronic and Electrical Engineering, Shanghai University of Engineering Science, Shanghai 201620, China
{xiong,M020119109}@sues.edu.cn
[2] Shanghai Key Laboratory of Multidimensional Information Processing, East China Normal University, Shanghai 200241, China

Abstract. Offline handwritten signatures play an important role in biometrics and document forensics, and it has been widely used in the fields of finance, judiciary and commerce. However, the skilled signature forgeries bring challenges and difficulties to personal privacy protection. Thus it is vital to discover micro but critical details between genuine signatures and corresponding skilled forgeries in signature verification tasks. In this paper, we propose an attention based Multiple Siamese Network (MSN) to extract discriminative information from offline handwritten signatures. MSN receives the reference and query signature images and their corresponding inverse images. The received images are fed to four parallel branches. We develop an effective attention module to transfer the information from original branches to inverse branches, which attempts to explore prominent features of handwriting. The weight-shared branches are concatenated in a particular way and formed into four contrastive pairs, which contribute to learn useful representations by comparisons of these branches. The preliminary decisions are generated from each contrastive pair independently. Then, the final verification result is voted from these preliminary decisions. In order to evaluate the effectiveness of proposed method, we conduct experiments on three publicly available signature datasets: CEDAR, BHSig-B and BHSig-H. The experimental results demonstrate the proposed method outperforms that of other previous approaches.

Keywords: Attention mechanism · Multiple siamese network · Offline signature verification

1 Introduction

The emergence of big data era has brought the rise in privacy concerns. Passwords are easy to guess, which creates a serious security threat for personal information. To tackle this problem, biometrics identification has been widely applied in various scenarios, since it is a promising replacement for conventional identification approaches. The biometric system takes into account inherent physiological or behavioral traits such as fingerprint, face, voice and

© Springer Nature Switzerland AG 2021
J. Lladós et al. (Eds.): ICDAR 2021, LNCS 12823, pp. 337–349, 2021.
https://doi.org/10.1007/978-3-030-86334-0_22

handwritten documents, and makes verification or identification decisions according to different tasks or objectives.

Handwriting carries rich information to reveal the identity of individuals, and it plays significant roles in human communication, perception, emotional behavior and so on. Signatures have been widely used in biometric systems to verify a person's identity. They are used in legal and financial fields such as contract agreement, bank checks, passports, receipts, identity certificates and many other applications. Besides the necessary characteristics of biometrics identification, signatures also have many good traits. For example, they are easy to access and readily accepted by people in daily life. Therefore, the researches about signature verification are very early. With the help of artificial intelligence technology, it is more convenient to build a automatic signature verification system.

Signature verification systems can be classified in two types according to the data acquisition means: online (dynamic) and offline (static). In online systems, signatures are collected as temporal sequences. The data such as positions, pressure, pen inclination and acceleration are recorded. In offline systems, the data are represented as static digital images. In both online and offline systems, the query signatures are judged as genuine samples or forgeries. The forgeries are commonly categorized into three types: random, simple and skilled forgeries. For random forgeries, the forger basically has no information about the forged object, and he/she has never seen the signature and does not even know the name of the forged person. In this case, the forged signature has a completely different shape and it contains very different semantic characteristics compared to genuine signature. For simple forgeries, the forger has basic information such as the name of the object being forged but not know about the writing pattern of signatures. The forgeries may be similar to the genuine signature under such circumstance. Skilled forgery means the forger not only knows the name of the object being forged, but also has the information of his/her signature, and even has practiced the writing pattern deliberately.

There exists highly similarities between genuine and forged signatures, and it is almost impossible to discriminate the difference for a person who has not been trained for handwriting verification, therefore it is a particularly challenging task. Many approaches have been proposed to solve the challenging problem. Various hand crafted features are used in the field. Ferrer et al. [1] used Local Binary Patterns (LBP) and statistical measures for automatic offline handwritten signature verification. Okawa [2] proposed a discriminative and robust feature extraction approach based on a Fisher Vector (FV) with fused "KAZE" features from both foreground and background offline signature images. Diaz et al. [3] proposed a complete framework to recover on-line Western signatures from image-based specimens. In Ref. [4], a parameter free, candidate graph mining method was introduced for offline signature coding and verification. In recent years, many deep learning based methods have been proposed. Convolutional Neural Network (CNN) has demonstrated its excellent capabilities in the fields of signature verification. Masoudnia et al. [5] combined the different but complementary advantages of different loss functions and proposed Multi-Loss Snapshot

Ensemble (MLSE). Li et al. [6] proposed the first black-box adversarial example attack against handwritten signature verification. Siamese network is a class of architecture that usually contains two weight-shared branches. It was first proposed by Bromley et al. [7] for verification of signatures written on a pen-input tablet. The siamese network tries to minimize the Euclidean distance between the feature representations and has been successfully used in face verification and signature verification. Dey et al. [8] designed a convolutional siamese network named Signet, and achieved good performance in writer independent feature learning. Wei et al. [9] proposed a novel inverse discriminative network (IDN) to resolve the sparse information issue in writer-independent handwritten signature verification. Mustafa et al. [10] proposed a two-channel CNN and fused user-independent CNN score with user-dependent SVM score to get verification results. Lin et al. [11] proposed to add dropout layers in the middle position of 2-Channel-2-Logit (2C2L) network to address the overfitting problems.

In this paper, we propose an attention based multiple siamese network for offline signature verification. The structure is regarded as an enhanced version of siamese network. It contains four weight-shared branches which two of them receive reference signatures and corresponding inverse images and the other two receive query signatures and corresponding inverse images. Both original reference and query signatures are gray scale images. Attention modules are used to connect the original and inverse signature branches, which can make the network focus on effective stroke details and suppress interference information. Then the model performance is improved with the help of attention modules. Furthermore, we propose contrastive pairs to learn useful representations by comparisons of branches. Specifically speaking, the features from four branches are grouped into four different pairs, then the contrastive pairs are fed into four classifiers and made final decisions by voting mechanism.

The remainder of this paper is organized as follows: Sect. 2 describes details of the proposed method; Sect. 3 provides experimental results and discussions, and Sect. 4 concludes this paper.

2   The Proposed Method

The network architecture of the proposed attention based multiple siamese network is illustrated in Fig. 1. The original reference and query signatures are images with white backgrounds and gray signature strokes. The images with black backgrounds are inverse signatures of original reference and query samples, respectively. The MSN contains four weight-shared branches. Two of them are original branches and the other two are inverse branches. These branches have the same structure. Each branch contains four convolutional modules, and the number of channels are 32, 64, 96 and 128, respectively. Every module contains two convolutional layers (the kernel size is 3×3 and the stride is 1) and a pooling layer (the kernel size is 2×2 and the stride is 2). Rectified Linear Units (ReLU) are utilized as the activation function in each module. There are eight attention modules in MSN. Between original and inverse branches, attention module plays a connecting role. In the forward propagation, it receives the output from convolutional module in the original branch and the output from the first convolutional layer in the inverse branch. Then the output of attention module are regarded as input of the second layer in the inverse branch convolutional module. The feature maps output from four branches are concatenated, forming four contrastive pairs. Then these pairs are fed into fully-connected layers through global average pooling layers and make classification decisions independently. The final verification results are obtained according to these decisions from voting.

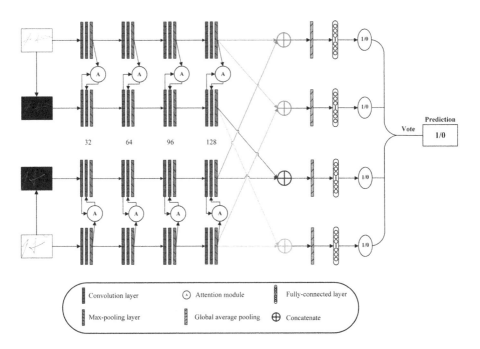

Fig. 1. Architecture of the proposed Multiple Siamese Network.

2.1   Pre-processing

For all signatures from training and testing datasets, we apply the same pre-processing strategy. In our research, signature images from different datasets have a variable size, for example, the size ranges from 153×258 to $819 \times 1,137$ in CEDAR. Our proposed model needs images with the same size as inputs. Therefore, all images are resized using linear interpolation. In signature images, there are large blanks around the foreground area which are useless. In order to reduce unnecessary calculations, we utilize a pixel search method to remove these margins. Besides, other pre-processing steps are also adopted. For example, we remove backgrounds while preserving the text. For convenience of network training, all original signature images are inverted using 255 minus the grayscale image matrix.

2.2   Attention Modules

In order to make the model focus on efficient and reliable stroke features in offline signature verification tasks, we introduce attention mechanism which may contribute to the feature learning in our network. Profiting from the special structure, MSN collects global information of the original image and combines them with features of the inverse image through attention modules, and discovers accurate stroke information effectively and quickly, thus guides the convolutional network focus on complement signature details.

The architecture of our redesigned attention module is inspired by [9,12–14]. The proposed attention modules are stacked to generate attention-aware features which can perform adaptively recalibration. Moreover, it can be used at any depth in the network. In the early layers, it enhances the quality of the shared lower-level feature representations. In later layers, it becomes specialised in a highly class-specific manner.

We adopt mixed attention mechanism in the module, which employs the residual learning method and is capable to capture crucial features. More specifically, spatial attention maps play an important role in deciding which area of signature images is informative. Channel attention maps are produced by exploiting the inter-channel relationship of features. Both global average-pooling and max-pooling method are utilized simultaneously. Average-pooling is to learn the extent of spatial information and aggregate them effectively. Max-pooling plays another significant role in gathering distinctive stroke features.

The architecture of designed attention modules is illustrated in Fig. 2. It contains both spatial and channel attention mechanisms. In the left side of the red dotted line, r is defined as the output from original branches. The feature map is resized to a fixed size using up-sampling operation which is based on nearest neighbor algorithm. Then a convolutional operation with sigmoid activation receives the resized feature maps. We defined g as the output after sigmoid activation, and o is the output of the first layer from convolutional modules in inverse branches. We multiply g and o, then make a element-wise addition which can be described as $g \cdot o + o$ to achieve desired spatial attention results. Subsequently, Global Average Pooling (GAP) and Global Max Pooling layers (GMP) are utilized to receive the spatial attention results. The architecture in the right side of the red dotted line in Fig. 2 can be regarded as channel attention mechanism which can be represented by:

$$W_c = \sigma(\text{FC}(\text{AvgPool}(g \cdot o + o)) + \text{FC}(\text{MaxPool}(g \cdot o + o))) \qquad (1)$$

where σ denotes the sigmoid function, W_c is defined as channel attention weights of c-th channel. The features are fed into the shared network which is composed of Fully-Connected layers (FC), then we sum the output features to generate the weight vector f through sigmoid activation. Finally, we get an attention mask $(g \cdot o + o) \times f$ which is fed into the second layer of convolutional modules in inverse branches.

MSN contains eight attention modules between original and inverse branches. With the help of attention mechanism, the important and effective stroke features are focused and strengthened. Figure 3 shows an example of feature maps output from attention modules. Row 1 represents the signature images after pre-processing, and following rows denote different visualization results of the output from different level attention modules between original and reverse branches. We compare the visualization results of proposed attention module with that of IDN [9]. Column 1 and 2 are the visualization results of IDN and proposed MSN, respectively. It can be clearly seen that the proposed attention modules contain more reliable features and focus on stroke information.

2.3 Contrastive Pairs

Our proposed MSN aims to learn useful representations by comparisons of reference and query signature samples. To achieve such a goal, we consider the ordered combination of original and inverse branches. More specifically speaking, four feature maps are generated from these branches. They are concatenated to four pairs which are inverse reference and original query signature, original reference and original query signature, original reference and inverse query signature, inverse reference and inverse query signature, respectively. Each pair is fed into FC layers through a GAP layer. The c-th channel value z after GAP layer is calculated from feature map p_c by:

$$z_c = \frac{1}{H \times W} \sum_{i=1}^{H} \sum_{j=1}^{W} p_c(i,j) \tag{2}$$

Each pair makes a two-class classification independently. The MSN is expected to make the same decisions for all pairs in spite of the background colors. In order to achieve this ambitious objective, we utilize a binary cross entropy based loss function to measure the performance of contrastive pairs.

$$Loss\,(X_i, Y_i) = -\sum_{i=1}^{4} w_i \left[y_i \log x_i + (1 - y_i) \log (1 - x_i) \right] \tag{3}$$

y_i denotes ground truth label, and it is binary variables. 0 indicates that reference and query samples are written by different person which means the query signature sample is forged. 1 indicates the query sample is genuine. x_i represents predicted probability results, and it ranges from 0 to 1. w_i is a hyper-parameter, and we set it in different values for different datasets.

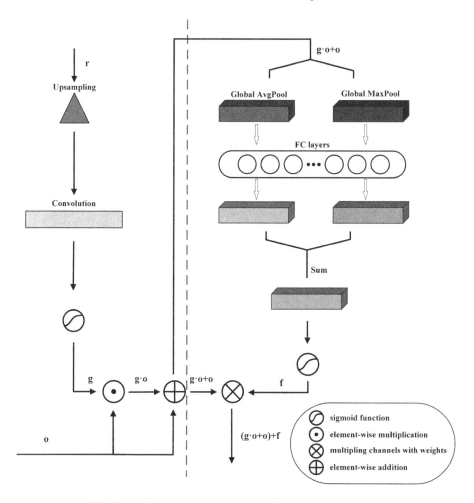

Fig. 2. Architecture of the proposed attention module.

$$Pre = \begin{cases} 1, N_p \geq 3 \\ 0, \ N_p < 3 \end{cases} \qquad (4)$$

We also design a voting mechanism for the final prediction results. N_p indicates the number of contrastive pairs being regarded as the same writer's sample. Pre has two values where 0 denotes the query signature is forged and 1 denotes both reference and query signatures belong to the same writer's handwriting.

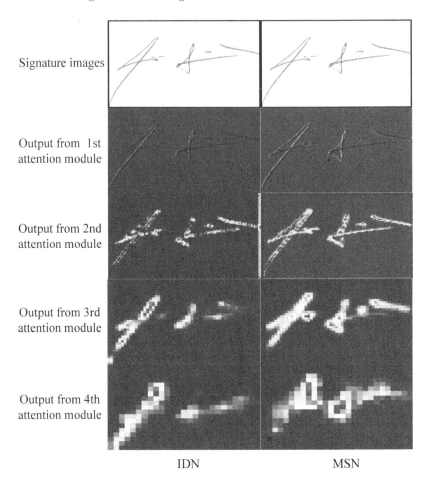

Signature images

Output from 1st attention module

Output from 2nd attention module

Output from 3rd attention module

Output from 4th attention module

IDN   MSN

Fig. 3. Feature maps visualization results output from attention modules of IDN and proposed MSN.

3   Experimental Results

3.1   Datasets

In order to evaluate the effectiveness of our method, we conduct experiments on several widely used public datasets: (1) CEDAR [15], (2) BHSig-B and (3) BHSig-H [16]. The brief introduction of three datasets is as follows.

CEDAR is an English signature dataset which contains 55 individuals' samples. Every writer are asked to sign 24 genuine signatures and 24 skilled forgeries in a predefined space of 22 in. Therefore, there are $55 \times 24 = 1,320$ genuine and 1,320 forged signature images. These signatures are scanned at 300 dpi in 8-bit gray scale and stored as PNG images.

BHSig-B is a Bengali dataset which contains 100 individuals' samples. 24 genuine signatures and 30 skilled forgeries are available for each writer, which results in $100 \times 24 = 2,400$ genuine and $100 \times 30 = 3,000$ forged signatures.

BHSig-H contains 160 individuals' samples which are written in Hindi. It consists of $24 \times 160 = 2,840$ genuine signatures and $30 \times 160 = 4,800$ skilled forgeries from 160 individuals altogether. Both BHSig-B and BHSig-H dataset are collected from individuals with different educational backgrounds and ages. The signatures are scanned in gray scale with 300 dpi resolution and stored in TIFF format.

3.2 Evaluation Metrics

In order to evaluate our proposed method, we applied several standard metrics: False Rejection Rate (FRR), False Acceptance Rate (FAR), Equal Error Rate (EER), Average Error Rate (AER) and Accuracy (Acc).

3.3 Experimental Settings

The experiments are performed under the framework of Pytorch (1.4.0), with NVIDIA-2080 for GPU acceleration, Inter (R) Core (TM) i7-9700k CPU and 16G memory. The operating system is Ubuntu 18.04 and the programming language for all methods is Python.

The method is designed for writer independent signature verification, and the datasets need to be divided into training and testing samples. In column 2 and 3 of Table 1, the number of writers used for training and testing are given. For CEDAR dataset, it contains 24 genuine signatures for each writer, thus there are $C_{24}^2 = 276$ sample pairs (genuine-genuine). By combining all the (genuine-forgery) signatures of each writer, we can get $24 \times 24 = 576$ pairs. 276 genuine-forgery pairs are randomly selected to avoid imbalanced data issue between different classes. Likewise, for BHSig-B dataset, we use 50 individuals' samples for training, and there are $2 \times C_{24}^2 = 552$ pairs for each individual. For BHSig-H dataset, we use 100 individuals' samples for training and 60 individuals' samples for testing. Thus, the dataset was split with $100 \times 2 \times C_{24}^2 = 55,200$ pairs of samples assigned to the training dataset and $60 \times 2 \times C_{24}^2 = 33,120$ pairs to the testing dataset. The column 4 and 5 in Table 1 give the number of positive and negative pairs of each writer used for training and testing in different datasets.

3.4 Results and Discussions

The structure of proposed MSN is similar to Inverse Discriminative Network (IDN) [9], thus we compare our proposed method with it. We conduct our experiments using two proposed approaches, respectively. The first approach aims to focus on efficient stroke features using our proposed attention module. The second approach considers using four contrastive pairs to concatenate the output of original and inverse branches, thus enhance the feature learning ability.

Table 1. Details of experimental protocol on different datasets.

Dataset	Train	Test	Positive pairs	Negative pairs
CEDAR	50	5	276	276 out of 576
BHSig-B	50	50	276	276 out of 720
BHSig-H	100	60	276	276 out of 720

Table 2 shows the comparison results with IDN on CEDAR dataset. $\mathcal{A}ttention^1$ denotes the attention module used in IDN. $\mathcal{A}ttention^2$ represents our proposed attention module. From row 1 and 2, it can be seen that our proposed attention module is able to focus on more efficient and reliable stroke features. In row 3 and 4, MSN employs four contrastive pairs in our experiments. MSN + $\mathcal{A}ttention^1$ and MSN + $\mathcal{A}ttention^2$ means that we utilize $\mathcal{A}ttention^1$ and $\mathcal{A}ttention^2$ in MSN, respectively. By comparing the results of MSN + $\mathcal{A}ttention^1$ and IDN + $\mathcal{A}ttention^1$, we can notice that our proposed MSN achieves the higher accuracy compared to IDN, which demonstrates that MSN is more capable of learning effective feature representations by comparisons of reference and query signature samples. We also conduct experiments to discover if our proposed $\mathcal{A}ttention^2$ can achieve better performance in MSN than $\mathcal{A}ttention^1$. As can be seen in Table 2, the accuracy of MSN + $\mathcal{A}ttention^2$ is higher than MSN + $\mathcal{A}ttention^1$, which demonstrate that our proposed $\mathcal{A}ttention^2$ is workable in MSN. To sum up, our proposed approaches are more effective than IDN on CEDAR dataset.

Table 2. Comparison with IDN on CEDAR Dataset.

Model	Acc	FAR	FRR	EER
IDN [9] + $\mathcal{A}ttention^1$	96.77	2.75	3.69	3.22
IDN [9] + $\mathcal{A}ttention^2$	97.28	3.98	1.45	2.71
MSN + $\mathcal{A}ttention^1$	97.93	**2.02**	2.10	2.06
MSN + $\mathcal{A}ttention^2$	**98.40**	3.18	**0**	**1.63**

∗ Note: $\mathcal{A}ttention^1$ denotes the attention module used in IDN, and $\mathcal{A}ttention^2$ represents our proposed attention module.

In order to evaluate the proposed approaches on BHSig-B and BHSig-H, we use MSN with our proposed attention module to conduct contrast experiments. These experiments are also to verify whether the query samples are genuine signatures or skilled forgeries. In Table 3, it shows that the proposed method achieves good performance on the two datasets. The system achieves the best performance on BHSig-B. Compared with IDN [9], the accuracy increases most on BHSig-B which is 2.16%, FRR and EER decrease most on BHSig-B which is 2.60% and 2.16%, respectively. It can be concluded from Table 3 that the performance of proposed method excels IDN [9] effectively.

Table 3. Comparison with IDN on BHSig-B and BHSig-H Dataset.

Dataset	Method	FAR	FRR	EER	Acc
BHSig-B	IDN [9]	12.16	9.04	10.59	89.40
	MSN + $\mathcal{A}ttention^2$	**10.42**	**6.44**	**8.43**	**91.56**
BHSig-H	IDN [9]	18.55	6.02	11.51	87.71
	MSN + $\mathcal{A}ttention^2$	**17.06**	**5.16**	**11.31**	**88.88**

∗ Note: We reproduced the network architecture of IDN and got the results.

We also compare the method with other approaches on the three datasets. Table 4 shows the comparative analysis on the BHSig-B dataset. It is clear from the table that the proposed MSN performs better than previous approaches which consider handcrafted or deep learning based features as feature extractors.

Table 4. Comparison with other approaches on the BHSig-B dataset.

Model	FRR	FAR	EER	Acc
Dey et al. [8]	13.89	13.89	-	86.11
Lin et al. [11]	-	-	11.92	88.08
Pal et al. [16]	33.82	33.82	33.82	66.18
Jadhav and Chavan [19]	-	-	-	90.36
Jain et al. [20]	-	-	-	76.03
MSN + $\mathcal{A}ttention^2$	**6.44**	**10.42**	**8.43**	**91.56**

Table 5 gives the evidence that the proposed method outperforms other approaches on CEDAR dataset. A possible reason for the higher performance on this dataset is the plenty number of signature samples for training. It is observed that the proposed method achieves performance improvement on all metrics than other approaches, which proves the superiority of our MSN.

Table 5. Comparison with other approaches on the CEDAR dataset.

Model	FRR	FAR	EER	AER
Hafemann et al. [17]	-	-	4.63	-
Kumar et al. [21]	8.33	8.33	-	8.33
Bhunia et al. [22]	-	-	-	1.64
Sharif et al. [23]	4.67	4.67	-	4.67
MSN + $\mathcal{A}ttention^2$	**0**	**3.18**	**1.63**	**1.59**

Table 6 depicts the comparison results on BHSig-H dataset. We receive 88.88% accuracy compared with previous approaches. It is easy to see that the proposed method achieves good performance and is capable of distinguishing reference and query samples effectively.

Table 6. Comparison with other approaches on the BHSig-H dataset(%).

Model	FRR	FAR	EER	Acc
Dey et al. [8]	15.36	15.36	-	84.64
Lin et al. [11]	-	-	13.34	86.66
Pal et al. [16]	24.47	24.47	24.47	75.53
Dutta et al. [18]	15.09	**13.10**	-	85.90
Jain et al. [20]	-	-	-	83.50
MSN + $\mathcal{A}ttention^2$	**5.16**	17.06	**11.31**	**88.88**

4   Conclusions

In the field of pattern recognition, offline signature verification task has been considered as a challenging problem since it is difficult to capture the small differences between genuine and forged samples. In this paper, we introduce attention based multiple siamese network to extract discriminative information from offline signature images. Attention modules are utilized to discover stroke details between original and inverse pairs. The discriminative information can be learned through contrastive pairs by comparing reference and query signature samples. Experiments on CEDAR, BHSig-B and BHSig-H dataset demonstrate our proposed method is effective in offline signature verification tasks.

In our future work, we would like to apply the proposed method into online signature verification and investigate more reliable feature learning approaches in cross-language verification tasks.

Acknowledgements. This work is jointly sponsored by the National Natural Science Foundation of China (Grant No.62006150), Shanghai Young Science and Technology Talents Sailing Program (Grant No. 19YF1418400), Shanghai Key Laboratory of Multidimensional Information Processing (Grant No. 2020MIP001), and Fundamental Research Funds for the Central Universities.

References

1. Ferrer, M.A., Vargas, J.F., Morales, A., Ordonez, A.: Robustness of offline signature verification based on gray level features. IEEE Trans. Inf. Forensic Secur. **7**(3), 966–977 (2012)
2. Okawa, M.: Synergy of foreground-background images for feature extraction: offline signature verification using fisher vector with fused kaze features. Pattern Recognit. **79**, 480–489 (2018)

3. Diaz, M., Ferrer, M.A., Parziale, A., Marcelli, A.: Recovering western on-line signatures from image-based specimens. In: ICDAR, pp. 1204–1209 (2017)
4. Zois, E.N., Zervas, E., Tsourounis, D., Economou, G.: Sequential motif profiles and topological plots for offline signature verification. In: CVPR, pp. 13245–13255 (2020)
5. Masoudnia, S., Mersa, O., Araabi, B.N., Vahabie, A., Sadeghi, M., Ahmadabadi, M.: Multi-representational learning for offline signature verification using multi-loss snapshot ensemble of CNNs. Expert Syst. Appl. **133**, 317–330 (2019)
6. Li, H., Li, H., Zhang, H., Yuan, W.: Black-box attack against handwritten signature verification with region-restricted adversarial perturbations. Pattern Recognit. **111**, 107689 (2021)
7. Bromley, J., et al.: Signature verification using a "Siamese" time delay neural network. Int. J. Pattern Recognit. Artif. Intell. **7**(4), 737–744 (1993)
8. Dey, S., Dutta, A., Toledo, J., Ghosh, S., Lladós, J., Pal, U.: Signet: Convolutional siamese network for writer independent offline signature verification, arxiv, 1707.02131 (2017)
9. Wei, P., Li, H., Hu, P.: Inverse discriminative networks for handwritten signature verification. In: CVPR, pp. 5757–5765 (2019)
10. Yilmaz, M.B., Öztürk, K.: Hybrid user-independent and user-dependent offline signature verification with a two-channel CNN. In: CVPRW, pp. 639–647 (2018)
11. Li, C., Lin, F., Wang, Z., Yu, G., Yuan, L., Wang, H.: DeepHSV: User-independent offline signature verification using two-channel CNN. In: ICDAR, pp. 166–171 (2019)
12. Hu, J., Shen, L., Albanie, S., Sun, G., Wu, E.: Squeeze-and-excitation networks. IEEE Trans. Pattern Anal. Mach. Intell. **42**(8), 2011–2023 (2020)
13. Wang, F., et al.: Residual attention network for image classification. In: CVPR, pp. 6450–6458 (2017)
14. Woo, S., Park, J., Lee, J., Kweon, I.: CBAM: convolutional block attention module. In: ECCV, pp. 3–19 (2018)
15. Kalera, M.K., Srihari, S., Xu, A.: Offline signature verification and identification using distance statistics. Int. J. Pattern Recognit. Artif. Intell. **18**(7), 1339–1360 (2004)
16. Pal, S., Alaei, A., Pal, U., Blumenstein, M.: Performance of an off-line signature verification method based on texture features on a large indic-script signature dataset. In: DAS, pp. 72–77 (2016)
17. Hafemann, L., Sabourin, R., Oliveira, L.: Learning features for offline handwritten signature verification using deep convolutional neural networks. Pattern Recognit. **70**, 163–176 (2017)
18. Dutta, A., Pal, U., Lladós, J.: Compact correlated features for writer independent signature verification. In: ICPR, pp. 3422–3427 (2016)
19. Jadhav, S.K., Chavan, M.K.: Symbolic representation model for off-line signature verification. In: ICCCNT, pp. 1–5 (2018)
20. Jain, A., Singh, S., Singh, K.P.: Signature verification using geometrical features and artificial neural network classifier. Neural Comput. Appl. **1**, 1–12 (2020)
21. Kumar, R., Sharma, J.D., Chanda, B.: Writer-independent off-line signature verification using surroundedness feature. Pattern Recognit. Lett. **33**(3), 301–308 (2012)
22. Bhunia, A., Alaei, A., Roy, P.: Signature verification approach using fusion of hybrid texture features. Neural Comput. Appl. **31**, 8737–8748 (2019)
23. Sharif, M., Khan, M., Faisal, M., Yasmin, M., Fernandes, S.L.: A framework for offline signature verification system: best features selection approach. Pattern Recognit. Lett. **139**, 50–59 (2020)

Attention to Warp: Deep Metric Learning for Multivariate Time Series

Shinnosuke Matsuo[1]([✉]), Xiaomeng Wu[2][iD], Gantugs Atarsaikhan[1],
Akisato Kimura[2], Kunio Kashino[2], Brian Kenji Iwana[1][iD],
and Seiichi Uchida[1][iD]

[1] Kyushu University, Fukuoka, Japan
shinnosuke.matsuo@human.ait.kyushu-u.ac.jp
[2] Communication Science Laboratories, NTT Corporation, Tokyo, Japan

Abstract. Deep time series metric learning is challenging due to the difficult trade-off between *temporal invariance* to nonlinear distortion and *discriminative power* in identifying non-matching sequences. This paper proposes a novel neural network-based approach for robust yet discriminative time series classification and verification. This approach adapts a parameterized attention model to time warping for greater and more adaptive *temporal invariance*. It is robust against not only local but also large global distortions, so that even matching pairs that do not satisfy the monotonicity, continuity, and boundary conditions can still be successfully identified. Learning of this model is further guided by dynamic time warping to impose temporal constraints for stabilized training and higher *discriminative power*. It can learn to augment the inter-class variation through warping, so that similar but different classes can be effectively distinguished. We experimentally demonstrate the superiority of the proposed approach over previous non-parametric and deep models by combining it with a deep online signature verification framework, after confirming its promising behavior in single-letter handwriting classification on the Unipen dataset.

Keywords: Attention model · Dynamic time warping · Metric learning · Signature verification

1 Introduction

Over the past two decades, time series classification and verification have been considered to be two of the most challenging problems in data mining. They have been applied to many applications, such as activity recognition, computational auditory scene analysis, cybersecurity, electronic health records, and handwritten biometric recognition [9]. The fundamental problem in these tasks is to define the distance between time series. This distance must be invariant to nonlinear temporal distortions, e.g., minute but diverse shifts, scaling, and noise, due to intra-class variations. It should also be able to capture temporal inconsistency so that time series of different classes can be optimally distinguished.

© Springer Nature Switzerland AG 2021
J. Lladós et al. (Eds.): ICDAR 2021, LNCS 12823, pp. 350–365, 2021.
https://doi.org/10.1007/978-3-030-86334-0_23

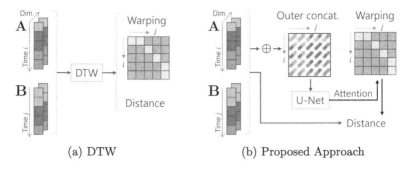

(a) DTW   (b) Proposed Approach

Fig. 1. Comparison between DTW and our approach. While DTW creates the warping path and distance with a non-parametric algorithm, our approach learns warping through an attention model where the attention (weights) form the warping path.

Dynamic Time Warping (DTW) [31] is the most widely used measure for this purpose [2,16,20,22,27]. It calculates the minimum temporal alignment cost between two time series by taking the summation of the distances between local elements that are optimally matched under monotonicity, continuity, and boundary conditions (Fig. 1a). By defining this minimum cost as a dissimilarity measure, DTW achieves invariance to temporal distortions and imposes strong constraints on the measure. However, DTW relies on handcrafted features and may be less effective when complex feature representations are required. Meanwhile, Hidden Markov Models (HMMs) [10,11] are often reported to be well-suited for time series classification thanks to their high adaptability to intra-class temporal variations. Similar to DTW, their limited complexity makes them difficult to take advantage of the information in the training data.

In recent years, more and more efforts have been made on neural network-based classification and metric learning [1,4–6,12,24,29,34,39,40] for discrete time series. In deep metric learning, a Siamese network or a triplet network is used to learn the distance metric from data, driving the distance to be small for *matching pairs* (data from the same class) and large for *non-matching pairs* (data from different classes). Some networks [6,24,29] produce a global feature representation for the time series, inflicting the heavy loss of useful temporal information. Others [1,4,34] encode the time series into a sequence of multidimensional feature vectors that are not explicitly invariant to nonlinear temporal distortions. Tallec and Ollivier [32] theoretically proved that the learnable gates in RNNs formally provide quasi-invariance to temporal transformations. Meanwhile, DECADE [5] and NeuralWarp [12] were proposed, in which the time series metric is defined as the (weighted) average of all the pairwise distances between deep feature sequences. These methods [5,12,32] offer robustness regarding temporal distortions but impose little constraints on the discrimination of non-matching time series. Wu et al. [39,40] proposed to incorporate the DTW into a Siamese network for simultaneous temporal invariance and discriminative power. However, these two approaches remain inefficient because DTW is a sequential algorithm and, unlike convolutional and fully connected layers, it is hard to take advantage of parallel GPU computing.

In this paper, we propose a novel deep model for time series metric learning. The model should be robust with regards to *temporal invariance* to distortion. Meanwhile, it should be able to impose temporal constraints on the learned metric for higher *discriminative power* in terms of non-matching pairs. To this end, we first propose an adaptation of a parameterized attention model to metric learning for greater and more adaptive temporal invariance. Given two time series, which can be raw signals or feature sequences extracted from a Siamese network, the attention model predicts a flexible decision for each pair of temporal locations regarding how proper it is to align the local data points at the two locations in time. The rectangular array of these decisions forms a warping matrix and is used for nonlinear distortion rectification (Fig. 1b). This proposal is inspired by the great success of attention models employed in image captioning and machine translation [3,21,36]. Metric learning drives the distance between a time series and a warped counterpart to be small for matching pairs and large for non-matching pairs.

We further propose a novel pre-training strategy to initialize the weights of the proposed model more effectively. The pre-training is guided by DTW so that temporal constraints can be imposed on the warping matrix for higher discriminative power. By applying the proposed approach to single-letter handwriting classification and online signature verification, we experimentally demonstrate our superior performance over conventional non-parametric and deep models for time series metric learning.

2 Proposed Approach

2.1 Overview

A schematic overview of the proposed network is shown in Fig. 2. Given two time series $\mathbf{A}$ and $\mathbf{B}$ as inputs, our network aims to output a distance that measures their dissimilarity. The time series can be raw signal sequences or deep feature sequences extracted from a Siamese network. The two inputs are first reshaped and concatenated to form an "outer concatenation" (see Sect. 2.2 for details), which is further passed to a fully convolutional network (FCN), i.e., the U-net in Fig. 2. The output of the FCN is a matrix $\mathbf{P}$ composed of flexible decisions indicating which local data points in $\mathbf{A}$ and $\mathbf{B}$ should be aligned in time. Row-wise softmax is carried out to generate two "warping paths" $\mathbf{P}_s$ and $\mathbf{P}_t$, which are used to warp $\mathbf{B}$ and $\mathbf{A}$, respectively. During testing, the time series distance is thus measured based on the inputs and their warped counterparts. In this network, the combination of outer concatenation, FCN, and row-wise softmax corresponds to the attention model adapted to time series metric learning.

We want the distance measure described above to be small for matching pairs and large for non-matching pairs during training. To this end, we define a contrastive loss [14] based on the distance, as shown in Fig. 2 and Sect. 2.2.

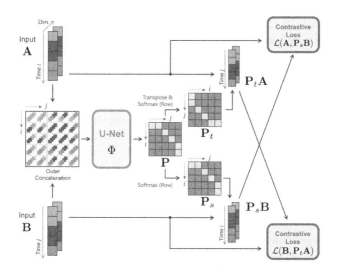

Fig. 2. Overview of our proposed network (training stage).

We initialize the weights of the network through pre-training, which is guided by DTW for improved discriminative power and is described in Sect. 2.3.

2.2   Attention to Warp

In the family of attention models, additive attention [3] and (scaled) dot-product attention [21,36] are the most widely used. In this study, we choose the parameterized additive attention mechanism and adapt it for time warping to ensure that the learned metric is explicitly invariant to temporal distortions.

Let $\mathbf{A} \in \mathbb{R}^{W \times K}$ and $\mathbf{B} \in \mathbb{R}^{W \times K}$ denote two multivariate time series, where W is the temporal length and K is the number of dimensions (variables). The time series can be a sequence of raw signals, e.g., an online signature, or a sequence of feature vectors, e.g., a deep feature sequence extracted from a Siamese network. We consider a function $\Phi : \mathbb{R}^{W \times K} \times \mathbb{R}^{W \times K} \mapsto \mathbb{R}^{W \times W}$ that produces a warping matrix $\mathbf{P} = \Phi(\mathbf{A}, \mathbf{B})$ such that $\mathbf{B}$ can be aligned in time with $\mathbf{A}$ in the form of $\mathbf{PB}$. Let $\mathbf{p}_i \in \mathbb{R}^{1 \times W}$ be the i-th row vector of $\mathbf{P}$ with $i \in [1, W]$. Suppose that $\mathbf{p}_i$ has been l^1-normalized. The warped time series $\mathbf{PB}$ can thus be understood as a sequence of K-dimensional vectors $\{\mathbf{p}_i\mathbf{B}\}$, each corresponding to a weighted average of all the K-dimensional vectors in $\mathbf{B}$. Based on this warping, we expect that each $\mathbf{p}_i\mathbf{B}$ best matches with the K-dimensional vector at the i-th time step of $\mathbf{A}$, so that intra-class temporal distortions can be accommodated.

The function Φ can be decomposed into $W \times W$ kernel functions denoted by φ. Let $\mathbf{a}_i \in \mathbf{R}^{1 \times K}$ and $\mathbf{b}_j \in \mathbf{R}^{1 \times K}$ be the vectors of $\mathbf{A}$ and $\mathbf{B}$ at the i-th and j-th time steps, respectively. Let p_{ij} be the j-th element in $\mathbf{p}_i$. The kernel function $p_{ij} = \varphi(\mathbf{a}_i, \mathbf{b}_j)$ thus transforms $\mathbf{a}_i$ and $\mathbf{b}_j$ into a flexible decision regarding whether the two vectors best match each other ($p_{ij} \approx 1$) or not ($p_{ij} \approx 0$). According to additive attention [3], φ can be defined by a neural network consisting of one or more fully connected layers, which takes the concatenation of $\mathbf{a}_i$ and $\mathbf{b}_j$ as the input.

In the case of the proposed approach, the attention model can be realized more efficiently. Let $\mathbf{A}$ be reshaped to a $W \times 1 \times K$ tensor and be *horizontally* stacked W times. Similarly, $\mathbf{B}$ is reshaped to a $1 \times W \times K$ tensor and is *vertically* stacked W times. They are then concatenated along their third axis, leading to a $W \times W \times 2K$ tensor. We call this operation an "outer concatenation" (analogous to the outer product) and denote it by $\mathbf{A} \oplus \mathbf{B}$. Instead of obtaining each p_{ij} individually, we can obtain the warping matrix $\mathbf{P}$ at once by the convolution between the outer concatenation $\mathbf{A} \oplus \mathbf{B}$ and the kernel function φ. Our attention model Φ can thus be defined by an FCN.

Backbone Network. In this paper, we adopt U-net [28], which was originally developed for biomedical image segmentation, as the FCN to define Φ. Since the input (outer concatenation) $\mathbf{A} \oplus \mathbf{B}$ of the FCN can be regarded as a $2K$-channel "image" of size $W \times W$, and the output (warping matrix) $\mathbf{P}$ as a single-channel "image" of the same size, it is appropriate to employ U-net as the FCN for direct "image-to-image" mapping. Compared to the 1×1 convolution described above, the U-net linearly increases the size of the receptive field by stacking multiple 3×3 convolution layers. The flexible decision p_{ij} thus depends not only on the local data points $\mathbf{a}_i$ and $\mathbf{b}_j$ but also on their neighborhood along the time axis. This enables the exploitation of beneficial contextual information. Note that for simplicity, we have assumed that the input time series $\mathbf{A}$ and $\mathbf{B}$ have the same length W, but our approach is not limited to this assumption since U-net can naturally handle inputs (outer concatenation $\mathbf{A} \oplus \mathbf{B}$) of arbitrary size.

Learning. Once the warping matrix $\mathbf{P}$ is output from the U-net, a row-wise softmax is conducted on $\mathbf{P}$ to ensure that each $\mathbf{p}_i \in \mathbf{P}$ is l^1-normalized and to make it peakier. $\mathbf{B}$ is then warped according to $\mathbf{P}_s\mathbf{B}$ to compensate for temporal distortions, where $\mathbf{P}_s$ is the output of the softmax. During training, a contrastive loss can thus be defined by Eq. (1), which is adapted from the extensively used conventional contrastive loss [14].

$$\mathcal{L}(\mathbf{A}, \mathbf{P}_s\mathbf{B}) = \begin{cases} \frac{1}{WK}\|\mathbf{A} - \mathbf{P}_s\mathbf{B}\|_{\mathrm{F}}^2 & \text{if } z = 1 \\ \max\left(0, \tau - \frac{1}{WK}\|\mathbf{A} - \mathbf{P}_s\mathbf{B}\|_{\mathrm{F}}^2\right) & \text{otherwise} \end{cases} \tag{1}$$

Here, τ is the margin of the hinge loss, and $z \in \{0, 1\}$ defines whether $\mathbf{A}$ and $\mathbf{B}$ are non-matching or matching, respectively. The distance between the two time series is defined by the Frobenius norm of the difference matrix $\mathbf{A} - \mathbf{P}_s\mathbf{B}$. Minimizing Eq. (1) drives this distance to be small for matching pairs and large for non-matching pairs.

To make the model symmetric, another warping path $\mathbf{P}_t$ is created by transposing $\mathbf{P}$ and then executing row-wise softmax. $\mathbf{A}$ is warped according to $\mathbf{P}_t\mathbf{A}$. A second contrastive loss can thus be defined between $\mathbf{B}$ and $\mathbf{P}_t\mathbf{A}$ as in Eq. (2).

$$\mathcal{L}(\mathbf{B}, \mathbf{P}_t\mathbf{A}) = \begin{cases} \frac{1}{WK}\|\mathbf{B} - \mathbf{P}_t\mathbf{A}\|_{\mathrm{F}}^2 & \text{if } z = 1 \\ \max\left(0, \tau - \frac{1}{WK}\|\mathbf{B} - \mathbf{P}_t\mathbf{A}\|_{\mathrm{F}}^2\right) & \text{otherwise} \end{cases} \tag{2}$$

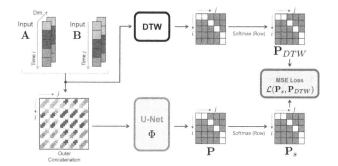

Fig. 3. Overview of our proposed network (pre-training stage).

Inference. During testing, the proposed model takes two times series as inputs and predicts a distance measuring their dissimilarity. This distance can be applied to a k-NN classifier for classification or a threshold classifier for verification. Specifically, we define the distance by Eq. (3), which is the average of the distances between $\mathbf{A}$ and $\mathbf{P}_s\mathbf{B}$ and between $\mathbf{B}$ and $\mathbf{P}_t\mathbf{A}$.

$$d(\mathbf{A}, \mathbf{B}, \mathbf{P}) = \frac{1}{2WK} \left(\|\mathbf{A} - \mathbf{P}_s\mathbf{B}\|_{\mathrm{F}}^2 + \|\mathbf{B} - \mathbf{P}_t\mathbf{A}\|_{\mathrm{F}}^2 \right) \tag{3}$$

2.3 Pre-training Guided by Dynamic Time Warping

In our experiments, we found that the widely-used He initialization [15] could not initialize the weights of the proposed model effectively on some datasets. Specifically, the contrastive losses in Eqs. (1) and (2) do not decrease as the training stage progresses. The reason might be because the attention model takes no temporal constraint into account and is limited for distinguishing non-matching pairs of time series.

In this paper, we incorporate temporal constraints in our model through pre-training guided by DTW. Specifically, we pre-train the U-net weights to be used in the proposed model with the guidance of DTW. As shown in Fig. 3, the input (outer concatenation $\mathbf{A} \oplus \mathbf{B}$ of the two time series) and output (matrix $\mathbf{P}$) of the U-net as well as the row-wise softmax remains the same as those shown in Fig. 2 and Sect. 2.2. The difference from the training stage lies in the loss function. To define the loss, we first apply DTW to the inputs $\mathbf{A}$ and $\mathbf{B}$ and obtain its warping path in the form of a binary matrix. A row-wise softmax is then applied to this matrix, which is similar to the same softmax for $\mathbf{P}$. Let $\mathbf{P}_{\mathrm{DTW}}$ denote the output of the softmax. Here, we want to make the attention model mimic DTW and force it to produce a warp matrix $\mathbf{P}_s$ that is as close to $\mathbf{P}_{\mathrm{DTW}}$ as possible. Therefore, we adopt the mean squared error (MSE) between the two matrices as the loss function of pre-training, as shown in Eq. (4).

$$\mathcal{L}(\mathbf{P}_s, \mathbf{P}_{\mathrm{DTW}}) = \frac{1}{W^2} \|\mathbf{P}_s - \mathbf{P}_{\mathrm{DTW}}\|_{\mathrm{F}}^2 \tag{4}$$

Another loss can also be defined based on the nearly symmetric warping matrix $\mathbf{P}_t$. In this paper, we do not do this for simplicity.

Note that mimicking DTW during pre-training is to guide the learning of our attention model so that its discriminative power can be improved. Our final purpose is not to mimic DTW but to learn warping through metric learning.

3   Experiments

To evaluate the performance of the proposed approach, we conducted experiments using Unipen [13], an online single-letter handwriting dataset, and MCYT-100 [26], an online signature dataset.

3.1   Online Single-Letter Handwriting Classification (Unipen)

Among the multiple subsets officially offered by the provider, Unipen 1a (numerical digits), Unipen 1b (uppercase alphabet), and Unipen 1c (lowercase alphabet) were selected for evaluation. For all three subsets, each time series is a sequence of 2D coordinates of the pen tip, resized to the same fixed temporal length 50. Some statistics are shown in Table 1.

For each class, two sets of 200 data were randomly sampled and used as validation and test data, respectively. The rest of the data were used as training data. When training the network, matching and non-matching pairs were created between all training data. Given a test time series, its distances from all training time series were calculated and a k-nearest neighbor (k-NN) classifier was used to determine its class label ($k = 3$ in this paper).

Baselines. The proposed approach was compared with the four subsequent baselines. In addition, the performance of the proposed approach without pre-training was also included in the comparison.

DTW. Given a test time series, its distances from the training time series are measured by applying DTW to their raw signals (2D coordinates). This baseline serves as a representative of a non-parametric (non-training) distance measure.

Support Vector Machine (SVM). For this baseline, we trained an SVM with a radial basis function (RBF) kernel for classification. The raw signals were used for training and testing.

Classification Network (CN). This is a convolutional neural network (CNN) that consists of eight convolutional layers and three fully connected (FC) layers. It was trained using a cross-entropy loss.

Siamese Network (SN). This network has almost the same architecture as CN, but was trained with contrastive loss. Its last FC layer produces a global feature representation rather than a probability distribution of classes. Similar to DTW and the proposed approach, SN uses a k-NN classifier for classification.

Implementation Details. The U-Net used in our model consists of an encoder and a decoder that are almost symmetrical, both containing seven convolution

Table 1. Classification accuracy (%) for Unipen.

Method	Unipen 1a	Unipen 1b	Unipen 1c
Proposed	**99.0**	**98.0**	**95.5**
w/o pre-training	98.4	97.3	90.7
DTW	98.4	96.0	94.1
SVM	98.2	93.9	94.1
CN	98.3	96.2	**95.5**
SN	98.6	97.2	**95.5**
#class	10	26	26
#training (per class)	756.2	232.3	391.2
#validation (per class)	200	200	200
#test (per class)	200	200	200

Table 2. Number of misclassifications between similar but different classes.

Method	Unipen 1b		Unipen 1c				
	J vs. T	U vs. V	f vs. t	g vs. y	h vs. k	h vs. n	v vs. w
Proposed	**7**	**13**	**10**	**6**	**6**	**13**	**2**
DTW	28	23	16	24	16	20	7

layers. The feature maps output by the encoder are concatenated with the corresponding layers of the decoder by skip connections. Further details can be found in the supplementary document[1]. Adam [17] was used as the optimizer to train the model, and the learning rate was set to 0.0001. The batch size was set to 512. The margin τ used in Eqs. (1) and (2) was set to one. The training was conducted for up to 20 epochs, and the best model was selected based on the performance measured by the validation data.

Results. The classification results when using Unipen are shown in Table 1. In all three subsets, the proposed approach outperformed or at least was comparable to all baselines. These results confirm the superiority of our deep model in terms of learnable time warping (compared to DTW) and high discriminative power (compared to SVM, CN, and SN). Although the improvement in accuracy is consistent, the range of improvement is not very large. This may be due to a large amount of training data and the short length and low dimensionality of the time series, which render this handwriting classification task a relatively easy classification problem. This also explains the small difference in accuracy between our models w/ and w/o pre-training. In Sect. 3.2, we apply our model to a more complex verification problem for further evaluation.

Examining the results of Unipen 1b, we found that the largest improvement achieved by the proposed approach lies in the discrimination between J (with a

---

[1] http://human.ait.kyushu-u.ac.jp/~matsuo/ICDAR2021_Appendix.pdf.

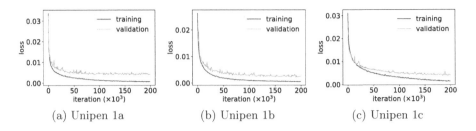

(a) Unipen 1a   (b) Unipen 1b   (c) Unipen 1c

Fig. 4. Training and validation losses.

horizontal serif) and T and between U and V. Table 2 compares DTW and the proposed approach in terms of the number of misclassifications between such similar but different alphabetic classes. Figure 4 shows the graphs of training and validation losses, which show good convergence and few signs of overfitting.

3.2   Online Signature Verification (MCYT-100)

We considered the most common type of signature verification, namely deciding whether a test signature is a genuine signature or a skilled forgery (a signature imitated by a forger) in relation to a claimed identity (subject). To this end, we adopted the dataset MCYT-100 [26] because it is one of the largest online signature datasets in the literature. MCTY-100 contains 100 subjects, each having 25 genuine signatures and 25 skilled forgeries. Each signature is a time series of 5D variables (2D coordinates, pressure, azimuth, and altitude angles of the pen tip). The temporal length ranges from a few hundred to a few thousand.

We tested the proposed approach with the following training protocol. The first 90/80/70/60/50% of subjects in MCYT-100 were used for training and the remaining subjects for testing. Therefore, the training and test sets are completely non-overlapping in terms of both classes (subjects) and time series (signatures). During testing, for each subject in the test set, the first five genuine signatures were used as reference signatures. The remaining genuine signatures and all the skilled forgeries were used as test signatures. The distances between each test signature and all the reference signatures of the claimed subject were computed and their average was used as the distance between the test signature and the claimed identity. This distance can be passed to a subject-independent threshold classifier for the final decision of signature authenticity. For each test signature, its distance to the corresponding five reference signatures is averaged. Based on this averaged distance, all test signatures (of all subjects) are then sorted and an Equal Error Rate (EER) is calculated, as in most related studies.

Implementation Details. For MCYT-100, we adopted the deep dynamic time warping (DDTW) [39] for time series feature learning. Specifically, a Siamese network with two raw signatures as input and two feature sequences as output was trained by DDTW following the training protocol described above. The temporal length W of each feature sequence is 256, and the number of dimensions K is 64. This network was then used to extract feature sequences from all

Table 3. EER(%) of online signature verification for MCYT-100.

Method	Percentage of training data				
	90%	80%	70%	60%	50%
Proposed	**0.50**	**2.00**	**2.33**	**2.13**	**2.20**
DTW [23,39]	4.00	3.00	4.17	4.37	4.60
w/ raw signatures as inputs	5.00	6.25	5.73	6.37	6.96
SN	5.50	6.80	6.27	7.33	8.40
w/ local embedding loss [39]	3.50	3.40	3.75	3.75	5.50
DDTW [39]	1.00	2.20	2.53	2.25	2.40
PSN [40]	**0.50**	2.50	2.40	2.50	4.50

training, reference, and test signatures. Afterward, we built the proposed deep model with two feature sequences as input time series and a distance as output and trained it in the way described in Sect. 2. During testing, this trained model was applied to the feature sequences of all reference and test signatures in order to calculate their distances. It is also possible to jointly learn feature representation and warping in an end-to-end manner, which will be our future work. Implementation details of DDTW can be found from its original paper [39].

The U-net for MCYT-100 is the same as Unipen, but the encoder and decoder each have eight convolution layers. This difference is due to the different size of outer concatenation. The learning rate was set to 0.0001. The batch size was set to 15. The ratio of matching to non-matching pairs per batch was fixed at 1 : 2.

Baselines. We compared our approach with the following four baselines.

DTW [23,39]. In this baseline, each signature was represented by a handcrafted feature sequence [23]. The distances between test and reference signatures were then measured with DTW. In addition, the performance of DTW that takes raw signatures as inputs was also included in the comparison.

SN. This SN has almost the same architecture as DDTW but was trained with conventional contrastive loss. Its output is not a feature sequence but a global feature vector produced by average pooling, and thus considers no temporal constraint. Another SN was also trained with the same local embedding loss as DDTW. It considers temporal constraint but offers limited temporal invariance.

DDTW [39]. The only difference between DDTW and our approach lies in the warping process. DDTW uses DTW for warping, while our approach warps signatures with the proposed deep attention model.

Prewarping Siamese Network (PSN) [40]. This approach prewarps two signatures with DTW for temporal invariance. A Siamese network with the warped signatures as inputs and a distance as output is then trained with the same local embedding loss as DDTW. Similar to DDTW, PSN takes both temporal invariance and temporal constraint into account.

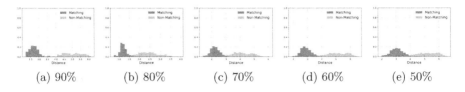

| (a) 90% | (b) 80% | (c) 70% | (d) 60% | (e) 50% |

Fig. 5. Distributions of distances measured by DDTW [39] for MCYT-100. The sub-captions indicate the percentage of training data.

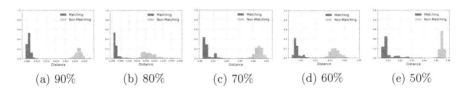

| (a) 90% | (b) 80% | (c) 70% | (d) 60% | (e) 50% |

Fig. 6. Distributions of distances measured by proposed approach for MCYT-100. The sub-captions indicate the percentage of training data.

Results. The EERs of our approach and the baselines are shown in Table 3. In general, our approach outperformed all baselines in all experiments. Our lower EER than DTW indicates our superiority in the ability to learn complex representations over handcrafted features. Our approach also significantly outperformed its competitors, which only consider either temporal invariance (SN) or temporal constraints (SN w/ local embedding loss). Our improvement over DDTW shows the greater effectiveness of our trainable attention model in terms of warping and distance measurement compared to the traditional DTW.

Figures 5 and 6 show the histograms of the distances of matching and non-matching pairs, corresponding to DDTW and the proposed approach, respectively. We can see that our approach shows a much smaller overlap than DTW. As described above, DDTW focuses only on learning time series features, while our approach further improves warping through learning to facilitate easier separation of the distances of matching and non-matching pairs.

We also tested our approach without pre-training, where He initialization [15] was used to initialize the model weights. However, the contrastive losses could not decrease as the training progresses. The reason may be due to the difficulty of the signature verification task, such as the small amount of training data and the subtle difference between genuine signatures and skilled forgeries. The failure of He initialization confirms the necessity of pre-training based on DTW.

Table 4 compares our performance with the state-of-the-art EERs previously reported for MCYT. Noted that similar to DDTW and PSN, our model is learning-based and so our superior accuracy was obtained using only the test set, which contains many fewer subjects. The comparison is thus biased toward our model relative to the state of the art. In the future, we shall consider training the network with a larger dataset and testing it on the whole of MCYT-100.

After training with 90% of the subjects, the EER of our approach was 0.5%, as shown in Table 3. Using similar experimental protocols, the EER of SynSig2Vec [18] was 1.7%, the EER of Li et al. [19] was 10.5%, and the accuracy of OSVNet [38]

Table 4. EERs (%) published on MCYT. #sub indicates the number of subjects in the test set. The number of reference signatures is five for all methods.

Method	#sub	Deep?	EER (%)
Yanikoglu and Kholmatov [41]	100	✗	7.80
Vivaracho-Pascual et al. [37]	280	✗	6.60
Faúndez-Zanuy [8]	280	✗	5.42
Nanni and Lumini [25]	100	✗	5.20
Cpalka et al. [7]	100	✗	4.88
Sae-Bae and Memon [30]	100	✗	4.02
Tang et al. [33]	100	✗	3.16
DDTW [39]	50	✓	2.40
PSN [40]	50	✓	4.50
Proposed approach	50	✓	**2.20**

was 93.0%. All of these previous studies are deep signature verification approaches. These results demonstrate the superiority of our approach over the state of the art. In this section, we have applied our approach only to skilled forgery verification, but it can be easily adapted to identify random forgeries. We will demonstrate this in the future. We also plan to test our approach on a larger dataset such as DeepSignDB [35].

3.3   Qualitative Analysis

Figure 7 shows examples of warping behavior of our approach on the Unipen dataset. In all examples, our network correctly classified the two time series into matching and non-matching pairs. Some interesting behaviors can be observed. For relatively easy matching pairs (first two rows in Fig. 7a), our network predicted warp paths that satisfied the monotonicity, continuity, and boundary conditions similar to DTW. When two letters of the same class but with completely different stroke orders came along (last two rows in Fig. 7a), our network could learn to adaptively break through the temporal constraints described above, defying the common wisdom of time warping. The correct classifications shown in the last two examples in Fig. 7a are not accidental, but because the training data includes such intra-class variation in stroke order (Fig. 8) and thanks to the great robustness of our attention model. The conventional DTW is obviously not capable of handling such challenging time series.

Figure 7b shows some challenging examples of non-matching pairs that are very similar in both appearance and stroke order. The warping suggested by DTW could not effectively distinguish these pairs. In comparison, if we focus on the pair of input handwriting $\mathbf{A}$ vs. warped "handwriting" $\mathbf{P}_s\mathbf{B}$ (or $\mathbf{B}$ vs. $\mathbf{P}_t\mathbf{A}$), we can observe a significant difference in appearance. That is, our network could learn to augment the inter-class variation through warping, so that similar but

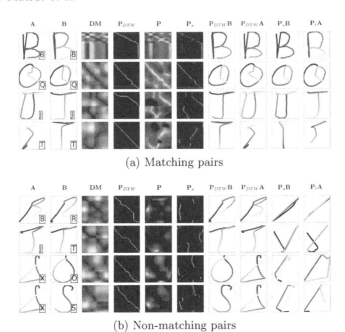

(a) Matching pairs

(b) Non-matching pairs

Fig. 7. Examples of input time series **A** and **B**, DTW distance matrix (DM), DTW warping path $\mathbf{P}_{\mathrm{DTW}}$, output **P** of U-net, our warping matrix $\mathbf{P}_s$, DTW-warped **B** and **A**, and our warped time series $\mathbf{P}_s\mathbf{B}$ and $\mathbf{P}_t\mathbf{A}$. Each time series starts with the darkest color. For DM, $\mathbf{P}_{\mathrm{DTW}}$, **P**, and $\mathbf{P}_s$, the brighter the color, the larger the value. (Color figure online)

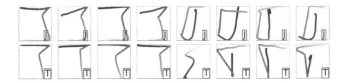

Fig. 8. Examples of Unipen training data of J (top) and T (bottom). Each time series starts with the darkest color and finishes with the lightest color. For both classes, the leftmost four letters have a totally different stroke order from the rightmost ones. (Color figure online)

different classes could be distinguished in a much easier manner. These results demonstrate the high discriminative power of our approach.

4   Conclusion

We proposed to learn a warping function to align the indices of time series using an attention model specialized for metric learning. This model is invariant not only to local or semi-local temporal distortions, but also to large global ones, so that even matching pairs that do not satisfy the monotonicity, continuity, and boundary conditions can still be successfully identified. We also proposed to pre-train our model with the guidance of DTW for stabilized learning and higher discriminative power.

The proposed pre-training can actually be accomplished using general time series data that do not have to be included in the training set. If this is true, our model would be relieved of the requirement for large-scale labeled data in the target (classification or verification) domain. This property would be beneficial for tasks, e.g., signature verification or writer identification, where only a small dataset is available. We will investigate this in future research.

References

1. Ahrabian, K., BabaAli, B.: Usage of autoencoders and Siamese networks for online handwritten signature verification. Neural Comput. Appl. **31**(12), 9321–9334 (2019)
2. Bagnall, A.J., Lines, J., Bostrom, A., Large, J., Keogh, E.J.: The great time series classification bake off: a review and experimental evaluation of recent algorithmic advances. Data Min. Knowl. Discov. **31**(3), 606–660 (2017)
3. Bahdanau, D., Cho, K., Bengio, Y.: Neural machine translation by jointly learning to align and translate. In: ICLR (2015)
4. Bromley, J., Guyon, I., LeCun, Y., Säckinger, E., Shah, R.: Signature verification using a Siamese time delay neural network. In: NIPS, pp. 737–744 (1993)
5. Che, Z., He, X., Xu, K., Liu, Y.: DECADE: a deep metric learning model for multivariate time series. In: Workshop on Mining and Learning from Time Series (2017)
6. Coskun, H., Tan, D.J., Conjeti, S., Navab, N., Tombari, F.: Human motion analysis with deep metric learning. In: ECCV, pp. 693–710 (2018)
7. Cpalka, K., Zalasinski, M., Rutkowski, L.: A new algorithm for identity verification based on the analysis of a handwritten dynamic signature. Appl. Soft Comput. **43**, 47–56 (2016)
8. Faúndez-Zanuy, M.: On-line signature recognition based on VQ-DTW. Pattern Recognit. **40**(3), 981–992 (2007)
9. Fawaz, H.I., Forestier, G., Weber, J., Idoumghar, L., Muller, P.: Deep learning for time series classification: a review. Data Min. Knowl. Discov. **33**(4), 917–963 (2019)
10. Fiérrez-Aguilar, J., Ortega-Garcia, J., Ramos, D., Gonzalez-Rodriguez, J.: HMM-based on-line signature verification: feature extraction and signature modeling. Pattern Recognit. Lett. **28**(16), 2325–2334 (2007)
11. Ge, X., Smyth, P.: Deformable Markov model templates for time-series pattern matching. In: KDD, pp. 81–90 (2000)
12. Grabocka, J., Schmidt-Thieme, L.: NeuralWarp: time-series similarity with warping networks. CoRR (2018)
13. Guyon, I., Schomaker, L., Plamondon, R., Liberman, M., Janet, S.: UNIPEN project of on-line data exchange and recognizer benchmarks. In: ICPR, pp. 29–33 (1994)
14. Hadsell, R., Chopra, S., LeCun, Y.: Dimensionality reduction by learning an invariant mapping. In: CVPR, pp. 1735–1742 (2006)

15. He, K., Zhang, X., Ren, S., Sun, J.: Delving deep into rectifiers: Surpassing human-level performance on ImageNet classification. In: ICCV, pp. 1026–1034 (2015)
16. Kholmatov, A., Yanikoglu, B.A.: Identity authentication using improved online signature verification method. Pattern Recognit. Lett. **26**(15), 2400–2408 (2005)
17. Kingma, D.P., Ba, J.: Adam: a method for stochastic optimization. In: ICLR (2015)
18. Lai, S., Jin, L., Lin, L., Zhu, Y., Mao, H.: SynSig2Vec: learning representations from synthetic dynamic signatures for real-world verification. In: AAAI, pp. 735–742 (2020)
19. Li, C., et al.: A stroke-based RNN for writer-independent online signature verification. In: ICDAR, pp. 526–532 (2019)
20. Lines, J., Bagnall, A.J.: Time series classification with ensembles of elastic distance measures. Data Min. Knowl. Discov. **29**(3), 565–592 (2015)
21. Luong, T., Pham, H., Manning, C.D.: Effective approaches to attention-based neural machine translation. In: EMNLP, pp. 1412–1421 (2015)
22. Marteau, P.: Time warp edit distance with stiffness adjustment for time series matching. IEEE Trans. Pattern Anal. Mach. Intell. **31**(2), 306–318 (2009)
23. Martinez-Diaz, M., Fiérrez, J., Krish, R.P., Galbally, J.: Mobile signature verification: Feature robustness and performance comparison. IET Biom. **3**(4), 267–277 (2014)
24. Mueller, J., Thyagarajan, A.: Siamese recurrent architectures for learning sentence similarity. In: AAAI, pp. 2786–2792 (2016)
25. Nanni, L., Lumini, A.: A novel local on-line signature verification system. Pattern Recognit. Lett. **29**(5), 559–568 (2008)
26. Ortega-Garcia, J., et al.: MCYT baseline corpus: a bimodal biometric database. IEE Proc. Vis. Image Sig. Process. **150**(6), 395–401 (2003)
27. Rakthanmanon, T., et al.: Searching and mining trillions of time series subsequences under dynamic time warping. In: KDD, pp. 262–270 (2012)
28. Ronneberger, O., Fischer, P., Brox, T.: U-net: convolutional networks for biomedical image segmentation. In: Navab, N., Hornegger, J., Wells, W.M., Frangi, A.F. (eds.) MICCAI 2015. LNCS, vol. 9351, pp. 234–241. Springer, Cham (2015). https://doi.org/10.1007/978-3-319-24574-4_28
29. Roy, D., Mohan, C.K., Murty, K.S.R.: Action recognition based on discriminative embedding of actions using Siamese networks. In: ICIP, pp. 3473–3477 (2018)
30. Sae-Bae, N., Memon, N.D.: Online signature verification on mobile devices. IEEE Trans. Inf. Forensics Secur. **9**(6), 933–947 (2014)
31. Sakoe, H., Chiba, S.: Dynamic programming algorithm optimization for spoken word recognition. IEEE Trans. Acoust. Speech Signal Process. **26**(1), 43–49 (1978)
32. Tallec, C., Ollivier, Y.: Can recurrent neural networks warp time? In: ICLR (2018)
33. Tang, L., Kang, W., Fang, Y.: Information divergence-based matching strategy for online signature verification. IEEE Trans. Inf. Forensics Secur. **13**(4), 861–873 (2018)
34. Tolosana, R., Vera-Rodríguez, R., Fiérrez, J., Ortega-Garcia, J.: Exploring recurrent neural networks for on-line handwritten signature biometrics. IEEE Access **6**, 5128–5138 (2018)
35. Tolosana, R., Vera-Rodríguez, R., Fiérrez, J., Ortega-Garcia, J.: DeepSign: deep on-line signature verification. IEEE Trans. Biom. Behav. Identity Sci. **3**(2), 229–239 (2021)
36. Vaswani, A., et al.: Attention is all you need. In: NIPS, pp. 5998–6008 (2017)
37. Vivaracho-Pascual, C., Faúndez-Zanuy, M., Pascual, J.M.: An efficient low cost approach for on-line signature recognition based on length normalization and fractional distances. Pattern Recognit. **42**(1), 183–193 (2009)

38. Vorugunti, C.S., Guru, D.S., Mukherjee, P., Pulabaigari, V.: OSVNet: convolutional Siamese network for writer independent online signature verification. In: ICDAR, pp. 1470–1475 (2019)
39. Wu, X., Kimura, A., Iwana, B.K., Uchida, S., Kashino, K.: Deep dynamic time warping: end-to-end local representation learning for online signature verification. In: ICDAR, pp. 1103–1110 (2019)
40. Wu, X., Kimura, A., Uchida, S., Kashino, K.: Prewarping Siamese network: Learning local representations for online signature verification. In: ICASSP, pp. 2467–2471 (2019)
41. Yanikoglu, B.A., Kholmatov, A.: Online signature verification using Fourier descriptors. EURASIP J. Adv. Sig. Process. (2009)

Document Forensics and Provenance Analysis

Customizable Camera Verification for Media Forensic

Huaigu Cao$^{(\boxtimes)}$ and Wael AbdAlmageed

Information Sciences Institute, University of Southern California, Marina del Rey, CA, USA
{hcao,wamageed}@isi.edu

Abstract. This paper presents our research work in camera verification. We expanded a convolutional network-based feature extraction/verification network to a multi-patch input and addressed the concerns over memory limitation and over-fitting issue. We have also made careful consideration for custom model training and provided strong results showing promising potential for real-world application of detecting scene text repurposing.

Keywords: Camera identification · PRNU · Media forensic

1 Introduction

1.1 Problem Overview

In media forensic, there has been extensive research in finding fake news using embedded imagery materials (photos or videos). Not only manipulated images can make fake news, but unaltered images can also be repurposed to make the news look real. Thus, when no evidence of manipulation can be found by using all scientific methods, one may also consider whether the materials were really taken from the event mentioned in the news.

There has been extensive research work on manipulation detection and repurposing detection aiming at verifying the digital integrity of the images, videos and textual content in the news. Besides these approaches, researchers have also come up with the idea of comparing the PRNU [1] of the reporter's camera with the original photograph in question. When successful, this approach can effectively tell if the photo was "borrowed" from an unspoken source (*e.g.*, some sort of online image search engine) rather than being taken by the reporter himself/herself. The PRNU of a camera is the abbreviation of the Photo Response Non-Uniformity Noise. The CCD or CMOS sensor has slightly different response from one pixel to another due to the limitation of the manufacture precision. The non-uniform response of the camera will lead to a fixed-pattern noisy 2D image when capturing a flatfield image (see Fig. 1). Identifying the PRNU pattern from a document image is easier. However, scene text images that appear more in media news tend to have complex background than document images which makes it a non-trivial problem to isolate the PRNU pattern from the image (see the example in Fig. 2).

© Springer Nature Switzerland AG 2021
J. Lladós et al. (Eds.): ICDAR 2021, LNCS 12823, pp. 369–379, 2021.
https://doi.org/10.1007/978-3-030-86334-0_24

G0070152_TL.JPG	01010059_TL.JPG	DSCN0014_TL.JPG	DSCN0002_TL.JPG
G0070153_TL.JPG	01010065_TL.JPG	DSCN0015_TL.JPG	DSCN0003_TL.JPG
G0070154_TL.JPG	01010068_TL.JPG	DSCN0016_TL.JPG	DSCN0004_TL.JPG
Camera 1	Camera 2	Camera 3	Camera 4

Fig. 1. The PRNU patterns from four cameras normalized using histogram equalization. Three instances of flatfield images were showed for each camera. Original images are flatfield images selected from the NIST MFC20 dataset.

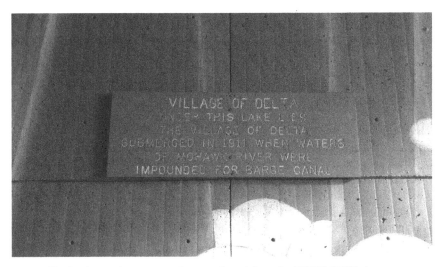

Fig. 2. A sample scene text image selected from the NIST MFC20 dataset.

1.2 Related Work

The PRNU pattern from a natural scene image can be extracted using the noise residual operator [2], homomorphic filter [3], image descriptors [4] or deep convolutional feature extractor [5, 6]. From any method, the PRNU is still contaminated heavily by noise and it requires sophisticated metric learning model [7] to compute the similarity between the image and the flatfield PRNU. Although camera verification is not a hot topic, established research work has been made focusing on representing the pattern and matching algorithm. Notably, NIST has collected the MFC training data of very good size and made the data publicly available through its open challenges in recent years (2018–2020) [8]. Several systems showed promising results within a range between 70% and 90% in AUC. The aforementioned performance looks encouraging. If it is possible to obtain more significant performance gain and further refine the steps of the application, it is very promising to meet the requirement of the real-world application.

1.3 Major Challenge and Motivation

- **Challenge in computation and memory requirement**
  One major challenge in camera verification is the huge input dimensions. The PRNU pattern can be thought of as the high frequency noise after a homomorphic filter is applied. Thus, we ideally need to feed the image in its original dimensions into a neural network. And this will be very demanding for the GPU memory given the prevailing dimensions of modern digital cameras.
  On the other hand, we have also experimented with the NIST MFC training data and trained a network with a CNN/dense-layers architecture and obtained competitive performance. In fact, the classifier we built only takes a 224x224 patch from the center of the image to verify, and makes the decision using the patch (As one can see from Fig. 3, the selected patch is a very small fraction of the entire image.) Surprisingly, we obtained a very competitive result: 85.8% AUC on the NIST MFC19 Eval set (the best team in the open challenge obtained 79.7%) and 77.5% on the MFC20 Eval set (only lower than one team with a 87.2% AUC). (Our deep learning classifier is stronger than the Siamese network [7] as the latter can only perform image-to-image matching.) Inspired by this result, we have been expecting to see significantly higher performance when more data are fed into the network.

Fig. 3. A sample image selected from the NIST MFC20 dataset with a 224 × 224 pixel region at the center of the image highlighted.

- **Ad-hoc creation of training data**

  Due to the way camera verification is used in media forensic, the analyst will not receive an off-the-shelf model and deploy it to predict the image-to-camera similarity. Instead, a real-world application requires the analyst to collect training data and feed the training data to the provided training pipeline and get the model. The risk of using a custom model is the performance will be unknown when designing the algorithm. Since it is very difficult to get a perfect guideline to instruct the analyst how to prepare the training set, we can only suggest very basic tips such as shooting at as many scenes as possible, avoiding taking multiple photos as the same scene, and including significant number of natural scenes rather than flatfield images to emulate the actual distribution of test data. A good example is showed in Fig. 4 where the photos from a camera in NIST MFC20 showed a reasonable representation and coverage of real-world data samples.

Fig. 4. Some of the sample images from camera PAR-1579 of NIST MFC20 dataset, showing a variety of scenes.

Besides, since each camera has its own model weights, even though the model may use a soft-max layer to predict a normalized confidence score, say, between 0 and 1, the same score can still mean different level of confidence as they are yielded by different models. A scientific way would be to align the score to its real confidence using validation data. It is also useful for the analyst to plot an ROC for each individual camera. Given a confidence score, one can also look up the raw data for the ROC to get the corresponding TPR/FPR values and provide a quantized interpretation of the camera verification score.

1.4 Overview of Presented Method

In the next two sections, we will present our customizable camera verification approach in detail including the network for extracting features, inferring the confidence score between a photo image and its claimed camera, and how the system obtained significant gains using multiple patches. Experimental results for each individual camera and extensive discussion are also presented.

2 Camera Verification Model

We adopted the VGG16 [9] architecture for feature extraction. The input layer takes a $224 \times 224 \times 3$ RGB image as the input. Following the last VGG16 layer, we appended a multi-layer classifier to classify positive (mismatched image and camera) vs. negative (matched image and camera) samples. The architecture of the classifier is showed in Fig. 5.

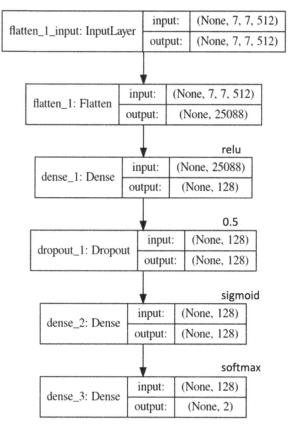

Fig. 5. Binary classifier architecture following the feature extraction stage.

When considering selecting multiple patches from the photo image, one effective way is to sample evenly from different locations of the image. We adopted a 3 row 3 column matrix of patches showed in Fig. 6 (a) and even more aggressively, 5 rows and 5 columns as showed in Fig. 6 (b). Note these patches are 224×224 each and they may overlap when the image is small enough.

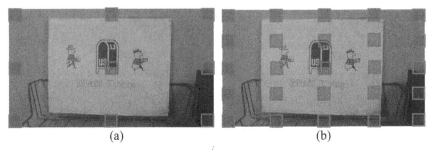

(a)   (b)

Fig. 6. Illustration of sampling multiple patches from an image. All these patches can be used to predict camera verification result.

Fig. 7. Examples of concatenated patches.

We could concatenate these patches for each photo to create a larger input image and feed it to the neural network (Fig. 7). However, there are at least three advantages to let the input layer take one patch at a time:

1) Most camera verification applications cannot afford collecting too many training images. Increasing the input dimensions exponentially will lead to the overfitting problem.
2) Increasing the input size exponential may also exceed the GPU memory.
3) When patches are not concatenated, it is more flexible to handle the situations such as cropped images.

The only thing that could be a concern when patches from different locations match each other very well. But the chance by which this happens is extremely low owing to the fact that the PRNU pattern is a sort of high frequency noise and it is unlikely to find two 224×224 spots that have well matched PRNU in camera sensors.

Here is how we handle multiple patches per photo in our system:

- Training:
  Each patch is treated as an independent training sample and no concatenation is performed. Each patch is rotated by 90, 180, and 270 degree to create three more training samples. The rotation-based augmentation is to handle all possible rotations of the camera.
- Inference:
  Inference using the trained model is performed on all patches from a photo and the maximum confidence score is taken as the confidence score of the photo.

3 Experimental Results

3.1 Dataset Overview

The NIST MFC20 [8] dataset for camera verification training has 106 cameras and 35695 natural scene images (previously released MFC19 and MFC18 can be considered subsets of MFC20.) The Eval set for MFC20 has 11288 image-camera pairs to verify. The Eval set for MFC19 has 8804 pairs to verify. Some of the training images have already been demonstrated in Fig. 4.

3.2 Single-Patch Open Set Verification Results

As we have mentioned above, to run the NIST evaluation protocol, we trained a model to predict the similarity score between each image-camera pair. And all the scores were collected to plot a global DET curve and AUC was also computed.

In this experiment, the classifier in Fig. 5 has a small modification such that the output layer has as many units as the number of cameras in the training set rather than 2. Thus, the training process was a closed set training on known cameras whereas the evaluation was performed as an open set problem [10] without requiring the negative sample to belong to any of the cameras in the training set. The system took the score corresponding to the camera as the similarity between the photo and the camera. Only the central 224×224 patch of each photo was used in training and evaluation. The AUC numbers are showed in Table 1. Our competitive results using only a 224×224 patch has inspired us to expand our system to a multi-patch customizable modeling approach.

Table 1. Camera verification results on the NIST MFC19 and MFC20 Eval sets.

Set	DET-AUC (best team in NIST open challenge)	DET-AUC (ours)
MFC19	79.7%	85.8%
MFC20	87.2%	77.5%

Table 2. A random selection of cameras from the MFC20 training set

Camera name	Number of photos (including rotated)
285540_Primary	132
50050172_Primary	156
50052808_Primary	60
PAR-1216_Primary	1548
PAR-1226_Primary	72
PAR-1579_Primary	160
PAR-1580_Primary	36
PAR-1581_Primary	180
PAR-1583_Primary	148
PAR-1589_Primary	176
PAR-2629_Primary	2260
PAR-2631_Primary	1216
PAR-3645_Primary	388
PAR-4335_Primary	648
Negative	136

3.3   Multi-patch Customizable Modeling Results

We randomly selected the following cameras showed in Table 2 from the MFC20 training set. A 64%: 16%: 20% ratio is adopted when creating the training: validation: test partition. We used the train set to train the model using the binary classification architecture, selected the best model within 50 epochs by monitoring the classification accuracy on the validation set, and applied the model to the test set.

Note the objective of a real-world application is to find repurposed image. Thus, it is important to define mismatched image-camera pairs as "positive". And the detection performance can be measured using the TPR values at given FPR. And these values can be computed from an ROC curve.

After we plotted the ROC curve for each camera, we obtained an AUC and TPR@0.2%FPR from each curve. These values are showed in Table 3. The average values of these two measurements are showed in Table 4. The result showed upgrading from a 1-patch-per-image model to a 9-patch-per-image model significantly improved the repurposing detection performance in both the average AUC and average TPR (FPR = 0.2%). Notably, the improvement in TPR value makes the application much more powerful in detection while maintaining the reliability of controlling false alarms.

We plotted the ROC curves for camera PAR-1589, the one that has the lowest performance when using one patch per image. First of all, the gain from using multiple patches can be visualized quite well using the ROC curves. And more importantly, all the TPR and FPR values in those curves have been linked to their corresponding confidence score. Thus, an analyst can retrieve the estimated TPR and FPR from the curves for each photo being examined and provide a quantized evidence for repurposing detection (Fig. 8).

Table 3. Repurposing detection performance using camera verification (performance of each camera).

Camera ID	Metric	Single-patch	9-patch	25-patch
285540_Primary	ROC-AUC	100.0%	100.0%	100.0%
	TPR@0.2%FPR	100.0%	100.0%	100.0%
50050172_Primary	ROC-AUC	89.8%	100.0%	100.0%
	TPR@0.2%FPR	28.6%	100.0%	100.0%
50052808_Primary	ROC-AUC	83.9%	85.7%	71.4%
	TPR@0.2%FPR	46.4%	57.1%	14.3%
PAR-1216_Primary	ROC-AUC	99.9%	92.0%	100.0%
	TPR@0.2%FPR	89.3%	53.6%	100.0%
PAR-1226_Primary	ROC-AUC	100.0%	100.0%	99.1%
	TPR@0.2%FPR	100.0%	100.0%	96.4%
PAR-1579_Primary	ROC-AUC	92.9%	100.0%	100.0%
	TPR@0.2%FPR	42.9%	100.0%	100.0%
PAR-1580_Primary	ROC-AUC	100.0%	100.0%	100.0%
	TPR@0.2%FPR	100.0%	100.0%	100.0%
PAR-1581_Primary	ROC-AUC	100.0%	99.2%	98.4%
	TPR@0.2%FPR	100.0%	85.7%	85.7%
PAR-1583_Primary	ROC-AUC	100.0%	100.0%	100.0%
	TPR@0.2%FPR	100.0%	100.0%	100.0%
PAR-1589_Primary	ROC-AUC	69.4%	99.1%	98.5%
	TPR@0.2%FPR	0.0%	89.3%	78.6%
PAR-2629_Primary	ROC-AUC	98.4%	99.0%	97.4%
	TPR@0.2%FPR	0.0%	57.1%	14.3%
PAR-2631_Primary	ROC-AUC	96.7%	99.9%	99.8%
	TPR@0.2%FPR	0.0%	82.1%	71.4%
PAR-3645_Primary	ROC-AUC	100.0%	100.0%	98.8%
	TPR@0.2%FPR	100.0%	100.0%	85.7%
PAR-4335_Primary	ROC-AUC	100.0%	100.0%	100.0%
	TPR@0.2%FPR	100.0%	100.0%	100.0%

Table 4. Repurposing detection performance using camera verification (Average of all cameras).

Metric	Single-patch	9-patch	25-patch
Avg ROC-AUC	95.1 ± 8.9%	98.2 ± 4.2%	97.4 ± 7.5%
Avg TPR@0.2%FPR	64.8 ± 43.0%	87.5 ± 18.1%	81.9 ± 30.1%

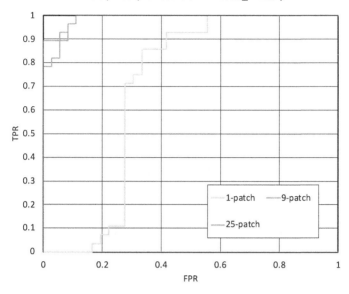

(a) Linear scale ROC

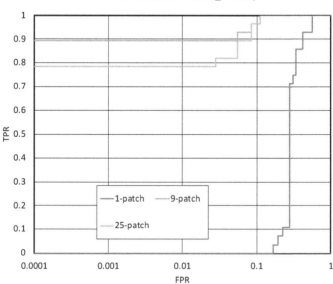

(b) Logarithm scale ROC

Fig. 8. ROC curves for camera PAR-1589.

4   Conclusion

To sum up, we have presented our research work in camera verification in this paper. We expanded a convolutional network-based feature extraction and verification network to multi-patch applications, addressed the concerns over memory limitation and overfitting issue. We have also made careful consideration for custom model training and provided strong results showing promising potential for real-world application to detecting scene text repurposing.

References

1. Akshatha, K.R., Karunakar, A.K., Anitha, H., Raghavendra, U., Shetty, D.: Digital camera identification using PRNU: a feature based approach. Digit. Investig. **19**, 69–77 (2016). ISSN 1742–2876. https://doi.org/10.1016/j.diin.2016.10.002. https://www.sciencedirect.com/science/article/pii/S1742287616300998
2. Goljan, J.F., Chen, M.: Defending against fingerprint-copy attack in sensor-based camera identification. IEEE Trans. Inf. Secur. Forensics **6**(1), 227–236 (2010)
3. Singh, R., Vatsa, M., Noore, A.: Improving verification accuracy by synthesis of locally enhanced biometric images and deformable model. Signal Process. **87**(11), 2746–2764 (2007)
4. Banerjee, S., Ross, A.: Impact of photometric transformations on PRNU estimation schemes: a case study using near infrared ocular images. In: 2018 International Workshop on Biometrics and Forensics (IWBF), Sassari, Italy, pp. 1–8 (2018). https://doi.org/10.1109/IWBF.2018.8401560
5. Cozzolino, D., Verdoliva, L.: Noiseprint: a CNN-based camera model fingerprint, CoRR, abs/1808.08396 (2018)
6. Athanasiadou, E., Geradts, Z., Van Eijk, E.: Camera recognition with deep learning. Forensic Sci. Res. **3**(3), 210–218 (2018). https://doi.org/10.1080/20961790.2018.1485198
7. Cozzolino, D., Marra, F., Gragnaniello, D., Poggi, G., Verdoliva, L.: Combining PRNU and noiseprint for robust and efficient device source identification. EURASIP J. Inf. Secur. **2020**(1), 1–12 (2020). https://doi.org/10.1186/s13635-020-0101-7
8. Fiscus, J., et al.: MediFor Challenge Evaluation Overview, p. 163 (2020). https://mig.nist.gov/MFC/Web/PIMeeting2020/NIST_MFC20_PIMeeting_All_Final_formated.pdf
9. . Simonyan, K., Zisserman, A.: Very deep convolutional networks for large-scale image recognition. In: 3rd International Conference on Learning Representations, ICLR 2015, San Diego, CA, USA, 7–9 May 2015
10. Bayar, B., Stamm, M.C.:Towards open set camera model identification using a deep learning framework. In: 2018 IEEE International Conference on Acoustics, Speech and Signal Processing (ICASSP), Calgary, AB, Canada, pp. 2007–2011 (2018). https://doi.org/10.1109/ICASSP.2018.8462383

Density Parameters of Handwriting in Schizophrenia and Affective Disorders Assessed Using the Raygraf Computer Software

Barbara Gawda[(⊠)] [iD]

Maria Curie-Sklodowska University, 20-031 Lublin, Poland
bgawda@wp.pl

Abstract. The present study aimed to examine fine motor impairments in individuals diagnosed with schizophrenia and affective disorders, manifested in their handwriting, in order to compare new global characteristic *morpheme and letter density of handwriting*. The handwriting samples were obtained from four groups: patients diagnosed with paranoid schizophrenia (n = 60), bipolar patients with depressive episode (n = 60), bipolar patients with manic episode (n = 60), and healthy controls (n = 60). The groups were matched in terms of intellectual level, sex, age, and education. The examined handwriting specimens were identical in terms of text content. The computer software Raygraf, developed by the Polish Forensic Association, was used in the examination of the structural graphical parameters associated with morpheme/letter density. The inter-group comparisons were made. The results showed significant differences in handwriting parameters between the clinical groups and the controls, however, there were no significant differences between the three different clinical groups. These results confirmed that individuals with schizophrenia, depression, and mania differ in terms of handwriting density features from the healthy controls. Logistic regression analysis and k-means clustering were performed to test whether any density parameters are predictive of pathological handwriting and allow to differentiate pathological and non-pathological writing. It has been found that the morpheme density parameter, which is the new structural feature of handwriting, can be of value in the differentiation of pathological handwriting movement, and can be successfully applied in forensic document examination.

Keywords: Forensic document examination · Handwriting · Pathological handwriting · Schizophrenia · Affective disorders

1 Introduction

For a long time, researchers have been interested in pathological writing patterns in the area of forensic document examination. The research was focused on psychotic and affective disorders. It has been found that fine motor impairments are associated with schizophrenia spectrum and affective disorders [1–5]. These

© Springer Nature Switzerland AG 2021
J. Lladós et al. (Eds.): ICDAR 2021, LNCS 12823, pp. 380–394, 2021.
https://doi.org/10.1007/978-3-030-86334-0_25

motor defects can be manifested in hand movement and measured with handwriting examination tools. It is expected that unusual forms of verbal, motor and cognitive-emotional behaviors of schizophrenic individuals would be reflected in handwriting [6]. Motor impairments associated with schizophrenia spectrum disorders include parkinsonism, akathisia, dystonia, and dyskinesia. Parkinsonism is characterized by muscle rigidity, tremor, postural abnormalities, bradykinesias, limited co-movements of upper limbs, and salivation. These motor disorders can be manifested in the graphical features of handwriting in a form of disturbances in fluency, rhythm, and pressure. Akathisia is associated with increased tension, restlessness, and muscular unease while dystonia is manifested by uneven muscle tension in various parts of the body, short-term/or longer-lasting muscular contractions such as torticollis, blepharospasms, trismus, or back arching [7]. Acute or tardive dyskinesias are a series of involuntary hyperkinetic movements e.g., chewing, sticking out the tongue, sideway jaw movements, jaw rotations, lip movements such as imitating gum chewing, teeth grinding, lip licking, smacking, puckering, and pursing, rapid eye blinking, chorea limb and torso movements [7]. There are also other motor disorders in the schizophrenia spectrum that are thought to be associated with obsessive-compulsive behaviors [8]. They take the form of intrusive, compulsive thoughts, repeated movements, rituals or unwanted behaviors, such as mannerisms. It was established that the disorder severity is reflected by the increased number of such behaviors and fine motor impairments [9].

There are many viewpoints regarding the etiology of these motor disorders. Firstly, they are considered to reflect acute or tardive (secondary) drug-induced extrapyramidal side effects [7,8]. On the other hand, these subtle motor impairments have been identified in individuals with schizotypal personality disorders and are also considered as innately related with schizophrenic disorders [10]. Supporting evidence to the latter comes from studies, where similar motor disorders have been found in first-degree biological relatives of schizophrenic patients [11]. What is more, such impairments have been recorded in about 7% of patients first-time diagnosed with schizophrenic disorder who are not yet on antipsychotic treatment [12] and in fully antipsychotic-naive patients [13–15]. Other dysfunctions, which can be also related with the subtle motor alteration impacting handwriting, such as eye movement dysfunction, have been found in 40 to 80% of schizophrenic patients and 25 to 40% of their first-degree relatives [16]. In a nutshell, two main hypotheses regarding the etiology of fine motor disturbances in schizophrenia can be formulated: the first assuming that motor impairments in schizophrenia are related to pharmacotherapy and the second, that motor fluency disorders are related to another brain mechanism such as motor automation dysfunction.

As for affective disorders, they are listed separately as depressive disorders and bipolar affective disorders in the DSM-V [17]. The typical syndromes of depressive disorders include depressed mood, anhedonia, and psychomotor slowing, while in manic disorders elevated mood and psychomotor acceleration are included. The reports related to motor symptoms accompanying affective disorders often excessively emphasize the fact that depression is associated with motor slowing, however, Tan argues that motor agitation should not be ignored [18]. Some studies have demonstrated motor slowing in patients diagnosed with depression, e.g., slower drawing, longer movement time, and longer motor response time [19–21]. Conversely, other studies failed to produce evidence related to motor slowing [22]. It was found that depressed subjects needed a similar amount of time for a drawing as the control group. Analyses of spatial properties and other characteristics of handwriting showed that elderly depressive persons differ from healthy controls [23].

Research on pathological handwriting in the area of forensic document examination is focused on searching differentiating patterns of pathological and non-pathological writing. This research uses post-factum methods in analyzing e.g. wills and anonymous letters, which means that they are already written and not in the process of being written. Thus, the use of global patterns in examining handwritten specimens in the post-factum analysis process can be valuable. In the last decades, different automatic systems have been proposed to support forensic handwriting examination. In addition to analyzing the writing process through kinetic analyses using the instruments such as the MovAlyzeR's software [2] or the ComPET developed at the University of Haifa [24], other tools for handwritten image analysis have been invented. The support tools for scanned handwritten document are as follows: FISH (Forensic Information System for Handwriting) identification system [25], WANDA (a technical and ergonomic update of the FISH writer identification system), CEDAR-FOX system described by Srihari et al. [26], or FHS (Forensic Handwriting Examiners) proposed by Parziale and colleagues [27]. These tools are focused on writer identification or/and authenticity of handwriting, i.e., they provide evidence to forged writing, altered, or modified writing. They are intended to compare handwritten documents and determine whether documents are produced by the same persons but not to analyze density patterns as we propose here by using the Raygraf program in this study. Our goal was to predict whether handwriting is pathological basing on the density pattern. The new global feature of handwriting proposed by forensic examiners is the handwriting density. The density coefficient is a characteristic feature of a handwriting sample that enables comparative testing in the document examination procedure. It is a structural property of writing. Handwriting density refers to spatial patterns and placing writing across a page. The forensic researchers and the Raygraf authors propose to specify morpheme and letter density. Morpheme is the smallest written unit of a letter [28]. Morpheme density coefficient is a structural parameter that is calculated as the total sample width divided by the ratio of the graphic elements' width and their number. Letter density coefficient is calculated as morpheme density coefficient divided by the letter number [28].

1.1 Hypothesis

Given the above findings, which suggest that fine motor impairments are manifested in the handwriting of patients with different clinical disorders, we conducted a study to examine the handwriting density parameters with the use of the Raygraf computer program. We aim to compare handwriting specimens, i.e. scanned documents, produced by persons in the clinical groups (patients diagnosed with paranoid schizophrenia, patients diagnosed with bipolar disorder related depression phase, patients diagnosed with bipolar disorder related manic episode) to the control group of persons with no diagnosed disorders. The main goal was to show whether density parameters are different in the clinical and control groups and whether these parameters would allow to differentiate pathological from non-pathological handwriting. The density-parameter-based approach to differentiate pathological vs. non-pathological handwriting was used in this manuscript for the first time.

2 Methods

2.1 Participants and Procedure

The examination of handwriting of individuals with schizophrenic and affective disorders was conducted with the use of the Raygraf computer program. This computer software was developed by a Polish team of forensic document examiners: Andrzej Łuszczuk and Krystyn Łuszczuk. Expert consultants for that forensic project development included forensic document examiners and scientists: Tadeusz Tomaszewski, Mieczysław Goc, and Kacper Gradoń. Raygraf is part of the GlobalGraf computer software package, distributed by the Polish Forensic Association [28]. The Raygraf program facilitates assessing structural-geometric features of handwriting, such as size of line segments, slope, angles, handwriting density, and pulse, including factors such as length of selected graphical elements, angles, width of graphical elements itself, or width of the spacing between elements [28]. Raygraf has been tested for validity and its accuracy in the assessment of the handwriting parameters was confirmed. This program is recommended by the Polish Forensic Association and is publicly available at the Polish Forensic Association (website https://kryminalistyka.pl).

Handwriting samples were collected from four groups in standardized conditions (participants wrote on white sheets the same text dictated at a medium speed; they wrote a 200-words long newspaper text about a construction company). The same dictation speed was guaranteed for all participants because the text was played from the recording. The sample included three clinical groups (diagnosis based on the DSM-IV-TR): individuals with schizophrenia (n = 60), depression (n = 60), mania (n = 60), and the control group of non-patients without disorders (n = 60). Mean age (and standard deviation) for the groups: controls M_{age} = 38.19(SD = 7.05), schizophrenia group M_{age} = 40.30 (SD = 12.02), mania group M_{age} = 41.76 (SD = 7.50), depression group M_{age} = 39.71 (SD = 9.68). Each group included 50% female and 50% male subjects. The participants from the control group were examined with the SCID-I/P, the same diagnostic tools as the clinical groups [29]. Means and standard deviations for modules A (affective disorders) and C (schizophrenia) of the SCID-I/P for groups: controls (M_A = 1.10, SD_A =.01, M_C) = .10, SD_C =.01), schizophrenia (M_A = .20, SD_A = .10, M_C, SD_C = 2.4), depression (M_A = 21.60, SD_A = 3.50, M_C = .50, SD_C = .03), mania (M_A = 18.94, SD_A = 2.30, M_C = .80,

$SD_C = .04$). All participants were right-handed. The groups were matched for demographic parameters, intelligence, as well as lack of visual, motor, and severe neurological impairments which was exclusion parameter for the experimental task. Intelligence was measured with the WAIS-R (controls: $M = 109.24$, $SD = 8.82$, schizophrenia group: $M = 107.61$, $SD = 7.63$, mania group: $M = 109.01$, $SD = 7.23$, depression group: $M = 110.06$, $SD = 7.00$). The experimental protocol 2017/01/PH has been approved by the local Ethics Committee at the University of Maria Curie-Sklodowska.

First, the researcher scanned 240 handwriting specimens/documents (60 from each group) and saved them in the Raygraf compatible file format - .jpg. First analysis step was to make the document format consistent, thus we performed image scaling to save them in the same format: dpi. Then, the analysis of a total of 2400 handwriting assessments was carried out: 10 assessments in each specimen (scanned document) were made, i.e., ten very same words were taken from each document as their content was exactly the same for each subject. The Raygraf program allows for some document elaboration, however, the words were manually extracted from the documents. The first parameter – total specimen width – was measured by marking the start and end points of each of the 10 assessments per specimen. The starting and ending points of a word are defined by the Raygraf authors; they are the furthest points to the left or right, irrespectively of a slant. Another parameter, writing density, was determined through measuring the width of every graphical element (letter or character), which was done by the researcher marking their starting and ending points (Fig. 1). Then, the researcher entered the number of letters/characters for each analyzed specimen. This allows for the morpheme density coefficient to be calculated. According to forensic experts, W_{gm} is an individual structural-geometric feature of handwriting and it does not refer to ink/pixel distribution but strokes on the surface [28]. The parameters included in the equations (1)–(3) are represented in Fig. 1. The example: S1 = 9 mm, S2 = 1 mm, S3 = 2 mm, S4 = .6 mm, S5 = 1.2 mm, S6 = 1 mm, S7 = 1.9 mm. Thus, Sm equals 16.7 mm (the sum of morpheme/graphical element/ widths from S1 to S7). Number of graphical elements i.e. morphemes = 7. Total width of specimen (Scp) = 19.6 mm. Then, we can calculate W_{gm} (2) = 19.6/(16.7 * 7), W_gm= .167. Then, W_{gl} (3) can be computed, number of letters in a word shown in Fig. 1 is 9. W_{gl} = .167/9. W_{gl} = .018.

Sum of widths of graphical elements/morphemes:

$$S_m = \sum S \tag{1}$$

Morpheme density coefficient:

$$W_{gm} = \frac{S_{cp}}{S_m * N_{morf}} \tag{2}$$

Letter density coefficient:

$$W_{gl} = \frac{W_{gm}}{N_{liter}} \tag{3}$$

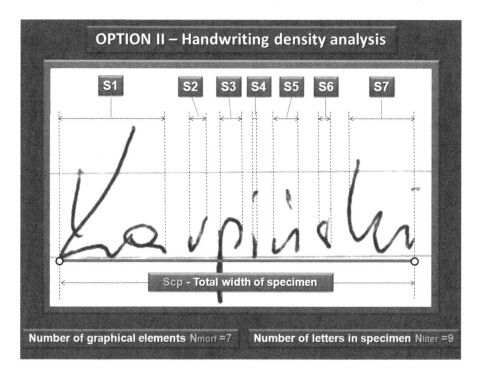

Fig. 1. Assessment of morpheme density (Łuszczuk et al. 2009–2011).

Width of the spaces between graphical elements is yet another parameter we can determine with the Raygraf program. It is measured by marking spaces between the elements in a specimen and comparing starting and ending points of individual graphical elements as in the Fig. 2 below. The example: number of spaces between graphical elements/morphemes is 6. Widths of the interspaces are as follows: O1 = .6 mm, O2 = .5 mm, O3 = .2 mm, O4 = .4 mm, O5 = .6 mm, O6 = .6 mm. Thus, the sum of the widths of interspaces (throughout O1 to O6) is 2.9 mm.

All the graphical parameters assessed in this study and identified with the use of the Raygraf program comprised the following component of density parameters: total specimen width, sum of the widths of the graphemes that make up the analyzed element (see Fig. 1), sum of the widths of interspaces (see Fig. 2), number of interspaces in a specimen, and two density coefficients:

- morpheme density coefficient which is "the quotient of the total width of the sample to the product of the sum of the widths of the graphical elements and their number in a given sample" (Eqs. 1 and 2),
- letter density coefficient defined as the quotient of the morpheme density coefficient to the number of letters/characters in a given specimen (Eq. 3).

The variables identified here are numeric. Before statistical analysis with the use of SPSS 26, variable distribution was analyzed and it has been found that the variables have normal distribution (descriptive statistics for variables are presented in Table 1).

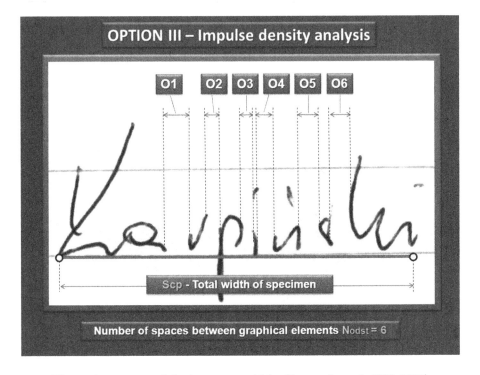

Fig. 2. Assessment of the interspace widths (Łuszczuk et al. 2009–2011)

3   Results

3.1   Comparisons Between Groups

One-way variance analysis was conducted to verify whether there are significant differences in the handwriting density parameters (and its components) between the four groups (patients with schizophrenia, patients with depression, patients with mania, and healthy controls). The analysis confirmed that there were significant differences between the control and clinical groups. The results indicated that the clinical groups significantly differ from healthy controls in all the following: total specimen width, sum of the width of graphemes, sum of the width of interspaces, number of interspaces, morpheme, and letter density coefficients (Table 1).

Next, *post hoc* multiple comparisons (Scheffe's test) were conducted to establish specific differences between schizophrenic patients, depressive patients, manic patients, and individuals from the control group (Table 1). This analysis revealed that the differences are found between the clinical and control groups, however, there were no significant differences in density parameters between patients with depression, with mania, and patients with schizophrenia. The *post hoc* multiple comparisons showed that both morpheme density and letter density coefficients are significantly lower in the clinical groups than in the control group. Then, total specimen width, sum of the grapheme widths, sum of the interspace widths, number handwriting interspaces are greater in the clinical groups than in healthy controls.

Table 1. Descriptive statistics for the graphical variables and one-way variance analysis between four groups: schizophrenia patients, depression, mania, control group.

Handwriting features	Groups	M	SD	Min.	Max.	F(3, 236)
Total width	Controls	12.92a	2.79	7.89	22.35	11.95***
	Schizophrenia	15.31b	3.10	10.46	21.89	
	Mania	15.34b	2.97	10.93	21.08	
	Depression	14.48b	3.13	9.41	20.63	
Width of graphemes	Controls	10.96a	2.80	5.08	21.59	10.24***
	Schizophrenia	12.80b	3.07	8.38	21.08	
	Mania	13.17b	2.94	9.39	21.08	
	Depression	11.21b	2.94	5.58	17.78	
Morpheme density coefficient	Controls	.56a	.17	.337	1.430	16.83***
	Schizophrenia	.44b	.10	.282	.676	
	Mania	.42b	.10	.299	.650	
	Depression	.45b	.08	.306	.635	
Letter density coefficient	Controls	.14a	.04	.084	.358	3.93**
	Schizophrenia	.11b	.02	.070	.169	
	Mania	.10b	.02	.075	.151	
	Depression	.14a	.15	.076	.940	
Sum of inter-spaces	Controls	1.52a	1.11	.25	5.08	12.66***
	Schizophrenia	2.76b	1.99	.25	9.65	
	Mania	2.84b	1.80	.25	7.62	
	Depression	2.51b	1.47	.51	6.10	
Number of interspaces	Controls	1.42a	.68	1.00	4.00	14.69***
	Schizophrenia	2.13b	.99	1.00	4.00	
	Mania	2.23b	.96	1.00	4.00	
	Depression	2.10b	1.02	1.00	4.00	

a, b - the same letters for means in columns indicate no significant differences (Scheffe's test); M – mean, SD - standard deviation, significance: $**p < .01$, $*** p < .001$.

3.2   Logistic Regression Analysis

The next step was to search whether any of these types of density parameters allowed to predict pathological/clinical group, i.e. pathological handwriting. To test this hypothesis, a logistic binary regression with *the enter method* based on the likelihood ratio was applied. The enter method means that the presented model is obtained in step 1 with all independent variables entered: morpheme density, letter density, total width, width of morphemes, width of interspaces, and number of interspaces. The dependent variable was binary: pathology vs. non-pathology (in SPSS value 1 stands for all clinical groups, value 0 - controls) while independent variables included all density parameters and their components. The Hosmer–Lemeshow test was not significant ($\chi^2 = 11.27$, $p = .19$) which means that the obtained model is well fitted to the empirical data. The Nagelkerke's pseudo $R^2 = .235$ means that the model explains 23.5% of the dependent variable variance. Summary of regression analysis is presented in Table 2. The values of Wald test for variables indicate that only morpheme density contributes significantly to the model ($p < .01$), while other variables do not. The B value for this variable is negative, which means that low morpheme density coefficient is associated with higher pathology. It is illustrated in Fig. 3. The results confirm that the morpheme density parameter can be a significant predictor for pathological handwriting (Table 2). Pathological handwriting refers to writing produced by clinical groups including schizophrenia and affective disorders.

Table 2. Summary of the logistic regression

Handwriting parameters	B	S.E.	Wald(1)	Sig.	Exp(B)
Morpheme density	−5.432	2.151	6.376	.010**	.004
Letter density	3.217	3.97	.364	.546ns	1.242
Total width	−.094	.136	.475	.491ns	.910
Width of graphemes	.270	.131	4.247	.059ns	1.310
Sum of interspaces	.361	.201	3.225	.073ns	1.435
Number of interspaces	.443	.306	2.103	.147ns	1.558
Constant	−.501	1.334	.146	.702ns	.601

S.E. – standard error, 2 Log. Likehood = 312.33, Cox & Snell $R^2 = .171$, Nagelkerke's psuedo $R^2 = .235$, Significance: **$p < .01$, ns – non-significant.

3.3   Clustering Analysis

Taking into account the findings that only W_{gm} (morpheme density coefficient) is a predictor for pathological writing, we aimed to precisely establish what type of W_{gm} is associated with pathological writing. Thus, we aimed to test whether there is any typology of W_{gm} (any types of persons who present the specific kind of morpheme density parameter) which allows to identify pathological handwriting. To test this hypothesis, k-mean clustering was used. This statistical method allows identifying relatively homogeneous groups of cases based on W_{gm}. The k-mean clustering analysis revealed six independent final cluster centers, i.e., six different morpheme density styles ($F(5, 234) = 15.10$, $p < .001$, see Table 3).

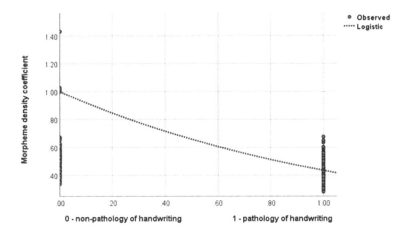

Fig. 3. Logistic binary regression plot.

Distances between them are shown in Table 4. Distances between cluster centers are computed using Euclidean distance. Each final cluster center differs from the other centers and presents a different configuration of morpheme density parameter. Table 4 shows the Euclidean distance between final clusters; greater distances correspond to greater dissimilarities. It shows that cluster centers 5 and 6 are the most dissimilar to other clusters and cluster centers 2 and 4 are the most similar to each other, i.e. include subjects with the most similar morpheme density coefficient.

Table 3. Final cluster centers: morpheme density parameter.

Number of cluster center	1	2	3	4	5	6
Morpheme density coefficient	1.430	.282	.541	.411	1.010	.645

To search density types associated with pathology, we performed Cramer's V statistics, as it measures associations between two nominal variables (pathology vs. non-pathology and six final clusters centers). Cramer's V statistics is used in the case of multivariate table (more than 2×2 contingency table). It showed that clusters 2 (W_{gm} values .282) and 4 (W_{gm} values .416) are positively associated with pathology (V = .410, $p < .001$), while clusters 1, 3, 5, and 6 are not (Fig. 4). It confirmed that morpheme density coefficient ranging between values .282 and .416 differentiates pathological handwriting (written by patients with schizophrenia and affective disorders) from non-pathological handwriting (written by healthy individuals).

Table 4. Distances between final cluster centers.

No of final cluster centers	1	2	3	4	5	6
1		.883	.433	.978	1.105	1.168
2	.883		.450	.095	.222	.813
3	.433	.450		.546	.672	.908
4	.978	.095	.546		.126	.823
5	1.105	.222	.672	.126		.852
6	1.168	.813	.908	.823	.852	

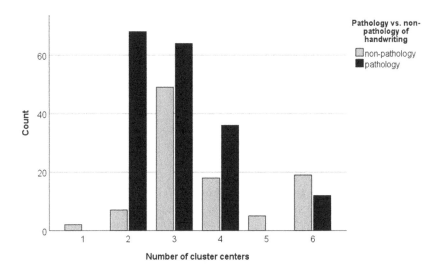

Fig. 4. Morpheme density coefficient in non-pathological and pathological groups.

4   Discussion

The comparative analyses of the handwriting density parameters produced by patients from different clinical groups showed significant differences between them and the control group. However, no differences were found in the handwriting density features between the three clinical groups in this study (patients diagnosed with paranoid schizophrenia, depression in bipolar disorder, and mania in bipolar disorder). The approach based on the handwriting density parameters has not been presented anywhere before in terms of differentiating pathological vs. non-pathological writing. Although density parameters have been previously examined in another study along with other structural-geometric handwriting characteristics; that research was focused on psychopathic personality disorder and no significant differences in any parameters between handwriting of individuals with psychopathic personality disorder and control group were evidenced [30].

In 2016, we have published results of a study related to schizophrenia disorder using traditional graphical-comparative method, in which we have not found specific distinctive patterns of handwriting in the analyzed group of outpatients [31]. Both the sample and method are different to the present work, where we focus on a new parameter named density of handwriting. Here, we found that all examined disorders are characterized by different density patterns than those observed among the controls. The most valuable parameter related to the handwriting density, which allows to differentiate handwriting produced by clinical patients and healthy individuals, is the morpheme density coefficient. In pathological groups, morpheme density coefficient is significantly lower than in healthy controls. Precise values of lowered morpheme density coefficient associated with pathologies such as schizophrenia or affective disorders have been established. These values are lower than .416. It suggests that handwriting gesture in clinical patients presents some fine motor disturbances and, in general, is less fluent or less integrated. The present findings are in line with studies on pathological handwriting indicating that patients with a schizophrenia spectrum disorder and bipolar disorder exhibit significant motor impairments such as decrease in velocity, acceleration and increase in the length, dysfluency in pressure [32]. Similarly to our study, no differences in handwriting parameters were found between patients diagnosed with schizophrenia and bipolar disorder in the study by Crespo and colleagues (parameters different than ours) [32]. This can be explained in two ways. First, all the subjects from the clinical groups experience a certain level of emotional tension, stress, and negative mood which can impact hand control. This results in impaired motor fluency and is manifested in the differences between the control and the clinical groups. Such emotional tension is shared by patients with different disorders and results in non-specific changes in handwriting [33]. There is evidence supporting this interpretation, for example, emotional arousal is associated with the movement expansiveness or its blockage, and with handwriting features [34]. The second explanation of the obtained results refers to the neurobiological mechanism as the potential basis for non-specific changes in graphism of various mental disorders. Some research found a similar neurobiological mechanism which is the basis for non-specific impairments of handwriting in mental disorders [35]. In general, patients with some mental disorders are characterized by more vigorous and less coordinated gesture. This seems to be associated with reduced visual-motor coordination and/or poorer visual-kinetic-motor integration. It has been documented that different mental disorders associated with cerebellar dysfunctions produce similar symptoms [36]. The dysfunctions of the cerebellum in affective disorders and schizophrenia can be manifested in similar properties of handwriting, precisely in similar density parameters, however, different from healthy persons' handwriting.

The density morpheme coefficient has been proposed here for the first time as a novel parameter potentially differentiating pathological handwriting, yet, in general our findings correspond to other results indicating occurrence of fine motor impairments in patients with schizophrenia spectrum disorders [37,38]. It has been documented that motor disorders such as parkinsonism, akathisia, dystonia, or dyskinesia can impact formation of handwriting gesture causing dysfluency of movement [1,2,5,7]. Less integrated handwriting resulted in low morpheme density and can be also associated with involuntary hyperkinetic movements or compulsive behaviors in schizophrenic patients [7,8].

Our findings are also consistent with data that indicate fine motor disturbances in patients with affective disorders and such disturbances can be assessed using handwriting kinematic analyses [4,21–23]. These tools have been found valuable in overseeing affective disorder pharmacotherapy. For instance, Mergl and associates stated that kinematic analysis of handwriting is a sensitive tool for monitoring the pharmaceutical effect on hand-motor function in depression [4,22].

Our analyses indicate that global parameters in pathological handwriting examination might be of a greater value than focus on single handwriting characteristics. There are inconsistencies that can be found between findings that base on single handwriting characteristics, such as stroke width. For instance, it has been documented that elderly patients with depression produced smaller stroke width in comparison with control groups [23]. Our research did not confirm such differences neither in the total specimen width nor in the morpheme width between depressive, manic, and control groups. We propose that global approach, such as morpheme density coefficient, will enable better differentiation between handwriting of clinical and non-clinical samples.

5 Conclusions

The evidence obtained in the study with the use of the Raygraf computer software indicates that there are significant differences in the handwriting density parameters between healthy individuals and clinical groups (patients diagnosed with schizophrenia, depression in bipolar disorder, and mania in bipolar disorder). However, significant differences in parameters such as total specimen width, sum of the width of morphemes, letter density coefficient, sum of the width of interspaces, and number of the handwriting interspaces between the clinical groups have not been identified. The results confirm that morpheme density coefficient might be of value in the differentiation of pathological handwriting from the handwriting produced by healthy persons and that mental disorders impact fine motor activity in general. The values of the morpheme density parameter mostly associated with pathological writing are in the range of .282 and .416.

References

1. Caligiuri, M.P., Teulings, H.L., Filoteo, V., Song, D., Lohr, J.: Quantitative measurement of handwriting in the assessment of drug-induced parkinsonism. Hum. Movement Sci. **25**(4–5), 510–522 (2006)
2. Caligiuri, M.P., Teulings, H.L., Dean, C.E., Niculescu, A.B., Lohr, J.: Handwriting movement analysis for monitoring drug-induced motor side effects in schizophrenia patients treated with risperidone. Hum. Movement Sci. **28**(5), 633–642 (2009)

3. Mavrogiorgou, P., et al.: Kinematic analysis of handwriting movements in patients with obsessive-compulsive disorder. J. Neurol. Neurosur. Psychiatry **70**, 605–612 (2001)

4. Mergl, R., et al.: Handmotor dysfunction in depression: characteristics and pharmacological effects. Clin. EEG Neurosci. **38**, 82–98 (2007)

5. Tigges, P., et al.: Digitized analysis of abnormal hand-motor performance in schizophrenic patients. Schizophr. Res. **45**, 133–143 (2000)

6. DSM-V: Diagnostic and Statistical Manual of Mental Disorders, 5th ed. APA, Washington DC (2013)

7. Gervin, M., Barnes, T.R.E.: Assessment of drug-related movement disorders in schizophrenia. Adv. Psychiatric Treat. **6**, 332–343 (2000)

8. Poyurovsky, M., et al.: Obsessive-compulsive disorder in hospitalized patients with chronic schizophrenia. Psychiatry Res. **102**, 49–57 (2001)

9. Tucha, O., Paul, G.M., Mecklinger, L., Eichhammer, D., Klein, H.K., Lange, K.W.: Handwriting in schizophrenia. A kinematic analysis of handwriting movements. Int. J. Forensic Doc. Examiners **6**(1), 1–3 (2003)

10. Cassady, S.L., Adami, H., Moran, M., Kunkel, R., Thaker, G.V.: Spontaneous dyskinesia in subjects with schizophrenia spectrum personality. Am. J. Psychiatry **155**, 70–75 (1998)

11. Egan, M.F., et al.: Relative risk in neurological signs in siblings of patients with schizophrenia. Am. J. Psychiatry **158**, 1827–1834 (2001)

12. Gervin, M., et al.: Spontaneous abnormal involuntary movements in first episode schizophrenia and schizophreniform disorder: baseline rate in a sample from Irish catchment area population. Am. J. Psychiatry **155**, 1202–1206 (1998)

13. Cuesta, M.J., et al.: Spontaneous parkinsonism is associated with cognitive impairment in antipsychotic-naive patients with first-episode psychosis: a 6-month follow-up study. Schizophrenia Bull. **40**(5), 1164–1173 (2014)

14. Gschwandtner, U., et al.: Fine motor function and neuropsychological deficits in individuals at risk for schizophrenia. Eur. Arc. Psy. Clin. N. **256**(4), 201–206 (2006)

15. Peralta, V., Campos, M.S., de Jalon, E.G., Cuesta, M.J.: DSM-IV catatonia signs and criteria in first-episode, drug-naive, psychotic patients: psychometric validity and response to antipsychotic medication. Schizophr. Res. **118**(1–3), 168–175 (2010)

16. Holzman, P.S.: Eye movements and the search for the essence of schizophrenia. Brain Res. Rev. **31**, 350–356 (2000)

17. Butcher, J.N., Hooley, J.M., Mineka, S.: Psychology of disorders in the DSM-V. GWP Sopot (2017)

18. Tan, Ü.: The psychomotor theory of human mind. Inter. J. Neurosci. **117**, 1109–1148 (2007)

19. Van Hoof, J.J.M., Hulstijn, W., Van Mier, J.I.A., Pagen, M.: Figure drawing and psychomotor retardation: preliminary results. J. Affect. Disorders **23**, 263–266 (1994)

20. Van Mier, J.I.A., Hulstijn, W.: The effects of motor complexity on imitation time in writing and drawing. Acta Psychol. **84**, 231–251 (1993)

21. Sabbe, B., Hulstijn, W., Van Hoof, J., Zitman, F.: Fine motor retardation and depression. J. Psychiatr. Res. **30**(4), 295–306 (1996)

22. Mergl, R., et al.: Kinematical analysis of handwriting movements in depressed patients. Acta Psychiat. Scand. **109**, 383–391 (2004)

23. Rosenblum, S., Werner, P., Dekel, T., Gurevitz, I., Heinik, J.: Handwriting process variables among elderly people with mild major depressive disorder: a preliminary study. Aging Clin. Exp. Res. **22**(2), 141–147 (2010)

24. Rosenblum, S., Parush, S., Epsztein, L., Weiss, P.L.: Process versus product evaluation of poor handwriting among children with Developmental Dysgraphia and ADHD. In: Teulings, H.F., Van Gemmert, A.W.A. (eds.) Proceeding of the 11th Conference of the International Graphonomics Society, pp. 169–173. Scottsdale Arizona (2003)

25. Franke, K., Schomaker, L.R.B., Veenhuis, C., Vuurpijl, L.G., van Erp, M., Guyon, I.: WANDA: a common ground for forensic handwriting examination and writer identification. ENFNEX news - Bull. Eur. Netw. Forensic Handwriting Experts **1**(04), 23–47 (2003)

26. Srihari, S.N., Srinivasan, B., Desai, K.: Questioned document examination using CEDAR-FOX. J. Forensic Doc. Examination **28**, 15–26 (2018)

27. Parziale, A., et al.: An interactive tool for forensic handwriting examination. In: 14th International Conference on Frontiers in Handwriting Recognition, pp. 440–445. IEEE (2014)

28. Łuszczuk, A., Łuszczuk, K., Tomaszewski, T., Goc, M.: GLOBALGRAF, computer programs GRAFOTYP, RAYGRAF, KINEGRAF, SCANGRAF - research financed by the ministry of science and higher education from the budget for research for the years 2009–2011 as a development project - program guide (electronic version). Polish Forensic Association, Warsaw University (2009–2011)

29. First, M.B., et al.: The structured clinical interview for the DSM-IV Axis I disorders (SCID-I/P) - research version. PTP Warsaw (2010)

30. Gawda, B.: The computational analyses of handwriting in individuals with psychopathic personality disorder. PLoS ONE **14**(12), e0225182 (2019)

31. Gawda, B.: Dysfluent handwriting in schizophrenic outpatients. Percept. Motor Skills **122**(2), 560–577 (2016)

32. Crespo, Y., Ibañez, A., Soriano, M.F., Iglesias, S., Aznarte, J.I.: Handwriting movements for assessment of motor symptoms in schizophrenia spectrum disorders and bipolar disorder. PLoS ONE **14**(3), e0213657 (2019)

33. Van Hoof, J.J.M., Hulstijn, W., Van Mier, J.I.A., Pagen, M.: Figure drawing and psychomotor retardation: preliminary results. J. Affect. Disorders **23**, 263–266 (1993)

34. Van Hoof, A.W.A., Van Galen, G.P.: Stress, neuromotor noise, and human performance: a theoretical perspective. J. Exp. Psychol. Hum. **23**, 1299–1313 (1997)

35. Baldaçara, L., Borgio, G.J., Lacerda, A., Jackowski, A.P.: Cerebellum and psychiatric disorders. Rev. Brasilian Psychiatry **30**(3), 281–289 (2008)

36. Chrobak, A.A., Siuda, K., Tereszko, A., Siwek, M., Dudek, D.: Mental disorders and structure, and functions of cerebellum - review of newest research. Psychiatria **11**(1), 15–22 (2014)

37. Walther, S., Morrens, M.: Editorial: psychomotor symptomatology in psychiatric illnesses. Frontiers Psychiatry **6**, 81 (2015)

38. Walther, S., Strik, W.: Motor symptoms and schizophrenia. Neuropsychobiology **66**(2), 77–92 (2012)

Pen-Based Document Analysis

Language-Independent Bimodal System for Early Parkinson's Disease Detection

Catherine Taleb[1]($\boxtimes$), Laurence Likforman-Sulem[2], and Chafic Mokbel[1]

[1] University of Balamand, Balamand El-Koura, Lebanon
`catherine.taleb@std.balamand.edu.lb`
[2] LTCI/Telecom Paris/Institut Polytechnique de Paris, Paris, France

Abstract. Parkinson's disease (PD) is a complex disorder characterized by several motor and non-motor symptoms that worsen over time, and that differ from person to another. In the early stages, when the symptoms are often incomplete, the diagnosis becomes difficult and at times, the subject may remain undiagnosed. This difficulty is a strong motivation for computer-based assessment tools that can aid in the early diagnosing and predicting the progression of PD. Handwriting's deterioration, vocal and eye movement impairments may be ones of the earliest indicators for the onset of the illness. A language independent model to detect PD at early stages by using multimodal signals has not been enough addressed. Due to the lack of multimodal and multilingual databases, database which includes online handwriting, speech signals, and eye movements recordings have been recently collected. After succeeding in building language independent models for PD early diagnosis using pure handwriting or speech, we propose in this work language independent models based on bimodal analyses (handwriting and speech), where both SVM and deep learning models are studied. Our experiments show that classification accuracy up to 100% can be obtained by our SVM model through handwriting/speech bimodal analysis.

Keywords: Parkinson's disease (PD) · 2D CNN · 1D CNN-BLSTM · SVM · 1D CNN-MLP · Handwriting · Speech · Data augmentation

1 Introduction

PD is a neurological disorder caused by a increased dopamine levels in the brain. This disease is characterized by motor and non-motor symptoms that worsen over time. The motor symptoms consist of tremor, rigidity, slowness of movement or bradykinesia, micrographia, and speech difficulty [20]. In advanced stages of PD, clinical diagnosis is clear-cut. However, in the early stages, when the symptoms are often incomplete or subtle, the diagnosis becomes difficult and at times, the subject may remain undiagnosed. Furthermore, there are no efficient and reliable methods capable of achieving PD early diagnosis with certainty [2]. The difficulty in early detection is a strong motivation for computer-based assessment tools/decision support tools/test instruments that can aid in

© Springer Nature Switzerland AG 2021
J. Lladós et al. (Eds.): ICDAR 2021, LNCS 12823, pp. 397–413, 2021.
https://doi.org/10.1007/978-3-030-86334-0_26

the early diagnosing and predicting the progression of PD [19]. Early detection of the disease could be hugely beneficial in order for the patient to have access to a therapy that will slow down the course of PD progression. Handwriting's deterioration and vocal impairment may be ones of the earliest indicators for the onset of the illness [21,22]. According to the reviewed literature, a language independent model to detect PD at early stages using multimodal signals has not been enough addressed. In our previous works [5], and [6], language-independent models for assessing the motor disorders in PD patients at early stages based on handwriting features have been developed; where two approaches were studied and compared: a classical feature extraction and classifier approach, and a deep learning approach. Also in [18] a language independent model based on pure speech analysis and SVM has been built. Approximately 97% classification accuracy was reached with both modalities and approaches. The main contribution of the present work is to build a language independent model for assessing the motor disorders in PD patients at early stages based on bimodal analysis (handwriting and speech), where two different approaches are studied and compared: handcrafted features and SVM, and deep learning.

The paper is organized as follows. In Sect. 2, we introduce our handwriting and speech datasets. In Sect. 3 an overview of related work is provided. Section 4 presents the bimodal system used for PD detection. We conducted several experiments that are described in Sect. 5. Conclusions and perspectives are drawn in Sect. 6.

2   Handwriting and Speech Datasets

Due to the lack of multimodal and multilingual PD database, a database (PDMultiMC) that includes handwriting tasks, speech samples, and eye movements recordings has been collected from PD patients attending an experienced neurologist, in two phases ("on-state" (1 h after taking L-dopa dosage for peak response to medication) and "off-state" (12 h after the last L-dopa medication dosage)), and from Healthy control (HC) subjects selected from our entourage and seen by a neurologist to make sure they do not have any neurological disease. 21 PD patients (16 Males and 5 Females), and another 21 HC subjects (5 Males and 16 Females) are included in PDMultiMC database. PD and HC subjects are matching for age, years of education, and hand dominance. This database includes samples in three languages: 31 Arabic, 9 French, and 2 English and will be released on the IAPR TC11 repository. Even though the language representation is not balanced, language independence is somehow provided by averaging the signal (speech or handwriting). This average can be considered as the summation of a certain noise (summation of the average of channel information and the average of linguistic information) with disease characteristics; where the noise will not interfere in the classification. The modified Hoehn and Yahr (mH&Y) scale was measured to show the presence and the severity of PD motor symptoms. The mean mH&Y of PD patients in our database is 1.81 ± 0.77; where around 95% of our PD patients show early to mild degree of disease severity. In this work, we focus on handwriting and speech modalities for early PD detection, where handwriting and speech samples are taken from HandPDMultiMC and SpeechPDMultiMC datasets (parts of PDMultiMC database). However, since our target is the early detection of the disease and since collecting large database at early stages is very difficult, and referring to [10] and [12], we make the assumption that mild stages can get nearer early stages one hour after taking L-dopa medication, since levodopa reduces the motor symptoms. For this reason, the seven handwriting tasks and the two speech tasks recorded for each of the 42 subjects (PD in their "on-state") are studied and analyzed. Participants were asked to complete handwriting and speech tasks; where these tasks were chosen in a manner to highlight as much as possible the differences between PD and control, and where neurologists were consulted in the process of selecting the tasks.

Handwriting samples were collected using Wacom intuos 5 tablet with a sample rate of 197 points/s. The trace of the pen tip (X-Y-Z coordinate), the pressure of the pen tip on the surface, the angles of the pen relative to the tablet (altitude and azimuth), and timestamp were collected per sample point, forming seven times series as the handwriting dynamic signals. The seven handwriting tasks and the captured handwriting dynamic signals for a given task are displayed in Fig. 1. The first three tasks (with different degrees of complexity) demand one continuous pen movement, similar to spiral and meander tasks, which emphasis hypokinesia and tremor. Since there is no consensus on which stroke type reflects better the disease, we have included different type of strokes in the cursive tasks. From the other side, since PD patients have difficulty in maintaining constant force in long tasks, we decided to include words repetition in the other 4 tasks.

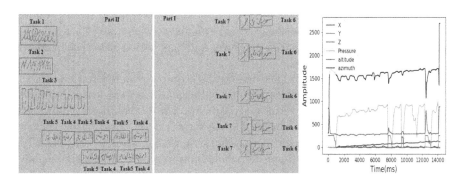

Fig. 1. The seven tasks segmented from the sheet filled by a PD subject (left figure) and the handwriting dynamic signals captured for task1 (right figure).

Speech wave signals were recorded using the internal microphone of the laptop (hp Elittbook 8570 w). Two channels sounds with 16-Bit depth and 44.1 KHz sampling rate were produced and saved. Two tasks were considered: 1) the participants were instructed to produce a sustained vowel 'a' (it is easy to be reproduced by elderly subjects and provides information about the phonatory and articulatory processes of speech production, and it was shown in [7] that the vowel 'a' reports the highest PD detection accuracy comparing to other vowels). 2) Participants were asked to read loudly a text appearing on the PC screen written with their familiar languages (such speech task permit judgments of speech rate, phrase length, voice quality, resonance, and precision of articulation, and also permits assessment of the prosodic features of speech [8]).

Many other complex tasks can more involve cognitive and functional issues and may better define the disease, but our main focus was on motor skills and the selected tasks can reflect most of the motor symptoms and can be easily reproduced by elderly people. Such tasks (except the sustained vowel) can also be used to assess dementia since they require linguistic skills, attention, and memory.

3   Single Modality PD Early Detection

3.1   Handwriting Modality

In literature, some works such as [4] focused on handwriting analysis for supporting early PD diagnosis where only patients with early to mild degrees of disease taken from PaHaW dataset [1] were studied. In our previous work [5] the main contribution was to find a feature selection approach for an improved PD early detection based on handwriting features suggested in [1]. Advanced language independent handwriting markers based on kinematic, stroke, pressure, entropy, and intrinsic features were extracted from the "on-paper" periods in each handwriting task (samples taken from HandPDMultiMC), forming a global feature vector of size 189. An SVM model was trained on these features, where a two-stage feature selection (FS) approach was applied. It was shown that handwriting can be a tool for PD early diagnosis with a 96.87% prediction accuracy when a set of kinematic, pressure, and correlation between kinematic and pressure features are used.

However, since hand-crafted features model required expert knowledge of the field, and since the database is small, pen-based features were learnt by means of deep learning where short term analysis is applied to avoid losing important information while applying global features extraction [6]. Two based learning models for end to end time series classification were proposed (the 2D CNN and the 1D CNN-BLSTM), where the whole handwriting dynamic signals have been studied so we can extract both in-air and on-surface features. Two new approaches were proposed to encode each handwriting dynamic signal into separate image for the 2D CNN model (spectrogram and modified Gramian Angular Field (GAF)), and compared to the approach proposed by Pereira et al. [3] that encode all the handwriting dynamic signals into single image. For the 1D CNN-BLSTM model, the raw signals are directly used. The number of handwriting dynamic signals k is a hyper-parameter varying between 1 and 7. Some pre-processing steps were applied: getting the same writing direction, and normalizing the X and Y coordinate. We have demonstrated the importance of both the 1D CNN-BLSTM, and the 2D CNN model with spectrograms as input in

PD detection (Fig. 2). These models have the ability to tackle the variation of information in time series either by explicitly considering the local short term information on the time axis of the non-stationary online handwriting signals or by dealing with raw time series directly. To cope with the limited data, and to improve our deep models, some data augmentation techniques (jittering, scaling, time-warping, and synthetic data generation) were applied on handwriting dynamic signals to generate new synthetic samples. We have found that combining both jittering and synthetic data augmentation techniques with the 1D CNN-BLSTM model yields 97.62% classification accuracy [6]. This diagnosis system has been validated on PaHaW database (closest to our database in term of handwriting dynamic signals compared to NewHandPD [3]), and it was found the importance of Z coordinates feature and the relevance of the results obtained (results are consistent on different datasets) [6].

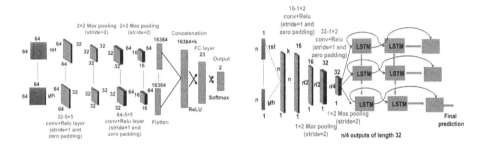

Fig. 2. The left model refers to a single-task 2D CNN architecture. This architecture takes as input k two dimensional representations of 1D time series. The right model refers to a single-task 1D CNN-BLSTM architecture on a multivariate time series of length n. The output of the 1D CNN is fed to a BLSTM as a sequence of length n/4.

3.2 Speech Modality

Several studies in literature focused on the early detection of PD based on speech analysis such as Rusz et al. [14] who studied early staged PD patients with mean H&Y of 2.20 ± 0.5. In [18], we have defined a language and task-independent acoustic feature set for assessing the motor disorders in PD patients, and have studied the influence of sampling rate and unvoiced sounds on the performance. Only phonation and articulation handcrafted features are studied, where the prosody features are excluded since they depend on the language spoken [9]. Some pre-processing steps were applied prior extracting the low level descriptors (LLD) features: removing silence at the start and the end of the speech, removing speech that does not refer to the subject and each spontaneous intervention introduced by the subject that was not directly related with the task, converting the 2 channels signal into mono signal, and down-sampling signal rate. The LLDs were extracted over 20 ms frames shifted by 10 ms from the processed voice signal using openSMILE toolkit. Global features were obtained from the z-scored LLD features by applying some statistical functions. This resulted in 220 global features per task. An SVM model was trained on these features, where the two-stage feature selection approach proposed in [5] was applied. Unvoiced sounds and sampling rate effects on classification performance of PD detection through voice analysis were studied. Our language independent SVM model for PD early diagnosis through voice analysis achieved 97.62% accuracy. It was found that the effect of sampling rate on PD classification may depend on task and features used. We have found that signals with low sampling rate (less than 16 KHz) can lose valuable information that can play a good role in PD detection, where a sampling rate of 24 KHz for sustained vowel 'a' and text reading (voiced sounds) tasks, and 16 KHz for text reading task (voiced and unvoiced sounds) are appropriate for the features analyzed. We have also demonstrated the importance of unvoiced frames in PD detection and the importance of MFCC coefficients to quantify the problems in speech articulation and to detect the disease [11].

4   Bimodal PD Early Detection

The main contribution of this work is to build a language independent bimodal system for assessing the motor disorders in PD patients at early stages based on handwriting and speech signals. Since there is no consensus on which modality is more appropriate to help on PD diagnosis in early stages; so combining and analyzing both signals may deliver a more accurate PD prediction. Two different learning approaches were applied: feature-based and deep learning approaches. Fusion of different modalities can be executed at different levels: data level, short term or global features level, or decision level as shown in Fig. 3. In this work, only global features and decision level fusion are applied since data-level and short term feature-level fusions are used when the multiple raw data even come from a same type of modality source, or are synchronized (which is not our case) [13].

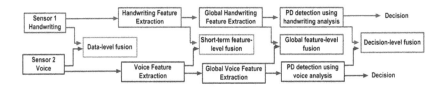

Fig. 3. Levels of bimodal fusion of handwriting with speech.

4.1   Feature-Based Approach

In this section, only global feature-level fusion is applied. Global feature-level fusion consists in combining two global features vectors, one for each modality. Each vector includes information of all tasks per subject. For each modality, the set of global and language-independent features defined in Sect. 3 (189 "on-paper" global handwriting features and the 220 global acoustic features) are extracted for each task then combined together to form a single feature vector. Two different methods are applied to combine the feature vectors in each modality: calculate the average feature vector across the different tasks, or concatenate the features vectors together. At a later stage, the features vectors obtained from the 2 modalities will be concatenated to form a global bimodal vector that will be used to detect PD using a SVM model with RBF kernel; where feature-level Min-Max is applied on each feature separately before classification. An overview of the SVM model trained on bimodal pre-engineered features is shown in Fig. 4.

For speech, all the pre-processing steps described in Sect. 3 are applied to the voice signal before extracting the LLD features, and speaker level z-normalization is applied to the LLD features to reduce the effects of variations that are not related to the disease (such as recording environment noise, speaking styles or accent etc.). Based on the results obtained in [18], the sustained vowel 'a' is sampled at 24 KHz while the text reading is sampled at 16 KHz and both voiced and unvoiced frames in the text reading are studied. The two-stage feature selection approach defined in [5] is also applied here, where the first stage consists of a pure statistical analysis of the data (where a sequence of significance levels between 0 and 1 were tested and the one with the best validation accuracy will be picked) and the second stage consists of applying a suboptimal approach that provides a kind of benchmark of the relevance of the features in the desired task.

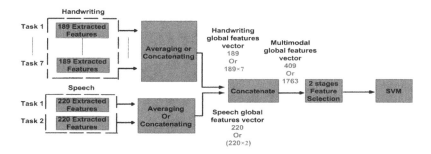

Fig. 4. SVM trained on bimodal pre-engineered features using global feature-level fusion.

4.2 Deep-Learning Approach

The core idea is to train our model with language-independent bimodal feature vector. For deep learning approach, to obtain language-independent feature vector, the model is trained on all the languages so the features will not be biased toward a specific language. Handwriting and audio pre-processing steps mentioned in Sect. 3 are applied in this section. For text reading task, the best sounds combination found in the first part of this work are also applied here.

However, the voice sampling rate for both tasks is set to 8 KHz to reduce computational time, and memory usage. For all the deep models, the number of hidden nodes is selected in a way to have a number of independent parameters smaller than the number of data points available.

2D CNN/2D CNN. One of the deep learning models studied is the 2D CNN model with spectrogram images (found in [6] and summarized in Fig. 2), which is applied in this work in both modalities; where spectrogram 2D representation for both raw handwriting signals and speech signals are obtained by applying Short Term Fourier Transform (STFT). Blackman windowing function is applied, where both the window length and the number of Fast Fourier Transform (FFT) points are set to 256 and the overlapping rate is 50%. Lanczos technique [18] is used to ensure that the number of input feature maps is identical for all subjects by resizing the spectrogram images to 64×64 pixels resolution. This model can be used for classification from a single image (voice case), or classification from k measurements, where each measurement is encoded into spectrogram image (handwriting case). The number of handwriting dynamic signals k is a hyper-parameter varying between 1 and 7. Fusion of both handwriting and voice modalities are executed here at two levels: global feature-level and decision-level.

Global Feature-Level Fusion. The aim here is to form one feature vector with information of all tasks per subject and per bio-signal. To do this, for each modality individual 2D CNNs are trained per task and the feature maps obtained by the convolutional layers (the output of the concatenated layer in Fig. 2) for each task are combined together whether by averaging (model M1) or by concatenating (model M2). The embeddings obtained from the 2 modalities are concatenated to form a bimodal vector per subject. The created feature vectors are then used to classify PD patients and HC subjects using fully connected layers as shown in Fig. 5.

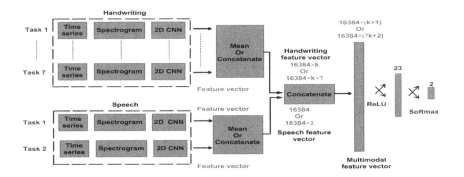

Fig. 5. Bimodal assessment using 2D CNN models and global feature-level fusion.

Decision-Level Fusion. For voice modality, two cases are studied: the first case (defined by case 1 and model M3) is when spectrogram is obtained for the whole audio signal (image input of size 64×64 pixels and 23 hidden nodes), and the second case (defined by case 2 and model M4) is when the audio signal is cropped into segments of size 4 s in order to increase the number of samples, and to keep the nonlinear variation over the time axis as shown in Fig. 6. Spectrogram is obtained for each segment separately (image input of size 249×129 pixels and 1024 hidden nodes). All the training segment slices images are considered independent training instances. Segmentation is also applied when predicting the label of a testing time series. No window slices referring to the same participant exist in training and test. To make the final prediction for each subject in the test set, the S probability vectors outputs of the 2D CNN models are considered as a Multivariate sequence of length S, and are used as input to a dynamic BLSTM to decide the final prediction. Two multilayer perceptron (MLPs) models (MLP1 and MLP2) are applied, where each one is used to combine the probability vectors (each of size 2) obtained by all tasks in each modality. At a later stage, another MLP model (MLP3) is used to combine the probability vectors provided by each of MLP1 and MLP2 (each of size 2) in order to get the final prediction (refer to Fig. 6).

1D CNN-BLSTM/1D CNN-MLP. The 1D CNN-BLSTM model with raw time series as input (see Fig. 2) is applied in this section. However, since we are working with long audio signals, we have decided to crop the audio signal into segments of fix length (4 s) and apply the 1D CNN-MLP model (shown in Fig. 7). This model is defined by M5. The number of handwriting dynamic signals k is a hyper-parameter varying between 1 and 7. Fusion of both handwriting and voice modalities are executed here at decision-level only since the feature maps in both modalities differ (the features map in handwriting modality is a sequence of length n/4 of vectors of size 32, where in voice modality the features map is a vector of length $32 \times n/4$), in addition in each modality n varies from one task to another and the number of audio segments of length 4s varies between tasks. Individual 1D CNN-BLSTMs are trained for each task in handwriting modality, where individual 1D CNN-MLPs followed by BLSTMs to make the final prediction are trained for each task in voice modality as shown in Fig. 7. Three different MLPs are also applied to get the final prediction.

1D CNN-BLSTM/2D CNN. The 1D CNN-BLSTMs and 2D CNNs models are combined. For voice, 2D CNNs are applied with the STFT as input, where for handwriting 1D CNN-BLSTMs are used with the raw signals as input. Fusion of both handwriting and voice modalities are also executed here at decision-level, and the number of handwriting dynamic signals k is a hyper-parameter varying between 1 and 7. For voice modality, spectrogram is even obtained for the whole audio signal (defined by model M6), or for each 4s segment as described previously (defined by M7). Individual 1D CNN-BLSTMs and 2D CNNs are trained for each task in handwriting and voice modalities respectively, where

MLPs are also applied to combine and obtain the final prediction (similar to Fig. 6, where the only difference is that for handwriting 1D CNN-BLSTMs are applied with the raw signals).

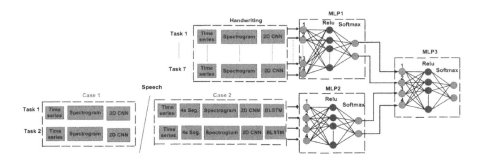

Fig. 6. Bimodal assessment using 2D CNN models and decision-level fusion.

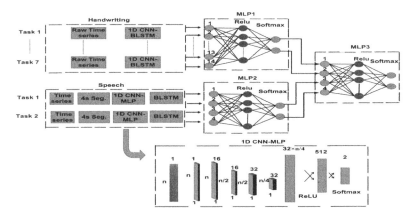

Fig. 7. Bimodal assessment using 1D CNN-BLSTM and 1D CNN-MLP models and decision-level fusion.

4.3   Data Augmentation

Based on the findings in [6], data augmentation applied to time series improves the 1D CNN-BLSTM model performance, and fails to improve the 2D CNN model performance. In addition, we have found the power of combining both jittering and synthetic data augmentation techniques with the 1D CNN-BLSTM model. Based on these findings, and after selecting the best bimodal systems, these techniques are applied. However, synthetic data generation is memory consuming method as long as we are working with long audio signals. For this reason, only jittering data augmentation method is applied with voice modality; where in handwriting modality, jittering is applied with 2D CNNs and the combination of jittering and synthetic data is applied with the 1D CNN-BLSTMs. For jittering, several values of noise intensity are studied in order to explore its effect on classification. In [6] we have found that the best results are obtained when the training data is augmented twice. This has also been adopted in this work.

5 Experiments and Results

In this work, bimodal assessment of PD is studied where both SVM model trained on handcrafted features and deep models are studied and compared, where the 42 subjects are divided into 3 folds, with the 66.66/33.33% (training/validation) proportion using stratified sampling method. Sequentially, one fold is validated using the classifier trained on the remaining 2 folds. The total accuracy is obtained by calculating the mean of all the folds accuracies. We have decided not to use a separate test set due to a low database size. As a result the validation set can be considered as test set. Starting with the SVM model based on global feature-level fusion defined in Sect. 4.1. The significance level with the best validation accuracy, the number of selected features and the performances obtained with one and two stage feature selection methods are shown in Table 1.

Table 1. Table of comparison between the performance obtained with one and two stage feature selection methods.

Method		1 stage FS	2 stages FS
Average	Acc (Sens, Spec) (%)	92.86 (95.24, 90.48)	100 (100, 100)
	Significance level	0.0933	
	# Selected features	125	22
Concatenation	Acc (Sens, Spec) (%)	97.62 (95.24, 100)	100 (100, 100)
	Significance level	0.0350	
	# Selected features	356	55

The highest classification accuracy obtained with average method is up to 100% for $N = 22$ features (15 handwriting and 7 acoustic). For concatenation method, also the highest classification accuracy obtained is up to 100% for $N = 55$ features (52 handwriting and 3 acoustic). Most of the selected features providing the best performance for both methods include kinematic, pressure, and correlation between kinematic and pressure features for handwriting modality, and MFCC coefficients for voice modality; agreed with the conclusion found in [5] and [18]. From a clinical point of view, acceleration and stroke size are regulated by the motor control of wrist and finger movement (mechanism inaccurate in PD). Moreover, pressure features can give further detailed information that cannot be obtained from kinematic features, hence, the significance to show the relationship between kinematic and pressure features [5]. From the other side PD affects the movement of the articulatory muscles resulting varying energy in frequency bands of speech signal, and MFCCs coefficients can compute these variations [18]. This can explain the frequent existence of such features in the selected set.

Moving to the bimodal assessment using deep learning, the three different combinations described in Sect. 4.2 were studied and compared. The results obtained are summarized in Table 2. For the 2D CNN model, decision-level fusion method performs better than feature-level fusion. This can be related to the fact that feature-level fusion is effective when time synchronized modalities are to be fused (fusion of speech and eye movements for example) [13]. The best models are selected and presented in Table 3; where the 'All-tasks' performances of both modalities beside the bimodal system are shown. Once the best models are selected, data augmentation techniques described in Sect. 4.3 are applied, where for jittering a random additive scalar is sampled from a Gaussian distribution with 0.3 STD for handwriting modality and 0.1 STD for voice modality. The new time series are either used directly with the 1D CNN-BLSTM or converted into spectrogram images with the 2D CNN. The 'All-tasks' performances of both modalities beside the bimodal system are shown in Table 4, and Task-wise accuracies in Table 5 respectively. Based on these results, we can see that the best handwriting features combination found is the same as the one found with handwriting modality [6], and it is clear how deep learned audio features has no effect on PD detection, and how the results obtained when applying deep learning to detect PD from raw speech signals are not satisfactory as the ones obtained with handwriting analysis. It is challenging to learn acoustic deep models from raw signals and especially when very few convolutional layers are used for acoustic feature extraction, which might be insufficient for building high-level discriminative features [15]. In our case, since we are working with small dataset, it will not be a good idea to enlarge our model. We investigate that it is more efficient to build the model using low-level audio descriptors instead of applying the raw audio waveforms directly.

Table 2. Bimodal classification of PD and control performance, where the seven models are defined in Sect. 4.

Model	Fusion method	Modality combination	Best handwriting features combination	Bimodal Acc (Sens, Spec) %
M1	Feature-level	Averaging	Z	78.57 (85.71,71.43)
M2		Concatenating	X+Y+Z+ Pre.+ Alt.+Azi	61.9 (61.9, 61.9)
M3	Decision-level	MLP	X+Y+Z+ Pre.+ Alt	**83.33** (87.71, 80.95)
M4			X+Y+Z+ Pre.+ Alt	**83.33** (87.71, 80.95)
M5			X+Y+Z+ Pre.+ Alt.+Azi	**88.1** (80.95, 95.24)
M6			X+Y+Z+ Pre.+ Alt.+Azi	**88.1** (80.95, 95.24)
M7			X+Y+Z+ Pre.+ Alt.+Azi	**88.1** (80.95, 95.24)

The log-spectrogram offers a rich representation of the temporal and spectral structure of the input signal. The use of log-spectrograms is thus studied. According to Table 4, the results show how the accuracy performance is improved from 52.38% to 71.43% after considering the log-spectrogram as input instead of the raw signal. However, the achieved results are still non-satisfactory compared to the results obtained with handwriting. A possible explanation of this behavior is that in spectrogram representations, it is difficult to separate simultaneous sounds since they all sum together into a distinct whole [16]. This means that a particular observed frequency in a spectrogram cannot be assumed to belong to a single sound. In addition, moving a sound vertically in a spectrogram might influence the meaning. Therefore, the spatial invariance that 2D CNNs provide might not perform as well for this form of data [16]. Finally, periodic sounds comprised of a fundamental frequency and a number of harmonics which are most often non-locally distributed on the spectrogram. Finding local features in spectrograms using 2D convolutions will be complicated in this case [16].

Table 3. Handwriting, audio and bimodal classification performance (Acc (Sen, Spec)) obtained with decision-level fusion method.

Model	Hand. Perf. (%)	Voice Perf. (%)	Bimodal Perf. (%)
M3	83.33 (87.71, 80.95)	54.76 (76.19, 33.33)	83.33 (87.71, 80.95)
M4	83.33 (87.71, 80.95)	52.38 (61.9, 42.86)	83.33 (87.71, 80.95)
M5	88.1 (80.95, 95.24)	54.76 (23.81, 85.71)	**88.1** (80.95, 95.24)
M6	88.1 (80.95, 95.24)	54.76 (76.19, 33.33)	**88.1** (80.95, 95.24)
M7	88.1 (80.95, 95.24)	52.38 (61.9, 42.86)	**88.1** (80.95, 95.24)

Table 4. Performance measures obtained after applying data augmentation and decision-level fusion, where the best handwriting features combination are the ones found in Table 2.

Model	Aug. Technique Hand/Voice	Hand. Perf. (%) Acc (Sens, Spec)	Voice Perf. (%) Acc (Sens, Spec)	Bimoda Perf. (%) Acc (Sens, Spec)
M3	Jitter/Jitter	83.33(87.71,80.95)	57.14(95.24,19.05)	83.33(87.71,80.95)
M4	Jitter/Jitter	83.33(87.71,80.95)	71.43(76.13,66.67)	85.71 (71.43, 100)
M5	Jitter+Syn./Jitter	**97.62**(95.24,100)	52.38(33.33,71.43)	**97.62** (95.24, 100)
M6	Jitter+ Syn./Jitter	**97.62**(95.24,100)	57.14(95.24,19.05)	**97.62** (95.24, 100)
M7	Jitter+ Syn./Jitter	**97.62**(95.24,100)	71.43(76.13,66.67)	95.24(95.24,95.24)

Nevertheless, cropping the audio signal into short segments and getting the log-spectrograms of each segment (short-term analysis) seems to be more effective than getting the log-spectrograms of the whole signal (global analysis) with 2D CNN (accuracy from 57.14% to 71.43%). We believe that this is due to the larger number of samples needed to train the 2D CNN, and the idea of maintaining the nonlinear variation over time axis. Data augmentation improves the 2D CNN model performance when audio signal segmentation is applied, and fails to improve the performance of the 1D CNN-MLP with raw audio signals and the 2D CNN without audio signal segmentation. In [6], we found that data augmentation does not improve the 2D CNN model with online handwriting spectrograms since the augmented time series are converted into spectrograms then normalized to a fixed dimension. While here since no normalization is applied on audio spectrograms, the model may benefit the most from the new generated signals.

Finally, from a quick analysis of the Task-wise accuracies presented in Table 5, text reading task performs better than the sustained phonation vowel 'a' in PD detection. We believe that the text reading task is richer in terms of acoustic and prosodic information, which makes them more convenient for automatic PD detection in contrast to maximum phonation time of vowel 'a' which contains less information [17]. The same conclusion was found in [18] when a SVM model was trained on pure handcrafted acoustic features.

In our opinion, it may be more efficient to build a deep model using some low level acoustic descriptors instead of using speech signal. From the other side, the acoustic handcrafted features proposed in this work have achieved good results in pure speech and in bimodal corpuses. In general, deep learning models are basically selected to avoid handcrafted features extraction that needs an expert knowledge of the field, or to achieve a better result by extracting deep features.

Table 5. Task-wise system and all-tasks system accuracies (in %) for various models and training schemes presented in Table 4.

Task	M3	M4	M5	M6	M7
Aug. technique	Jitter	Jitter	Jitter/Syn.	Jitter/Syn.	Jitter/Syn.
Repetitive letter 'l'	**69.05**	**69.05**	59.52/47.62	59.52/47.62	59.52/47.62
Triangular wave	71.43	71.43	**80.95**/78.57	**80.95**/78.57	**80.95**/78.57
Rectangular wave	61.9	61.9	71.43/**76.19**	71.43/**76.19**	71.43/**76.19**
Repetitive "Monday"	59.52	59.52	**78.57**/76.19	**78.57**/76.19	**78.57**/76.19
Repetitive "Tuesday"	**71.43**	**71.43**	57.14/59.52	57.14/59.52	57.14/59.52
Repetitive "Name"	52.38	52.38	**57.14**/50	**57.14**/50	**57.14**/50
Repetitive "Family Name"	**71.43**	**71.43**	69.05/64.29	69.05/64.29	69.05/64.29
Aug. technique	Jitter	Jitter	Jitter	Jitter	Jitter
Sustained vowel 'a'	52.38	**66.67**	52.38	52.38	**66.67**
Text reading	54.76	**73.81**	54.76	54.76	**73.81**
All tasks	83.33	85.71	**97.62**	**97.62**	95.24

6   Conclusions

The main contribution of this work is to build a language independent bimodal system for PD early diagnosis by combining both handwriting and speech signals, where this combination may be more appropriate to help on PD diagnosis in early stages. Both SVM and deep learning models are studied and compared in this work. For text reading task, the best combination (only voiced or the combination of voiced and unvoiced sounds) found in our previous work [18] is applied in this work with both approaches (SVM and deep learning). However, the best sampling rates found in [18] are only applied with the SVM, where a low sampling rate value (8 KHz) was applied with deep models (for memory usage problem). The results obtained with the SVM model are better than the ones obtained with deep learning. We have found how SVM with the combination of information from both handwriting and speech modalities deliver a more accurate PD early prediction (accuracy up to 100% that needs to be confirmed on larger scaled data) than pure handwriting (96.87% accuracy [5]) and speech (97.62% accuracy [18]) analyses.

The observations and conclusions obtained in this work are many. We have found that decision-level fusion method performs better than feature-level fusion in case of combining non-synchronized signals. In addition we have noticed how it is challenging to learn acoustic deep models from raw signals and especially when very few convolutional layers are used for acoustic feature extraction. Feeding a CNN with 2D spectrograms performs better than the raw signals but the results are still non-satisfactory. Higher number of training data samples and preservation of the signal nonlinear variation over the time axis improve the performance. Text reading task performs better than the sustained phonation vowel 'a' due to the existence of acoustic and prosodic information. Data augmentation methods applied on voice signals may increase deep learning model performance when the raw signals are converted into 2D spectrograms and the non-linearity over time axis is preserved. Deep models with the combination of handwriting and speech modalities deliver same PD prediction accuracy as pure handwriting analysis, and more accurate PD prediction than pure speech analysis. Since we believe that it may be more efficient to build deep models with some low level acoustic descriptors as inputs instead of raw speech signals in case we are working with small database, as a future work it will important to build such model to approve the effectiveness of speech analysis in PD detection. Despite the encouraging results obtained, there are still some works to do before putting our PD detection bimodal model into clinical use due to the fact that we have few subjects, in comparison with the real world where we would have thousands of subjects, but the findings in this work can form a solid basis to a future stage of research that needs to involve a much larger set of patients. For this reason, the observations and conclusions obtained in this work and the relevance of our system should be validated on a larger scaled database in future work (since for the time being we can not find any another public bimodal database (handwriting and speech) that can be used to validate our diagnosis system).

References

1. Drotar, P., Mekyska, J., Rektorova, I., Masarova, L., Smekal, Z., Faundez-Zanuy, M.: Decision support framework for Parkinsons disease based on novel handwriting markers. IEEE Trans. Neural Syst. Rehabil. Eng. 1 (2015). https://doi.org/10.1109/tnsre.2014.2359997
2. Weiner, W.J., Shulman, L.M., Lang, A.E.: Parkinsons Disease: A Complete Guide for Patients and Families. Johns Hopkins University Press, Baltimore (2013)

3. Pereira, C.R., et al.: Handwritten dynamics assessment through convolutional neural networks: an application to Parkinson's disease identification. Artif. Intell. Med. **87**, 67–77 (2018)
4. Impedovo, D., Pirlo, G., Vessio, G.: Dynamic handwriting analysis for supporting earlier Parkinson's disease diagnosis. Information **9**, 247 (2018)
5. Taleb, C., Likforman-Sulem, L., Khachab, M., Mokbel, C: Feature selection for an improved Parkinson's disease identification based on handwriting. In: 2017 1st International Workshop on Arabic Script Analysis and Recognition (ASAR), Nancy, France (2017)
6. Taleb, C., Likforman-Sulem, L., Mokbel, C., Khachab, M.: Detection of Parkinson's disease from handwriting using deep learning: a comparative study. Evol. Intell. (2020). https://doi.org/10.1007/s12065-020-00470-0
7. Orozco-Arroyave, J.R., et al.: Characterization methods for the detection of multiple voice disorders: neurological, functional, and laryngeal diseases. IEEE J. Biomed. Health Inform. **19**, 1820–1828 (2015)
8. Duffy, J.: Motor speech disorders: clues to neurologic diagnosis. In: Parkinson's Disease and Movement Disorders: Diagnosis and Treatment Guidelines for the Practicing Physician, pp. 35–53 (2000). https://doi.org/10.1016/j.wocn.2017.01.009
9. Pinto, S., Chan, A., Guimarães, I., Rothe-Neves, R., Sadat, J.: A cross-linguistic perspective to the study of dysarthria in Parkinson's disease. J. Phon. **64**, 156–167 (2017)
10. Shalash, A.S., et al.: Non-motor symptoms as predictors of quality of life in Egyptian patients with Parkinson's disease: a cross-sectional study using a culturally adapted 39-item Parkinson's disease questionnaire. Front. Neurol. **9**, 357 (2018). https://doi.org/10.3389/fneur.2018.00357
11. Khan, T.: Running-speech MFCC are better markers of Parkinsonian speech deficits than vowel phonation and diadochokinetic (2014). http://www.diva-portal.org/smash/record.jsf?pid=diva2:705196. Accessed 10 May 2021
12. Mazuel, L., et al.: Proton MR spectroscopy for diagnosis and evaluation of treatment efficacy in Parkinson disease. Radiology **278**, 505–513 (2016)
13. Dumas, B., Lalanne, D., Oviatt, S.: Multimodal interfaces: a survey of principles, models and frameworks. In: Lalanne, D., Kohlas, J. (eds.) Human Machine Interaction. LNCS, vol. 5440, pp. 3–26. Springer, Heidelberg (2009). https://doi.org/10.1007/978-3-642-00437-7_1
14. Rusz, J., et al.: Imprecise vowel articulation as a potential early marker of Parkinson's disease: effect of speaking task. J. Acoust. Soc. Am. **134**, 2171–2181 (2013). https://doi.org/10.1121/1.4816541
15. Dai, W., Dai, C., Qu, S., Li, J., Das, S.: Very deep convolutional neural networks for raw waveforms. In: 2017 IEEE International Conference on Acoustics, Speech and Signal Processing (ICASSP), New Orleans, LA, USA (2017)
16. Lonce, W.: Audio spectrogram representations for processing with convolutional neural networks. In: Proceedings of the 1st International Workshop on Deep Learning for Music, Anchorage, AK, USA (2017)
17. Pompili, A., et al.: Automatic detection of Parkinson's disease: an experimental analysis of common speech production tasks used for diagnosis. In: Ekštein, K., Matoušek, V. (eds.) TSD 2017. LNCS (LNAI), vol. 10415, pp. 411–419. Springer, Cham (2017). https://doi.org/10.1007/978-3-319-64206-2_46
18. Taleb, C.: Parkinson's disease detection by multimodal analysis combining handwriting and speech signals (Unpublished doctoral dissertation). Telecom Paris, France (2020)

19. Jeancolas, L., et al.: Automatic detection of early stages of Parkinsons disease through acoustic voice analysis with mel-frequency cepstral coefficients. In: 2017 International Conference on Advanced Technologies for Signal and Image Processing (ATSIP) (2017). https://doi.org/10.1109/atsip.2017.8075567
20. Sharma, R.K., Gupta, A.K.: Voice analysis for telediagnosis of Parkinson disease using artificial neural networks and support vector machines. Int. J. Intell. Syst. Appl. **7**, 41–47 (2015)
21. Little, M., Mcsharry, P., Hunter, E., Spielman, J., Ramig, L.: Suitability of dysphonia measurements for telemonitoring of Parkinsons disease. IEEE Trans. Biomed. Eng. **56**, 1015–1022 (2009)
22. Rosenblum, S., Samuel, M., Zlotnik, I., Schlesinger, I.: Handwriting as an objective tool for Parkinson's disease diagnosis. J. Neurol. **260**, 2357–2361 (2013)

TRACE: A Differentiable Approach to Line-Level Stroke Recovery for Offline Handwritten Text

Taylor Archibald$^{(\boxtimes)}$ (iD), Mason Poggemann, Aaron Chan (iD), and Tony Martinez

Brigham Young University, Provo, UT, USA
`taylor.archibald@byu.edu, martinez@cs.byu.edu`
`https://axon.cs.byu.edu`

Abstract. Stroke order and velocity are helpful features in the fields of signature verification, handwriting recognition, and handwriting synthesis. Recovering these features from offline handwritten text is a challenging and well-studied problem. We propose a new model called TRACE (Trajectory Recovery by an Adaptively-trained Convolutional Encoder). TRACE is a differentiable approach that uses a convolutional recurrent neural network (CRNN) to infer temporal stroke information from long lines of offline handwritten text with many characters and dynamic time warping (DTW) to align predictions and ground truth points. TRACE is perhaps the first system to be trained end-to-end on entire lines of text of arbitrary width and does not require the use of dynamic exemplars. Moreover, the system does not require images to undergo any pre-processing, nor do the predictions require any post-processing. Consequently, the recovered trajectory is differentiable and can be used as a loss function for other tasks, including synthesizing offline handwritten text.

We demonstrate that temporal stroke information recovered by TRACE from offline data can be used for handwriting synthesis and establish the first benchmarks for a stroke trajectory recovery system trained on the IAM online handwriting dataset.

Keywords: Handwriting · Stroke recovery · Deep learning

1 Introduction

Handwriting is prevalent in both the physical and digital world. When handwriting is captured by a digital device, such as a pen-based computer screen, it is referred to as *online* handwriting data. At a minimum, these data include the location of the pen tip or stylus when touching the screen through time [17]. On the other hand, *offline* handwriting data refers to digital images of handwriting inscribed on some physical medium.

While online data can be readily rendered as an image, the reverse process is much more difficult, as offline data lack a temporal component and often contain artifacts inherent to the writing medium or digitization process. Consequently, online handwriting data can make many tasks easier or more accurate [16],

© Springer Nature Switzerland AG 2021
J. Lladós et al. (Eds.): ICDAR 2021, LNCS 12823, pp. 414–429, 2021.
https://doi.org/10.1007/978-3-030-86334-0_27

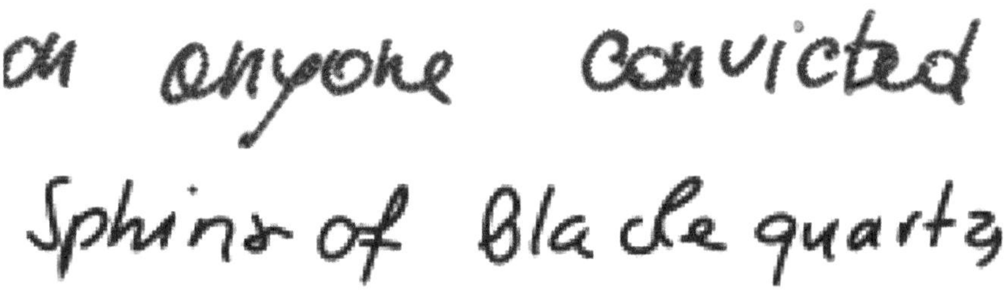

Fig. 1. TRACE recovery and synthesis. (1) is a visualization of strokes recovered by TRACE from an offline handwriting image. Blue arrows indicate the predicted direction and orange points indicate the beginning of a new stroke. (2) is an example of a synthetically generated image that mimics the style of (1), and demonstrates how strokes recovered from offline data can be used for other tasks. (Color figure online)

including handwriting recognition, signature verification, writer identification, and handwriting synthesis (see Fig. 1). While capturing handwriting online is becoming increasingly common, offline handwriting data collection can be easier in many instances and offline handwriting recognition remains an important challenge.

We propose a novel, differentiable model for stroke recovery called TRACE (**T**rajectory **R**ecovery by an **A**daptively-trained **C**onvolutional **E**ncoder). Our model is based on a CRNN that outputs a series of predicted stroke points, the number of which is proportional to the width of the original image. These predictions are then aligned using a Dynamic Time Warping algorithm (DTW) and compared against the ground truth (GT) stroke points to calculate a loss, from which the network is updated. The most important contribution TRACE offers is it extends prior trajectory recovery deep learning approaches to work on arbitrarily long lines of text. We provide the first trajectory recovery benchmarks for the IAM online handwriting database (IAM-On) [12] and the IAM offline handwriting database (IAM-Off) [13]. Additionally, we demonstrate that strokes recovered by TRACE can be used to synthesize handwriting in the manner of a given style.

2   Related Work

Traditional, explicit methods to stroke recovery broadly split the problem into two phases: *local examination*, where strokes are analyzed for startpoints, end-points, loops, ambiguous zones, and other complications, and *global reconstruction*, where the strokes are reconstructed based on features and observations derived from local examination [15]. A drawback of these approaches is that they often require handcrafted rules for each script. Moreover, they often rely on image preprocessing, including image skeletonization, which is sensitive to noise [15] and discards data that may be useful for determining the trajectory. With the advent of deep learning however, new approaches to trajectory recovery offer possible resolutions to these challenges.

Bhunia et al. [11] were the first to publish an end-to-end deep learning model for stroke recovery. Using an encoder-decoder style network, Bhunia et al. employed a CNN followed by an LSTM [9]. Benefits of using an LSTM include that it can have an arbitrary number of states and can learn long-term dependencies. They demonstrate its effectiveness on single stroke characters with square images, although encoding the image as a finite length vector and their chosen loss function limit the applicability of this model to more difficult tasks, including wider, multistroke images. Hung Tuan Nguyen et al. extend this approach to recover multiple strokes in a single image and add an attention mechanism in [14], though they only demonstrate it for single Japanese kanji characters.

In [22], Zhao et al. use a CNN and dynamic energy prediction network for Chinese single-character recognition. In [19], Sumi et al. use a Cross Variational Autoencoder to translate from offline to online characters and vice versa. The process involves learning a shared latent space between online and offline representations of characters, by iteratively passing online data to one encoder and offline data to another encoder. Representations encoded in this latent space are then decoded into both an online and offline data representation, where the difference between the reconstruction and ground truth is used to tune the network, although it is only demonstrated for single characters.

However, these efforts have focused on trajectory recovery for single characters and require significant revisions to process variable width images consisting of many characters.

3   Method

Our method broadly extends the ideas in [11], but we propose two additional methods to better align the prediction points with ground truth points, while also modifying the architecture to support encoding arbitrarily wide images, predict when a new stroke begins, and predict relative coordinates.

3.1   Loss

Our loss is broadly the distance between a sequence T of target (x, y) coordinates and the sequence of our predictions P according to some mapping.

One goal of our loss function is it should favor stroke point combinations that could have generated the original image. This means that all predicted stroke points should lie somewhere on the original strokes, and they should collectively cover the entirety of the original strokes. We also desire for the model to accurately predict both the order the strokes were originally written as well as the direction of each stroke.

In [11], the authors employed an L_1 loss, or the Manhattan distance from the predicted stroke points to the ground-truth points, where each predicted point is mapped to the GT point with the same index in the sequence. A potential issue with this approach is a set of stroke points that accurately reconstructs the original image might still incur a very large loss if the intervals between stroke points do not align well with the GT. Moreover, if the stroke has a small loop, the network might learn to exploit the loss by tending to place stroke points in the middle of the loop to minimize distance to all possible points in the loop if it cannot infer the direction of the stroke.

Our goal is then to find a better mapping from the predicted points and GTs. Since many different sequences of points can define the same function, we do not constrain our investigation to a bijective mapping, matching each point in our prediction sequence to precisely one target. Rather, we consider many to many mappings that favor ensuring every predicted point is near a GT point, every GT point is near a predicted point, and the order of the predicted points mirrors that of the GT points.

Formally, for all $t \in T$, we wish to minimize the distance between t and the nearest $p \in P$, p^*. The constraint,

$$\min_{p \in P} ||t_i - p||, \tag{1}$$

ensures that our prediction spans the entire GT function r.

Similarly, we wish to minimize the distance between any $p \in P$ and the nearest point $t^* \in T$, i.e.,

$$\min_{t \in T} ||p_i - t||. \tag{2}$$

This ensures that each p is proximate to the original function, or that each predicted point lies on the original stroke.

Finally, we wish to find a mapping that preserves the order each point appears within the stroke. Specifically, for i indexing the sequence of points in some stroke and s indexing some mapping, our mapping should require monotonicity

$$i_{s-1} \leq i_s \tag{3}$$

and continuity

$$i_s - i_{s-1} \leq 1. \tag{4}$$

A fast, dynamic programming algorithm that satisfies these constraints is dynamic time warping (DTW) [18]. DTW is robust to translations and dilations along the "time" dimension. DTW is typically computed by first computing a cumulative cost matrix, where each cell is the cumulative minimum cost needed to reach that cell. The last cell then provides the cost of the optimal alignment, while the optimal alignment can be solved with backward induction. To further reduce temporal complexity, we employ a uniform warping window on the cost matrix.

For our experiments, we prefer the L_1 loss, since L_2 penalizes outliers more and tends to produce more conservative predictions that fail to cover the corresponding GT strokes (e.g., it may predict only points in the middle of a crossbar in the letter "t"). However, a solution that predicts stroke points that span the entire crossbar but in the wrong direction might be a preferable for many tasks.

3.2   Adaptive Ground Truth

In many instances, it is impossible to infer from the image the direction a stroke
was drawn, or the order in which the strokes were drawn. A writer may cross a
"t with a left-to-right or right-to-left stroke. Similarly, a writer may cross a "t
or dot an "i" immediately or upon the completion of a word or sentence.

This presents a challenge for deep learning systems, since the loss function
provides its most salient feedback if the system correctly predicts the order and
direction of each stroke. However, for many applications, it is more important
for the system to predict strokes that reproduce the image of the stroke in high
fidelity, while being more invariant to the direction and order of the original
strokes. A potential weakness of employing DTW as our mapping algorithm is
that it enforces continuity and monotonicity on the entire sequence of strokes.
While monotonicity and continuity should be enforced within each stroke, we
wish to relax this requirement for a set of strokes to ensure that pathological
strokes do not inhibit the ability of the system to make predictions that faithfully
reconstruct the image.

Since we wish to preserve sequences of strokes while simultaneously minimizing the number of pathological model updates induced by reversed or out-of-
order strokes, we employ a method for permuting stroke order and direction during training. Because each instance will be assessed by the model many times, we
can adopt an iterative stroke reordering approach. Specifically, for each update,
we perform at most one alteration to the GT strokes sequence, either swapping
the order of adjacent strokes or inverting the sequence of points of a single stroke.
We can identify candidate strokes by those with the greatest total or average
DTW loss:

$$\Delta(P, T) := [\delta(P_i, T_j)]_{ij} \in \mathbb{R}^{n \times m}. \tag{5}$$

To select a candidate to swap, we perform a softmax on this loss for each
stroke to compute the sample probability of that stroke being altered:

$$\text{Softmax}(x_i) = \frac{\exp(x_i)}{\sum_j \exp(x_j)}. \tag{6}$$

After sampling a stroke, we compute what the loss would have been had the
GT stroke been altered by one of three transformations (if applicable): revers-
ing the direction of the stroke, swapping that stroke with the next stroke, or
swapping that stroke with the previous stroke. After performing this transformation, we recompute a new alignment and cost for this stroke. The stroke that
yielded the lowest loss is saved as the GT for subsequent epochs, and the loss
is computed relative to this adapted GT. As the number of iterations increases
and each training instance is trained on multiple times, the model converges to
a solution that better reconstructs the original images.

To improve efficiency, we only recompute the DTW alignment for some window around the affected stroke, and we do not need to recompute every possible
cost, as the cost matrix prior to the affected stroke has not changed. Because we

desire to preserve the actual stroke ordering and directions as much as possible, we employ the adaptive GT method to fine-tune the network after it has largely converged.

Even with these modifications, the temporal complexity of our loss function is still $O(N * k)$, where N is the number of points and k is the window size. To improve performance during training, the model can also be pretrained using a smaller DTW window and fine-tuned later with larger windows.

3.3   Encoder-Decoder Network

We employ a CRNN of the variety commonly used in handwriting recognition, depicted in Fig. 2. Unlike [11,22], and [19], our model uses a variable-length encoder, which allows it to perform robust predictions for potentially arbitrarily wide images.

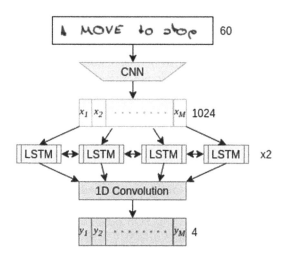

Fig. 2. The network architecture for TRACE. The input image for the CNN is 60 pixels high with an arbitrary width. The resulting feature maps are approximately the same width as the input and 1024 feature maps. This is passed into a 2-layer, bi-directional LSTM, followed by a 1D convolution. The result is a sequence of stroke point predictions.

We start with an 11-layer CNN that expects the input images to be to 60 pixels tall, 1 channel, and any possible width, to handle handwriting segments of varying length. The output is a matrix of variable width, rescaled proportionally to the length of the width of the original image and 1024 feature maps. We use 3×3 kernels for convolution, and both 2×2 and 2×1 windows for MaxPool operations (Fig. 3).

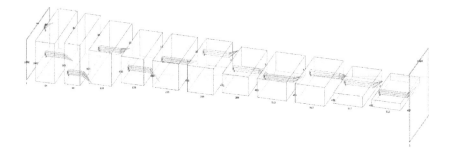

Fig. 3. Our CNN architecture. We use primarily 3×3 kernels to perform the series of layer operations: Conv, MaxPool, Conv, MaxPool, Conv, Conv, MaxPool, Conv, Conv, MaxPool, Conv. The output is a matrix of variable width and a height of 1024.

For each step, the CRNN predicts a relative coordinate (x, y) from the last position, whether the point is the start of a new stroke (SOS token), and whether the point represents the end of a sequence (EOS token).

If the model is trained to predict relative coordinates with no other constraints, the resulting strokes tend to be reasonably accurate in isolation. However, collectively, these strokes often do not align well with the input image. This is due, in part, because pen-up movements are infrequent, and thus more difficult to learn, but simultaneously have a disproportionately large role ensuring the prediction is aligned to the original image. To achieve a kind of global consistency, we compute the cumulative sum of these relative coordinate predictions. We then employ an L_1 loss to compute the difference between these summed predictions and the GT absolute coordinates.

For predicting SOS tokens, we first compute the DTW alignment between the predicted stroke and GT. Once the alignment is computed, we consider the first predicted point that matches to a GT SOS point to be the corresponding SOS (i.e., when multiple predictions match to a GT SOS). We employ a cross-entropy loss with class weights due to the class imbalance between SOS and non-SOS points.

We similarly employ a cross-entropy loss when predicting the EOS. To mitigate the class imbalance issue for EOS tokens, we duplicate the EOS GT stroke point 20 times and append it to the end of the GT stroke sequence. The model is thus trained to predict an EOS token for all successive points after the first EOS stroke point (in contrast to the SOS process described above). This approach allows for the model to learn a smoother transition from non-EOS points to EOS points, while also mitigating the class imbalance issue.

For training, we used the ADAM optimizer [10] with a batch size of 32, a learning rate of 0.0001, and a learning rate schedule that decreased the learning rate at a rate of .96 every 180,000 training instances.

3.4   Data

Since our goal is to reconstruct offline handwritten strokes, ideally we would have a set of offline images with corresponding online GT data. However, since these data are comparatively more difficult to collect, we adopt an approach of approximating offline data by rendering online data as images and degrading them. For our experiments, we train our model on the training and validation sets from the IAM online handwriting database (IAM-On) [12], a corpus of trajectory data collected from 221 different writers and 10,426 lines. To validate the model, we test it on both IAM-On test set, as well as the entire IAM offline handwriting database (IAM-Off) [13], which is composed of 13,353 lines by 500 different writers.

Each line in IAM-On contains a series of positional coordinates (x, y) as well as a time coordinate t. Since our model outputs predictions that are proportional to the width of the image, we resample the GT so that the number of stroke points is proportional to the width of the image. Moreover, because we are more concerned with recovering the shape of the strokes than the velocity, the points are resampled with respect to the cumulative stroke distance, so that points within a particular stroke are equidistant. While the model can be used to predict velocity as well by keeping the GT strokes parameterized by time, we prefer to use distance, as generally fewer points are needed to faithfully reconstitute the original strokes. Because of this reparameterization, TRACE is not technically recovering online handwriting data in our experiments (since it recovers this reparameterized data), though we have no reason to suspect TRACE would not work similarly well if it were trained on data parameterized by time.

Each set of strokes is then rendered as an image to be processed by the CNN. To better mimic offline data, a series of augmentations and degradations are applied to each image. These include varying stroke width and contrast, as well as applying random grid warping [20], random Gaussian noise, blurring, and other distortions [2]. We further supplement these data with 200,000 synthetic samples drawn from a generative model for online handwriting based on the one described in [7] trained on the IAM-On training and validation data.

4   Experiments

We evaluate the success of our model using quantitative metrics for how well it recovers strokes from online data and offline data. We establish the first baseline performance for stroke recovery for lines of text on both the IAM-On and IAM-Off data. We also demonstrate how it can be used in online handwriting synthesis, a potential downstream task.

4.1   Online Evaluation

We first consider how successfully the model is able to recover stroke trajectory information from online data rendered as images. Since the GT strokes are

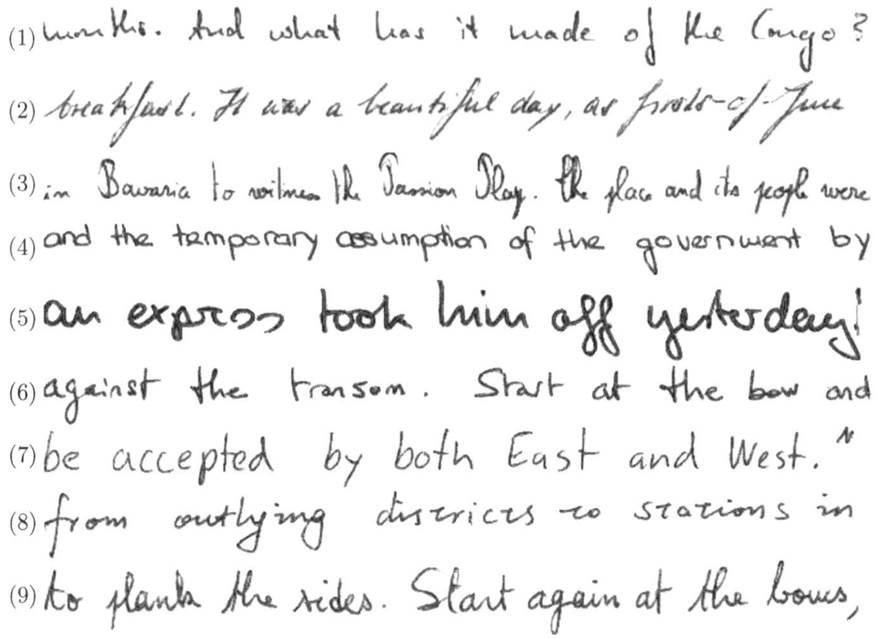

(1) months. And what has it made of the Congo?

(2) breakfast. It was a beautiful day, as first-of-June

(3) in Bavaria to witness the Passion Play. The place and its people were

(4) and the temporary assumption of the government by

(5) An express took him off yesterday!

(6) against the transom. Start at the bow and

(7) be accepted by both East and West."

(8) from outlying districts to stations in

(9) to plank the sides. Start again at the bows,

Fig. 4. Random sample of IAM-offline stroke reconstructions

Table 1. Average DTW distance

Distance metric	Average DTW loss	
	(equidistant GT)	(original GT)
L_1	0.03060	0.03452
L_2	0.02423	0.02745

known, we use the DTW distance score between the GT stroke points and our predicted points (i.e., the cumulative sum of the relative points, as in the loss function). Table 1 reports L_1 and L_2 average DTW scores, both for the actual GT (where points are sampled as a function of time), as well as a resampled version where points are sampled as a function of cumulative stroke distance, as TRACE was trained to predict equidistant stroke points and ignore velocity. Additionally, because TRACE predicts more points than were in the original GT and predicting more points tends to decrease average DTW distances, we resample the predictions to have the same number of points as the original GTs. The DTW distances are scaled so that the distance from the lowest stroke point to the highest stroke point has a unit distance of 1.

Table 2. Online NN distance

Type	Average NN distance (L_2)	
	(equidistant GT)	(original GT)
GT to nearest prediction	0.01662	0.01751
Prediction to nearest GT	0.01405	0.01615

We also report the average distance to the nearest neighbor (NN distance). In this case, we measure the distance between each predicted point and the nearest GT point and vice versa. Measuring each predicted point to the nearest GT is a measure akin to precision, and measures the extent to which the predicted points lie somewhere on the GT. Conversely, measuring the distance between each GT point to the nearest prediction resembles recall, and measures how well the predictions cover the entire space of GT points. Table 2 suggests that TRACE is slightly better at ensuring the predicted stroke points are near GT stroke points than it is at ensuring every GT stroke point is near a predicted point, which is supported by the observation that TRACE does not dot every "i" or cross every "t". As with DTW loss, using equidistant GT points decreases error.

4.2   Offline Evaluation

Because IAM-Off does not have ground truth strokes, we cannot compute the DTW distance. Instead, we consider the average NN distance from each predicted point to the nearest GT pixel (as opposed to the nearest stroke point), and define a GT pixel on the image as one that has an intensity of less than 127.5 on a scale from 0 to 255, which creates many more GT points for comparison than in the online experiment. This result is reported in Table 3.

Table 3. Offline NN distance

Type	Average NN distance (L_2)
Prediction to nearest GT	9.311×10^{-4}

Table 4. LPIPS metric

Model	LPIPS metric
White image (baseline)	0.423
TRACE without DTW	0.296
TRACE	**0.106**

In Table 4, we report the Learned Perceptual Image Patch Similarity (LPIPS) [21] metric between the offline image and an image of the reconstructed strokes using a 2 pixel stroke width. Specifically, we report the Alex-lin variant LPIPS metric for TRACE, an identical model but trained without DTW loss, and a baseline of all white images.

Qualitative results of the process can be observed in Fig. 4, which shows a random sample of GT offline images with the recovered strokes overlaid in red. TRACE tends to do very well in predicting neat, well-spaced handwriting, and generally better when the handwriting stands in high contrast to the background. TRACE struggles somewhat with predicting punctuation, as well as isolated strokes, or strokes that are often not drawn consecutively with respect to the rest of the character (as in the dot in an "i" or the cross in a "t"). Moreover, it sometimes fails to accurately predict stroke extremities, or strokes that approach too near the top of the line (as in the "P" in "Passion" on line 3).

Figure 5 shows it is partially robust to anomalies as it successfully resumes after an aberrant, scribbled out marking.

Fig. 5. TRACE achieves robust performance despite the presence of an anomalous marking.

4.3   Synthesis Evaluation

While recovering handwritten stroke trajectory may be of interest for its own sake, often it is considered an intermediate step for solving other handwriting tasks, including handwriting recognition, handwriting synthesis, and writer verification. While there are many possible ways to incorporate recovered strokes into these systems, we demonstrate one way it can be done for handwriting synthesis.

One of the first handwriting synthesis models is described in [7]. In this model, an LSTM is used to parameterize a Mixture Density Network (MDN), which can then be iteratively sampled to predict each successive stroke point. Data can then be passed to the LSTM to prime the model to mimic a particular handwriting style.

The success of this particular model depends on the input data being structured as a sequence of points, rather than as an image, as in the case of offline handwriting. In our case, we use recovered strokes from offline data to variously train or prime the model to synthesize offline handwriting styles.

Using offline handwriting in this system is useful not only for cases when the target handwriting style has not been captured online, but also because online and offline handwriting styles are not the same. Factors that contribute to these differences include the friction between the writing surface and implement, the responsiveness and sensitivity of the digital screen, and any changes to the way a person holds his or her hand when writing on a digital surface (e.g., to not touch the screen with the side of his or her hand). Moreover, strokes recovered from offline text can be used as a way to augment the training data of these kinds of systems, which improves generalization.

Figure 6 demonstrates the ability of the synthesis system to mimic the style of an offline sample. Note that all synthetic texts are rendered with the same stroke width and consequently mimic only the rough shape of the original input and not, e.g., line quality. For synthesizing experiments, we synthesize English pangrams, sentences that include every letter of the alphabet at least once. Our test sentences include

– Sphinx of black quartz, judge my vow.
– The five boxing wizards jump quickly.
– How vexingly quick daft zebras jump.

As Fig. 6 shows, more common letters and n-grams generally appear to produce better results.

4.4   Synthetic Evaluation

Figure 7 compares text generated by a system trained only with online data, another with only converted offline data, and another one trained on both, all being primed with a converted offline sample. While all three achieve various success in this task generally, the system is more prone to produce worse or degenerate samples when primed with a style it has not been trained on, as noted in [7].

Note that the inability of the synthesis system trained only on native online data to synthesize text when seeded with converted offline data may be due, in part, to either the imperfections that arise from the conversion process or inherent distinctions between online and offline handwriting styles. When the system is trained on offline data, and particularly when it is trained on other samples of a particular author, the system produces better synthetic samples.

Thus, TRACE enables seeding the synthesis model with an offline sample to mimic an offline handwriting style, while also augmenting the set of training data available to the synthesis model, enabling it to produce better synthetic handwriting samples.

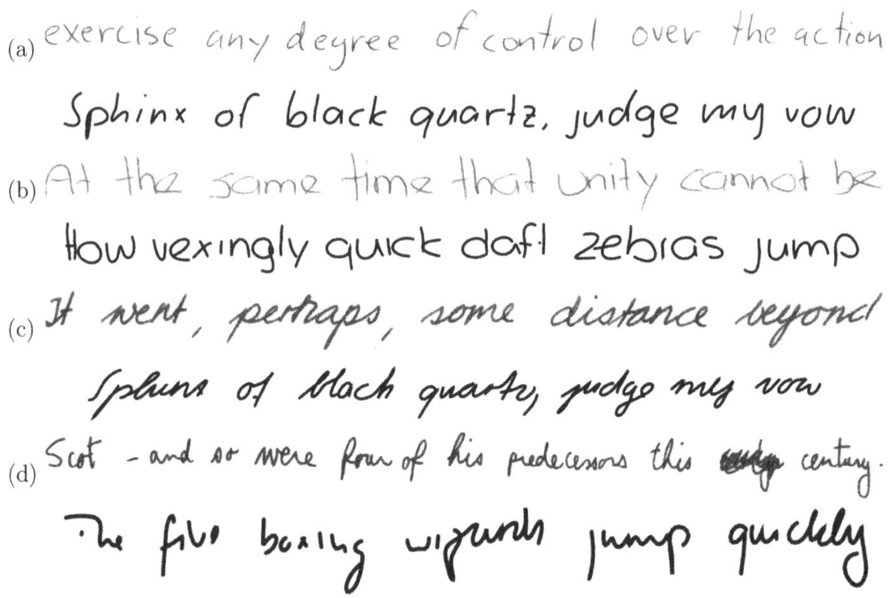

(a) exercise any degree of control over the action

Sphinx of black quartz, judge my vow

(b) At the same time that unity cannot be

How vexingly quick daft zebras jump

(c) It went, perhaps, some distance beyond

Sphinx of black quartz, judge my vow

(d) Scot - and so were four of his predecessors this ~~early~~ century.

The five boxing wizards jump quickly

Fig. 6. Each pair of lines above constitutes (1) an offline image used to prime the synthetic text model and (2) an image of synthetic text generated by the model in the style of the offline image.

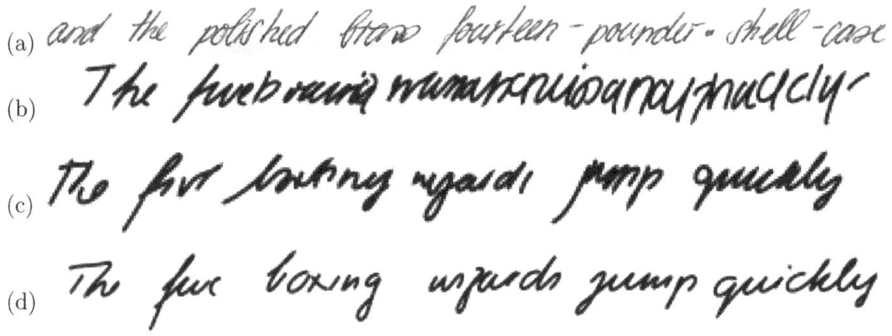

(a) and the polished brass fourteen-pounder-shell-case

(b) The five boxing wizards jump quickly

(c) The five boxing wizards jump quickly

(d) The five boxing wizards jump quickly

Fig. 7. Comparison of synthesized text based on different training data. (a) is a sample of offline handwriting data used to prime the handwriting synthesis model. (b), (c), and (d) are results from models trained on only online data, only offline data, and both online and offline data, respectively. The synthetic text is the pangram "The five boxing wizards jump quickly."

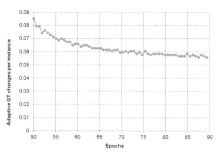 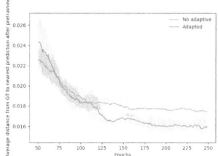

(a) The percentage of instances that undergo a GT adaptive change converges.

(b) Adapting GTs on the training set improves NN loss on test set.

Fig. 8. Adaptive GT performance by epoch

4.5   Adaptive GT Ablation Study

To demonstrate how the adaptive GT can improve the system, we first train the network for 50 epochs with the original GTs to have the system first learn the GT stroke orders and directions, before fine-tuning it with adaptive GTs to improve the system's ability to handle anomalous strokes.

Once the network has been pretrained, we employ the adaptive GT algorithm described in Sect. 3. Figure 8a shows how the number of swaps and changes is small to begin with, as a change to the GT is helpful for fewer than 1 in 10 instances. After 40 additional epochs, the number of changes has started to converge with 35% fewer changes per instance than initially. Figure 8b shows how the average NN loss on 5 runs converges to a lower loss than without the use of adaptive GTs on the test set.

5   Conclusion

In this work, we have proposed TRACE, a novel method to recover stroke trajectories from offline data, and, in effect, rendering it in an online format. TRACE works well on wide images composed of many characters and strokes and is completely differentiable. We have demonstrated that it can be used to enable online handwriting synthesis to work with offline data.

A possible future application of TRACE is it can be used as a loss function for synthesizing an offline handwriting image directly (as opposed to synthesizing strokes only) and could supplement approaches presented in [4] and [6]. Another possibility would be using it to augment training data for online recognition systems.

To improve the stroke recovery method, one direction might be to use a generative model, such as training a mixture density network as in [7,8], or an invertible neural network [1,5]. This would afford the model greater ability to model uncertainty and provide multimodal solutions. Another possible method to improve performance is to reparameterize input strokes as Bézier curves, as in [3].

Acknowledgements. We would like to thank Chris Tensmeyer for his suggestions and feedback.

References

1. Ardizzone, L., Kruse, J., Rother, C., Köthe, U.: Analyzing inverse problems with invertible neural networks. In: International Conference on Learning Representations (2019). https://openreview.net/forum?id=rJed6j0cKX
2. Breuel, T.M.: Tutorial on OCR and layout analysis. Technical report (2018)
3. Carbune, V., et al.: Fast multi-language LSTM-based online handwriting recognition. Int. J. Doc. Anal. Recognit. (IJDAR) **23**(2), 89–102 (2020). https://doi.org/10.1007/s10032-020-00350-4
4. Davis, B., Tensmeyer, C., Price, B., Wigington, C., Morse, B., Jain, R.: Text and style conditioned GAN for generation of offline handwriting lines. In: 31st British Machine Vision Conference, BMVC (2020)
5. Dinh, L., Sohl-Dickstein, J., Bengio, S.: Density estimation using Real NVP. In: 5th International Conference on Learning Representations, ICLR 2017 - Conference Track Proceedings, May 2016. http://arxiv.org/abs/1605.08803
6. Fogel, S., Averbuch-Elor, H., Cohen, S., Mazor, S., Litman, R.: Scrabblegan: semi-supervised varying length handwritten text generation. In: Proceedings of the IEEE/CVF Conference on Computer Vision and Pattern Recognition, pp. 4324–4333 (2020)
7. Graves, A.: Generating sequences with recurrent neural networks. arXiv preprint arXiv:1308.0850 (2013)
8. Graves, A.: Stochastic Backpropagation through Mixture Density Distributions, July 2016. http://arxiv.org/abs/1607.05690
9. Hochreiter, S., Schmidhuber, J.: Long short-term memory. Neural Comput. **9**(8), 1735–1780 (1997)
10. Kingma, D.P., Ba, J.L.: Adam: a method for stochastic optimization. In: 3rd International Conference on Learning Representations, ICLR 2015 - Conference Track Proceedings. International Conference on Learning Representations, ICLR, December 2015. https://arxiv.org/abs/1412.6980v9
11. Kumarbhunia, A., et al.: Handwriting trajectory recovery using end-to-end deep encoder-decoder network. In: Proceedings - International Conference on Pattern Recognition, vol. 2018-August, pp. 3639–3644. Institute of Electrical and Electronics Engineers Inc. (2018). https://doi.org/10.1109/ICPR.2018.8546093
12. Liwicki, M., Bunke, H.: IAM-OnDB - an on-line English sentence database acquired from handwritten text on a whiteboard. In: Proceedings of the International Conference on Document Analysis and Recognition, ICDAR, vol. 2005, pp. 956–961 (2005). https://doi.org/10.1109/ICDAR.2005.132
13. Marti, U.V., Bunke, H.: The IAM-database: an English sentence database for offline handwriting recognition. Int. J. Doc. Anal. Recogn. **5**(1), 39–46 (2002)
14. Nguyen, H.T., Nakamura, T., Nguyen, C.T., Nakawaga, M.: Online trajectory recovery from offline handwritten Japanese kanji characters of multiple strokes. In: 2020 25th International Conference on Pattern Recognition (ICPR), pp. 8320–8327. IEEE (2021)

15. Nguyen, V., Blumenstein, M.: Techniques for static handwriting trajectory recovery. In: Proceedings of the 8th IAPR International Workshop on Document Analysis Systems - DAS 2010, pp. 463–470. ACM Press, New York (2010). https://doi.org/10.1145/1815330.1815390. http://portal.acm.org/citation.cfm?doid=1815330.1815390

16. Plamondon, R., Privitera, C.M.: The segmentation of cursive handwriting: an approach based on off-line recovery of the motor-temporal information. IEEE Trans. Image Process. **8**(1), 80–91 (1999). https://doi.org/10.1109/83.736691

17. Plamondon, R., Srihari, S.N.: On-line and off-line handwriting recognition: a comprehensive survey. IEEE Trans. Pattern Anal. Mach. Intell. **22**(1), 63–84 (2000)

18. Sakoe, H., Chiba, S.: Dynamic programming algorithm optimization for spoken word recognition. IEEE Trans. Acoust. Speech Signal Process. **26**(1), 43–49 (1978)

19. Sumi, T., Iwana, B.K., Hayashi, H., Uchida, S.: Modality conversion of handwritten patterns by cross variational autoencoders. In: 2019 International Conference on Document Analysis and Recognition (ICDAR), pp. 407–412. IEEE (2019)

20. Wigington, C., Stewart, S., Davis, B., Barrett, B., Price, B., Cohen, S.: Data augmentation for recognition of handwritten words and lines using a CNN-LSTM network. In: Proceedings of the International Conference on Document Analysis and Recognition, ICDAR, vol. 1, pp. 639–645. IEEE Computer Society (2017). https://doi.org/10.1109/ICDAR.2017.110

21. Zhang, R., Isola, P., Efros, A.A., Shechtman, E., Wang, O.: The unreasonable effectiveness of deep features as a perceptual metric. In: Proceedings of the IEEE Conference on Computer Vision and Pattern Recognition, pp. 586–595 (2018)

22. Zhao, B., Yang, M., Tao, J.: Pen tip motion prediction for handwriting drawing order recovery using deep neural network. In: Proceedings - International Conference on Pattern Recognition, vol. 2018-August, pp. 704–709. Institute of Electrical and Electronics Engineers Inc. (2018). https://doi.org/10.1109/ICPR.2018.8546086

Segmentation and Graph Matching for Online Analysis of Student Arithmetic Operations

Arnaud Lods[1,2]($\boxtimes$) (iD), Éric Anquetil[2] (iD), and Sébastien Macé[1]

1 Learn&Go, Rennes, France
{arnaud.lods,sebastien.mace}@learn-and-go.com
2 Univ Rennes, CNRS, IRISA, 35000 Rennes, France
{arnaud.lods,eric.anquetil}@irisa.fr

Abstract. This paper is based on a research project aiming at improving learning long arithmetic operations in primary school using pen-based tablets. The goal is to automatically analyze a student's handwritten answer by comparing it to an expected answer and to provide immediate feedback. This comes down to find any mistake made such as a calculus mistake, missing carry over or symbol misalignment. We use the correspondence obtained by the Graph Edit Distance (GED) computed between both the student and expected answers. In order to reduce graph sizes to overcome the computational complexity of the GED on large graphs, we present a new semantic graph of line segmentation. We propose a backtracking process to correct potential early mis-recognition mistakes for non-corresponding vertices. We evaluate the improvement on the analysis performances for an increasing number of backtracks on an in-house dataset composed of 400 handwritten operations.

Keywords: Arithmetical operation analysis · Graph edit-distance · On-line hand-drawn structured document recognition

1 Introduction

Through the use of Intelligent Tutors, we are able to provide students with immediate feedback to help them correct their mistakes. For mathematics some tutors were proposed using keyboard interfaces [1,2], but with the improvement of pen-tablet devices such systems can be enhanced to transfer solving mathematical problem from numerical devices to paper back and forth. The online handwritten input is a list of sequences of points in the 2D space. A long arithmetic operation is a 2D structure represented by a set of related handwritten mathematical symbols. Figure 1 displays an example of such input. Given a long arithmetic problem we can represent the expected answer to confront it to the handwritten answer made by the student. The goal is to produce adapted

With the support from the LabCom **ScriptAndLabs** founded by the **ANR** ANR-16-LVC2-0008-01. With the support from the ANRT.

© Springer Nature Switzerland AG 2021
J. Lladós et al. (Eds.): ICDAR 2021, LNCS 12823, pp. 430–444, 2021.
https://doi.org/10.1007/978-3-030-86334-0_28

feedback for the student, which depends on the nature of its mistakes. Given the learning context, mistakes can be a wrong instruction recopy, misaligned symbols, a forgetting of symbols, calculus mistakes, an excess of symbols... (see Fig. 5) Due to the complexity of the input, there is a need to devise a recognition system to transform the handwriting into an adapted representation. The comparison between this representation and the expected answer also needs to be computed in no more than a couple of seconds to present "immediate" feedback to the students.

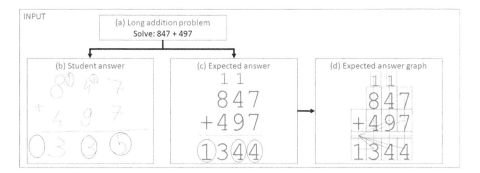

Fig. 1. Input of the system. (a) A long arithmetic problem is given to the student. (b) His answer is handwritten using a pen-tablet device. (c) An expected answer is created based on the problem. (d) A graph-based representation is generated. The dissimilarities we are looking for as an output of the analysis system are circled.

To represent both the handwritten input and the expected answer, we build a graph-based representation commonly used to represent mathematical expression and their symbols relationships [3]. We can produce corresponding representations to compute a comparison between the two graphs using the Graph edit Distance [4]. It is a popular and general graph similarity computation that searches the best vertices and edges correspondence between a pair of graphs. The dissimilarities found between graphs can be used to produce the analysis. For instance corresponding vertices with different labels might highlight calculus mistakes. However long arithmetic operation graphs easily reach 15+ vertices. To overcome the computational complexity of the GED on such graphs, we propose the use of a new graph segmentation. We transform the initial graph of symbols into a graph of lines by clustering symbols. This greatly reduces the size of graphs while keeping their structural identity. We propose to use the GED results by introducing backtracking steps into the system. Non-corresponding vertices between graphs can be re-evaluated to create several new recognition hypotheses to find a better fit to the expected answer.

1.1   Related Works

Handwritten mathematical expression (HME) recognition is a widely researched subject with the popular CROHME competition [3]. The most recent and best performing systems use end-to-end neural network by transforming the set of online strokes to an image. A CNN is used to extract a features map encoded and decoded by a Recurrent Neural Network with attention based model [5]. Those systems are limited to the translation of an HME to a recognized expression, reducing the relationships between symbols to the bare minimum to construct the LATEX expression. Other works focus on sequential solutions to build graphs, such as visibility graphs [6,7], representing mathematical symbols by vertices and relationships by edges. Those graphs are then parsed using a grammar to produce a Symbol Layout Tree containing only relationships to build a valid mathematical expression. The lower recognition results are mainly due to the missing context on the recognition step introduced by attention model in neural network. However they create a complete representation of the 2D structure of the mathematical expression in regards to relationships between symbols. In order to produce explicit feedback to the students, we need such an explicit representation of the structure of the operation with the complete relationships between symbols.

Given the graph representation of a student answer, it is then possible to compare it to another generated expected answer with the same architecture. The common method used for exact graph comparison are based on the Graph-Edit Distance, initially introduced in [8], which searches for the best pair-by-pair transformation from one graph to another. Several algorithms on the exact computation of the GED are presented in [4], and though several improvement using Depth-First Search [9], using sub-graph isomorphism [10] or through a mathematical solver [11] are proposed, it is yet challenging to compute this GED on graphs larger than 16 vertices without reaching the fixed time-out of 100 s. In practice we are able to compute the GED in a couple of seconds for smaller sizes of graphs. Other much faster methods rely on an inexact computation of the GED [12,13]. Though efficient, an incorrect matching between symbols could induce unexpected dissimilarities between the student answer and the expected answer, and thus produce incorrect feedback. Other methods focus on graph sizes reduction by clustering graph nodes such as in [14]. In [15] they use this graph clustering algorithm to create hierarchical graph representation in order to embed graphs to a vectorial space while keeping most of the important structural information.

To overcome the computational complexity of the GED on larger graphs, we present a similar approach to graph clustering with an adapted graph segmentation for arithmetical operation to reduce the sizes of graphs while keeping the structural information of the operation. Furthermore, to avoid early mis-recognition made by our sequential solution and to take advantage of our knowledge of the expected answer, we propose to use the results of the GED computation to develop a backtracking process. By selecting strokes with an incorrect fit with the expected answer, we can call back the recognition step to produce new segmentation hypotheses on these specific strokes in order to look for a better fit to the expected answer.

The rest of the paper is structured as follows. We present our graph representation and construction with an overview of our system in Sect. 2.1 before introducing the sub-segmentation of graphs in Sect. 2.2. We present the backtracking solution to correct potential recognition mistakes in Sect. 2.3. The experimental results on analysis accuracy and the quantity of operations timing-out on an in-house Handwritten Arithmetical Operation (HAO) dataset are detailed in Sect. 3. We discuss future improvements in Sect. 4.

2   Proposed Method

2.1   Overview

In the following sections we present our contributions. We first present an overview of the system before presenting the graph of lines segmentation and backtracking steps.

The Fig. 2 presents an overview of our system. (1) A graph is first constructed given a list of strokes as input: several classifiers are used namely for the strokes segmentation into symbols, the symbols classification and the symbols relationships classification. Using the classification rates of the previous classifiers, we will be able to use previous computation on pair of strokes or symbols to create new hypotheses. The segmentation features and classifier are based on [16]. The symbol classifier uses a standard VGG [17] architecture. The relationship features and classifier, computed using learned fuzzy landscapes [18] for mathematical relationship, are based on our previous work [19]. (2) Once the graph containing the symbols and their mathematical relationships is built, it is once again segmented using those same relationships to create a new graph of lines (see Sect. 2.2). This line segmentation is also applied to the expected answer graph.

(3) We can compute the GED using the DFS algorithm [9] to produce a first correspondence between our graphs of lines. Each corresponding pair of lines, represented each by a small subset of symbols, can later be matched to obtain a symbol-to-symbol correspondence. (5) If a pair of lines has a perfect match, it is possible to fix each corresponding pair of symbols, reducing the search space of the GED applied on the graph of symbols. (4) If any dissimilarities are detected between lines, we can backtrack to create new recognition hypotheses. By looking for new recognition hypotheses on non-corresponding strokes, we may correct mis-recognition by finding a better fit to the expected answer (see Sect. 2.3).

2.2   Graph of Lines Segmentation

In [4] the authors have shown that computing the exact GED on graphs larger than 16 vertices in less than 100 s is still challenging. To propose a system able

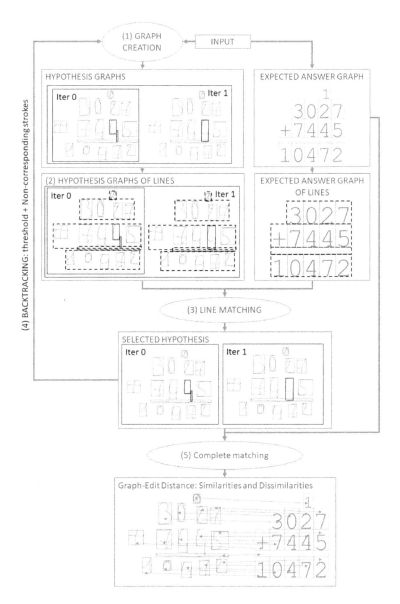

Fig. 2. Overview of our system. (1) From the input, a hypothesis graph and the expected answer graph are produced. (2) Those graphs are segmented into graphs of lines, (3) and then matched with the DFS algorithm. (4) If dissimilarities arise, the process is iterated with a lower threshold for non-corresponding strokes to create new hypotheses. (5) Once the steps are completed, the best hypothesis is kept and matched on the symbols level. In this example, the first iteration produce an incorrect segmentation on the highlighted digit 4. The corresponding line doesn't have a perfect match, thus we backtrack to the recognition process with a lower threshold. The second iteration create a new segmentation hypothesis on this digit, with a better fit to the expected answer (no dissimilarities arise).

to produce immediate feedback for the students, our goal is to reduce this time-out to a couple of seconds at most. To overcome this computational problem, we propose to create a graph representation with a reduced number of vertices each containing more knowledge while keeping the structural identity of the operation.

An arithmetical operation is composed of numbers defined by the instruction, the expected result and single digits identified either as carry over or report. Those digits can have a variety of placement and be misaligned, however numbers are always written and aligned from left to right and each symbol constituting the number are temporally close one to another. As specified earlier, each operation is represented by a graph in which a vertex is a mathematical symbol and each edge represents a mathematical relationship between a pair of symbols. Using the relationships defined earlier, we can cluster those vertices into a sub-set of vertices, each one representing a line of the operation.

Algorithm 1. Graph of lines creation from a set of symbols and corresponding relationships

---

   INPUT: V (set of symbols), E (set of relationships), t (temporal gap defined to reject symbols)

   $V_{lines} = \{\}, E_{lines} = \{\}$

   // We check every vertex in the graph

   while V is not empty **do**

      // The nearest neighbor with Left or Right relationship are used to cluster the symbol vertex into a line vertex

      $search \leftarrow (V[0], Right), (V[0], Left)$

      $line \leftarrow \{V[0]\}$

      // Until no new neighbor with the corresponding relationship are found, we extend the neighbor search to new vertices

      while $search! = \phi$ **do**

         $v, r \leftarrow search[0]$

         for $v', r' \in E[v]$ **do**

            if $r' == r \&\& gap(v, v') < t$ **then**

               $search+ = (v', r')$

               $line+ = v'$

            end if

         end for

         $pop(search[0]), pop(V[v])$

      end while

      $V_{lines}+ = line$

   end while

   Each vertex is segmented into a line, and relationships between lines are deduced from relationships between vertex inside the lines

   return $V_{lines}, inherit\_E(V_{lines}, V, E)$

Figure 3 shows the resulting Algorithm 1 applied on an addition. (a) A vertex is randomly selected, and linked to its neighbor with *Right* and *Left* relationships. Those vertices are then selected and the search is extended on both sides with the corresponding edges. (b) Once no more vertices are found, a single vertex, containing the previously selected vertices, can be created. (c) The process is repeated until all vertices are checked to produce the complete graph of lines. To avoid clustering instruction vertices with their related carry over or report, a time constraint is considered. Given a vertex and a neighbor candidate with the correct relationship, if the temporal gap between them is too high, they are not considered belonging to the same cluster and thus the neighbor is rejected. The edges between clustered vertices one to another are inherited from the known relationships between symbols vertices. An expected answer graph of lines can also be generated with the same rules. This new representation contains less vertices, each vertex contains more information and the structural information between them is inherited from the graph of symbols representation.

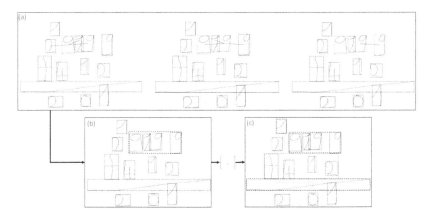

Fig. 3. An addition with 14 vertices written with the corresponding long arithmetic problem: 999 + 412 = 1411. (a) A random vertex is selected and linked to its neighbor with Right and Left relationships. The highlighted edges between the 9 digits and the 1 and 2 digit are ignored because of the huge temporal gap between those vertices (carry over). The process is repeated for each neighbor until no new neighbor are found to create the line (b). The process is iterated until all vertices belongs to a line. (c) The resulting graph of lines contains 7 vertices.

Given this graph of lines representation, we can apply the DFS on this new pair of graphs. To compute the cost of replacing one line by another, we transform those sub-graphs into strings by using the left-right order inside each sub-graph. We can apply a Levenshtein distance to compute this cost. If two lines have a cost similarity of 0, we can fix them by adding this knowledge to the graph of symbols. To avoid an imperfect match between numbers due to the malleability of the operators position, the insertion cost of an operator is fixed to 0. If at least a vertex has an incorrect label or is missing, other vertices belonging to the same cluster won't be directly matched.

The complete DFS on the pair of graph of symbols can be computed with the knowledge of fixed vertices with an updated cost (Eq. 1). If at least one of the vertex was found and the id does not match, then an infinite cost is applied. Thanks to the updated cost for matched vertices, a large part of the search tree will be ignored. The DFS is computed much faster on initially larger graphs with the same resulting correspondence.

$$d(v_i, v_j) = \begin{cases} \text{if } found \text{ and } f_i! = f_j & \infty \\ \text{else if } \mu_i \in [\text{-}, +] \text{ and } \mu_j \notin [\text{-}, +] & 50 \\ \text{else if } \mu_i! = \mu_j & 10 \\ \text{else} & 0 \end{cases} \tag{1}$$

2.3 Backtracking Recognition

Thanks to the line segmentation we are able to compute the DFS faster with the updated cost of corresponding symbols to non-corresponding symbols, thus considering only a sub-set of correspondence. However in addition to mistakes from the students, it is also possible to make wrong assumption on the recognition steps, either due to a wrong segmentation or by missing relationships with misaligned symbols. Thus vertices which should belong to the same cluster are split into different lines. In [19] we proposed to create multiple hypotheses using a fixed segmentation threshold before computing the matching. However on operations with many overlapping strokes, it is likely to create an exponential number of new hypotheses, many of whom will not fit the expected answer and could have been avoided. Moreover it is needed to compute a partial matching with each hypothesis to select the best fit, thus reaching a time-out if too many hypotheses are created.

We propose to include a backtracking step to take advantage of the resulting correspondence, detailed in the Algorithm 2. Figure 4 displays a backtracking example where a mis-segmentation on a digit can be revised by computing new segmentation hypotheses on the set of unmatched strokes. In the previous Section, we obtain a correspondence between graph of lines, which yield a cost transformation from lines to lines. If any lines from the expected answer is not perfectly matched with a transformation cost of 0, then non-corresponding strokes are labeled as such in the initial set of strokes. The recognition step is once again called with an increased threshold τ to create new hypotheses. Only the labeled pair of non-corresponding strokes are re-evaluated. If new hypotheses or relationships are created from one step to another, then the lines segmentation and matching is once again applied on each hypothesis. By reducing the search space to non-corresponding strokes, we can greatly reduce the number of new hypotheses that might be generated.

When multiple graphs are generated after a new recognition steps, each hypothesis is segmented into a graph of lines and matched to the expected answer. The hypothesis with the lowest transformation cost is kept for further analysis while others are discarded. If the end condition is not reached (either a number of backtracking steps reached or all lines returning a perfect correspondence), then the recognition process is applied on non-corresponding strokes with an increased threshold τ. Previous segmentation hypotheses will be discarded, only new segmentation hypotheses as well as the current best will be kept for the new selection. Once the end condition is reached, we can compute the DFS on the remaining unmatched symbols on the best hypothesis selected.

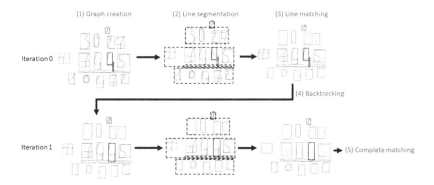

Fig. 4. Backtracking applied over an addition. The highlighted digit 4 is mis-segmented on the first iteration, resulting in an imperfect line matching. New segmentation hypotheses is computed on these un-matched strokes. The resulting new hypothesis has a better lines matching cost and new symbols can be preemptively matched. This hypothesis is kept for the DFS symbols-to-symbols.

3   Experiments

The dataset of handwritten arithmetical operation (HAO) presented in Table 1 is composed of both additions and substractions written by primary school students (age 6 to 9) on pen-based tablet. Samples from this dataset are displayed in Fig. 5. An operation is considered incorrect when an expected symbol is missing (number, operator or carry over) or when an expected symbol is matched with an incorrect label. The result of an analysis is based on these mistakes. Symbols in excess are not counted as mistakes due to the noisy input devices but they are expected to not be matched to any symbol from the expected answer.

A separate training set of 200 operations was indiscriminately input by both adults and students and was used to train the segmentation and relation classifiers presented earlier. For the symbol classifier, a much larger training set from the CROHME 2019 competition [3] was used. The test set of 200 operations is exclusively composed of handwritten operation written by primary school students. 110 out of 200 operations contains at least one mistake. No writer is found in both the training and testing set.

Algorithm 2. Recognition backtracking through lines matching

INPUT: S (set of strokes), τ (classification threshold), G_{EA} (expected answer graph), V_{fixed} (set of matched vertices)

// Generate a set of hypothesis given the threshold and previous found vertices

$\mathcal{G} \leftarrow reco(S, V_{fixed}, \tau)$

$G_{EA_{line}} \leftarrow line(G_{EA})$

cost $= \{\}$, path $= \{\}$

// For each hypothesis, apply the line segmentation and compute the correspondence with the expected answer

for $G \in \mathcal{G}$ **do**

    $G_{line} \leftarrow line(G)$

    $path, cost+ = DF - GED(G_{line}, G_{EA_{line}})$

end for

$path, cost = min(path, cost)$

// For each corresponding line in the best hypothesis, if its transformation cost is 0, keep the vertices

$V = V_{fixed}$

for $(line, line_{EA}) \in path$ **do**

    path_line, cost_line $\leftarrow$ Levenstein$(line, line_{EA})$

    if $cost\_line == 0$ **then**

        $V+ = line$

    end if

end for

// If all lines are matched, or if last iteration is reached, we return the correspondence symbol to symbol

if $(V == G_{EA_{line}} \| \tau == 0.2)$ **then**

    return DF-GED(best_hyp, G_{EA}, EP, V)

end if

// Otherwise, we backtrack with a greater threshold to generate more hypotheses while keeping the previous found vertices

return $backtrack(S, \tau + 0.05, G_{EA}, V)$

Table 1. Description of the HAO dataset. The size of the training and test set as well as the number of independent writers are reported. The number of mistakes contained through all operations and expected to be detected is also quantified. A single operation may contain multiple mistakes (see Fig. 5).

	Training set	Test set
Size	200	200
Writers	28	24
Symbols	2908	3285
Incorrect symbols label	–	83
Missing symbols	–	122

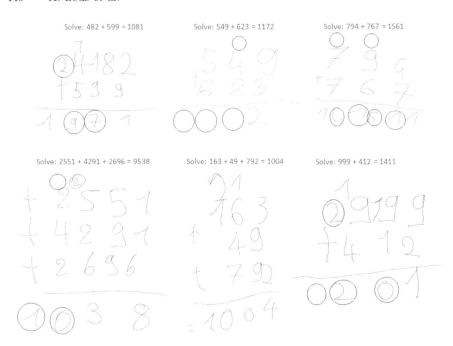

Fig. 5. Samples from the HAO dataset with their related long arithmetic problem. Students mistakes we are looking for are circled.

Evaluations are conducted on a machine with an Intel i5-8250U processor with 8GB of RAM. A time-out of 5 s was set for the computation on each operation. If the computation reaches a time-out, the last matching obtained is used for the analysis. We evaluate the improvement of the new graph segmentation in regards to the analysis score and quantity of operation reaching the time-out for different initial sizes of graphs. For the backtracking method, we consider a backtracking threshold value $\tau = 0.05\%$ with 5 backtracking iterations. The accuracy is computed on the detected dissimilarities between the operation and the expected answer compared to the ground truth (see Fig. 8). If the dissimilarities detected are different, then the analysis result is considered incorrect.

We compare 4 different systems. (1) The DFS applied on the first graph of symbols hypothesis. (2) As proposed previously in [19], a partial matching applied on sub-graphs to match part of the operation before computing the DFS on the complete graph of symbols. (3) The DFS applied subsequently on the first graph of lines hypothesis then on the graphs of symbols with the fixed vertices. (4) The DFS applied in the complete backtracking process on several hypotheses and on the last selected graph of symbols hypothesis and its fixed vertices.

The goal is to evaluate the analysis score improvement of using the new graph of lines representation over different sizes of graphs in relation to the number of operations reaching a time-out. The results on the analysis accuracy are reported in the Fig. 6. The analysis score are improved using the iterative process compared to previous methods. The backtracking allows for the correction of some mis-recognition from the system as we notice an improvement on the accuracy using the backtracking solution. The graph segmentation doesn't generate incorrect correspondence between graphs. We report the quantity of operations reaching a time-out in the Fig. 7 for each method. We notice a much lower number of operations reaching the time out of 5 s by using the DFS on the graph of lines. The lines segmentation greatly improves the number of operations solved before reaching the time-out, which is in direct relation with operations in which an incorrect analysis was yielded before. The best process can match 162 out of 200 operations under 5 s, as opposed to the results previously reported in [19] of 134 operations.

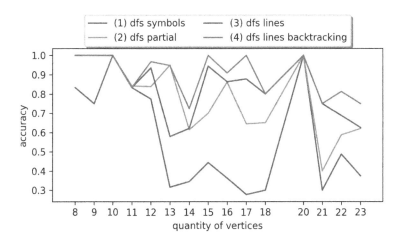

Fig. 6. Accuracy for each graph size. The accuracy is computed on the resulting analysis: if dissimilarities found were expected, the analysis is correct

Figure 8 displays two operations, the resulting recognition results from the backtracking process as well as the ground-truth which was previously annotated to evaluate the experiments. On the first operation, we are able to correct a mis-segmentation by keeping a new recognition hypothesis after a single iteration. However for the second operation, by searching a new recognition hypothesis with a better fit, we generated an hypothesis which yield an incorrect analysis. No hypothesis produces a correct matching on the result line because both segmentation do not correspond to the expected answer, and the system ends up selecting the incorrect segmentation hypothesis. Thanks to the lines matching, the DFS is computed before the time-out but an incorrect analysis is produced compared to the ground-truth. It would be necessary to have a balance between

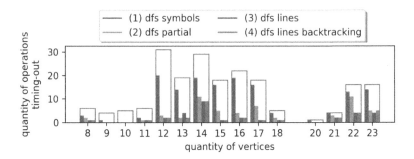

Fig. 7. Quantity of operations reaching time-out for each graph size confronted. The purple indicator represents the quantity of operations in the dataset for each graph size.

the fitting cost of the operation and the deterioration of the recognition results to avoid too far-fetched recognition hypothesis.

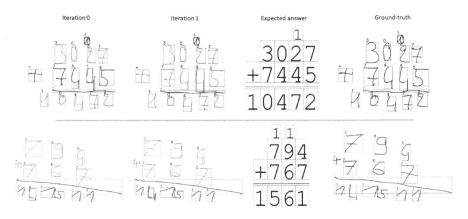

Fig. 8. Top: An operation with an initial incorrect segmentation, corrected thanks to the backtracking process on the second iteration. Bottom: incorrect segmentation on the highlighted 4 digit on the second iteration resulting in a similar high line matching cost, resulting in an incorrect hypothesis selected.

4   Conclusion and Future Works

We tackle the problematic of the analysis of on-line handwritten arithmetical operation in the context of primary school students long arithmetic teaching. We use a graph-based representation to transform the handwritten input. We propose to use the knowledge of the expected answer and the correspondence between this answer and the student answer using Graph-Edit distance to correct recognition mistakes. To speed up the matching process and reduce the size of the search space, we propose a sub-segmentation to transform graphs of symbols to graphs of lines using the relative positioning and temporality between symbols. The structural identify of the operation is preserved while greatly reducing the graphs sizes and each vertex also contains more information. This allows for a faster yet correct matching between graphs. We introduce a backtracking step to produce several hypotheses for non-corresponding strokes. By doing so we produce new recognition hypotheses to look for a better fit to the expected answer.

The experiments on a dataset of 400 arithmetical operations show that we're able to improve the analysis results by correcting recognition mistakes through the backtracking steps. Those steps can be quickly computed thanks to the lines segmentation. The number of new hypothesis created doesn't increase exponentially by limiting the new recognition hypotheses to non-corresponding strokes. However relying on the correspondence with the expected answer might influence the system to select an incorrect recognition hypothesis with a better fit to the expected answer. To avoid those mistakes, it would be advised in future works to combine both the matching costs of hypotheses as well as the classifier scores for each hypothesis to avoid selecting a hypothesis degrading too much the recognition results.

References

1. Xin, Y.P., Tzur, R., Hord, C., Liu, J., Park, J.Y., Si, L.: An intelligent tutor-assisted mathematics intervention program for students with learning difficulties. Learn. Disabil. Q. **40**(1), 4–16 (2017)
2. Huang, X., Craig, S.D., Xie, J., Graesser, A., Hu, X.: Intelligent tutoring systems work as a math gap reducer in 6th grade after-school program. Learn. Individ. Differ. **47**, 258–265 (2016)
3. Mahdavi, M., Zanibbi, R., Mouchere, H., Viard-Gaudin, C., Garain, U.: ICDAR 2019 CROHME + TFD: competition on recognition of handwritten mathematical expressions and typeset formula detection. In: 2019 International Conference on Document Analysis and Recognition (ICDAR), pp. 1533–1538 (2019)
4. Blumenthal, D.B., Gamper, J.: On the exact computation of the graph edit distance. Pattern Recognit. Lett. **134**, 46–57 (2020). Applications of Graph-based Techniques to Pattern Recognition
5. Zhang, J., Du, J., Dai, L.: Track, attend, and parse (tap): an end-to-end framework for online handwritten mathematical expression recognition. IEEE Trans. Multimedia **21**(1), 221–233 (2018)
6. Hu, L., Zanibbi, R.: MST-based visual parsing of online handwritten mathematical expressions. In: 2016 15th International Conference on Frontiers in Handwriting Recognition (ICFHR), pp. 337–342. IEEE (2016)
7. Zhang, T., Mouchère, H., Viard-Gaudin, C.: A tree-BLSTM-based recognition system for online handwritten mathematical expressions. Neural Comput. Appl. **32**(9), 4689–4708 (2018). https://doi.org/10.1007/s00521-018-3817-2
8. Sanfeliu, A., Fu, K.-S.: A distance measure between attributed relational graphs for pattern recognition. IEEE Trans. Syst. Man Cybern. **3**, 353–362 (1983)
9. Abu-Aisheh, Z., Raveaux, R., Ramel, J.-Y., Martineau, P.: An exact graph edit distance algorithm for solving pattern recognition problems. In: 4th International Conference on Pattern Recognition Applications and Methods 2015, Lisbon, Portugal, January 2015

10. Blumenthal, D.B., Gamper, J.: Exact computation of graph edit distance for uniform and non-uniform metric edit costs. In: Foggia, P., Liu, C.-L., Vento, M. (eds.) GbRPR 2017. LNCS, vol. 10310, pp. 211–221. Springer, Cham (2017). https://doi.org/10.1007/978-3-319-58961-9_19

11. Lerouge, J., Abu-Aisheh, Z., Raveaux, R., Héroux, P., Adam, S.: New binary linear programming formulation to compute the graph edit distance. Pattern Recogn. **72**, 254–265 (2017)

12. Fischer, A., Plamondon, R., Savaria, Y., Riesen, K., Bunke, H.: A hausdorff heuristic for efficient computation of graph edit distance. In: Fränti, P., Brown, G., Loog, M., Escolano, F., Pelillo, M. (eds.) S+SSPR 2014. LNCS, vol. 8621, pp. 83–92. Springer, Heidelberg (2014). https://doi.org/10.1007/978-3-662-44415-3_9

13. Bai, Y., Ding, H., Bian, S., Chen, T., Sun, Y., Wang, W.: Simgnn: a neural network approach to fast graph similarity computation. In: Proceedings of the Twelfth ACM International Conference on Web Search and Data Mining, pp. 384–392 (2019)

14. Girvan, M., Newman, M.E.: Community structure in social and biological networks. Proc. Natl. Acad. Sci. **99**(12), 7821–7826 (2002)

15. Dutta, A., Riba, P., Lladós, J., Fornés, A.: Hierarchical stochastic graphlet embedding for graph-based pattern recognition. Neural Comput. Appl. **32**(15), 11579–11596 (2020)

16. Hu, L., Zanibbi, R.: Line-of-sight stroke graphs and parzen shape context features for handwritten math formula representation and symbol segmentation. In: 2016 15th International Conference on Frontiers in Handwriting Recognition (ICFHR), pp. 180–186. IEEE (2016)

17. Simonyan, K., Zisserman, A.: Very deep convolutional networks for large-scale image recognition, arXiv preprint arXiv:1409.1556 (2014)

18. Delaye, A., Anquetil, E.: Fuzzy relative positioning templates for symbol recognition. In: 2011 12th International Conference on Document Analysis and Recognition (ICDAR), pp. 1220–1224. IEEE (2011)

19. Lods, A., Anquetil, E., Macé, S.: Graph edit distance for the analysis of children's on-line handwritten arithmetical operations. In: 2020 17th International Conference on Frontiers in Handwriting Recognition (ICFHR), pp. 337–342. IEEE (2020)

Applying End-to-End Trainable Approach on Stroke Extraction in Handwritten Math Expressions Images

Elmokhtar Mohamed Moussa[1,2(✉)], Thibault Lelore[1],
and Harold Mouchère[2]

[1] MyScript SAS, Nantes, France
{elmokhtar.mohamed.moussa,thibault.lelore}@myscript.com
[2] LS2N, University of Nantes, UMR, 64004 Nantes, France
harold.mouchere@univ-nantes.fr

Abstract. In this paper, we propose a novel end-to-end system to extract strokes from offline math expressions. Using a multi-task neural network we simultaneously predict the location of the pen and the pen state. Our approach is based on a recent state-of-the-art image-to-sequence method limited to small fixed-sizes images. We generalize it to large and multi-symbol images without preprocessing steps such as skeletonization or binarization. This architecture allows an end-to-end training. A curriculum learning strategy have been used to address the complexity of the images. We achieve comparable results to the state of the art on the UNIPEN English character dataset considering the next point prediction. We propose a stroke level metrics that allows us to measure the stroke reconstruction. Experiments show the advantages and limitations of the adopted Image-to-Sequence method when scaling up to large and complex images such as math equations.

Keywords: Stroke extraction · End-to-end trainable system · Handwritten mathematical expressions

1 Introduction

Traditionally, handwriting recognition approaches are split in two categories: off-line systems which use as input an image of a scanned document; and on-line systems which use the pen trajectory recorded with e-pen or touch-sensitive surfaces. In most of real use cases, there is no choice in the input modality. Finger traces on a smartphone are on-line data, ancient documents are always digitized as images. However, each modality has its own advantages. The on-line signal keeps the dynamic of the pen trace which is useful for most of the recognition algorithms. In addition, on-line signals facilitate editing to the users. Image based systems have the advantage to better take into account local 2D context or the global layout of a full document. Multi-modal systems try to keep advantages from both worlds [18,19] but they need the two modalities.

© Springer Nature Switzerland AG 2021
J. Lladós et al. (Eds.): ICDAR 2021, LNCS 12823, pp. 445–458, 2021.
https://doi.org/10.1007/978-3-030-86334-0_29

Very few datasets provide both modalities natively [17]. Rasterization or on-line world to the off-line conversion is a straightforward problem. This is very useful to produce training pairs for multi-modal systems from on-line data [3] or to produce new training image samples (data augmentation) [10]. In this last case, rendering realistic raster images from on-line signals usually involves adding background noise, variation in pen tip width to simulate movement speed and different artifacts [7]. The reverse operation, off-line to on-line conversion, is mainly used in two different contexts.

The first one is the vectorization which attempts to model a line drawing image as a set of mathematical primitives (polygons, parametric curves, *etc.*) associated to vector elements found in vector image format such as SVG. Application can be for technical drawing vectorization [6] and 2D animation [7]. In this case, retrieving temporal information is less relevant. The ordering between the different primitives is not an interest here and parametric curves have no drawing direction.

The second one is pen trajectory recovery from image of handwritten documents focus on retrieving the original temporal information. It is a crucial step for many applications, such as handwriting recognition and signature verification. The availability of temporal information in online systems often makes them better performing than their offline analogue [13]. In 2019, the Competition on Recognition of Online Handwritten Mathematical Expressions (CROHME) [12] included for the first time a offline recognition task. It has sparked since a great interest for offline to online conversion [4].

Recovering the stroke structure (or skeleton) is one of the initial steps in most document image analyses and understanding systems. It plays a key role in document processing since its performance affects quite critically the degree of success in a subsequent character segmentation and recognition. Degradation appear frequently and can be due to several reasons which range from the acquisition source type to environmental conditions [11]. Sketches and rough pencil drawings can add difficulties to the process when multiple overlapped lines should be merged into a single line [16]. Skeletonization is a ubiquitous step in classical approaches [13] of on-line to off-line conversion. Those approach suffer from the aforementioned limitations of skeletonization.

Based on the work of Zhao *et al.* [20], we propose a fully convolutional neural and multitask network, based on U-Net [15] to predict the pen location and the skeleton. A final fully connected layer is added for pen state classification. Pen state prediction enable the reconstruction of strokes and to halt the iteration framework (thanks to *END* state). Training data is generated using variable width strokes [7] making the system robust to stroke width variations.

1.1 Related Works

Pen Trajectory Recovery. Over the years, researchers have proposed many methods to recover the temporal information. Usually they follow similar steps: topology extraction, local ambiguous regions detection (junctions, double traced strokes, *etc.*). The ambiguities are then resolved using handcrafted

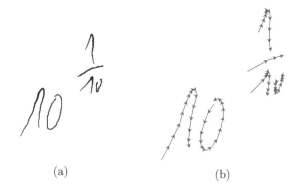

Fig. 1. Offline image (a) rendered from a math expression online signal (b).

heuristics. Existing approach often fall into three categories: recognition-based, topology-based and tracking-based. Recognition-based methods [5] were initially addressed to drawing composed of regular shapes (diagrams, engineering drawing, *etc.*). They detect those shapes using geometric primitives. By nature, this approach is not well-suited to handwriting and limit the user to a limited graphical vocabulary. Topology-based approaches build a representation from topological information in original image (skeleton, contour, *etc.*) and express the pen trajectory recovery as a global or local optimization problem. Qiao *et al.* proposed a weighted graph approach to recover the pen trajectory by finding the best matching paths [14]. They achieve good performance on English characters. Tracking-based approach iteratively estimates the relative direction of the pen. Bhunia *et al.* applied sequence-to-sequence modeling with an end-to-end Convolutional-BLSTM network obtaining excellent results on Tamil, Telugu and Devanagari characters [2]. However, Their approach is limited to single isolated characters. Although the mentioned scripts are closely related, they train a separate model for each script. Zhao *et al.* [20] proposed an image-to-sequence iterative framework to generate pen trajectories with a CNN followed by fully connected layers to predict the pen position at each time step. They obtain good results on Chinese and English handwriting datasets. However, these approaches suffer from the same drawback: the complexity of the model is directly dependent on the offline image resolution. Small resolutions (such as 28×28 or 64×64) can be sufficient for characters level applications but will lead to illegible images in the case of larger content like math equations. Using higher resolutions implies a quadratic increase of the total number of parameters in the model, stemming from the fully connected layers. Their method also relies on the skeleton to guide the prediction and to end the iteration framework. Skeletonization of real-world offline image often results in noisy and incomplete skeletons.

Line Drawing Vectorization is an essential step for 2D animations and sketching. Vectorization focus on converting the drawing images to vector graphics.

Artist usually start by sketching up a first version of their work with a pen and paper and then manually vectorize and finalize their work digitally. Vectorization of rough and complex real-world sketches is a difficult task. Multiple overlapping lines should be merged into a single line, noisy background and non-essential lines (*e.g.* construction lines) need to be cleaned. Simo-Serra *et al.* [16] proposed a fully convolutional simplification network augmented with a discriminator network to clean high resolution sketches. Guo *et al.* [7] presented a two-phase method using two networks to vectorize clean line drawings. A first multi-task CNN extract the skeleton and junction images. The skeleton is subdivided to many lines by removing junctions. A second CNN reconstruct the line connectivity around junctions. They achieve state of the art on the public *Quick, Draw!* [9] dataset. Nonetheless, their approach is limited to junction of valence 3 to 6 of small size 32×32.

2   Proposed Method

2.1   Overview

Our approach is based on Zhao *et al.* image to sequence method [20]: we design a pen trajectory prediction network to model pen position frames and infer the pen trajectory from handwriting offline images by iteratively predicting the next pen position with the said network. Our approach goes beyond limitations of the original method by being applicable to arbitrary image resolution, reconstructing the different strokes using pen states classification and eliminating the need for a pre-processing skeletonization step. The Fig. 2 shows the inputs and outputs of our model. The input of our network consists of fives images, the previous and current images $F_{i-1}, F_i \in \{-2, -1, 0\}^{h \times w}$, the grayscale offline image $I \in [0, 1]^{h \times w}$. We also provide the pixels coordinates as two images $I_X, I_Y \in [0, 1]^{h \times w}$. In fact, the network has a receptive field size of 32×32, the spatial clues we provide are global information that can help improve the network decisions in the local regions. The network outputs three images, the full skeleton I_S of the input image, all the stroke end points I_E, and the next pen position I_{POS}^{i+1}. Furthermore, we also predict a pen state in {Down, Up, End}. The redundancy in the different outputs allows to guide the training of the main task (the pen position and pen state). We train our network on synthetic off-line images with a variable stroke-width generated from the on-line signals *c.f.*Sect. 3.1. Contrary to [20], the network learns the skeleton image and can handle different image resolutions.

2.2   Network Architecture

Motivated by the successful application of fully convolutional neural networks (FCNNs) in the recent work of [16] for sketch simplification, we adapt our model from the U-Net architecture [15]. U-Net is a FCNN used for image segmentation in biomedical applications. It consists of a downsampling and an upsampling path. The downsampling path is a succession of 3×3 convolutions followed by

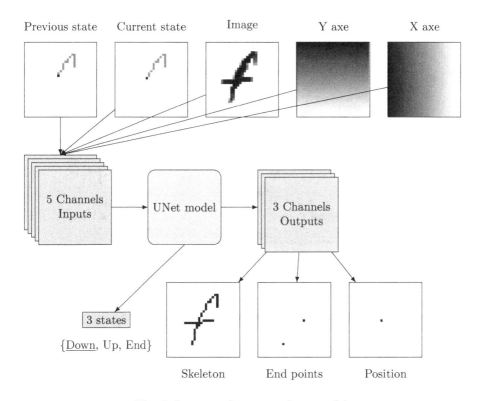

Fig. 2. Inputs and outputs of our model.

ReLU and a 2×2 max pooling layers. It can be seen as an encoder, that encodes the input in a small hidden feature map H. The upsampling path decodes the feature map H to the original resolution using up convolutions with a stride of 2×2. The high-resolution information is reused thanks to shortcut connections from the downsampling layers to the upsampling layers.

As shown in Fig. 3, the input frame images (F_{i-1}, F_i), the offline image I and the coordinate images I_X, I_Y are encoded to a small size hidden feature map $H \in \mathbb{R}^{448}$:

$$H = Encoder(F_{i-1}, F_i, I, I_X, I_Y).$$

The upsampling path decodes the feature map H to a map $O \in R^{h \times w \times 28}$ at the original resolution and outputs the target skeleton image $\hat{I}_S$, the stroke ends positions $\hat{I}_E$ and locate the next pen position $\hat{I}_{pos}^{i+1}$. The Eqs. (1) to (3) explain how these images are computed.

$$O = \mathrm{Decoder}(H) \tag{1}$$

$$\hat{I}_S, \hat{I}_E = \sigma(\mathrm{conv}_2(O)) \tag{2}$$

$$\hat{I}_{pos}^{i+1} = \sigma(\mathrm{conv}_1(O) \odot I_S) \tag{3}$$

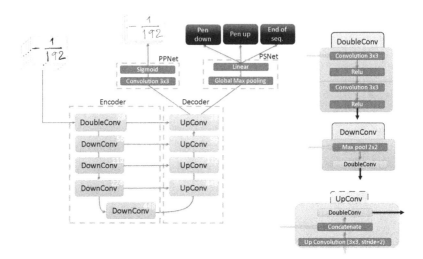

Fig. 3. Network architecture. The number of filters used in each convolution is shown on every block. The encoder and decoder are shared by PSNet and PPNet.

The $conv_1$ and $conv_2$ functions are classical convolution layers with respectively 1 and 2 kernels. The σ function is the sigmoid function. The product with the skeleton I_S before the output of the next position allows to constraint the next position to be on the predicted skeleton. This simplifies the task of the last $conv_1$ operation. A similar operation is done in [20] but as a post-processing step, using the skeleton extracted separately.

The encoder and decoder define the pen position prediction network PPNet. We modify U-Net by adding third path, a pen state classification network PSNet with one fully connected layer. We aggregate the decoder output O to a fixed size vector with global max pooling and input it to the classification layer:

$$\hat{P}_{i+1} = \text{PSNet}(O). \tag{4}$$

This allows the classification network to have a complete view of the input image whereas the decoder has a fixed size receptive field to make it prediction. Fortunately, the max pooling layers exponentially increase the receptive field. The pen can take three different state values. In addition to the standard *pen down* and *pen up*, we define an *end* state indicating that the scripter has finished writing. We consider that an *end* is a *pen up*, which implies multi-label classification. The *end* state is necessary, it's a stopping condition for the iterative framework used in the inference. Checking if every pixel from the skeleton has been visited (as done in [20]) is insufficient, as a scripter can draw certain pixels multiple times (pixels at junctions and double traced segments).

2.3   Loss Functions

We train our network with a multi-task loss composed of a binary-cross entropy loss of the outputed pen state $\hat{P}$, a soft-F1 loss of the predicted skeleton $\hat{I}_S$ and for the predicted end of stroke positions $\hat{I}_E$ and L2 loss of the predicted pen position $\hat{I}_{pos}$.

$$\mathcal{L} = \mathcal{L}_P + \mathcal{L}_S + \mathcal{L}_E + \mathcal{L}_{POS}$$

$$\mathcal{L}_P(\hat{P}, P) = \frac{1}{3} \sum_{y \in \{down, up, end\}} -(P_y \log(\hat{P}_y) + (1 - P_y) \log(1 - \hat{P}_y))$$

$$\mathcal{L}_S(\hat{I}_S, I_S) = \frac{1}{h \times w} \sum_{h,w} \text{softF1}(\hat{I}_S, I_S) \tag{5}$$

$$\mathcal{L}_E(\hat{I}_E, I_E) = \frac{1}{h \times w} \sum_{h,w} \text{softF1}(\hat{I}_E, I_E)$$

$$\mathcal{L}_{POS} = \frac{1}{h \times w} \sum_{h,w} \left\| I_{pos} - \hat{I}_{pos} \right\|_2^2$$

Where $I_S, I_E, I_{pos} \in \{0, 1\}^{h \times w}$ are the ground-truth skeleton, next pen position and strokes ends images.

3   Experimental Results

This section describes the used data sets and their preparation, the training protocol with curriculum learning and finally the proposed evaluations using the stroke level metrics.

3.1   Dataset Preparation

We use the online data from the CROHME 2019 [12] dataset to create a synthetic offline dataset of handwritten math expressions. We also use the isolated symbols of UNIPEN online handwriting [8] to allow a fair comparison with [20]. We apply the same prepossessing steps on both datasets.

The following preprocessing steps are applied to the online data:

- Symbol-wise normalization: the online signals are recorded with different resolutions and written in different handwriting sizes. To reduce handwriting size variation, we normalize the writing size to an average stroke diagonal size (so about a symbol size) of 32 pixels.
- Rasterization: We raster the online signals to gray-scale images with four stroke thickness selection strategies as in [7]. We vary the strokes-widths between 1 and 3 pixels.
- Frames generation: Each frame indicates the position of the pen. Similar to [20], we set background pixels to −2, already drawn pixels that are on the pen trajectory to −1. We encode the current pen position with 0 value.

The CROHME 2019 dataset provides 9,993 math equations for the training set, 986 and 1199 for the validation and test sets respectively (see Table 1). To reduce computational time, the validation set is reduced to a smaller subset of 40 randomly selected equations. The Unipen dataset only contains small images and thus allows comparison of both systems. The CROHME dataset contains larger and more complex images with much more strokes per samples.

Table 1. Dataset split between training, validation and test sets. And the corresponding total frame images.

Dataset	Number	Training	Validation	Test
CROHME	# of equations	9,993	40	1,119
	# of strokes	137K	539	17K
	# of frames	6M	25K	770K
Unipen	# of characters	8,000	500	2,000
	# of strokes	12K	722	2,852
	# of frame	407K	26K	102K

3.2   Training

During training we follow a curriculum learning [1] strategy. We start by selecting frame images of one symbol (image resolution of 4096 pixels, *e.g.* 64×64) and equally sub-sample each pen state class. We compute the total f1-score of the substasks with Eq. (6) on the validation set regularly during training. When the validation f1-score has not improved over 10 evaluation steps, the image resolution selection threshold is doubled. We stop at an image resolution of $202,752$ pixels (*e.g.* 256×792). As all images in a mini-batch should have the same size, each image is padded with white background inside each mini-batch. To optimize the mini-batch content, each one is built with images of about the same size. Its means that mini-batches with small images contain more samples than one with large images.

$$\mathcal{L}_C(\hat{I}_S, I_S) = \frac{1}{h \times w} \sum_{h,w} \left[\text{softF1}(\hat{I}_E, I_E) + \text{softF1}(\hat{I}_S, I_S) \right.$$
$$\left. + \text{softF1}(\hat{I}_{pos}, I_{pos}) \right] + \sum_{y \in \{down, up, end\}} \text{F1}(\hat{P}_y, P_y) \tag{6}$$

The Fig. 4 illustrates the training progression showing the F1 score of the validation dataset which always contains the same images, and the F1 score of the training set which is updated with new larger images every time the validation F1-score converges. As the validation set contains simple (small images with few strokes) and complex (large images with numerous strokes), the f1 score is progressing after each update of the training set. However the training score remains stable as complex images are smoothly introduced.

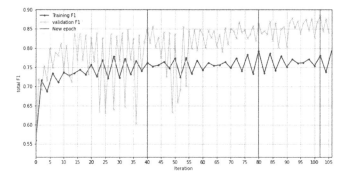

Fig. 4. Network training and validation learning curve. The different phases of curriculum learning are illustrated with the green vertical lines. (Color figure online)

We use *Adam* optimizer with a learning rate of $1e^{-3}$. The maximum batch size is set to 10. The network is trained on a single NVIDIA GeForce RTX 2080 Ti 11 GB GPU, taking 18 h to be completed.

3.3 Evaluation and Metrics

In this section, we provide a comprehensive performance analysis of our network architecture. First we evaluate each subtasks such as skeletonization, next frame prediction and pen state classification. We then assess the efficiency of their combination in the iteration framework using an end-to-end image-to-sequence scenario. We also compare our results with [20] on the Unipen dataset.

Skeletonization, Pen Location and State Prediction. The output of PPNet jointly predicts the position of the next point, the structure of the skeleton as well as the end point of the strokes, in three separate output layers. We define the position of the next point as the pixel with the highest activation in the dedicated layer and compute the TOP-1 error as an evaluation metric for the pen position prediction. We choose the F1 score metric for the skeleton and end points extraction. To evaluate PSNet, we compute the per-class f1-score for pen state classification. The evaluation results on UNIPEN and CROHME datasets are listed in Table 2.

In comparison, [20] achieves a pen position top-1 error rate of 1.2% on UNIPEN. No public implementation of their work is available. We implemented and trained their neural network on UNIPEN. Some meta parameters were not available in the paper, thus we optimized them on the validation dataset. We obtained a 4.6% error rate. This result should be compared with the 15.5% of error obtained by our network on the same dataset in Table 2.

Table 2. Evaluation results of PPNet and PSNet on the Unipen and CROHME test datasets. The table shows the pen position TOP-1 error, the skeleton and end points f1-score and the per-class f1-score for pen state classification.

	Pen position error	Skeleton	End points	Pen state		
				Down	Up	End
CROHME	0.100	0.993	0.754	0.990	0.770	0.870
UNIPEN	0.155	0.939	0.6691	0.990	0.682	0.789

Stroke Extraction. We adapt the iteration framework [20] to extract strokes with our network. At every iteration, in addition to the next point prediction, the pen state output is used to recover the strokes. The next point is constrained to be on the skeleton by multiplying the heat map I_{POS}^{i+1} by the predicted skeleton I_S. We evaluate the proposed stroke extraction algorithm on UNIPEN and CROHME. We use the stroke intersection over union SIoU from [7] defined as

$$\text{SIoU} = \frac{1}{n} \sum_{i=1,\dots,n} \max_{j=1,\dots,m} \frac{P_i \cap \hat{P}_j}{P_i \cup \hat{P}_j}, \tag{7}$$

with n being the number of strokes in the ground-truth online signal and m the stroke number in the predicted one. A groundtruth stroke P_i is matched with the predicted stroke $\hat{P}_j$ with the highest IoU. We add a new metric, SIoU 75% which is the rate of strokes for which the SIoU is greater than 75%.

Table 3 compare these metrics for the two systems on UNIPEN dataset. The method from [20] does not provide pen up information. Thus, we consider that a pen up state has been reached if the next pen positions is not 8-connected to the current position.

Table 3. Stroke extraction evaluation and comparison on UNIPEN dataset.

Method	SIoU	SIoU 75%	Number of parameters
Zhao et al. [20]	0.3587	0.4673	17.8 Millions
Ours	0.568	0.283	5.9 Millions

We can observe that our approach has a better SIoU than the other system. However both have quite low results, in the best case, there is only 56.8% of matching between the predicted strokes and the ground-truth strokes. The SIoU75% shows that the reasons are different for both systems. It seems that our approach over segment the strokes (only 28.3% of the strokes match at more than 75%) and the system from Zhao et al. seems to merge strokes. We can also notice that our system has 3 times less parameters that the other one.

To better understand the behavior of the proposed system on large images, we study in Table 4 on different subsets of the CROHME data set depending of the number of strokes in the original ink. We can observe that a small decrease of SIoU for ink between 10 to 20 strokes, but the results are globally stable arround 53.2%. The low value of the SIoU75% shows that we over segment most of the strokes as only 22% of the stroke match more than 75% of the original stroke. However, we note surprisingly that the lowest result is for small expression (17.7% for less than 5 strokes). We think that this is because the system well succeeded in long straight strokes (as fraction bar, integral, equals, ...) which are rare in small expressions.

Table 4. Matching rate of predicted stroke on the CROHME data set, considering different sizes of ink.

	[1,5]	]5,10]	]10,15]	]15,20]	]20,25]	All
SIoU	0.510	0.511	0.498	0.498	0.535	0.532
SIoU75%	0.177	0.178	0.192	0.217	0.229	0.220

Figure 5 presents results of complete inference on two samples with different sizes. We can see in the original images the used variability in the pen styles (width and gray level). The second line in the figure shows the ground-truth strokes. We can see in the last line that the predicted strokes correctly follow the true skeleton. We can notice that the strokes end at the extremity of the skeleton and no symbol or part of symbol is missing. However, we globally observe an over segmentation of the ink. The system well predicts long strait lines (fraction bars, equal parts) and smooth curves (parts of the α, θ or t). The difficulties rise in the complex parts at crossing strokes (in $alpha$, θ symbols), at inflection points (in 7, 1, or tan symbols).

4   Discussions

One of the main difficulties we faced in this work was the absence of absolute ground-truth for the strokes order. In the case of equations, the great diversity which exists in the way of writing a single number, but also the various corrections made *a posteriori* on the beginning of an equation (*e.g.* the prolongation of a fraction bar) means that the order of the strokes captured by the user does not always correspond to a logical and semantically valid visual order. One solution to overcome this issue would be to use an online recognition system to evaluate if the produced ink is recognizable. We think that it would not completely overcome this problem because these systems generally put a rather strong prior knowledge on the dynamics and the temporality of the entries, and for the moment none produce a perfect recognition. Moreover, the non-differentiable nature of this approach prevents the backpropagation from using the stroke order tolerance of the online math recognition engine to improve the stroke extraction. That is why we proposed the SIoU and SIoU75% metrics which evaluate the strokes independently of their global order. However, the stroke direction (order of points in

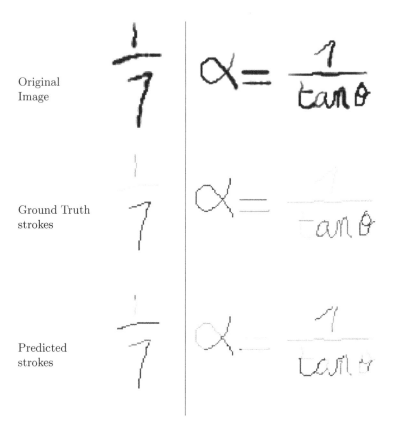

Fig. 5. Stroke extraction visualisation for a simple and a more complex image. Each stroke is drawn with a different color (better view in color). (Color figure online)

a stroke) should be evaluated, but the over/sub-segmentation of strokes makes it difficult. A metric allowing this evaluation still needs to be proposed (maybe based on DTW distance).

The proposed approach very well succeed to produce the skeleton and constraining the next pen position to be on the skeleton was an interesting proposal. However, the pen state strategy proposed too much pen-up. This decision needs local and global context. Indeed, the same local configuration (*e.g.* crossing points) can be solved differently depending on the symbol level context. On the one hand, the global pooling layer allows a concurrency between all candidate points of the image for this type of decision. On the other hand, it brings confusion on difficult points. The network favors solving obvious regions (*e.g.* straight lines) of the image before ambiguous regions (*e.g.* junctions), resulting in an over segmentation of the strokes. Furthermore, the network learns a short temporal transition $(t \rightarrow t+1)$ from a temporal context limited to the previous state $t-1$ therefore, by design it doesn't necessarily learn to model longer term temporal order.

5   Conclusion

In this paper, we presented image-to-sequence approach based on FCNN network to extract strokes from off-line images. The network simultaneously predicts the next pen position, skeleton image, stroke end points and pen state at a given time step. The iterative framework from [20] is complemented with the pen state information enabling stroke extraction. To the best of our knowledge, we are the first to tackle stroke extraction from arbitrary image resolution with a neural network approach. The skeletonization and stroke end points extraction shows good results that encourage a convolutional approach. However, we show the limitations of the Image-to-sequence with a FCNN approach on this type of problem, the network over segments the strokes because of a lack of mid-term target. A solution can be to model longer temporal transition and to provide a longer temporal context combining an attention based approach with a recurrent framework, keeping the CNN backbone.

References

1. Bengio, Y., Louradour, J., Collobert, R., Weston, J.: Curriculum learning. In: Proceedings of the 26th Annual International Conference on Machine Learning - ICML 2009, Montreal, Quebec, Canada, pp. 1–8. ACM Press (2009). https://doi.org/10.1145/1553374.1553380
2. Bhunia, A.K., et al.: Handwriting trajectory recovery using end-to-end deep encoder-decoder network. In: 2018 24th International Conference on Pattern Recognition (ICPR), pp. 3639–3644, August 2018. https://doi.org/10.1109/ICPR.2018.8546093
3. Bluche, T., Louradour, J., Messina, R.: Scan, attend and read: end-to-end handwritten paragraph recognition with MDLSTM attention. In: 2017 14th IAPR International Conference on Document Analysis and Recognition (ICDAR), vol. 01, pp. 1050–1055, November 2017. https://doi.org/10.1109/ICDAR.2017.174
4. Chan, C.: Stroke extraction for offline handwritten mathematical expression recognition. IEEE Access **8**, 61565–61575 (2020). https://doi.org/10.1109/ACCESS.2020.2984627
5. Doermann, D., Intrator, N., Rivin, E., Steinherz, T.: Hidden loop recovery for handwriting recognition. In: Proceedings Eighth International Workshop on Frontiers in Handwriting Recognition, pp. 375–380, August 2002. https://doi.org/10.1109/IWFHR.2002.1030939
6. Egiazarian, V., et al.: Deep vectorization of technical drawings. In: Vedaldi, A., Bischof, H., Brox, T., Frahm, J.-M. (eds.) ECCV 2020. LNCS, vol. 12358, pp. 582–598. Springer, Cham (2020). https://doi.org/10.1007/978-3-030-58601-0_35
7. Guo, Y., Zhang, Z., Han, C., Hu, W., Li, C., Wong, T.T.: Deep line drawing vectorization via line subdivision and topology reconstruction. Comput. Graph. Forum **38**(7), 81–90 (2019). https://doi.org/10.1111/cgf.13818
8. Guyon, I., Schomaker, L., Plamondon, R., Liberman, M., Janet, S.: UNIPEN project of on-line data exchange and recognizer benchmarks. In: Proceedings of the 12th IAPR International Conference on Pattern Recognition, vol. 3 - Conference C: Signal Processing (Cat. No. 94CH3440-5), vol. 2, pp. 29–33 (1994). https://doi.org/10.1109/ICPR.1994.576870

9. Ha, D., Eck, D.: A Neural Representation of Sketch Drawings, p. 16 (2018)

10. Le, A.D., Indurkhya, B., Nakagawa, M.: Pattern generation strategies for improving recognition of handwritten mathematical expressions. Pattern Recogn. Lett. **128**, 255–262 (2019). https://doi.org/10.1016/j.patrec.2019.09.002

11. Lelore, T., Bouchara, F.: FAIR: a fast algorithm for document image restoration. IEEE Trans. Pattern Anal. Mach. Intell. **35**(8), 2039–2048 (2013). https://doi.org/10.1109/TPAMI.2013.63

12. Mahdavi, M., Zanibbi, R., Mouchere, H., Viard-Gaudin, C., Garain, U.: ICDAR 2019 CROHME + TFD: competition on recognition of handwritten mathematical expressions and typeset formula detection. In: 2019 International Conference on Document Analysis and Recognition (ICDAR), pp. 1533–1538, September 2019. https://doi.org/10.1109/ICDAR.2019.00247

13. Nguyen, V., Blumenstein, M.: Techniques for static handwriting trajectory recovery: a survey. In: Proceedings of the 9th IAPR International Workshop on Document Analysis Systems, DAS 2010, pp. 463–470. Association for Computing Machinery, New York, June 2010. https://doi.org/10.1145/1815330.1815390

14. Qiao, Y., Nishiara, M., Yasuhara, M.: A framework toward restoration of writing order from single-stroked handwriting image. IEEE Trans. Pattern Anal. Mach. Intell. **28**(11), 1724–1737 (2006). https://doi.org/10.1109/TPAMI.2006.216

15. Ronneberger, O., Fischer, P., Brox, T.: U-Net: convolutional networks for biomedical image segmentation. In: Navab, N., Hornegger, J., Wells, W.M., Frangi, A.F. (eds.) MICCAI 2015. LNCS, vol. 9351, pp. 234–241. Springer, Cham (2015). https://doi.org/10.1007/978-3-319-24574-4_28

16. Simo-Serra, E., Iizuka, S., Ishikawa, H.: Mastering sketching: adversarial augmentation for structured prediction. ACM Trans. Graph. **37**(1), 11:1–11:13 (2018). https://doi.org/10.1145/3132703

17. Viard-Gaudin, C., Lallican, P.M., Knerr, S., Binter, P.: The IRESTE on/off (IRONOFF) dual handwriting database. In: Proceedings of the Fifth International Conference on Document Analysis and Recognition, ICDAR 1999 (Cat. No. PR00318), pp. 455–458, September 1999. https://doi.org/10.1109/ICDAR.1999.791823

18. Vinciarelli, A., Perone, M.: Combining online and offline handwriting recognition. In: Proceedings of Seventh International Conference on Document Analysis and Recognition, pp. 844–848, August 2003. https://doi.org/10.1109/ICDAR.2003.1227781

19. Zhang, J., Du, J., Dai, L.: Track, attend, and parse (TAP): an end-to-end framework for online handwritten mathematical expression recognition. IEEE Trans. Multimedia **21**(1), 221–233 (2019). https://doi.org/10.1109/TMM.2018.2844689

20. Zhao, B., Yang, M., Tao, J.: Pen tip motion prediction for handwriting drawing order recovery using deep neural network. In: 2018 24th International Conference on Pattern Recognition (ICPR), pp. 704–709, August 2018. https://doi.org/10.1109/ICPR.2018.8546086

A Novel Sigma-Lognormal Parameter Extractor for Online Signatures

Jianhuan Huang(ID) and Zili Zhang(✉)(ID)

College of Computer and Information Science,
Southwest University, Chongqing, China
hjh526400@email.swu.edu.cn, zhangzl@swu.edu.cn

Abstract. Online signature analysis can be widely applied in e-security and health. The latest method combines the Sigma-Lognormal model and visual feedback to extract the kinematic and spatial parameters of online signatures, but the model still does not perform well in complex handwriting signatures. Inaccurate parameters cannot reveal health information about users and cannot correctly reconstruct the online signature. This paper presents a novel Sigma-Lognormal parameter extractor for this drawback. On the one hand, this extractor estimates the parameters of pen-up and optimizes the parameters without the stroke midpoint. On the other hand, the extractor dynamically corrects the salient point position deviation by the velocity minimum point and velocity intersection point of adjacent strokes. The new extractor solves the parameter distortion caused by the ignored pen-ups and the hidden time deviation. The experiments demonstrate the accuracy and robustness of our method on multiple databases and verifiers, and the results show that the performance of the new extractor is better than the state-of-the-art method.

Keywords: Online signature · Motor equivalence theory · Visual feedback mechanism · Sigma-Lognormal model

1 Introduction

An online signature is acquired by a handheld computer device. This device contains various information during handwriting, including position, angle, pressure, timestamp, etc. In the applications of e-security and health, the data are generally analyzed by pattern recognition and signal processing techniques to reveal the identity or health state of the user [8].

The velocity profile of the online signature is an important issue in various theories. Plamondon et al. pointed out that kinematic signals should be preferred over kinetic signals and have the most discriminative signature verification space [21]. Based on velocity, many models can reproduce human movements, such as the multicomponent sinusoidal model [3], beta-elliptic model [4], equilibrium point model [19], or behavioral model [19]. These models are formalized by various mathematical functions, such as the exponential function, the Gaussian function, the beta function, and the spline function. Among them, kinematic

© Springer Nature Switzerland AG 2021
J. Lladós et al. (Eds.): ICDAR 2021, LNCS 12823, pp. 459–473, 2021.
https://doi.org/10.1007/978-3-030-86334-0_30

theory based on the Sigma-Lognormal model provides the best performance in explaining the human motor control of rapid and automated human movements [18]. The theory models the velocity of a neuromuscular system by the vector sum of lognormal functions, which is widely used in security and health fields [8]. These include the single reference signature system (SRSS) [5], the generation of duplicated signatures [1], the improvement of forgery detection [11], the development of biomedical tools for the diagnosis of neurodegenerative diseases [20,24], and the analysis of the development and acquisition process of handwriting skills [2,6].

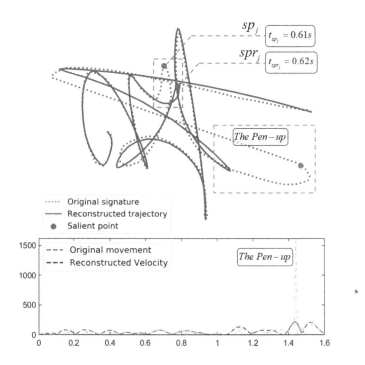

Fig. 1. Reconstructed signature example by iDeLog. Reconstructed distortions are marked by dotted boxes. Green dotted line frames the deformation at the end of the reconstructed stroke. Blue dotted line marks the deformation of the pen-up that was not captured by the tablet. (Color figure online)

The Sigma-lognormal model decomposes the complex movement into a linear combination of lognormal strokes and extracts six parameters from each stroke. The first automatic Sigma-Lognormal parameter extractor named ScriptStudio was presented in 2009 [16]. The application was developed to implement the robust XZERO algorithm. As the sigma lognormal parameter estimator is velocity-based, the errors in the estimation are propagated over the whole signal. To avoid accumulating too many spatial deviations in the long signals, an improvement was published in 2015 [15]. It chopped the original signal into short pieces, and estimated the parameters. In 2018, Ferrer et al. presented iDeLog [9] to erase spatial deviations that accumulate during the estimation by a visual feedback mechanism [17]. It can estimate the parameters of the whole signal without chopping. To date, iDeLog has achieved the highest performance available. However, there is still parameter distortion in the result of the extractor. Figure 1 shows a signature reconstructed by iDeLog. Two deformed trajectories in Fig. 1 are highlighted with dotted boxes. In the green dotted box, the reconstructed salient point spr_j deviates from the correct position sp_j, because of the 0.01 s time difference between spr_j and sp_j. In the blue dotted box, the reconstructed trajectory misses a stroke due to the ignored Pen-up.

The main contribution of this paper is the design of the novel Sigma-Lognormal parameter extractor. In the method, we take the pen-up as an independent stroke for parameter estimation and dynamically correct the spatial deviation between the original and reconstructed signatures. The new extractor excellently ensures the accuracy of the parameters and solves the deformation of the reconstructed trajectory. The procedure proposed in this paper considers the kinematic significance of handwritten signatures. This procedure improves the parameter extraction of duplicated signatures in three aspects. We estimate the Sigma-Lognormal parameters of the pen-up to fill the gap of pen-up processing. Then, we propose a new construction model of the stroke in visual feedback to optimize the parameters of stroke and pen-up. These two improvements solve the mismatch of pen-ups on the sigma-lognormal model. We use the reconstructed velocity minima point and intersection point to dynamically correct the salient point position deviation caused by the hidden time deviation. The proposed method, which fits the parameter extraction of complex signatures, is assessed with multiple databases and verifiers. The experimental results show that the reconstruction of the new extractor is less deformed than the baseline.

The remainder of the paper is organized as follows: the following section contains a brief description of the motor equivalence theory, visual feedback mechanism, and deformation problem analysis. Section 3 presents a novel sigma-lognormal parameter extractor and details its three key parts. Experimental results and a comparison with the state-of-the-art method are discussed in Sect. 4. Finally, the conclusion is drawn in Sect. 5.

2   Related Work

This section contains a brief description of the motor equivalence theory and the visual feedback hypothesis.

2.1   Motor Equivalence Theory

Motor equivalence theory shows that the brain decomposes a handwriting movement in two steps. In the first step, the brain stores the movement as a series of action plans. Second, the brain sends a series of motor commands to certain muscles to execute a given action plan. The Sigma-Lognormal model [16] provides an analytical representation of motor equivalence theory. Figure 2 shows a signature that is parsed by the model. The action plan is marked as a red dotted arc in Fig. 2 and refers to an elementary neuromuscular command in motor equivalence theory. The executed action plan will form a stroke. Before the current action plan is nearly completed, the next action plan will be executed simultaneously. Thus, the stroke will not reach the action plan endpoint "virtual target point" instead forming a "salient point" on the trajectory close to the virtual target point. The salient point is a stroke endpoint, and its corresponding velocity value is a local minimum.

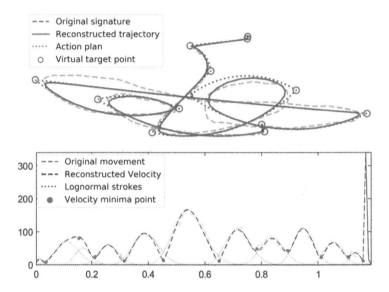

Fig. 2. Reconstructed signature example by iDeLog. In the Sigma-Lognormal model, stroke is an elementary movement of the signature. Its action plan is an arc, and its velocity lobe is a lognormal function.

Based on the correspondence between the trajectory salient point and the velocity minimum point [16], the velocity of stroke can be approximated by a unimodal curve v_{O_j} between two velocity minima points as:

$$v_{O_j}(t) = \begin{cases} v_O(t) & t_{\min_{j-1}} \leq t \leq t_{\min_j} \\ 0 & t < t_{\min_{j-1}}, t_{\min_j} < t \end{cases}, \tag{1}$$

where t is the time, $v_O(t)$ is the original velocity of the signature, $1 \leq j \leq N$, and N is the number of strokes. $t_{\min_0}$ is the starting time of the signature, $t_{\min_N}$ is the ending time of the signature, and $t_{\min_j}$ is not only the time of the jth velocity minima but also the time of the jth salient point. Hence, the jth stroke can be segmented by the times $t_{\min_{j-1}}$ and $t_{\min_j}$.

Each stroke has six Sigma-Lognormal parameters. The kinematic parameters of the jth stroke $\{t_{0_j}, \mu_j, \sigma_j\}$ are estimated from the velocity $v_{O_j}(t)$, where t_{0_j} is the time of stroke occurrence, μ_j is the stroke time delay, and σ_j is the stroke response time. The velocity profile of the jth stroke can be expressed as a lognormal function $|\overrightarrow{v_j}(t)|$:

$$|\overrightarrow{v_j}(t)| = \frac{D_j}{\sigma_j(t-t_{0_j})\sqrt{2\pi}} \cdot \exp\left(\frac{[\ln(t-t_{0_j})-\mu_j]^2}{-2\sigma_j^2}\right), \tag{2}$$

where D_j is the length of stroke, which can be obtained by the following rough calculation:

$$D_j = \int_{t=t_{\min_{j-1}}}^{t_{\min_j}} v_{O_j}(t)\,dt. \tag{3}$$

and t_{0_j}, μ_j, and σ_j are determined by fitting the original velocity peak $v_{O_j}(t)$ with the lognormal function (2), which can be expressed as:

$$t_{0_j}, \mu_j, \sigma_j = \underset{\hat{t}_{0_j}, \hat{\mu}_j, \hat{\sigma}_j}{\arg\min} \int_{t=t_{\min_{j-1}}}^{t_{\min_j}} \left||\overrightarrow{v_j}(t)| - v_{O_j}(t)\right|^2 dt, \tag{4}$$

where $\hat{t}_{0_j}$, $\hat{\mu}_j$, and $\hat{\sigma}_j$ and are the initial estimations by the robust XZERO algorithm.

The spatial parameters $\{D_j, \theta_{s_j}, \theta_{e_j}\}$ are estimated from the action plan of the jth stroke, and the angular position of the jth stroke can be approximated by:

$$\phi_j(t) = \theta_{s_j} + \frac{\theta_{e_j}-\theta_{s_j}}{D_j} \cdot \int_0^t |\overrightarrow{v_j}(\tau)|d\tau. \tag{5}$$

where θ_{s_j} is the starting angle of the action plan, θ_{e_j} is the ending angle of the action plan, and D_j is the length of the action plan. The action plan is a directed arc that starts from virtual target point tp_{j-1}, passes through the stroke midpoint mp_j, and finally reaches tp_j. tp_j is located on an extension line from the bottom midpoint $(sp_{j-1} + sp_{j+1})/2$ to vertex sp_j of triangle which is formed by salient points sp_{j-1}, sp_j and sp_{j+1}. The direction of the action plan can be determined by the vector cross product $\overrightarrow{tp_{j-1}mp_j} \times \overrightarrow{tp_{j-1}tp_j}$. If the action plan is clockwise, θ_{s_j} is the direction angle of the tangent vector $(-y_{tp_{j-1}}+y_{O_j}, x_{tp_{j-1}}-x_{O_j})$, where O_j is the center of the action plan. θ_{e_j} equals θ_{s_j} minus the vectorial angle $\angle tp_{j-1}O_jtp_j$, and D_j is the arc length. If the action plan is counterclockwise, θ_{s_j} is the direction angle of $(y_{tp_{j-1}}-y_{O_j}, -x_{tp_{j-1}}+x_{O_j})$, and θ_{e_j} equals θ_{s_j} plus $\angle tp_{j-1}O_jtp_j$.

2.2 Visual Feedback Mechanism

The study on visual feedback shows that it does not slow down the open-loop movement of skilled handwriting [17]. As shown in Fig. 3, the visual feedback based method corrects the spatial difference between the estimated and observed movements by moving the virtual target point. This ensures that the stroke still has a smooth and unimodal velocity curve after optimization, which means that the method maintains the kinematic property of the stroke.

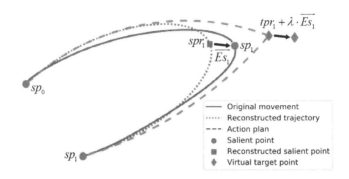

Fig. 3. Visual feedback mechanism that moves the virtual target point according to the error between the observed and reconstructed salient points to optimize the spatial parameters

The trajectory can be reconstructed by the sigma-lognormal parameters. Based on (2) and (5), the velocity of movement on the X-axis and Y-axis can be calculated by the Sigma-Lognormal function:

$$\overrightarrow{v}(t) = \begin{bmatrix} v_x(t) \\ v_y(t) \end{bmatrix} = \sum_{j=1}^{N} \begin{bmatrix} |\overrightarrow{v_j}(t)| \cdot \cos(\phi_j(t)) \\ |\overrightarrow{v_j}(t)| \cdot \sin(\phi_j(t)) \end{bmatrix}. \tag{6}$$

Then, the coordinates of the reconstructed trajectory are calculated by following the variable upper limit integral of the velocity component:

$$\begin{cases} x_r(t) = \int_0^t v_x(\tau)d\tau \\ y_r(t) = \int_0^t v_y(\tau)d\tau \end{cases}. \tag{7}$$

The coordinate of the jth reconstructed virtual target point $[x_{tpr_j}, y_{tpr_j}]^T$ can be calculated by (6) and (7) when N equals j. The visual feedback mechanism is moving tpr_j according to the difference between the observed and reconstructed j^{th} salient points. This optimization procedure can be expressed as:

$$\begin{bmatrix} \hat{x}_{tpr_j} \\ \hat{y}_{tpr_j} \end{bmatrix} = \begin{bmatrix} x_{tpr_j} \\ y_{tpr_j} \end{bmatrix} + \lambda \cdot \begin{bmatrix} x(t_{min_j}) - x_r(t_{min_j}) \\ y(t_{min_j}) - y_r(t_{min_j}) \end{bmatrix}, \tag{8}$$

where t_{min_j} is the time of the jth salient point. λ is a constant that controls the movement of the virtual target point tpr_j. When $\lambda = 1$, the system optimizes each stroke only once, and tpr_j will be moved directly to the final position. When

---

Algorithm 1. The Novel Sigma-Lognormal Extractor

Input: On-line samples of an observed signature: $(x, y, P, timestamp)$
Output: $(t_{0_j}, \mu_j, \sigma_j, D_j, \theta_{s_j}, \theta_{e_j})$

1: $t_{penup_i} = $ PenUpCheck(x,y,P,timestamp)
2: $v_0 = $ VelocityProfile(x,y)
3: $t_{min_j}, N = $ StrokeSegment(v_0, t_{penup_i})
4: //Estimate the jth stroke parameters;
5: **for** $j = 0$ **to** $N - 1$ **do**
6:    **if** The jth stroke is a pen-up **then**
7:       //Section 3.1;
8:       $(t_{0_j}, \mu_j, \sigma_j) = $ TimestampBasedEstimation($x, y, timestamp, t_{min_j}$)
9:       $(D_j, \theta_{s_j}, \theta_{e_j}) = $ StraightLineApproximation(x, y, t_{min_j})
10:    **else**
11:       //Section 2.1;
12:       $(t_{0_j}, \mu_j, \sigma_j) = $ RX0Algorithm($x, y, timestamp, t_{min_j}$)
13:       $(t_{0_j}, \mu_j, \sigma_j) = $ CurveFitting($v_0, t_{min_j}, t_{0_j}, \mu_j, \sigma_j$)
14:       $(D_j, \theta_{s_j}, \theta_{e_j}) = $ ArcApproximation(x, y, t_{min_j})
15:    **end if**
16: **end for**
17: $(x_r, y_r) = $ ReconstructTrajectory($t_{0_j}, \mu_j, \sigma_j, D_j, \theta_{s_j}, \theta_{e_j}$)
18: $t_{min_j} = $ LocateReconsturctedSalientPoint($t_{0_j}, \mu_j, \sigma_j, D_j, \theta_{s_j}, \theta_{e_j}$) //Section 3.2;
19: //Optimize the jth stroke parameters
20: **for** $j = 0$ **to** $N - 1$ **do**
21:    $(x_{tpr_j}, y_{tpr_j}) = $ MovingVirtualtargetPoint($x, y, t_{min_j}, x_r, y_r, t_{r_j}$) //Section 2.2
22:    $(D_j, \theta_{s_j}, \theta_{e_j}) = $ ActionPlanConstruct($x_{tpr_j}, y_{tpr_j}, \theta_{s_j}, \theta_{e_j}$)//Section 3.3
23: **end for**
24: **return** $(t_{0_j}, \mu_j, \sigma_j, D_j, \theta_{s_j}, \theta_{e_j})$

$0 < \lambda < 1$, the system will move tpr_j by λ times the spatial deviation, update the spatial deviation and move tpr_j again, until tpr_j is close enough to the final position.

Finally, as mentioned in Sect. 2.1, the method constructs the action plan by the new position of tpr_j and updates the spatial parameters from the new action plan.

3 Method

We present a new extractor to solve the problem of parameter distortion in a complex signature. This extractor is summarized in Algorithm 1, and the following subsections detail its three key parts.

3.1 Parameter Estimation of Pen-Up

A previous study described the full signature with one Sigma-lognormal model. It has ignored some pen-ups that are not captured by the tablet. This forms a joined-up writing of pen-up and normal stroke. Because only a stroke with more than three sampling points can be recognized by the system, the pen-up has only the starting and ending points. The shape of joined-up writing cannot be approximated as one arc, which leads to the distortion shown in the green dotted box of Fig. 1. To solve this problem, the pen-up parameters should be estimated as independent strokes (Algorithm 1:8).

The pen-ups are separated from the signature by pressure, timestamp, or button status. We set the velocity of pen-up endpoints as the zero-crossings $v_2(t_2)$ and $v_4(t_4)$. We estimate the parameters of pen-up by timestamps t_2 and t_4 instead of the unavailable values v_2 and v_4. The stroke response time σ_j of the pen-up is approximately estimated by solving the following nonlinear equation [16]:

$$\frac{t_2 - t_3}{t_4 - t_2} = \frac{e^{-a_2} - e^{-a_3}}{e^{-a_4} - e^{-a_2}} \tag{9}$$

where t_3 is the median of t_2 and t_4. a_2, a_3 and a_4 are temporary variables calculated as

$$\begin{cases} a_2 = \frac{3}{2}\sigma_j^2 + \sigma_j\sqrt{\frac{\sigma_j^2}{4} + 1} \\ a_3 = \sigma_j^2 \\ a_4 = \frac{3}{2}\sigma_j^2 - \sigma_j\sqrt{\frac{\sigma_j^2}{4} + 1} \end{cases} \tag{10}$$

The rest of the kinematic parameters μ_j and t_{0_j} can be calculated as:

$$\mu_j = \ln\left(\frac{t_4 - t_2}{e^{-a_4} - e^{-a_2}}\right), \tag{11}$$

$$t_{0_j} = t_2 - e^{\mu_j - 2a_2}, \tag{12}$$

We assumes that the curvature of the pen-up is zero so that the pen-up angles θ_{s_j} and θ_{e_j} are equal to the slope of the pen-up, and D_j is the length of the pen-up.

The segmentation and parameter estimation of the pen-up ensure that the adjacent stroke will not be distorted by the pen-up. However, the estimated pen-up parameters are still not sufficiently accurate and need to be optimized.

3.2   Salient Point Correction

Before parameter optimization, there is a problem that needs to be solved. As shown in Fig. 1, the time of reconstructed salient point spr_j is different from sp_j. Thus, locating sp_j by the time of spr_j is not correct enough. We proposed a method to locate the reconstructed salient point before optimization. The method locates spr_j by the time of the reconstructed velocity minima point. It can ensure the accuracy of the estimated spatial deviation in the visual feedback mechanism.

In addition, the time difference between sp_j and spr_j means that there is error in the estimated parameters. This error may cause a great overlap of two

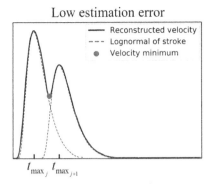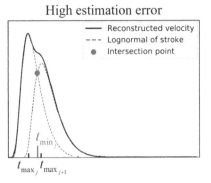

Fig. 4. High estimation error causes the disappearance of the velocity minima point. We use the intersection point of lognormals to locate spr_j when velocity minima fail.

lognormals, and the velocity minima point of reconstructed velocity may be hidden by the faster lognormal neighbor [22]. As Fig. 4 shows, we use the intersection point of lognormals to locate spr_j, if there is no velocity minimum between t_{max_j} and $t_{max_{j+1}}$. The time t_{min_j} is calculated as:

$$t_{min_j} = \arg\min_t |\overrightarrow{v_{j+1}}(t) - \overrightarrow{v_j}(t)|, t_{max_j} < t < t_{max_{j+1}}, \tag{13}$$

where t_{max_j} is the time of the jth lognormal peak point, which equals $t_0 + e^{\mu_j - \sigma_j^2}$.

The method (Algorithm 1:18) dynamically corrects the position spr_j by the reconstructed velocity minimum point and intersection point. If there is no velocity minimum and intersection point in the area $(t_{max_j}, t_{max_{j+1}})$, we use the time of sp_j to locate spr_j. The method ensures that virtual target point tpr_j can be moved to the right position in the visual feedback mechanism.

3.3 Stroke Construction in Optimization

The parameters of the ignored pen-ups lead to a new problem, that is, how to optimize such parameters. The current optimization needs the stroke midpoint to construct the action plan after moving the virtual target point, but the pen-up only contains two points without a midpoint.

To optimize the pen-up parameters, we propose a new method (Algorithm 1:25) to construct the action plan by the known central angle instead of the midpoint. Because the action plan is approximated by an arc in the model, the central angle ensures that the shape of the constructed action plan is the same as the original action plan, where the central angle $d\theta_j$ equals $\hat{\theta}_{e_j}$ minus $\hat{\theta}_{s_j}$.

As shown in Fig. 5, the clockwise action plan has a negative central angle $d\theta_j$, a center O_j, a starting point A_j, an ending point C_j, and a midpoint B_j of

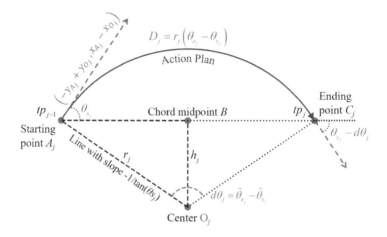

Fig. 5. After moving the virtual target point, the method constructs the action plan by central angle $d\theta_j$ instead of midpoint and updates the spatial parameters from the new arc.

the chord A_jC_j. According to the right-angled triangle $\Delta A_jB_jO_j$, the length of $\overrightarrow{OB}$ can be obtained by:

$$h_j = \frac{\left|\overrightarrow{A_jC_j}\right|/2}{\tan\left(d\theta_j/2\right)} = \sqrt{\left(x_{O_j} - x_{B_j}\right)^2 + \left(y_{O_j} - y_{B_j}\right)^2}, \tag{14}$$

where $[x_{O_j}, y_{O_j}]$ is the coordinate of the center O_j. It can be established by the following two equations:

$$y_{O_j} - y_{B_j} = \frac{-\left(x_{O_j} - x_{B_j}\right)}{\arctan\left(y_{C_j} - y_{A_j}/x_{C_j} - x_{A_j}\right)}, \tag{15}$$

$$\arctan\left(\frac{y_{C_j} - y_{O_j}}{x_{C_j} - x_{O_j}}\right) - \arctan\left(\frac{y_{A_j} - y_{O_j}}{x_{A_j} - x_{O_j}}\right) = d\theta_j, \tag{16}$$

where Eq. (15) is based on the direction of the vector $\overrightarrow{O_jB_j}$, and Eq. (16) is according to the turned angle from $\overrightarrow{O_jA_j}$ to $\overrightarrow{O_jC_j}$. The center $[x_{O_j}, y_{O_j}]$ can be obtained by substituting $[x_{A_j}, y_{A_j}]$, $[x_{B_j}, y_{B_j}]$, and $d\theta_j$ into Eqs. (14), (15), and (16).

Then, according to the estimated center coordinate and central angle, the new starting angle θ_{s_j} of the clockwise action plan can be obtained from the vector $(-y_{A_j} + y_{O_j}, x_{A_j} - x_{O_j})$. The new ending angle θ_{e_j} is equal to θ_{s_j} plus the signed number $d\theta_j$. The new arc length D_j is $r_j \cdot |d\theta_j|$. The center coordinate of the counterclockwise action plan can also be estimated by Eqs. (14), (15), and (16). The spatial parameter calculation method of the counterclockwise action plan can be found in Sect. 2.1.

4   Experiments

We tested our extractor on four databases to compare the performance of our work and the baseline which is iDelog. Performance experiments and automatic signature verifier experiments aim to validate the new Sigma-lognormal parameter extractor in several ways.

These four databases only acquired the sampling points of pen-downs, but not of pen-ups. The first database is a publicly available SUSIG-Blind database [12], which contains 2610 signatures from 87 users, with 20 genuine and 10 skilled forged signatures per user. These signatures are captured by a WACOM tablet. The second is the SCUT-MMSIG Tablet Subcorpus database [14], which includes 57 users with 20 genuine and 10 skilled forged signatures. The signatures were acquired by a low-cost UGEE EX05 pen tablet. The third database is Chinese online signatures from SigComp2011 [13]. This database consists of 21 real signatures and 34 skilled forged signatures for 10 users. The collection device is a WACOM table, and the captured pressure has a rounding error. The fourth is a proprietary database with 600 genuine signatures from 12 users.

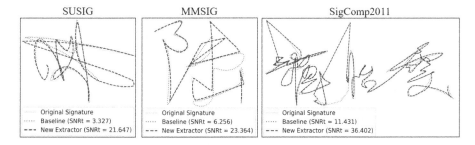

Fig. 6. Comparison of the reconstruction effect between the baseline and new extractor. From left to right are the signatures from the SUSIG, MMSIG and SigComp2011 databases. (Color figure online)

4.1   Performance Experiments

The experimental results are quantified by the number of lognormal strokes and the signal-to-noise ratio (SNR) between the reconstructed and observed movement. A higher SNR value indicates better reconstruction quality. The SNR of trajectory is denoted as SNR_t

$$SNR_t = 10 \cdot \log_{10} \left(\frac{\int_{t=0}^{T} \left(x'_O(t)^2 + y'_O(t)^2 \right) dt}{\int_{t=0}^{T} \left(dx(t)^2 + dy(t)^2 \right) dt} \right), \qquad (17)$$

where $[x'_O(t), y'_O(t)]$ is the zero-mean coordinate of the original signature, and $[dx(t), dy(t)]$ is the difference between the observed and reconstructed trajectories. As shown in Fig. 6, the blue line is closer to the green line than the red line, and the proximity can be quantified as the high SNR_t of the reconstruction from the new extractor. The SNR of velocity is written as SNR_v:

$$SNR_v = 10 \cdot \log_{10} \left(\frac{\int_{t=0}^{T} v_0(t)^2 dt}{\int_{t=0}^{T} (v_0(t) - v_r(t))^2 dt} \right). \tag{18}$$

The extractor performance in the whole database can be calculated by the quantification of each reconstruction. The mean values $\overline{SNR_t}$ and $\overline{SNR_v}$ are the fitness qualities between the reconstructed and original databases. The ratio of the mean $\overline{SNR_t}$ to the number of strokes $NbLog$ expresses the extraction consistency of the method [7]. The $\overline{SNR_t}$ and $\overline{SNR_t}/Nblog$ cannot be big enough. The standard deviation σ_{SNR_t} and σ_{SNR_v} show the robustness of the extractor. A lower σ_{SNR} means better performance of the method.

Table 1. Comparison results of reconstruction fitness quality, robustness, and extraction consistency.

Database	Procedure	$\overline{SNR_t}$	σ_{SNR_t}	$\frac{SNR_t}{NbLog}$	$\overline{SNR_v}$	σ_{SNR_v}	$\frac{SNR_v}{NbLog}$
SUSIG-blind	Baseline	18.82	6.05	0.75	13.22	5.14	0.53
	New extractor	25.77	3.23	0.98	14.76	1.96	0.56
SCUT-MMSIG	Baseline	13.29	6.13	0.86	8.75	3.57	0.56
	New extractor	22.99	2.65	1.36	12.48	1.07	0.73
SigComp2011	Baseline	25.42	3.88	0.41	16.16	3.36	0.25
	New extractor	31.41	3.85	0.35	13.13	1.78	0.14
proprietary	Baseline	22.29	3.24	0.2	10.33	4.16	0.08
	New extractor	28.81	3.14	0.2	11.83	0.51	0.09

Resampling the trajectory 200 Hz can further improve the performance [16], but in this study, we did not preprocess any more than baseline. Table 1 shows the performance comparison between the new extractor and baseline. The new extractor has better trajectory fitness quality and robustness in all the databases. In addition, the velocity fitness quality and extraction consistency of the new extractor are not as good as the baseline performance with the SigComp2011 database. The pressure rounding error of this database causes incorrect segmentation of pen-ups and requires extra preprocessing.

4.2   Automatic Signature Verifications Experiments

In our experiment, we use two automatic signature verifications (ASVs) to check the feature similarity degree between the observed and reconstructed signatures. The protocol was followed. That is, the first 5 genuine signatures of the set from a specific user were used to enroll a template. The remaining genuine signatures and the forged signatures from the user were used to test the performance of ASVs at different threshold levels.

The equal error rate (EER) is a special value when the FAR is equal to the FRR, where the FAR is the false acceptance rate of forgery signatures, and the FRR is the false rejection rate of genuine signatures. The smaller the difference in $EERs$ between the reconstructed and original databases, the more features the extractor keeps.

The first ASV is a dynamic time warping-based verifier (DTW) [10]. It verifies the writer of online signatures by calculating the minimum Euclidean distance of two variable-length sequences. The $EERs$ of the DTW verifier reflect the similarity of the local features between the original and reconstructed databases.

The second ASV is a Manhattan distance-based verifier (MD) [23]. It verifies the signature by calculating the Manhattan distance of two histograms. These histograms are the global feature sets that are obtained by statistical attributes.

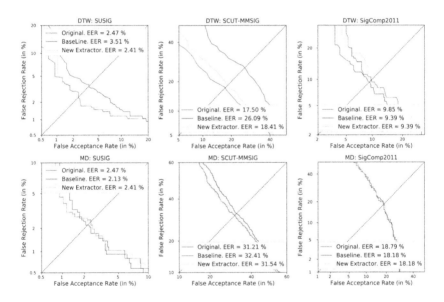

Fig. 7. DET curves of the DTW and MD with the original, the baseline reconstructed and new approach reconstructed databases. The databases are SUSIG-blind, SCUT-MMSIG, and SigComp2011.

Figure 7 shows the detection error tradeoff (DET) curve which takes FAR and FRR as the axis. In the DTW results, the DET curves of the new extractor are closer to the original than the baseline, and the EER values for the new extractor are closer to the original than the baseline. Our work retains more local features in reconstructed data than the baseline. In the MD, the global feature similarity degree of the new extractor has a slight increase over the baseline since the two methods both have no preprocessing.

5   Conclusion

This paper presents a novel sigma-lognormal parameter extractor that carries out valuable innovation with three methods. First, the method estimates the parameters of neglected pen-ups. Then, the extractor dynamically locates the position of reconstructed salient points before visual feedback. After that, in visual feedback, we establish a new model to construct the action plan without the stroke midpoint. Based on these improvements, the new algorithm solves the reconstruction deformation from the ignored pen-ups and the time deviation. Compared with the state-of-the-art extractor, the new method has a higher signal-to-noise ratio and better reconstruction effect on four databases. The results of the ASV experiment show that the new algorithm keeps the features from the original signature better than the baseline. In summary, the new extractor is expected to further apply the Sigma-lognormal model in more complex situations. Furthermore, the time deviation between observed and reconstructed salient points may be the breakthrough to optimizing the kinematic parameters. It eliminates the trade-off problem between trajectory and velocity. Finally, the new extractor is expected to further apply the Sigma-Lognormal model in more complex situations.

References

1. Bhattacharya, U., Plamondon, R., Dutta Chowdhury, S., Goyal, P., Parui, S.K.: A sigma-lognormal model-based approach to generating large synthetic online handwriting sample databases. Int. J. Doc. Anal. Recognit. (IJDAR) **20**(3), 155–171 (2017). https://doi.org/10.1007/s10032-017-0287-5
2. Carmona-Duarte, C., Ferrer, M.A., Parziale, A., Marcelli, A.: Temporal evolution in synthetic handwriting. Pattern Recogn. **68**, 233–244 (2017)
3. Choudhury, H., Prasanna, S.M.: Representation of online handwriting using multi-component sinusoidal model. Pattern Recogn. **91**, 200–215 (2019)
4. Dhieb, T., Ouarda, W., Boubaker, H., Halima, M.B., Alimi, A.M.: Online Arabic writer identification based on beta-elliptic model. In: 2015 15th International Conference on Intelligent Systems Design and Applications (ISDA), pp. 74–79. IEEE (2015)
5. Diaz, M., Fischer, A., Ferrer, M.A., Plamondon, R.: Dynamic signature verification system based on one real signature. IEEE Trans. Cybern. **48**(1), 228–239 (2016)
6. Djeziri, S., Guerfali, W., Plamondon, R., Robert, J.: Learning handwriting with pen-based systems: computational issues. Pattern Recogn. **35**(5), 1049–1057 (2002)
7. Duval, T., Rémi, C., Plamondon, R., Vaillant, J., O'Reilly, C.: Combining sigma-lognormal modeling and classical features for analyzing graphomotor performances in kindergarten children. Hum. Mov. Sci. **43**, 183–200 (2015)
8. Faundez-Zanuy, M., Fierrez, J., Ferrer, M.A., Diaz, M., Tolosana, R., Plamondon, R.: Handwriting biometrics: applications and future trends in e-security and e-health. Cogn. Comput. **12**(5), 940–953 (2020)
9. Ferrer, M.A., Diaz, M., Carmona-Duarte, C., Plamondon, R.: iDeLog: iterative dual spatial and kinematic extraction of sigma-lognormal parameters. IEEE Trans. Pattern Anal. Mach. Intell. **42**(1), 114–125 (2018)

10. Fischer, A., Diaz, M., Plamondon, R., Ferrer, M.A.: Robust score normalization for DTW-based on-line signature verification. In: 2015 13th International Conference on Document Analysis and Recognition (ICDAR), pp. 241–245. IEEE (2015)
11. Gomez-Barrero, M., Galbally, J., Fierrez, J., Ortega-Garcia, J., Plamondon, R.: Enhanced on-line signature verification based on skilled forgery detection using sigma-lognormal features. In: 2015 International Conference on Biometrics (ICB), pp. 501–506. IEEE (2015)
12. Kholmatov, A., Yanikoglu, B.: SUSIG: an on-line signature database, associated protocols and benchmark results. Pattern Anal. Appl. **12**(3), 227–236 (2009)
13. Liwicki, M., et al.: Signature verification competition for online and offline skilled forgeries (sigcomp2011). In: 2011 International Conference on Document Analysis and Recognition, pp. 1480–1484. IEEE (2011)
14. Lu, X., Fang, Y., Kang, W., Wang, Z., Feng, D.D.: SCUT-MMSIG: a multimodal online signature database. In: Zhou, J., et al. (eds.) CCBR 2017. LNCS, vol. 10568, pp. 729–738. Springer, Cham (2017). https://doi.org/10.1007/978-3-319-69923-3_78
15. Martín-Albo, D., Plamondon, R., Vidal, E.: Improving sigma-lognormal parameter extraction. In: 2015 13th International Conference on Document Analysis and Recognition (ICDAR), pp. 286–290. IEEE (2015)
16. O'Reilly, C., Plamondon, R.: Development of a sigma-lognormal representation for on-line signatures. Pattern Recogn. **42**(12), 3324–3337 (2009)
17. Pertsinakis, M.: Effect of visual feedback on the static and kinematic characteristics of handwriting. J. Forensic Doc. Exam. **27**, 5–21 (2017)
18. Plamondon, R., Alimi, A.M., Yergeau, P., Leclerc, F.: Modelling velocity profiles of rapid movements: a comparative study. Biol. Cybern. **69**(2), 119–128 (1993)
19. Plamondon, R., O'reilly, C., Galbally, J., Almaksour, A., Anquetil, É.: Recent developments in the study of rapid human movements with the kinematic theory: applications to handwriting and signature synthesis. Pattern Recognit. Lett. **35**(SI), 225–235 (2014)
20. Plamondon, R., O'Reilly, C., Ouellet-Plamondon, C.: Strokes against stroke–strokes for strides. Pattern Recogn. **47**(3), 929–944 (2014)
21. Plamondon, R., Parizeau, M.: Signature verification from position, velocity and acceleration signals: a comparative study. In: 9th International Conference on Pattern Recognition, pp. 260–261. IEEE Computer Society (1988)
22. Rubin, D.B.: Inference and missing data. Biometrika **63**(3), 581–592 (1976)
23. Sae-Bae, N., Memon, N.: Online signature verification on mobile devices. IEEE Trans. Inf. Forensics Secur. **9**(6), 933–947 (2014)
24. Van Gemmert, A., Plamondon, R., O'Reilly, C.: Using the sigmalognormal model to investigate handwriting of individuals with Parkinson's disease. In: Proceedings of 16th Biennial Conference on the International Graphonomics Society, Nara Japon, pp. 119–122 (2013)

Human Document Interaction

Near-Perfect Relation Extraction from Family Books

George Nagy$^{(\boxtimes)}$ (iD)

Rensselaer Polytechnic Institute, Troy, NY 12180, USA
nagy@ecse.rpi.edu

Abstract. Precise sequence constraints are proposed to accelerate information extraction from a class of "semi-structured" documents that includes hundreds of thousands of digitized genealogical records. While Named Entity Recognition (NER) and Named Relation Recognition (NRR) on free-running text lack universally applicable solutions, under these constraints generalized template-matching can accomplish both. Interactive information extraction is demonstrated on three digitized books. The book-text tokens are first labeled according to their role (e.g. Head, Spouse, or Birthdate), then pairs of labeled entities are combined into labeled relations (e.g. <Head [Spouse]>, or <Spouse [Birthdate]>). Accurate NRR is ensured by high-precision (>99%) NER. On semi-structured text the proposed NRR algorithm produces only valid relations from correctly labeled entities. About three hours of user interaction and a few minutes of laptop run time suffice to produce database- or ontology-ready input from a new book.

Keywords: Information extraction · Text analysis · Language models

1 Introduction

In response to the need for less laborious recovery of genealogical facts from printed family records, we present a model of semi-structured text that leads to simple and accurate information extraction by generalized template matching. Our program processes Unicode text with no formatting other than line breaks. The user needs to specify only a few exemplary templates and a list of the relations to be extracted.

Semi-structured family books typically contain interspersed *sign phrases* (like *born, died, spouse, son of*, or *dau. of*) and *value phrases* (*Henrietta Mills Hyde* or *1828*). A value may be located several lines away from its sign and span a variable configuration of tokens. The *GreenEx* collection of python modules scours the text for data needed to populate genealogical databases and ontologies.

The first pass over the text, *Named Entity Recognition* (NER), attaches each label (HEAD, SPOUSE, BIRTHDATE...) to its value. The second pass, *Named Relation Recognition* (NRR), aka *Semantic Relation Extraction,* links the labeled values into binary relations like <HEAD[CHILD]: Henry Hyde [Henrietta Mills Hyde]> or, from a subsequent record of Henry's daughter Henrietta's family, <SPOUSE [MARRIAGE-DATE]: Hyde Charles Smith Shelton, [1848]>. We tested the method on the 18th century

© Springer Nature Switzerland AG 2021
J. Lladós et al. (Eds.): ICDAR 2021, LNCS 12823, pp. 477–491, 2021.
https://doi.org/10.1007/978-3-030-86334-0_31

Kilbarchan parish register [1], the early 20th century Miller funeral home records [2], and The Ely Ancestry that spans three centuries [3] (illustrated in the Appendix).

The contributions we demonstrate are (1) a set of constraints imposed on text by the desired relations, and (2) a generalized template matching algorithm that extracts the specified relations from any text subject to the constraints. The experimental results confirm that the permissive semi-structure constraints obeyed by three diverse family books suffice for fast and accurate relationship extraction.

The next section is a review of relevant prior work. Section 3 defines the proposed semi-structure constraints. The following sections present Sect. 4. entity labeling, Sect. 5. relation extraction, Sect. 6. experimental results and Sect. 7. conclusions.

2   Prior Work

We discuss here only properties of relations of interest in information extraction. Among the pioneering achievements before 2000 were The Acquisition of Hyponyms [4], the NYU Proteus System (later extended to news, scientific papers and patents) [5], and Snowball for finding patterns in plain text [6].

A popular survey of relation extraction up to 2006 is Bach's and Badaskar's [7]. Their taxonomy, based on the amount of required human interaction and of relevant data, has stood the test of time. Zettlemoyer offers a lively introduction to both NER and NRR [8]. Dependency tree methods derive distance measures from grammatical relations between tokens [9]. The value of extending seed lists with unlabeled data is demonstrated in [10]. Joint NER-NRR was initiated in 2006 [11]. *Distant supervision* combines the advantages of supervised and unsupervised approaches by exploiting linguistic resources only indirectly related to the searched text [12, 13]. It is the focus of an authoritative CSUR review [14]. Approaches based on neural networks came along a little later [15]. Chen and Gu review NRR research up to 2019, catalog the shortcomings of existing methods, and compare their probabilistic joint NER-NRR to other methods on several biomedical benchmarks [16]. Many papers address only binary relations because in NER useful n-ary (n > 2) relations can often be factorized into binary relations [17]. *NLP-Progress*, a website that tracks developments in natural language processing, lists test data and competitions by language, model, and application [18].

A category-based language model is compared with a probabilistic finite-state machine model for labeling family roles in handwritten 17th Century Catalan marriage records in [19]. With a large fraction (6/7) of the 173-pages used for training, and seven-fold cross-validation, both methods yielded 70–80% Precision and Recall. Information extraction was also the topic of a 2017 ICDAR competition. Using neural networks and conditional random fields, the winning team (from Harbin Institute of Technology) achieved a remarkable character error rate of ~8% on the same database [20], but 100 of the 125 pages had to be manually labeled for training and validation. Combining NER/NR with HWR or OCR appears to be unique to the DAR community.

Our example-based approach is similar in spirit to end-user-provided training examples for scanned business documents [21]. Literals and semantic tags were anticipated in [22]. The effects of OCR errors on information retrieval were discussed in [23]. Recent shifts in the very nature of documents were reviewed in [24]. The popular Stanford

Named Entity Recognizer [25] failed on our books because it depends on probabilistic sentence analysis, but for tokenization we use the Natural Language Tool Kit (NLTK) that it spawned. Preceding the rapid rise of deep learning methods, rule-based extraction like ours was favored over machine learning in [26]. It remains to be seen whether a machine learning approach applied to semi-structured text can match generalized template matching with respect to user time, minimal training data, and accuracy.

The work closest to ours is that of the BYU and FamilySearch research team, which has access to 460,000 digitized publications of genealogical interest [27]. They describe in [28] a pipeline based on conceptual modeling for possible integration into Family-Search's Family Tree. In [29] they present a thorough review of recent research on information extraction for genealogical purposes as well as some experiments on the same books as we used. The BYU team also proposed automated discovery of errors (like inconsistent dates) in sources of data [30]. They reported recovering family information from obituaries [31] and from lists abstracted from family books [32].

GreenEx was initiated to accelerate the construction of character-level REGEX templates at BYU [33]. The first versions of GreenEx lacked floating extracts, format variants and auto-suggestion routines and failed to exceed a Figure-of-Merit of 0.95 [34, 35]. (*Green* is for pattern recognition programs that never waste a user action, from [36]). Generalizing template matching yielded much better (~99%) entity recognition with fewer templates [37]. The development of accurate entity extraction on semi-structured genealogical records laid the foundation for relation extraction. We found no formal examination of semi-structure in the literature, and no comparable experiments.

3 Structure Constraints

Informally, *semi-structured documents* are lists of quasi-repetitive records where some tokens can be designated as signs or values of the information to be extracted. The, *sign* (a semiotic term) is also called *marker, query,* or *search phrase,* and *value* is *extract* or *target.* The subject of a record is called *Head.* Each f*amily record* is a sequence of phrases (*factoids*) about the Head's family. Figure 1 includes two family records (in lines 9–16 and 17–23) that we will use as a running example. Each record begins with a sign for the Head (here a six-digit record identifier) and extends to the first token of the next Head sign. In other books, the sign for Head is the name itself at the beginning of a line.

We define a binary relation R of *type* Label_1 [Label_2] as a pair of values (v_1 [v2]). An *instance* of a relation, from lines 10 and 11 of Fig. 1, is:

$R_{\text{SPOUSE [BIRTHDATE]}} = \ <\text{SPOUSE [BIRTHDATE]};$ *Charles Smith Shelton* [*1819*]>

GreenEx reports every instance of $R_{\text{SPOUSE [BIRTHDATE]}}$.

We need some notation for the signs and values of R in semi-structured text. A sequence number (*SeqNo*) N, ranging from 1 to the number of word tokens in the book, is assigned to each word token. Let *s1, s2, v1,* and *v2* denote the signs and values designated to extract instances of relation R. Let $N(s1)$ denote the SeqNo of the first token of *s1,* and $N(s2)$ that of *s2.* Let *N0* be the SeqNo of the first token of the family record that contains *v1,* and *N1* the SeqNo of this record's last token. Figure 2 shows plausible signs and values for the genealogical factoids of a record in Fig. 1.

1.	THE ELY ANCESTRY. 423
2.	SEVENTH GENERATION.
3.	b. 1826, d. 1857, son of Daniel Havens and Desire Holmes; m. 3rd,
4.	1862, Herbert Post, Marion, Ala., who was b. 1827, son of Truman Post
5.	and Betsy Atwater. Their children:
6.	1. Robert Alexander, b. 1846; m. 1869, Katherine Pierce Parker.
7.	2. Julia Hyde, b. 1855.
8.	3. Etta Hyde, h. 1856, d. 1857.
9.	243357. Henrietta Mills Hyde (127 St. James Place, Brooklyn, N.
10.	Y.), b. 1826, dau. of Julia Ely and Zabdial Hyde; m. 1848, Charles
11.	Smith Shelton, Madura, India (missionary), b. 1819, d. 1879, son of
12.	George Shelton and Betsy Wooster. Their children :
13.	1. Fanny Arabella, b. 1850; m. 1874, Arthur Harry Bissell.
14.	2. Julia Elizabeth, b. 1851 ; m. 1878, Chas. J. Van Tassel.
15.	3. Charles Henry, b. 1854, M.D., 288 Fourth St., Jersey City, N. responds.
16.	4. Henry Hyde, b. 1858.
17.	243358. Aurelia Carrington Hyde (127 St. James Place, Brooklyn,
18.	N. Y), b. 1828, dau. of Julia Ely and Zabdial Hyde; m. 1848, Edward
19.	Chauncey Halsey (Directory), 165 Warren St., Brooklyn, N. Y., who
20.	was b. 1825. Their children :
21.	1. Eleanor Shelton, b. 1850.
22.	2. Adeline Sanford, b. 1852.
23.	3. Edward Carrington, b. 1854.
24.	243359. Zabdiel Sterling Hyde (care E. Goddard & Sons, 1020

Fig. 1. Part of a page of OCR'd text from the Ely Ancestry (line numbers and highlight for record # 243357 added)

243357. *Henrietta Mills Hyde* (127 St. James Place, Brooklyn, N. Y.), **b.** *1826*, **dau. of** *Julia Ely* and *Zabdial Hyde*; **m.** *1848*, *Charles Smith Shelton*, Madura, India (missionary), **b.** *1819*, **d.** *1879*, **son of** *George Shelton* **and** *Betsy Wooster*. Their **children** :
1. *Fanny Arabella*, **b.** *1850*; **m.** *1874*, *Arthur Harry Bissell*.
2. *Julia Elizabeth*, **b.** *1851* ; **m.** *1878*, *Chas. J. Van Tassel*.
3. Charles Henry, **b.** *1854*, M.D., 288 Fourth St., Jersey City, N. J., responds.
4. *Henry Hyde*, **b.** *1858*.

Fig. 2. Family record for Henrietta Mills Hyde, from the page shown in Fig. 1. Potential signs and values are colored green (bold) and red (italicized) respectively. (Color figure online)

Then the structure constraints on the text that suffice for a relation to be extractable are:

1. $N0 \leq min(N(s1), N(s2), N(v1), N(v2))$, and $max(N(s1), N(s2), N(v1), N(v2)) \leq N1$ (*every sign and value participating in a relation must be in the same family record*)
2. $N(s1) \leq N(v1)$ and $N(s2) \leq N(v2)$ (*a value cannot precede its sign*)
3. $N(s1) \leq N(v1) \leq N(s2) < N(v2)$ (*signs and values alternate, except for collocations indicated by equality for "$\leq$", where the sign and value share some or all tokens*)

4. If a relation *R1* associates *v1* and *v2*, then there cannot be a relation *R2* of the same type as *R1* between *v1* and *v2* (*relations of the same type cannot be nested*)

The following examples would violate these constraints:

Constraint #1:   243357. Henrietta Mills Hyde b. 243357. 1826 Abel
Constraint #2:   243357. Henrietta Mills Hyde 1826 b. 243357.
Constraint #3:   43357. b. Henrietta Mills Hyde 1826 243357.
Constraint #4:   Henrietta Mills Hyde Fanny Arabella b. 1850 **b. 1826**

Constraint #1 implies that extracting inter-record relations requires further processing. #2 can be obviated with a reverse second pass. According to #3, either *m. 1848, Charles Smith Shelton* or *m. Charles Smith Shelton 1848* would be acceptable, with *m* serving in either case as the sign for both the *marriage date* 1848 and the *spouse* Charles Smith Shelton. This constraint occasionally requires some ingenuity in formulating the appropriate template (e.g. for twins with a single birth date). We have never seen a violation of #4.

The family record of Henrietta Mills Hyde (Fig. 2) is semi-structured with respect to every relation listed in Table 3 of Sect. 6. Semi-structure is a substitute for sentence structure to aid human comprehension.

4 First Level Template Matching

Generalized template matching achieves high precision entity recognition with few templates by (1) application-oriented (but not document-specific) *word tagging* based on alphanumeric format, (2) substituting common alternative noun and date configurations (*format variants*) for the ones specified in the template, and (3) extending the search for the value corresponding to a sign (*floating templates*) [37].

Recall is further improved by *auto-suggestion routines* that scan the book for tokens that should have been labeled but were not. Common causes of unlabeled tokens are unusual word configurations (like *John, in adultery 1675* instead of the expected *John, born 1675*), typesetting errors (often in punctuation), and OCR errors (*I* instead of *1*) that affect tagging. When the inconsistent text segments are displayed on a clickable form, the user can add a template that will correct the current error as well as similar errors elsewhere in the text [37].

Adding half-a-dozen templates (for each book) based on the suggestion routines raised recall by about a half percent with insignificant change in precision. Half percent is not negligible at Recall >98%. However, the effectiveness of adding templates gradually decreases: eventually each new template will correct only one or two errors.

When all recognizable tokens have been labeled, GreenEx assembles consecutive same-label tokens and their locations into *extract groups*. Then the extract groups of each family, bracketed by HEAD labels, are collected into *labeled family records*. The book-length list of labeled family records (e.g. Fig. 3) is the input to relation extraction.

...HEAD:,578,10,9,3,Henrietta,Mills,Hyde,B_DATE:,578,11,9,1,1826,PARENT1:,
578,11,23,2,Julia,Ely,PARENT2:,578,11,37,2,Zabdal,Hyde,M_DATE:,578,11,54,1,
1848,SPOUSE:,578,11,60,3,Charles,Smith,Shelton,B_DATE:,578,12,47,1,1819,
D_DATE:,578,12,56,1,1879,PARENT1:,578,13,1,2,George,Shelton,PARENT2:,
578,13,20,2,Betsy,Wooster,CHILD:,578,14,4,2,Fany,Arabella,B_DATE:,578,14,23,1,
1850,M_DATE:,578,14,32,1,1874,SPOUSE:,578,14,38,3,Arthur,Harry,Bissell,
CHILD:,578,15,4,2,Julia,Elizabeth,B_DATE:,578,15,24,1,1851,M_DATE:,578,15,34
,1,1878,SPOUSE:,578,15,40,5,Chas,.,J.,Van,Tsel,CHILD:,578,16,4,2,Charles,Henry,
B_DATE:,578,16,22,1,1854,CHILD:,578,17,4,2Henry,Hyde,B_DATE:,578,17,19,1,
1858,HEAD:, ...

Fig. 3. Labeled family record of Henrietta Mills Hyde, including book coordinates: page, line, character and length (number of tokens)

5 Relation Extraction

Named Relation Recognition has been intensively studied for thirty years without finding a universal solution. The complexity of natural language requires complex language models, many training examples, or external resources for avoiding misses and errors. Our main point is that the task is much easier for text semi-structured with respect to the desired relations because every relation to be extracted is fully defined by the *labels* of the participating entities. Figure 4 shows pseudo-code for binary relation extraction from labeled family records.

In our notation, A[B] stands for a unique tuple within a family. Therefore relations like HEAD[CHILD] and CHILD[BIRTHDATE] must be understood as HEAD [FIRST-CHILD] or HEAD[SECOND-CHILD] and FIRST-CHILD[BIRTHDATE] or SECOND-CHILD[BIRTHDATE]. (N-ary relations are also restricted to *unique* tuples, and can therefore always be decomposed into dyads. Some authors exclude such tuples from the definition of n-ary (n > 2) relations.)

GreenEx extracts all the specified relations from one labeled family record at a time. In contrast to the NER pass, there is no need for tagging, format variants, provisions for line-ends and page breaks, or interactive template construction. The signs and values can only be the algorithmically assigned labels that define the relations. For a desired <Label_X [Label_Y]> relation, the program just loops over the label groups in each family to locate a Label_X group and the next Label_Y group. The values of these label groups constitute the sought relation. Therefore errors in relations can occur only when one of the constituent tag phrases was mislabeled or unlabeled in the NER pass.

Constraint #1 is satisfied by limiting the search for a value to the current family record. Restarting the search at the current sign satisfies *Constraint* #2. Halting the search before the next identical sign satisfies both #3 *and* #4. The stopping rules convert potential Precision errors in relation extraction (due to labels missed in the NER pass) to Recall errors. Therefore if the labels are correct, and the labeled text satisfies the constraints imposed by the specified relations, then only valid relations are extracted.

Twenty-six types of binary relations that can be extracted from our family books, under the semi-structure constraints on the current labels, are listed in Table 3. Shown below are some relations extracted from our running example. (We envy mathematicians, whose notation for a binary relation is typically (a,b). In the first relation below, a is *Henrietta,Mills,Hyde* and b is *Charles,Smith,Shelton*). The following examples of extracted relations show relation type, book locations (page, line, token) of the values, and the values themselves. The book locations are *attributes* of the relations.

Two examples of an instance of a user-specified binary relation:

<HEAD   [SPOUSE]   57810   9,   578   11   60   Henrietta,Mills,Hyde [Charles,Smith,Shelton]>

   <CHILD [SPOUSE] 578 15 4, 578 15 40 Julia,Elizabeth [Chas,.,J.,Van Tassel]>

```
Function Grelex (FamilyRecords, DesiredTuples)
   Input:  FamilyRecords, DesiredTuples
   Output: ExtractedRelations

Convert FamilyRecords to Families      % Family Records is a book-length list
                                          of labeled groups of tokens
For Family in Families:      % Family is a list of Groups, each [Label, Page, Line,
      Offset, Value], in a single Family; Families is a book-length list of Family(s)
   For XGroup in Family            % restrict search to this Family
      Excerpt XLabel, XValue, XID from XGroup
                           % XID_ is Page, Line, Offset of XGroup's value
      For Tuple in DesiredTuples:  % for every specified relation
         Excerpt LeftLabel, RightLabel from Tuple  % set the two search arguments
         StopGroup ← Stopper(LeftLabel, RightLabel)
                           % StopGroup is a context-dependent label
         If XLabel =LeftLabel     % if an entity label matches left label of this tuple
            For YGroup in Family from XGroup to StopGroup :
                           % search forward to find a match for the right label
               If YLabel = RightLabel   % if a match for right label is found
                  Excerpt YLabel, YValue, YID from YGroup
                  Relation ← [[XLabel, YLabel, XID YID],[XValue],[YValue]]
                           % extract this instance of the desired relation
                  Append Relation to ExtractedRelations
               End If YLabel = RightLabel
            End  For YGroup
         End  If XLabel =LeftLabel
      End  For Tuple
   End  For XGroup
End  For Family
Return(ExtractedRelations)     % ExtractedRelations is a list of attributed relations
         of the form Label_1[Label_2], , ID_1, ID_2, Value_1[ Value_2]
```

Fig. 4. Simplified pseudocode for extracting binary relations from family records. Code for stopping rules with provisions for multiples (e.g. spouses, children, twins) and alternatives (birth-date, christening date) not shown.

The derivation of a decomposable n-ary relation from its constituent binary relations is straightforward. A recursive GreenEx routine factors each n-ary relation into binary relations, e.g. A[B[C]] into A[B] and B[C], or A[B,C] into A[B] and A[C]. The program then fills the slots of the n-ary relation with the elements of the already extracted binary relations. Many important applications (e.g. populating relational databases and Resource Description Framework RDF triples) require only binary relations. The experiment below is confined to attributed binary relations. Two examples of an instance of a user-specified n-ary relation:

<HEAD [CHILD, [B_DATE]] 578 10 9, 578 15 40, 578 15 24; Henrietta,Mills,Hyde [Julia,Elizabeth [1803]]>

<HEAD, [SPOUSE [[PARENT1], [PARENT2]]] 400 13 08, 400 14 32, 400 15 27, 400 15 48; Henrietta,Mills,Hyde [Charles,Smith,Shelton [[George,Shelton], [Betsy,Wooster]]]>

6 Experimental Results

Table 1 shows the results of processing the three books, and the accuracy on the manually labeled test data. The tokens labeled "NONE", like addresses, occupations, military ranks, officiating clergy, and the names of informants, were excluded from the labeled family records. They were included as an additional class in the precision and error calculations. In two of the books Precision is 99.9%, and in the third it is 99.7%. We note, however, that first-stage labeling failure of a single shared value could cause missing several relations. Table 2 summarizes the results of relation extraction. Table 3 displays the number of relations extracted from each book by relation type.

We did not find any instance of the 26 relations listed that failed to obey the constraints. All missed relations in the ground-truthed pages were due to OCR errors or misprints (but some were caught by the suggestion routines). Extracting almost all of

Table 1. Data characteristics and accuracy

	Kilbarchan	Miller	Ely
Pages processed	139	389	301
Specified labels	8	10	8
Tokens assigned specified labels	39203	91633	39440
Tokens labeled NONE	34077	131462	101485
Test Set with ground truth			
Pages	6	6	6
Tokens (including "NONE")	3126	3842	3423
Precision	0.999	0.999	0.997
Recall	0.981	0.991	0.992
F-measure	**0.990**	**0.996**	**0.994**

the desired information from a new book takes less than three hours of interactive template construction and only a few minutes runtime on a 2.4-GHz Dell Optiplex 7010 with 8-GB RAM running Python 3.6 with IDLE under Windows 10.

Table 2. Summary of relation extraction

	Kilbachan	Miller	Ely
Types to extract	26	26	26
Types found	15	17	16
Families	2615	4186	1219
Extracted Groups	14152	38146	20781
Extracted Relations	**11487**	**25633**	**18780**

Table 3. Number of extracted relations of each type

Relation	Kilbarchan	Miller	Ely
HEAD[CHILD]	4076	837	3448
HEAD[TWINS]	47	0	0
HEAD[SPOUSE]	2102	1730	1306
HEAD[B_DATE]	6	2849	1128
HEAD[M_DATE]	963	0	1020
HEAD[D_DATE]	0	4119	374
HEAD[BU_DATE]	0	3524	0
HEAD[B_PLACE]	0	2798	0
HEAD[M_PLACE]	134	0	0
HEAD[BU_PLACE]	0	4053	0
HEAD[PARENT1]	0	2824	1203
HEAD[PARENT2]	0	2726	1210
SPOUSE[B_DATE]	3	9	1003
SPOUSE[M_DATE]	12	0	136
SPOUSE[D_DATE]	0	8	360
SPOUSE[B_PLACE]	0	10	0
SPOUSE[M_PLACE]	12	0	0
SPOUSE[PARENT1]	0	9	1015
SPOUSE[PARENT2]	0	7	1028
CHILD[B_DATE]	1143	43	3282
CHILD[C_DATE]	2873	0	0
CHILD[D_DATE]	0	82	831
CHILD[SPOUSE]	34	5	819
CHILD[M_DATE]	38	0	617
TWINS[B_DATE]	6	0	0
TWINS[C_DATE]	38	0	0
TOTAL	**11487**	**25633**	**18780**

7 Conclusion

What we learned from the experiments is the unexpected simplification of Named Entity Recognition and Named Relation Extraction enabled by appropriate characterization of semi-structured text. The constraints listed in Sect. 3 proved just tight enough to allow generalized template recognition to yield much higher precision and recall on both tasks than reported by others on free-flowing test data, and loose enough to fit the three diverse books recommended to us for testing. These books differed not only from each other because of purpose and date, but also internally because they were compiled over several lifetimes by many authors.

Particularly gratifying was the discovery that template matching enables error-free binary relation extraction on correctly labeled semi-structured text. This is assured because template matching just maps the extracted entity labels into a highly redundant list of relations without using or introducing any external information. No such claim can be made for free-flowing text.

Template matching is linear in the length of the input, so checking text compliance with the rules directly would be only slightly faster than running GreenEx. Template construction is necessarily book-specific. For example, in Miller, *m* indicates *mother*, but in Kilbarchan and Aly it points to *married*. Fortunately, we were able to show in earlier papers that customizing the system to each book (with appropriate computer help and a user-friendly interface) requires surprisingly little human interaction. We expect, but have not proved, that the skill level required is within reach of most current users of genealogical software.

No machine learning was tried or compared. With all F-scores $\geq 99\%$, what could any comparison on the same data prove? Avoiding the laborious preparation of training data is the main point of the proposed approach. No comparison with statistical classifiers, including deep learning, can contest that. From the perspective of genealogists it seems more urgent and useful to determine what fraction of the plethora of family books obeys the postulated semi-structure constraints.

The most serious potential error at the NER stage is a missed Head. This could assign a Child, a Parent, or a Spouse to the preceding Head, and consequently yield some incorrect relations. Although there were some OCR errors on Heads in our test set, they were all caught by the auto-suggestion routines.

The results could be filtered by genealogy-specific checks to detect missing names (every person with some attribute must have one), missing birthdates (in Ely, every Head has one), more than two parents, inconsistent birth, marriage and death dates, and other definite or suspect genealogical inconsistencies. We don't, however, have any dependable method for automatic correction of detected errors. We expect the most significant advance to come from larger scale projects that combine results from multiple genealogical sources covering the same community.

Only part of the simple tagging routine in GreenEx is specific to English family books. Therefore the proposed method could perhaps be extended to other semi-structured books of historical interest like city directories and product or merchandise catalogs, and to other languages and scripts. As it stands, the only contribution claimed is a simple and effective method of named relation extraction from family books.

Acknowledgment. I am grateful to Emeritus BYU Professor David E. Embley and his colleagues for the digitized family books and for their sustained interest, advice and critique. The cogent suggestions of the three ICDAR reviewers prompted appropriate revisions.

Appendix I. Sample of Text from the Kilbarchan Parish Register

```
Parish of Kilbarchan.
Adame, Robert, par., and Issobell Adame, par. of Loch-
winnoch, in Pennell 1679 m- 2I Mar. '678
A daughter, 30 Mar. 1679.
Adam, William, par., and Elizabeth Alexander, par. of Paisley
m. Paisley, 15 May 1650
Adamson, Alexander, in Kilbarchan, and Mary Aitken p. 12 Feb. 1763
Mary, born 16 Oct. 1763.
David, born 1 May 1765.
Aird, William, and Margaret Aitken, in Auchincloigh
Margaret, 9 Feb. 1707.
Aitken (Akin), and Elspa Orr m. 18 Dec. 1693
Aitken, Allan, and Mary Aitken
Agnes, 10 May 1741.
Aiken, David, and Janet Stevenson m. 29 Sept. 1691
Aitkine, Thomas, and Geills Ore m. 21 Dec. 1661
W. Richard Allasone and Ninian Aitkine.
Aikine, James, and Jean Allason, in Ramferlie, 1696 in Kaimhill
m. 23 Jan. 1679
John, 28 Nov. 1679.
William, 28 Aug. 1681.
Isobel, 12 July 1691.
Thomas, 19 Jan. 1696.
Allan, 19 July 1698.
Elizabeth, 26 May 1701.
Aitken, James, in Sandholes, and Mary Henderson p. 10 July 1741
John, 26 Dec. 1742.
James, born 28 Sept. 1744.
Robert, born 12 May 1747.
Matthew, born 11 April 1749.
William, born 22 April 1756.
Aitken, James, and Janet Moodie
Elizabeth, 25 July 1742.
Aitken. James, in Abbey par. of Paisley, and Janet Lyle, par.
p. 7 June 1755
Aitken, James, in Kilbarchan, and Jane Lindsay
Jane, born 4 July 1755.
Margaret, born 9 Sept. 1757.
John, born 21 Oct. 1762.
James, born 26 Oct. 1764.
Janet, born 14 April 1768.
Aitken, James, in Lochermiln, and Janet Gardner 1760 in Barbush
Mary, born 15 May 1758.
Christian, born 1 May 1760.
Janet, born 19 Jan. 1764.
Robert, born 28 May 1766.
Jane, born 28 Aug. 1768.
```

Aitken, James, par., and Janet Houstoun, par. of Houstoun p. 11 Jan. 1772
Aitkine, John, par., and Janet Muire, par. of Paisley
m. Paisley, 22 Oct. 1650
Akine, John, 1655 in Todhils
William, 15 Oct. 1652.
James and William, 9 April 1654.
Jonet, 1 July 1655.
Margaret, 1 May 1659.

Appendix II. Sample of Text from the Miller Funeral Home Records

ABERNATHY, ELMER d 4 April 1924 252 Bellevernon Ave BD Greenville Cem 6 Apr
1924 b 5 Oct 1863 age 60-5-29 pd by ELLEN ABERNATHY
ACCETTE, FRANK d 16 Oct 1942 Friday 3:15p.m. Greenville Dke Co OH BD Oct 1942
Abbottsville Cem Dke Co OH b 20 April 1897 Montreal Canada age 45-5-26
f JOSEPH ACCETTE m AGNES QUEIRLLON waiter in restaurant sp & informant
LENA ACCETTE 405 Central Ave physician Dr Mills religion Catholic
War record: enlisted 21 Feb 1918 disch 19 Aug 1919 World War I Canadian
Expeditionary Force Army 2nd Depot Batt C.O. Reg . Services Catholic
Church clergy Father Gnau
ADAMS, ADAM DANIEL d 30 Aug 1931 Miami Valley Hosp Dayton OH BD Castine Cem
2 Sept 1931 b 18 May 1872 Preston Co WV age 59-3-12 f COLEMAN AD.AMS
Barber Co WV m RACHAEL BOWMAN Barber Co WV married farmer
ADAMS, ANNA E. 2215 Rustic Road Dayton OH d 24 Aug 1942 Monday 2:45a.m. Dayton
Montgomery Co OH BD 26 Aug 1942 Frankli n Cem OH b 7 Feb 1856 Franklin OH
age 86-6-17 f DAVID ADAMS single housekeeper informant Mrs LOUIS MEYERS
2215 Rustic Road Dayton OH physician Dr Sacks clergy Rev Jones Dayton OH
services Baptist Church in Franklin OH
ADKINS, HESTER d 6 Nov 1925 Weaver's Station BD Fort Jefferson Cem 8 Nov 1925
age 84-3-9 chg to RILEY ADKINS, pd by JAMES A. ADKINS
ADKINS, JAMES ALEXANDER d 2 Sept 1944 Wayne Hosp Greenville OH BD 4 Sept 1944
Fort Jefferson Cem Dke Co OH b 19 June 1872 Vandalia IL age 72-2-13
f RILEY ADKINS Dke Co OH m HESTER McCOOL retired rural mail carrier
sp CORA ADKINS 65 years sisters Mrs MARY VIETS Dayton & Mrs CLATE RIEGLE
Fort Jefferson
AIKEY, JACOB CLARENCE d 2 Oct 1937 N.W. of Pikeville 1~ mile BD Oakland Cem
5 Oct 1937 b 14 Dec 1855 Union Co PA age 81-9-18 f THOMAS AIKEY Maine
m ALVINA KATHERMAN married retired farmer sp LYDIA
AIKEY, LYDIA ANN d 27 Aug 1925 7~ mile N.E. of Greenville BD Oakland Cem 30
Aug 1925 age 60-9-23 chg to JACOB AIKEY
ALBRIGHT, ADAM C. d 28 June 1920 Piqua OH hosp BD Abbottsville Cem 1 July 1920
age 72-7-27
ALBRIGHT, CARL ROLAND d 21 June 1917 VanBuren Twp BD Abbottsville Cem 23 June
1917 b VanBuren Twp age 11-2-10 f ALLEN ALBRIGHT m ANNA WEAVER
ALBRIGHT, CATHARINE d 10 May 1930 4 mile S.W. BD Greenville Mausoleum 13 May
1930 age 94-5-20 pd by DAYTON & CHAS ALBRIGHT
ALBRIGHT, ESTHER R. d 1 Jan 1946 113 Sherman St Dayton OH BD Abbottsville Cem
Dke Co OH 3 Jan 1946 b 22 July 1863 Butler Co OH age 82-6-9 f THOMAS
BENTON MORRIS Butler Co OH m ANGELINE HARROD Hamilton Co OH housekeeper
widow sp WINFIELD S. ALBRIGHT 1 daughter Mrs HENRY RANCH 4 sons HENDER-
SON of Greenville WILBUR of Greenville GEO of Dayton ELBERT of Dayton
12 grandchildren 2 brothers ARTHUR MORRIS Venice OH & SAM MORRIS Harrison
OH 2 sisters Miss ELLA MORRIS Greenvi lle OH & Mrs ADA HARP Tulsa OK
1

Appendix III. Sample of Text from the Ely Ancestry

```
422 THE ELY ANCESTRY.
SEVENTH GENERATION.
243331- James Joseph Ernest Ely, son of Elisha Mills Ely and
Catherine Elizabeth Boode; m. Anna Horloff. Their children:
1. Alphonse.
2. August.
3. Alice.
4. Alfred.
243332. Alphonso Ethelbert Mills Ely, Palmyra, Mo., b. 1821, son
of Elisha Mills Ely and Catherine Elizabeth Boode; m. 1841, Drusilla
Pinkston, Palmyra, Mo., who was b. 1820, dau. of Peter Pinkston and
Abig-ail Davis. Their children :
1. Laura Ann Catherine, b. 1842.
2. Emma McLellan, b. 1850.
3. Alphonse Ethelbert Mills, b. 1852.
4. Mary Bailey, b. 1855.
5. Ophelia Goldburg, b. 1861.
243351. Elizabeth Plummer Hyde, b. 1814, d. 1855, dau. of Julia
Ely and Zabdial Hyde; m. 1834, Robert McClay Henning (243362X),
who was b. 1812, d. 1875, son of James Gordon Henning and Alicia
Courtney Spinner. Their children:
1. James Spencer, b. 1835.
2. Julia Ely, b. 1837; m. 1854, Robert Pearce (Mrs. Julia Ely Pearce, 58
St. John's PL, Brooklyn).
3. Edwin Courtney, b. 1838.
4. Henrietta Mills, b. 1841. (Her son is Dr. Chas. H. Shelton, 288 Fourth
St., Jersey City).
5. Elizabeth Alicia, b. 1843.
6. Robert McClay, b. 1847.
7. James Woodruff, b. 1850.
243353. Edwin Clark Hyde (13 Warren St., St. Louis), b. 1819,
son of Julia Ely and Zabdial Hyde ; m. 1844, Elizabeth Ann Peake (Gor-
don), who was b. 1816, dau. of Henry Peake and Isabella Herring
( Snyder) . Their children :
1. Henrietta Mills, b. 1845, d. 1848.
2. Susan Isabella, b. 1847.
3. Samuel Peake, b. 1850.
4. Annie Carroll, b. 1851, d. 1857.
5. Allen Withers, b. 185S, d. 1856.
243356. Julia Ely Hyde, Marion, Perry Co., Ala., b. 1824, 'dau. of
Julia Ely and Zabdial Hyde; m. 1844, Alexander Clark Bunker, who
(vas b. 1822, d. 1846, son of Thomas Bunker and Sally Raymond; m.
2nd, 1854, Washington Holmes Havens, St. Francisville, Mo., who was
```

References

1. Grant, F.J. (ed.): Index to the Register of Marriages and Baptisms in the PARISH OF KILBARCHAN, pp. 1649–1772. J. Skinner & Company, Ltd, Edinburgh (1912)
2. Miller Funeral Home Records, 1917–1950, Greenville, Ohio. Darke County Ohio Genealogical Society, Greenville (1990)
3. Vanderpoel, G.: The Ely Ancestry: Lineage of RICHARD ELY of Plymouth, England. The Calumet Press, New York (1902)

4. Hearst, M.A.: Automatic acquisition of hyponyms from large text corpora. In: Proceedings of the Fourteenth International Conference on Computational Linguistics, Nantes, France (1992)

5. Grishman, R., Sterling, J., Macleod, C.: Description of the Proteus system as used for MUC-3. In: Proceedings of the Third Message Understanding Conference, San Diego, CA, pp. 183–190 (1991)

6. Agichtein, E., Gravano, L.: Snowball: extracting relations from large plain-text collections. In: Fifth ACM Conference on Digital Libraries, San Antonio, TX (2000)

7. Bach, N., Badaskar, S.: A Review of Relation Extraction (2006). https://www.cs.cmu.edu/~nbach/papers/A-survey-on-Relation-Extraction.pdf

8. Zettlemoyer, L.: Relation Extraction (2013). https://docplayer.net/31229549-Relation-extraction-luke-zettlemoyer-cse-517-winter-2013.html

9. Culotta, A., Sorensen, J.: Dependency tree kernels for relation extraction. In: Proceedings of the 42nd Annual Meeting of the Association for Computational Linguistics, Barcelona, pp. 423–429 (2004)

10. Talukdar, P.P., Brants, T., Liberman, M., Pererira, F.: A context pattern induction method for named entity extraction. In: Proceedings of the 10th Conference on Computational Natural Language Learning (CoNLL-X), New York City, pp. 141–148 (2006)

11. Choi, Y., Brock, E., Cardie, C.: Joint extraction of entities and relations for opinion recognition. In: Proceedings of the 2006 Conference on Empirical Methods in Natural Language Processing (EMNLP 2006), pp. 431–439, July 2006

12. Zettlemoyer, L.: Advanced Relation Extraction (2013). https://cs.nyu.edu/courses/spring17/CSCI-GA.2590-001/DependencyPaths.pdf

13. Min, B., Grishman, R., Wan, L., Wang, C., Gondek, D.: Distant supervision for relation extraction with an incomplete knowledge base. In: Human Language Technologies: Conference of the North American Chapter of the Association of Computational Linguistics, pp. 777–782 (2013)

14. Smirnova, A., Cudré-Mauroux, P.: Relation extraction using distant supervision: a survey. ACM Comput. Surv. (2018). Article no. 106

15. Cai, R., Zhang, X., Wang, H.: Bidirectional recurrent convolutional neural network for relation classification. In: Proceedings of the 54th Annual Meeting of the Association for Computational Linguistics, Berlin, Germany, pp. 756–765 (2016)

16. Chen, J., Gu, J.: Jointly extract entities and their relations from biomedical text. IEEE Access **7**, 162818–16227 (2019)

17. McDonald, R., Pereira, F., Kulick, S., Winters, S., Jin, Y., White, P.: Simple algorithms for complex relation extraction with applications to biomedical IE. In: Proceedings of the 43rd Annual Meeting on Association for Computational Linguistics, Ann Arbor, MI, pp. 491–498 (2005)

18. NLP-Progress. http://nlpprogress.com/. Accessed 15 Mar 2020

19. Romero, V., Fornes, A., Vidal, E., Sanchez, J.A.: Using the MGGI methodology for category-based language modeling in handwritten marriage licenses books. In: Proceedings of the 15th International Conference on Frontiers in Handwriting Recognition (2016)

20. Fornes, A., et al.: ICDAR 2017 competition on information extraction in historical handwritten records. In: Proceedings of the 14th IAPR International Conference on Document Analysis and Recognition (2017)

21. Schuster, D., et al.: Intellix – end-user trained information extraction for document archiving. In: Proceedings of the International Conference on Document Analysis and Recognition, Washington (2013)

22. Sutherland, S.: Learning information extraction rules for semi-structured and free text. Mach. Learn. **34**, 232–272 (1999)

23. Taghve, K., Nartker, T.A., Borsack, J.: Information access in the presence of OCR errors. In: Proceedings of the ACM Hardcopy Document Processing Workshop, Washington, DC, pp. 1–8 (2004)
24. Nagy, G.: Disruptive developments in document recognition. Pattern Recogn. Lett. (2016). https://doi.org/10.1016/j.patrec.2015.11.024
25. Finkel, J.R., Grenager, T., Manning, C.: Incorporating non-local information into information extraction systems by Gibbs sampling. In: Proceedings of the 43rd Annual Meeting of the Association for Computational Linguistics (ACL 2005), pp. 363–370 (2005)
26. Chiticariu, L., Li, Y., Reiss, F.R.: Rule-based Information Extraction is Dead! Long Live Rule-based Information Extraction Systems! Seattle, Washington, USA, pp. 827–832 (2013)
27. Family History Archives. https://www.familysearch.org/blog/en/family-history-books/. Accessed 21 Mar 2020
28. Embley, D.W., Liddle, S.W., Eastmond, S., Lonsdale, D.W., Woodfield, S.N.: Conceptual modeling in accelerating information ingest into family tree. In: Cabot, J., Gómez, C., Pastor, O., Sancho, M. (eds.) Conceptual Modeling Perspectives. Springer, Cham, pp. 69–84 (2017). https://doi.org/10.1007/978-3-319-67271-7_6
29. Embley, D.W., Liddle, S.W., Lonsdale, D.W., Woodfield, S.N.: Ontological document reading. An experience report. Int. J. Concept. Model. **13**(2), 133–181 (2018)
30. Woodfield, S.N., Seeger, S., Litster, S., Liddle, S.W., Grace, B., Embley, D.W.: Ontological deep data cleaning. In: Trujillo, J.C., et al. (eds.) ER 2018. LNCS, vol. 11157, pp. 100–108. Springer, Cham (2018). https://doi.org/10.1007/978-3-030-00847-5_9
31. Schone, P., Gehring, J.: Genealogical indexing of obituaries using automatic processes. In: Proceedings of the Family History Technical Workshop (FHTW 2016), Provo, Utah, USA (2016). https://fhtw.byu.edu/archive/2016
32. Packer, T.L., Embley, D.W.: Unsupervised training of HMM structure and parameters for OCRed list recognition and ontology population. In: Proceedings of the 3rd International Workshop on Historical Document Imaging and Processing, Nancy, France, pp. 23–30 (2015)
33. Kim, T.: A green form-based information extraction system for historical documents. MA thesis, Brigham Young University, Provo, Uta (2017)
34. Embley, D.W., Nagy, G.: Green interaction for extracting family information from OCR'd books. In: Proceedings of the Document Analysis Systems Workshop (DAS 2018), Vienna (2018). https://doi.org/10.1109/DAS.2018.58
35. Embley, D.W., Nagy, G.: Extraction Rule Creation by Text Snippet Examples, Family History Technology Workshop, Provo, UT (2018)
36. Nagy, G.: Estimation, learning, and adaptation: systems that improve with use. In: Proceedings of the Joint IAPR International Workshop on Structural, Syntactic, and Statistical Pattern Recognition, Hiroshima, Japan, pp. 1–10 (2012)
37. Nagy, G.: Green information extraction from family books. SN Comput. Sci. **1**, 23 (2020)

Estimating Human Legibility in Historic Manuscript Images - A Baseline

Simon Brenner[(✉)] , Lukas Schügerl, and Robert Sablatnig

Institute of Visual Computing and Human-Centered Technology,
TU Wien, 1040 Vienna, Austria
sbrenner@cvl.tuwien.ac.at
https://cvl.tuwien.ac.at

Abstract. For accessing degraded historic text sources, humanist research increasingly relies on image processing for the digital restoration of written artifacts. A problem of these restoration approaches is the lack of a generally applicable objective method to assess the results. In this work we motivate the need for a quality metric for historic manuscript images, that explicitly targets human legibility. Reviewing previous attempts to evaluate the quality of manuscript images or the success of text restoration methods, we can not find a satisfying solution: either the approaches have a limited applicability, or they are insufficiently validated with respect to human perception. In order to establish a baseline for further research in this area, we test several candidates for human legibility estimators, while proposing an evaluation framework based on a recently published dataset of expert-rated historic manuscript images.

Keywords: Legibility estimation · Historic manuscripts · Document image quality · Human perception · Evaluation framework

1 Introduction

Written heritage is a valuable resource for humanist research. However, the physical artifacts preserved may be in a condition that prohibits the direct accessing of the text. Addressing this problem, a research field dedicated to the digital restoration of such degraded sources based on specialized imaging and image processing methods has ensued [1,11,15,34,39]. In general, corresponding approaches aim at producing output images in which a text of interest is maximally legible for a human observer. We note that an inherent problem of this research field is the absence of a generally applicable objective metric for the property of human legibility. Consequently, the evaluation of proposed text restoration approaches is commonly based on expert ratings, the demonstration on selected examples or case studies [11,34,39]. This practice is unfavorable for the research field: it does not allow for an automated evaluation on large public datasets, such that an objective comparison of different approaches is impeded.

© Springer Nature Switzerland AG 2021
J. Lladós et al. (Eds.): ICDAR 2021, LNCS 12823, pp. 492–506, 2021.
https://doi.org/10.1007/978-3-030-86334-0_32

Another use case where a quantitative legibility metric would be beneficial is automated parameter tuning for producing an optimally readable version of a specific document [13,43,46], where the meaning of 'parameter' can range from the choice of a processing algorithm up to fine tuning of method-specific constants. For a meaningful application in the area of ancient text restoration we propose the following requirements that must be fulfilled by a quality metric:

1. **Rank correlation to human legibility.** The transcription of heavily degraded manuscripts is performed by experienced scholars and currently inconceivable using automated methods. It is thus the usefulness of an image for a human reader that ultimately defines its quality; machine readability and aesthetical properties are subordinate. A quality metric should thus be able to order images with respect to legibility the same way a human reader would.

2. **Applicability to non-binarized images.** Binarization is a typical preprocessing step for optical character recognition (OCR), but not desirable for human observers. For historical handwritten documents, the risk of information loss due to binarization artifacts is to high [53], especially when they are heavily degraded.

3. **Robustness to script and language.** Considering the variety of scripts and languages found in ancient documents, it would be desirable for a quality metric to being largely agnostic of them. This is especially relevant for rare scripts of which not many samples are available.

4. **Independence from reference.** In order to make a judgement, a quality metric should require no information other than the image being judged. This extends the applicability of the metric from dedicated datasets to arbitrary images of unknown documents.

To the best of our knowledge, none of the previous attempts to quantitatively assess document image quality or the success of text restoration methods meet all of these requirements (see Sect. 2). Partly this can be attributed to the previous absence of a suitable ground truth dataset of subjectively rated manuscript images, the like of which usually employed for the development and validation of general image quality assessment (IQA) methods [38,45,48]. Recent research addresses this problem: Shahkolaei et al. publish a dataset of 177 RGB images of historic manuscripts, subjectively rated for quality on a full-page basis [43]. The SALAMI dataset [3] consists of 250 grayscale images of manuscript regions, with corresponding spatial maps of mean legibility based on a study conducted with 20 experts of philology and paleography. This dataset is specifically targeted at measuring legibility.

This paper establishes a baseline for the problem of estimating human legibility in historic manuscript images. In Sect. 2 previous attempts to objectively evaluate text restoration methods or the quality of historic document images are reviewed. Section 3 is dedicated to experimentally testing candidate legibility estimators for their correlation with human perception: after defining the evaluation framework and the ground truth dataset, we describe the specific candidate estimators being tested and the results obtained. Section 4 concludes this work with summarizing observations.

2   Related Work and Potential Legibility Estimators

In the following, previous approaches to quantitatively assess the quality of manuscript images or the success of legibility enhancement methods are reviewed. We regard legibility estimation as a variation of image quality assessment (IQA), where *quality* is defined in terms of text legibility. In the subsequent discussion, we thus adhere to terminology prevalent in IQA literature [33, 35] and coarsely classify the approaches with respect to the additional information necessary to make a quality judgment: Full/reduced reference approaches require a reference image or a reduced representation, such as a ground truth (GT) segmentation, for operation. No-reference or blind methods on the other hand operate on the input image only. Such methods might be trained and tested using reference information, but do not require this information in production mode. This means that they are applicable to arbitrary images and usable in a wide range of scenarios where no reference is available.

2.1   Full/Reduced Reference Approaches

Giacometti et al. created a multispectral image dataset of manuscript patches before and after artificial degradation [15]. This allows the quantitative assessment of digital restoration methods by comparing the results to the non-degraded originals - a full reference approach [33]. As for historical manuscripts an image of its non-degraded state is usually not available, the application is limited to expensively created test datasets.

Arsene et al. have specifically evaluated the application of dimensionality reduction methods for text restoration based on multispectral manuscript images [1]. Their evaluation is based on both expert ratings (ranking of the result images done by 7 scholars) and cluster separability metrics (Davies-Bouldin Index [6] and Dunn-Index [10]). The clusters tested for separability are defined as manually selected pixels from foreground and background. In their experiments, the cluster separability metrics could correctly identify the images variants rated best by the human judges; apart from that, cluster scores and human assessments did not correlate well.

Shaus et al. introduce the metric of *potential contrast* and suggest its application as a quality metric for images of ostraca [12, 44]. The metric measures the contrast achievable between foreground and background under arbitrary grayscale transformations. Similarly to the previous method, it relies on user-defined foreground and background pixels. The correspondence to human assessments is not evaluated.

For restoration approaches producing binary images, evaluation is typically done by quantitative comparison to a GT segmentation [47]. This comparison is straight forward with standard error metrics such as F-score or peak signal-to-noise ratio [40]. Datasets with GT segmentations of historic manuscripts have been published, with both RGB images [14, 40] and multispectral images [19, 20] as inputs.

Another quantitative metric for evaluating document image enhancement methods is the performance of Optical Character Recognition (OCR) on the enhanced images [21,30]. This approach addresses the property of *legibility* more directly than the approaches mentioned before and is a reasonable choice when the purpose of text restoration lies in the subsequent automated transcription, rather than in the delivery to a human observer. The correspondence between machine legibility and human legibility, however, is not straightforward [37]; a fact that is exploited for example by *CAPTCHA* authentication methods [49]. A further problem with this method is the strong dependence of OCR systems on the scripts and languages they have been trained on [26]; this is especially problematic when considering ancient documents using rare historical scripts.

2.2 No-Reference/Blind Approaches

For the general IQA problem (i.e., estimating human perceived image quality), several blind approaches have been proposed, mostly based on natural scene statistics and machine learning [31,33]. General IQA metrics such as BRISQUE [35] are used as reference implementations for specialized document IQA methods, showing comparable performance [13,43]. Ye et al. propose a general IQA method based on filter learning and specifically demonstrate it on the assessment of document image quality [51]. Garg and Chaudhury later use the same method for automated parameter tuning for document image enhancement [13]. Both publications validate their document quality estimates via OCR performance on machine written documents. The correlation of the predictions with human legibility or perceived quality is not addressed.

We find a similar situation in dedicated *document IQA* literature [50]: Quality is mostly defined in terms of OCR accuracy and the proposed methods are trained and tested accordingly [27,28,32,50]. Document IQA thus amounts to OCR performance prediction. We have argued in Sect. 2.1 that OCR performance cannot be directly used as a human legibility estimator; similarly, OCR performance *predictors* cannot be directly used as human legibility estimators [37].

Few approaches explicitly target the estimation of human legibility or perceptual quality in documents. Stommel and Frieder propose an automatic legibility estimation for binarized historical documents, intended for unsupervised parameter tuning of binarization algorithms [46]. The method relies on OCR features computed for small image patches. Based on these features and subjective ratings (good, medium or bad legibility) an SVM classifier is trained. Applied to a whole document in a sliding window manner, the method produces a spatial map of legibility.

Obafemi and Agam propose a document IQA estimator based on classical feature extraction and a neural network classifier. The method operates on binary images of segmented characters. They show that their system can be trained to estimate both human ratings and OCR accuracy; however, it must be trained for each of these use cases separately. The approach is demonstrated on machine written text only.

Shahkolaei et al. created a dataset for quality assessment of subjectively rated ancient document images [43]. The dataset is based on 177 RGB images of Arabic manuscript pages. Mean opinion scores for the quality of the pages were obtained in a study where 28 students of technical subjects rated the images in a pair comparison mode. The images were compared with respect to not further specified 'quality' on a full-page basis. This dataset is used to train and test a novel method for objective quality assessment of degraded manuscript images, which uses support vector regression (SVR) on Gabor filter responses [43].

Lastly, we would like to address a class of methods that to the best of our knowledge was not used for legibility estimation before, but is worth investigating. While the final transcription accuracy of OCR systems was argued to be problematic for the purpose, early stages of OCR pipelines could be useful for blind legibility estimation: before text can be transcribed, it must be localized in the input image. In the domain of document image processing, text line detection and layout analysis are typical processing stages [4,16,17]. When working with natural images (i.e., trying to detect text 'in the wild'), the problem is referred to as scene text detection [29,36,54]. For both cases, systems that are largely independent of script and language are described [7,8,17,42]. We hypothesize that e.g., the amount of detected text lines in a given image area (possibly weighted by corresponding detection confidences) could be an implicit indicator for human legibility. In Sect. 3, the correlation of popular layout analysis and scene text detection methods with human legibility is tested.

2.3   Roundup

Table 1 gives an overview of the approaches discussed in this section. All of the approaches in the full/reduced reference category were previously used to estimate handwritten document images or evaluate restoration approaches; however, none of them was shown to correlate to human legibility. Furthermore, the dependence on reference information limits the applicability and thus contradicts the fourth requirement given in Sect. 1.

In the no-reference category we only disqualify the approaches that operate on binary images (see second requirement in Sect. 1). The other approaches are in principle valid candidates for a legibility estimator; yet their correspondence to human legibility and robustness to script and language remains to be shown.

3   Experiments

This section describes the experiments conducted to test the correlation of candidate legibility estimators with human legibility. First, the SALAMI dataset [3], which serves as the ground truth and is an integral part of our methodology, is briefly described. Subsequently, the test framework and the tested candidate methods are defined. The section concludes with the experimental results obtained.

Table 1. Overview of related work and approaches to legibility estimation.

Full/reduced reference approaches			
Description	Publication	Reference	Limited applicability
Comparison to non-degraded reference	Giacometti et al. 2017 [15]	Image the object before artificial degradation	Dedicated datasets only
Cluster separability	Arsene et al. 2018 [1]	Pixel labels	–
Potential contrast	Shaus et al. 2017 [44]	Pixel labels	–
Comparison to GT binarization	Multiple (see e.g. [40,47])	GT binarization	Binarized documents
OCR performance	Multiple (e.g. [21,30])	GT transcription	Known alphabets/languages
No-reference/blind approaches			
Description	Publication	Ground truth	Limited applicability
General blind IQA	Multiple (e.g. [2,35,51]	Human perceived quality	–
Blind document IQA	Multiple (e.g. [27,28,32,50])	OCR performance on machine written documents	–
Handcrafted features + SVM	Stommel and Frieder 2011 [46]	Human legibility rating (good, medium, bad)	Binarized documents
Handcrafted features + NN	Obafemi and Agam 2012 [37]	Human quality rating	Binarized and segmented characters
Gabor filter responses + SVR	Shahkolaei et al. 2018 [43]	Human quality rating	–
Text detection	Multiple (e.g. [29,36,54])	labeled text areas	–

3.1   The SALAMI Dataset

The SALAMI dataset [3] was created as a ground truth for the development of objective legibility estimation methods. It is based on 50 manuscript regions of 60×60 mm sampled from 48 different manuscripts and 8 language families. Depending on line height and layout, they contain 1–17 lines of text; image regions with multiple layers of text (palimpsests) and image regions without text are not part of the dataset. The source imagery consists of multispectral images acquired during several research projects, using different imaging devices and protocols [9,20,22]. To unify spatial resolution and number of spectral layers, the source images were re-sampled to 720×720 pixels and reduced to 5 images per region via principal component analysis. Figure 1 shows an example. With 5 image variants for each of the 50 manuscript regions the SALAMI dataset contains 250 test images.

Legibility scores for the images were obtained in a study conducted with 20 scholars of philology and paleography. All participants were experienced in reading original manuscripts and consisted of pre-doctoral researchers, post-doctoral researchers and professors. In the study, the participants were required to mark all visible text areas with rectangular bounding boxes, which are then rated individually on a five point scale. The scale labels refer to the percentage of text within the selected region that is deemed clear enough to read: score 1 corresponds to 0–20% readable, score 1 corresponds to 20–40% readable, etc.;

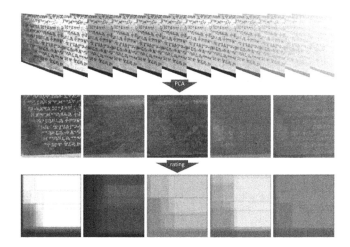

Fig. 1. Creation of the SALAMI dataset. The original multispectral image (top) is reduced to 5 principal components, that constitute the input images of the dataset (middle). Mean legibility maps (bottom) are obtained in an expert rating study.

background regions receive score 0. This way, the ambiguities of ordinal scales (with labels like "Bad", "Poor", "Fair", etc.) are avoided and scores on a true interval scale are obtained [52].

The experiments produced one spatial map of legibility for each image and each participant. For the final dataset, the score maps of all the participants are averaged. See Fig. 1 for an example of input images with the corresponding mean legibility maps. Statistical analyses of the subjective ratings show excellent inter-rater agreement and a low impact of participant traits on the scores obtained [3].

3.2   Test Framework

Our experiments aim at testing the outputs of candidate legibility estimators for rank correlation with the subjective human legibility scores of the SALAMI dataset, which we report in the form of Spearman Rank Correlation Coefficients (SRC) [41]. Evaluation takes place with respect to the following dimensions:

Spatial Accuracy. The correlations are evaluated on several spatial levels: The rating of the whole image (720×720 pixels/60×60 mm), as well as the rating of sub-squares of 360, 180, 90 and 45 pixels side-length are evaluated. The ground truth legibility scores are obtained by averaging the legibility maps of the SALAMI dataset across the corresponding regions.

Absolute Legibility vs. Single Content Ranking. The ability to estimate absolute legibility is measured by correlating the estimates with the ground truth across all manuscript regions. As the SALAMI dataset contains 5 image versions for each manuscript region, it can also be used to test the ability of an estimator to correctly rank different versions of the same content. This is done by computing the SRC separately for each manuscript region. This per-content correlation is a weaker property than the overall correlation (as the former is implied by the latter); however, it can be sufficient for many applications, including the comparison of different enhancement algorithms and automated parameter optimization.

All Areas vs. Text Areas Only. Lastly, we compare the performance on the whole images with the performance on text areas only. These test cases correspond to different application scenarios and pose different challenges. The former case represents the most general situation where no additional information about the content is available; here the estimators must be able to handle background areas and assign minimal legibility to them. The latter case is relevant for scenarios, where the location of text is known a priori; here the estimator must not handle background areas and only the discriminatory power between text legibility levels is relevant. Within the SALAMI dataset, we define text areas as areas with a mean legibility score of at least 1 in at least one of the 5 image variants; following this definition, text areas make up ca. 80% of the dataset.

Some of the evaluated methods are trained on the SALAMI dataset. In these cases, we resort to a leave-one-out cross-validation approach: For each of the 50 manuscript regions represented in the dataset an own model is trained, with the 5 corresponding images excluded from training. This approach is chosen in order to cope with a relatively small dataset.

3.3   Tested Candidate Estimators

The SALAMI dataset consists of input images paired with spatial maps of subjective legibility, without additional information. Therefore, we can only test blind/no-reference candidate estimators (see Sect. 2.2) within our framework. In the following, the specific metrics and methods that are tested for correlation with subjective legibility scores are defined.

Simple Image Statistics. For a basic reference, the correlation of legibility scores to image statistics with plausible impact on legibility is tested:

- *RMS contrast:* Equivalent to the standard deviation of intensities: $\sqrt{\frac{1}{n}\sum_{i=1}^{n}(x_i - \mu)^2}$, where n is the number of pixels in the area of interest, x_i is the intensity of pixel i and μ is the mean intensity.
- *entropy:* Shannon entropy of intensities: $-\sum_{i=1}^{n} P(x_i) \log_2 P(x_i)$, where n is the number of pixels in the area of interest and $P(x_i)$ is the probability of pixel i having the intensity x_i, based on a histogram of intensities.
- *edge intensity:* Edge intensity images are computed as magnitudes of gradients resulting from convolution with horizontal and vertical Sobel operators. The edge intensity images are then averaged over the area of interest.

Text Detection and Layout Analysis. This class of methods detects text in natural scenes or documents. To obtain legibility scores from detected text regions, first a *detection map* is created: depending on the specific method, pixels inside of detected regions are set to the associated detection confidences or to one, if no detection confidence is available; background pixels are set to zero. The legibility score of a given area of interest is then defined as the mean of the corresponding detection map. None of the methods were trained on the SALAMI dataset. The following approaches are tested:

- *Neumann and Matas 2012* [36]: A scene text detector based on cascade classification of extremal regions, trained on segmented characters and non-character images. Binary detection maps of multiple input layers [36] are summed up.
- *Zhou et al. 2017 (EAST)* [54]: A scene text detector based on a Convolutional Neural Network (CNN), trained on the ICDAR 2015 dataset [24]. In the resulting detection map, regions are weighted by their confidences.
- *Liao et al. 2017 (TextBoxes)* [29]: A CNN-based scene text detector, trained on the ICDAR 2013 dataset [25]. In the resulting detection map, regions are weighted by their confidences.
- *Grüning et al. 2019* [17]: A CNN-based text line detector for layout analysis, as implemented in the Transkribus plattform [23]. We evaluate both on text line level and text region level. Detection maps are binary.

(Document) Image Quality Assessment. Blind IQA methods estimate quality scores for a given input image, which can be directly used as legibility estimates. Generic "image quality", especially with respect to natural images, might prioritize image properties irrelevant for human legibility. We thus evaluate the selected methods once trained on the datasets used in the original publications, and once trained on the SALAMI dataset with its legibility scores. The following approaches are tested:

- *Mittal et al. 2012 (BRISQUE)* [35]: A popular general IQA metric appearing as a reference implementation in document IQA approaches [13,43]. The original version is trained on the LIVE dataset [45].
- *Shahkolaei et al. 2018* [43]: A dedicated document IQA method for historic manuscript images. The original version is trained on the authors' own dataset [43].

A Dedicated CNN. Finally, we train a CNN for legibility estimation from scratch, in order to obtain a baseline for dedicated deep-learning based legibility estimation:

- *CNN regression:* The network architecture is based on an 18 layer ResNet [18], equipped with an extra linear layer for regression. Considering the leave-one-out cross-validation approach mentioned at the beginning of the section, 50 models are trained for 20 epochs each on 240×240 pixel patches of images of the SALAMI dataset. For both training and inference, the patches are extracted from the original images in a sliding window manner with a stride of 10 pixels. Inference results are thus directly interpretable as legibility maps compatible with our evaluation framework.

3.4   Results

Figure 2 shows the SRCs of tested candidate estimators with the legibility scores of the SALAMI dataset for different spatial resolution levels (window sizes). The performance of most of the tested candidates is comparable to the performance of simple image statistics such as the *RMS contrast*. Exceptions are the dedicated *CNN regression*, as well as the method by Shahkolaei et al. [43] when trained on the SALAMI dataset. The EAST scene text detector [54] out of the box performs reasonably well on large images, without any training on the SALAMI dataset. No statistically significant differences were observable when comparing the results for the entire dataset with the results for text areas only.

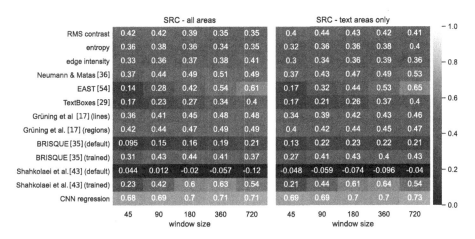

Fig. 2. Spearman rank correlation coefficients between tested methods and ground truth (SALAMI dataset [3]), given separately for different window sizes. Results for the entire dataset are shown on the left, results for only text areas on the right.

Results of the single content ranking tests (i.e. the SRCs are computed for each manuscript region separately, as described in Sect. 3.2) are shown as box plots in Fig. 3. From this experiment, additional insights are gained:

1. The interquartile ranges indicate a high variability in performance between different manuscript regions. The *CNN regression*, which shows the best overall correlation, also appears to be most invariant to the input region.
2. Comparing the medians to the overall SRCs (Fig. 2), no general trend is observable: some of the tested methods show higher correlations in the single content situation, others show lower correlations. Notable in this context are

the results of *RMS contrast* and *entropy* on the largest window size, where they are only surpassed by *Neumann & Matas* (which is also performing unusually well here) and the *CNN regression*.

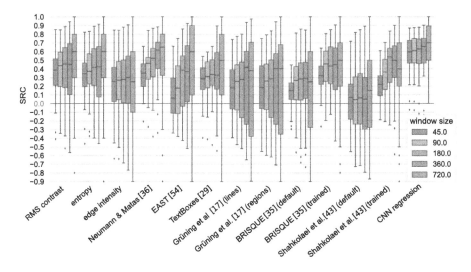

Fig. 3. Boxplots of Spearman rank correlation coefficients computed for each manuscript region separately, grouped by different window sizes. The boxplots show the quartiles of SRCs observed in the different regions, outliers are shown as dots.

4 Conclusion

In this paper, we motivate the need for a quality metric for historic manuscript images that explicitly targets human legibility. Reviewing previous approaches to evaluate the quality of manuscript images or the success of text restoration methods, we find that presently no metric suitable for this purpose exists: either the approaches have a limited applicability (specialized dataset, binary images, machine written documents) or they are insufficiently validated with respect to domain expert ratings. In order to establish a baseline for further research in this area, we propose an evaluation framework for legibility estimators that is based on correlation with the subjective legibility scores of the novel SALAMI dataset [3]. Several possible candidates for legibility estimators are then evaluated based on this framework.

We find that few of the tested methods surpass basic image statistics such as *RMS contrast*. The work of *Shahkolaei et al.* [43] is conceptually most compatible with our definition of a valid legibility metric. Interestingly, the results are close to random when applied to the salami dataset in its originally trained form; when trained on the SALAMI dataset, however, the performance increases drastically. This behavior might result from the fundamental differences in the study designs used to create their dataset and the SALAMI dataset (pair comparisons of whole pages by technical students vs. spatially resolved absolute ratings by domain experts). Also the *BRISQUE* [35] IQA metric shows a significant gain in correlation after training on the SALAMI dataset, suggesting that generic image quality does not directly translate to human legibility. However, even after training the performance stays in the range of simple image statistics.

The legibility estimates generated via text detection and layout analysis show mediocre performance. However, the translation from text detections to legibility scores adopted for this study is rather crude and is worth further elaboration in future work.

Unsurprisingly, the best results are obtained with a CNN regression trained from scratch on the SALAMI dataset. The problem with using CNN predictions as a quality metric, however, lies in the lack of decision transparency [5]; when pursuing a CNN path in future work, this aspect must be given the same attention as the improvement of correlation to human perception.

References

1. Arsene, C.T.C., Church, S., Dickinson, M.: High performance software in multi-dimensional reduction methods for image processing with application to ancient manuscripts. Manuscript Cult. **11**, 73–96 (2018)
2. Bosse, S., Maniry, D., Muller, K.R., Wiegand, T., Samek, W.: Deep neural networks for no-reference and full-reference image quality assessment. IEEE Trans. Image Process. **27**(1), 206–219 (2018)
3. Brenner, S., Sablatnig, R.: Subjective assessments of legibility in ancient manuscript images - the SALAMI dataset. In: Del Bimbo, A., et al. (eds.) ICPR 2021. LNCS, vol. 12667, pp. 68–82. Springer, Cham (2021). https://doi.org/10.1007/978-3-030-68787-8_5
4. Bukhari, S.S., Kadi, A., Jouneh, M.A., Mir, F.M., Dengel, A.: anyOCR: an open-source OCR system for historical archives. In: 2017 14th IAPR International Conference on Document Analysis and Recognition (ICDAR), vol. 01, pp. 305–310, November 2017
5. Chattopadhay, A., Sarkar, A., Howlader, P., Balasubramanian, V.N.: Grad-CAM++: generalized gradient-based visual explanations for deep convolutional networks. In: 2018 IEEE Winter Conference on Applications of Computer Vision (WACV), pp. 839–847, March 2018
6. Davies, D.L., Bouldin, D.W.: A cluster separation measure. IEEE Trans. Pattern Anal. Mach. Intell. **1**(2), 224–227 (1979)
7. Dey, S., et al.: Script independent approach for multi-oriented text detection in scene image. Neurocomputing **242**, 96–112 (2017)
8. Diem, M., Kleber, F., Sablatnig, R.: Text line detection for heterogeneous documents. In: 2013 12th International Conference on Document Analysis and Recognition, pp. 743–747, August 2013
9. Diem, M., Sablatnig, R.: Registration of ancient manuscript images using local descriptors. In: Digital Heritage, Proceedings of the 14th International Conference on Virtual Systems and Multimedia, pp. 188–192 (2008)
10. Dunn, J.C.: A fuzzy relative of the ISODATA process and its use in detecting compact well-separated clusters. J. Cybern. **3**(3), 32–57 (1973)

11. Easton, R.L., Christens-Barry, W.A., Knox, K.T.: Spectral image processing and analysis of the Archimedes Palimpsest. In: European Signal Processing Conference (Eusipco), pp. 1440–1444 (2011)
12. Faigenbaum-Golovin, S., et al.: Multispectral images of ostraca: acquisition and analysis. J. Archaeol. Sci. **39** (2012)
13. Garg, R., Chaudhury, S.: Automatic selection of parameters for document image enhancement using image quality assessment. In: 2016 12th IAPR Workshop on Document Analysis Systems (DAS), pp. 422–427, April 2016
14. Gatos, B., Ntirogiannis, K., Pratikakis, I.: ICDAR 2009 document image binarization contest (DIBCO 2009). In: 2009 10th International Conference on Document Analysis and Recognition (ICDAR), pp. 1375–1382, July 2009
15. Giacometti, A., et al.: The value of critical destruction: evaluating multispectral image processing methods for the analysis of primary historical texts. Digit. Scholarsh. Hum. **32**(1), 101–122 (2017)
16. Grüning, T., Labahn, R., Diem, M., Kleber, F., Fiel, S.: READ-BAD: a new dataset and evaluation scheme for baseline detection in archival documents. In: 2018 13th IAPR International Workshop on Document Analysis Systems (DAS), pp. 351–356, April 2018
17. Grüning, T., Leifert, G., Strauß, T., Michael, J., Labahn, R.: A two-stage method for text line detection in historical documents. Int. J. Doc. Anal. Recogn. (IJDAR) **22**(3), 285–302 (2019)
18. He, K., Zhang, X., Ren, S., Sun, J.: Deep residual learning for image recognition. In: 2016 IEEE Conference on Computer Vision and Pattern Recognition (CVPR), pp. 770–778. IEEE, Las Vegas, June 2016
19. Hedjam, R., Nafchi, H.Z., Moghaddam, R.F., Kalacska, M., Cheriet, M.: ICDAR 2015 contest on MultiSpectral text extraction (MS-TEx 2015). In: Proceedings of the International Conference on Document Analysis and Recognition, ICDAR, pp. 1181–1185 (2015)
20. Hollaus, F., Brenner, S., Sablatnig, R.: CNN based binarization of MultiSpectral document images. In: Proceedings of the International Conference on Document Analysis and Recognition, ICDAR, pp. 533–538 (2019)
21. Hollaus, F., Diem, M., Sablatnig, R.: Improving OCR accuracy by applying enhancement techniques on MultiSpectral images. In: Proceedings - International Conference on Pattern Recognition, pp. 3080–3085 (2014)
22. Hollaus, F., Gau, M., Sablatnig, R.: Multispectral image acquisition of ancient manuscripts. In: Ioannides, M., Fritsch, D., Leissner, J., Davies, R., Remondino, F., Caffo, R. (eds.) EuroMed 2012. LNCS, vol. 7616, pp. 30–39. Springer, Heidelberg (2012). https://doi.org/10.1007/978-3-642-34234-9_4
23. Kahle, P., Colutto, S., Hackl, G., Mühlberger, G.: Transkribus - a service platform for transcription, recognition and retrieval of historical documents. In: 2017 14th IAPR International Conference on Document Analysis and Recognition (ICDAR), vol. 04, pp. 19–24, November 2017
24. Karatzas, D., et al.: ICDAR 2015 competition on robust reading. In: 2015 13th International Conference on Document Analysis and Recognition (ICDAR), pp. 1156–1160, August 2015
25. Karatzas, D., et al.: ICDAR 2013 robust reading competition. In: 2013 12th International Conference on Document Analysis and Recognition (ICDAR), pp. 1484–1493, August 2013
26. Leifert, G., Labahn, R., Sánchez, J.A.: Two semi-supervised training approaches for automated text recognition. In: 2020 17th International Conference on Frontiers in Handwriting Recognition (ICFHR), pp. 145–150 (Sep 2020)

27. Li, H., Zhu, F., Qiu, J.: CG-DIQA: no-reference document image quality assessment based on character gradient. In: 2018 24th International Conference on Pattern Recognition (ICPR), pp. 3622–3626, August 2018

28. Li, H., Zhu, F., Qiu, J.: Towards document image quality assessment: a text line based framework and a synthetic text line image dataset. In: 2019 International Conference on Document Analysis and Recognition (ICDAR), pp. 551–558, September 2019

29. Liao, M., Shi, B., Bai, X., Wang, X., Liu, W.: TextBoxes: a fast text detector with a single deep neural network. In: Thirty-First AAAI Conference on Artificial Intelligence, February 2017

30. Likforman-Sulem, L., Darbon, J., Smith, E.H.: Enhancement of historical printed document images by combining total variation regularization and non-local means filtering. Image Vis. Comput. **29**(5), 351–363 (2011)

31. Liu, X., Van De Weijer, J., Bagdanov, A.D.: RankIQA: learning from rankings for no-reference image quality assessment. In: 2017 IEEE International Conference on Computer Vision (ICCV), pp. 1040–1049. IEEE, October 2017

32. Lu, T., Dooms, A.: A deep transfer learning approach to document image quality assessment. In: 2019 International Conference on Document Analysis and Recognition (ICDAR), pp. 1372–1377, September 2019

33. Manap, R.A., Shao, L.: Non-distortion-specific no-reference image quality assessment: a survey. Inf. Sci. **301**, 141–160 (2015)

34. Mindermann, S.: Hyperspectral Imaging for Readability Enhancement of Historic Manuscripts. Master's thesis, TU München (2018)

35. Mittal, A., Moorthy, A.K., Bovik, A.C.: No-reference image quality assessment in the spatial domain. IEEE Trans. Image Process. **21**(12), 4695–4708 (2012)

36. Neumann, L., Matas, J.: Real-time scene text localization and recognition. In: 2012 IEEE Conference on Computer Vision and Pattern Recognition (ICPR), pp. 3538–3545. IEEE, Providence, June 2012

37. Obafemi-Ajayi, T., Agam, G.: Character-based automated human perception quality assessment in document images. IEEE Trans. Syst. Man Cybern. Part A Syst. Hum. **42**(3), 584–595 (2012)

38. Ponomarenko, N., et al.: Image database TID2013: peculiarities, results and perspectives. Sig. Process. Image Commun. **30**, 57–77 (2015)

39. Pouyet, E., et al.: Revealing the biography of a hidden medieval manuscript using synchrotron and conventional imaging techniques. Anal. Chim. Acta **982**, 20–30 (2017)

40. Pratikakis, I., Zagoris, K., Karagiannis, X., Tsochatzidis, L., Mondal, T., Marthot-Santaniello, I.: ICDAR 2019 competition on document image binarization (DIBCO 2019). In: 2019 International Conference on Document Analysis and Recognition (ICDAR), pp. 1547–1556 (Sep 2019)

41. Puth, M.T., Neuhäuser, M., Ruxton, G.D.: Effective use of Spearman's and Kendall's correlation coefficients for association between two measured traits. Anim. Behav. **102**, 77–84 (2015)

42. Ryu, J., Koo, H.I., Cho, N.I.: Language-independent text-line extraction algorithm for handwritten documents. IEEE Signal Process. Lett. **21**(9), 1115–1119 (2014)

43. Shahkolaei, A., Nafchi, H.Z., Al-Maadeed, S., Cheriet, M.: Subjective and objective quality assessment of degraded document images. J. Cult. Herit. **30**, 199–209 (2018)

44. Shaus, A., Faigenbaum-Golovin, S., Sober, B., Turkel, E.: Potential contrast - a new image quality measure. Electron. Imaging **2017**(12), 52–58 (2017)

45. Sheikh, H.R., Sabir, M.F., Bovik, A.C.: A statistical evaluation of recent full reference image quality assessment algorithms. IEEE Trans. Image Process. **15**(11), 3441–3452 (2006)

46. Stommel, M., Frieder, G.: Automatic estimation of the legibility of binarised historic documents for unsupervised parameter tuning. In: 2011 International Conference on Document Analysis and Recognition (ICDAR), pp. 104–108, September 2011

47. Sulaiman, O.: Nasrudin: degraded historical document binarization: a review on issues, challenges, techniques, and future directions. J. Imaging **5**(4), 48 (2019)

48. Virtanen, T., Nuutinen, M., Vaahteranoksa, M., Oittinen, P., Häkkinen, J.: CID2013: a database for evaluating no-reference image quality assessment algorithms. IEEE Trans. Image Process. **24**(1), 390–402 (2015)

49. Xu, X., Liu, L., Li, B.: A survey of CAPTCHA technologies to distinguish between human and computer. Neurocomputing **408**, 292–307 (2020)

50. Ye, P., Doermann, D.: Document image quality assessment: a brief survey. In: 2013 12th International Conference on Document Analysis and Recognition (ICDAR), pp. 723–727, August 2013

51. Ye, P., Kumar, J., Kang, L., Doermann, D.: Real-time no-reference image quality assessment based on filter learning. In: 2013 IEEE Conference on Computer Vision and Pattern Recognition (CVPR), pp. 987–994, June 2013

52. Ye, P., Doermann, D.: Combining preference and absolute judgements in a crowdsourced setting. In: Proceedings of International Conference on Machine Learning, pp. 1–7 (2013)

53. Yousefi, M.R., Soheili, M.R., Breuel, T.M., Kabir, E., Stricker, D.: Binarizationfree OCR for historical documents using LSTM networks. In: 2015 13th International Conference on Document Analysis and Recognition (ICDAR), pp. 1121–1125, August 2015

54. Zhou, X., et al.: EAST: an efficient and accurate scene text detector. In: 2017 IEEE Conference on Computer Vision and Pattern Recognition (CVPR), pp. 2642–2651. IEEE, Honolulu, HI, July 2017)

A Modular and Automated Annotation Platform for Handwritings: Evaluation on Under-Resourced Languages

Chahan Vidal-Gorène[1,2]([⊠]) [iD], Boris Dupin[2], Aliénor Decours-Perez[2], and Thomas Riccioli[2]

[1] École Nationale des Chartes - Université Paris, Sciences and Lettres,
65 rue de Richelieu, 75002 Paris, France
`chahan.vidal-gorene@chartes.psl.eu`
[2] Calfa, MIE Bastille, 50 rue des Tournelles,
75003 Paris, France
{`boris.dupin,alienor.decours,thomas.riccioli`}`@calfa.fr`
`http://www.chartes.psl.eu`, `https://calfa.fr`

Abstract. There is today several approaches for automatic handwritten document analysis. HTR achieve in particular convincing results both in layout analysis and text recognition, but also in more up-to-date requests like name entity-recognition, script identification or manuscript datation. These systems are trained and evaluated with large open and specialized databases. Manual annotation and proofreading of handwritten documents is a key step to train such systems. However, it is a time-consuming task, especially when the formats required by the systems display considerable variations, or when the interfaces do not manage several level of information. We propose a new modular and collaborative interface online, ready-to-use, for multilevel annotation and quick-view solution for handwritten and printed documents, including for right-to-left languages. This interface undertakes the creation of customized projects, and the management, the conversion and the export of data in the different formats and standards of the state-of-the-art. It includes automated tasks for layout analysis and text lines extraction with high level fine-tuning capacities. We present this new interface through the case study of the creation of a database for Armenian, an under-resourced language with specific paleographical issues.

Keywords: HTR · OCR · Historical documents · Layout analysis · Text line extraction · Crowdsourcing · Dataset · Under-resourced language · Armenian

1 Introduction

Manual annotation of handwritten documents is a key step for any recognition process, layout analysis or relevant information extraction. It is a time-consuming task that often requires high linguistic and paleographic proficiency,

© Springer Nature Switzerland AG 2021
J. Lladós et al. (Eds.): ICDAR 2021, LNCS 12823, pp. 507–522, 2021.
https://doi.org/10.1007/978-3-030-86334-0_33

especially in the case of an ancient and under-resourced language like Classical Armenian for which the annotation can only be achieved by a small number of people. This specificity is not limited to Armenian and despite the growing number of datasets created in recent years (READ datasets, cBAD datasets, etc. [6, 13, 30]), very few are dealing with under-resourced and Oriental languages [5, 17, 18]. The current dynamic of digitization and the promotion of IIIF standard increase access to the manuscripts, even for the least resourced languages. Most labeling tools (see *infra* Sect. 2) include collaborative work or automated tasks. Nevertheless, these interfaces remain specialized in a very specific type of document or annotation project, or they require technical expertise to create models dedicated to a specific project. These limitations result in the multiplication of interfaces, each dedicated to one purpose.

We undertook the creation of a database of Armenian handwritings, from which we are presenting a sample centered on old manuscripts (from 10th to 20th century). To build this database, we have created an open access web-based, multi-level and modular labeling platform: Calfa Vision[1]. It offers modular architecture to initiate and monitor annotation projects: projects of layout analysis, of text recognition or requiring more in-depth description (name entity, datation, etc.). This article presents on the one hand a state of the art of current labeling tools and approach for handwritings annotation, and on the other hand the description of the proposed interface, with evaluation of generic models it includes on various datasets and impact of fine-tuning for collaborative annotation. The final part of the paper is dedicated to the description of the built database and the evaluation of the last campaign of collaborative annotation.

2   Related Work

There are many interfaces for manuscripts annotation. For instance, HTR systems like OCRopus [3] and its forks incorporate basic interface for line-by-line transcription and an export function for images of line and the related ground-truth into a text file. We can cite three interfaces that are particularly used and meet specific needs, given the detection and recognition approach considered. First, the python interface *labelme* [32] enables to draw manually a great number of baselines, bounding-boxes or polygons and to generate a JSON file with the coordinates of the drawn geometric shapes. Second, the *GraphManuscribble* interface developed by the University of Freiburg facilitates the layout annotation through crowdsourcing with a graph-based approach [8]. Third, the *Transkribus* interface [16] has been used in recent years for ICFHR and ICDAR, in particular READ datasets and cBAD datasets, and generates an XML file with all information necessary to both the description of the document structure and the textual information (text-region, line, baseline, transcription, etc.) based on the annotation. Besides *Transkribus* performs automatic labeling for line with P2PaLA [24] and CITlab [21], text prediction and enables collaborative annotation. Three platforms are currently being developed. Two of them

---

[1] https://vision.calfa.fr.

are open source, collaborative and web-based: (i) *OCR4all* [25] in collaboration with the *OCR-D* project [23] includes different OCR architectures (notably tesseract). It is specifically dedicated to printed and incunabula recognition. And (ii) *escriptorium* [19], framework that incorporates pre-trained layout analysis models and offers the opportunity to train a HTR model, dedicated to a given annotation project. Google is currently working on expanding its OCR system to manuscripts with a game-oriented platform [15]. There are other platforms for different more precise tasks, like *tranScriptorium* [9], *DigiPal* [4] or *Archetype* projects, for a collaborative paleographical description of manuscripts; overall, many institutions propose their own interface (e.g. the Smithsonian Institution).

Regarding the databases, there is none for handwritten Armenian that combines descriptive information on the content (script, datation), on the layout and on the related ground-truth. For ancient and under-resourced languages, besides the databases presented at ICDAR and ICFHR, several reference databases exist, like the BADAM dataset [18] for Arabic scripts, the GRPOLY dataset [10] for Old Greek polytonic documents (printed and handwritten), or the DIVA-HisDB dataset [28] of the University of Freiburg for medieval manuscripts (in Latin, Italian, Greek, Alemannic and German). In view of its quick and easy labeling and the possibility to infer a polygon for the whole line, the baseline annotation is favored [7, 13].

3   Common Issues for Under-Resourced Languages: Example on Armenian

General Considerations: The platforms described in Sect. 2 have proved to be effective on a wide-ranging set of handwritten and printed documents, encompassing various layouts and languages. They are focused on the annotation of documents (layout and HTR). The automatization of annotation requires either the use of generic models, often specialized on the Latin scripts and layouts (e.g. for the reading order, the baseline prediction, etc.), or the creation of model from scratch which is resource and time-consuming. The approach with polygons associated with lines, when chosen, equally suffers from the focus on Latin writings, and proves to be inconsistent with the image size, with the line height or the character size (e.g. for instance, it is the case of the Arabic vocalization [31] or Armenian capital scripts). Therefore, processing specific layouts (e.g. complex layout composed of vertical writing, curved lines, etc.) requires to create dedicated datasets [18] from scratch, as is the case for Arabic scripts with the need for a specific baseline for multiple columns versification [18].

For the purposes of processing the Armenian language, we are dealing with several types of layout. In ancient manuscripts, the most common layouts are one or two columns spread with marginalia, and catchword at the end of folios. In gospels particularly, the margins are often ornamented with floral or zoomorphic illuminations. There are very complex layouts (vertical writing, curved lines, etc.) for scientific manuscripts.

The main difference with the Latin lays in the notion of baseline. In *erkat'agir* script, a monumental Armenian script [20], the baseline is the line upon which the letter lies (a lower baseline as in Latin). For the *bolorgir* script, a bicameral script, the baseline becomes the starting point for stems. Thus, the letters are often aligned by the top, on the mean line (as in Hebrew), especially in the latest manuscripts (see Fig. 1). Managing with mean line and topline is a major issue encountered by several language and to which we answer with the platform to build the dedicated dataset.

Joining Letters – The Armenian letters are all separated from one another. Due to the gradual cursive introduction of *nōtrgir* and its numerous prolongements and ligatures, copyists start to join the letters together to create new independent graphemes, or ligatures in-between lines (see Fig. 3).

Scriptio Continua and Text Justification – The Armenian manuscripts, except the latest ones, are written in *scriptio continua*, it means that, as for the manuscripts of Latin and Greek Antiquities, the text is continuous. The spaces are not in between words, but are solely used by copyists for aesthetic purposes (to justify the text). *Scriptio continua* raises issues for word-segmentation. In the same vein, hyphenation is arbitrary, and according to the current research, it follows no formal rule. The text justification is almost systematic, and failing spaces, the letters are either stretched to bridge the gaps of the text (e.g. Fig. 3, image i, line 5), or the text is written above or below the line. This last solution causes difficulties in particular for baseline detection and for matching content to the line of the text. The same can be observed for other languages [18].

Abbreviations – Abbreviations in Armenian are an unprecedented challenge. While there is a wide variety more or less codified, the Armenian language introduces notably ideograms that transform a whole word into a drawing (see Fig. 1). There is a very wide range of ideograms that cover a very large lexicon (the dataset contains 18 ideograms). Therefore, the platform enables to create dictionaries specific to a project, this functionality will be extended to incorporate non-unicode characters in the future.

The list of characteristics specific to a language or another could not be exhaustive. Hence, the issues raised by these languages often require to create manually the data from scratch, even leading to new informatics developments, regardless of the type of project considered. The platform we are describing further in the article relies on a modular architecture for project definition, and on the fine-tuning of models pre-trained on other under-resourced languages, in order to meet the needs of the scientific community more rapidly and to achieve the building of appropriate datasets.

Fig. 1. (i) Example of abbreviations by contraction (1) and ideograms (2 and 3), LOB Or 13941 (British Library); (ii) Difference between traditional baseline (red, below) and predicted baseline/Armenian baseline (green, above), W545 (Walters Art Museum) (Color figure online)

4   Proposed Interface

The crowdsourcing approach is relevant to annotate large databases, when the goals are clearly defined and some simple and tailored tools are provided [2,11]. We propose a simple project management interface to create and monitor an annotation project, private or public, with semi-automated annotation of documents (see Fig. 5). The platform design is modular in order to allow for the creation of projects other than HTR-focused. It can just as well be used for creating ground-truth for HTR training purposes as for assisted manuscript transcription. The interface is developped especially for Oriental languages (Arabic scripts, Hebrew, Syriac, Armenian, etc.), that are presenting specific difficulties for analysis and transcription, and that we had to overcome in order to process efficently a language such as Armenian. The aim is to foster easily and quickly the creation of multi-level customizable data. The highlight is set on internal fine-tuning to quickly have robust models for languages on which latin-based models are less appropriate.

4.1   Project Definition and Management

The user has at his disposal a management and monitoring interface for all of his annotation projects (personnal or shared). The user can either use the templates provided by the platform (for OCR/HTR or for name-entity recognition purposes) with different kinds of pre-trained models, or define a new one for customized tasks (e.g. labeling of characters for a paleographical description, see *infra* Sect. 6.1). The interface enables collaborative work, with unlimited users by projects. To that end, the project manager has at his disposal numerous statistics to monitor the progress and the built database. For his part, a user involved in an annotation project has access to statistics on his labels as compared to the other users also working on it.

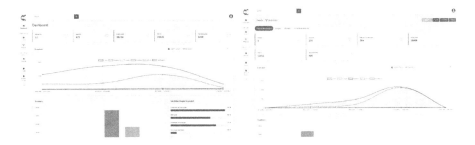

Fig. 2. (i) Main project management dashboard and (ii) dashboard of a project open to crowdsourcing

The main dashboard (see Fig. 2, i) as well as the project dashboard (see Fig. 2, ii) compiles in particular the latest actions (including third party modifications)

and the tasks completed or ongoing for every image (e.g. list of pending automatic predictions). The project manager thus defines the missions and objectives of annotation, visible to all users involved.

4.2   Import/Export

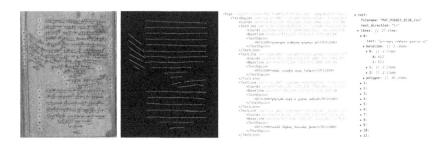

Fig. 3. Original labeled image (predictions and seamcarves corrected by a user), and some customizable inputs/outputs (non exhaustive list): mask of baseline, pageXML file and a customized JSON file. Manuscript: MAF65 (BVMM-IRHT)

Data provided as input can take three forms: (i) a single image; (ii) an image with a pre-existing ground truth and (iii) a link to a IIIF manifest, that imports all metadata from the manuscript. Pre-existing ground truth consists in: (a) an image - plain text pair, (b) an image - labeled image pair (mask of baseline or of region), (c) a HTML file (e.g. *OCRopus* output), (d) an image - pageXML/ALTO pair, (e) an image - customized JSON pair, and (f) a text file. When importing a HTML file, generated by *OCRopus* or by one of its forks, the import infers the baselines associated to the bounding-box. The images are compressed for quick on-screen display, whereas the automatic processing and the exports retain both original format and definition. The platform accepts PDF, TIFF, PNG and JPG formats. The platform enables to import files one at a time or altogether through ZIP file.

Results can be exported according to needs, using the same process as for input: image - text pair, image of a line - text or image - mask pair, pageXML format or other XML formats (ALTO, etc.), JSON format defined by the user. The output feature can also be used to convert data from one format to another, because if the research in layout analysis incorporates today baselines annotation, many older datasets do not have this information and may appear obsolete. To convert them into new standards and formats is a laborious task. Only the project manager has the ability to edit the metadata associated to the images, to export the annotations, to lock the project (once the proofreading is complete) or to launch automatic analysis for one or several images simultaneously.

4.3   Default Annotation Interface

The interface has a classical set of options and layout for annotation projects (see Fig. 4). Display is customizable according to the user's preferences (either image-text display [left-right, right-left, top-bottom] or the text over the image). The focus is on providing an easy-to-use interface and few default options, for the volunteers to get familiar with. In case of numerous labels, the image align with the text considered. The interface accepts RTL languages and enables the correction of predicted or uploaded data, including the correction of baselines and polygons. The first default level of annotation is the text-region, for any given OCR/HTR project. The text-region can be defined by (i) a tag (table, image, footnote, etc.) and (ii) the surrounding polygon. The tags can be customized through the management project settings. The second level of annotation is the text line, that can be defined by (i) the baseline, (ii) the surrounding polygon and (iii) the text (see Fig. 4). A text line must be located within a text-region, and can not be created outside of one. Automatic detection is available for text-region and text lines (see *infra* Sect. 5). Two different geometric objects are available to draw: the polygonal lines (for baselines) and the polygons (for the surrounding polygons of text-region and text lines). The surrounding polygons can also be drawn with a box. Only the project manager has at his disposal more advanced tools like image enhancement settings and automated tasks.

Fig. 4. (i) Annotation interface with identified text-regions and baseline editing and (ii) editing of polygons associated to the text lines and their predicted transcripts. Manuscript: Ber.Or.Quar.805 (Staatsbibliothek zu Berlin - PK)

5   Automatic Annotation and Fine-Tuning Impact

The platforms offers a two-phased semi-automatic annotation: (i) text-region and baselines detection, and (ii) text lines extraction. The first step consists in detecting the surrounding polygons of the text-region in an image, as well as the baselines of the text lines within the text-region defined. We have been using the following pipeline: classification of the pixels image and mask vectorization. The classification of the pixels image – or semantic segmentation – has been achieved

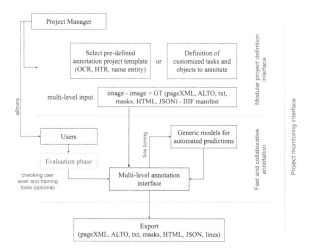

Fig. 5. Detailed operation blueprint of Calfa Vision platform

through a U-net [26] – architecture derived from the Fully Convolutional Network [27]. Nowadays, this type of network is conventional in layout analysis [7,12,22].

We therefore trained several models on the same U-net architecture, depending on the tasks intended and the type of target documents. The vectorization consists in extracting the coordinates of the wanted geometric objects from the obtained mask. For the text-region in particular, we are using the algorithm proposed by [29]. For the baselines, we are undertaking morphological operations to extract and to simplify their skeleton. The second step consists in infering the polygons of the text lines from the baselines. Our approach is based on a seam carve [1]. The predictions are achieved without prior binarization, besides the users do not need to enhance their images. The automatic annotation has been divided into two steps to ensure the full control of the process by the user and to correct effectively the potential mistakes in detection. The user can also choose to simply draw the regions for which he wish to obtain baseline prediction.

The detection models for text-regions and baselines have been specifically trained with the database described in this paper (see *infra* Sect. 6.1) and with various handwritten documents in Arabic scripts (800 pages, notably in Arabic Maghrebi), in Hebrew (500 pages), in Greek (300 pages), in Syriac (250 pages) and Georgian (250 pages), from Libraries we are in partnership with. The training dataset comprises a mix of layouts and also includes some printed documents (newspapers), but is mainly focused on handwritten documents from the Middle Ages up to the mid-20th century. These documents display numerous artifacts and contents in the margins, notably curving text lines. The aim is to propose on the platform several polyvalent models, suited for several projects and in particular focused on under-resourced languages that are under-recognized by state-of-the-art systems (see *supra* Sect. 2).

Fig. 6. Prediction of regions and baselines for different scripts and layouts. [From left to right] (i) Testamentum vetus-1290, f. 27v (Staatsbibliothek zu Berlin - PK, *Hebrew*), (ii) Bulac ARA 609, f. 20v (BULAC, BINA, *Maghrebi Arabic*), (iii) Ber. Sachau 168, f. 58v (Staatsbibliothek zu Berlin - PK, *Syriac*), (iv) Ber. Or. Quart 337, f. 18v (Staatsbibliothek zu Berlin - PK, *Armenian*), (v) MAF54, f. 149v-150r (BVMM-IRHT, *Armenian*), (vi) Safi (Arabic) and (vii) MM-74, f. 303a (Matenadaran, *Armenian*)

The default analysis models (text-regions and baselines) running on the platform are robust on a wide variety of layouts and languages (see Fig. 6). However our strategy relies on the ability of these models to be quickly fine-tuned within an annotation project, in order to meet the needs of the user. For a given project, batch of recently proofread or annotated data can be defined in order to customize the default models of Calfa Vision. The process is imperceptible to the user who sees a substantial gain in the analysis throughout the annotation process (see Table 1 and Fig. 7). The tasks targeted by this strategy are text-region detection, baseline detection and so, line extraction. The fine-tuning approach is already implemented at the OCR and HTR level, whose efficency has been proven for the processing of Latin scripts incunabula [25]. In an annotation project, this approach facilitates data creation, as was the cas with the internal Calfa Vision dataset annotation as well as the RASAM dataset for Arabic Maghrebi scripts (voir *infra* Sect. 6).

We assessed the baseline and region predictions on several classical datasets (see Table 1), first with the Calfa Vision default model and secondly with a fine-tuning strategy. From the perspective of an annotation project of the datasets cBAD simple track (ICDAR 2017), BADAM and READ (ICFHR 2016), we evaluate the benefits of fine-tuning to annotate them. For each competition, the fine-tuning process consists in training successively the Calfa Vision default model with new batches of 50 random samples taken from training dataset of

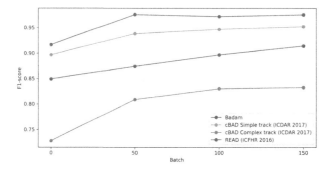

Fig. 7. Evolution of fine-tuning for baseline predictions

the competition. Evaluation is performed on the respective testing sets of each competition. We also assess more precisely this gain and the impact of fine-tuning as part of a dedicated project on Arabic Maghrebi scripts [31].

For baseline detection, we have measured the precision, the recall and the F1 as defined in [14]. The cBAD dataset, the complex track in particular, presents several very different documents (tables, maps, etc.), dissimilar to the target of our training dataset. Nevertheless, Fig. 7 shows that a 50 images batch is sufficient enough to achieve a first accurate model, and with a 150 images batch, the outcomes achieved are consistent with those of the state of the art. BADAM is a good example of the model capacity to be fine-tuned efficiently. This dataset presents indeed great complexity due to the numerous curved or vertical lines. The authors have in particular defined another kind of baseline to manage with verses in Arabic manuscripts, definition very far from the common baseline used in other datasets. As shown in Fig. 6 on several oriental manuscripts[2], the model remains effective on various layouts, with or without scenery. This strategy allows to quickly fit models according to annotator needs in his annotation project.

The targeted datasets don't always have semantic annotation of text-regions, thus limiting the evaluation of the text-region detection model. For the dataset introduced in this paper, we obtain the following results for the text-region detection, with four classic classes (title, paragraph, marginalia and page number): 0.9702% of mean accuracy and 0.9542% of mean IU. The same fine-tuning process allows to go from an UI mean of 0.8935% on all the regions of cBAD2019, to 0.9245% in the first batch of 50 images (addition of new classes) and to 0.9453% in the third batch.

Text-region and baseline predictions could be not entirely relevant for a given set of documents. As for projects including less images variety, as it was for cBAD2019, our first experiments show an appropriate specialization with a new batch of 25 images (see Fig. 6, image vi for which user has added two classes

---

[2] Staatsbibliothek zu Berlin - PK (Testamentum vetus-1290, Sachau 168, Ber. Or. Quart 337), Bibliothèque Virtuelle des Manuscrits Médiévaux (BVMM) – IRHT-CNRS (MAF54), Digitized heritage collections of the Bulac (ARA. 609) and Matenadaran.

Table 1. Evaluation of the platform fine-tuning impact for baselines prediction on various datasets (May 2021)

	Precision (%)	Recall (%)	F1-score (%)
cBAD simple track (ICDAR 2017)			
BYU	0.878	0.907	0.892
DMRZ	0.973	0.970	0.971
dhSegment	0.943	0.939	0.941
ARU-Net	0.977	0.980	0.978
BADAM	0.944	0.966	0.954
Vision (default)	0.8945	0.8991	0.8967
Vision (fine-tuned: 150)	0.9511	0.9538	0.9525
cBAD complex track (ICDAR 2017)			
BYU	0.773	0.820	0.796
DMRZ	0.854	0.863	0.859
Vision (default)	0.7430	0.7131	0.7278
Vision (fine-tuned: 150)	0.7937	0.8761	0.8329
READ (ICFHR 2016)			
Vision (default)	0.7265	0.9353	0.8178
Vision (fine-tuned: 150)	0.9649	0.9868	0.9757
BADAM			
BADAM	0.941	0.901	0.924
Vision (default)	0.8353	0.8641	0.8495
Vision (fine-tuned: 150)	0.9132	0.8575	0.8844

[footnote and running title], allowing the model to predict these two new classes). This functionality is being assessed to automate the fine-tuning process in the future. Meanwhile, default models are regularly updated.

As of now, we propose online text prediction for under-resourced languages, four in particular: Armenian, Georgian, Syriac and Arabic Maghrebi. This option is restricted for our partners for now.

6 Crowdsourcing Campaigns for Dataset Creation

Three types of default annotation projects are proposed: OCR, HTR and name-entity recognition projects. However, the user can create his own annotation project, by defining the objects, their characteristics (e.g. text, coordinates, etc.) and their hierarchy. Many variations can thus be defined for manuscripts classification, language identification in a paragraph, script / copyist identification, simple region and related text annotation (without lines), etc.

There are numerous case studies, we are presenting three of them that held crowdsourcing campaigns recently on the Calfa Vision platform. The main case study we will present is the creation of a dataset for paleographical analysis of Armenian manuscripts, this dataset is included in the training data described earlier.

6.1   Armenian Paleographical Dataset

The number of remaining Armenian manuscripts worldwide is estimated to 31,000. A specimen of this Armenian database is accessible online on the platform and contains 550 labeled images, comprehensive samples from five digital collections of Armenian handwritten documents and one private collection.

The 550 images come from 49 manuscripts that have been copied by more than 60 different hands. They originate from the following collections, from institutions we are partnering with and who are involved in digitization process: the Musée Arménien de France-IRHT CNRS (16 manuscripts), the Matenadaran (15), the Walters Art Museum (10), the Congregation of the Mekhitarists Fathers of Venice (7), and the British Library (1). Images offer different resolutions (from 200 to 600 DPI). The images have been selected to cover the written works from the 10th to the 20th century, and to present every type of writing examined in Armenian paleography [20]. For each manuscript, we chose a dozen pages with various layouts and, whenever possible, at least two different hands or writing tools (e.g. fragments used as flyleaves, colophon pages, etc.).

The remaining Armenian written production is mostly comprised of religious or liturgical writings, that were created with great care. Nevertheless, the selected manuscripts display different types of works: illuminated prayer books, hymnals and religious texts; apocryphical hagiographies; synaxarium; song books; collections of poetry; medieval historical chronicles; late thematic compilations of texts and fragments (miscellanea). The dataset encompasses many difficulties with table structures or curved/vertical lines (see Fig. 6, image vii). In fact, it shares similarities with other Oriental languages datasets [18]. The annotations have been achieved with the platform described above, through crowdsourcing and checked by experienced users. A five-level annotation was implemented in a customized project: (i) at the character level (see Fig. 8 allowing sorting by date or by type of writing); (ii) at the pixel level for baseline (each visual stand-alone entity was labeled separately); (iii) at the line level; (iv) at the text level, transcription made in accordance with the layout of the text, namely the *scriptio continua*; and (v) at the layout of the page level (text-regions and illuminations only). An ideogram dictionary (see *supra* Sect. 3) has been designed to match the text, however some creative ligatures could have been not so rigorously transcribed. These ligatures are either considered as an independant class defined afterwards in postprocessing, or decomposed into the letters comprised.

To date, the public database holds 254,794 characters, 714 marginal illuminations and initials, and 22,414 lines. The assessment was conducted on 110 images. The annotation campaign has been supported by the Calouste Gulbenkian Foundation. The Armenian language raises various problems of recognition (layout and character recognition) that we have described briefly in Sect. 3. The challenge of the Armenian characters recognition lies more in the distinctiveness of its scripts, and the deterioration of its manuscripts than in the layout, overall less-elaborated than in the Latin or Arabic world.

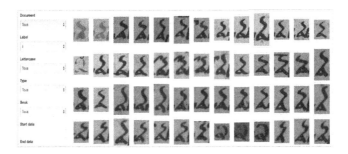

Fig. 8. Creation of a project for paleographical description

6.2   Oriental Historical Manuscripts Datasets

From December 2020 to January 2021, a crowdsourcing campaign took place on the platform for the annotation of medieval manuscripts in Oriental languages (Syriac, Armenian and Arabic) and of Oriental newspapers. The purpose of this campaign has consisted in checking and correcting the automatic predictions (layout and baseline) and in typing the corresponding transcription. The campaign gathered 24 users, divided in 13 projects (10 projects initially planned and 3 created by the users). In total, 1,585 images have been annotated, adding up to 74,385 text lines. The corrected lines have been used, at regular intervals, to specialize the automatic layout analysis models for each project. The short-term format (two months) was preferred to maintain the users' commitment.

With a similar approach, from January 2021 to April 2021 a hackathon was held for the creation of an open dataset for Arabic Maghrebi. It was co-organized by GIS MOMM and the BULAC Libary. The dataset, results and benefits of fine-tuning are evaluated [31] et show a significant gain in annotation (fine-tuning accurate from 50 checked images for layout analysis and 20 checked images for HTR). We get 96.27% of accuracy for layout analysis models from the first batch. This dataset is further described in a separate paper [31].

7   Conclusion

We have presentend a new modular and automated annotation platform, ready-to-use without any computing knowledge, for printed and handwritten documents. It allows the creation of various kind of projects for documents and materials descriptions according to projects needs. Oriental and ancient languages, and the ability to read manuscripts, require extensive academic studies, apparently conflicting with crowdsourcing with often unexperienced volunteers. The interface enables the involvement of users no matter their experience whilst ensuring a system with great versatility, and highly adaptable, to cover the different needs of the scientific community engaged in HTR related research. The aim is to quickly provide resources for under-resourced languages also poorly represented in Digital Humanities. The models presented are polyvalent, robust to various types of documents, and keep on improving thanks to the corrections of the users. We show that our pre-trained models can be quickly fine-tuned to match with a project specification and results on classical datasets are consistent with those of the state of the art. The fine tuning strategy can be considerably improved with an active learning process, consisting of re-training the models with the worst predicted samples. This is a strategy that we plan to explore in future developments. We intend to increase the number of default templates proposed, by adding new dedicated and fine-tuned models, notably for keyword spotting, name-entity recognition and manuscripts classification. The use of external technologies (via API) is ongoing, to allow user to integrate its own character recognition engine. We have tested this interface in particular through the creation of the first database for Armenian manuscripts.

Acknowledgement. Images used as examples in Figs. 3, 4 and 6 come from various open Digital Libraries, that are the Staatsbibliothek zu Berlin - PK, the Bibliothèque Virtuelle des Manuscrits Médiévaux (BVMM) – IRHT-CNRS, the Walters Art Museum and the British Library. We especially thank the Matenadaran, the Musée Arménien de France, the Mekhitarist Congregation of Venice and the BULAC library for providing us HD reproductions of manuscripts for this research.

References

1. Arvanitopoulos, N., Süsstrunk, S.: Seam carving for text line extraction on color and grayscale historical manuscripts. In: 2014 14th International Conference on Frontiers in Handwriting Recognition, pp. 726–731 (2014)
2. BnF: Réalisation d'une étude d'usages des utilisateurs de la plateforme experimentale correct. Technical report. ACM 248, Bibliothèque Nationale de France (2015)
3. Breuel, T.M.: The OCRopus open source OCR system. In: Yanikoglu, B.A., Berkner, K. (eds.) Document Recognition and Retrieval XV, vol. 6815, pp. 120–134. International Society for Optics and Photonics, SPIE (2008)
4. Brookes, S., Stokes, P.A., Watson, M., De Matos, D.M.: The DigiPal project for European scripts and decorations. Essays Stud. **68**, 25–59 (2015)
5. Clausner, C., Antonacopoulos, A., McGregor, N., Wilson-Nunn, D.: ICFHR 2018 competition on recognition of historical arabic scientific manuscripts - RASM2018. In: 2018 16th International Conference on Frontiers in Handwriting Recognition (ICFHR), pp. 471–476 (2018)
6. Diem, M., Kleber, F., Fiel, S., Grüning, T., Gatos, B.: cBAD: ICDAR2017 competition on baseline detection. In: 2017 14th IAPR International Conference on Document Analysis and Recognition (ICDAR), vol. 01, pp. 1355–1360, November 2017

7. Diem, M., Kleber, F., Sablatnig, R., Gatos, B.: cBAD: ICDAR2019 competition on baseline detection. In: 2019 International Conference on Document Analysis and Recognition (ICDAR), pp. 1494–1498 (2019)
8. Garz, A., Seuret, M., Simistira, F., Fischer, A., Ingold, R.: Creating ground-truth for historical manuscripts with document graphs and scribbling interaction. In: 2016 12th IAPR Workshop on Document Analysis Systems (DAS), pp. 126–131 (2016)
9. Gatos, B., et al.: Ground-truth production in the Transcriptorium project. In: 2014 11th IAPR International Workshop on Document Analysis Systems, pp. 237–241 (2014)
10. Gatos, B., et al.: GRPOLY-DB: an old Greek polytonic document image database. In: 2015 13th International Conference on Document Analysis and Recognition (ICDAR), pp. 646–650 (2015)
11. Granell, E., Romero, V., Martínez-Hinarejos, C.D.: Multimodality, interactivity, and crowdsourcing for document transcription. Comput. Intell. **34**, 398–419 (2018)
12. Grüning, T., Leifert, G., Strauß, T., Labahn, R.: A two-stage method for text line detection in historical documents. Int. J. Doc. Anal. Recogn. (IJDAR) (2018)
13. Grüning, T., Labahn, R., Diem, M., Kleber, F., Fiel, S.: READ-BAD: a new dataset and evaluation scheme for baseline detection in archival documents. In: 2018 13th IAPR International Workshop on Document Analysis Systems (DAS), pp. 351–356 (2018)
14. Grüning, T., Labahn, R., Diem, M., Kleber, F., Fiel, S.: READ-BAD: a new dataset and evaluation scheme for baseline detection in archival documents. arXiv:1705.03311 [cs] (2017)
15. Ingle, R.R., Fujii, Y., Deselaers, T., Baccash, J., Popat, A.C.: A scalable handwritten text recognition system. In: 2019 International Conference on Document Analysis and Recognition (ICDAR), pp. 17–24. IEEE (2019)
16. Kahle, P., Colutto, S., Hackl, G., Mühlberger, G.: Transkribus - a service platform for transcription, recognition and retrieval of historical documents. In: 2017 14th IAPR International Conference on Document Analysis and Recognition (ICDAR), vol. 04, pp. 19–24 (2017)
17. Kassis, M., Abdalhaleem, A., Droby, A., Alaasam, R., El-Sana, J.: VML-HD: the historical Arabic documents dataset for recognition systems. In: 2017 1st International Workshop on Arabic Script Analysis and Recognition (ASAR), pp. 11–14 (2017)
18. Kiessling, B., Ezra, D.S.B., Miller, M.T.: BADAM: a public dataset for baseline detection in Arabic-script manuscripts. In: Proceedings of the 5th International Workshop on Historical Document Imaging and Processing, HIP 2019, pp. 13–18. Association for Computing Machinery (2019)
19. Kiessling, B., Tissot, R., Stokes, P., Ezra, D.S.B.: eScriptorium: an open source platform for historical document analysis. In: 2019 International Conference on Document Analysis and Recognition Workshops (ICDARW), vol. 2, pp. 19–19. IEEE (2019)
20. Kouymjian, D., Stone, M., Lehmann, H.: Album of Armenian Paleography. Aarhus University Press (2002)
21. Leifert, G., Strauß, T., Grüning, T., Labahn, R.: CITlab ARGUS for Historical Handwritten Documents (2016)
22. Lombardi, F., Marinai, S.: Deep learning for historical document analysis and recognition-a survey. J. Imaging **6**(10), 110 (2020)

23. Neudecker, C., et al.: OCR-D: an end-to-end open source OCR framework for historical printed documents. In: Proceedings of the 3rd International Conference on Digital Access to Textual Cultural Heritage, pp. 53–58 (2019)

24. Quirós, L.: Multi-Task Handwritten Document Layout Analysis. arXiv:1806.08852 [cs] (2018)

25. Reul, C., et al.: OCR4all-an open-source tool providing a (semi-) automatic OCR workflow for historical printings. Appl. Sci. **9**(22), 4853 (2019)

26. Ronneberger, O., Fischer, P., Brox, T.: U-net: convolutional networks for biomedical image segmentation. arXiv:1505.04597 [cs] (2015). arXiv: 1505.04597

27. Shelhamer, E., Long, J., Darrell, T.: Fully convolutional networks for semantic segmentation. arXiv:1605.06211 [cs] (2016)

28. Simistira, F., Seuret, M., Eichenberger, N., Garz, A., Liwicki, M., Ingold, R.: DIVA-HisDB: a precisely annotated large dataset of challenging medieval manuscripts. In: 2016 15th International Conference on Frontiers in Handwriting Recognition (ICFHR), pp. 471–476 (2016)

29. Suzuki, S., be, K.: Topological structural analysis of digitized binary images by border following. Comput. Vis. Graph. Image Process. **30**(1), 32–46 (1985)

30. Sánchez, J.A., Romero, V., Toselli, A.H., Vidal, E.: ICFHR2016 competition on handwritten text recognition on the READ dataset. In: 2016 15th International Conference on Frontiers in Handwriting Recognition (ICFHR), pp. 630–635 (2016)

31. Vidal-Gorène, C., Lucas, N., Salah, C., Decours-Perez, A., Dupin, B.: RASAM - a dataset for the recognition and analysis of scripts in arabic maghrebi. In: Barney Smith, E.H., Pal, U. (eds.) ICDAR 2021. LNCS, vol. 12916 (2021). https://doi.org/10.1007/978-3-030-86198-8_19

32. Wada, K.: labelme: image polygon annotation with Python (2016). https://github.com/wkentaro/labelme

Reducing the Human Effort in Text Line Segmentation for Historical Documents

Emilio Granell[1]([✉])[iD], Lorenzo Quirós[1][iD], Verónica Romero[2][iD], and Joan Andreu Sánchez[1][iD]

[1] Pattern Recognition and Human Language Technology Research Center, Universitat Politècnica de València, Valencia, Spain
{emgraro,loquidia,jandreu}@prhlt.upv.es
[2] Departament d'Informàtica, Universitat de València, Valencia, Spain
veronica.romero@uv.es

Abstract. Labeling the layout in historical documents for preparing training data for machine learning techniques is an arduous task that requires great human effort. A draft of the layout can be obtained by using a document layout analysis (DLA) system that later can be corrected by the user with less effort than doing it from scratch. We research in this paper an iterative process in which the user only supervises and corrects the given draft for the pages automatically selected by the DLA system with the aim of reducing the required human effort. The results obtained show that similar DLA quality can be achieved by reducing the number of pages that the user has to annotate and that the accumulated human effort required to obtain the layout of the pages used to train the models can be reduced more than 95%.

Keywords: Document layout analysis · Text line segmentation · Human effort reduction · Historical document

1 Introduction

The state of the art in document digitalization has increased the interest in preserving and providing access to handwriting historical documents. A large number of institutions worldwide, such as libraries, museums and archives, have been performing mass digitization of paper documents for more than twenty years, and the results are stored in digital libraries, usually web-accessible to the general public. Among the digitized documents, it is worth highlighting the large collections of historical documents containing records of quotidian activities. These historical record documents contain only limited information when considered individually, but provide an intriguing look into the historic life when considered as a complete collection. Examples of these kind of documents are birth, marriage, and death records, notarial and court records, border and census forms, medical forms, and ship logs, among many other collections.

The paper-to-image conversion ensures long term preservation of documents, which in some cases were in advanced state of degradation with a high risk of loss

© Springer Nature Switzerland AG 2021
J. Lladós et al. (Eds.): ICDAR 2021, LNCS 12823, pp. 523–537, 2021.
https://doi.org/10.1007/978-3-030-86334-0_34

of information. However, unless the textual information contained in the scanned images is digitally extracted, accessibility to the document contents is limited if not impossible. Moreover, in order to make these digital documents really useful their textual contents need to be transcribed, so they can be provided for searching, indexing and querying them.

However, manually transcribing these documents is a slow and expensive process which by no means can be afforded for large collections of historical manuscripts.

To alleviate this problem, Handwritten Text Recognition (HTR) technology can be used. For instance, approaches referred to as "segmentation free off-line HTR" [10,27,28] are available. These approaches aim at recognizing all text elements in a line (words and characters) as a whole, without any prior segmentation of the image into these elements. The only segmentation process required in these documents is in text lines.

Therefore, given that the input of the HTR systems are the text line images to be transcribed, some previous steps of *document layout analysis* (DLA) where the text lines in the page images are detected and extracted is required. This process is known as Text Line Segmentation (TLS) and it is widely accepted that can be solved by solving the baseline detection problem [23]. TLS or baseline detection is particularly hard in historical handwritten text, due to document page images containing text lines with different orientations, slope variations through the length of the line, touching and overlapping characters from adjacent lines and varying inter-line and inter-word spacing issues.

TLS in printed documents can be easily resolved, even when the documents present warping or some type of degradation [15]. However, this problem can be very difficult when handwritten documents are considered, specially historical documents due to the lack of strict layout rules and degradation problems [15].

Different techniques have been proposed in the literature to solve the TLS problem, including vertical projection profiles and Hough transformation [9,15, 17], dynamic programming to search for optimal paths between overlapping text lines [16], making use of HTR training information [4] and natural language inspired proposals [5,8]. In the recent years, end-to-end systems, based on neural networks, where all the main document components (text areas and text lines) are detected at the same time have become popular [6,18,20]. In this case the main drawback is the amount of data needed to train a good model, and therefore efficient techniques have to be developed to alleviate the necessary human effort.

This paper presents a semi-supervised TLS scenario, where in every iteration the user is asked to manually correct some pages in order to increment the training material. A study about the theoretical reduction of the human effort in this correction process is carried out. This study has been carried out in two historical record collections. On one hand we have used the *Oficio de Hipotecas de Girona* (OHG) dataset [22], which is composed by notarial records. On the other hand, the *HisClima* database [19] has been used. This database was compiled from the log book of a ship and is composed mainly by handwritten tables.

The rest of the paper is structured as follows: recent related works are reviewed in the next section (Sect. 2); the proposed DLA process is detailed in Sect. 3; the experimental framework is described in Sect. 4; the performed experiments and the obtained results are reported in Sect. 5; the conclusions and future work lines are drawn in Sect. 6; finally, the information necessary to reproduce this work can be found in Sect. 7.

2   Related Work

Automatically detecting text lines (baselines) from document images has been long studied. However, most researchers today are focusing on boosting the detection rate paying less attention to the human effort needed to generate the ground-truth required by most of the systems to train their models or to correct any error left by the system.

Nevertheless, some works pointed out the cost of generating a big grand-truth and propose new datasets generated semi-automatically or as the union of other smaller datasets [11,29], while others focus on how to reuse data already labeled for a different dataset or a different task [26].

Having the user in mind, other papers pointed out how to reduce the cost of labeling new data. For instance in [1,2,7] a set of user interfaces were developed to facilitate the task to the users, while [23] pointed out that the cost of text-line labeling can be significantly reduced by annotate baselines instead of the complete polygon that surrounds a text-line. In [12] an adversarial approach is proposed to extend the training procedure to unlabeled samples using a loss function that takes into account a confidence map. In this work, we estimate a similar confidence measure, while keeping the user in the loop to review the selected samples.

On the other hand, in [20,21] experiments were carried out to determine the effect of increasing the number of images in the training set when the new pages comes from a new batch generated by an user. Which is a more realistic approach than the common academic setup (where new images are randomly selected).

In this work, we analyze a more realistic semi-supervised scenario, where we keep the user in the loop in order to estimate the best ratio between user effort and effectiveness of the baseline detection system. In effect, many active learning strategies have been proposed to select the unlabeled samples that best improves the model [25,30]. Moreover, as we also want to minimize the user effort per page, in this work we will focus on the use of the *most* confident samples, while uncertainty sampling [14] and other strategies are left to future works.

3   Proposed Document Layout Analysis Process

Labeling the layout of historical documents by hand and from scratch is an arduous task. Therefore, instead of labeling lots of pages manually we propose an iterative process with computational assistance with the aim of reducing the human effort in two ways: reducing the number of pages that needs to be labeled

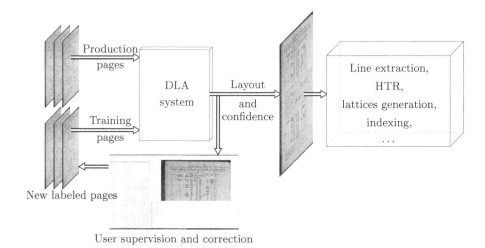

Fig. 1. Proposed iterative semi-supervised document layout analysis process.

to train a good DLA model, and reducing the effort of labeling each page by offering a draft that the user must check and correct if necessary.

Figure 1 shows a representation of our proposal. On one hand we have a set of labeled pages (training pages) that will be used to train an initial DLA model. On the other hand we have a set of unlabeled pages (production pages) that will be used to refine the model.

In this process a DLA model initialized with a few pages (manually labeled) is iteratively re-trained adding new labeled pages to the training set. These new pages are selected according to a confidence measure provided by the DLA system, and then reviewed and corrected by the user, helped by a graphical tool [13]. This way, the human effort required to obtain the layout for a collection of historical documents can be considerably reduced.

Then, the obtained layout can be used in later stages to, for example, extract the text lines, perform automatic handwritten text recognition, generate word graphs and/or content indexing.

On this work we simulate the user review of the drafts by substituting the draft by the ground-truth available for each dataset. Note that in such case, the evaluation of the human effort reduction is computed automatically as we describe below.

3.1   Document Layout Analysis System

The DLA system[1] uses a model to detect the baselines to generate the layout draft and a second model to estimate the confidence in the generated draft.

---

[1] P2PaLA: https://github.com/lquirosd/P2PaLA.

Baseline Detection Model. We used the technique that was presented in [20] for baseline detection, not just because the competitive results presented but because it can be trained very fast from scratch. Equally important is to notice that this technique provides a way to estimate the confidence of the results obtained at page level as we explain in the next section.

The technique is proposed as a two stages process. First, following the formulation in [20] for a single task—baseline detection in our case—our problem is a binary classification problem under a random variable y (background: $y = 0$, baseline: $y = 1$) for each input image x, where the conditional distribution $p(y \mid x)$ is estimated under a naive Bayes assumption for each pixel in the image by M-net, the Unet-like Neural Network presented in [20].

Under this assumption, the classification problem can be computed element by element as:

$$y^\star_{i,j} = \arg\max_{y \in \{0,1\}} \mathcal{M}_{i,j,y}(x), \quad 1 \leq i \leq w, 1 \leq j \leq h \tag{1}$$

where $\mathcal{M}(\cdot)$ is the output of the latest layer of M-net, and w and h are the width and height of the input image respectively.

Then, in the second stage, a set of contours are extracted from the previously classified image $(y^\star)$ as the set of connected components on it. Finally for each contour one baseline is extracted as the simplified digital curve that best fits all the bottom black pixels in the binarized image of x inside the contour.

Confidence Estimation Model. In general DLA confidence estimation is a complex problem, because most state-of-the-art systems rely on independence assumptions that prevent it of compute the confidence at page level, directly from the model used to detect the baselines. For instance, in the system used in the experiments described in this paper, each pixel is assumed independent on the classification stage and then in a second stage the baselines are detected.

Although a pixel level confidence can be computed, it does not make sense for our problem because the input of the system is the whole page. Instead, a page level confidence is preferred.

Given a confidence level for a page, the best pages can be selected to improve the baseline detection model, or the user can be directed to review those pages that helps the most to improve the baseline detection model.

This confidence can be estimated by means of random variable $z \in \{0, 1\}$. This is $z = 0$ if the generated hypothesis do not belong to the input image x and $z = 1$ if the generated hypothesis belongs to x. Then the conditional probability distribution $p(z \mid x, y)$ can be interpreted as the confidence that the hypothesis y belong to the image x.

In this work we estimate the confidence at page level by means of an Adversarial Neural Network. This Adversarial Neural Network is trained in parallel to the main network to estimate if the hypotheses generated by the baseline detection model are equal to the ground-truth or not.

Formally, lets $\mathcal{A}(\cdot)$ the output of the adversarial network and $\mathcal{M}(\cdot)$ the output of the main network. The adversarial network is trained by minimizing the following loss function:

$$\mathcal{L}_A(\boldsymbol{X}, \boldsymbol{Y}) = \frac{-1}{2N} \sum_{n=1}^{N} \log \mathcal{A}(\boldsymbol{x}_n, \boldsymbol{y}_n) + \log(1 - \mathcal{A}(\boldsymbol{x}_n, \arg\max_y \mathcal{M}(\boldsymbol{x}_n))) \quad (2)$$

where N is the batch size. Finally the confidence is estimated for each page as:

$$\mathrm{p}(z \mid \boldsymbol{x}, \boldsymbol{y}) = \mathcal{A}(\boldsymbol{x}, \arg\max_y \mathcal{M}(\boldsymbol{x})) \quad (3)$$

4   Experimental Framework

Extracting useful conclusion from the proposed method introduced in this paper could be highly conditioned by the task. Therefore, we considered two very different tasks. In one task the layout is not very difficult since the lines can be easily detected. In the other task, tables with text and numbers mixed with running text is considered. This section presents these two historical manuscripts used in the experiments, and the evaluation metrics.

4.1   Datasets

Oficio de Hipotecas de Girona (OHG). The Oficio de Hipotecas de Girona (OHG) collection is composed of hundreds of thousands of notarial deeds from the XVIII and XIX centuries (1768–1862), and it is provided by the Centre de Recerca d'Història Rural from the Universitat de Girona (CRHR)[2]. Sales, redemption of censuses, inheritance and matrimonial chapters are among the most common documentary typologies in the collection. This collection is divided in batches of 50 pages each, digitized at 300ppi in 24 bit RGB color, available as TIF images along with their respective ground-truth layout in PAGE XML format. OHG pages exhibit a relatively low complex layout, with 40 text lines per page in average. Examples are depicted in Fig. 2.

We use a publicly available portion of 350 pages from the collection [22], divided randomly into training, validation, production and test sets, 5 pages, 15 pages, 280 pages and 50 pages, respectively. Splits are available online along with the source code used in this experiments (see Sect. 7).

Hisclima Database. The HisClima database is a freely available handwritten text database [19] compiled from a log book of a ship called Jeannette, which sailed the Arctic ocean from July of 1880 until February of 1881.

This logbooks documents are composed by two different kind of pages, some pages containing tables and other ones containing descriptive text. The Jeannette

---

2 http://www2.udg.edu/tabid/11296/Default.aspx.

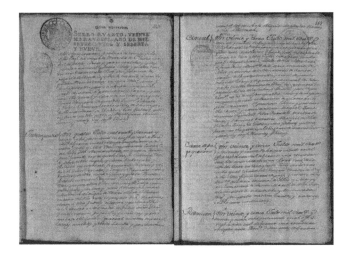

Fig. 2. Two pages of the OHG dataset.

log book follows this structure and it is composed of 419 pages, 208 correspond to table pages and the other 211 correspond to descriptive text.

Given that the most relevant information for researchers of this kind of documents is included in the tables, that contain a lot of numerical data, we focus only on table pages. These pages are divided into two parts, the upper part is for registering the information in the AM period of each day and the bottom part registers the information referred to the PM period of each day (see Fig. 3).

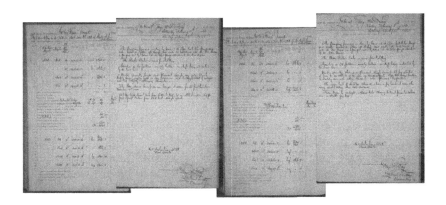

Fig. 3. Example of ship logs with annotation about the weather conditions.

These documents entail some challenges related with different areas such as layout analysis, handwritten recognition and information extraction. One of the main difficulties for automatic layout analysis come from the fact that the information included in a cell sometimes is replaced with quotation marks (") when the data is the same as the data in the same column in the previous row. This quotation marks are very short and sometimes they are difficult to automatically detect. Another layout difficulties that can be found are: data related with a cell that is really written in the upper and lower cells, crossed out column names, words written between cells, and different number of rows completed in every table.

The HisClima database has been endowed with two different types of annotations. First, a layout analysis of each page was done to indicate blocks, columns, rows, and lines. Second, the text was completely transcribed by an expert paleographer, including relevant semantic information.

With respect to the layout analysis, in the table regions the lines were marked at cell level, resulting in a total of 3 525 lines.

In this paper, the partitions defined in [24] have been used. In these partitions, the 208 pages of the HisClima database has been divided into three shuffled subsets: one composed by 143 pages for training (here further divided randomly into 5 pages for training and 138 for production), another of 15 pages for validation, and the last one composed by 50 pages for testing the trained system.

4.2   Evaluation Metrics

We report precision (P), recall (R) and its harmonic mean (F1) measures as defined specifically for the baseline detection problem in [11]. Tolerance parameters are set to default values in all experiments (see [11] for details about measure definition, tolerance values and implementation).

Precision represents the probability that a random section of a baseline in the hypothesis exits as part of a baseline in the ground-truth. Correspondingly, Recall represents the probability that a random section of a baseline in the ground-truth, will also be predicted to exist in the hypothesis.

The F1 score corresponds to the harmonic mean between precision and recall:

$$F1 = 2 \cdot \frac{P \cdot R}{P + R} \tag{4}$$

In a similar way, we can estimate the human effort necessary to correct the recognized layout by means of the false discovery rate (FDR $= 1 - P$) and the miss rate (MR $= 1 - R$). The false discovery rate represents an upper-bound of the percentage of the recognized baselines that needs to be corrected, while the miss rate represents an upper-bound of the percentage of the baselines that were not recognized and need to be added to the layout. Then, we can define their harmonic mean to estimate the total human effort:

$$\text{Effort} = 2 \cdot \frac{\text{FDR} \cdot \text{MR}}{\text{FDR} + \text{MR}} \tag{5}$$

It is important to notice that FDR and MR are upper-bounds of the user effort. Hence, both should be considered pessimistic.

Furthermore, the statistical significance of the experimental results is esti-mated by means of confidence intervals of probability 95% ($\alpha = 0.025$) calculated using the bootstrapping method with $10,000$ repetitions [3].

5   Experimental Results

In the experiments, an initial DLA model (called touchstone model) will be iteratively refined by adding new samples to the training set in each iteration. In an initial effort, the user labels 20 pages: 15 from the validation set and only 5 from the training set. The rest of unlabeled production pages of the set will be processed in each iteration and according to the confidence measured will be selected to be supervised by the user and added to the training set for the next iteration.

In order to limit the user workload, in each iteration between 1 and 5 pages are selected. This selection is made by ordering the pages that remain to be processed (those that have already been supervised are not processed again) from highest to lowest confidence and those that exceed a threshold are selected. The threshold is dynamic, it starts at 0.9 and every time 5 pages are selected, the confidence value of the fifth page (the one with the lowest confidence) is taken as the new threshold.

The performance of the DLA system is compared in two modes: semisuper-vised and unsupervised. In the semisupervised mode the layout of the selected pages is corrected by the user before to be added to the training set, while in the unsupervised mode the obtained layout is added as it is.

The model obtained in each iteration is evaluated with a test set of 50 pages that are never used in any other part of the process.

5.1   Touchstone and Oracle Results

In a preliminar experiment the initial models were obtained as a touchstone. Additionally, DLA models were trained by using the full training set labelled in order to estimate the best performance that can be achived for each corpus, we call this later model 'oracle'.

As can be observed the obtained results for both corpora (Table 1) five pages is not enough to obtain good results. Using all the pages of the training set gives almost perfect results in the case of OHG. But, in the case of HisClima, although they are not that good, they, also show an improvement of around 30% when more pages are added to the training set.

Regarding the human effort necessary to correct the obtained layout for the test set, as can be observed in Table 2, the touchstone models reduce the expected effort to less than 30% respect to labeling it from scratch. However, the oracle models reduce the expected effort to less than 10% in the case of HisClima and to less than 1% en the case of OHG. Which means that we have a room for improvement of about 20% on HisClima and 29% on OHG.

Table 1. Quality for the test set given by the DLA touchstone and oracle models.

Measure	HisClima		OHG	
	Touchstone	Oracle	Touchstone	Oracle
Precision	0.61 [0.56, 0.65]	0.90 [0.84, 0.94]	0.41 [0.37, 0.45]	0.98 [0.97, 0.99]
Recall	0.77 [0.76, 0.79]	0.85 [0.84, 0.87]	0.82 [0.76, 0.86]	0.98 [0.98, 0.99]
F1	0.66 [0.60, 0.70]	0.85 [0.80, 0.89]	0.53 [0.49, 0.58]	0.98 [0.98, 0.99]

Table 2. Human effort required to correct the drafts given by the DLA touchstone and oracle models for the test partition.

Measure	HisClima		OHG	
	Touchstone	Oracle	Touchstone	Oracle
FDR	0.39 [0.35, 0.44]	0.10 [0.06, 0.16]	0.59 [0.56, 0.63]	0.02 [0.01, 0.03]
MR	0.23 [0.21, 0.25]	0.15 [0.13, 0.16]	0.18 [0.14, 0.24]	0.02 [0.01, 0.02]
Effort	0.27 [0.25, 0.29]	0.09 [0.08, 0.10]	0.26 [0.21, 0.31]	0.01 [0.01, 0.02]

5.2   Iterative Results

In the following experiment an iterative process will be followed in order to estimate the minimum human effort required to achieve similar DLA results than the oracle models.

As can be observed in Fig. 4, the width of the confidence intervals gives us an idea of the difficulty of the task, specifically in the case of HisClima. In this case, the confidence intervals for F1 (see Fig. 4(a)) overlap from the third iteration for both the semi-supervised and unsupervised processes with the results obtained for the oracle model. However, from the tenth iteration, it can be observed how the results continue to improve in the semi-supervised process while the unsupervised process tends to converge to a value of F1 around 0.72.

The semi-supervised process reaches the F1 values of the oracle model (that was trained using 143 pages) in the iteration 20, using 61 pages to train the model, the initial 5 and 56 corrected by the user. This means that similar results can be obtained using 57.3% less pages.

In the case of OHG (see Fig. 4(b)), from the eleventh iteration, significant differences are observed between the semi-supervised and the unsupervised processes. The unsupervised process tends to converge to a value of F1 around 0.82, while the semi-supervised process reaches the F1 values of the oracle model (trained with 280 pages) in the iteration 54, using only 101 pages (64.6% less pages), the initial 5 and 96 corrected by the user.

For both corpora, Table 3 presents the detail of the obtained quality results, Table 4 the effort required to correct the drafs given by the DLA models in the optimal iterations (20 for HisClima and 54 for OHG), for the test partitions, and Table 5 presents an overview of the number of pages used in the training sets.

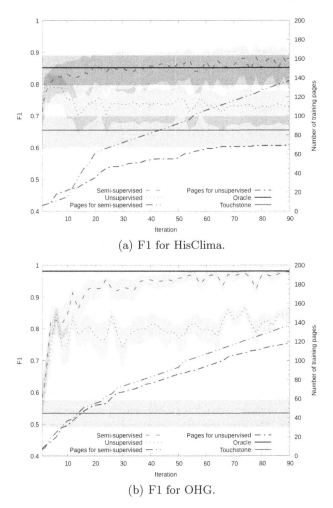

(a) F1 for HisClima.

(b) F1 for OHG.

Fig. 4. Quality measures for the test partitions of both corpora.

It has been found that similar results can be obtained with a smaller number of labeled pages for training. Moreover, by using the iterative process, the human effort required to obtain the labels of additional training pages is considerably reduced. In Fig. 5 the estimated human effort to correct the selected pages per iteration and accumulated is presented. For both corpora the same trend can be observed, as the model is refined with new supervised samples it offers better results and therefore the human effort to correct new pages is reduced.

Taking into account that labeling from scratch the entire production partitions represents 100% of the human effort, we can make the following assumptions. In the case of HisClima, the effort accumulated in the iteration 20 in correcting the 56 pages is only 4.1% of the effort required to label all the 138

Table 3. Quality for the test set given by the DLA models in the optimal iteration.

Measure	HisClima	OHG
	Iteration 20	Iteration 54
Precision	0.91 [0.85, 0.95]	0.96 [0.94, 0.98]
Recall	0.85 [0.84, 0.87]	0.98 [0.97, 0.98]
F1	0.86 [0.81, 0.90]	0.97 [0.95, 0.98]

Table 4. Human effort required to correct the drafts given by the DLA models in the optimal iteration for the test partition.

Measure	HisClima	OHG
	Iteration 20	Iteration 54
FDR	0.09 [0.05, 0.15]	0.04 [0.02, 0.07]
MR	0.14 [0.13, 0.15]	0.02 [0.02, 0.03]
Effort	0.07 [0.06, 0.08]	0.02 [0.01, 0.02]

pages of the production set (and as mentioned before the qualitative results are the same). However, if we do not use the drafts the estimated effot for labelling the selected 56 out of 138 pages from scratch is 40.6%. Therefore, when using the drafts provided by our proposal, a reduction in effort of 95.9%[3] is achieved.

For OHG the accumulated effort in the iteration 54 for correcting the 96 out of 280 pages is reduced to 1.5% of the effort required to label the entire production set, while the estimate effort required for labelling the selected pages whithout the help of the draft is 34.3%. As a result, a reduction in effort of 98.5%[4] can be achieved respect to labeling all the 280 pages from scratch.

Table 5. Overview of the number of pages used in the training set. **Best** represent the optimal iteration, the number 20 for HisClima and the number 54 for OHG.

	HisClima			OHG		
	Touchstone	Best	Oracle	Touchstone	Best	Oracle
Labelled pages	5	5	143	5	5	285
Semisupervised pages	–	56	–	–	96	–
Total number of pages	5	61	143	5	101	285

---

[3] According to a consulted paleographer, the estimated human work time required for labelling HisClima from scratch is around 20 minutes per page and only 5 minutes in the case of OHG. This means that our proposal can save more than 44 hours of human work in the case of HisClima.

[4] It represents a saving of around 23 hours of human work for labelling OHG.

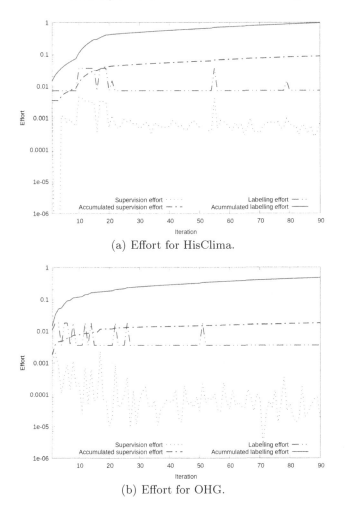

(a) Effort for HisClima.

(b) Effort for OHG.

Fig. 5. Human effort to correct the layout of the selected pages of both corpora.

6   Conclusions and Future Work

Labeling the layout of historical documents is an arduous task that requires great human effort. A draft of the layout can be obtained by using a DLA system that later can be corrected by the user with less effort than doing it from scratch.

In this work we have proposed an iterative process in which the user should only supervise and correct the given draft for the selected pages.

In the experiments, carried out with two different historical manuscripts, it was observed that is not necessary to have a large number of labeled pages to train DLA models. In addition, the proposed iterative approach allows to considerably reduce the human effort necessary to obtain the layout of the selected pages by correcting a draft that, as the model is refined with new samples, contains fewer errors and therefore requires less effort.

Concretely, we obtained models capable of offering a quality similar to that obtained with the models trained with all the training samples, but using only 42.7% and 35.4% of the available pages for each one of the two manuscripts used on the experimentation. Moreover, the expected user effort is reduced in more than 95% in both cases, when the drafts are used to help the user in the process.

Future work lines include the experimentation with other datasets, tests with real users, and the exploration of new assistive and interactive strategies such as selecting the pages with less confidence to check if the DLA model improves when correcting the most difficult samples and how it affects human effort.

7   Reproducibility

We hope this work will help to boost research on how to reduce the human effort to fix DLA problems and related research areas. For this reason, the source code used in the experiments presented in this paper is freely available on https://github.com/PRHLT/iterative-dla-icdar2021, along with the subsets used as training, validation, test and production for each dataset.

Acknowledgement. The authors thank Alejandra Macián Fuster for her collaboration as paleographer. Work partially supported by the Universitat Politècnica de València under grant FPI-II/900, by EU JPICH project "HOME—History Of Medieval Europe" (PEICTI Ref. PCI2018-093122), by the Generalitat Valenciana under the EU-FEDER Comunitat Valenciana 2014-2020 grant IDIFEDER/2018/025 "Sistemas de fabricación inteligente para la indústria 4.0", and by Ministerio de Ciencia, Innovación y Universidades project DocTIUM (Ref. RTI2018-095645-B-C22).

References

1. Acuna, D., Ling, H., Kar, A., Fidler, S.: Efficient interactive annotation of segmentation datasets with polygon-RNN++. In: CVPR, pp. 859–868 (2018)
2. Andriluka, M., Uijlings, J.R., Ferrari, V.: Fluid annotation: a human-machine collaboration interface for full image annotation. In: ACM MM, pp. 1957–1966 (2018)
3. Bisani, M., Ney, H.: Bootstrap estimates for confidence intervals in ASR performance evaluation. In: ICASSP, vol. 1, pp. 409–412 (2004)
4. Blouche, T., Moysset, B., Kermorvant, C.: Automatic line segmentation and ground-truth alignment of handwritten documents. In: ICFHR, pp. 667–672 (2014)
5. Bosch, V., Toselli, A.H., Vidal, E.: Semiautomatic text baseline detection in large historical handwritten documents. In: ICFHR, pp. 690–695 (2014)
6. Chen, K., Seuret, M., Hennebert, J., Ingold, R.: Convolutional neural networks for page segmentation of historical document images. In: ICDAR, pp. 965–970 (2017)
7. Dutta, A., Zisserman, A.: The via annotation software for images, audio and video. In: MM, vol. 27. ACM (2019)
8. Fernandez, F.C., Terrades, O.R.: Handwritten line detection via an EM algorithm. In: ICDAR, pp. 718–722 (2013)

9. Gatos, B., Louloudis, G., Stamatopoulos, N.: Segmentation of historical handwritten documents into text zones and text lines. In: ICFHR, pp. 464–469 (2014)
10. Graves, A., Liwicki, M., Fernández, S., Bertolami, R., Bunke, H., Schmidhuber, J.: A novel connectionist system for unconstrained handwriting recognition. IEEE TPAMI **31**(5), 855–868 (2009)
11. Grüning, T., Labahn, R., Diem, M., Kleber, F., Fiel, S.: READ-BAD: a new dataset and evaluation scheme for baseline detection in archival documents. CoRR (2017). http://arxiv.org/abs/1705.03311
12. Hung, W.C., Tsai, Y.H., Liou, Y.T., Lin, Y.Y., Yang, M.H.: Adversarial learning for semi-supervised semantic segmentation. In: BMVC (2018)
13. Kahle, P., Colutto, S., Hackl, G., Mühlberger, G.: Transkribus - a service platform for transcription, recognition and retrieval of historical documents. In: ICDAR, vol. 04, pp. 19–24 (2017)
14. Lewis, D.D., Gale, W.A.: A sequential algorithm for training text classifiers. In: SIGIR, pp. 3–12 (1994)
15. Likforman-Sulem, L., Zahour, A., Taconet, B.: Text line segmentation of historical documents: a survey. IJDAR **9**, 123–138 (2007)
16. Liwicki, M., Indermuhle, E., Bunke, H.: On-line handwritten text line detection using dynamic programming. In: ICDAR, vol. 1, pp. 447–451 (2007)
17. Louloudis, G., Gatos, B., Pratikakis, I., Halatsis, C.: Text line and word segmentation of handwritten documents. Pattern Recogn. **42**(12), 3169–3183 (2009)
18. Oliveira, S.A., Seguin, B., Kaplan, F.: dhSegment: a generic deep-learning approach for document segmentation. In: ICFHR (2018)
19. PRHLT: Hisclima dataset, October 2020. https://doi.org/10.5281/zenodo.4106887
20. Quirós, L.: Multi-task handwritten document layout analysis. arXiv e-prints, 1806.08852 (2018). https://arxiv.org/abs/1806.08852
21. Quirós, L., Bosch, V., Serrano, L., Toselli, A.H., Vidal, E.: From HMMs to RNNs: computer-assisted transcription of a handwritten notarial records collection. In: ICFHR, pp. 116–121 (2018)
22. Quirós, L., et al.: Oficio de Hipotecas de Girona. A dataset of Spanish notarial deeds (18th Century) for Handwritten Text Recognition and Layout Analysis of historical documents (2018). https://doi.org/10.5281/zenodo.1322666
23. Romero, V., Sánchez, J.A., Bosch, V., Depuydt, K., de Does, J.: Influence of text line segmentation in handwritten text recognition. In: ICDAR, pp. 536–540 (2015)
24. Romero, V., Sánchez, J.A.: The HisClima database: historical weather logs for automatic transcription and information extraction. In: ICPR (2020)
25. Settles, B., Craven, M.: An analysis of active learning strategies for sequence labeling tasks. In: EMNLP, pp. 1070–1079 (2008)
26. Studer, L., et al.: A comprehensive study of imagenet pre-training for historical document image analysis. In: ICDAR, pp. 720–725 (2019)
27. Toselli, A.H., et al.: Integrated handwriting recognition and interpretation using finite-state models. IJPRAI **18**(4), 519–539 (2004)
28. Voigtlaender, P., Doetsch, P., Ney, H.: Handwriting recognition with large multidimensional long short-term memory recurrent neural networks. In: ICFHR, pp. 228–233 (2016)
29. Zhong, X., Tang, J., Yepes, A.J.: Publaynet: largest dataset ever for document layout analysis. In: ICDAR, pp. 1015–1022. IEEE (2019)
30. Zhu, X.: Semi-supervised learning literature survey. Technical report, TR-1530, University of Wisconsin-Madison (2008)

DSCNN: Dimension Separable Convolutional Neural Networks for Character Recognition Based on Inertial Sensor Signal

Fan Peng, Zhendong Zhuang, and Yang Xue[✉]

School of Electronic and Information Engineering,
South China University of Technology, Guangzhou, China
{202020112442,eezhuangzhendong}@mail.scut.edu.cn, yxue@scut.edu.cn

Abstract. Most existing researches on air-writing recognition based on inertial sensors only employ acceleration and angular velocity data to recognize characters. The main characteristics of acceleration and angular velocity data are concentrated in two parts, namely, the change of signal characteristics over time and the correlation between different dimensions of signal. However, existing models usually cannot effectively extract these two types of features. In this paper, we propose a dimension separable convolutional neural networks (DSCNN) that uses one-dimensional convolution and group convolution to fully mine the signal characteristics that change over time, the unique mutual information between different dimensions, and the global characteristics of the signal. Based on DSCNN, we design a model suitable for the character recognition of the sensor signal and conduct experiments on the three databases: DB1, DB2 and DB3. The experimental results show that our method compared with the previous methods not only has a certain improvement in the case of mixed-user, but also achieves a high accuracy under the condition of user-independent, which proves that the model has good generalization performance. As the average time of single-character recognition is very short regardless of GPU or CPU conditions, our method also has good real-time performance.

Keywords: Air-writing recognition · Inertial sensors · Dimension separable · One-dimensional convolution · Group convolution

1 Introduction

With the rapid development of modern science and technology, smart devices with various functions are continuously created and integrated into people's lives. On this basis, human-computer interaction (HCI) technology has gradually matured. Air-writing recognition is an important part of HCI. It refers to writing meaningful characters in the air, and then using a computer or smart

© Springer Nature Switzerland AG 2021
J. Lladós et al. (Eds.): ICDAR 2021, LNCS 12823, pp. 538–552, 2021.
https://doi.org/10.1007/978-3-030-86334-0_35

device to recognize the written characters and convert them into corresponding instructions to control the machine [2].

There are many methods to realize air-writing recognition in today's researches. These methods are based on different types of signals, including video signals [7], WiFi signals [9], radar signals [4,16], and inertial sensor signals [1,3,5,22,23]. Compared with other methods, the inertial sensor-based method has higher degree of freedom, lower energy consumption, and lower environmental requirements, which can be used in more complex scenarios [17]. The inertial sensor signal consists of three-dimensional acceleration and three-dimensional angular velocity. The acceleration describes the speed variation of inertial sensors while the angular velocity describes angle variation.

In order to study air-writing recognition based on inertial sensor signal, many models have been proposed. In [3], Christoph et al. proposed a HMM to model continuous inertial sensor signal and realize air-writing recognition. In [22], Xu et al. used Baum-Welch algorithm to train CHMM to enhance its performance in air-writing recognition. However, each state in HMM is only related to its previous state, which makes it impossible to make full use of the relationships of whole signal. To solve this problem, LSTM was employed to capture the historical context and the correlation along time [1,23]. Although LSTM can concern more sequence variant information, but it can't well capture other two key information contained in sequence data: local dependence and mutual relationship.

To better extract the above two key information, a numbers of 2D CNN methods have been proposed in past few years [13,26]. Gholami et al. [10] spliced all the signals together along time and employed the two-dimensional convolution to extract feature. However, if signals come from multiple sensors, such as accelerometer and gyroscope, the characteristics forms of different sensors are always different. And the two-dimensional convolution uses the same weights among different dimensions of signal which will interact with each other. It weakens the ability of the models to capture mutual relations. Therefore, in this paper, inspired by [8,20,25], we propose a new method to improve the ability to extract local dependence and mutual relationship from initial signals.

In the field of action recognition, data from different sensors are processed separately to achieve good performance in the task [19,21]. Ding et al. [8] proposed a multi-stream CNN model to process acceleration and angular velocity separately to extract features. Then the features extracted from the two models were spliced together for classification. Although it avoids the mutual influence between different type of data, multi-stream CNN has two problems. First, it utilizes multiple network branches to extract individual features of different types of sensors, and the structure of these networks is very complex. Second, even if the data from the same sensor is fed to each sub-net, the mutual information between each pair of axes is different, and the local weight sharing distribution will affect feature extraction.

TextCNN adopted one-dimensional convolution to recognize text sequences, which avoided local parameter sharing and was more suitable for two-dimensional sequences recognition [25]. Then Guo et al. proposed a multi-channel TextCNN model to improve text classification with weighted word embeddings [12]. Some researchers also tried to use one-dimensional convolution for the recognition of 3D signatures, and achieved good results [11]. It shows that one-dimensional convolution can extract spatial features well. The multi-resolution convolution proposed by Sun et al. [20] used group convolution to divide input channels into several subset channels, and extracted features of different resolutions from convolution on each subset, which was equivalent to using group convolution to divide the data into multiple branches and extracted different features.

Considering all factors mentioned above, we combine the idea of TextCNN and group convolution and propose a dimension separable convolutional neural networks (DSCNN) for air-writing recognition.

First, the two-dimensional convolution is replaced by one-dimensional convolution. As local parameter sharing is avoided, DSCNN can effectively extract the change of signal characteristics over time and the correlation among different dimensions of signal.

Secondly, on the basis of one-dimensional convolution, the convolution operation is upgraded to group convolution, and the grouping function of group convolution is used to realize the ability of the network to extract multi-level features. We design a single-dimensional convolution module, a single-sensor convolution module, and an overall signals convolution module, which can fully consider the unique characteristics of single-dimensional signals and single-sensor signals, and extract the overall features to ensure that multi-level features are extracted. Furthermore, our network uses a single model to achieve multi-branch feature extraction, which reduces the complexity of the network compared to multi-stream networks [8,19,21].

Thirdly, from the recognition results of the three databases, we have reached a recognition rate of over 95% in the case of mixed-user. The network has also achieved very good results under the condition of user-independent, which proves that our model has good generalization performance. The average time to recognize a single character in the three databases is 2 ms and 4 ms under GPU and CPU conditions respectively, so our model has good real-time performance.

2   Our Proposed Approach

The pipeline of our approach is shown in Fig. 1. Firstly, the original inertial sensor data with variable-length are preprocessed into fixed length, then the data are fed to our DSCNN. The architecture of DSCNN mainly consists three modules, which are single-dimensional signal convolution module, single-sensor signal convolution module and overall signal convolution module. As shown in Fig. 1, the signal group [Ax, Ay, Az, Gx, Gy, Gz] are the input data collected by accelerometer and gyroscope. In the network, the single-dimensional signal convolution module will extract the unique features [Fax, Fay, Faz, Fgx, Fgy, Fgz] of each single dimension of signal. Then, the single-sensor signal convolution module combines the characteristics of the acceleration signal [Fax, Fay, Faz] and the characteristics of the gyroscope [Fgx, Fgy, Fgz] as [Fa] and [Fg], which are the combination characteristics on a single sensor. It can further extract the

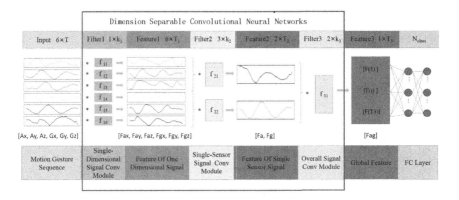

Fig. 1. The pipeline of our approach to character identification based on inertial sensor data.

features of each single sensor. Finally, the model extracts the features of whole signals with the overall signal convolution module. Using the softmax fully connected (FC) layer, the network finally obtains the probability of each category. Our network makes full use of the characteristics of group convolution and one-dimensional convolution. The specific implementation details are as follows.

2.1 One-Dimensional Convolution

The characteristics of acceleration and angular velocity data are concentrated on the change of signal characteristics over time and the correlation between different dimensions of signal. General 2D convolution [15] is a feature extractor that uses the same parameters for different dimensions of signal. The weights of 2D convolutional kernel used to extract between different adjacent signal groups are the same. For example, the weights used to calculate the correlation among Ax, Ay, Az are the same as those among Ay, Az and Gx, as showed in the upper part of Fig. 2. As signals of different sensors have different characteristic forms, weight sharing will cause signals to influence each other.

So we choose one-dimensional convolution as the feature extraction method. The width of the convolution kernel is exactly the same as the number of the signal's dimensions. The convolution kernel will only slide along the time axis. So during convolution, each dimension of signal has its own parameters to extract features. The lower part of Fig. 2 shows that the first dimensional signal are fixed to the parameters in the blue box during convolution. The parameters in the blue box fully extract the features of the first dimensional signal during sliding horizontally. Other dimensional signals are also processed in the same way with their own parameters to ensure that each dimensional signal can be extracted separately during convolution. Not only that, the final output in the green box is also the result of integrating the features extracted from each dimension. So 1D convolution is very conducive to the feature extraction of inertial sensor signals.

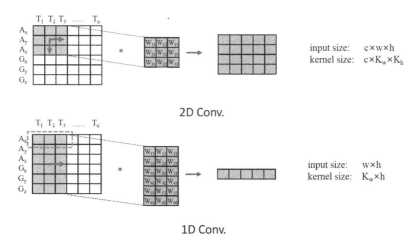

Fig. 2. Illustration of 1D Conv and 2D Conv. The upper part of the figure is 2D Conv. When we use it to extract features of the input signal, the convolution kernel can slide horizontally and vertically on the signal, and the parameters between the data are fully shared. The lower part of the figure is 1D Conv. When the size of input data is $w \times h$, the size of the convolution kernel is $K_w \times h$, and the convolution kernel can only slide horizontally on the data. (Color figure online)

2.2   Group Convolution

Group convolution was firstly proposed in AleNet [14]. When we group data in several parts, parallel training can be easily realized, which is good for GPU acceleration. Furthermore, researchers also proposed to use group convolution to reduce the amount of calculation of the model [6]. This is essentially grouping the data and then using different convolution kernels to extract features of each part. This method not only reduces the amount of calculation during convolution by splitting the data, but also creates an independent extraction of input features. Group convolution can extract the individual characteristics of each group.

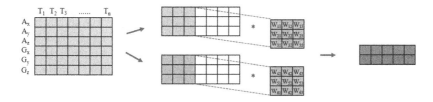

Fig. 3. Illustration of group convolution.

We believe that multi-level feature extraction can be achieved through reasonable data division. Suppose the size of input data is N × L, N is the dimensions of the signal and L is the length of the signal on the time axis. First, we divide the data into N groups to extract the features of each group of signals. Then, we divide the data into M groups according to the number of sensors to extract the features on each inertial sensor. Finally, we integrate the data into one group, and extract the overall characteristics of the signal. As shown in the Fig. 3, the input data consists of three-dimensional acceleration signal and three-dimensional angular velocity signal. When we extract the characteristics on a single sensor, we divide it into two groups according to the signal category, and use one-dimensional convolution to extract the features respectively. After obtaining unique features of the two groups, we combine the two parts of data to get output.

2.3   Multi-level Features Extraction

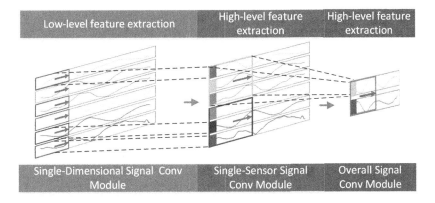

Fig. 4. Illustration of Multi-level features extraction. The colored boxes represent the convolution of the data to extract characteristics. The red arrows represent the direction in which the convolution kernel moves. (Color figure online)

The extraction of multi-level features can be regarded as the sequential composition of two methods of extracting features. One is to directly extract characteristics from raw signal with low-level semantics, and the other is to extract the unique mutual information between different dimensions of signal from features with high-level semantics. Our data are inertial sensor signals related to dynamic motion gestures. Different kinds of signals and different axes of the same signals have different response positions to the same action. As features are scattered in the raw signal, extracting correlation between different dimensions of signal directly in raw signal will be affected by the receptive field. In order to solve this problem, we adopt a multi-level method to extract features. As Fig. 4 shows, we first extract the characteristics on each axis of the inertial sensor signal separately. Then we extract the characteristics of each single inertial sensor to get the features with high-level semantics, the characteristics of which are more concentrated. Finally, we extract correlation from features with high-level semantics to get the overall characteristics of the signal. The receptive field of each feature point of high-level semantic information is very large, and the feature is compressed over time. So it is less affected by different response positions and we can easier to capture the characteristics of whole signal.

As for characteristics extraction of each axis of the signal, the output of each single-dimensional extracted feature is not affected by other axes of the signal. And we passes through those features to rectified linear unit (relu) independently. So the influence of the non-linear is also reduced. For each single-dimensional signal, it can build a feature extractor with more mathematical representation ability.

Based on the above advantages, we believe that multi-level features are more friendly to air-writing recognition based on inertial signals.

2.4   DSCNN Architectures

Table 1. Dimension separable convolution architectures for $N \times L$ signals.

DSCNN		Form of feature grouping
13 Conv1D stride = 2 group = N 13 Conv1D stride = 1 group = N	Channel = $N \times 32$	$N \times (32 \times L/2)$
5 Conv1D stride = 2 group = N/3 5 Conv1D stride = 1 group = N/3 5 Conv1D stride = 1 group = N/3	Channel = $N/3 \times 24$	$N/3 \times (24 \times L/4)$
13 Conv1D stride = 1 group = 1 5 Conv1D stride = 2 group = 1 5 Conv1D stride = 1 group = 1 5 Conv1D stride = 1 group = 1	Channel = 12	$1 \times (12 \times L/8)$

Table 1 shows the architectures of the dimension separable network we designed. Compared with Y. Ding's multi-stream network [8], which needs multiple branch models to process different types of sensor signals, our network only needs a single model. Signals of different sensor are processed separately, which reduces the complexity of the network and simplifies the structure of the network. Furthermore, our network can extract multiple features from a single model, and the network design is more flexible. Our model can not only distinguish different types of sensor signals, but also achieve three different feature extraction levels at the same time. The advantages of our network is also reflected in the adjustment of the adaptive network framework. As described in Table 1, when the dimensions of input signals are N × L, the network design will be adaptively adjusted according to the values of N and L. Both the number of channels and the number of neurons in the network are represented by variables N and L. When the dimensions of input data change, the architecture of network can adjust instantly based on the input data. So the network has strong adaptability, not only for air-writing recognition tasks based on inertial signals, but also for other tasks based on inertial sensor signals.

3   Experiments

3.1   Databases

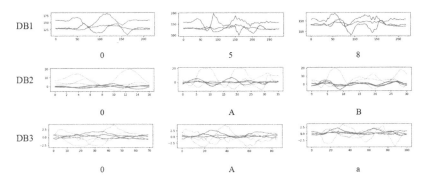

Fig. 5. The waveforms of inertial sensor data of digit '0', '5', '8', lowercase letter 'a', uppercase letter 'A' and 'B' belong to three different databases. The green waveforms represent accelerations, and the orange ones represent angular velocities. (Color figure online)

We used three databases to verify the effectiveness of our network, including DB1, DB2 [22] and DB3 [5]. DB1 contains 40 collectors (all males) who wrote 1130 3D acceleration sequences of 10 Arabic numerals 100 Hz sampling. DB2 consists of 49 collectors (31 males and 18 females) who wrote 14530 3D acceleration and 3D angular velocity 15 Hz sampling. DB3 contains 22 collectors (17 males and 5 females) who wrote uppercase, lowercase letters and numbers 60 Hz sampling. It includes 8571 sets of data, each set of data contains 3D acceleration and 3D angular velocity sequences. Figure 5 shows the waveforms of the inertial sensor signal of each database. We can find that the signal of DB1 are very long, while the DB2's are very short. This is caused by the different sampling rate of the signal. As can be seen from the figure, a longer signal can better reflect the details of the signal changes.

3.2 Data Preprocessing

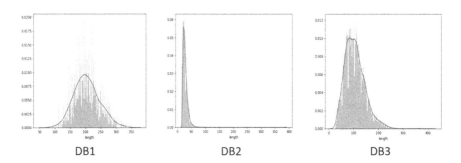

DB1   DB2   DB3

. **Fig. 6.** The length distributions of three databases.

Since the CNN model requires the input data with the same size, we need to unify the length of the inertial sensor signals. We have drew the distribution map of the length of the three databases. As shown in Fig. 6, we find that the data in the first and third databases are generally longer, and the data length of the second database is shorter, so we divide the data into two groups, the first and third databases are the first group, the rest are the second group. In the process of unifying the length, we have to consider two aspects. The length of data should be able to fully contain the valid characteristics of various data for easy recognition. And we should minimize the amount of network calculation to save computing resources.

Therefore, according to the distribution of the data length, we unify the length of the first group to 256 and the length of the second group to 64. If the length of data is less than the unified length, we use the zero padding to ensure that the characteristics and distribution of the data will not be changed. If the length is longer than the unified length, we resample the original data to shorten the length of the data. Furthermore, we use moving average filter (MA) to filter out the noise in the data, and use Z-score normalization to speed up the gradient descent to find the optimal solution.

3.3 Experiments and Analysis

We use the five-fold cross-validation method in our experiments. There are three division criteria for cross-validated data, which are the category of writers, the category of characters, and the total number of data. Our experiments are divided into two forms: mixed-user and user-independent. In the mixed-user experiment, it is necessary to ensure that the data of each fold are similar in the three standards above. In the user-independent experiment, under the premise of ensuring that the types of characters and the total number of data contained in each fold are as the same as possible, the writers contained in each fold are independent and not crossed, which means the data of each writer can only exist in a single fold. Comparing with the two experiments, the second one can better reflect the generalization performance of the model.

In order to verify the effectiveness of one-dimensional convolution and group convolution, we modify the design of DSCNN to get TextCNN and CNN. We set the group in DSCNN to 1 for TextCNN, which means that all one-dimensional group convolutions are replaced by one-dimensional convolutions. For CNN, we change the size of the input data from (b, c, w) to (b, 1, w, c). Then, we replace each one-dimensional convolution with a kernel size of k with a two-dimensional convolution with a kernel size of (k, 3). And we set the stride of the second convolution to (2, 2), and the others to (s, 1), where s is the original one-dimensional convolution step.

Table 2. Evaluation of our model on DB3 database.

Algorithm	Accuracy (%)
CHMM [22]	95.91
LSTM [23]	94.75
BLSTM [24]	99.27
CNN [1]	99.26
LSTM [1]	99.32
CNN	98.89
TextCNN	99.34
DSCNN	**99.40**

A. DB3 Dataset Evaluation. We first verify the effectiveness of our algorithm on a public dataset DB3 [5] used by many researchers. Table 2 shows a detailed comparison. Xu and Xue performed an extensive analysis of this dataset [22]. They considered both user-dependent and user-independent cases and followed CHMM based approach of which the accuracy is 95.91% in the mixed-user condition. Then they designed a model based on the long and short-term memory method, but did not improve the recognition performance on this dataset [23]. In that contemporary time, Yana and Onoye combined the BLSTM and CNN features and achieved 99.27% maximum accuracy [24]. Furthermore, Alam et al. proposed a LSTM-based network which achieved 99.32% accuracy [1]. In this paper, the proposed DSCNN outperforms previous work, achieving 99.40% accuracy which is the highest so far. The result fully proved the effectiveness of our network for air-writing recognition.

B. Overall Assessment Results. In order to compare with the original experimental results based on the CHMM and LSTM models, we use CNN, TextCNN

Table 3. Evaluation results on three databases.

Indicators (%)	Databases of multi-complexity dynamic motion gestures					
	DB1		DB2		DB3	
	Independent	Mixed	Independent	Mixed	Independent	Mixed
CHMM [22]	65.64 ± 8.74	72.04 ± 1.19	82.29 ± 4.68	87.52 ± 0.84	62.69 ± 2.91	95.91 ± 0.47
LSTM [23]		90.60 ± 1.87		93.96 ± 0.40		94.75 ± 0.31
CNN	83.74 ± 5.65	91.21 ± 1.75	85.71 ± 3.62	96.29 ± 0.42	88.06 ± 1.84	98.89 ± 0.32
TextCNN	87.93 ± 4.38	93.54 ± 1.29	88.35 ± 3.73	97.59 ± 0.20	90.83 ± 1.20	99.34 ± 0.25
DSCNN	**90.68 ± 2.45**	**95.40 ± 1.30**	**89.43 ± 3.53**	**97.84 ± 0.16**	**91.55 ± 1.13**	**99.40 ± 0.22**

and DSCNN to conduct experiments on three databases. The results show that the proposed DSCNN model achieves the best results (Table 3). The recognition accuracy of TextCNN has been higher than that of all original methods, and DSCNN has further improved the performance on this basis. We find that both networks reach a high accuracy in the case of mixed-user. DSCNN achieves accuracy of 95.40%, 97.84% and 99.40% respectively on complicated air-writing databases DB1, DB2 and DB3 while TextCNN achieves 93.54%, 97.59% and 99.34%. We think there are two main reasons. First, One dimensional convolution is suitable for feature extraction of inertial sensor signal. Second, as the reason of mixed-user, the data of training set and test set are similar, which reduces the difficulty of the task. However, in the case of user-independent, the results of DSCNN are significantly higher than TextCNN's. DSCNN achieves the state of the art accuracy, reaching 90.68%, 89.43% and 91.55% on these three databases respectively. The recognition task is more difficult in the condition of user-independent, and the performance of the model is more distinguished. We believe that, compared with CNN, TextCNN solves the problem of mutual information extraction caused by local weight sharing. Therefore, the recognition performance is improved. Compared with TextCNN, DSCNN can extract more levels of features. A more complete feature extraction process improves the recognition ability of the model.

To explain why DSCNN's performance is better than previous methods, we use the Grad-CAM [18] method to generate coarse localization maps for TextCNN and DSCNN respectively, which can highlight the important regions of the inertial sensor signal for character recognition. As Fig. 7 shows, most of the signal corresponding to TextCNN is purple, which means that TextCNN will ignore these areas when recognizing characters. In the signal corresponding to DSCNN, there are few purple parts, which means that DSCNN can pay attention to the entire signal when recognizing characters. It confirms that the features extracted by DSCNN model can cover the signals as complete as possible. And it makes those confusing characters easier to distinguish. So DSCNN can bring certain improvements compared with TextCNN.

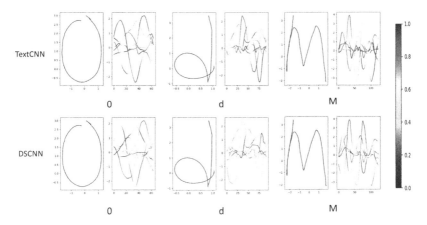

Fig. 7. Grad-CAM: localizes class-discriminative regions for TextCNN and DSCNN. The left side of each part is the trajectory corresponding to the character, and the right side is the six-dimensional inertial sensor signal. The different colors of the signal indicate the different importance of this part in the process of character recognition. From red to purple, the importance of the regions decrease gradually. (Color figure online)

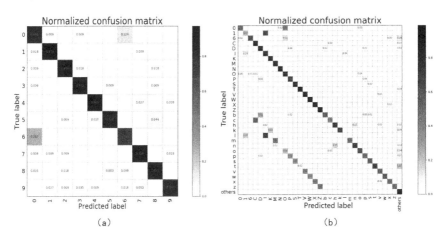

Fig. 8. Confusion matrix of DSCNN on (a) DB1 and (b) DB3. As for DB3, we compress the categories of the confusion matrix. We keep the confusing character categories while merging the remaining categories into 'others' category. In this way, we can more clearly analyze which characters are confused in the model.

In order to analysis erroneous cases, we give the confusion matrix of DB1 and DB3. As Fig. 8 shows, the confusion between the number '0' and the number '6' in DB1 is the main reason for the low accuracy. The strokes of the two numbers are partly the same, which may cause the model to confuse them. In future research, the two numbers should be distinguished in order to achieves better results. On DB3, we found that in user-independent condition, many uppercase and lowercase letters with similar strokes would be confused with each other, such as 'k' and 'K', 's' and 'S'. Some numbers would also be confused with letters, such as '0', 'o' and 'O', '1', 'l' and 'I'. Everyone's writing habits are different. Some people write uppercase letters in a similar way to the lowercase letters of others. At the same time, the database has only 8571 samples, which is small. In order to train a better model, we should try to expand the database and solve the problem of confusion between signals.

Table 4. Real-time performance evaluation of DSCNN model.

Indicators (ms)	DB1		DB2		DB3	
	Independent	Mixed	Independent	Mixed	Independent	Mixed
GPU	1.89	1.88	2.31	2.34	2.40	2.44
CPU	4.03	3.78	4.13	3.99	4.66	4.61

C. Real-Time Performance Evaluation. Our model includes the complete processes of character recognition. For each received signal, it first adjusts the data to a fixed length, then filters it with MA and normalizes it with Z-score, and finally feeds the data into the network for recognition. These include all the steps of character recognition in practical application, so we further tested the time of the model to recognize a single character to evaluate its real-time performance. The experiments are performed on a 64bit PC with Intel Core i7-4790 CPU 3.60 GHz × 8 and a single GeForce GTX TITAN X GPU. Due to the complexity of the actual environment, we not only tested under GPU condition, but also under CPU condition. From the Table 4, the average time to recognize a single character on the three databases is about 2 ms under GPU condition, while it is about 4 ms under CPU condition. The recognition time of this model in both cases is very short, so it can well meet the real-time requirements.

4   Conclusion

In this paper, we present a dimensional separable convolution neural network for character recognition based on inertial sensor signal. The network flexibly separates dimensions of the signals. It can fully extract the features of single dimension of signal, the unique features of the single sensor and the overall characteristics of all the sensors to recognize the characters. Compared with the multi-branch network, DSCNN is a single model, which can flexibly extract multi-level features, and has less complexity than previous models. The experimental results in three databases show that our method has a great improvement in recognition performance compared with the previous methods under the conditions of user-mixed and user-independent, which proves the effectiveness of our proposed method. High accuracy in the user-independent condition also proves that the model has good generalization performance. Our model has a short recognition time regardless of GPU or CPU conditions, which proves that the model has good real-time performance. In addition, DSCNN has strong self-adaptive ability, and can automatically adjust the network structure according to the dimensions of input data. So it is also suitable for many other types of recognition tasks. In this paper, we mainly discuss the effectiveness of dimensional separable convolution in air-writing recognition based on inertial sensors. Exploring the DSCNN network in other tasks is a future direction that can be further advanced.

Acknowledgements. This research is supported in part by NSFC (No. 61771199) and Guangdong Basic and Applied Basic Research Foundation (No. 2021A1515011870).

References

1. Alam, M., Kwon, K.C., Abbass, M.Y., Imtiaz, S.M., Kim, N., et al.: Trajectory-based air-writing recognition using deep neural network and depth sensor. Sensors **20**(2), 376 (2020)
2. Amma, C., Gehrig, D., Schultz, T.: Airwriting recognition using wearable motion sensors. In: Proceedings of the 1st Augmented Human international Conference, pp. 1–8 (2010)
3. Amma, C., Georgi, M., Schultz, T.: Airwriting: hands-free mobile text input by spotting and continuous recognition of 3D-space handwriting with inertial sensors. In: 2012 16th International Symposium on Wearable Computers, pp. 52–59. IEEE (2012)
4. Arsalan, M., Santra, A.: Character recognition in air-writing based on network of radars for human-machine interface. IEEE Sens. J. **19**(19), 8855–8864 (2019)
5. Chen, M., AlRegib, G., Juang, B.H.: 6DMG: a new 6D motion gesture database. In: Proceedings of the 3rd Multimedia Systems Conference, pp. 83–88 (2012)
6. Chollet, F.: Xception: deep learning with depthwise separable convolutions. In: Proceedings of the IEEE Conference on Computer Vision and Pattern Recognition, pp. 1251–1258 (2017)
7. Choudhury, A., Sarma, K.K.: A CNN-LSTM based ensemble framework for in-air handwritten assamese character recognition. Multimedia Tools Appl. 1–36 (2021)
8. Ding, Y., Xue, Y.: A deep learning approach to writer identification using inertial sensor data of air-handwriting. IEICE Trans. Inf. Syst. **102**(10), 2059–2063 (2019)
9. Fu, Z., Xu, J., Zhu, Z., Liu, A.X., Sun, X.: Writing in the air with WiFi signals for virtual reality devices. IEEE Trans. Mob. Comput. **18**(2), 473–484 (2018)
10. Gholami, M., Rezaei, A., Cuthbert, T.J., Napier, C., Menon, C.: Lower body kinematics monitoring in running using fabric-based wearable sensors and deep convolutional neural networks. Sensors **19**(23), 5325 (2019)
11. Ghosh, S., Ghosh, S., Kumar, P., Scheme, E., Roy, P.P.: A novel spatio-temporal siamese network for 3D signature recognition. Pattern Recognit. Lett. **144**, 13–20 (2021)
12. Guo, B., Zhang, C., Liu, J., Ma, X.: Improving text classification with weighted word embeddings via a multi-channel textcnn model. Neurocomputing **363**, 366–374 (2019)

13. Ha, S., Yun, J.M., Choi, S.: Multi-modal convolutional neural networks for activity recognition. In: 2015 IEEE International Conference on Systems, Man, and Cybernetics, pp. 3017–3022. IEEE (2015)
14. Krizhevsky, A., Sutskever, I., Hinton, G.E.: Imagenet classification with deep convolutional neural networks. Adv. Neural. Inf. Process. Syst. **25**, 1097–1105 (2012)
15. Lecun, Y., Bengio, Y.: Convolutional networks for images, speech, and time-series. In: The Handbook of Brain Theory and Neural Networks (1995)
16. Leem, S.K., Khan, F., Cho, S.H.: Detecting mid-air gestures for digit writing with radio sensors and a CNN. IEEE Trans. Instrum. Meas. **69**(4), 1066–1081 (2020)
17. Livingston, L.M.M., Deepika, P., Benisha, M.: An inertial pen with dynamic time warping recognizer for handwriting and gesture recognition. IJETT **35**(11), 506–510 (2016)
18. Selvaraju, R.R., Cogswell, M., Das, A., Vedantam, R., Parikh, D., Batra, D.: Gradcam: visual explanations from deep networks via gradient-based localization. In: IEEE International Conference on Computer Vision (2017)
19. Simonyan, K., Zisserman, A.: Two-stream convolutional networks for action recognition in videos. arXiv preprint arXiv:1406.2199 (2014)
20. Sun, K., et al.: High-resolution representations for labeling pixels and regions. arXiv preprint arXiv:1904.04514 (2019)
21. Tu, Z., et al.: Multi-stream CNN: learning representations based on human-related regions for action recognition. Pattern Recogn. **79**, 32–43 (2018)
22. Xu, S., Xue, Y.: Air-writing characters modelling and recognition on modified CHMM. In: 2016 IEEE International Conference on Systems, Man, and Cybernetics (SMC), pp. 001510–001513. IEEE (2016)
23. Xu, S., Xue, Y.: A long term memory recognition framework on multi-complexity motion gestures. In: 2017 14th IAPR International Conference on Document Analysis and Recognition (ICDAR), vol. 1, pp. 201–205. IEEE (2017)
24. Yana, B., Onoye, T.: Fusion networks for air-writing recognition. In: Schoeffmann, K., et al. (eds.) MMM 2018. LNCS, vol. 10705, pp. 142–152. Springer, Cham (2018). https://doi.org/10.1007/978-3-319-73600-6_13
25. Zhang, Y., Wallace, B.: A sensitivity analysis of (and practitioners' guide to) convolutional neural networks for sentence classification. arXiv preprint arXiv:1510.03820 (2015)
26. Zhao, B., Lu, H., Chen, S., Liu, J., Wu, D.: Convolutional neural networks for time series classification. J. Syst. Eng. Electron. **28**(1), 162–169 (2017)

Document Synthesis

DocSynth: A Layout Guided Approach for Controllable Document Image Synthesis

Sanket Biswas[1]([✉])[iD], Pau Riba[1][iD], Josep Lladós[1][iD], and Umapada Pal[2][iD]

[1] Computer Vision Center and Computer Science Department,
Universitat Autònoma de Barcelona, Bellaterra, Spain
{sbiswas,priba,josep}@cvc.uab.es
[2] CVPR Unit, Indian Statistical Institute, Kolkata, India
umapada@isical.ac.in

Abstract. Despite significant progress on current state-of-the-art image generation models, synthesis of document images containing multiple and complex object layouts is a challenging task. This paper presents a novel approach, called DocSynth, to automatically synthesize document images based on a given layout. In this work, given a spatial layout (bounding boxes with object categories) as a reference by the user, our proposed DocSynth model learns to generate a set of realistic document images consistent with the defined layout. Also, this framework has been adapted to this work as a superior baseline model for creating synthetic document image datasets for augmenting real data during training for document layout analysis tasks. Different sets of learning objectives have been also used to improve the model performance. Quantitatively, we also compare the generated results of our model with real data using standard evaluation metrics. The results highlight that our model can successfully generate realistic and diverse document images with multiple objects. We also present a comprehensive qualitative analysis summary of the different scopes of synthetic image generation tasks. Lastly, to our knowledge this is the first work of its kind.

Keywords: Document synthesis · Generative Adversarial Networks · Layout generation

1 Introduction

The task of automatically understanding a document is one of the most significant and primary objectives in the Document Analysis and Recognition community. Nowadays, especially in business processes, paper scanned and digitally born documents coexist. There is a big variability in real-world documents coming from different domains (forms, invoices, letters, etc.). Modern Robotic Process Automation (RPO) tools in paperless offices have a compelling need for managing automatically the information of document workflows, which can integrate both reading and understanding. According to the standardized recommendations of the Office Document Architecture (ODA) [5], a document representation could be expressed by formalisms that obey two crucial aspects.

© Springer Nature Switzerland AG 2021
J. Lladós et al. (Eds.): ICDAR 2021, LNCS 12823, pp. 555–568, 2021.
https://doi.org/10.1007/978-3-030-86334-0_36

The first one considers a document as an image for printing or displaying, while the second one considers its textual and graphical representation for interpreting its layout and logical structure.

The layout structure of a document is fundamentally represented by layout objects (e.g. text or graphic blocks, images, tables, lines, words, characters and so on) while the logical structure describes the semantic relationship between conceptual elements (e.g. company logo, signature, title, body or paragraph region and so on). The recognition of document layout has been one of the most challenging problems for decades. The understanding of layout is a necessary step towards the extraction of information. Business intelligence processes require the extraction of information from document contents at large scale, for subsequent decision-making actions. Many examples can be found in different X-tech areas: fin-tech (analyze sales trends based on intelligent reading of invoices), legal-tech (determine if a clause of a contract has been violated), insurance-tech (liability from accident statement understanding). Document layout syntactically describes the whole document, and therefore allows to give context to the individual components (named entities, graphical symbols, key-value associations). Thus, performance in information extraction is boosted when it is driven by the layout. As in many other domains, the deep learning revolution has open new insights in the layout understanding problem. Consequently, there is a need for annotated data to supervise the learning tasks. Having big amounts of data is not always possible in real scenarios. In addition to the manual effort to annotate layout components, such types of images have privacy restrictions (personal data, corporate information) which prevent companies and organizations to disclose it. Data augmentation strategies are a good solution. Among the different strategies for augmenting data, synthetic generation of realistic images is one of the most successful.

This work discusses a research effort which is intended to develop a synthetic generation tool called DocSynth for rendering realistic printed documents with plausible layout objects desired by the user. A simple illustration of this task is as shown in Fig. 1. The proposed model is able to generate samples given a single reference layout image. Thus, it can generate training data with one single sample per class and can be adapted to few-shot settings for document classification tasks. This automatic document image synthesizer could provide a possible solution to manage all papers as well as electronic documents with a centralized platform manipulated by the user. In fact, this practical application has the potential to improve visual search and information retrieval engines. Usually, in retrieval task the user wants to index in a repository of documents (real ones). Instead one could create variations of the query sample to improve retrieval performances of the model.

While classical computer graphics techniques have been used in modeling for example geometry, projections, surface properties and cameras, the more recent computer vision techniques rely on the quality of designed machine learning approaches to learn from real world examples to generate synthetic images. Explicit reconstruction and rendering of document properties (both graphical and textual) in the form of complex layout objects is a hard task from both computer graphics and vision perspective. To this end, traditional image-based rendering approaches tried to overcome these issues, by using simple heuristics to combine a captured imagery. But applying these heuristical approaches for synthesizing images with complex document layouts generate artifacts and does not provide an optimal solution. Neural rendering brings the promise of addressing the problem of both reconstruction and rendering by using deep generative models like Generative Adversarial Networks (GANs) and Variational Auto encoders (VAEs) and to learn complex mappings from captured images to novel images. They help to combine physical knowledge, e.g., mathematical models of projection and geometry, with learned components to yield new and powerful algorithms for controllable image generation.

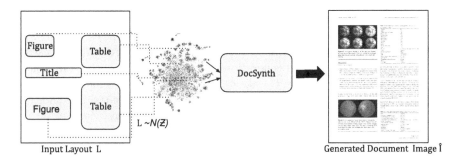

Fig. 1. Illustration of the Task: Given an input document layout with object bounding boxes and categories configured in an image lattice, our model samples the semantic and spatial attributes of every layout object from a normal distribution, and generate multiple plausible document images as required by the user.

The main contributions of this work are as follows.

1. A new model is proposed for synthetic document image generation guided by the layout of a reference sample.
2. Qualitative and quantitative results on the PubLayNet dataset [18], demonstrate our model's capability to generate complex layout documents with respect to spatial and semantic information of object categories.
3. Also this work addresses the layout-guided document image synthesis task with an analytical understanding as the first of its kind in the document analysis community.

The rest of this paper is organized as follows: in Sect. 2 we review the relevant literature. Section 3 describes the main methodological contribution of the work. In Sect. 4 we provide a quantitative and qualitative experimental analysis. Finally, Sect. 5 draws the main contributions and outlines future perspectives.

2   Related Work

The analysis of structural and spatial relations between complex layouts in documents has been a significant challenge in the field of Document Analysis and Recognition. Extracting the physical and logical layout in documents is a required step in tasks such as Optical Character Recognition for document image transcription, document classification, or information extraction. The reader is referred to [1] for a comprehensive survey on the state of the art on document layout analysis.

As it has been introduced in Sect. 1 the main objective of this work is to construct a generative neural model to construct visually plausible document images given a reference layout. The strategy for augmenting data and its corresponding ground truth by synthetic images automatically generated has gained interest among the Computer Vision community. Since they were proposed by Goodfellow in 2014, Generative Adversarial Networks (GANs) [3] and subsequent variants have been a successful method to generate realistic images, ranging from handwritten digits to faces and natural scenes. A step forward which is a scientific challenge in the controlled generation of images in terms of the composition of objects and their arrangement. Lake et al. in [9] suggested a hierarchical generative model that can build whole objects from individual parts, it is shown to generate Omniglot characters as a composition of the strokes. Zhao et al. [16] proposed a model that can generate a set of realistic images with objects in the desired locations, given a reference spatial distribution of bounding boxes and object labels. Our work has been inspired in this work.

Preserving the reliable representation of layouts has shown to be very useful in various graphical design contexts, which typically involve highly structured and content-rich objects. One such recent intuitive understanding was established by Li et al. [10] in their LayoutGAN, which aims to generate realistic document layouts using Generative Adversarial Networks (GANs) with a wireframe rendering layer. Zheng et al. [17] used a GAN-based approach to generate document layouts but their work focused mainly on content aware generation, that primarily uses the content of the document as an additional prior. To use more highly structured object generation, it is very important to focus operate on the low dimensional vectors unlike CNN's. Hence, in the most recent literature, Patil et al. [13] has come up with a solution called 'READ' that can make use of this highly structured positional information along with content to generate document layouts. Their recursive neural network-based resulting model architecture provided state-of-the-art results for generating synthetic layouts for 2D documents. But their solution could not be applicable for document image level analysis problems. Kang et al. [6] actually exploited the idea to generate synthetic data at the image level for handwritten word images.

Summarizing, state-of-the-art generative models are still unable to produce plausible yet diverse images for whole page documents. In this work we propose a direction to condition a generative model for whole page document images with synthesized variable layouts.

3   Method

In this section we describe the contributions of the work. We first formally formulate the problem, and introduce the basic notation. Afterwards, we describe the proposed approach, the network structure, its learning objectives, and finally the implementation details.

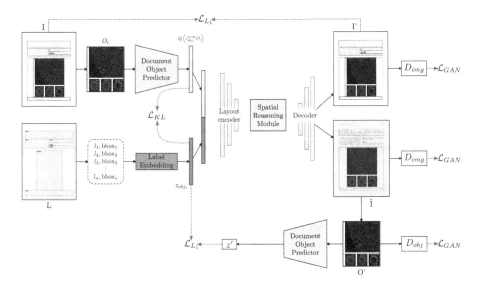

Fig. 2. Overview of our DocSynth Framework: The model has been trained adversarially against a pair of discriminators and a set of learning objectives as depicted.

3.1   Problem Formulation

Let us start by defining the problem formally. Let X be an image lattice (e.g. of size 128×128) and I be a document image defined on the lattice. Let $L = \{(\ell_i, \text{bbox}_i)_{i=1}^{n}\}$ be a layout which contains n labeled object instances with defined class categories $\ell_i \in O$ and bounding boxes of these instances represented by top-left and bottom-right coordinates on the canvas, $\text{bbox}_i \subset X$, and $|O|$ is the total number of document object categories (e.g. table, figure, title and so on). Let Z_{obj} be the overall sampled latent estimation comprising every object instance O_i in the layout L, which can be represented as $Z_{obj} = \{\mathbf{z}_{obj_i}\}_{i=1}^{n}$. The latent estimation have been sampled randomly for the objects from the standard prior Normal distribution $\mathcal{N}(0, 1)$ under the i.i.d. setup.

The layout-guided document image synthesis task can be codified as learning a generator function G which can map a given document layout input (L, Z_{obj}) to the generated output image $\tilde{I}$ as shown in Eq. 1.

$$\tilde{I} = G(L, Z_{obj}; \Theta_G) \tag{1}$$

where Θ_G represents the parameters of the generation function G which needs to be learned by our model. Primarily, a generator model $G(.)$ is able to capture the underlying conditional data distribution $p(\tilde{I}|L, Z_{obj}; \Theta_G)$ present in a higher dimensional space, equivariant with respect to spatial locations of document layout objects bbox$_i$.

The proposed model in this work for the above formulated task investigates three different challenges: (1) Given the user provides the input document layout L, is the model capable of synthesizing plausible document images while preserving the object properties conditioned on L? (2) Given the user provides the input document layout L, can the model generate multiple variable documents using different style $\mathbf{z}_{obj_i}$ of objects while retaining the object configuration ℓ_i, bbox $_i$ in the input layout? (3) Given a tuple (L, Z_{obj}), is the generator capable of generating consistent document images for different $(\tilde{L}, \tilde{Z}_{obj})$ where a user can add an object to existing layout L or just modify the location or label of existing objects?

Handling such complexities using deep generative networks is difficult due to the difficulty of sampling the posterior elements. This work focuses to tackle the problem by designing a single generator model $G(\cdot)$ that tries to provide an answer to the above mentioned research questions.

3.2   Approach

To build on the layout-guided synthetic document generation pipeline, we aim to explain our proposed approach in two different parts: Training and Inference.

Training: The overall training pipeline of our proposed approach is illustrated in Fig. 2. Given an input document image I and its layout $L = \{(\ell_i, \text{bbox}_i)_{i=1}^n\}$, the proposed model creates a category label embedding e_i for every object instance O_i in the document. A set of object latent estimations $Z_{obj} = \{\mathbf{z}_{obj_i}\}_{i=1}^n$ are sampled from the standard prior normal distribution $\mathcal{N}(0, 1)$, while another set of object latent estimations $Z_{obj}^{crop} = \left\{\mathbf{z}_{obji}^{crop}\right\}_{i=1}^n$ are sampled from the posterior distribution $Q\left(\mathbf{z}_{obji}^{crop} \mid O_i\right)$ conditioned on the features received from the cropped objects O_i of input image I as shown in Fig. 2 in the document object predictor. This eventually allows us to synthesize two different datasets: (1) A collection of reconstructed images I' from ground-truth image I during training by mapping the input (L, Z_{obj}^{crop}) through the generator function G. (2) A collection of generated document images $\tilde{I}$ by mapping (L, Z_{obj}) through G, where the generated images match the original layout but exhibits variability in object instances which is sampled from random distribution. To allow consistent mapping between the generated object $\tilde{O}$ in $\tilde{I}$ and sampled Z_{obj}, the latent estimation is regressed by the document object predictor as shown in Fig. 2. The training is done with an adversarial approach by including two discriminators to classify the generated results as real or fake at both image-level and object-level.

Inference: During inference time, the proposed model synthesizes plausible document images from the layout L provided by the user as input and the object latent estimation Z_{obj} sampled from the prior $\mathcal{N}(0,1)$ as illustrated in the Fig. 1.

3.3 Generative Network

The proposed synthetic document image generation architecture consists of mainly three major components: two object predictors E and E', a conditioned image generator H, a global layout encoder C, an image decoder K and an object and image discriminator denoted by D_{obj} and D_{img} respectively.

Object Encoding: Object latent estimations Z_{obj}^{crop} are first sampled from the ground-truth image I with the object predictor E. They help to model variability in object appearances, and also to generate the reconstructed image I'. The object predictor E predicts the mean and variance of the posterior distribution for every cropped object O_i from the input image. To boost the consistency between the generated output image $\tilde{I}$ and its object estimations, the model also has another predictor E' which infers the mean and variances for the generated objects O' cropped from $\tilde{I}$. The predictors E and E' consist of multiple convolutional layers with two dense fully-connected layers at the end.

Layout Encoding: Once the object latent estimation $\mathbf{z}_i \in \mathbb{R}^n$ has been sampled from the posterior or the prior distribution $\left(\mathbf{z}_i \in \left\{ Z_{obj}^{crop}, Z_{obj} \right\} \right)$, the next step is to construct a layout encoding denoted by F_i with the input layout information $L = \{(\ell_i, \text{ bbox }_i)_{i=1}^n\}$ as provided by the user for every object O_i in the image I. Each feature map F_i should contain the disentangled spatial and semantic information corresponding to layout L and appearance of the objects O_i interpolated by latent estimation $\mathbf{z}_i$. The object category label ℓ_i is transformed as a label embedding $e_i \in \mathbb{R}^n$ and then concatenated with the latent vector $\mathbf{z}_i$. The resultant feature map F_i for every object is then filled with the corresponding bounding box information bbox_i to form a tuple represented by $< \ell_i, \mathbf{z}_i, \text{bbox}_i >$. These feature maps encoding this layout information are then fed to a global layout encoder network C containing multiple convolutional layers to get downsampled feature maps.

Spatial Reasoning Module: Since the final goal of the model is to generate plausible synthetic document images with the encoded input layout information, the next step of the conditioned generator H would be to generate a good hidden feature map h to fulfill this objective. The hidden feature map h should be able to perform the following: (1) encode global features that correlate an object representation with its neighbouring ones in the document layout (2) encode local features with spatial information corresponding to every object (3) should invoke spatial reasoning about the plausibility of the generated document with respect to its contained objects.

To meet these objectives, we choose to define the spatial reasoning module with a convolutional Long-Short-Term Memory(conv-LSTM) network backbone. Contrary to vanilla LSTMs, conv-LSTMs replace the hidden state vectors with feature maps instead. The different gates in this network are also encoded by convolutional layers, which also helps to preserve the spatial information of the contents more accurately. The conv-LSTM encodes all the object feature maps F_i in a sequence-to-sequence manner, until the final output of the network gives a hidden layout feature map h.

Image Reconstruction and Generation: Given the hidden layout feature map h already generated by the spatial reasoning module, we move towards our final goal for the task. An image decoder K with a stack of deconvolutional layers is used to decode this feature map h to two different images, I' and $\tilde{(I)}$. The image I' is reconstructed from the input image I using latent estimation Z_{obj}^{crop} conditioned on its objects O. The image $\tilde{(I)}$ is the randomly generated image using Z_{obj} directly sampled from the prior $\mathcal{N}(0,1)$. Both these images retain the same layout structure as mentioned in the input.

Discriminators: To make the synthetic document images look realistic and its objects noticeable, a pair of discriminators D_{img} and D_{obj} is adopted to classify an input image as either real or fake by maximizing the GAN objective as shown in Eq. 2. But, the generator network H is being trained to minimize $\mathcal{L}_{GAN}$.

$$\mathcal{L}_{GAN} = \underset{x \sim p_{rcal}}{\mathbb{E}} \log D(x) + \underset{y \sim p_{fake}}{\mathbb{E}} \log(1 - D(y)) \tag{2}$$

While the image discriminator D_{img} is applied to input images I, reconstructed images I' and generated sampled images $\tilde{I}$, the object discriminator D_{obj} is applied at the object-level to assess the quality of generated objects O' and make them more realistic.

3.4   Learning Objectives

The proposed model has been trained end-to-end in an adversarial manner with the generator framework and a pair of discriminators. The generator framework, with all its components help to minimize the different learning objectives during training phase. Our GAN model makes use of two adversarial losses: image adversarial loss $\mathcal{L}_{\text{GAN}}^{\text{img}}$ and object adversarial loss $\mathcal{L}_{\text{GAN}}^{\text{obj}}$. Four more losses have been added to our model, including KL divergence loss $\mathcal{L}_{\text{KL}}$, image reconstruction loss $\mathcal{L}_1^{\text{img}}$, object reconstruction loss $\mathcal{L}_1^{\text{obj}}$ and auxiliary classification loss $\mathcal{L}_{\text{AC}}^{\text{obj}}$, to enhance our synthetic document generation network.

The overall loss function used in our proposed model can be defined as shown in Eq. 3:

$$\mathcal{L}_G = \lambda_1 \mathcal{L}_{\text{GAN}}^{\text{img}} + \lambda_2 \mathcal{L}_{\text{GAN}}^{\text{obj}} + \lambda_3 \mathcal{L}_{\text{AC}}^{\text{obj}} + \lambda_4 \mathcal{L}_{\text{KL}} + \lambda_5 \mathcal{L}_1^{\text{img}} + \lambda_6 \mathcal{L}_1^{\text{obj}} \tag{3}$$

3.5   Implementation Details

In order to stabilise training for our generative network, we used the Spectral-Normalization GAN [12] as our model backbone. We used conditional batch normalization [2] in the object predictors to better normalize the object feature maps. The model has been adapted for 64×64 and 128×128 image sizes. The values of the six hyperparameters λ_1 to λ_6 are set to 0.01, 1, 8, 1,1 and 1, respectively. These values have been set experimentally. The Adam optimizer [7] was to train all the models with batch size of 16 and 300,000 iterations in total. For more finer details, we will make our code publicly available.

4   Experimental Validation

Extensive experimentation was conducted to evaluate our adapted DocSynth framework. Since this work introduces the first fundamental approach towards the problem of layout-guided document image synthesis, we try to conduct some ablation studies that are important for proposing our model as a superior baseline for the task. Also, we try to analyse our obtained results with the model, both qualitatively and quantitatively. All the code necessary to reproduce the experiments is available at github.com/biswassanket/synth_doc_generation using the PyTorch framework.

4.1   Datasets

We evaluate our proposed DocSynth framework on the PubLayNet dataset [18] which mainly contains images taken from the PubMed Central library for scientific literature. There are five defined set of document objects present in this dataset: text, title, lists, tables and figures. The entire dataset comprises 335,703 images for training and 11,245 images for validation.

4.2   Evaluation Metrics

Plausible document images generated from layout should fulfill the following conditions: (1) They should be realistic (2) They should be recognizable (3) They should have diversity. In this work, we have adapted two different evaluation metrics for evaluating our rendered images for the problem.

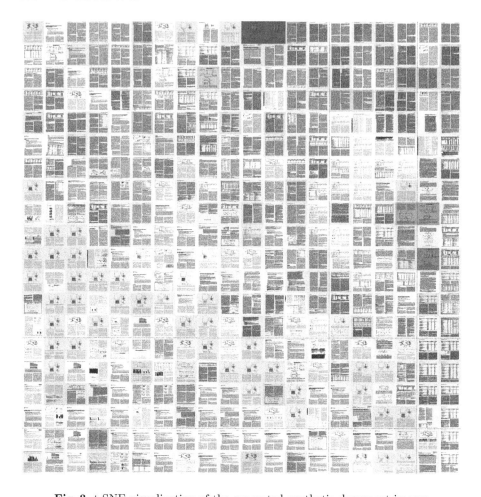

Fig. 3. t-SNE visualization of the generated synthetic document images

Fréchet Inception Distance (FID): The FID metric [4] is a standard GAN performance metric to compute distances between the feature vectors of real images and the feature vectors of synthetically generated ones. A lower FID score denotes a better quality of generated samples and more similar to the real ones. In this work, the Inception-v3 [14] pre-trained model were used to extract the feature vectors of our real and generated samples of document images.

Diversity Score: Diversity score calculates the perceptual similarity between two images in a common feature space. Different from FID, it measures the difference of an image pair generated from the same input. The LPIPS [15] metric actually used the AlexNet [8] framework to calculate this diversity score. In this work, we adapted this perceptual metric for calculating diversity of our synthesized document images.

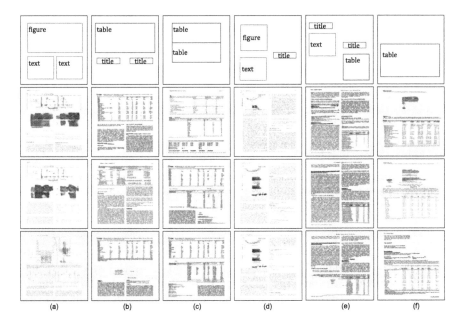

Fig. 4. Examples of diverse synthesized documents generated from the same layout: Given an input document layout with object bounding boxes and categories, our model samples 3 images sharing the same layout structure, but different in style and appearance.

4.3   Qualitative Results

Generating Synthetic Document Images: In order to highlight the ability of our proposed model to generate diverse realistic set of document images, we present in Fig. 3 a t-SNE [11] visualization of the different synthetic data samples with plausible layout content and variability in overall style and structure. Different clusters of samples correspond to particular layout structure as observed in the figure. We observe that synthetic document samples with complex layout structures have been generated by our model. From these examples, it is also observed that our model is powerful enough to generate complex document samples with multiple objects and multiple instances of the same object category. All the generated samples from the model shown in Fig. 3 have a dimension of 128×128. In this work, we propose two final model baselines, for generated images of 64×64 and 128×128 dimension.

Controllable Document Synthesis: One of the most intriguing challenges in our problem study was the controllable synthesis of document images guided by user specified layout provided as an input to our DocSynth generative network. We proposed a qualitative analysis of the challenge in two different case studies.

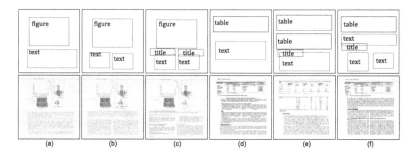

Fig. 5. Examples of synthesized document images by adding or removing bounding boxes based on previous layout: There are 2 groups of images (a)–(c) and (d)–(f) in the order of adding or removing objects.

In the first case study as shown in Fig. 4, it shows a diverse set of generated documents from a single reference layout as specified by the user. In real life scenario, documents do have the property to exhibit variability in appearance and style while preserving the layout structure. The generated samples in this case obey the spatial constraints of the input bounding boxes, and also the generated objects exhibit consistent behaviour with the input labels.

In the second case study as shown in Fig. 5, we demonstrate our model's ability to generate documents with complex layouts by starting with a very simple layout and then adding a new bounding box or removing an existing bounding box from the input reference layout. From these results we can clearly infer that new objects can be introduced in the images at the desired locations by the user, and existing objects can be modified as new content is added.

4.4 Quantitative Results

Table 1 summarizes the comparison results of the FID and Diversity Score of our proposed model baselines for both 128×128 and 64×64 generated documents. For proper comparison, we have compared the performance scores of the model generated images with the real images. The model generated images are quite realistic as depicted by the performance scores for both FID and Diversity scores. The performance scores obtained for 64×64 image generation model are slightly better compared to those obtained for 128×128 images.

4.5 Ablation Studies

We demonstrate the importance of the key components in our model by creating some ablated models trained on the PubLayNet dataset [18]. The following studies clearly illustrate the importance of these elements for solving the task.

Table 1. Summary of the final proposed model baseline for synthetic document generation

Method	FID	Diversity score
Real images (128×128)	30.23	0.125
DocSynth (128×128)	**33.75**	**0.197**
Real images (64×64)	25.23	0.115
DocSynth (64×64)	**28.35**	**0.201**

Spatial Reasoning Module: As already discussed, the spatial reasoning module comprising conv-LSTM to generate the hidden feature map h is one of the most significant components in our model. For generating novel realistic synthetic data, we compare our model results with conv LSTM over vanilla LSTM and also modifying its number of layers for exhaustive analysis (Table 2).

Table 2. Ablation Study based on different Spatial Reasoning backbones used in our model

Reasoning backbone	FID
No LSTM	70.61
Vanilla LSTM	75.71
conv-LSTM ($k = 1$)	37.69
conv-LSTM ($k = 2$)	36.42
conv-LSTM ($k = 3$)	**33.75**

5   Conclusion

In this work, we have presented a novel approach to automatically synthesize document images according to a given layout. The proposed method, is able to understand the complex interactions among the different layout components to generate synthetic document images that fulfill the given layout. Despite the low resolution of the generated images, we believe that this work supposes the first step towards the generation of whole synthetic documents whose contents are related to the context of the page. Indeed, other applications arise from this synthetic generation besides generating realistic images which opens a large variety of future research lines.

The future scope will be mainly focused on two research lines. Firstly, high resolution documents with understandable content is the final goal for any synthetic document generator, therefore, we plan to extend our model towards this end. Secondly, exploiting the generated data for supervision purposes, can improve the performance on tasks such as document classification, table detection or layout analysis.

Acknowledgment. This work has been partially supported by the Spanish projects RTI2018-095645-B-C21, and FCT-19-15244, and the Catalan projects 2017-SGR-1783, the CERCA Program/Generalitat de Catalunya and PhD Scholarship from AGAUR (2021FIB-10010).

References

1. Binmakhashen, G.M., Mahmoud, S.A.: Document layout analysis: a comprehensive survey. ACM Comput. Surv. (CSUR) **52**(6), 1–36 (2019)
2. De Vries, H., Strub, F., Mary, J., Larochelle, H., Pietquin, O., Courville, A.: Modulating early visual processing by language. arXiv preprint arXiv:1707.00683 (2017)
3. Goodfellow, I.J., et al.: Generative adversarial networks. In: Advances in Neural Information Processing Systems, pp. 2672–2680 (2014)
4. Heusel, M., Ramsauer, H., Unterthiner, T., Nessler, B., Hochreiter, S.: GANs trained by a two time-scale update rule converge to a local nash equilibrium. arXiv preprint arXiv:1706.08500 (2017)
5. Horak, W.: Office document architecture and office document interchange formats: current status of international standardization. Computer **18**(10), 50–60 (1985)
6. Kang, L., Riba, P., Wang, Y., Rusiñol, M., Fornés, A., Villegas, M.: GANwriting: content-conditioned generation of styled handwritten word images. In: Vedaldi, A., Bischof, H., Brox, T., Frahm, J.-M. (eds.) ECCV 2020. LNCS, vol. 12368, pp. 273–289. Springer, Cham (2020). https://doi.org/10.1007/978-3-030-58592-1_17
7. Kingma, D.P., Ba, J.: Adam: a method for stochastic optimization. arXiv preprint arXiv:1412.6980 (2014)
8. Krizhevsky, A., Sutskever, I., Hinton, G.E.: Imagenet classification with deep convolutional neural networks. Adv. Neural. Inf. Process. Syst. **25**, 1097–1105 (2012)
9. Lake, B.M., Salakhutdinov, R., Tenenbaum, J.B.: Human-level concept learning through probabilistic program induction. Science **350**(6266), 1332–1338 (2015)
10. Li, J., Yang, J., Hertzmann, A., Zhang, J., Xu, T.: Layoutgan: generating graphic layouts with wireframe discriminators. arXiv preprint arXiv:1901.06767 (2019)
11. Van der Maaten, L., Hinton, G.: Visualizing data using t-SNE. J. Mach. Learn. Res. **9**(11) (2008)
12. Miyato, T., Kataoka, T., Koyama, M., Yoshida, Y.: Spectral normalization for generative adversarial networks. arXiv preprint arXiv:1802.05957 (2018)
13. Patil, A.G., Ben-Eliezer, O., Perel, O., Averbuch-Elor, H.: Read: recursive autoencoders for document layout generation. In: Proceedings of the IEEE Conference on Computer Vision and Pattern Recognition Workshops, pp. 544–545 (2020)
14. Szegedy, C., Vanhoucke, V., Ioffe, S., Shlens, J., Wojna, Z.: Rethinking the inception architecture for computer vision. In: Proceedings of the IEEE Conference on Computer Vision and Pattern Recognition, pp. 2818–2826 (2016)
15. Zhang, R., Isola, P., Efros, A.A., Shechtman, E., Wang, O.: The unreasonable effectiveness of deep features as a perceptual metric. In: Proceedings of the IEEE Conference on Computer Vision and Pattern Recognition, pp. 586–595 (2018)
16. Zhao, B., Yin, W., Meng, L., Sigal, L.: Layout2image: image generation from layout. Int. J. Comput. Vision **128**, 2418–2435 (2020)
17. Zheng, X., Qiao, X., Cao, Y., Lau, R.W.: Content-aware generative modeling of graphic design layouts. ACM Trans. Graph. **38**(4), 1–15 (2019)
18. Zhong, X., Tang, J., Yepes, A.J.: Publaynet: largest dataset ever for document layout analysis. In: Proceedings of the International Conference on Document Analysis and Recognition, pp. 1015–1022 (2019)

Font Style that Fits an Image – Font Generation Based on Image Context

Taiga Miyazono, Brian Kenji Iwana$^{(\boxtimes)}$ ⓘ, Daichi Haraguchi,
and Seiichi Uchida ⓘ

Kyushu University, Fukuoka, Japan
{iwana,uchida}@ait.kyushu-u.ac.jp

Abstract. When fonts are used on documents, they are intentionally selected by designers. For example, when designing a book cover, the typography of the text is an important factor in the overall feel of the book. In addition, it needs to be an appropriate font for the rest of the book cover. Thus, we propose a method of generating a book title image based on its context within a book cover. We propose an end-to-end neural network that inputs the book cover, a target location mask, and a desired book title and outputs stylized text suitable for the cover. The proposed network uses a combination of a multi-input encoder-decoder, a text skeleton prediction network, a perception network, and an adversarial discriminator. We demonstrate that the proposed method can effectively produce desirable and appropriate book cover text through quantitative and qualitative results. The code can be found at https://github.com/Taylister/FontFits.

Keywords: Text generation · Neural font style transfer · Book covers

1 Introduction

Fonts come in a wide variety of styles, and they can come with different weights, serifs, decorations, and more. Thus, choosing a font to use in a medium, such as book covers, advertisements, documents, and web pages, is a deliberate process by a designer. For example, when designing a book cover, the title design (i.e., the font style and color for printing the book title) plays an important role [14,25]. An appropriate title design will depend on visual features (i.e., the appearance of the background design), as well as semantic features of the book (such as the book content, genre, and title texts). As shown in Fig. 1 (a), given a background image for a book cover, typographic experts determine an appropriate title design that fits the background image. In this example, yellow will be avoided for title design to keep the visual contrast from the background; a font style that gives a hard and solid impression may be avoided by considering the impression from the cute rabbit appearance. These decisions about fonts are determined by the image context.

This paper aims to generate an appropriate text image for a given context image to understand their correlation. Specifically, as shown in Fig. 1 (b), we

© Springer Nature Switzerland AG 2021
J. Lladós et al. (Eds.): ICDAR 2021, LNCS 12823, pp. 569–584, 2021.
https://doi.org/10.1007/978-3-030-86334-0_37

Fig. 1. (a) Our question: is there any title design that fits a book cover? (b) Our approach: if there is a specific design strategy, we can realize a neural network that learns the correlation between the cover image and the title design and then generate an appropriate title design.

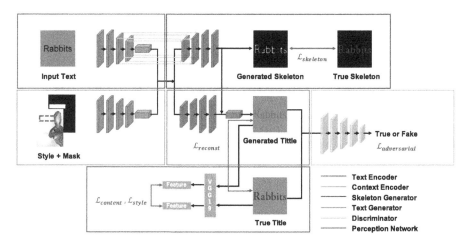

Fig. 2. Overview of the proposed framework.

attempt to generate a title image that fits the book cover image by using a neural network model trained by 104,295 of actual book cover samples. If the model can generate a similar title design to the original, the model catches the correlation inside it and shows the existence of the correlation.

Achieving this purpose is meaningful for two reasons. First, its achievement shows the existence of the correlation between design elements through objective analysis with a large amount of evidence. It is often difficult to catch the correlation because the visual and typographic design is performed subjectively with a huge diversity. Thus, the correlation is not only very nonlinear but also very weak. If we prove the correlation between the title design and the book cover image through our analysis, it will be an important step to understand the "theory" behind the typographic designs. Secondly, we can realize a design support system that can suggest an appropriate title design from a given background image. This system will be a great help to non-specialists in typographic design.

In order to learn how to generate a title from a book cover image and text information, we propose an end-to-end neural network that inputs the book cover image, a target location mask, and a desired book title and outputs stylized text. As shown in Fig. 2, the proposed network uses a combination of a Text Encoder, Context Encoder, Skeleton Generator, Text Generator, Perception Network, and Discriminator to generate the text. The Text Encoder and Context Encoders encode the desired text and given book cover context. The Text Generator use skeleton-guided text generation [29] to generate the text, and the Perception Network and adversarial Discriminator refine the results.

The main contributions of this paper are as follows:

- We propose an end-to-end system of generating book title text based on the cover of a book. As far as we know, this paper presents the first attempt to generate text based on the context information of book covers.
- A novel neural network is proposed, which includes a skeleton-based multi-task and multi-input encoder-decoder, a perception network, and an adversarial network.
- Through qualitative and quantitative results, we demonstrate that the proposed method can effectively generate text suitable given the context information.

2   Related Work

Recently, font style transfer using neural networks [1] has become a growing field. In general, there are three approaches toward neural font style transfer, GAN-based methods, encoder-decoder methods, and NST-based methods. A GAN-based method for font style transfer uses a conditional GAN (cGAN). For example, Azadi et al. [2] used a stacked cGAN to generate isolated characters with a consistent style learned from a few examples. A similar approach was taken for Chinese characters in a radical extraction-based GAN with a multi-level discriminator [10] and with a multitask GAN [28]. Few-shot font style transfer with encoder-decoder networks have also been performed [33]. Wu et al. [29] used a multitask encoder-decoder to generate stylized text using text skeletons. Wang et al. [26] use an encoder-decoder network to identify text decorations for style transfer. In Lyu et al. [18] an autoencoder guided GAN was used to generate isolated Chinese characters with a given style. There also have been a few attempts at using NST [6] to perform font style transfer between text [1,7,21,24].

An alternative to text-to-text neural font style transfer, there have been attempts to transfer styles from arbitrary images and patterns. For example, Atarsaikhan et al. [1] proposed using a distance-based loss function to transfer patterns to regions localized to text regions. There are also a few works that use context information to generate text. Zhao et al. [32] predicted fonts for web pages using a combination of attributes, HTML tags, and images of the web pages. This is similar to the proposed method in that the context of the text is used to generate the text.

Yang et al. [30] stylized synthetic text to become realistic scene text. The difference between the proposed method and these methods is that we propose inferring the font style based on the contents of the desired medium and use it in an end-to-end model to generate text.

3   Automatic Title Image Generation

The purpose of this study is to generate an image of the title text with a suitable font and color for a book cover image. Figure 2 shows the overall architecture of the proposed method. The network consists of 6 modules: Text Encoder, Context Encoder, Text Generator, Skeleton Generator, Perception Network, and Discriminator. The Text Encoder and Context Encoder extracts text and styles from the input text and style cover. The Generator generates the stylized text suitable for the book cover input, and the Skeleton Generator creates a text skeleton to guide the Generator. The Perception Network and Discriminator help refine the output of the Text Generator.

3.1   Text Encoder

The Text Encoder module extracts character shape features from an image I_t of the input text. These features are used by the Text Generator and the Skeleton Generator to generate their respective tasks. As shown in Fig. 3, the Text Encoder input is an image of the target text rendered on a background with a fixed pixel value.

The encoder consists of 4 convolutional blocks with residual connections [9]. The first block consists of two layers of 3×3 convolutions with stride 1. The subsequent blocks contain a 3×3 convolutional layers with stride 2 and two convolutional layers with stride 1. Leaky Rectified Linear Units (Leaky ReLU) [19] are used as the activation function for each layer. The negative slope is set to 0.2. There are skip-connections between second and third convolutional blocks and the convolutional blocks of the Skeleton Generator, which is described later.

3.2   Context Encoder

The Context Encoder module extracts the style features from the cover image S_c and the location mask image S_m. Examples of the covers S_c and the location mask S_m are shown in Fig. 3. The cover image S_c provides the information about the book cover, such as color, objects, layout, etc.) for the Context Encoder to predict the style of the font and the location mask S_m provides target location information. It should be noted that the cover image S_c has the text inpainted, i.e., removed. Thus, the style is inferred solely based on the cover and not on textual cues.

The Context Encoder input is constructed of S_c and S_m concatenated in the channel dimension. Also, the Context Encoder structure is the same as the Text Encoder, except that the input is only 2 channels (as opposed to 3 channels,

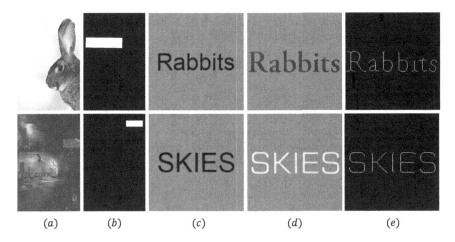

(a) (b) (c) (d) (e)

Fig. 3. Training data example. From left to right: style cover image, style cover mask image, input text image, true title image, true skeleton image.

RGB, for the Text Encoder). As the input of the generators, the output of the Context Encoder is concatenated in the channel dimension with the output of the Text Encoder.

3.3   Skeleton Generator

In order to improve the legibility of the generated text, a skeleton of the input text is generated and used to guide the generated text. This idea is called the Skeleton-guided Learning Mechanism [29]. This module generates a skeleton map, which is the character skeleton information of the generated title image. As described later, the generated skeleton map is merged into the Text Generator. By doing so, the output of the Text Generator is guided by the skeleton to generate a more robust shape.

This module is an upsampling CNN that uses four blocks of convolutional layers. The first block contains two standard convolutional layers, and the three subsequent blocks have one transposed convolutional layer followed by two standard convolutional layers. The transposed convolutional layers have 3 × 3 transposed convolutions at stride 2, and the standard convolutional layers have 3 × 3 convolutions at stride 1. All the layers use Batch Norm and LeakyReLU. In addition, there are skip-connections between the Text Encoder and after the first layer of the second and third convolutional blocks.

To train the Skeleton Generator, the following skeleton loss $\mathcal{L}_{\text{skeleton}}$ is used to train the combined network:

$$\mathcal{L}_{\text{skeleton}} = 1 - \frac{2\sum_{i}^{N}(T_{\text{skeleton}})_{i=1}(O_{\text{skeleton}})_i}{\sum_{i=1}^{N}(T_{\text{skeleton}})_i + \sum_{i=1}^{N}(O_{\text{skeleton}})_i}, \tag{1}$$

where N is the number of pixels, T_{skeleton} is true skeleton map, and O_{skeleton} is the output of Skeleton Generator. This skeleton loss is designed based on DiceLoss [20].

3.4   Text Generator

The Text Generator module takes the features extracted by Text Encoder and Style Encoder and outputs an image of the stylized title. This output is the desired result of the proposed method. It is an image of the desired text with the style inferred from the book cover image.

The Text Generator has the same structure as the Skeleton Generator, except with no skip-connections and with an additional convolutional block. The additional convolutional block combines the features from the Text Generator and the Skeleton Generator. As described previously, the Skeleton Generator generates a skeleton of the desired stylized text. To incorporate the generated skeleton into the generated text, the output of the Skeleton Generator is concatenated with the output of the fourth block of the Text Generator. The merged output is further processed through a 3×3 convolutional layer with a stride of 1, Batch Normalization, and Leaky ReLU activation. The output O_t of the Text Generator has a tanh activation function.

The Text Generator is trained using a reconstruction loss $\mathcal{L}_{\text{reconst}}$. The reconstruction loss is Mean Absolute Error between the generated output O_t and the ground truth title text T_t, or:

$$\mathcal{L}_{\text{reconst}} = |T_t - O_t|. \tag{2}$$

While loss guides the generated text to be similar to the original text, it is only one part of the total loss. Thus, the output of the Text Generator is not strictly the same as the original text.

3.5   Perception Network

To refine the results of the Text Generator, we use a Perception Network. Specifically, the Perception Network is used to increase the perception of the generated images [13,29]. To do this, the output of the Text Generator is provided to a Perception Network, and two loss functions are added to the training. These loss functions are the Content Perceptual loss $\mathcal{L}_{\text{content}}$ and the Style Perceptual loss $\mathcal{L}_{\text{style}}$. The Perception Network is a VGG19 [22] that was pre-trained on ImageNet [5].

Each loss function compares differences in the content and style of the features extracted from the Perception Network when provided the generated title and the ground truth title images. The Content Perceptual loss minimizes the distance between the extracted features of the generated title images and the original images, or:

$$\mathcal{L}_{\text{content}} = \sum_{l \in \mathcal{F}} |P_l(T_t) - P_l(O_t)|, \tag{3}$$

where P is the Perception Network, $P_l(\cdot)$ is the feature map of P's l-th layer given the generated input image O_t or ground truth image T_t, and $\mathcal{F}$ is the set of layers used in these loss. In this case, $\mathcal{F}$ is set to the relu1_1, relu2_1, relu3_1, relu4_1, and relu5_1 layers of VGG19. The Style Perceptual loss compares the texture and local features extracted by the Perception Network, or:

$$\mathcal{L}_{\text{style}} = \sum_{l \in \mathcal{F}} |\Psi_l^P(T_t) - \Psi_l^P(O_t)|, \tag{4}$$

where $\Psi_l^P(\cdot)$ is a Gram Matrix, which has $C_l \times C_l$ elements. Given input $I \in \{O_t, T_t\}$, $P_l(I)$ has a feature map of shape $C_l \times H_l \times W_l$. The elements of Gram Matrix $\Psi_l^P(I)$ are given by:

$$\Psi_l^P(I)_{c,c'} = \frac{1}{C_l H_l W_l} \sum_{h=1}^{H_l} \sum_{w=1}^{W_l} P_l(I)_{h,w,c} P_l(I)_{h,w,c'}, \tag{5}$$

where c and c' are each element of C_t. By minimizing the distance between the Gram Matrices, a style consistency is enforced. In other words, the local features of both images should be similar.

3.6 Discriminator

In addition to the Perception Network, we also propose the use of an adversarial loss $\mathcal{L}_{\text{adversarial}}$ to ensure that the generated results are realistic. To use the adversarial loss, we introduce a Discriminator to the network architecture. The Discriminator distinguishes between whether the input is a real title image or a fake image. In this case, we use the true tile image T_t as the real image and the Text Generator's output O_t as the fake image.

The Discriminator in the proposed method follows the structure of the DCGAN [17]. The Discriminator input goes through 5 down-sampling 5×5 convolutional layers with stride 2 and finally a fully-connected layer. The output is the probability that the input is a true title image. Except for the output layer, the LeakyReLU function is used. At the output layer, a sigmoid function is adopted. The following adversarial loss is used to optimize the entire generation model and the Discriminator:

$$\mathcal{L}_{\text{adversarial}} = \mathbb{E}_{I_t, T_t}[\log D(I_t, T_t)] + \mathbb{E}_{I_t, S}[\log\{1 - D(I_t, G(I_t, S))\}], \tag{6}$$

where G is the whole generation module, D is the discriminator, S is the style condition(S_c, S_m), T_t is the true title image, and I_t is the input title text.

3.7 Title Image Generation Model

As we have explained, the proposed network can generate title text suitable for a style. As a whole, the Text Encoder receives an input text image I_t and the Context Encoder receives a style image (S_c, S_m), and the Skeleton Generator

| (a) Detected Text | (b) Text Mask | (c) Extracted Text |

Fig. 4. Mask generation and title extraction.

outputs a skeleton map O_{skeleton} and the Text Generator outputs a title image O_t. This process is shown in Fig. 2. The training is done with alternating adversarial training with the Discriminator and the rest of the modules in an end-to-end manner. The Text Generator, Skeleton Generator, Text Encoder, and Context Encoder are trained using a total loss $\mathcal{L}_{\text{total}}$ through the Text Generator, where

$$\mathcal{L}_{\text{total}} = w_1 \mathcal{L}_{\text{reconst}} + w_2 \mathcal{L}_{\text{skeleton}} + w_3 \mathcal{L}_{\text{adversarial}} + w_4 \mathcal{L}_{\text{content}} + w_5 \mathcal{L}_{\text{style}}. \quad (7)$$

Variables w_1 to w_5 are weights for each of the losses.

4 Experimental Setup

4.1 Dataset

In the experiment, as shown in Fig. 3, to train the network, we need a combination of the full book cover image without text, a target location mask, a target plain input text, the original font, and the skeleton of the original words. Thus, we use a combination of the following datasets and pre-processing to construct the ground truth.

We used the Book Cover Dataset [12]. This dataset consists of 207,572 book cover images[1]. The size of the book cover image varies depending on the book cover, but for this experiment, we resize the images to 256 × 256 pixels in RGB.

To ensure that the generated style is only inferred by the book cover and not any existing text, we remove all of the text from the book covers before using them. To remove the text, we first use Character Region Awareness for Text Detection (CRAFT) [4] to detect the text, then cut out regions defined by the detected bounding-boxes with dilated with a 3 × 3 kernel. CRAFT is an effective text detection method that uses a U-Net-like structure to predict character proposals and uses the affinity between the detected characters to generate word-level bounding-box proposals. Then, Telea's inpainting [23] is used to fill the removed text regions with the plausible background area. The result is images of book covers without the text.

For the title text, CRAFT is also used. The text regions found by CRAFT are recognized using the scene text recognition method that was proposed by Baek et al. [3]. Using the text recognition method, we can extract and compare

---

[1] https://github.com/uchidalab/book-dataset.

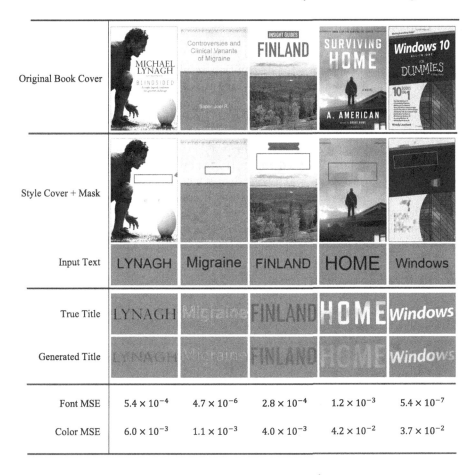

Original Book Cover					
Style Cover + Mask					
Input Text	LYNAGH	Migraine	FINLAND	HOME	Windows
True Title	LYNAGH	Migraine	FINLAND	HOME	Windows
Generated Title	LYNAGH	Migraine	FINLAND	HOME	Windows
Font MSE	5.4×10^{-4}	4.7×10^{-6}	2.8×10^{-4}	1.2×10^{-3}	5.4×10^{-7}
Color MSE	6.0×10^{-3}	1.1×10^{-3}	4.0×10^{-3}	4.2×10^{-2}	3.7×10^{-2}

Fig. 5. Successful generation results.

the text regions to the ground truth. Once the title text is found, we need to extract the text without the background, as shown in Fig. 4. We generate a text mask (Fig. 4b) using a character stroke separation network to perform this extraction. The character stroke separation network is based on pix2pix, [11] and it is trained to generate the text mask based on the cropped detected text region of the title. By applying the text mask to the detected text, we can extract the title text without a background. The background is replaced with a background with pixel values (127.5, 127.5, 127.5). Since the inputs are normalized between $[-1, 1]$, the background represents a "0" value. Moreover, a target location mask (Fig. 3b) is generated by masking the bounding-box of the title text. The plain text input is generated using the same text but in the font "Arial." Finally, the skeleton of the text is obtained using the skeletonization method of Zhang et al. [31].

4.2   Implementation Details

In this experiment, for training the proposed model on high-quality data pairs, only images where the character recognition results [3] of the region detected by CRAFT (Fig. 4a) and the image created by the character stroke separation network (Fig. 4c) match were used. As a result, our proposed model has been trained on 195,922 title text images and a corresponding 104,925 book cover images. Also, 3,702 title text images and 1,000 accompanying book cover images were used for the evaluation. The training was done end-to-end using batch size 8 for 300,000 iterations. We used Adam [15] as optimizer, and set the learning coefficient $lr = 2 \times 10^{-4}$, $\beta_1 = 0.5$, and $\beta_2 = 0.999$. We also set $w_1 = 1$, $w_2 = 1$, $w_3 = 1.0 \times 10^{-2}$, $w_4 = 1$, and $w_5 = 1.0 \times 10^3$ for the weighting factors of the losses. The weights are set so that the scale of each loss value is similar. For w_3, we set a smaller value so that Discriminator does not have too much influence on the generation module's training and can generate natural and clear images [16].

4.3   Evaluation Metrics

For quantitative evaluation of the generate title images, we use standard metrics and non-standard metrics. For the standard metrics, Mean Absolute Error (MAE), Peak Signal-to-Noise Ratio (PSNR), and Structural Similarity (SSIM) [27] are used. We introduce two non-standard metrics for our title generation task, Font-vector Mean Square Error (Font MSE) and Color Mean Square Error (Color MSE). Font MSE evaluates the MSE between font style vectors of the original (i.e., ground-truth) and the generated title images. The font style vector is a six-dimensional vector of the likelihood of six font styles: serif, sans, hybrid, script, historical, and fancy. The font style vector is estimated by a CNN trained with text images generated by SynthText [8] with 1811 different fonts. Color MSE evaluates the MSE between three-dimensional RGB color vectors of the original and the generated title images. The color vector is given as the average color of the stroke detected by the character stroke separation network. These evaluations are only used in the experiment where the target text is the same as the original text. It should be noted that the three standard evaluations MAE, PSNR, and SSIM, can only be used when the target text is the same as the original text. However, we can use Font MSE and Color MSE, even when the generated text is different because they measure qualities that are common to the generated text and the ground truth text.

5   Experimental Results

5.1   Qualitative Evaluation

This section discusses the relationship between the quality of the generated images and the book cover images by showing various samples generated by the network and the corresponding original book cover images. Figure 5 shows the samples where the network generates a title image close to the ground truth

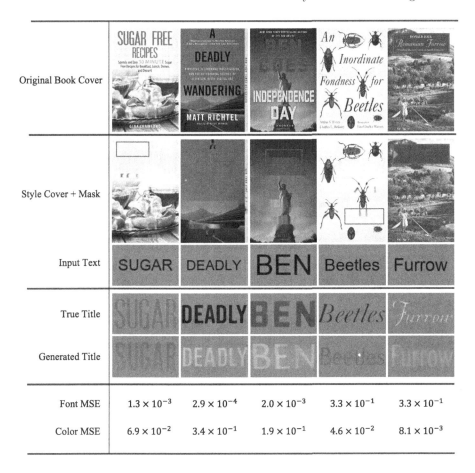

Original Book Cover					
Style Cover + Mask					
Input Text	SUGAR	DEADLY	BEN	Beetles	Furrow
True Title					
Generated Title					
Font MSE	1.3×10^{-3}	2.9×10^{-4}	2.0×10^{-3}	3.3×10^{-1}	3.3×10^{-1}
Color MSE	6.9×10^{-2}	3.4×10^{-1}	1.9×10^{-1}	4.6×10^{-2}	8.1×10^{-3}

Fig. 6. Failure results.

successfully, that is, with smaller Font MSE and Color MSE. From the figure, we can see that the strokes, ornaments, and the size of the text are reproduced. Especially, the first example shows the serif font is also reproducible even if the input text is always given as a sans-serif image.

Among the results in Fig. 5, the "Windows" example clearly demonstrates that the proposed method can predict the font and style of the text given the book cover and the target location mask. This is due to the book cover being a recognizable template from the "For Dummies" series in which other books with similar templates exist in the training data. The figure demonstrates that the proposed method effectively infers the font style from the book cover image based on context clues alone.

Figure 6 shows examples where the generator could not successfully generate a title image close to the ground truth. The first, second, and third images in Fig. 6 show examples of poor coloring in particular. The fourth and fifth images

Fig. 7. Other generation results. The top and bottom rows show the results selected by FontMSEs and ColorMSEs, respectively. The left four columns show lower MSE (i.e., successful) cases and the remaining two show the higher MSE (i.e., failure) cases. Each result is comprised of the true (top) and generated (middle) titles and the whole book cover image (bottom).

show examples where the proposed method could not predict the font shape. For the "SUGAR" and "BEN" images, there are no clues that the color should be red. In the "DEADLY" book cover image, one would expect light text on the dark background. However, the original book cover used dark text on a dark background. For the "Beetles" and "Furrow" examples, the fonts are highly stylized and difficult to predict.

Figure 7 shows several additional results including successful and failure cases. Even in this difficult estimation task from a weak context, the proposed method gives a reasonable style for the title image. The serif for "Make" and the thick style for "GUIDEBOOK" are promising. We also observe that peculiar styles, such as very decorated fonts and vivid colors, are often difficult to recover from the weak context.

Finally, in Fig. 8, we show results of using the proposed method, but with text that is different from the ground truth. This figure demonstrates that we can produce any text in the predicted style.

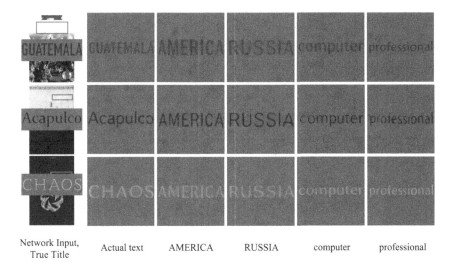

| Network Input, True Title | Actual text | AMERICA | RUSSIA | computer | professional |

Fig. 8. Generation results using text that is different from the original text.

5.2 Ablation Study

To measure the importance of the components of the proposed method, quantitative and qualitative ablation studies are performed. The network of the proposed method consists of six modules and their associated loss functions. Therefore, we measure the effects of the components. All experiments are performed with the same settings and the same training data.

The following evaluations are performed:

- **Proposed**: The evaluation with the entire proposed model.
- **Baseline**: Training is performed only using the Text Encoder and Text Generator with the reconstruction loss $\mathcal{L}_{reconst}$.
- **w/o Context Encoder**: The proposed method but without the Context Encoder. The results are expected to be poor because there is no information about the style to learn from.
- **w/o Skeleton Generator**: The proposed method but with no guidance from the Skeleton Generator and without the use of the skeleton loss $\mathcal{L}_{skeleton}$.
- **w/o Discriminator**: The proposed method but without the Discriminator and the adversarial loss $\mathcal{L}_{adversarial}$.
- **w/o Perception Network**: The proposed method but without the Perception Network and the associated losses $\mathcal{L}_{content}$ and $\mathcal{L}_{style}$.

The quantitative results of the ablation study are shown in Table 1. The results show that Proposed has the best results in all evaluation methods except one, Color MSE. For Color MSE, w/o Perception Network performed slightly better. This indicates that the color of text produced by the proposed method without the Perception Network was more similar to the ground truth. However, as shown in Fig. 9, the Perception Network is required to produce reasonable

Table 1. Quantitative evaluation results.

Method	MAE ↓	PSNR ↑	SSIM ↑	Font MSE↓	Color MSE↓
Baseline	0.041	20.49	0.870	0.174	0.102
Proposed	**0.035**	**21.58**	**0.876**	**0.062**	0.064
w/o Context Encoder	0.036	20.79	0.872	0.126	0.093
w/o Discriminator	**0.035**	21.46	0.814	0.105	0.081
w/o Perception Network	0.061	19.39	0.874	0.085	**0.058**
w/o Skeleton Generator	**0.035**	21.09	0.874	0.112	0.080

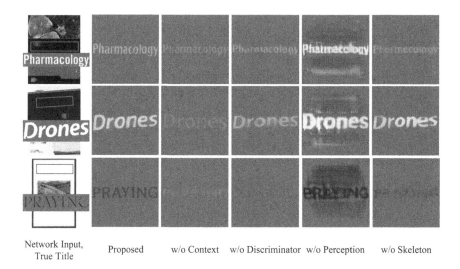

Network Input,   Proposed   w/o Context   w/o Discriminator w/o Perception   w/o Skeleton
True Title

Fig. 9. Results of the ablation study.

results. In the figure, the colors are brighter without the Perception Network, but there is also a significant amount of additional noise. This is reflected in the results for the other evaluation measures in Table 1.

Also from Fig. 9, it can be observed how important each module is to the proposed method. As seen in Table 1, the Font MSE and Color MSE are much larger for w/o Context Encoder than the proposed method. This is natural due to knowing the style information being provided to the network. There are no hints such as color, object, texture, etc. Thus, as shown in Fig. 9, w/o Context Encoder only generates a basic font with no color information. This also shows that the book cover image information is important in generating the title text. A similar trend can be seen with w/o Discriminator and w/o Skeleton Network. The results show that the Discriminator does improve the quality of the font and the Skeleton Network ensures the structure of the text is robust.

6   Conclusion

In this study, we proposed a method of generating the design of text based on context information, such as the location and surrounding image. Specifically, we generated automatic book titles for given book covers using a neural network. The generation of the title was achieved by extracting the features of the book cover and the input text with two encoders, respectively, and using a generator with skeletal information. In addition, an adversarial loss and perception network is trained simultaneously to refine the results. As a result, we succeeded in incorporating the implicit universality of the design of the book cover into the generation of the title text. We obtained excellent results quantitatively and qualitatively and the ablation study confirmed the effectiveness of the proposed method. The code can be found at https://github.com/Taylister/FontFits. In the future, we will pursue the incorporation of the text back onto the book cover.

Acknowledgement. This work was in part supported by MEXT-Japan (Grant No. J17H06100 and Grant No. J21K17808).

References

1. Atarsaikhan, G., Iwana, B.K., Uchida, S.: Guided neural style transfer for shape stylization. PLOS ONE **15**(6), e0233489 (2020)
2. Azadi, S., Fisher, M., Kim, V., Wang, Z., Shechtman, E., Darrell, T.: Multi-content GAN for few-shot font style transfer. In: CVPR (2018)
3. Baek, J., et al.: What is wrong with scene text recognition model comparisons? Dataset and model analysis. In: ICCV (2019)
4. Baek, Y., Lee, B., Han, D., Yun, S., Lee, H.: Character region awareness for text detection. In: CVPR (2019)
5. Deng, J., Dong, W., Socher, R., Li, L.J., Li, K., Fei-Fei, L.: ImageNet: a large-scale hierarchical image database. In: CVPR (2009)
6. Gatys, L., Ecker, A., Bethge, M.: A neural algorithm of artistic style. J. Vis. **16**(12), 326 (2016)
7. Gomez, R., Biten, A.F., Gomez, L., Gibert, J., Karatzas, D., Rusinol, M.: Selective style transfer for text. In: ICDAR (2019)
8. Gupta, A., Vedaldi, A., Zisserman, A.: Synthetic data for text localisation in natural images. In: CVPR (2016)
9. He, K., Zhang, X., Ren, S., Sun, J.: Deep residual learning for image recognition. arXiv preprint arXiv:1512.03385 (2015)
10. Huang, Y., He, M., Jin, L., Wang, Y.: RD-GAN: few/zero-shot Chinese character style transfer via radical decomposition and rendering. In: Vedaldi, A., Bischof, H., Brox, T., Frahm, J.-M. (eds.) ECCV 2020, Part VI. LNCS, vol. 12351, pp. 156–172. Springer, Cham (2020). https://doi.org/10.1007/978-3-030-58539-6_10
11. Isola, P., Zhu, J.Y., Zhou, T., Efros, A.A.: Image-to-image translation with conditional adversarial networks. In: CVPR (2017)
12. Iwana, B.K., Raza Rizvi, S.T., Ahmed, S., Dengel, A., Uchida, S.: Judging a book by its cover. arXiv preprint arXiv:1610.09204 (2016)

13. Johnson, J., Alahi, A., Fei-Fei, L.: Perceptual losses for real-time style transfer and super-resolution. In: Leibe, B., Matas, J., Sebe, N., Welling, M. (eds.) ECCV 2016, Part II. LNCS, vol. 9906, pp. 694–711. Springer, Cham (2016). https://doi.org/10.1007/978-3-319-46475-6_43

14. Jubert, R.: Typography and graphic design: from antiquity to the present. Flammarion (2007)

15. Kingma, D.P., Ba, J.: Adam: a method for stochastic optimization. arXiv preprint arXiv:1412.6980 (2014)

16. Ledig, C., et al.: Photo-realistic single image super-resolution using a generative adversarial network. In: CVPR (2017)

17. Li, J., Jia, J., Xu, D.: Unsupervised representation learning of image-based plant disease with deep convolutional generative adversarial networks. In: CCC (2018)

18. Lyu, P., Bai, X., Yao, C., Zhu, Z., Huang, T., Liu, W.: Auto-encoder guided GAN for Chinese calligraphy synthesis. In: ICDAR (2017)

19. Maas, A.L., Hannun, A.Y., Ng, A.Y.: Rectifier nonlinearities improve neural network acoustic models. In: ICML (2013)

20. Milletari, F., Navab, N., Ahmadi, S.A.: V-net: fully convolutional neural networks for volumetric medical image segmentation. In: 3DV (2016)

21. Narusawa, A., Shimoda, W., Yanai, K.: Font style transfer using neural style transfer and unsupervised cross-domain transfer. In: Carneiro, G., You, S. (eds.) ACCV 2018. LNCS, vol. 11367, pp. 100–109. Springer, Cham (2019). https://doi.org/10.1007/978-3-030-21074-8_9

22. Simonyan, K., Zisserman, A.: Very deep convolutional networks for large-scale image recognition. arXiv preprint arXiv:1409.1556 (2014)

23. Telea, A.: An image inpainting technique based on the fast marching method. J. Graph. Tools **9**(1), 23–34 (2004)

24. Ter-Sarkisov, A.: Network of steel: neural font style transfer from heavy metal to corporate logos. In: ICPRAM (2020)

25. Tschichold, J.: The New Typography: A Handbook for Modern Designers, vol. 8. University of California Press, California (1998)

26. Wang, W., Liu, J., Yang, S., Guo, Z.: Typography with decor: intelligent text style transfer. In: CVPR (2019)

27. Wang, Z., Bovik, A.C., Sheikh, H.R., Simoncelli, E.P.: Image quality assessment: from error visibility to structural similarity. IEEE Trans. Image Process. **13**(4), 600–612 (2004)

28. Wu, L., Chen, X., Meng, L., Meng, X.: Multitask adversarial learning for Chinese font style transfer. In: IJCNN (2020)

29. Wu, L., et al.: Editing text in the wild. In: ACM ICM (2019)

30. Yang, X., He, D., Kifer, D., Giles, C.L.: A learning-based text synthesis engine for scene text detection. In: BMVC (2020)

31. Zhang, T., Suen, C.Y.: A fast parallel algorithm for thinning digital patterns. Commun. ACM **27**(3), 236–239 (1984)

32. Zhao, N., Cao, Y., Lau, R.W.: Modeling fonts in context: font prediction on web designs. Comput. Graph. Forum **37**(7), 385–395 (2018)

33. Zhu, A., Lu, X., Bai, X., Uchida, S., Iwana, B.K., Xiong, S.: Few-shot text style transfer via deep feature similarity. IEEE Trans. Image Process. **29**, 6932–6946 (2020)

Bayesian Hyperparameter Optimization of Deep Neural Network Algorithms Based on Ant Colony Optimization

Sinda Jlassi[1], Imen Jdey[1,2(✉)], and Hela Ltifi[1,2]

[1] Faculty of Sciences and Techniques of Sidi Bouzid, University of Kairouan, Kairouan, Tunisia
Jlassi.Sinda@fstsbz.u-kairouan.tn, {imen.jdey,
hela.ltifi}@fstsbz.rnu.tn
[2] REsearch Groups in Intelligent Machines, University of Sfax, National School of Engineers (ENIS), BP 1173, 3038 Sfax, Tunisia

Abstract. Within this paper we proposed a new method named BayesACO, to improve the convolutional neural network based on neural architecture search with hyperparameters optimization. At its essence BayesACO in first side uses Ant Colony Optimization (ACO) to generate the best neural architecture. In other side, it uses bayesian hyperparameters optimization to select the best hyperparameters. We applied this method on Mnist and FashionMnist datasets. Our proposed method proven competitive results with other methods of convolutional neural network optimization.

Keywords: Convolutional neural network · Neural architecture search · Hyperparameters optimization · Ant colony optimization · Bayesian hyperparameters optimization

1 Introduction

Deep learning is a new area of machine learning research which was introduced with the aim of bringing machine learning closer to its main goal artificial intelligence. These are algorithms inspired by the structure and functioning of the brain they can learn several levels of representation [28–30].

The performance of many deep learning methods is highly sensitive to many decisions including choosing the right neural structures, training procedures and methods of hyperparameters optimization; in order to get the desired result. This is a problem whether for new users or even experts. Therefore, Automated Machine Learning (AutoML) can improve performance while saving a lot time and money. AutoML fields aims to make these decisions in data-driven, objective and automated manner. The most helpful and remarkable method of AutoML is the Neural Architecture Search (NAS) method [9].

We choose one of the latest NAS method that uses the ant colony optimization algorithm to design its structure with minimal weights [7]. Where the general concept

© Springer Nature Switzerland AG 2021
J. Lladós et al. (Eds.): ICDAR 2021, LNCS 12823, pp. 585–594, 2021.
https://doi.org/10.1007/978-3-030-86334-0_38

of ant colony optimization algorithms, inspired by original ant demeanor [17], is a combination of prerequisites about a promising solution structure with basic in-formation about the priori obtained network structure [4]. It is branched into several types, which are of interest to us in this study, namely Ant Colony System (ACS), where a group of ant team up to explore good solutions for treatment providers, using an indirect form of pheromone mediated communication deposited at the edges of the travelling salesman problem (TSP) diagram while building solutions [8].

The hyperparameters optimization has a significant impact on the performance of the neural network. Many technologies have been applied successfully. The most common ones are grid search, random search, and Bayesian optimization [2]. Grid search, searches all possibilities. So, it takes a lot of time devoted to the search of hyperparameters. Whereas, random search is based on Grid search, with the aim of creating a network of excessive parameter values, to randomly choose a combination of them. Therefore, this process automatically takes a long time, so it cannot converge with global optimum or guarantee a stable and competitive result. These two methods need all the possible values for each parameter. On the other hand, Bayesian optimization just needs the order of values. The Bayesian optimization is looking for a global optimum with minimal stages [12].

In this work, we are interested in modeling a new optimization process for a convolutional neural network model. To verify the feasibility of our work we compared it with others previous methods of optimization, such as Deepswarm [7], Udeas [2], and LDWPSO [3].

The remainder of the paper is organized as follows: Sect. 2 presents the study background concerning the deep learning, the ant colony optimization and the bayesian hyperparameter optimization; Sect. 3 introduces our proposed method; Sect. 4 provides an evaluation of our method; Sect. 5 concludes the paper and explores possible future directions.

2   Study Background

2.1   Deep Learning

Deep learning supports computer models composed of different processing layers to explore representations of data with different advanced levels these technics have dramatically afflicted the technical level of various domains [28, 29].

A deep neural network, within deep learning, includes many categories, including convolutional neural networks. So convolutional refers to a computation, and it is a specialized type of linear operation. They are simply neural networks that use convolution rather than repeating the general matrix in at least one of their layers. Its first appearance was in the 1990s, when models were developed to recognize handwritten numbers and were of high performance [18]. Even CNN structures have seen increased development, with ImageNet's challenge error rate dropping below 4%. When developing AlexNet researchers from 8 layers to 152 layers [19]. CNN has also excelled in other computer vision tasks, such as human action recognition [20], object localization [21, 22], pedestrian detection [23], and face recognition [24]. CNN subsequently demonstrated that it was effective in natural language and speech processing, and achieved excellent results in stratification [25], sentence modeling [26], and speech recognition [27].

2.2 Ant Colony Optimization

The algorithm to improve the ant colony mainly aims to find the path closest to the target [8]. By secreting the so-called pheromone, in order to leave a trace of the following ants, and according to the nature of the ants, it follows the smell of the most recent and most pheromone in terms of quantity, which means the ant that took the shortest path towards the target. The choice is between each node and another with the rule of choosing a pseudo random procedure (based on probability). The environment altered in two various modes, one to update the local track, when the other to update the global track as shown in Fig. 1. The amounts of pheromone are updated in the edges, where ants move between the contract, with the aim of building a new solution. Automatically all ants apply updating pheromone offline. Global Trail Update comes when all nodes complete through the shortest path by updating the edges that have the best ingenuity on their way [4].

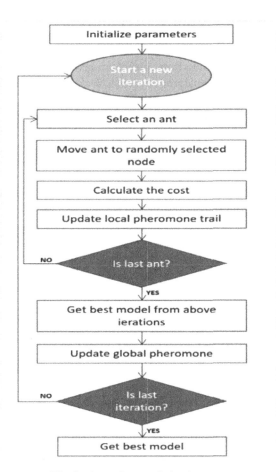

Fig. 1. Ant colony optimization steps

The contact weights are adjusted according to the number of ants, so different combinations of contact weight values are determined. For an independent Ant Colony Optimization (ACO) training application and ACO-BP hybrid training for forward neural networks training for categorization of patterns [10]. Through the global research of the ant colony, weights and bias of artificial neural network (ANN) and deep neural network (DNN) models were modified to achieve optimum performance in the prediction of capital cost [1]. ACO is used to form an ant clan that uses pheromone information to collectively search for the best neurological structure [7]. Is also used for recurrent neural network to develop the structure [11].

2.3 Bayesian Hyperparameter Optimization

We can say that the bayesian hyperparameters optimization algorithm is repeated $t - 1$ times [12]. Within this loop, it will increase the acquisition function and then update the pre-distribution. With each cycle, the pre-distribution is constantly updated, and based on the new setting, the dot at whom the acquisition function is incremented and collected to the training data set is organize. The entire process is duplicated until the maximum number of duplicates is reached or the difference between the current value and the optimum value obtained so far is less than a predetermined threshold [12]. Using bayesian optimization to control data, compare models for an ideal network [13]. Gaussian Process Model (GP) is an algorithm for integrating learning performance. Its performance is affected by some options like kernel type and handling super parameters. The advance is that, the algorithms take into account the variable cost of learning algorithm experiments, which can benefit from multiple cores of parallel experiments [14]. Where the only available standards are artificial test functions that do not represent practical applications. To alleviate this problem, a library of standards was introduced from the pre-eminent application to improve hyperparameters [15]. In [16], the authors define a new kernel for conditional parameter distances that clearly includes information about the relevant parameters in a particular structure, to link the collected performance data for different architectures. In the case of searching for structures that have parameters of different values. For example, we might want to research neural network architectures with an unknown number of layers.

The central idea of BO is to optimize the hyperparameters of the neural network. In this work, we suggest an improvement, that we called BayesACO.

3 Proposed Algorithm

The parameters embedded in our model are internal to the neural network, assessed automatically or learned from learning samples. Therefore, hyperparameters, which are external parameters determined by the neural network, have a great influence on the accuracy of the neural network, hence it is hard to detect these values. Our algorithm takes place in two phases as shown in Fig. 2:

- Initial phase: Neural architecture searches with ANT COLONY OPTIMIZATION to obtain the best weight that form the convolutional neural network structure which improves the weight of the CNN model in order to get optimal performance.

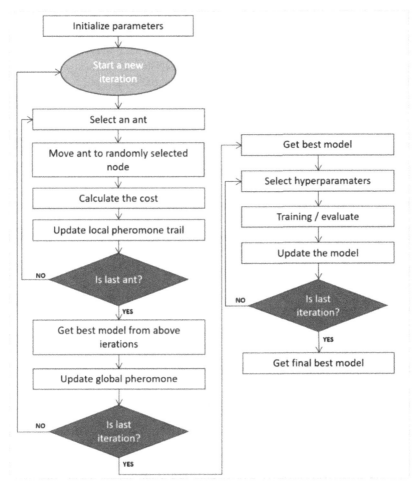

Fig. 2. BayesACO workflow

- Second phase: Using Bayesian Hyperparameters Optimization enables the optimization of certain number of hyperparameters. Thus, we can find better hyperparameters in less period of time because they are reflected on the best set of hyperparameters to an evaluation based on the former experiences.

3.1 Bayesian Hyperparameter Optimization

Throughout this phase, we are going to apply the Bayesian Hyperparameters Optimization. To begin first, we have to specify which function is to be optimized. Then we start with selecting the hyperparameters of the model in a random way. Afterwards, we train the model by evaluating and updating it until it gets the best performance. This stage is repeated with certain number of iterations which is specified by the user in a manner that each iteration depends on the previous one.

4   Application

Along this section, we will deal with hyperparameter and parameter optimization. Evaluating hyperparameters and model structure in CNN to get the best performance as possible is performed on the Mnist and FashionMnist datasets.

Table 1 shows the optimization parameters of our methodology. The number of filters in the convolution Node is optimized between 32 and 128, and the kernel size between 1 and 5 the learning rate of Dropout Node is between 0.1 and 0.3 the "stride" of Pooling Node between 2 and 3, and the type which max or average the size of Dense Node can be 64 or 128, and their activation function is ReLU or Sigmoid validation split between 0.0 and 0.3, batch size which 32, 64 or 128 and epochs number which 5, 10 or 20.

Table 1. Optimization parameters of BayeACO.

Parameter	Optimization value
filter_count	{32, 64, 128}
kernel_size	{1, 3, 5}
rate	{0.1, 0.3}
pool_type	{max, average}
stride	{2, 3}
output_size	{64, 128}
activation	{ReLU, Sigmoid}
validation split	{0.1, 0.3}
batch size	{32, 64, 128}
epochs	{5, 10, 20}

The main parameters of BayesACO use for optimization with mnist. The ant count is 16 and the maximum number of depths is 10. The epochs number is 15, the batch size equals 64, and the learning rate is equal to 0.1.

The first model that we present in Fig. 3 is composed of two convolutional layers and two fully connected layers and one max pooling, dropout and flatten layer.

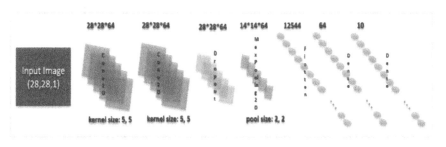

Fig. 3. The best architecture discovered with Mnist

From Fig. 4 The accuracy of training and testing increases with the number of epochs, this reflects that with each epoch the model learns more information. If the precision is reduced then we will need more information to teach our model and therefore we must increase the number of epochs and vice versa. Likewise, the learning and validation error decreases with the number of epochs. We also notice that the total misclassified images are 57 images, an error rate of 0.57% and the total well classified images is 9943 an accuracy rate of 99.43%.

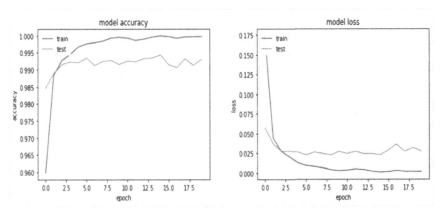

Fig. 4. Accuracy and Error for mnist model Fashion-MNIST dataset

The initial parameters of BayesACO use for optimization with fashionmnist. The ant count is 16 and the maximum number of depth is 15. The epochs number is 20, the batch size equals 64, and the learning rate is equal to 0.1.

The second model that we present in Fig. 5 is composed of three layers of convolution and two layers of averagepooling and a dropout and fully connected layer.

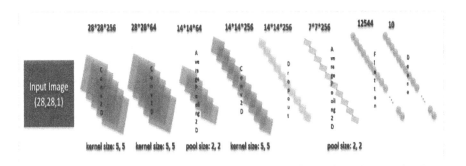

Fig. 5. The best architecture discovered with FashionMnist

Beyond this, we compare the differences between the results of our final method and other methods of improving the convolutional neural network such as Deepswarm, Udeas and LDWPSO.

The variations in results of the algorithms obviously indicate the effectiveness of our proposed methodology in term of cost which has been defined as the value of the test accuracy as it is clarified in the Table 2.

Table 2. Results of optimization methods.

Method	Accuracy	
	Mnist	FashionMnist
Lenet5	99%	81.6%
Resnet18	99.2%	92.1%
XgBoost	95.8%	89.8%
Deepswarm [7]	99.22%	93.4%
LDWPSO [3]	98.95%	–
uDeas [2]	99.1%	–
BayesACO	99.43%	93.8%

We can conclude that our result with the mnsit database is 3.63% higher compared to the xgboost architecture, and higher by 0.48 with the ldwpso optimization method and with the fashionmnsit database the precision obtained is 12.4% higher compared to the lenet5 architecture, and higher by 8.71% with the bayesian optimization method.

5   Conclusion

In this paper, we are interested in integrating the bayesian optimization of hyperparameters in the stages of an existing neural architecture search. This system was developed to optimize the convolutional neural network.

The combination of the hyperparameters optimization with the neural architecture search allows reducing human intervention because the process of extracting the network will become fully automated. Thus, we gain time and give us more accurate results.

As perspectives, we think it is important to run the proposed method on other databases. To evaluate the method in terms of time compared to other competing methods, and to develop this approach using more advanced techniques than those which already exist to obtain better results.

References

1. Zhang, H., et al.: Developing a novel artificial intelligence model to estimate the capital cost of mining projects using deep neural network-based ant colony optimization algorithm. Res. Policy **66**, 101604 (2020)
2. Yoo, Y.J.: Hyperparameter optimization of deep neural network using univariate dynamic encoding algorithm for searches. Knowl.-Based Syst. **178**, 74–83 (2019)

3. Serizawa, T., Fujita, H.: Optimization of convolutional neural network using the linearly decreasing weight particle swarm optimization. arXiv:2001.05670 (2020)

4. Katiyar, S., Ibraheem, N., Ansari, A.Q.: Ant colony optimization: a tutorial review. In: National Conference on Advances in Power and Control, pp. 99–110 (2015)

5. Wu, J., Chen, X.Y., Zhang, H., Xiong, L.D., Lei, H., Deng, S.H.: Hyperparameter optimization for machine learning models based on bayesian optimization. J. Electr. Sci. Technol. **17**(1), 26–40 (2019)

6. Andonie, R.: Hyperparameter optimization in learning systems. J. Membr. Comput., 1–13 (2019)

7. Byla, E., Pang, W.: DeepSwarm: optimising convolutional neural networks using swarm intelligence. In: Zhaojie, J., Yang, L., Yang, C., Gegov, A., Zhou, D. (eds.) Advances in Computational Intelligence Systems: Contributions Presented at the 19th UK Workshop on Computational Intelligence, Portsmouth, UK, 4–6 September, 2019, pp. 119–130. Springer International Publishing, Cham (2020). https://doi.org/10.1007/978-3-030-29933-0_10

8. Dorigo, M., Gambardella, L.M.: Ant colony system: a cooperative learning approach to the traveling salesman problem. IEEE Trans. Evol. Comput. **1**(1), 53–66 (1997)

9. Hutter, F., Kotthoff, L., Vanschoren, J. (eds.): Automated Machine Learning. TSSCML, Springer, Cham (2019). https://doi.org/10.1007/978-3-030-05318-5

10. Mavrovouniotis, M., Yang, S.: Training neural networks with ant colony optimization algorithms for pattern classification. Soft Comput. **19**(6), 1511–1522 (2014)

11. Desell, T., Clachar, S., Higgins, J., Wild, B.: Evolving deep recurrent neural networks using ant colony optimization. In: Ochoa, G., Chicano, F. (eds.) Evolutionary Computation in Combinatorial Optimization, pp. 86–98. Springer International Publishing, Cham (2015). https://doi.org/10.1007/978-3-319-16468-7_8

12. Zhang, X., Chen, X., Yao, L., Ge, C., Dong, M.: Deep neural network hyperparameter optimization with orthogonal array tuning. In: Advances in Neural Information Processing, Vancouver, BC, Canada, pp. 287–295 (2019)

13. MacKay, D.J.C.: Probable networks and plausible predictions—a review of practical Bayesian methods for supervised neural networks. Netw. Comput. Neural Syst. **6**(3), 469–505 (1995)

14. Snoek, J., Larochelle, H., Adams, R.P.: Practical Bayesian optimization of machine learning algorithms. In: Advances in Neural Information Processing Systems, Lake Tahoe, Nevada, pp. 2951–2959 (2012)

15. Eggensperger, K., et al.: Towards an empirical foundation for assessing Bayesian optimization of hyperparameters. In: NIPS Workshop on Bayesian Optimization in Theory and Practice, 10 December 2013

16. Swersky, K., Duvenaud, D., Snoek, J., Hutter, F., Osborne, M.A.: Raiders of the lost architecture: Kernels for Bayesian optimization in conditional parameter spaces. arXiv:1409.4011 (2014)

17. Dorigo, M., Birattari, M., Stutzle, T.: Ant colony optimization. IEEE Comput. Intell. Mag. **1**(4), 28–39 (2006)

18. Lecun, Y., Bottou, L., Bengio, Y., Haffnern, P.: Gradient-based learning applied to document recognition. In: Proceedings of the IEEE (1998)

19. He, K., Zhang, X., Ren, S., Sun, J.: Identity mappings in deep residual networks. In: European Conference on Computer Vision (2016)

20. Ji, S., Xu, W., Yang, M., Yu, K.: 3D convolutional neural networks for automatic human action recognition. US Patent 8,345,984 (2013)

21. He, K., Zhang, X., Ren, S., Sun, J.: Delving deep into rectifiers: surpassing human-level performance on imagenet classification. In: Proceedings of the IEEE (2015)

22. Ren, S., He, K., Girshick, R., Sun, J.: Faster R-CNN: towards real-time object detection with region proposal networks. Adv. Neural Inf. (2015)

23. Angelova, A., Krizhevsky, A., Vanhoucke, V., Ogale, A., Ferguson, D.: Real-time pedestrian detection with deep network cascades (2015)
24. Schroff, F., Kalenichenko, D., Philbin, J.: Facenet: a unified embedding for face recognition and clustering. In: Proceedings of the IEEE (2015)
25. Kim, Y.: Convolutional neural networks for sentence classification. arXiv:1408.5882. (2014)
26. Kalchbrenner, N., Grefenstette, E., Blunsom, P.: A convolutional neural network for modelling sentences. arXiv preprint (2014)
27. Abdel-Hamid, O., Mohamed, A., Jiang, H., Deng, L., Penn, G., Yu, D.: IEEE/ACM Transactions on Audio, Speech, and Language Processing (2014)
28. Bengio, Y.: Learning deep architectures for AI. Found. Trends Mach. Learn. 2(1), 1–127 (2009)
29. Deng, L., Yu, D.: Deep learning: methods and applications. Found. Trends Signal Process. 7(3–4), 197–387 (2014)
30. Jdey, I., Bouhlel, M.S., Dhibi, M.: Comparative study of two decisional fusion techniques: dempester Shafer theory and fuzzy integral theory in radar target recognition. Fuzzy Sets Syst. 241, 68–76 (2014)

End-to-End Approach for Recognition of Historical Digit Strings

Mengqiao Zhao[1], Andre Gustavo Hochuli[2] (ID), and Abbas Cheddad[1(✉)] (ID)

[1] Faculty of Computing, Blekinge Institute of Technology, 371 79 Karlskrona, Sweden
mezh18@student.bth.se, abbas.cheddad@bth.se
[2] Pontifical Catholic University of Parana (PPGIa/PUCPR),
R. Imaculada Conceição 1155, Curitiba, PR 80215-901, Brazil
aghochuli@ppgia.pucpr.pr

Abstract. The plethora of digitalised historical document datasets released in recent years has rekindled interest in advancing the field of handwriting pattern recognition. In the same vein, a recently published data set, known as ARDIS, presents handwritten digits manually cropped from 15.000 scanned documents of Swedish churches' books that exhibit various handwriting styles. To this end, we propose an end-to-end segmentation- free deep learning approach to handle this challenging ancient handwriting style of dates present in the ARDIS dataset (4-digits long strings). We show that with slight modifications in the VGG-16 deep model, the framework can achieve a recognition rate of 93.2%, resulting in a feasible solution free of heuristic methods, segmentation, and fusion methods. Moreover, the proposed approach outperforms the well-known CRNN method (a model widely applied in handwriting recognition tasks).

Keywords: Handwriting digit string recognition · Segmentation-free · Historical document processing

1 Introduction

Due to the rapid growth of document storage in modern society, handwriting recognition has become an important research branch in the field of machine learning and pattern recognition. In the context of handwriting digit strings recognition (HDSR), there are several application scenarios such as postcode recognition, bank checks, document indexing [3] and word spotting [1,4,14].

Most state-of-the-art approaches focus on the segmentation of connected components, followed by training a model to classify each component. However, the performance collapses when two or more digits are touching. Besides, variant approaches based on multiple segmentation algorithms have been proposed. For such algorithms, to generate a potential segmentation cut, a heuristic analysis based on background and foreground information, contour, shape, or a combination of these is required to be implemented [26]. The over-segmentation strategy

© Springer Nature Switzerland AG 2021
J. Lladós et al. (Eds.): ICDAR 2021, LNCS 12823, pp. 595–609, 2021.
https://doi.org/10.1007/978-3-030-86334-0_39

is frequently used to optimize the segmentation of touching components by over segmenting the string. Although over-segmentation increases the probability of obtaining a plausible segmentation cut, it also increases the computational cost since the hypothesis space expands exponentially with the increase of the segmentation cuts. However, the use of heuristic to deal with touching digits has shown that performance degrades by the presence of noise, fragments, and the lack of context in digit strings, such as a lexicon and the string length.

To better address these issues and to take advantage of deep learning models, segmentation-free methods came to the surface, offering unique capabilities to the research community in the domain [2,17,19,23]. Along this line, the proposed approaches rely on implicit segmentation through a deep model or a set of them. Recently, Hochuli et al. [18] evaluated object recognition models into the HSDR context. For that, a digit string is considered a sequence of objects. The advantage is that these models can encode the background, the shape, and the neighbourhood of digits efficiently. Nonetheless, the annotation of digit bounding boxes (ground-truth) is a drawback when synthetic data are not available. Finally, sequence-to-sequence approaches were proposed resulting in feasible end-to-end models [30] for word strings' recognition. The primary strategy is to split the input string into fragments to feed a recurrent model (RNN), and then, a transcription layer determines the resulting string. As stated in [18,23], due to the lack of context, this approach did not achieve outstanding results in HDSR; however, it produces a good trade-off between data annotation, training complexity and accuracy.

An essential aspect of the strategies mentioned above is that most of them have been proposed for modern handwritten digit strings recognition. There is still a lack of approaches in the context of historical (ancient) document recognition. The challenges are different from modern ones, such as paper texture deterioration, noise, ancient handwriting style, ink failure, bleed-through, and the lack of data [20]. Remarkably, the performance of the modern approaches applied to historical document context is a matter of discussion.

Recently, Kusetogullari et al. [20] released to the public the ARDIS dataset, composed of historical handwriting digit strings extracted from the Swedish church record. Their comprehensive analysis reveals that a model trained with modern isolated digits (MNIST, USPS, etc.) fails by a fair margin to correctly encode the isolated digit from ARDIS due to its unique characteristics. However, a comprehensive analysis using the historical digit strings of ARDIS dataset is missing. In light of this, we propose to assess two state-of-the-art approaches by adapting slight modifications into their architecture to better fit the ARDIS digit strings characteristics. Moreover, we introduce data augmentation techniques to represent the classes more efficiently. This work's resulting analysis eventually proposes a baseline for the ARDIS strings dataset and pin-points, which efforts are needed to implement feasible solutions for historical digit strings recognition. Moreover, it highlights the research gaps for further investigations.

The rest of this document is organized as follows: The related work is presented in Sect. 2. The assessed approaches are described in Sect. 4, and then

Sect. 5 provides a discussion around the experiments. Section 6 presents the conclusion and future directions.

2   Related Work

We surveyed the state-of-art and divided the related work into two main categories: (a) Segmentation-based and (b) Segmentation-free approaches. The related work is discussed in the following paragraphs.

Segmentation-Based Approaches: By segmenting the connected components as much as possible, we attain the concept of over-segmentation, which is the most commonly used strategy. It maximizes the probability of generating the optimal segmentation point. However, as mentioned earlier, it also increases the hypothesis space, resulting in a higher computational cost to classify all the candidates compared with a strategy based on single segmentation.

An implicit filter is proposed by Vellasques et al. [32] to reduce the computational cost of over-segmentation, in which a Support Vector Machine (SVM) classifier is used to determine whether the cut produces reliable candidates. The proposed filter succeeded in eliminating up to 83% of unnecessary segmentation cuts in their experimental results.

In Roy et al. [27], a segmentation-based approach is devised to segment out destination address block for postal applications; a review of postal and check processing applications is warranted in [16].

Besides the strategies mentioned above, several segmentation algorithms were proposed in the last decade. Ribas et al. [26] assessed most of them considering their performance, computational cost, touching types, and complexity. This characterization aimed to identify the limitations of the algorithms based on a given pair of touching digits. Moreover, the work reveals that most of the heuristic segmentation strategies are biased towards the characteristic of the dataset's characteristic under scrutiny; thus, a suitable method that works for all touching types is impractical.

Segmentation-Free Approaches: To the best of our knowledge, the first attempt along this line was proposed by Matan et al. [22]. A convolutional neural network (CNN) based model is displaced from left to right over the input. The proposed approach is termed SDNN (Space Displacement Neural Network), which reported 66% of correct classification on 3000 images of ZIP Codes. LeCun et al. [21], stated that SDNN is an attractive technique but has not yielded better results than heuristic segmentation methods.

Years later, Choi and Oh [7] presented a modular neural network composed of 100 sub-networks trained to recognize 100 classes of touching digits (00..99). The recognition rate of 1374 pairs of digits extracted from the NIST database reaches 95.3%. A similar concept was presented by Ciresan [10], in which 100-class CNN was trained with 200,000 images reporting a recognition rate of 94.65%.

An image-based sequence recognition was proposed by Shi et al. [29]. The end-to-end framework, named Convolutional Recurrent Neural Network

(CRNN), naturally handles sequences in arbitrary lengths without character segmentation or horizontal scale normalization. The approach achieved outstanding performance on recognising the scene text (text in the wild) and music scores.

To make handwriting digit recognition less dependent on segmentation algorithms, Hochuli et al. [19] proposed a segmentation-free framework based on a dynamic selection of classifiers. The authors postulate that a set of convolutional neural networks trained to (a) predict the size of touching components and (b) specific-task models to recognize up to three touching digits performs better than if the digits were segmented. However, this algorithm's generalisation to other datasets needs further verification since there is a lack of diversity of the used datasets in the experimental protocol.

Cheng et al. [5] proposed a strategy based on the improved VGG-16 model to overcome the lack of texture features in handwriting digit recognition. The model was examined on the extended MNIST dataset, eventually achieved a high accuracy of 99.97%, which indicates that this VGG-based model has a robust feature extraction ability than traditional classifiers and can meet the requirements of handwriting digits classification and recognition. Besides, a VGG-like model with multiple sub classifiers was built to recognize CAPTCHAs. Although the CAPTCHA images for the test are featured by a lot of noise and touching digits, the model accuracy reached 98.26% without any pre-segmentation.

End-to-end approaches are frequently proposed in recent years, including those tackling writer identification [8] and document analysis and recognition [24, 25]. Recently, approaches based on object recognition models have been exploited with the HDSR task [17,18]. Considering that a string is a sequence of objects, these models can efficiently encode the background, shape, and neighbourhood of digits, providing an end-to-end solution for the problem. Additionally, they reduce the restrictions imposed on the number of touching components or string length. However, the annotation of each digit bounding box (ground-truth) is a bottleneck when synthetic data are not applicable.

3 ARDIS Dataset

The Arkiv Digital Sweden (ARDIS) Dataset comprises historical handwriting digit strings extracted from Swedish church document images written by different priests from 1895 to 1970. The dataset is fully annotated [20], including the digits bounding-boxes [17]. The sub-datasets of ARDIS are exemplified in Fig. 1.

The dataset (I) is composed of 4-digit strings that represent the year of a record. Most of the samples were cropped with the size of 175 x 95 pixels from the document image and stored in its pristine RGB colour space. The dataset (I) contains 75 classes mapping to different years. However, due to insufficient samples, classes later than 1920 were not considered in this work. The class distribution used in this study is shown in Fig. 2.

The historical digit strings pose several challenges to classification, including variations in terms of variability of handwriting styles, touching digits, ink

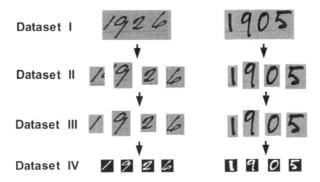

Fig. 1. Examples from the different representations of ARDIS subsets [20].

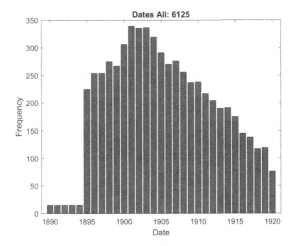

Fig. 2. Class distribution of ARDIS Strings[20]. Classes later than 1920 were not considered in this work due to an insufficient number of samples.

failure, and noisy handwriting. Figure 3 demonstrates some of the aforementioned challenges. Moreover, as observed in Fig. 2, the classes are not equally distributed.

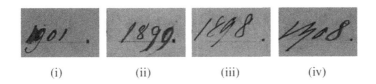

|(i)|(ii)|(iii)|(iv)|

Fig. 3. Challenging samples in ARDIS Dataset (I) representing noisy handwritten digits (i) (ii), ink failure (iii) and touching digits (iv).

The dataset (II) comprises cropped digits from the original digit strings, containing artefacts and fragments of the neighbour digits. In dataset (III), the isolated digits were manually cleaned. For completeness, a uniform distribution of each digit's occurrences was ensured in the dataset (III), resulting in 7600 de-noised digit images in RGB colour space. For dataset (IV), the isolated digits are normalised and binarised.

3.1   Synthetic Data

Recently, data augmentation techniques in handwriting digit were proposed to generate a synthetic training set [2, 19, 30]. In order to improve the data representation of ARDIS strings (Dataset I), we propose the creation of synthetic data by permuting and concatenating several single digits from dataset III. Figure 4 depicts two synthetic samples. Although both digit strings belong to the label "1987", the representation (e.g., style) is remarkably different.

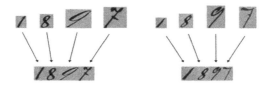

Fig. 4. Synthetic strings creation from isolated digits of dataset III.

We combine synthetic data and real data from up to 1000 samples for each class during the training phase to balance the distribution of classes. The number of real and synthetic data are summarized in Table 1. Considering the limited amount of data in the ARDIS data set, it is relatively reasonable (closer to standard practice) to retain about 43% of the real data for testing $(2651/(3474 + 2651))$ and use the remaining 57% of the real data set for training. It is worth mentioning that the synthetic data was used only for training, which represents 88.8% of the training samples. In total, 31000 and 2651 images are used for training and testing, respectively. All the data used in this work are already publicly available through the ARDIS Website[1] by the authors of [20].

4   Approaches for Historical Handwriting Digit String Recognition

As stated by [26] and [19], the segmentation problem has been overcome by segmentation-free approaches in the recent advances in deep learning models. In light of this, we propose to evaluate two segmentation-free approaches on the ARDIS dataset to tackle the task of historical handwriting digit string

---

[1] https://ardisdataset.github.io/ARDIS/.

Table 1. Training protocol using synthetic and real data

Label	Training samples		Test samples
	Synthetic	Real	
1890	1000	0	15
1891	1000	0	15
1892	1000	0	15
1893	1000	0	15
1894	1000	0	15
1895	875	125	100
1896	846	154	100
1897	846	154	100
1898	825	175	100
1899	833	167	100
1900	794	206	100
1901	761	239	100
1902	765	235	100
1903	764	236	100
1904	781	219	100
1905	809	191	100
1906	830	170	100
1907	824	176	100
1908	844	156	100
1909	863	137	100
1910	862	138	100
1911	883	117	100
1912	896	104	100
1913	910	90	100
1914	909	91	100
1915	925	75	100
1916	955	45	100
1917	962	38	100
1918	983	17	100
1919	981	19	100
1920	1000	0	76
Total	27526	3474	2651

recognition. The first approach (Sect. 4.1) is based on the well-known VGG-16 model. The second one (Sect. 4.2) is based on a sequence-to-sequence model.

Motivations on the Choice of Models: Given the VGG-16 model's decent performance in other character recognition tasks, we consider it the baseline of this experiment to evaluate against other alternative models. The VGG network model [31] was proposed by the Visual Geometry Group (VGG) at Oxford University. When first created, the focus of this network was to classify materials by their textural appearance and not by their colour. Due to the excellent generalisation performance of VGG-Net, its pre-trained model on the ImageNet dataset is widely used for feature extraction problems [9,13] such as: object candidate frame (object proposal) generation [15], fine-grained object localization, image retrieval [34], image co-localization [35], etc. On the other hand, our new approach is based on modifying the concept of CRNN [29]. The CRNN is mainly used for end-to-end recognition of indefinite length text sequence. It does not require pre-segmentation on long continuous text.

4.1   Specific-Task Classifiers

In this new approach, we adapted the well-known VGG-16 model to the context of the ARDIS dataset, which is composed of 4-digit strings. Instead of using a dynamic selection of classifiers [19], we proposed to parallelize the classification task by adding four dense layers (classifiers) on the bottom of the architecture. The final architecture is depicted in Fig. 5. With this simple modification, we produce an end-to-end pipeline avoiding both the heuristic segmentation and fusion methods.

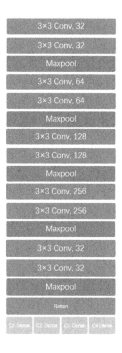

Fig. 5. Structure of the 4 classifiers CNN framework

The rationale here is that each specific-task classifier (C_1, C_2, C_3, and C_4) should determine the ten classes (0..9) for each digit of the 4-digit string. The prediction of input digit string is defined as follows:

Let $\mathcal{C}(x) = \max_{0 \leq i \leq 9} p^i(x)$ be the probability produced by the digit classifier (10-classes). Then, an input digit x is assigned to the class ω_j ($j = 0...9$) according to Eq. 1.

$$P(\omega_j | x) = \max(\mathcal{C}) \tag{1}$$

Considering that the input image I contains $n = 4$ digits, the most probable interpretation of the written amount M is given by Eq. 2.

$$P(M|I) = \prod_{i=1}^{n} P_i(\omega_j | x_i) \tag{2}$$

4.2 CRNN

A Convolutional Recurrent Neural Networks (CRNN) [12,30,33] is a sequence-to-sequence model that can be trained from end to end. The pipeline for such a network is depicted in Fig. 6a. First, convolutional layers extract features from an input image, and then a sequence of feature vectors is extracted from feature maps.

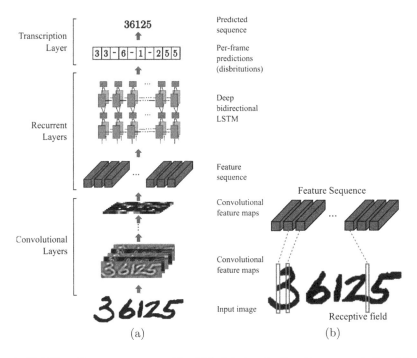

Fig. 6. CRNN architecture proposed by [30]: (a) the pipeline from convolutional layers to transcription layer and (b) the receptive field for each feature vector.

Since each region of the feature map is associated with a receptive field in the input image, each vector in the sequence is a descriptor of this image field, as illustrated in Fig. 6b. Next, this sequence is fed to the recurrent layers, which are composed of a bidirectional Long-Short Term Memory (LSTM) [28] network, producing a per-frame prediction from left to right of the image. Finally, the transcription layer determines the correct sequence of classes to the input image by removing the repeated adjacent labels and the blanks, represented by the character '-'. This solution is well suited when the past and future context of a sequence contributes to recognising the whole input. With the aid of contextual information, such as a lexicon, this approach achieves high text recognition performance. The application of this solution to handwriting digits is a matter of discussion since we have fewer classes (0..9) as compared to words, but there is no lexicon to mitigate possible confusion.

To address the context of historical digit string recognition, we propose a modification of the Recurrent Layers Architecture. Due to the lack of data, we replaced the LSTM with a Gated Recurrent Unit (GRU) [6]. Since the latter has fewer parameters, besides reducing training time, the vanishing gradient's impact is minimised. Moreover, we combined two identical GRU Layers to process the feature maps from left to right and vice-versa, and then, the output of both GRUs are merged. It is worth mentioning that the feature maps fed to the GRU are reshaped to vectors to provide a sequence of information. Further, another two identical GRUs repeat the process; however, their outputs are concatenated. Finally, a fully connected layer determines the class probabilities, and the connectionist temporal classification (CTC) layer determines the final prediction. The architecture and characterisation of our modified CRNN approach are depicted in Fig. 7.

5   Experiments

In this section, we assess all reported models in the context of HDSR. We also added to the comparison the native VGG16 model (i.e., without our modification). All metrics used to measure the performance are described in Sect. 5.1. The training protocol is presented in Sect. 5.2. Finally, the results are discussed in Sect. 5.3.

5.1   Evaluation Metrics

To better assess the performance of the proposed approaches, besides the well-known accuracy and F1-score, we propose using the Normalized Levenshtein Distance (NLD) and the Average Normalized Levenshtein Distance (ANLD).

The Normalized Levenshtein Distance (NLD) [11] describes how close a predicted string is from the ground truth by eliminating the influence of string length on performance measurement. The NLD can be defined as follow:

$$NLD(a_T, a_R) = \frac{LD(a_T, a_R)}{|a_T|} \tag{3}$$

where $|a_T|$ presents the length of the string a_T and LD represents the Levenshtein distance, which refers to the minimum number of editing operations of each character (insert, delete, and substitute) to convert from one string to another. A zero value indicates a correct prediction.

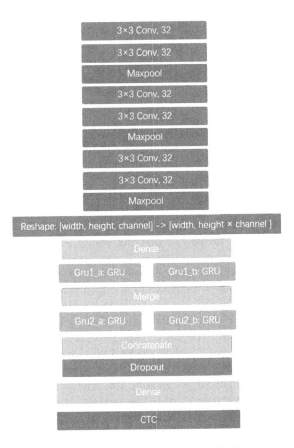

Fig. 7. CRNN architecture for HDSR.

Complementary to this, the Average Normalized Levenshtein Distance (ANLD) is a soft metric to evaluate the model performance:

$$ANLD = \frac{\sum_{i=1}^{T} NLD\left(a_T^i, a_R^i\right)}{T} \tag{4}$$

where T indicates the number of evaluated strings. Lower values represent a good performance.

5.2   Training

The models were fine-tuned using the data described in Table 1, which comprises real and synthetic data that sums up to 1000 training samples for each one of 31 classes (number of years in the selected period). All the classifiers were trained up to 100 epochs, and the over-fitting was prevented through early-stopping when no convergence occurs after ten epochs. The training parameters are summarized in Table 2.

Table 2. Training parameters

Frameworks	Batch size	Loss function	Reg.	Opt.	Learning rate	Batch norm
Specific-task	32	Cat. crossentropy	Dropout (0.25)	Adam	10^{-3}	TRUE
CRNN	128	CTC	Dropout (0.25)	Adadelta	10^{-3}	FALSE
VGG-16	32	Cat. crossentropy	L2 $(5*10^{-4})$	Adam	10^{-5}	FALSE

5.3   Results and Discussion

The performance of the evaluated models in this work is reported in Table 3. As described in Sect. 4.1, we modified the last layer of VGG-16 architecture by adding four classifiers instead of one. As stated in Table 3, this modification achieved the best performance since it can encode the string by implicit segmenting the digits with the specific-task classifiers. Comprehensive analysis reveals that the whole image's information is difficult to encode in several cases by only one classifier (VGG-16) due to the handwriting variability. For example, let us assume that we have the following strings "1890" and "1819". From a computational perspective, the global representation poses challenges in discriminating the digit '0' and '9' that mislead the classifier. However, for the specific-task approach, the information can be implicitly segmented according to each classifier's domain space. Regarding this, the models C_1 and C_2 exhibit a reduced complexity when compared to the models C_3 and C_4 since the former two classifiers need to discriminate fewer classes ([1,8,9]).

Table 3. Comparisons among the three frameworks in terms of different metrics

Frameworks	Accuracy	F1-score	ANLD
VGG-16	36.2	38.5	28.7
CRNN	85.0	83.5	5.8
Specific-task	93.2	90.8	2.3

Regarding the CRNN, we believe that the model suffers due to a lack of context. Contrary to the word recognition, the CTC layer missed the prediction since there is no lexicon to mitigate some confusions, such as fragment recognition and repeated labels. The issue is quite similar to the over-segmentation.

6   Conclusion and Future Work

In this work, we explored the recognition of historical digit strings. Based on this context, an image-based dataset containing 31 classes representing handwriting years ranging from 1890 to 1920 is utilised.

To this end, we proposed to evaluate three models implementing end-to-end solutions. The use of synthetic data was employed to overcome the lack of data.

The proposed approach that combines four specific-task classifiers achieved outstanding results. This promising performance can be explained by the implicit segmentation of the input string made by the domain space of each specific-task classifier. On the other hand, the approach based on a single classifier suffers due to the handwriting variability represented in a global perspective. Regarding the CRNN approach, it suffers from the lack of lexicon, as also stated in [18].

For future endeavours, once we have context information about the first and second digits, a reduced number of classifiers for the specific-task approach could be examined. Also, we will investigate a dynamic approach on VGG-16 that can implicitly handle different length of strings.

Acknowledgments. This project is supported by the research project "DocPRE-SERV: Preserving and Processing Historical Document Images with Artificial Intelligence", STINT, the Swedish Foundation for International Cooperation in Research and Higher Education (Grant: AF2020-8892).

References

1. Almazan, J., Gordo, A., Fornés, A., Valveny, E.: Word spotting and recognition with embedded attributes. IEEE Trans. Pattern Anal. Mach. Intell. **36**(12), 2552–2566 (2014). https://doi.org/10.1109/TPAMI.2014.2339814
2. Aly, S., Mohamed, A.: Unknown-length handwritten numeral string recognition using cascade of PCA-SVMNET classifiers. IEEE Access **7**, 52024–52034 (2019). https://doi.org/10.1109/ACCESS.2019.2911851
3. Cecotti, H.: Active graph based semi-supervised learning using image matching: application to handwritten digit recognition. Pattern Recogn. Lett. **73**, 76–82 (2016)
4. Cheddad, A.: Towards query by text example for pattern spotting in historical documents. In: 2016 7th International Conference on Computer Science and Information Technology (CSIT), pp. 1–6 (2016). https://doi.org/10.1109/CSIT.2016. 7549479
5. Cheng, S., Shang, G., Zhang, L.: Handwritten digit recognition based on improved vgg16 network. In: Tenth International Conference on Graphics and Image Processing (ICGIP 2018), vol. 11069, p. 110693B. International Society for Optics and Photonics (2019)
6. Cho, K., van Merrienboer, B., Gulcehre, C., Bougares, F., Schwenk, H., Bengio, Y.: Learning phrase representations using RNN encoder-decoder for statistical machine translation. In: Conference on Empirical Methods in Natural Language Processing (EMNLP 2014) (2014)

7. Choi, S.M., Oh, I.S.: A segmentation-free recognition of handwritten touching numeral pairs using modular neural network. Int. J. Pattern Recogn. Artif. Intell. **15**(06), 949–966 (2001)
8. Cilia, N., De Stefano, C., Fontanella, F., Marrocco, C., Molinara, M., Di Freca, A.S.: An end-to-end deep learning system for medieval writer identification. Pattern Recogn. Lett. **129**, 137–143 (2020). https://doi.org/10.1016/j.patrec.2019.11.025, https://www.sciencedirect.com/science/article/pii/S0167865519303460
9. Cimpoi, M., Maji, S., Vedaldi, A.: Deep filter banks for texture recognition and segmentation. In: 2015 IEEE Conference on Computer Vision and Pattern Recognition (CVPR), pp. 3828–3836 (2015). https://doi.org/10.1109/CVPR.2015.7299007
10. Ciresan, D.: Avoiding segmentation in multi-digit numeral string recognition by combining single and two-digit classifiers trained without negative examples. In: 2008 10th International Symposium on Symbolic and Numeric Algorithms for Scientific Computing, pp. 225–230. IEEE (2008)
11. Diem, M., et al.: ICFHR 2014 competition on handwritten digit string recognition in challenging datasets (HDSRC 2014). In: 2014 14th International Conference on Frontiers in Handwriting Recognition, pp. 779–784. IEEE (2014)
12. Dutta, K., Krishnan, P., Mathew, M., Jawahar, C.V.: Improving CNN-RNN hybrid networks for handwriting recognition. In: 2018 16th International Conference on Frontiers in Handwriting Recognition (ICFHR), pp. 80–85, August 2018. https://doi.org/10.1109/ICFHR-2018.2018.00023
13. Gao, B.B., Wei, X.S., Wu, J., Lin, W.: Deep spatial pyramid: the devil is once again in the details. arXiv preprint arXiv:1504.05277 (2015)
14. Gao, Y., Mishchenko, Y., Shah, A., Matsoukas, S., Vitaladevuni, S.: Towards data-efficient modeling for wake word spotting. In: ICASSP 2020–2020 IEEE International Conference on Acoustics, Speech and Signal Processing (ICASSP), pp. 7479–7483 (2020). https://doi.org/10.1109/ICASSP40776.2020.9053313
15. Ghodrati, A., Diba, A., Pedersoli, M., Tuytelaars, T., Gool, L.V.: DeepProposals: hunting objects and actions by cascading deep convolutional layers. Int. J. Comput. Vis. **124**(2), 115–131 (2017). https://doi.org/10.1007/s11263-017-1006-x
16. Gilloux, M.: Document analysis in postal applications and check processing. In: Doermann, D., Tombre, K. (eds.) Handbook of Document Image Processing and Recognition, pp. 705–747. Springer, London (2014). https://doi.org/10.1007/978-0-85729-859-1_22
17. Hochuli, A.G., Britto, A.S., Barddal, J.P., Sabourin, R., Oliveira, L.E.S.: An end-to-end approach for recognition of modern and historical handwritten numeral strings. In: 2020 International Joint Conference on Neural Networks (IJCNN), pp. 1–8 (2020). https://doi.org/10.1109/IJCNN48605.2020.9207468
18. Hochuli, A.G., Britto, A.S., Jr., Saji, D.A., Saavedra, J.M., Sabourin, R., Oliveira, L.S.: A comprehensive comparison of end-to-end approaches for handwritten digit string recognition. Expert Syst. Appl. **165**, 114196 (2021)
19. Hochuli, A.G., Oliveira, L.S., Britto, A., Jr., Sabourin, R.: Handwritten digit segmentation: Is it still necessary? Pattern Recogn. **78**, 1–11 (2018)
20. Kusetogullari, H., Yavariabdi, A., Cheddad, A., Grahn, H., Hall, J.: ARDIS: a Swedish historical handwritten digit dataset. Neural Comput. Appl. **32**, 16505–16518 (2019)
21. LeCun, Y., Bottou, L., Bengio, Y., Haffner, P.: Gradient-based learning applied to document recognition. Proc. IEEE **86**(11), 2278–2324 (1998)
22. Matan, O., Burges, C.J., LeCun, Y., Denker, J.S.: Multi-digit recognition using a space displacement neural network. In: Advances in Neural Information Processing Systems, pp. 488–495 (1992)

23. Neto, A.F.D.S., Bezerra, B.L.D., Lima, E.B., Toselli, A.H.: HDSR-Flor: a robust end-to-end system to solve the handwritten digit string recognition problem in real complex scenarios. IEEE Access **8**, 208543–208553 (2020)

24. Neudecker, C., et al.: OCR-D: an end-to-end open source OCR framework for historical printed documents. In: Proceedings of the 3rd International Conference on Digital Access to Textual Cultural Heritage, pp. 53–58. DATeCH2019, Association for Computing Machinery, New York (2019). https://doi.org/10.1145/3322905.3322917

25. Palm, R.B., Laws, F., Winther, O.: Attend, copy, parse end-to-end information extraction from documents. In: 2019 International Conference on Document Analysis and Recognition (ICDAR), pp. 329–336 (2019). https://doi.org/10.1109/ICDAR.2019.00060

26. Ribas, F.C., Oliveira, L., Britto, A., Sabourin, R.: Handwritten digit segmentation: a comparative study. Int. J. Doc. Anal. Recogn. (IJDAR) **16**(2), 127–137 (2013)

27. Roy, K., Vajda, S., Pal, U., Chaudhuri, B.B., Belaid, A.: A system for Indian postal automation. In: Eighth International Conference on Document Analysis and Recognition (ICDAR 2005), vol. 2, pp. 1060–1064 (2005).https://doi.org/10.1109/ICDAR.2005.259

28. Schuster, M., Paliwal, K.K.: Bidirectional recurrent neural networks. IEEE Trans. Sig. Process. **45**(11), 2673–2681 (1997). https://doi.org/10.1109/78.650093

29. Shi, B., Bai, X., Yao, C.: An end-to-end trainable neural network for image-based sequence recognition and its application to scene text recognition. IEEE Trans. Pattern Anal. Mach. Intell. **39**(11), 2298–2304 (2017). https://doi.org/10.1109/TPAMI.2016.2646371

30. Shi, B., Bai, X., Yao, C.: An end-to-end trainable neural network for image-based sequence recognition and its application to scene text recognition. IEEE Trans. Pattern Anal. Mach. Intell. **39**(11), 2298–2304 (2016)

31. Simonyan, K., Zisserman, A.: Very deep convolutional networks for large-scale image recognition. In: International Conference on Learning Representations (2015)

32. Vellasques, E., Oliveira, L.S., Britto, A., Jr., Koerich, A.L., Sabourin, R.: Filtering segmentation cuts for digit string recognition. Pattern Recogn. **41**(10), 3044–3053 (2008)

33. Voigtlaender, P., Doetsch, P., Ney, H.: Handwriting recognition with large multidimensional long short-term memory recurrent neural networks. In: 2016 15th International Conference on Frontiers in Handwriting Recognition (ICFHR). pp. 228–233, October 2016. https://doi.org/10.1109/ICFHR.2016.0052

34. Wei, X.S., Luo, J.H., Wu, J., Zhou, Z.H.: Selective convolutional descriptor aggregation for fine-grained image retrieval. Trans. Img. Proc. **26**(6), 2868–2881 (2017)

35. Wen, W., Wu, C., Wang, Y., Chen, Y., Li, H.: Learning structured sparsity in deep neural networks. In: Proceedings of the 30th International Conference on Neural Information Processing Systems, NIPS 2016, pp. 2082–2090. Curran Associates Inc., Red Hook (2016)

Generating Synthetic Handwritten Historical Documents with OCR Constrained GANs

Lars Vögtlin[✉], Manuel Drazyk, Vinaychandran Pondenkandath, Michele Alberti, and Rolf Ingold

Document Image and Voice Analysis Group (DIVA), University of Fribourg, Fribourg, Switzerland
{lars.vogtlin,manuel.drazyk,vinaychandran.pondenkandath, michele.alberti,rolf.ingold}@unifr.ch

Abstract. We present a framework to generate synthetic historical documents with precise ground truth using nothing more than a collection of unlabeled historical images. Obtaining large labeled datasets is often the limiting factor to effectively use supervised deep learning methods for Document Image Analysis (DIA). Prior approaches towards synthetic data generation either require human expertise or result in poor accuracy in the synthetic documents. To achieve high precision transformations without requiring expertise, we tackle the problem in two steps. First, we create template documents with user-specified content and structure. Second, we transfer the style of a collection of unlabeled historical images to these template documents while preserving their text and layout. We evaluate the use of our synthetic historical documents in a pre-training setting and find that we outperform the baselines (randomly initialized and pre-trained). Additionally, with visual examples, we demonstrate a high-quality synthesis that makes it possible to generate large labeled historical document datasets with precise ground truth.

Keywords: OCR · CycleGAN · Synthetic data · Historical documents

1 Introduction

Large labeled datasets play a major role in the significant performance increases seen in DIA and computer vision over the last decade. These datasets – often containing millions of labeled samples – are typically used to train deep neural networks in a supervised setting, achieving state-of-the-art performance in tasks such as text line segmentation [2], Optical Character Recognition (OCR) [4] or layout analysis [24]. However, such methods are much more challenging to train in settings where no labeled data is available. The size of labeled datasets is limited to a few hundred or thousand samples – as is often the case with historical documents [7,14].

L. Vögtlin and M. Drazyk—Both authors contributed equally to this work.

© Springer Nature Switzerland AG 2021
J. Lladós et al. (Eds.): ICDAR 2021, LNCS 12823, pp. 610–625, 2021.
https://doi.org/10.1007/978-3-030-86334-0_40

     (a)   (b)   (c)

Fig. 1. Inputs for the second step of our framework and the output of the network. (a) represents the style template for our output document. (b) shows a source document that was generated using LATEX. (c) shows the corresponding transformed version of the template image (b). The transformation between (b) and (c) preserves overall structure and content.

Common strategies to deal with limited labeled data include (1) transfer-learning, (2) synthesizing artificial data, or (3) unsupervised learning. In (1) typical procedure is to train a deep neural network on similar data and then fine-tune this network on the small labeled target dataset. The success depends on having datasets similar enough to the target dataset to perform pre-training. (2) has been an active area of DIA research. Baird [3], Kieu et al. [16] and Seuret et al. [25] focus on degrading real document images using defect models to augment datasets. Other tools such as DocEmul [5] and DocCreator [14] aim to create synthetic document images using a combination of user-specified structure, background extraction, degradation methods, and other data augmentation approach. However, such approaches still require human expertise in designing appropriate pipelines to generate realistic documents. When large unlabeled datasets are available for the target task, a common practice is to use unsupervised learning methods such as autoencoders [20] to learn representations. However, recent work [1] shows that autoencoders trained for reconstruction are not useful for this task. Another possibility is to use unlabeled data in a Generative Adversarial [10,32] setting to synthesize artificial data that looks similar in appearance to the unlabeled data.

More recent work in document image synthesis has used deep learning, and Generative Adversarial Network (GAN) based approaches. But these approaches [12,15,23,29] result in various issues: the produced data matches the overall visual style of historical documents but fail to produce meaningful textual content; they require paired datasets, which defeats the purpose of using unlabeled data; only create text of fixed length.

In this paper, we present a framework to generate historical documents without relying on human expertise or labeled data[1]. We approach this problem in two steps. First, we create template document images that contain user-specified content and structure using LaTeX[2]. Second, using the user-specified template documents and a collection of unlabeled historical documents, we learn a mapping function to transform a given template document into the historical style while preserving the textual content and structure present in the template document.

We evaluate the usefulness of our synthetically generated images by measuring the performances of a deep learning model on the downstream task of OCR. Specifically, we measure the performances of this model when (1) trained only on the target dataset (St. Gall [9]), (2) pre-trained on a similar dataset (IAM Handwriting database [19]) and then fine-tuned on the target dataset and finally when (3) pre-trained on our synthetic images and then fine-tuned on the target dataset. This will allow us to compare against a standard supervised baseline as well as a reasonable transfer learning baseline. Our empirical experiments show that, the model pre-trained on our synthetic images (see point 3 above) is outperforming the supervised and transfer learning baselines by 38% and, respectively, 14% lower Character Error Rate (CER).

Main Contribution

This paper extends the existing work on synthetic document generation by providing a general framework that can produce realistic-looking historical documents with a specific style and textual content/structure. We introduce a two-step CycleGAN based process that leverages two Text Recognizer (TR) networks to condition the learning process. This additional signal let us overcome the main limitations of previous work and enable us to obtain significantly better performance measured on a robust set of benchmarks.

2   Datasets

In this work, we use three datasets: the user-specified template document dataset (source domain dataset; see Sect. 2.1); a dataset of real unlabeled historical documents (target domain dataset) whose style which we want to learn in the transformation function; a dataset of real labeled historical documents (evaluation dataset) with transcription ground truth that we use to evaluate our methods.

2.1   Source Domain Dataset

We create a collection of template documents with user-specified content and structure as Pondenkandath et al. [23]. Our template document images are generated based on the specifications from LaTeX files; they define the layout, font,

---

[1] https://github.com/DIVA-DIA/Generating-Synthetic-Handwritten-Historical-Documents.

[2] This can be done with any other word processing tool such as MS Word.

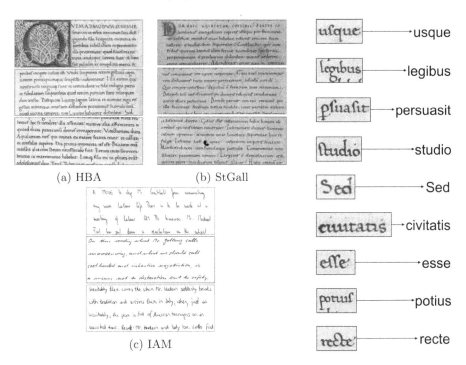

(a) HBA   (b) StGall

(c) IAM

Fig. 2. We use the HBA dataset (a) as the target historical style, the Saint Gall dataset (b) for evaluating our synthetic data in a pre-training setting and the IAM Handwriting Database (c) as a pre-trained baseline.

Fig. 3. Samples from the hand annotated subset of the HBA dataset used for validation purposes.

size, and content (see Fig. 1). As a text, we use the *Bellum Gallicum* [8] with a one or two-column layout. Additionally, we populated each document with different decorative starting letters. The advantage of this technique is that we have very precise ground truth, which is the transcription of the document and the exact position of the word on the page. This dataset contains 455 document images with a resolution of 2754×3564.

2.2 Target Domain Dataset

The target domain dataset refers to the collection of historical documents whose style we aim to learn in the transformation function. To create this dataset, we use the historical documents present in the Historical Book Analysis Competition (HBA) 1.0 dataset [21]. The HBA dataset comprises 11 books, where 5 are manuscripts and 6 books are printed, containing 4436 scanned historical document images. These books were published between the 13th and 19th centuries and are written in different scripts and languages. We use one book of this dataset; the handwritten Latin book "Plutarchus, Vitae illustrium viro-rum". This book contains 730 colored pages with a resolution of 6158×4267 (see Fig. 2a) from which we filtered out 120 pages (blank, binding, and title pages), leaving us with 600 pages that contain only text. To validate the best model for our downstream evaluation task, we hand-labeled 350 individual word crops from this book.

2.3   Evaluation Dataset

As part of the evaluation process, we use two different datasets. Our evaluation protocol involves pre-training a Handwritten Text Recognition (HTR) model using synthetic data generated using our method, and then evaluating it in a fine-tuning setting on the St. Gall dataset [9] (see Fig. 2b). The Saint Gall dataset includes 40 pages of labeled historical handwritten manuscripts containing 11'597 words and 4'890 word labels. Each image has a resolution of 3328×4992 with quality of 300 dpi.

To compare our synthetic data pre-training against pre-training on a real handwritten dataset, we pre-train an HTR model on the IAM Handwriting Database [19] (see Fig. 2c). This HTR model (pre-trained on the IAM Hand-writing Database) is then evaluated similarly in a fine-tuning setting on the St. Gall dataset. The IAM Handwriting Database contains 1'539 handwritten scanned pages with 115'320 words and 6'625 unique words. The word images are normalized and in black-white colorization.

3   Method

Our method uses a CycleGAN formulation, along with HTR models, to further constrain the synthesis process. To train the CycleGAN, we use unpaired col-lections of user-specified template images (source domain) and real historical images (target domain). The source domain documents specify the content and overall structure, and the target domain documents exemplify the style we want in our final synthetic historical documents.

Pondenkandath et al. [23] have shown that using only the CycleGAN formu-lation with the source and target domain datasets is enough to produce synthetic documents that appear stylistically similar to the target domain. However, they do not contain the content or structure specified in the source domain docu-ments. We add a loss term using HTR models that aim to read user-specified content from the synthesized historical documents to address this issue. After completing training, we obtain a generator that transforms any given template image to a corresponding synthetic historical version.

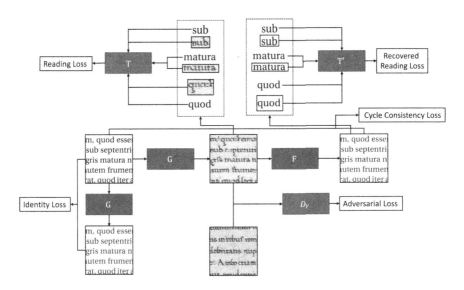

Fig. 4. The CycleGAN architecture presented in [32] with two additional TRs T and T' and the five different loss terms.

3.1 Model Architecture

Our model architecture is based on the CycleGAN formulation. It uses the cycle-consistency loss to transform an image from a given source domain to a target domain in a bi-directional fashion. This architecture introduces two main challenges. First, generating text in the target domain that is human-readable at the character and word levels is difficult due to the under-constrained nature of the CycleGAN architecture for our task. Second, CycleGANs are prone to emergent GAN steganography [31]; where the generators in a GAN can learn to hide information from the discriminator within the synthesized image and use it for perfect reconstruction.

To tackle the first problem of generating human-readable text, we introduce two HTR models T and T' to our architecture (see Fig. 4). We aim to adjust for the under-constrained nature of the CycleGAN by adding additional loss terms based on this HTR model. We adopt the bi-directional Long Short-Term Memory (LSTM) and Connectionist Temporal Classification (CTC) based HTR architecture used by the winners of the text recognition competition at ICFHR'18 [26].

The first HTR model T evaluates the quality of the characters or words produced by transforming a source domain template image to the target historical domain. To do this, it takes as input the synthetic images produced by the source-to-target generator G as well as the textual content and location information from the template document images. The second HTR model T' evaluates the quality of the reconstructed source domain documents (produced by the target-source generator F) by comparing the reconstructed image against the same textual content and location information as T.

The second challenge is overcoming the tendency of CycleGANs to hide information about the input within the generated synthetic image [6]. This tendency arises naturally due to cooperation between the generators and is potentially exacerbated by the presence of the HTR models. To minimize the cyclic consistency loss as well as the loss introduced by the TRs, the generator G attempts to hide information that can be effectively decoded by generator F to produce good reconstructions, as well as information that allows the TRs to recover the textual content. These results in synthetic documents that do not satisfy the constraints of our synthesis process, yet produce very low reconstruction losses and HTR losses. In some of our preliminary experiments, the generator places the encoded template document into the target document by adding or subtracting the encoded value from each pixel. The influence on the image is so small that it is nearly impossible for humans to detect, and it is even challenging to be detected by the style discriminator. Allowing the CycleGAN to cheat prevents it from learning the correct mapping from the target domain back to the source domain, negatively affecting the style representation learned by the GAN.

To prevent the CycleGAN from creating this hidden embedding, we add Gaussian noise to the synthetic document images. This low-frequency noise disturbs the encoded message of the generator, making it much harder to cheat by using steganography. This noise effectively prevents the network from cheating, as a much stronger signal would be needed, which would manipulate the appearance of the image in a way that is more easily detected by the human eye as well as the style discriminator, and thus would achieve a much lower performance score.

3.2 Loss Functions

We train with a loss objective that consists of five different loss terms (see Fig. 4). The identity loss, the adversarial loss, and the cycle consistency loss are the loss terms presented in the original CycleGAN paper [32]. To solve the readability problem described in Sect. 3.1, we introduce two additional loss terms using the HTRs system, the reading loss, and the recovered reading loss. The identity, adversarial, and cycle consistency loss are calculated in both directions, but the reading loss terms are just calculated once per cycle.

Formally, we aim to learn mappings between two domains X and Y with the help of N training samples $x \in X$ and M samples $y \in Y$. Each document image x is composed of pairs of its word images and the corresponding word text (ground truth) $x = ((x_1, z_1), (x_2, z_2), ..., (x_n, z_n))$ where $n = |x|$ and $|x|$ is the amount of words in a document.

The data distributions are denoted as $x \sim p_{data}(x)$ and $y \sim p_{data}(y)$. We also define a projection function α where $\alpha_1(x)$ refers to the first and $\alpha_2(x)$ to the second element of the tuple. The transformation functions of generators G (source-target) and F (target-source) are denoted respectively by $g : X \to Y$ and $f : Y \to X$. Additionally, we have two adversarial discriminators D_x and

D_y. The task of D_x is to distinguish the images of $\{x\}$ and $\{f(y)\}$, and in the same fashion D_y learns to differentiate between $\{y\}$ and $\{g(x)\}$.

Identity Loss. This loss term [28,32] is used to regularize both generators to function as identity mapping functions when provided with real samples of their respective output domains. Zhu et al. [32] observed that in the absence of this identity loss term, the generators G and F were free to change the tint between the source and target domains even without any need to do it. The identity loss is defined as follows:

$$\mathcal{L}_{\text{identity}}(G, F) = \mathbb{E}_{x \sim p_{\text{data}}(x)}[\|G(\alpha_1(x)) - \alpha_1(x)\|_1]$$
$$+ \mathbb{E}_{y \sim p_{\text{data}}(y)}[\|F(y) - y\|_1]. \tag{1}$$

Adversarial Loss. The adversarial loss [10] shows how well the mapping function g can create images $g(x)$ which looks similar to images in the domain Y, while the discriminator D_y aims to tries to distinguish between images from $g(x)$ and real samples from Y. g tries to minimize this objective against D_y, which tries to maximize it, i.e. $min_g max_{D_y} \mathcal{L}_{\text{GAN}}(g, D_Y, X, Y)$. As we use a CycleGAN, this loss is applied twice, once for g and its discriminator D_y, as well as for f and the discriminator D_x.

$$\mathcal{L}_{\text{GAN}}(g, D_Y, X, Y) = \mathbb{E}_{y \sim p_{\text{data}}(y)}[\log D_Y(y)]$$
$$+ \mathbb{E}_{x \sim p_{\text{data}}(x)}[\log(1 - D_Y(g(\alpha_1(x))))]. \tag{2}$$

Cycle Consistency Loss. The cycle consistency loss [32] further restricts the freedom of the GAN. Without it, there is no guarantee that a learned mapping function correctly maps an individual x to the desired y. Hence, for each pair $(x_i, z_i) \in x$ the cycleGAN should be able to bring the image x_i back into the original domain X, i.e. $x_i \rightarrow g(x_i) \rightarrow f(g(x_i)) \approx x_i$. As the nature of the cycleGAN is bidirectional the reverse mapping must also be fulfilled, i.e. $y \rightarrow f(y) \rightarrow g(f(y)) \approx y_i$.

$$\mathcal{L}_{\text{cyc}}(g, f) = \mathbb{E}_{x \sim p_{\text{data}}(x)}[\|f(g(\alpha_1(x))) - \alpha_1(x)\|_1]$$
$$+ \mathbb{E}_{y \sim p_{\text{data}}(y)}[\|g(f(y)) - y\|_1]. \tag{3}$$

Reading Loss and Recovered Reading Loss. As described in Sect. 3.1 and shown in Fig. 4, we use the reading loss to ensure that the GAN produces readable images, i.e. images containing valid Latin characters. The TRs T and T' are trained with a CTC loss [11,18], which is well suited to tasks that entail challenging sequences alignments.

$$\mathcal{L}_{CTC}(\mathbf{x}, \mathbf{y}) = -\ln p(\mathbf{x}|\mathbf{y}). \tag{4}$$

To calculate the reading loss, the template word text z_i and the corresponding transformed word image $G(x_i)$ is passed to the TRs T and T'. The loss evaluates the mapping g to our target domain Y at a character level.

This discriminator evaluates the readability of the reconstructed image. Hence, its input is a word text from the source domain z_i and the respective reconstruction $f(g(x_i))$. As above, we calculate the CTC-loss on a word level x_i. Since the documents all have a different length, the per word losses for each document are summed up and divided by the length of the document $|x|$.

The two reading loss terms are combined to form the overall reading loss defined as

$$\mathcal{L}_{\text{reading}}(g, f) = \mathbb{E}_{x \sim p_{\text{data}}(x)} \left[\frac{\sum_{v, w \in s(g, x)} \mathcal{L}_{CTC}(\alpha_2(v), w)}{|x|} \right]$$
$$+ \left[\frac{\sum_{v, w \in s(f(g, x))} \mathcal{L}_{CTC}(\alpha_2(v), w)}{|x|} \right] \quad (5)$$

where $s(h, u) = \{(u_i, h(u_i)) | i = 1, ..., |u|\}$ and h represents the transformation function and u all word image and ground truth pair of a document.

Combined Loss. The different loss term are weighted with $\lambda_{\text{cyc}} = 10$, $\lambda_{\text{read}} = 1$, and $\lambda_{\text{id}} = 5$ as suggested by Zhu et al. [32] and Touvron et al. [30] and summed up to form the overall loss objective:

$$\mathcal{L}_{\text{total}}(g, f, D_X, D_Y) = \mathcal{L}_{\text{GAN}}(g, D_Y, X, Y) + \mathcal{L}_{\text{GAN}}(f, D_X, Y, X)$$
$$+ \lambda_{cyc} \times \mathcal{L}_{\text{cyc}}(g, f) + \lambda_{id} \times \mathcal{L}_{\text{Identity}}(g, f)$$
$$+ \lambda_{read} \times \mathcal{L}_{\text{reading}}(g, f). \quad (6)$$

The combined loss is used in a min-max fashion, the generator tries to minimize it, and the discriminators aim to maximize it:

$$g^*, f^* = \arg \min_{g, f} \max_{D_x, D_Y} \mathcal{L}_{\text{total}}(g, f, D_X, D_Y). \quad (7)$$

4   Experimental Setup

Model Architecture. To achieve the goal of learning a transformation from source domain X to target domain Y using unpaired collections of images, we use an architecture based on the CycleGAN [32] framework. The generators G and F are each 24 layers deep Convolutional Neural Network (CNN) architectures with 11.3 million parameters. The discriminators D_x and D_y are based on the PatchGAN architecture [13], and have 5 layers and 2.7 million parameters each. Our Text Recognizer (TR) networks T and T' are based on the winning HTR model from the ICFHR2018 competition [26]. Both these networks contain 10 convolutional and batch normalization layers followed by 2 bi-directional LSTM layers for a total of 8.3 million parameters. For all architectures, we apply the preprocessing steps (e.g. resizing) as suggested in their respective publications. The data gets min-max normalized.

Task. The first step in our two-stage method is to create the source domain dataset images as described in Sect. 2.1. The structure and content of these documents are specified using LaTeX. In the second step, we use the source domain dataset files along with a collection of unlabeled historical document images (see Sect. 2.2) to train our CycleGAN and TR networks. In the training process, we learn a mapping function g that transforms source domain documents to the target domain as well as a mapping function f, which works in the other direction. The TR networks are trained simultaneously to recover the user-specified content from $g(x)$ and $f(g(x))$. After completing training, we use the generator G to transform document images from the source domain to the target domain while preserving content and structure.

Pre-processing. Due to GPU memory constraints, we use to train our models using image patches of size 256×256. These image patches are randomly cropped from the document images and fed into the CycleGAN architecture. The TR networks T and T' receive individual words cropped (128×32) from $g(x)$ and $f(g(x))$ respectively. Additionally we add Gaussian Noise to $g(x)$ as described in Sect. 3.1.

Training Procedure. We train the CycleGAN and TR components of our system simultaneously. The models are trained for 200 epochs using the Adam optimizer [17] with a learning rate of 2×10^{-4} and a linear decay starting at 100 epochs. The optimizer uses 5×10^{-5} weight decay and 0.5, 0.999 beta values for the generators and discriminators, respectively. We use a batch size of 1 to facilitate the varying amount of words per patch that is fed to T and T'.

Evaluation Procedure. We evaluate the quality of the synthetic historical documents produced with our method qualitatively and quantitatively. We first evaluate the synthetic historical documents produced qualitatively with a visual inspection, highlighting the successfully transformed and key limitations of the produced synthetic documents.

We use synthetic historical documents produced with our method in a pre-training setting to provide a quantitative evaluation. We generate 70'000 synthetic words in the historical style of the target domain dataset and use these words to train a new TR network called $\mathcal{R}_{\text{syn}}$. We then fine-tune $\mathcal{R}_{\text{syn}}$ using various subsets (10%, 20%, 50%, and 100%) of the training data from the St. Gall dataset (see Sect. 2.3) and evaluate its text recognition performance on the test set. As baselines, we compare $\mathcal{R}_{\text{syn}}$ against $\mathcal{R}_{\text{base}}$ and $\mathcal{R}_{\text{IAM}}$. $\mathcal{R}_{\text{base}}$ is randomly initialized and then trained directly on the St. Gall dataset in a similar manner as $\mathcal{R}_{\text{syn}}$. $\mathcal{R}_{\text{IAM}}$ is pre-trained on the IAM Handwriting Database (see Sect. 2.3) and fine-tuned on the St. Gall dataset.

To determine the best performing pre-trained models of $\mathcal{R}_{\text{syn}}$ and $\mathcal{R}_{\text{IAM}}$, we train both networks until convergence and select the best performing model based on validation score from the hand-labeled subset of HBA (see Fig. 3 and

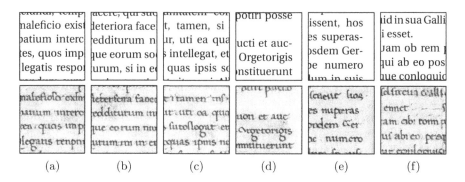

(a)   (b)   (c)   (d)   (e)   (f)

Fig. 5. Examples of template documents (upper row) and their corresponding synthetic historical images (bottom row). We can see that most characters are accurately transformed to the historical style while remaining readable. Notable exceptions include 'o' → 'a' and 's' → 'n'.

validation split of the IAM Handwriting Database. The performance of these three models is compared on the test split of the St. Gall dataset using the Character Error Rate (CER) and Word Error Rate (WER) metrics [22].

5   Results

We use two ways to evaluate the results of our generative model: a qualitative visual inspection and a qualitative evaluation. We use a qualitative human-based approach to evaluate the output from a visual perspective and a qualitative approach to measure the influence of our generated data on a downstream text recognition task.

5.1   Visual Analysis

As we can see in Fig. 5, the synthetic historical documents generated using our method achieve a high degree of similarity to documents from the target domain (see Sect. 2.2). The two primary goals of our approach were to preserve structure and content during the transformation of the source domain document into the target domain.

From Fig. 1, we can observe that the generator preserves the location of the text from the source domain to the target domain, resulting in the overall structure in the synthetic document matching the input document structure. In most cases, the transformation preserves the number of characters, words, and lines from the source document. However, we observe that on rare occasions, our approach results in synthetic documents where two letters in the source document are combined into a single letter (*legatis* in Fig. 5a) or a single letter is expanded into multiple letters (*rem* in Fig. 5f). We can also see from Fig. 1c that our approach is not very effective at transforming the large decorative characters at

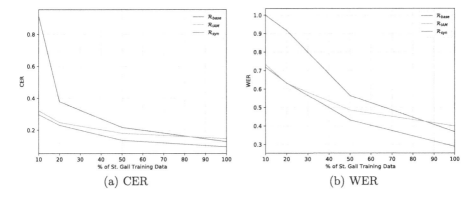

(a) CER   (b) WER

Fig. 6. We can see that the network pre-trained with synthetic data (in green) outperforms the two baselines (orange and blue) in all categories and for both metrics CER and WER. (Color figure online)

the beginning of paragraphs. The color of these decorative characters is transformed to the historical style, but they appear slightly distorted. This effect can be viewed as a side-effect of our training procedure, which does not emphasize transforming the decorative elements apart from the general style discrimination provided by D_x and D_y. Additionally, we see artifacts where the patches are stitched together because they are generated individually with 10% overlap and then combined by averaging.

Considering the preservation of textual content, our approach successfully transforms most individual characters to the style of the target domain dataset. Individual words are readable and require some effort to distinguish from real historical image samples – even to expert eyes. However, our approach struggles with the transformation of certain letters. From Fig. 5a, we can see that the character 'o' is mistransformed into an 'a'. However, the shape and appearance of these two letters are very similar and often hard to distinguish. Our approach also has problems transforming the letter 's'. This character is sometimes transformed into the character 'n', for e.g., in the word *superas* in Fig. 5e, the first 's' is transformed into 'n', however the second 's' is correctly preserved. Despite these small mistakes, we can observe that overall the method produces a very faithful transformation of the source document into the target historical style while preserving content and structure.

5.2   Quantitative Evaluation

In Fig. 6 we visualize the empirical results of our experiments where we compare our proposed approach against a purely supervised method and a transfer learning baseline method, with respect to the fraction of labels used in the target dataset. This way, we can assess the performances of those methods in the conditions of arbitrarily (and here, controlled) small datasets. We recall that small datasets are the common scenario in this domain, as opposed to more mainstream computer vision domains. As expected, with a small amount of data, the pre-trained methods ($\mathcal{R}_{\mathrm{syn}}$ and $\mathcal{R}_{\mathrm{IAM}}$) vastly outperform the baseline ($\mathcal{R}_{\mathrm{base}}$). This margin decreases as we train on large proportions of training data from St. Gall, however, $\mathcal{R}_{\mathrm{syn}}$ consistently achieves the lowest CER (see Fig. 6a), and is narrowly beat by $\mathcal{R}_{\mathrm{IAM}}$ only when considering the WER at the 20% subset (see Fig. 6b). On average, $\mathcal{R}_{\mathrm{syn}}$ has a 38% lower CER and a 26% lower WER compared to the model trained only on the St. Gall dataset, and a 14% lower CER and 10% lower WER compared to the model pre-trained on the IAM Handwriting Database.

Interestingly, when using the entire training set, $\mathcal{R}_{\mathrm{base}}$ achieves a lower error rate than $\mathcal{R}_{\mathrm{IAM}}$, which could be attributed to stylistic differences between the IAM Handwriting Database and the St. Gall dataset. Similar to observations from Studer et al. [27], the benefits of pre-training on a different domain could decrease when more training data is available from the actual task. Therefore, the stylistic similarity of the synthetic historical images and documents from the St. Gall dataset could explain the lower error rates of $\mathcal{R}_{\mathrm{syn}}$ compared to $\mathcal{R}_{\mathrm{base}}$.

6   Conclusion

We presented a two-step framework for generating synthetic historical images that appear realistic. The two steps are (1) creating electronic user-defined datasets (e.g., with LaTeX) for which the text content is known, and then feed it to step (2) where we use an improved CycleGAN based deep learning model to learn the mapping to a target (real) historical dataset. Differently from previous works in the field, our approach leverages two Text Recognizer (TR) networks to constrain the learning process further to produce images from which the text can still be read. The outcome of the process is a model capable of synthesizing a user-specified template image into historical-looking images. The content is known, i.e., we have the perfect ground truth for all the synthetic data we generate. These synthetic images—which come with a OCR ground truth—can then be used to pre-train models for downstream tasks. We measured the performances of a standard deep learning model using images created with our approach as well as other existing real historical datasets. We show that our approach consistently outperforms the baselines through a robust set of benchmarks, thus becoming a valid alternative as a source dataset for transfer learning. This work extends the already conspicuous work on the field of synthetic document generation. It distinguishes itself for providing the ground truth and high-quality synthetic historical images. Finally, the images generated with our methods are still distinguishable from real genuine ones due to small imperfections. Therefore we envisage that further work would improve upon our open-source implementation.

Acknowledgment. The work presented in this paper has been partially supported by the HisDoc III project funded by the Swiss National Science Foundation with the grant number 205120_169618. A big thanks to our co-workers Paul Maergner and Linda Studer for their support and advice.

References

1. Alberti, M., Seuret, M., Ingold, R., Liwicki, M.: A pitfall of unsupervised pre-training (2017). arXiv: 1703.04332
2. Alberti, M., Vögtlin, L., Pondenkandath, V., Seuret, M., Ingold, R., Liwicki, M.: Labeling, cutting, grouping: an efficient text line segmentation method for medieval manuscripts. In: 2019 International Conference on Document Analysis and Recognition (ICDAR), pp. 1200–1206. IEEE (2019)
3. Baird, H.S.: Document Image Defect Models. In: Baird, H.S., Bunke, H., Yamamoto, K. (eds.) Structured Document Image Analysis, pp. 546–556. Springer, Heidelberg (1992). https://doi.org/10.1007/978-3-642-77281-8_26
4. Bluche, T., Louradour, J., Messina, R.: Scan, attend and read: end-to-end handwritten paragraph recognition with MDLSTM attention. In: 2017 14th IAPR International Conference on Document Analysis and Recognition (ICDAR), vol. 01, pp. 1050–1055 (2017)
5. Capobianco, S., Marinai, S.: DocEmul: a toolkit to generate structured historical documents. In: 2017 14th IAPR International Conference on Document Analysis and Recognition (ICDAR), vol. 01, pp. 1186–1191 (2017)
6. Chu, C., Zhmoginov, A., Sandler, M.: CycleGAN, a master of steganography (2017)
7. Clausner, C., Pletschacher, S., Antonacopoulos, A.: Aletheia - an advanced document layout and text ground-truthing system for production environments. In: 2011 International Conference on Document Analysis and Recognition, pp. 48–52 (2011)
8. Edwards, H.J.: Caesar: The Gallic War. Harvard University Press Cambridge, Cambridge (1917)
9. Fischer, A., Frinken, V., Fornés, A., Bunke, H.: Transcription alignment of Latin manuscripts using hidden Markov models. In: Proceedings of the 2011 Workshop on Historical Document Imaging and Processing, HIP 2011, pp. 29–36. Association for Computing Machinery (2011)
10. Goodfellow, I.J., et al.: Generative Adversarial Networks (2014)
11. Graves, A., Fernández, S., Gomez, F., Schmidhuber, J.: Connectionist temporal classification: labelling unsegmented sequence data with recurrent neural networks. In: Proceedings of the 23rd International Conference on Machine Learning, ICML 2006, pp. 369–376. Association for Computing Machinery (2006)
12. Guan, M., Ding, H., Chen, K., Huo, Q.: Improving handwritten OCR with augmented text line images synthesized from online handwriting samples by style-conditioned GAN. In: 2020 17th International Conference on Frontiers in Handwriting Recognition (ICFHR), pp. 151–156 (2020)
13. Isola, P., Zhu, J.Y., Zhou, T., Efros, A.A.: Image-to-image translation with conditional adversarial networks. In: Proceedings of the IEEE Conference on Computer Vision and Pattern Recognition (CVPR) (2017)
14. Journet, N., Visani, M., Mansencal, B., Van-Cuong, K., Billy, A.: DocCreator: a new software for creating synthetic ground-truthed document images. J. Imaging **3**(4), 62 (2017)

15. Kang, L., Riba, P., Wang, Y., Rusiñol, M., Fornés, A., Villegas, M.: GANwriting: content-conditioned generation of styled handwritten word images. In: Vedaldi, A., Bischof, H., Brox, T., Frahm, J.-M. (eds.) ECCV 2020. LNCS, vol. 12368, pp. 273–289. Springer, Cham (2020). https://doi.org/10.1007/978-3-030-58592-1_17
16. Kieu, V.C., Visani, M., Journet, N., Domenger, J.P., Mullot, R.: A character degradation model for grayscale ancient document images. In: Proceedings of the 21st International Conference on Pattern Recognition (ICPR2012), pp. 685–688 (2012)
17. Kingma, D.P., Ba, J.: Adam: A Method for Stochastic Optimization (2017)
18. Li, H., Wang, W.: Reinterpreting CTC training as iterative fitting. Pattern Recog. **105**, 107392 (2020)
19. Marti, U.V., Bunke, H.: The IAM-database: an English sentence database for offline handwriting recognition. IJDAR **5**(1), 39–46 (2002)
20. Masci, J., Meier, U., Cireşan, D., Schmidhuber, J.: Stacked convolutional auto-encoders for hierarchical feature extraction. In: Honkela, T., Duch, W., Girolami, M., Kaski, S. (eds.) ICANN 2011. LNCS, vol. 6791, pp. 52–59. Springer, Heidelberg (2011). https://doi.org/10.1007/978-3-642-21735-7_7
21. Mehri, M., Héroux, P., Mullot, R., Moreux, J.P., Coüasnon, B., Barrett, B.: HBA 1.0: a pixel-based annotated dataset for historical book analysis. In: Proceedings of the 4th International Workshop on Historical Document Imaging and Processing, HIP 2017, pp. 107–112. Association for Computing Machinery (2017)
22. Märgner, V., Abed, H.E.: Tools and metrics for document analysis systems evaluation. In: Doermann, D., Tombre, K. (eds.) Handbook of Document Image Processing and Recognition, pp. 1011–1036. Springer, London (2014). https://doi.org/10.1007/978-0-85729-859-1_33
23. Pondenkandath, V., Alberti, M., Diatta, M., Ingold, R., Liwicki, M.: Historical document synthesis with generative adversarial networks. In: 2019 International Conference on Document Analysis and Recognition Workshops (ICDARW), vol. 5, pp. 146–151 (2019)
24. Scius-Bertrand, A., Voegtlin, L., Alberti, M., Fischer, A., Bui, M.: Layout analysis and text column segmentation for historical Vietnamese steles. In: Proceedings of the 5th International Workshop on Historical Document Imaging and Processing, HIP 2019, pp. 84–89. , Association for Computing Machinery (2019)
25. Seuret, M., Chen, K., Eichenbergery, N., Liwicki, M., Ingold, R.: Gradient-domain degradations for improving historical documents images layout analysis. In: 2015 13th International Conference on Document Analysis and Recognition (ICDAR), pp. 1006–1010 (2015)
26. Strauß, T., Leifert, G., Labahn, R., Hodel, T., Mühlberger, G.: ICFHR2018 competition on automated text recognition on a READ dataset. In: 2018 16th International Conference on Frontiers in Handwriting Recognition (ICFHR), pp. 477–482 (2018)
27. Studer, L., et al.: A comprehensive study of imagenet pre-training for historical document image analysis. In: 2019 International Conference on Document Analysis and Recognition (ICDAR), pp. 720–725 (2019)
28. Taigman, Y., Polyak, A., Wolf, L.: Unsupervised Cross-Domain Image Generation (2016)
29. Tensmeyer, C., Brodie, M., Saunders, D., Martinez, T.: Generating realistic binarization data with generative adversarial networks. In: 2019 International Conference on Document Analysis and Recognition (ICDAR), pp. 172–177 (2019)
30. Touvron, H., Douze, M., Cord, M., Jégou, H.: Powers of layers for image-to-image translation (2020). arXiv:2008.05763

31. Zhang, K.A., Cuesta-Infante, A., Xu, L., Veeramachaneni, K.: SteganoGAN: high capacity image steganography with GANs (2019). arXiv:1901.03892
32. Zhu, J.Y., Park, T., Isola, P., Efros, A.A.: Unpaired image-to-image translation using cycle-consistent adversarial networks. In: Proceedings of the IEEE International Conference on Computer Vision (ICCV) (2017)

Synthesizing Training Data
for Handwritten Music Recognition

Jiří Mayer[(✉)] and Pavel Pecina

Institute of Formal and Applied Linguistics, Charles University,
Prague, Czech Republic
{mayer,pecina}@ufal.mff.cuni.cz

Abstract. Handwritten music recognition is a challenging task that could be of great use if mastered, e.g., to improve the accessibility of archival manuscripts or to ease music composition. Many modern machine learning techniques, however, cannot be easily applied to this task because of the limi'ted availability of high-quality training data. Annotating such data manually is expensive and thus not feasible at the necessary scale. This problem has already been tackled in other fields by training on automatically generated synthetic data. We bring this approach to handwritten music recognition and present a method to generate synthetic handwritten music images (limited to monophonic scores) and show that training on such data leads to state-of-the-art results.

Keywords: Handwritten music recognition · Synthetic training data generation

1 Introduction

Handwritten music recognition (HMR) is the automatic process of converting handwritten sheet music to a machine-readable format. Having the music in digital form lets us easily preserve, analyze, search through, modify, engrave, or play the music [5]. While printed music recognition is still far from a solved problem, HMR has additional difficulties to contend with, due to the vast variability in handwriting style [3] and limited availability of data for training [12]. In addition to the low variability of graphical style, recognition of printed music has easier access to synthetic training data that can be generated using engraving tools such as Lilypond[1] or Verovio[2]. This has already been reflected in the literature: for instance, Deep-Scores [26] consists of 300k synthetic images of printed music scores for performing symbol classification, image segmentation, and object detection; the PrIMuS dataset [6] provides more than 80k printed images of music staves with ground truth readily available for end-to-end HMR learning.

Acquisition of training data for HMR, however, requires an expensive and slow manual process [12], which involves either transcription of existing sheet music into a symbolic representation (producing annotations) or handwriting

---

[1] http://lilypond.org/.
[2] https://www.verovio.org/.

© Springer Nature Switzerland AG 2021
J. Lladós et al. (Eds.): ICDAR 2021, LNCS 12823, pp. 626–641, 2021.
https://doi.org/10.1007/978-3-030-86334-0_41

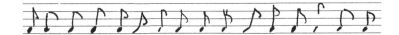

Fig. 1. Examples of eighth notes sampled from the MUSCIMA++ dataset. The handwriting style between writers varies greatly. Noteheads vary in shape and size, stems vary in slant and size, and sometimes the stem does not connect to the head.

music symbols on sheets of paper based on existing music representation (producing images). Both approaches require capturing the high variability of handwriting styles and therefore engagement of many writers (see Fig. 1).

The existing datasets for HMR are scarce. Most handwritten music datasets contain only individual symbols. The only resource containing entire staves is CVC-MUSCIMA [8], which comprises of 1,000 scanned sheets of music. While the variability of handwriting styles in this dataset is rather rich, the variability of music content is very limited (about 110 unique melody staves) and certainly not sufficient for end-to-end learning. This situation forces many researchers in the field to resort to data augmentation or transfer learning [2]. Despite all this, nobody has yet tried to create synthetic training data for HMR.

Synthetic data is data generated by a computer simulation of a real-world process. Training on such data is being used in machine learning where original (authentic) data is not available in sufficient amounts and the data generation process can be simulated by a computer program (e.g., image classification [20], natural scene text recognition [14], or handwritten text recognition [16]).

In this paper, we present *Mashcima* – an engine designed to generate realistic images of handwritten music for training HMR models. It exploits symbol masks extracted from the MUSCIMA++ dataset [12] (a richly annotated subset of CVC-MUSCIMA) that are rearranged and placed on one staff according to a music annotation in a newly proposed encoding adopted from the PrIMuS dataset [6] (see Fig. 2). *Mashcima* is highly configurable and customizable to alter the resulting visual style of the image. It can, for instance, mix handwriting of multiple people or generate images in the style of only one author. A non-trained human reported difficulties distinguishing a real-world sample from a well-synthesized one (see Fig. 3). The presented version of *Mashcima* generates only monophonic scores (music with one voice, without chords, and with musical symbols spanning only one staff). Polyphonic music presents challenges with regards to music representation and will be addressed in future versions.

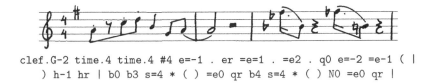

```
clef.G-2 time.4 time.4 #4 e=-1 . er =e=1 . =e2 . q0 e=-2 =e-1 ( |
     ) h-1 hr | b0 b3 s=4 * ( ) =e0 qr b4 s=4 * ( ) NO =e0 qr |
```

Fig. 2. An example of a synthetic image generated from the given annotation.

Fig. 3. Comparison of a real-world image (top) to a synthetic image (bottom) containing the same music. The top image is taken from the CVC-MUSCIMA dataset. The synthetic image is generated by *Mashcima* using symbols produced by a single writer.

The evaluation is conducted using a model based on a Convolutional Recurrent Neural Network (CRNN) with the Connectionist Temporal Classification (CTC) loss function, similar to the model proposed for printed music recognition by Calvo-Zaragoza and Rizo [6]. The model, trained on synthetic data produced by *Mashcima*, is evaluated on an unseen subset of MUSCIMA++ and a fully independent sample of real-world sheet music. A quantitative and qualitative comparison with previously published state-of-the-art results indicates the superior performance of our approach. The complete code of *Mashcima* and the experiments is available on GitHub.[3]

This text continues with an overview of related work (Sect. 2), followed by the proposal of the music encoding used by *Mashcima* (Sect. 3) and a detailed description of the *Mashcima* system (Sect. 4). Section 5 then presents our experiments, including details of the HMR model, training configurations, results, and analysis. Section 6 concludes the text and outlines our future work.

2   Related Work

Optical Music Recognition (OMR) aims at converting images of musical scores into a computer-readable form [5,9,22]. Traditionally, the research has mainly focused on printed music in the common Western music notation. Recognition of printed music is less challenging compared to handwritten music due to the enormous differences in the variability of printing/handwriting styles.

Until recently, most of the approaches to printed OMR employed the traditional pipeline consisting of several recognition steps (preprocessing, music object detection, notation assembly, and encoding) [5]. In 2017, van der Wel et al. [28] presented the first end-to-end OMR approach based on Convolutional Neural Networks (CNNs) and sequence-to-sequence models, although limited to monophonic music scores only. In 2018, Calvo-Zaragoza et al. [6] presented an end-to-end model (also limited to monophonic scores) based on CRNN [21] and CTC [11] that did not require alignment of graphical symbols and ground truth.

---

[3] https://github.com/Jirka-Mayer/Mashcima.

Fig. 4. Symbol masks and their relationships form a notation graph in MUSCIMA++.

The initial attempts to HMR focused on particular stages of the traditional OMR pipeline including preprocessing [4], staff removal [19], symbol detection and recognition [27]. In 2019, Baró et al. [2] published the first baseline for full HMR (not limited to monophonic music) based on CRNN and transfer learning. The model was pretrained on the PrIMuS dataset (printed notation) and fine-tuned on the MUSCIMA++ data (handwritten notation). Calvo-Zaragoza et al. [7] used CRNN models for HMR of mensural notation. The attention mechanism used in HMR of historical music was explored by Baró et al. [1]. Pacha et al. [18] attempted to reconstruct full notation graphs.

There are only two papers focusing on synthetic data generation for OMR: DeepScores [26] and PrIMuS [6], both synthesising printed scores.

2.1 Datasets

This section provides an overview of datasets exploited in our work. For a more complete overview, see the webpage by Alexander Pacha[4].

CVC-MUSCIMA [8] is a dataset originally designed for two tasks: writer identification and staff removal. It contains 20 pages of music, each manually transcribed by 50 different writers (1,000 pages total). It contains ground truth for staff removal but not for music recognition. We manually annotated a small subset of the dataset to be used for evaluation in our experiments (see Sect. 5).

MUSCIMA++ [12] is a subset (140 pages) of CVC-MUSCIMA enriched with detailed information about the placement and relationships of individual music symbol primitives in the form of a notation graph (MuNG, see Fig. 4) which allows recovering pixel masks of music symbols (e.g., notes) and their positions with respect to staff lines. We harvest this dataset to collect samples of the pixel masks to be used for rendering synthetic images of handwritten scores.

PrIMuS [6] is a dataset of 87,678 music incipits taken from the RISM catalog[5] containing over 1.2M records from various musical sources. Each PrIMuS incipit is encoded in five machine-readable formats (including an agnostic encoding format) and engraved in a raster image (Fig. 5). This data was sourced for real-world music annotations to produce the synthetic data for our experiments.

Fig. 5. An original image of an incipit from the PrIMuS dataset (rendered by Verovio).

3   Music Representation

Several formats have been proposed for representing music in a machine-readable form, differing in nature and purpose: e.g., MIDI [24] is designed for music playback, MusicXML [10] is primarily for music notation editing and engraving, MEI [23] captures various notation semantics, MuNG [12] was recently proposed for precise annotation of symbols in (handwritten) scores and their relationships.

In this work, we employ our own encoding similar to the agnostic encoding used in the PrIMuS dataset [6]. The PrIMuS agnostic encoding represents each staff as a sequence of tags corresponding to graphical symbols in the score, following the relative left-to-right and top-down ordering. Each graphical symbol is represented as a tag (without any predefined musical meaning) and position in the staff (line/space). Note beams are vertically sliced and slurs are represented by opening and closing tags surrounding a subsequence of symbols.

Such an agnostic encoding presents several advantages for HMR: It is very simplistic and describes only the way a piece of music looks (not what it means) and does not enforce any implicit constraints, such as proper rhythm (e.g., number of beats per measure). This implies that machine learning models aiming to produce this graphical-level encoding as output avoid the issue of large-distance dependencies that arise when interpreting music notation, such as clefs influencing the meaning of notes on the entire line, and thus have an easier task. Once this agnostic encoding is recovered, the musical semantics can be computed deterministically. There is no explicit alignment to the corresponding image (musical symbols and their absolute positions), which allows easy and relatively fast annotation of music scores from scratch. A single annotation can represent multiple scores (containing the same music but with a different appearance, which is typical in handwritten music). This is also important for generating synthetic data where multiple diverse images can be rendered for a single input. Most importantly, the sequential nature of the agnostic encoding allows feasible training of neural network models (e.g., using CTC) and straightforward evaluation based on string edit distance (e.g., Symbol Error Rate, see Sect. 5).

The PrIMuS agnostic encoding, however, was not designed for direct manual annotation (typing by hand). The tags have rather long names, which hinders readability. To simplify and speed up the manual annotation of the evaluation data used in our experiments, we designed our own *Mashcima* encoding, although not principally different from the PrIMuS agnostic encoding (see Fig. 6).

Tag names in the *Mashcima* encoding are considerably shorter and visually similar to the musical symbols they represent. Notes are represented by a one-letter symbol signaling their duration and a pitch number. The pitch 0

```
clef.G-2 b0 b3 time.C/ h2 . q3 | h4 h4 | h4 . q1 |
        q0 . s=1 =s2 q1 q-1 q0 | h-1 * q-1 | h-2
```

clef.G-L2 accidental.flat-L3 accidental.flat-S4 metersign.C/-L3 note.half-L4 dot-S3
note.quarter-S4 barline-L1 note.half-L5 note.half-L5 barline-L1 note.half-L5 dot-S5
        note.quarter-S3 barline-L1 note.quarter-L3 dot-S2 note.beamedRight2-S3
        note.beamedLeft2-L4 note.quarter-S3 note.quarter-S2 note.quarter-L3
        barline-L1 note.half-S2 dot-S2 note.quarter-S2 barline-L1 note.half-L2

Fig. 6. Image of an incipit from the PrIMuS dataset produced by *Mashcima*. Below
the image are the *Mashcima* encoding and the PrIMuS agnostic encoding of the image.

corresponds to notes on the center staff line. This makes the pitch space symmetrical and the annotation less prone to errors. Even pitches correspond to notes
on staff lines while odd pitches correspond to spaces. Rests are represented by
the letter r instead of a pitch number, slurs are denoted by round brackets.

Up until this point, the two encodings are equivalent. The differences are
the following: *Mashcima* distinguishes between duration dots *, ** and staccato
dots . while the PrIMuS agnostic supports only duration dots. The PrIMuS
agnostic encodes multimeasure rest digits and time signature digits the same
way; our encoding treats them differently (but does not support numeric multimeasure rests, only symbolic ones). *Mashcima* also introduces the ? token for
unsupported symbols. The current version of the synthesizer does not support
grace notes, fermatas, and tuplets (they will be added in future versions).

The complete description of the *Mashcima* encoding is available on GitHub,
including a (deterministic) tool to convert the PrIMuS agnostic encoding to the
Mashcima encoding which is used in our experiments.

4   Synthetic Data Generation

The aim of *Mashcima* is to produce synthetic images mimicking handwritten
music scores to train a system for recognition of such real-world images. This
is achieved by reusing images of individual music symbols extracted from scans
of real (authentic) handwritten scores and rearranging them onto a blank image
given a prescription specified in the *Mashcima* encoding. We describe all stages
of the process: 1) acquisition of music symbol masks, 2) obtaining ground-truth
annotations, 3) symbol placement on a staff, and 4) image rendering.

4.1   Acquisition of Music Symbol Masks

Mashcima collects authentic instances of music symbols from MUSCIMA++ in
the form of pixel masks (image pixels belonging to the given symbol) in binarized
images of handwritten scores. The masks are obtained for the entire symbols

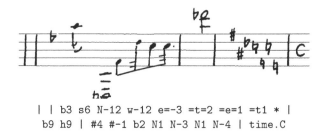

| | b3 s6 N-12 w-12 e=-3 =t=2 =e=1 =t1 * |
b9 h9 | #4 #-1 b2 N1 N-3 N1 N-4 | time.C

Fig. 7. An image synthesized from a pseudo-random annotation (denoted as *random*).

from the masks of their components (e.g., an eighth-note is extracted with its notehead, stem, and flag) through the music notation graphs and stored in a repository to be sampled from during later stages of the process. A total of 38,669 symbol instances were collected. Each symbol instance is then assigned anchor points (center of notehead, center of accidental, stem top) to control positioning on the rendered image. The acquisition can be constrained to the symbols of one writer to mimic the handwriting of a single person.

4.2   Obtaining Ground-Truth Annotations

Mashcima renders images based on music annotations in the *Mashcima* encoding. Two principled ways of obtaining such annotations are supported.

The first resort is to reuse existing annotations of real music. We leverage the converter from PrIMuS agnostic encoding to *Mashcima* encoding and use the PrIMuS dataset [6] as a source of such annotations. The entire dataset contains 87,678 music incipits, but only 64,127 of them (approx. 73%) can be successfully converted (due to symbols not supported by the current version of *Mashcima*, i.e., tuplets and multi-measure rests). Since PrIMuS contains no staccato dots, yet they are fairly abundant during evaluation, we converted about one half of the duration dots to staccato dots. Although this damages the musical meaning of (some) annotations, it allows recognition of staccato dots which would not be possible otherwise (adding training data without musical meaning actually helps the model, see Sect. 5). Apart from PrIMuS, additional annotations can be obtained, e.g., from RISM (the source of PrIMuS) and possibly other sources.

The second option is to generate the annotations by a stochastic process with only minimal constraints for valid token sequences (e.g., beams have to be connected, slur tags have to be paired). This approach is not expected to produce valid music, but this is not disqualifying. The amount of data that can be generated this way is basically unlimited and training on such noisy data can eventually improve the robustness of the model and its overall performance.

In our approach, most values (e.g., pitch) are drawn independently from a uniform distribution. For every image, the algorithm generates 5–15 token groups, where each group may be a key signature, a simple note, a beamed group, a barline, etc. The group distribution is chosen so that larger groups

Fig. 8. Groups with bounding boxes. Large crosses mark symbol origins, smaller crosses mark where a stem ends and a beam attaches. A key signature has no anchor point.

(e.g., beamed groups) are less likely to appear. Each note (within a group) gets a random pitch and a set of ornaments and accidentals. Lastly, slurs are added (without crossing each other). An example of the result is displayed in Fig. 7.

4.3   Symbol Placement

Symbol placement begins by positioning an empty staff drawn from the symbol repository to an empty image. The staff is aligned horizontally, so the horizontal dimension of the image can serve directly as a temporal dimension and the vertical one for pitch. A table that maps pitch values to vertical pixel offset is precomputed for each staff/image. Symbols are then positioned by their anchor points onto the proper pitch offset with no variation. The random symbol selection already varies the vertical position well enough.

The input token sequence is then clustered into token groups of symbols that act as a single unit (e.g., a note with its accidentals and ornaments is a single group), however, many tokens are left alone in their group (rests, barlines). A mask is randomly chosen from the symbol repository for every symbol in the group. Stem orientation is determined based on the note pitch (with randomization around the center). All symbols within a group are positioned relative to each other (accidentals are placed in front of the notes, duration dots behind, staccato dots below). Ornaments are positioned with some random noise. The groups are then placed from left to right onto the staff with randomized padding.

When the placement is settled, the stem length for beamed notes is adjusted by scaling the stem masks vertically. This is the only place where masks are distorted. The beam is then positioned as a simple straight line of a fixed width (no masks used). Similarly, the slur positions are determined and then drawn as parabolic arcs. Each note has three anchor points where a slur can end: in front, after, and below.

4.4   Image Rendering

With all symbols positioned and adjusted, they are all drawn onto the staff image (without any distortion and scaling). The order of drawing does not matter since the image is binarized. An example of a final image with symbol bounding boxes and anchor points is shown in Fig. 8. The synthesizer is able to draw three staves onto one image, which further helps with mimicking a real cropped image (barlines can be rendered across multiple staves, see Fig. 9).

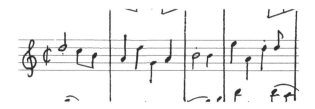

Fig. 9. Three staves synthesized in one image mimicking the look of a cropped image.

Limitations. The approach of simple positioning of masks cannot be applied to beams and slurs, so the system here fails to mimic the real world accurately (Fig. 10). There is also a lot of complexity regarding slur placement, that is not fully implemented in the current version of the synthesizer. The evaluation presented in the next section reveals that many errors are indeed related to slurs.

5   Experiments

This section presents the experiments evaluating how the proposed system is useful for synthesising HMR training data. We employ a state-of-the-art model trained on synthetic data generated by *Mashcima* and evaluate on real music sheets from MUSCIMA++ and a completely independent piece of written music.

5.1   Model Architecture

The neural network model employed in our experiments is inspired by the architecture proposed by Scheidl [25] for handwritten text recognition (the architecture used by Baró et al. [2] is not an alternative since it requires pixel-perfect annotation alignment that is not available in our data). The model is based on Convolutional Recurrent Neural Network (CRNN) with the Connectionist Temporal Classification (CTC) loss function [11, 21]. In our model, the convolutional block is slightly modified by adding one layer and shifting the pooling parameters to preserve the temporal resolution. Inspired by Calvo-Zaragoza et al. [6], we also add dropout to the Bidirectional Long Short-Term Memory (BLSTM) layer to improve convergence.

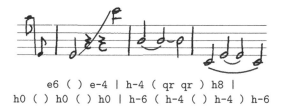

```
e6 ( ) e-4 | h-4 ( qr qr ) h8 |
hO ( ) hO ( ) hO | h-6 ( h-4 ( ) h-4 ) h-6
```

Fig. 10. Example of an incipit with misplaced slurs. Slurs do not affect positioning of neighbouring symbols and may improperly intersect them. Nested slurs are allowed.

Table 1. Model architecture overview.

Layer	Shape	Note
Input	w × 64 × 1	
Convolution	w × 64 × 16	Kernel 5 × 5
Max pooling	w/2 × 32 × 16	Stride 2,2
Convolution	w/2 × 32 × 32	Kernel 5 × 5
Max pooling	w/4 × 16 × 32	Stride 2,2
Convolution	w/4 × 16 × 64	Kernel 5 × 5
Max pooling	w/4 × 8 × 64	Stride 1,2
Convolution	w/4 × 8 × 128	Kernel 3 × 3
Max pooling	w/4 × 4 × 128	Stride 1,2
Convolution	w/4 × 4 × 128	Kernel 3 × 3
Max pooling	w/4 × 2 × 128	Stride 1,2
Convolution	w/4 × 2 × 256	Kernel 3 × 3
Max pooling	w/4 × 1 × 256	Stride 1,2
Reshape	w/4 × 256	
BLSTM	w/4 × 256 + w/4 × 256	Dropout
Concatenate	w/4 × 512	
Fully connected	w/4 × num_classes+1	No activation function
CTC	≤w/4	

Our architecture is described in Table 1. The first convolutional layer accepts images with the height of 64 pixels and a variable width. The convolutional block then squishes the dimensions until the height is equal to 1. The next block is a bidirectional recurrent layer with dropout. Finally, CTC produces the output tokens in the *Mashcima* encoding. In each experiment, the model is trained using the adaptive learning rate optimizer (Adam) [15] with the learning rate equal to 0.001, $\beta_1 = 0.9$, $\beta_2 = 0.999$, and $\varepsilon = 10^{-8}$. For evaluation, the beam search decoding algorithm is used with a beam width of 100, while during training the greedy algorithm is applied (the beam width set to 1) [13].

5.2 Data

All training and validation data used in experiments are synthetic, generated by *Mashcima*. We experiment with two types of such data, generated from the two types of annotations described in Sect. 4.2: i) *realistic* data synthesized from the 64,127 PrIMuS incipit annotations converted to the *Mashcima* encoding, ii) *random* data of the same size, synthesized from pseudo-randomly generated annotations. In both cases, 1,000 instances are randomly selected and used as validation data and the remaining 63,127 instances comprise the training data.

Two datasets are used for evaluation: The *in-domain* set is an unseen subset of 17 pages from MUSCIMA++ containing monophonic music written by 6 writers. The number of pages per writer ranges from 1 to 5. Some pages contain

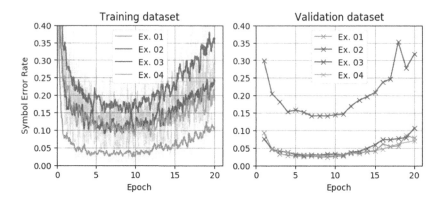

Fig. 11. SER on the training and validation sets throughout the training process: the training SER shown for each batch and smoothed over 11 values, the validation SER shown after each epoch (Ex. 4 has half as many epochs, each with twice as much data).

identical music but written by different writers.[6] This dataset contains 115 staves with 5,840 symbols (tokens) in total (50 symbols per staff on average). Images of these pages were split to single staves, each was manually annotated using the *Mashcima* encoding. This evaluation set is completely unseen – not only does the evaluation set contain no symbols from the training data, but there is also no intersection in terms of handwriting style and music content. However, we still (conservatively) consider this set in-domain (similar to the training data).

The *out-of-domain* set is created from a handwritten score of Cavatina by J. Raff for saxophone on three pages, which were scanned, binarized by a fixed threshold, and broken to images of single staves, each manually annotated in the *Mashcima* encoding (no additional preprocessing was performed). This evaluation set contains 35 staves and 1,831 symbols in total. We consider this evaluation set out-of-domain since there is no relation to the training data at all.

5.3   Experiment Design

We design four experiments to assess the effect of the two types of synthetic training data and their combination. In Experiment 1, the model is trained on the complete *realistic* data (63,127 instances). In Experiment 2, training is done on the *random* data of the same size. In Experiment 3, the model is trained on the same amount of data, while one half is a random sample from the *realistic* data and one half is sampled from the *random* data. Experiment 4 exploits both sets together doubling the training data size w.r.t. the previous experiments.

For quantitative evaluation, we employ Symbol Error Rate (SER) averaged over all evaluation instances. SER is the Levenshtein edit distance [17] normalized by the length of the ground-truth annotation. It was proposed for the evaluation of CRNN models for OMR in [6]. SER processes annotations at the token level with insertion, deletion, and substitution as elementary operations.

---

[6] Writers(pages): 13(2, 3, 16); 17(1); 20(2, 3, 16); 34(2, 3, 16); 41(2, 3, 16); 49(3, 5, 9, 11).

Table 2. Experiment results (%) averaged over three runs with standard deviation.

Ex.	Training	Validation	Evaluation (SER)	
			In-domain	*Out-of-domain*
1	Full *realistic*	*Realistic*	34.2 (±0.4)	59.2 (±1.3)
2	Full *random*	*Random*	28.0 (±1.4)	58.7 (±1.6)
3	1/2 *realistic* + 1/2 *random*	*Realistic*	**25.1 (±1.2)**	**49.2 (±2.0)**
4	Full *realistic* + full *random*	*Realistic*	25.6 (±0.9)	51.9 (±2.6)

All models are trained for a maximum of 20 epochs, keeping the model with the lowest validation SER (usually from the 6th to 10th epoch, see Fig. 11).

5.4   Results and Analysis

Results of our experiments are displayed in Table 2. Overall, the results are very optimistic, especially given the fact that training is done on synthetic data only. The best SER on the *in-domain* evaluation set is achieved by Experiment 3 and is as low as 25.1%. Vaguely interpreted, this indicates that only about 25% of all symbols in the *in-domain* evaluation set are misrecognized.

Another interesting finding comes from the comparison of Experiments 1 and 2 on the *in-domain* evaluation set. The results surprisingly indicate that training on the *random* data is more effective than training on the *realistic* data (SER drops from 34.2% to 28.0%). The extensive variability of music symbols occurring chaotically in the *random* data (unconstrained by any musical/notation rules) seems more beneficial than learning from real melodies and valid notations.

However, further comparison of Experiments 2 and 3 shows that the effect of training on the *realistic* data should not be neglected – if half of the *random* data in Experiment 2 is replaced by the same amount of the *realistic* data (Experiment 3), the SER on the *in-domain* evaluation set improves substantially (decreases from 28.0% to 25.1%). This suggests that seeing music symbols in their natural/meaningful configurations (think about clefs placed only at the beginning of scores, etc.) might be positive for learning.

Experiment 4 allows a direct comparison with all previous experiments and reveals additional findings. In comparison with Experiment 1, it quantifies the contribution of the *random* data added on top of the *realistic* data (-8.6% SER), which is much larger than the benefit of the *realistic* data added on top of the *random* data (−2.4% SER, cf. Experiment 2). In comparison with Experiment 3, it illustrates the effect of doubling the size of the training data, which is negligible (+0.5% SER). This suggests that enlarging the amount of training data of the same nature would not probably bring any major performance gains.

By comparing the *in-domain* and *out-of-domain* results, the model performs significantly worse on the *out-of-domain* data. This may be due to the fact that the *out-of-domain* data was preprocessed differently from the MUSCIMA++ data used for generating the training data. Cavatina contains faded text written onto

Table 3. Results achieved by Baró et al. [2] and our model from Experiment 4 on two sheets from MUSCIMA++. All scores are SER (%), however, not directly comparable.

Page	Writer	Baró et al. [2]			Our method	
		Rhythm	Pitch	Combined	#symbols	SER
1	17	52.8	34.9	59.2	304	28.1
3	13	22.6	17.5	27.0	334	11.6

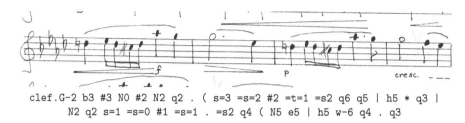

```
clef.G-2 b3 #3 N0 #2 N2 q2 . ( s=3 =s=2 #2 =t=1 =s2 q6 q5 | h5 * q3 |
    N2 q2 s=1 =s=0 #1 =s=1 . =s2 q4 ( N5 e5 | h5 w-6 q4 . q3
```

Fig. 12. Example from the *out-of-domain* evaluation data (SER on this staff is 49%).

the piece by a permanent marker, that cannot be fully removed by thresholding. Staff lines are also far thicker than in CVC-MUSCIMA and the model oftentimes confuses them with beams (also rendered as straight lines of the same width). Lastly, after binarization, the image is overall noisier (see Fig. 12).

5.5   Comparison with Other Works

Baró et al. [2] is the only work we can compare with. However, their method was not limited to monophonic music – the model was based on CRNN and did not use the CTC loss function, therefore the temporal resolution of their output is an order of magnitude higher than ours (since CTC collapses repeated tokens). Although their evaluation data overlaps in two pages (1 and 3) with our *in-domain* set and their evaluation was also done by SER, our scores (although appearing much higher, see Table 3) cannot be fairly compared to theirs.

The only possible direct comparison of their and our method is qualitative. Baró et al. [2] illustrated the effectiveness of their model on a certain image from MUSCIMA++ (page 3, writer 13). We recognized the same image by our model from Experiment 4 and display the results in Fig. 13 for direct comparison. The output of Baró et al. [2] contains four misrecognized notes and two missing slur endings (SER = 12.5%). Our model makes one error in note recognition and misses one slur beginning (causing the whole slur to disappear, SER = 4.2%).

Fig. 13. Qualitative comparison of our results to Baró et al. [2]. The first three staves are taken from Baró et al. [2]. The first staff is the input image from CVC-MUSCIMA (page 03, writer 13). The second staff is produced by PhotoScore. The third staff is the output of Baró et al. [2]. The last staff is the results of our model (Ex. 4) manually rendered by MuseScore (https://musescore.org/) followed by the output annotation in the *Mashcima* encoding.

6 Conclusion

We introduced the *Mashcima* system for creating synthetic handwritten music scores. It allows generating vast amounts of training data for HMR. Currently, HMR is mainly held back by the lack of training data. We were able to achieve state-of-the-art results by training on synthetic data generated by a relatively simple system. The MUSCIMA++ dataset is a step in the right direction. It provides rich annotations of music on various levels, from which any (simpler) encoding can be derived. However, its size is not sufficient for large experiments, which is understandable, given how much work such annotations take to produce. In *Mashcima*, MUSCIMA++ serves extremely well as a source of symbols for synthetic data generation. It is yet to be discovered what is the optimal mixture of random and realistic data to achieve even better results.

The *Mashcima* system still has limitations. Although it produces very real-world-looking data, the images are similar in style to those in the MUSCIMA++ dataset. The synthesized images are born binarized as they would have gone through the same binarization process as the original data from MUSCIMA++. Furthermore, the symbols and handwriting styles in the synthesized images are limited to the ones present in MUSCIMA++. In fact, *Mashcima* only enlarges existing datasets by reusing the existing symbols in new configurations and does not produce entirely new data. Altogether, this reduces the effectiveness of a model trained on such data and applied to data from a different source. This problem could be tackled by increasing the data variability, e.g., by additional distortion of symbol masks and/or collecting additional masks from other sources.

Our future work on *Mashcima* includes a more realistic rendering of slurs and beams, rendering of unsupported symbols, adding noise, and additional distortions of symbols and entire images. Ultimately, we aim at handling polyphonic music which would require changes in the HMR model architecture. This could be realized by producing more complex data with segmentation masks and relationship graphs as in MUSCIMA++. This would provide extensive amounts of data for many different HMR approaches to be trained and compared.

Despite the limitations of this first version, we believe the *Mashcima* system for handwritten musical score synthesis presents a valuable first-of-its-kind contribution to the field of OMR.

Acknowledgment. This work described in this paper has been supported by the Czech Science Foundation (grant no. 19-26934X), CELSA (project no. 19/018), and has been using data provided by the LINDAT/CLARIAH-CZ Research Infrastructure (https://lindat.cz), supported by the Ministry of Education, Youth and Sports of the Czech Republic (project no. LM2018101). The authors would like to thank Jan Hajič jr. for his valuable comments.

References

1. Baró, A., Badal, C., Fornés, A.: Handwritten historical music recognition by sequence-to-sequence with attention mechanism. In: 17th International Conference on Frontiers in Handwriting Recognition (ICFHR), Dortmund, Germany, pp. 205–210 (2020)
2. Baró, A., Riba, P., Calvo-Zaragoza, J., Fornés, A.: From optical music recognition to handwritten music recognition: a baseline. Pattern Recogn. Lett. **123**, 1–8 (2019)
3. Baró, A., Riba, P., Fornés, A.: Towards the recognition of compound music notes in handwritten music scores. In: 15th International Conference on Frontiers in Handwriting Recognition (ICFHR), Shenzhen, China, pp. 465–470 (2016)
4. Calvo-Zaragoza, J., Castellanos, F., Vigliensoni, G., Fujinaga, I.: Deep neural networks for document processing of music score images. Appl. Sci. **8**(5), 654 (2018)
5. Calvo-Zaragoza, J., Hajič, J., Jr., Pacha, A.: Understanding optical music recognition. ACM Comput. Surv. **53**(4), 77 (2020)
6. Calvo-Zaragoza, J., Rizo, D.: End-to-end neural optical music recognition of monophonic scores. Appl. Sci. **8**(4), 606 (2018)
7. Calvo-Zaragoza, J., Toselli, A., Vidal, E.: Handwritten music recognition for mensural notation with convolutional recurrent neural networks. Pattern Recogn. Lett. **128**, 115–121 (2019)
8. Fornés, A., Dutta, A., Gordo, A., Lladós, J.: CVC-MUSCIMA: a ground truth of handwritten music score images for writer identification and staff removal. Int. J. Doc. Anal. Recogn. **15**, 243–251 (2011)
9. Fornés, A., Sánchez, G.: Analysis and recognition of music scores. In: Doermann, D., Tombre, K. (eds.) Handbook of Document Image Processing and Recognition, pp. 749–774. Springer, London (2014). https://doi.org/10.1007/978-0-85729-859-1_24

10. Good, M.: MusicXML: An internet-friendly format for sheet music. In: Proceedings of the XML Conference, Orlando, FL, USA, pp. 3–4 (2001)
11. Graves, A., Fernández, S., Gomez, F., Schmidhuber, J.: Connectionist temporal classification. In: Proceedings of the 23rd International Conference on Machine Learning (ICML), Pittsburgh, PA, USA, pp. 369–376 (2006)
12. Hajič, J., Jr., Pecina, P.: The MUSCIMA++ dataset for handwritten optical music recognition. In: 14th IAPR International Conference on Document Analysis and Recognition (ICDAR), Kyoto, Japan, pp. 39–46 (2017)
13. Hwang, K., Sung, W.: Character-level incremental speech recognition with recurrent neural networks. In: IEEE International Conference on Acoustics. Speech and Signal Processing (ICASSP), Lujiazui, Shanghai, China, pp. 5335–5339 (2016)
14. Jaderberg, M., Simonyan, K., Vedaldi, A., Zisserman, A.: Synthetic data and artificial neural networks for natural scene text recognition (2014)
15. Kingma, D., Ba, J.: Adam: A method for stochastic optimization. In: 3rd International Conference on Learning Representations (ICLR), San Diego, USA (2014)
16. Krishnan, P., Jawahar, C.: Generating synthetic data for text recognition (2016)
17. Levenshtein, V.: Binary codes capable of correcting spurious insertions and deletions of ones. Probl. Inf. Transm. **1**, 8–17 (1965)
18. Pacha, A., Calvo-Zaragoza, J., Hajič, J., Jr.: Learning notation graph construction for full-pipeline optical music recognition. In: 20th International Society for Music Information Retrieval Conference (ISMIR), Delft, Netherlands, pp. 75–82 (2019)
19. Pacha, A., Choi, K.Y., Eidenberger, H., Ricquebourg, Y., Coüasnon, B., Zanibbi, R.: Handwritten music object detection: open issues and baseline results. In: 13th IAPR Interantional Workshop on Document Analysis Systems (DAS), Vienna, Austria, pp. 163–168 (2018)
20. Peng, X., Sun, B., Ali, K., Saenko, K.: Learning deep object detectors from 3D models. In: IEEE International Conference on Computer Vision (ICCV), Santiago, Chile, pp. 1278–1286 (2015)
21. Puigcerver, J.: Are multidimensional recurrent layers really necessary for handwritten text recognition? In: 14th IAPR International Conference on Document Analysis and Recognition (ICDAR), Kyoto, Japan, pp. 67–72 (2017)
22. Rebelo, A., Fujinaga, I., Paszkiewicz, F., Marçal, A., Guedes, C., Cardoso, J.: Optical music recognition: State-of-the-art and open issues. Int. J. Multimed. Inf. Retr. **1**, 173–190 (2012)
23. Roland, P.: The music encoding initiative (MEI). In: First International Conference on Musical Application Using XML, Milan, Italy, pp. 55–59 (2002)
24. Rothstein, J.: MIDI: A Comprehensive Introduction, vol. 7. AR Editions, Inc. (1992)
25. Scheidl, H.: Handwritten text recognition in historical documents. Master's thesis, Vienna University of Technology (2018)
26. Tuggener, L., Elezi, I., Schmidhuber, J., Pelillo, M., Stadelmann, T.: DeepScores - A dataset for segmentation, detection and classification of tiny objects. In: 24th International Conference on Pattern Recognition (ICPR), Beijing, China, pp. 3704–3709 (2018)
27. Tuggener, L., Elezi, I., Schmidhuber, J., Stadelmann, T.: Deep watershed detector for music object recognition. In: Proceedings of the 19th International Society for Music Information Retrieval Conference (ISMIR), Paris, France, pp. 271–278 (2018)
28. van der Wel, E., Ullrich, K.: Optical music recognition with convolutional sequence-to-sequence models. In: Proceedings of the 18th International Society for Music Information Retrieval Conference (ISMIR), Suzhou, China, pp. 731–737 (2017)

Towards Book Cover Design via Layout Graphs

Wensheng Zhang$^{(\boxtimes)}$, Yan Zheng, Taiga Miyazono, Seiichi Uchida[ORCID],
and Brian Kenji Iwana[ORCID]

Kyushu University, Fukuoka, Japan
{zhang.wensheng,taiga.miyazono,yan.zheng}@human.ait.kyushu-u.ac.jp,
{uchida,iwana}@ait.kyushu-u.ac.jp

Abstract. Book covers are intentionally designed and provide an introduction to a book. However, they typically require professional skills to design and produce the cover images. Thus, we propose a generative neural network that can produce book covers based on an easy-to-use layout graph. The layout graph contains objects such as text, natural scene objects, and solid color spaces. This layout graph is embedded using a graph convolutional neural network and then used with a mask proposal generator and a bounding-box generator and filled using an object proposal generator. Next, the objects are compiled into a single image and the entire network is trained using a combination of adversarial training, perceptual training, and reconstruction. Finally, a Style Retention Network (SRNet) is used to transfer the learned font style onto the desired text. Using the proposed method allows for easily controlled and unique book covers.

Keywords: Generative model · Book cover generation · Layout graph

1 Introduction

Book covers are designed to give potential readers clues about the contents of a book. As such, they are purposely designed to serve as a form of communication between the author and the reader [9]. Furthermore, there are many aspects of the design of a book cover that is important to the book. For example, the color of a book cover has shown to be a factor in selecting books by potential readers [13], the objects and photographs on a book cover are important for the storytelling [24], and even the typography conveys information [10,35]. Book covers [17,28] and the objects [21] on book covers also are indicators of genre.

While book cover design is important, book covers can also be time-consuming to create. Thus, there is a need for easy-to-use tools and automated processes which can generate book covers quickly. Typically, non-professional methods of designing book covers include software or web-based applications. There are many examples of this, such as Canva [6], fotor [11], Designhill [8], etc. These book cover designers either use preset templates or builders where the user selects from a set of fonts and

© Springer Nature Switzerland AG 2021
J. Lladós et al. (Eds.): ICDAR 2021, LNCS 12823, pp. 642–657, 2021.
https://doi.org/10.1007/978-3-030-86334-0_42

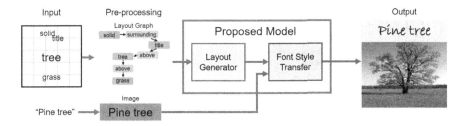

Fig. 1. Overview of generating a book cover using a layout graph.

images. The issue with these methods is that the design process is very restrictive and new designs are not actually created. It is possible for multiple authors to use the same images and create similar book covers.

Recently, there has been an interest in machine learning-based generation. However, there are only a few examples of book cover-based generative models. In one example, the website deflamel [7] generates designs based on automatically selected background and foreground images and title font. The images and font are determined based on a user-entered description of the book plus a "mood." The use of Generative Adversarial Networks (GAN) [12] have been used to generate books [27,28]. Although, in the previous GAN-based generation methods, the created book covers were uncontrollable and generate gibberish text and scrambled images.

The problem with template-based methods is that new designs are not created and the problem with GAN-based methods is that it is difficult to control which objects are used and where they are located. Thus, we propose a method to generate book covers that addresses these problems. In this paper, we propose the use of a *layout graph* as the input for users to draw their desired book cover. The layout graph, as shown in Fig. 1, indicates the size, location, positional relationships, and appearance of desired text, objects, and solid color regions. The advantage of using the layout graph is that it is easy to describe a general layout for the proposed method to generate a book cover image from.

In order to generate the book cover image, the layout graph is provided to a generative network based on scene graph-based scene generators [2,19]. In Fig. 2, the layout graph is fed to a Graph Convolution Network (GCN) [32] to learn an embedding of the layout objects (i.e. text objects, scene objects, and solid regions). This embedding is used to create mask and bounding-box proposals using a mask generator and box regression network, respectively. Like [2], the mask proposals are used with an appearance generator to fill in the masks with contents. The generated objects are then aggregated into a single book cover image using a final generator. These generators are trained using four adversarial discriminators, a perception network, and L1 loss to a ground truth image. Finally, the learned text font is transferred to the desired text using a Style Retention Network (SRNet) [37].

The main contributions of this paper are summarized as follows:

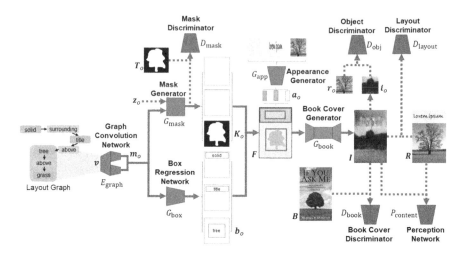

Fig. 2. The layout generator.

- As far as the authors know, this is the first instance of using a fully generative model for creating book cover images while being able to control the elements of the cover, such as size, location, and appearance of the text, objects, and solid regions.
- We propose a method of using a combination of a layout graph-based generator and SRNet to create user-designed book cover images.

Our codes are shown at https://github.com/Touyuki/Cover_generation.

2   Related Work

2.1   Document Generation

There are many generative models for documents. For example, automatic text and font generation is a key task in document generation. In the past, models such as using interpolation between multiple fonts [5,36] and using features from examples [34] have been used. More recently, the use of GANs have been used for font generation [1,14] and neural font style transfer [3] has become an especially popular topic in document generation.

There have also been attempts at creating synthetic documents using GANs [4,31] and document layout generation using recursive autoencoders [30]. Also, in a similar task to the proposed method, Hepburn et al. used a GAN to generate music album covers [16].

However, book cover generation, in particular, is a less explored area. Lucieri et al. [28] generated book covers using a GAN for data augmentation and the website Booksby.ai [27] generated entire books, including the cover, using GANs.

However, while the generated book covers have features of book covers and have the feel of book covers, the objects and text are completely unrecognizable and there is little control over the layout of the cover.

2.2   Scene Graph Generation

The proposed layout graph is based on scene graphs for natural scene generation. Scene graphs are a subset of knowledge graphs that specifically describe natural scenes, including objects and the relationships between objects. They were originally used for image retrieval [20] but were expanded to scene graph-based generation [19]. In scene graph generation, an image is generated based on the scene graph. Since the introduction of scene graph generation, there has been a huge boom of works in the field [38]. Some examples of scene graph generation with adversarial training, like the proposed method, include Scene Graph GAN (SG-GAN) [23], the scene generator by Ashual et al. [2], and PasteGAN [26]. These methods combine objects generated by each node of the scene graph and use a discriminator to train the scene image as a whole. As far as we know, we are the first to propose the use of scene graphs for documents.

3   Book Cover Generation

In this work, we generate book covers using a combination of two modules. The first is a Layout Generator. The Layout Generator takes a *layout graph* and translates it into an initial book cover image. Next, the neural font style transfer method, SRNet [37], is used to edit the generated placeholder text into a desired book cover text or title.

3.1   Layout Generator

The purpose of the Layout Generator is to generate a book cover image including natural scene objects, solid regions (margins, headers, etc.), and the title text. To do this, we use a layout graph-based generator which is based on scene graph generation [2,19]. As shown in Fig. 2, the provided layout graph is given to a comprehensive model of an embedding network, four generators, four discriminators, and a perceptual consistency network. The output of the Layout Generator is a book cover image based on the layout graph.

Layout Graph. The input of the Layout Generator is a layout graph, which is a directed graph with each object o represented by a node $\mathbf{n}_o = (\mathbf{c}_o, \mathbf{l}_o)$, where $\mathbf{c}_o$ is a class vector and l_o is the location vector of the object. The class vector contains a 128-dimensional embedding of the class of the object. The location vector $\mathbf{l}_o$ is a 35-dimensional binary vector that includes the location and size of the object. The first 25 bits of $\mathbf{l}_o$ describe the location of the object on a 5×5 grid and the last 10 bits indicate the size of the desired object on a scale of 1 to 10.

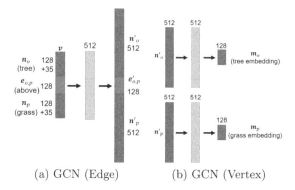

(a) GCN (Edge)   (b) GCN (Vertex)

Fig. 3. Illustration of the Graph Convolution Network. The red boxes are vertex vectors, the blue is the edge vector, the yellow is a hidden layer, and the arrows are full connections. (Color figure online)

The edges of the layout graph are the positional relations between the objects. Each edge $\mathbf{e}_{o,p}$ contains a 128-dimensional embedding of six relationships between every possible pairs of nodes o and p. The six relationships include "right of", "left of", "above", "below", "surrounding" and "inside".

Graph Convolution Network. The layout graph is fed to a GCN [32], E_{graph}, to learn an embedding $\mathbf{m}_o$ of each object o.

Where a traditional Convolutional Neural Network (CNN) [25] uses convolutions of shared weights across an image, a GCN's convolutional layers operate on graphs. They do this by traversing the graph and using a common operation on the edges of the graph.

To construct the GCN, we take the same approach as Johnson et al. [19] which constructs a list of all of the nodes and edges in combined vectors $\mathbf{v}$ and then uses a multi-layer perceptron (MLP) on the vectors, as shown in Fig. 3. Vector $\mathbf{v}$ consists of a concatenation of an edge embedding $\mathbf{e}_{o,p}$ and the two adjacent vertices o and p and vertex embeddings $\mathbf{n}_o$ and $\mathbf{n}_p$. The GCN is consists of two sub-networks. The GCN (Edge) network in Fig. 3a takes in vector $\mathbf{v}$ and then performs the MLP operation. The output is then broken up into temporary object segments $\mathbf{n}'_o$ and $\mathbf{n}'_p$ and further processed by individual GCN (Vertex) networks for each object. The result of GCN (Vertex) is a 128-dimensional embedding for each object, which is used by the subsequent Box Regression Network and Mask Generator.

Mask Generator and Discriminator. The purpose of the Mask Generator is to generate a mask of each isolated object for the Appearance Generator. The Mask Generator is based on a CNN. The input of the Mask Generator is the object embedding $\mathbf{m}_o$ learned from the GCN and the output is a 32×32 shape mask of the target object. This mask is only the shape and does not include size information. Furthermore, since the Mask Generator creates detailed masks, a variety of shapes should be used. To do this, a 64-dimensional random vector $\mathbf{z}_o$ is concatenated with the object embedding $\mathbf{m}_o$ before being given to the Mask Generator.

In order to produce realistic object masks, an adversarial Mask Discriminator D_{mask} is used. The Mask Discriminator is based on a conditional Least Squares GAN (LS-GAN) [29] with the object class s_o as the condition. It should be noted that the object class $\mathbf{s}_o$ is different than the 128-dimensional class vector $\mathbf{c}_o$ in the layout graph. The GAN loss $\mathcal{L}_{mask}^{D}$ is:

$$\mathcal{L}_{\mathrm{mask}}^{D} = [\log D_{\mathrm{mask}}(\mathbf{T}_o, \mathbf{s}_o)] + \mathbb{E}_{\mathbf{z}_o \sim \mathcal{N}(0,1)^{64}}[\log(1 - D_{\mathrm{mask}}(G_{\mathrm{mask}}(\mathbf{m}_o, \mathbf{z}_o), \mathbf{s}_o))], \tag{1}$$

where G_{mask} is the Mask Generator and $\mathbf{T}_o$ is a real mask. Accordingly, the Mask Discriminator D_{mask} is trained to minimize $-\mathcal{L}_{mask}^{D}$.

Box Regression Network. The Box Regression Network generates a bounding box estimation of where and how big the object should be placed in the layout. Just like the Mask Generator, the Box Regression Network receives the object embedding $\mathbf{m}_o$. The Box Regression Network is an MLP that predicts the bounding box $\mathbf{b}_o = \{(x_0, y_0), (x_1, y_1)\}$ coordinates for each object o.

To generate the layout, the outputs of the Mask Generator and the Box Regression Network are combined. In order to accomplish this, the object masks from the Mask Generator are shifted and scaled according to bounding boxes. The shifted and scaled object masks are then concatenated in the channel dimension and used with the Appearance Generator to create a layout feature map F for the Book Cover Generator.

Appearance Generator. The objects' appearances that are bound by the mask are provided by the Appearance Generator G_{app}. The Appearance Generator is a CNN that takes real images of cropped objects of $(64 \times 64 \times 3)$ resolution and encodes the appearance into a 32-dimension appearance vector. The appearance vectors $\mathbf{a}_o$ represent objects within the same class and changing the appearance vectors allows the appearance of the objects in the final generated result to be controlled. This gives the network to provide a variety of different object appearances even with the same layout graph. A feature map $\mathbf{F}$ is created by compiling the appearance vectors to fill the masks that were shifted and scaled by the bounding boxes.

Book Cover Generator. The Book Cover Generator G_{book} is based on a deep Residual Network (ResNet) [15] and it generates the final output. The network has three parts. The first part is the contracting path made of strided convolutions which encodes the features from the feature map $\mathbf{F}$. The second part is a series of 10 residual blocks and the final part is an expanding path with transposed convolutions to upsample the features to the final output image $\mathbf{I}$.

Perception Network. In order to enhance the quality of output of the Book Cover Generator a Perception Network is used. The Perception Network P_{content} is a pre-trained very deep convolutional network (VGG) [33] that is only used to establish a perceptual loss $\mathcal{L}^P_{\text{content}}$ [18]. The perceptual loss:

$$\mathcal{L}^P_{\text{content}} = \sum_{u \in \mathcal{U}} \frac{1}{u} \left| P^{(u)}_{\text{content}}(I) - P^{(u)}_{\text{content}}(R) \right| \tag{2}$$

is the content consistency between the extracted features of the VGG network P_{content} given the generated layout image I and a real layout image R. In Eq. (2), u is a layer in the set of layers $\mathcal{U}$ and $P^{(u)}_{\text{content}}$ is a feature map from P_{content} at layer u.

Layout Discriminator. The Layout Discriminator D_{layout} is a CNN used to judge whether the layout image $\mathbf{I}$ appears realistic given the layout $\mathbf{F}$. In this way, through the compound adversarial loss $\mathcal{L}_{\text{layout}}$, the generated layout will be trained to be more indistinguishable from images of real layout images $\mathbf{R}$ and real layout feature maps $\mathbf{Q}$. The loss $\mathcal{L}_{\text{layout}}$ is defined as:

$$\begin{aligned}
\mathcal{L}_{\text{layout}} = {} & \log D_{\text{layout}}(\mathbf{Q}, \mathbf{R}) + \log(1 - D_{\text{layout}}(\mathbf{Q}, \mathbf{I})) \\
& + \log(1 - D_{\text{layout}}(\mathbf{F}, \mathbf{R})) + \log D_{\text{layout}}(\mathbf{Q}', \mathbf{R})
\end{aligned} \tag{3}$$

where $\mathbf{Q}'$ is a second layout with the bounding box, mask, and appearance attributes taken from a different, incorrect ground truth image with the same objects. This is used as a poor match despite having the correct objects. The aim of the Layout Discriminator is to help the generated image $\mathbf{I}$ with ground truth layout $\mathbf{Q}$ to be indistinguishable from real image $\mathbf{R}$.

Book Cover Discriminator. The Book Cover Discriminator is an additional discriminator that is used to make the generated image look more like a book. Unlike the Layout Discriminator, the Book Cover Discriminator only compares the generated image $\mathbf{I}$ to random real book covers $\mathbf{B}$. Specifically, an adversarial loss:

$$\mathcal{L}_{\text{book}} = \log D_{\text{book}}(\mathbf{B}) + \log(1 - D_{\text{book}}(\mathbf{I})), \tag{4}$$

where D_{book} is the Book Cover Discriminator, is added to the overall loss.

Object Discriminator. The Object Discriminator D_{obj} is another CNN used to make each object images look real. $\mathbf{i}_o$ is an object image cut from the generated image by the generated bounding box and $\mathbf{r}_o$ is a real crop from the ground truth image. The object loss $\mathcal{L}_{\text{obj}}$ is:

$$\mathcal{L}_{\text{obj}} = \sum_{o=1}^{O} \log D_{\text{obj}}(\mathbf{r}_o) - \log D_{\text{obj}}(\mathbf{i}_o). \tag{5}$$

Fig. 4. Solid regions.

Training. The entire Layout Generator with all the aforementioned networks are trained together end-to-end. This is done using a total loss:

$$
\mathcal{L}_{\text{total}} = \lambda_1 \mathcal{L}_{\text{pixel}} + \lambda_2 \mathcal{L}_{\text{box}} + \lambda_3 \mathcal{L}_{\text{content}}^{P} + \lambda_4 \mathcal{L}_{\text{mask}}^{D} + \lambda_5 \mathcal{L}_{\text{obj}}^{D} \\
+ \lambda_6 \mathcal{L}_{\text{layout}}^{D} + \lambda_7 \mathcal{L}_{\text{book}}^{D} + \lambda_8 \mathcal{L}_{\text{mask}}^{P} + \lambda_9 \mathcal{L}_{\text{layout}}^{P},
\tag{6}
$$

where each λ is a weighting factor for each loss. In addition to the previously described losses, Eq. (6) contains a pixel loss $\mathcal{L}_{\text{pixel}}$ and two additional perceptual losses $\mathcal{L}_{\text{mask}}^{P}$ and $\mathcal{L}_{\text{layout}}^{P}$. The pixel loss $\mathcal{L}_{\text{pixel}}$ is the L1 distance between the generated image $\mathbf{I}$ and the ground truth image $\mathbf{R}$. The two perceptual losses $\mathcal{L}_{\text{mask}}^{P}$ and $\mathcal{L}_{\text{layout}}^{P}$ are similar to $\mathcal{L}_{\text{content}}^{P}$ (Eq. 2), except instead of a separate network, the feature maps of all of the layers of discriminators D_{mask} and D_{layout} are used, respectively.

3.2 Solid Region Generation

The original scene object generation is designed to generate objects in natural scenes that seem realistic. However, if we want to use it in book cover generation we should generate more elements that are unique to book covers, such as the solid region and the title information (Fig. 4).

We refer to *solid regions* as regions on a book with simple colors. They can be a single solid color, gradients, or subtle designs. As shown in Fig. 8, they are often used for visual saliency, backgrounds, and text regions. Except for some text information, usually, there are no other elements in these regions. To incorporate the solid regions into the proposed model, we prepared solid regions as objects in the Layout Graph. In addition, the solid regions are added as an object class to the various components of the Layout Generator as well as added to the ground truth images $\mathbf{R}$ and layout feature maps $\mathbf{Q}$. To make sure we can generate realistic solid regions, in our experiment, we used solid regions cut from real book covers.

3.3 Title Text Generation

Text information is also an important part of the book covers. It contains titles, sub-titles, author information, and other text. In our experiment, we only consider the title text.

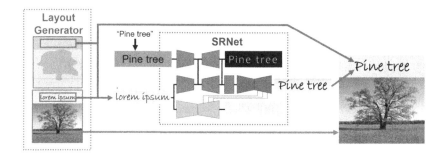

Fig. 5. The process of the SRNet.

Unlike other objects, like trees, the text information cannot be random variations and has to be determined by the user. However, the text still needs to maintain a style and font that is suitable for the book cover image.

Thus, we propose to generate the title text in the image using a placeholder and use font style transfer to transfer the placeholder's font to the desired text. Figure 5 shows our process of transferring the font style to the title text. To do this, we use SRNet [37]. SRNet is a neural style transfer method that uses a skeleton-guided network to transfer the style of text from one image to another. In SRNet, there are two inputs, the desired text in a plain font and the stylized text. The two texts are fed into a multi-task encoder-decoder that generates a skeleton image and a stylized image of the desired text. Using SRNet, we can generate any text using the style learned by the Layout Generator and use it to replace the placeholder.

To train the Layout Generator, we use a placeholder text, "Lorem Ipsum," to represent the title. Similar to the solid region object, the title object is also added as an object class. For the ground truth images $\mathbf{R}$, a random font, color, and location are used. However, the purpose of the Book Cover Discriminator D_{book} is to ensure that the combination is realistic as books.

4   Experimental Results

4.1   Dataset

To train the proposed method two datasets are required. The first is the Book Cover Dataset[1]. This dataset is made of book cover images and is used to train the Book Cover Discriminator.

For the second dataset, a natural scene object dataset with semantic segmentation information is required. For this, we use 5,000 images from the COCO[2] dataset. For the ground truth images and layouts, random solid regions and titles are added. The cropped parts of COCO are used with the Mask Discriminator

---

[1] https://github.com/uchidalab/book-dataset.
[2] https://cocodataset.org/.

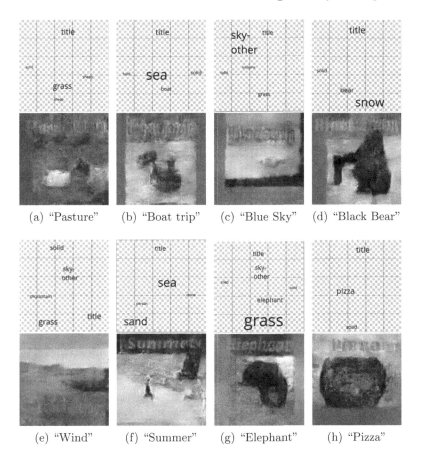

(a) "Pasture"   (b) "Boat trip"   (c) "Blue Sky"   (d) "Black Bear"

(e) "Wind"   (f) "Summer"   (g) "Elephant"   (h) "Pizza"

Fig. 6. Results with different layouts.

and the Object Discriminator, and modified images of COCO are used for the Layout Discriminator and Perception Network. All of the images are resized to 128×128.

4.2 Settings and Architecture

The networks in the Layout Generator are trained end-to-end using Adam optimizer [22] with $\beta = 0.5$ and an initial learning rate of 0.001 for 100,000 iterations with batch size 6. For the losses, we set $\lambda_1, \lambda_4, \lambda_6, \lambda_7 = 1$, $\lambda_2, \lambda_3, \lambda_8, \lambda_9 = 10$, and $\lambda_5 = 0.1$. The hyperparameters used in the experiments are listed in Table 1. For SRNet, we used a pre-trained model[3].

---

[3] https://github.com/Niwhskal/SRNet.

Table 1. The architecture of the networks.

Network	Layers	Activation	Norm.
GCN (Edge)	FC, 512 nodes	ReLU	
	FC, 1,152 nodes	ReLU	
GCN (Vertex)	FC, 512 nodes	ReLU	
	FC, 128 nodes	ReLU	
Box regression network	FC, 512 nodes	ReLU	
	FC, 4 nodes	ReLU	
Mask generator	Conv. (3×3), 192 filters, stride 1	ReLU	Batch norm.
	Conv. (3×3), 192 filters, stride 1	ReLU	Batch norm.
	Conv. (3×3), 192 filters, stride 1	ReLU	Batch norm.
	Conv. (3×3), 192 filters, stride 1	ReLU	Batch norm.
	Conv. (3×3), 192 filters, stride 1	ReLU	Batch norm.
Appearance generator	Conv. (4×4), 64 filters, stride 2	LeakyReLU	Batch norm.
	Conv. (4×4), 128 filters, stride 2	LeakyReLU	Batch norm.
	Conv. (4×4), 256 filters, stride 2	LeakyReLU	Batch norm.
	Global average pooling		
	FC, 192 nodes	ReLU	
	FC, 64 nodes	ReLU	
Book cover generator	Conv. (7×7), 64, stride 1	ReLU	Inst. norm.
	Conv. (3×3), 128 filters, stride 2	ReLU	Inst. norm.
	Conv. (3×3), 256 filters, stride 2	ReLU	Inst. norm.
	Conv. (3×3), 512 filters, stride 2	ReLU	Inst. norm.
	Conv. (3×3), 1,024 filters, stride 2	ReLU	Inst. norm.
($\times 10$ residual blocks)	Conv. (3×3), 1,024 filters, stride 1	ReLU	Inst. norm.
	Conv. (3×3), 1,024 filters, stride 1	ReLU	Inst. norm
	T. conv. (3×3), 512 filters, stride 2	ReLU	Inst. norm.
	T. conv. (3×3), 256 filters, stride 2	ReLU	Inst. norm.
	T. conv. (3×3), 128 filters, stride 2	ReLU	Inst. norm.
	T. conv. (3×3), 64 filters, stride 2	ReLU	Inst. norm.
	Conv. (7×7), 3 filters, stride 1	Tanh	
Mask discriminator	Conv. (3×3), 64 filters, stride 2	LeakyReLU	Inst. norm.
	Conv. (3×3), 128 filters, stride 2	LeakyReLU	Inst. norm.
	Conv. (3×3), 256 filters, stride 1	LeakyReLU	Inst. norm.
	Conv. (3×3), 1 filters, stride 1	LeakyReLU	
	Ave. Pooling (3×3), stride 2		
Layout discriminator	Conv. (4×4), 64 filters, stride 2	LeakyReLU	
	Conv. (4×4), 128 filters, stride 2	LeakyReLU	Inst. norm.
	Conv. (4×4), 256 filters, stride 2	LeakyReLU	Inst. norm.
	Conv. (4×4), 512 filters, stride 2	LeakyReLU	Inst. norm.
	Conv. (4×4), 1 filter, stride 2	Linear	
	Conv. (4×4), 64 filters, stride 2	LeakyReLU	
	Conv. (4×4), 128 filters, stride 2	LeakyReLU	Inst. norm.
	Conv. (4×4), 256 filters, stride 2	LeakyReLU	Inst. norm.
	Conv. (4×4), 512 filters, stride 2	LeakyReLU	Inst. norm.
	Conv. (4×4), 1 filter, stride 2	Linear	
	Ave. pooling (3×3), stride 2		
Book cover discriminator	Conv. (4×4), 64 filters, stride 2	LeakyReLU	
	Conv. (4×4), 128 filters, stride 2	LeakyReLU	Batch norm.
	Conv. (4×4), 256 filters, stride 2	LeakyReLU	Batch norm.
	Conv. (4×4), 512 filters, stride 2	LeakyReLU	Batch norm.
	Conv. (4×4), 512 filters, stride 2	LeakyReLU	Batch norm.
	Conv. (4×4), 1 filter, stride 2	Sigmoid	
Object discriminator	Conv. (4×4), 64 filters, stride 2	LeakyReLU	
	Conv. (4×4), 128 filters, stride 2	LeakyReLU	Batch norm.
	Conv. (4×4), 256 filters, stride 2	LeakyReLU	Batch norm.
	Global Average Pooling		
	FC, 1024 nodes	Linear	
	FC, 174 nodes	Linear	
Perception network	Pre-trained VGG [33]		
Font style transfer	Pre-trained SRNet [37]		

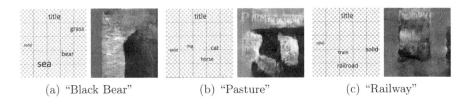

(a) "Black Bear"   (b) "Pasture"   (c) "Railway"

Fig. 7. Example of poor results.

4.3   Generation Results

Examples of generated book covers are shown in Fig. 6. We can notice that not only the object images can be recognizable, but also the solid regions make the results resemble book covers. In addition, for most of the results, the generated titles are legible. While not perfect, these book covers are a big step towards book cover generation. We also shows some images with poor quality in Fig. 7. In these results the layout maps are reasonable, but the output is still poor. This is generally due to having overlapping objects such as "grass" on the "title" or objects overlapping the solid regions.

4.4   Creating Variations in Book Covers

As mentioned previously, the advantage of using a layout graph is that each node contains information about the object, location, and appearance embedding. This allows for the ease of book cover customization using an easy to use interface. Thus, we will discuss some of the effects of using the layout graph to make different book cover images.

Location on the Solid Region. Along with the scene objects, the title text and the solid region can be moved on the layout graph. Figure 8 shows examples of generated book covers with the same layout graph except for the "Solid" nodes. By moving the "Solid" node to have different relationship edges with other nodes, the solid regions can be moved. In addition, multiple "Solid" nodes can be added to the same layout graph to construct multiple solid regions.

Variation in the Appearance Vector. Due to each node in the layout graph containing its own appearance vector, different variations of generated book covers can be created from the same layout graph. Figure 9 shows a layout graph and the effects of changing the appearance vector of individual nodes. In the figure, only one node is changed and all the rest are kept constant. However, even though only one element is being changed in each sub-figure, multiple elements are affected. For example, in Fig. 9 (c) when changing the "Grass" node, the generated grass area changes and the model automatically changes the "Solid" and "Sky" regions to

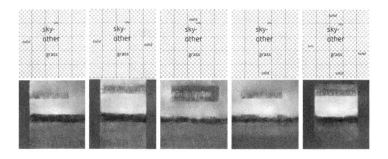

Fig. 8. Examples of moving or adding solid region nodes.

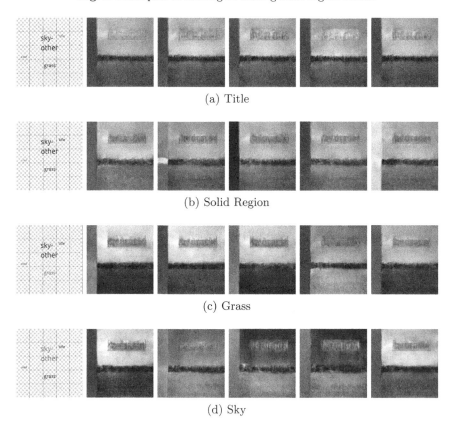

(a) Title

(b) Solid Region

(c) Grass

(d) Sky

Fig. 9. Examples of the effect of changing the appearance vector for different nodes. Each sub-figure changes the appearance vector for the respective node and keeps all other nodes constant.

match the appearance of the "Grass" region. As it can be observed from the figure, the solid bar on the left is normally contrasted from the sky and the grass. This happens because each node is not trained in isolation and the discriminators have a global effect on multiple elements and aim to generate more realistic compositions.

4.5   Effect of Text Style Transfer

The SRnet is used to change the placeholder text to the desired text in the generated image. In Fig. 10, we show a comparison of book covers before and after using SRNet. As can be seen from the figure, SRNet is able to successfully transfer the font generated by the Layout Generator and apply it to the desired text. This includes transferring the color and font features of the placeholder. In addition, even if the title text is short like "Sheep" or "Color," SRNet was able to still erase the longer placeholder text. However, "Winter Day" appears to erroneously overlap with the solid region, but that is due to the predicted bounding box of the text overlapping with the solid region. Thus, this is not a result of a problem with SRNet, but with the Box Regression Network.

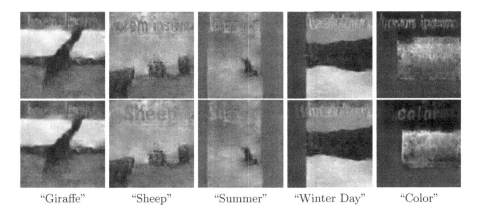

   "Giraffe"   "Sheep"   "Summer"   "Winter Day"   "Color"

Fig. 10. Using SRNet to change the placeholder title into a target text. The top row is the output before SRNet and the bottom is after SRNet.

5   Conclusion

We proposed a book cover image generation system given a layout graph as the input. It comprises an image generation model and a font style transfer network. The image generation model uses a combination of a GCN, four generators, four discriminators, and a perception network to a layout image. The font style transfer network then transfers the style of the learned font onto a replacement with the desired text. This system allows the user to control the book cover elements and their sizes, locations, and appearances easily. In addition, users can write any text information and fonts fitting the book cover will be generated. Our research is a step closer to automatic book cover generation.

Acknowledgement. This work was in part supported by MEXT-Japan (Grant No. J17H06100 and Grant No. J21K17808).

References

1. Abe, K., Iwana, B.K., Holmer, V.G., Uchida, S.: Font creation using class discriminative deep convolutional generative adversarial networks. In: ACPR (2017)
2. Ashual, O., Wolf, L.: Specifying object attributes and relations in interactive scene generation. In: ICCV (2019)
3. Atarsaikhan, G., Iwana, B.K., Narusawa, A., Yanai, K., Uchida, S.: Neural font style transfer. In: ICDAR (2017)
4. Bui, Q.A., Mollard, D., Tabbone, S.: Automatic synthetic document image generation using generative adversarial networks: application in mobile-captured document analysis. In: ICDAR (2019)
5. Campbell, N.D.F., Kautz, J.: Learning a manifold of fonts. ACM Tran. Graph. **33**(4), 1–11 (2014)
6. Canva: Canva - the free book cover maker with stunning layouts. https://www.canva.com/create/book-covers/. Accessed 15 Feb 2021
7. Deflamel Corp.: deflamel. https://deflamel.com/. Accessed 15 Feb 2021
8. Designhill.com: Designhill - book cover maker. https://www.designhill.com/tools/book-cover-maker. Accessed 15 Feb 2021
9. Drew, N., Stemberge, P.: By Its Cover: Modern American Book Cover Design. Princeton Architectural Press, Princeton (2005)
10. EL-Sakran, T., Ankit, A.: Representing academic disciplines on academic book covers. Int. J. Pedagogical Innov. **6**(02), 151–163 (2018)
11. Everimaging Limited: fotor - book cover maker. https://www.fotor.com/design/book-cover. Accessed 15 Feb2021
12. Goodfellow, I., et al.: Generative adversarial nets. In: NeurIPS (2014)
13. Gudinavičius, A., Šuminas, A.: Choosing a book by its cover: analysis of a reader's choice. J. Doc. (2018)
14. Hayashi, H., Abe, K., Uchida, S.: GlyphGAN: style-consistent font generation based on generative adversarial networks. Knowl. Based Sys. **186**, 104927 (2019)
15. He, K., Zhang, X., Ren, S., Sun, J.: Deep residual learning for image recognition. In: CVPR (2016)
16. Hepburn, A., McConville, R., Santos-Rodriguez, R.: Album cover generation from genre tags. In: MUSML (2017)
17. Iwana, B.K., Raza Rizvi, S.T., Ahmed, S., Dengel, A., Uchida, S.: Judging a book by its cover. arXiv preprint arXiv:1610.09204 (2016)
18. Johnson, J., Alahi, A., Fei-Fei, L.: Perceptual losses for real-time style transfer and super-resolution. In: Leibe, B., Matas, J., Sebe, N., Welling, M. (eds.) ECCV 2016. LNCS, vol. 9906, pp. 694–711. Springer, Cham (2016). https://doi.org/10.1007/978-3-319-46475-6_43
19. Johnson, J., Gupta, A., Fei-Fei, L.: Image generation from scene graphs. In: CVPR (2018)
20. Johnson, J., et al.: Image retrieval using scene graphs. In: CVPR (2015)
21. Jolly, S., Iwana, B.K., Kuroki, R., Uchida, S.: How do convolutional neural networks learn design? In: ICPR (2018)
22. Kingma, D.P., Ba, J.: Adam: a method for stochastic optimization. arXiv preprint arXiv:1412.6980 (2014)
23. Klawonn, M., Heim, E.: Generating triples with adversarial networks for scene graph construction. In: AAAI (2018)
24. Kratz, C.A.: On telling/selling a book by its cover. Cultur. Anthropol. **9**(2), 179–200 (1994)

25. LeCun, Y., Bottou, L., Bengio, Y., Haffner, P.: Gradient-based learning applied to document recognition. Proc. IEEE **86**(11), 2278–2324 (1998)
26. Li, Y., Ma, T., Bai, Y., Duan, N., Wei, S., Wang, X.: PasteGAN: a semi-parametric method to generate image from scene graph. In: NeurIPS (2019)
27. Loose, M.T., Refsgaard, A.: Booksby.ai - about. https://booksby.ai/about/. Accessed 15 Feb 2021
28. Lucieri, A., et al.: Benchmarking deep learning models for classification of book covers. SN Comput. Sci. **1**(3), 1–16 (2020)
29. Mao, X., Li, Q., Xie, H., Lau, R.Y., Wang, Z., Paul Smolley, S.: Least squares generative adversarial networks. In: CVPR (2017)
30. Patil, A.G., Ben-Eliezer, O., Perel, O., Averbuch-Elor, H.: READ: recursive autoencoders for document layout generation. In: CVPR Workshops (2020)
31. Rusticus, D., Goldmann, L., Reisser, M., Villegas, M.: Document domain adaptation with generative adversarial networks. In: ICDAR (2019)
32. Scarselli, F., Gori, M., Tsoi, A.C., Hagenbuchner, M., Monfardini, G.: The graph neural network model. IEEE Trans. Neural Netw. **20**(1), 61–80 (2009)
33. Simonyan, K., Zisserman, A.: Very deep convolutional networks for large-scale image recognition. arXiv preprint arXiv:1409.1556 (2014)
34. Suveeranont, R., Igarashi, T.: Example-based automatic font generation. In: Smart Graphics (2010)
35. Tschichold, J.: The New Typography: A Handbook for Modern Designers, vol. 8. University of California Press, Berkeley (1998)
36. Uchida, S., Egashira, Y., Sato, K.: Exploring the world of fonts for discovering the most standard fonts and the missing fonts. In: ICDAR (2015)
37. Wu, L., et al.: Editing text in the wild. In: ACM ICM (2019)
38. Xu, P., Chang, X., Guo, L., Huang, P.Y., Chen, X., Hauptmann, A.G.: A survey of scene graph: generation and application. EasyChair Preprint (2020)

Graphics Recognition

Complete Optical Music Recognition via Agnostic Transcription and Machine Translation

Antonio Ríos-Vila[1]([✉])(iD), David Rizo[1,2](iD), and Jorge Calvo-Zaragoza[1](iD)

[1] U.I. for Computing Research, University of Alicante, Alicante, Spain
{arios,drizo,jcalvo}@dlsi.ua.es
[2] Instituto Superior de Enseñanzas Artísticas de la Comunidad Valenciana (ISEA.CV), Valencia, Spain

Abstract. Optical Music Recognition workflows currently involve several steps to retrieve information from music documents, focusing on image analysis and symbol recognition. However, despite many efforts, there is little research on how to bring these recognition results to practice, as there is still one step that has not been properly discussed: the encoding into standard music formats and its integration within OMR workflows to produce practical results that end-users could benefit from. In this paper, we approach this topic and propose options for completing OMR, eventually exporting the score image into a standard digital format. Specifically, we discuss the possibility of attaching Machine Translation systems to the recognition pipeline to perform the encoding step. After discussing the most appropriate systems for the process and proposing two options for the translation, we evaluate its performance in contrast to a direct-encoding pipeline. Our results confirm that the proposed addition to the pipeline establishes itself as a feasible and interesting solution for complete OMR processes, especially when limited training data is available, which represents a common scenario in music heritage.

Keywords: Optical Music Recognition · Graphics recognition · Machine translation

1 Introduction

Music represents a valuable component of our cultural heritage. Most music has been preserved in printed or handwritten music notation documents. In addition to the efforts to correctly maintain the documents that inherently get damaged over time, huge efforts are being done to digitize them. The same way Optical Character Recognition (OCR) and Handwritten Text Recognition (HTR) are successfully being applied to extract the content from text images, Optical Music Recognition (OMR) systems are applied to encode the music content that appears in score sheets [6].

© Springer Nature Switzerland AG 2021
J. Lladós et al. (Eds.): ICDAR 2021, LNCS 12823, pp. 661–675, 2021.
https://doi.org/10.1007/978-3-030-86334-0_43

Specifically, OMR joins the knowledge from areas like computer vision, machine learning and digital musicology to perform the recognition and the digitising of music scores. Despite of being sometimes addressed as "OCR for music" [3], its two-dimensional nature, along with many ubiquitous contextual dependencies among symbols, differentiates OMR from other document recognition areas, such as OCR and HTR [5]. To illustrate this, we could use a simple example: A note is identified, graphically speaking, at a specific position in the staff. However, its interpretation could change depending on multiple factors, such as the clef position, the accidentals that may be present nearby, the key signature, or just a bar line cancelling previous alterations. Indeed, there is also a required temporal interpretation that does not depend on that specific position in the staff, but on the shape of the note (as each glyph represents a different duration of the note during interpretation).

Most of the existing OMR literature is framed within a multi-stage workflow, with steps involving image pre-processing—including staff-line detection and removal [10]—symbol classification [20] and notation assembly [19]. Advances in Deep Learning (DL) lead the image processing steps to be replaced with neural approaches such as Convolutional Neural Networks (CNN). But more importantly, DL brought alternatives to these traditional multi-stage workflows. On the one hand, we have the segmentation-based approach, where the complex multi-stage symbol isolation workflows have been replaced for region-based CNN that directly recognize symbols in a music staff [21,29]. On the other hand, there are end-to-end approaches. Specifically, we find solutions based on Convolutional Recurrent Neural Networks (CRNN) that come in varying configurations: the so-called Sequence to Sequence (*Seq2Seq*) architecture [2] ones, and also those trained using the Connectionist Temporal Classification (CTC) loss function [8].

Typically, these DL-based approaches cover all the processes that involve the transcription of an input image, which is usually a music staff, into a sequence that represents the recognized glyphs and positions of the symbols in the given score. Even obtaining such descriptive sequences, these results cannot be used by an end-user or reproduced in a music tool or visualizer, as there exists the need to recover music semantics as well. This last step to achieve an actual digital score is the so-called encoding process, where the graphical recognition outputs, without specific musical meaning, are converted into a standard semantic encoding. Unfortunately, this step is hardly addressed in the DL-based OMR literature, due to the focus the community has given to the challenges of the previous steps require.

In this paper, we research how to complete the OMR process, starting from a music-staff image as an input and producing a semantic standard encoding sequence as output. As a novelty, we introduce the use of Machine Translation to perform this last step of parsing a purely visual representation extracted from a graphic recognition process and exporting it in a standard musical encoding document.

The rest of the paper is organized as follows: in Sect. 2 we discuss why we approach the OMR encoding step with Machine Translation techniques, instead of hand-crafted rule-based heuristic approaches. In Sect. 3, we describe the specific implemented systems used to perform our experiments. In Sect. 4, we explain our experimentation environment for the sake of replicability; in Sect. 5 we present and discuss the obtained results regarding the comparison between different alternatives; and, we conclude our work in Sect. 6.

2 Machine Translation for Encoding in Optical Music Recognition

We discuss in this section how to approach the encoding step of an OMR system, as it is an issue that has not been fully solved in previous works. We remind the reader that the encoding step consists of the production of a symbolic music notation format from the symbol recognition phase in the previous OMR step, which typically works at the image level. This means eventually producing a score encoded in a standard digital format from a collection of musical glyphs located in a two-dimensional space. From now on, we will denote the output from the graphical process as *agnostic encoding*; while the music standard format is referred to as *semantic encoding*. These terms are becoming common in OMR literature [7,22].

A usual approach in most commercial systems to convert from agnostic encoding to semantic encoding is laying the task on a rule-based translation system. This has been proved to be a challenging task in complex scores [5,13]. This approach also has significant issues in both extrapolability for different notation types, and scalability terms, as it requires the redesign of the rules or systems when the source and/or the target encoding vary. This scalability issue also appears when moving into more complex music domains, such as polyphonic scores, where the task of designing rules which adapt to these documents may become unfeasible. As we can observe, this is hardly maintainable when complexity both on the document type and the notations scale. This situation leaves us to look for more sophisticated models in the Machine Translation community.

One simple approach could be to apply Statistical Machine Translation (SMT) techniques [16], which are data-driven approaches for Machine Translation where several independent models are combined to produce both a translation unit and a language model to convert a source language sequence into a target language one. These combinations allow SMT to provide balanced predictions in accuracy and fluency, as they implement mechanisms to deal with translation issues such as word reordering. Another benefit they bring is that, currently, there exist standard and well-known toolkits to perform SMT, such as Moses [17]. However, during preliminary experimentation with these techniques [26], we observed that, despite offering interesting results with few labeled data, they do not produce flexible models where the input can have structural errors. This is a significant drawback in our case, as we cannot expect the graphical recognition step of the OMR pipeline to be completely accurate. In addition to this issue, SMT techniques also require an additional feature engineering process for both the source and the target languages, as we are dealing with data-driven models which might not get their best results by just inputting raw sequences. This preprocessing requirement implies an additional workload that may become unfeasible if the considered encodings got extended.

For all the above, we decided to implement Neural Machine Translation (NMT). As other knowledge areas, the Machine Translation community has also moved towards DL techniques to perform automatic translation between languages. These neural models typically need more training data than SMT. However, they produce models that have proven to be more robust in the musical context [26], one aspect that we discuss further below. Another benefit of integrating these systems into the OMR pipeline is that they share technological features with the previously performed steps, so it is possible to easily produce a complete system that includes both the recognition steps and the translation one, which can be packaged to be served in practical applications. Therefore, given the advantages that this approach offers, we propose to tackle the encoding step via Machine Translation techniques, specifically with NMT.

2.1   Target Encoding Format

One relevant goal of this work is to showcase a suitable music notation format to be used as the target semantic encoding of the NMT process. We have analyzed which format is more beneficial in terms of exportability and later compatibility with musicology tools, which are the target destinations of our pipeline outputs.

The first options that may be considered are the most extended semantic encodings in music information retrieval and musicology contexts: MEI [14] and MusicXML [11], which represent music score components and metadata in XML languages. These can be understood as analogous markup-based encoding languages such as TEI [4] in the text recognition context. Despite being comprehensive formats, these semantic representations have a significant drawback in a Machine Translation context: they are highly verbose. This means that the target language would require a huge number of tokens for even small music excerpts, thereby making the automatic encoding task unnecessarily complicated.

Previous works on this area have proposed alternatives to tackle this issue [25], such as using *Humdrum **kern* [15]. This is a robust and widely-used semantic encoding for many musicological projects. Its benefits for our purpose lie in a simple vocabulary, a sequential-based format, and in its compatibility with dedicated musicology software like Verovio Humdrum Viewer [28], which brings the possibility of automatically converting into other formats.

For all the above, we selected **kern as our target semantic encoding language. An example of the convenience of this format over other XML-based ones, like MEI, is shown in Fig. 1.

3   Methodology

We define a complete-pipeline OMR as a process that eventually exports the written notation in a standard digital format. Our methodology assumes an

(a) Example music excerpt

```
... <music> ↵ <body> ↵ <mdiv> ↵ <score> ↵
<scoreDef xml:id="scoredef-0000000430793170" key.sig="2s" meter.sym="common">
<scoreDef xml:id="scoredef-0000000430793170" key.sig="2s" meter.sym="common">
<staffGrp xml:id="staffgrp-0000000321535565">
<staffGrp xml:id="staffgrp-0000000321535565">
<staffDef xml:id="staffdef-0000000979385103" clef.shape="G"
 clef.line="2" n="1" lines="5" />
</staffGrp> ↵ </scoreDef> ↵ <section xml:id="section-0000002102168953"> ↵
<measure xml:id="measure-0000000817881159" right="single">
<staff xml:id="staff-0000000752632627" n="1">
<layer xml:id="layer-0000001525105800" n="1">
<note xml:id="note-0000000088370008" dur="2" oct="5" pname="d" tie="i" />
<beam xml:id="beam-0000000838622227">
<note xml:id="note-0000001323524379" dur="16" oct="5" pname="d" tie="t" />
<note xml:id="note-0000000788593928" dur="16" oct="5" pname="e" />
<note xml:id="note-0000001776562259" dur="16" oct="5" pname="d" />
<note xml:id="note-0000000069259125" dur="16" oct="5" pname="c" />
</beam> ...
```

(b) MEI representation of the music excerpt ('↵' represents the end-of-line character.)

```
**skern ↵ *clefG2 ↵ *k[f#c#] ↵ *met(C) ↵2dd[ ↵ 16dd] ↵ 16ee ↵ 16dd ↵ 16cc#
```
(c) **kern representation of the music excerpt ('↵' represents the end-of-line character.)

Fig. 1. Example of MEI and **kern representations of a simple music excerpt, show-casing the different verbosity between formats.

initial pre-process to divide a full-page score into a sequence of staves, much in the same way as HTR typically assumes a previous text-line segmentation [27]. This is not a strong assumption as there exist specific layout analysis algorithms for OMR that decompose the image into staves [24].

Our OMR pipeline is divided here into a two-step procedure that first recovers the graphical information and then performs a proper encoding. Formally, let $\mathcal{X}$ be the input image space of single-staff sections, and Σ_a and Σ_s be denoted as the alphabet of *agnostic* symbols and the alphabet of *semantic* symbols, respectively. Then, the OMR system becomes a *graphical recognition* $f_g : \mathcal{X} \to \Sigma_a$ followed by a *translation process* $f_t : \Sigma_a \to \Sigma_s$. An overview of the methodology is illustrated in Fig. 2.

Additionally, a *direct encoding* approach $f_d : \mathcal{X} \to \Sigma_s$ will be proposed as a baseline for our experimentation, in order to demonstrate the benefits of the two-step strategy.

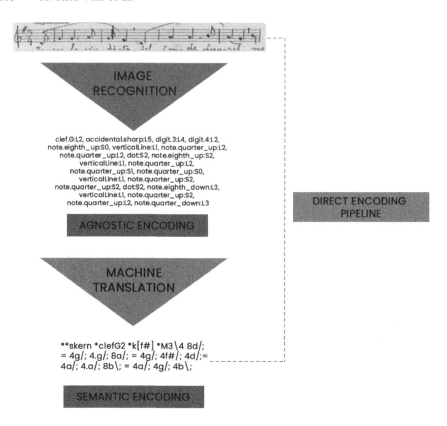

Fig. 2. Overview of the procedures proposed for complete OMR, receiving a staff-section image as input and predicting a semantic music encoding sequence as output.

3.1   Graphical Recognition

The graphical recognition step f_g takes an input image and produces a sequence of agnostic symbols. Given an input staff-section image $x \in \mathcal{X}$, f_g seeks for the sequence $\hat{s}$ such that

$$\hat{s} = \arg\max_{\mathbf{s} \in \Sigma_a^*} P(\mathbf{s} \mid x). \tag{1}$$

To implement this step, we consider the state-of-the-art model OMR, which consists of a CRNN model trained with a CTC loss function. We follow the configuration specified in [6].

The convolutional block learns relevant features of the image and the recurrent block interprets them as a sequence of frames. The model eventually computes the probability of each symbol appearing in each input frame. To approximate $\hat{s}$ of Eq. 1, we resort to a *greedy* decoding algorithm that takes the most likely symbol per frame, concatenates consecutive frames with the same symbol, and then removes the 'blank' symbol introduced to train the model with the CTC loss function [12].

3.2   Translation Process

The graphical recognition produces a discrete sequence of agnostic symbols, where just the shape and the position within the staff are encoded (graphical features). As discussed above, this is insufficient to retrieve meaningful music information, so we need an additional step to obtain a semantic output.

Given a sequence of agnostic symbols, $\mathbf{s} \in \Sigma_a^*$, the translation step f_t can be expressed as seeking the sequence $\hat{\mathbf{t}}$, such that

$$\hat{\mathbf{t}} = \arg\max_{\mathbf{t} \in \Sigma_s^*} P(\mathbf{t} \mid \mathbf{s}). \tag{2}$$

We compute this probability by means of NMT. Given the novelty of this approach in the context of OMR, we consider two alternatives, whose effectiveness will be analyzed empirically.

The first approach is a Seq2Seq model with Attention mechanisms, hereafter referred to as *Seq2Seq-Attn*. This model is an encoder-decoder approach based on Recurrent Neural Networks (RNN), where the first part (the encoder) creates an embedding representation of the input sequence, usually known as the context vector, and the decoder produces, from this context vector and the previously predicted tokens, the next token of the translated sequence. Specifically, we resort to an advanced strategy which implements attention mechanisms: the "Global Attention" approach, proposed by Luong et al. [18], where the previously mentioned context vector is regulated by an attention matrix, whose scoring regulators are given by the scalar product between the encoder and the decoder outputs.

The second considered model is the *Transformer* [30], that currently represents the backbone of state-of-the-art NMT. This model implements an encoder-decoder architecture, such as the previously described system, that replaces all the recurrent layers by attention-based ones, which are referred to in the literature as the multi-headed attention units. These units are not only able to compute faster the training process (as they are easily parallelizable) but have also proven to obtain better quality context vectors and translation decodings, which allows them to learn relevant grammatical and semantic information from the input sequences themselves.

In both cases, the specific configuration is set as done in our previous work [26], where good results for processing music encoding formats were attained.

3.3   Direct Encoding

A direct encoding performs a function $f_d : \mathcal{X} \to \Sigma_s$. Formally, given an input staff-section image $x \in \mathcal{X}$, it seeks for a sequence $\hat{\mathbf{t}}$ such that

$$\hat{\mathbf{t}} = \arg\max_{\mathbf{t} \in \Sigma_s^*} P(\mathbf{t} \mid x) \tag{3}$$

As far as we know, there is no single-step complete OMR system in the literature. In our case, we decided to implement the CRNN-CTC model used for image recognition (Sect. 3.1), but modifying the output alphabet to be that of the semantic output.

This implementation establishes a good comparison baseline, as it is the easiest and simplest model to implement and reduces the number of steps to one.

4   Experimental Setup

In this section, we present our experimental environment to evaluate the OMR pipelines. We detail the corpora used to perform and the evaluation process considered to obtain the results presented in Sect. 5.

4.1   Corpora

Two corpora of music score images, with varying features in printing style, have been used to assess and discuss the performance of the different pipelines.

The first considered corpus is the "Printed Images of Music Staves" (PrIMuS) dataset; specifically, the camera-based version [7]. It consists of 87, 678 music incipits[1] from the RISM collection [1]. They consist of music scores in common western modern notation, rendered with Verovio [23] and extended with synthetic distortions to simulate the imperfections that may be introduced by taking pictures of sheet music in a real scenario, such as blurring, low-quality resolutions, and rotations.

The second considered corpus is a collection of four groups of handwritten score sheets of popular Spanish songs taken from the 'Fondo de Música Tradicional IMF-CSIC' (FMT),[2] that is a large set of popular songs manually transcribed by musicologists between 1944 and 1960.

The characterization of these corpora can be found in Table 1, while representative examples are shown in Fig. 3 and Fig. 4 for PrIMuS and FMT, respectively, along with agnostic and semantic annotations.

4.2   Evaluation Process

One issue that one may find when performing OMR experiments is to correctly evaluate the performance of a proposed model, as music notation has specific features to take into account. However, OMR does not have a standard evaluation protocol [6]. In our case, it seems convenient to use text-related metrics to approach the accuracy of the predictions. Despite not considering specific music features, in practical terms, we are dealing with text sequences.

---

[1] Short sequence of notes, typically the first ones, used for identifying a melody or musical work.

[2] https://musicatradicional.eu.

Table 1. Details of the considered corpora.

	PrIMuS	FMT
Engraving	Printed	Handwritten
Size of the corpus (staves)	87,678	872
Agnostic vocabulary size ($|\Sigma_a|$)	862	266
Semantic vocabulary size ($|\Sigma_s|$)	1,421	206
Running symbols (agnostic)	2,520,245	18,329
Running symbols (semantic)	2,425,355	18,616

(a) Staff-section image.

```
clef.G-L2, accidental.sharp-L5, accidental.sharp-S3, accidental.sharp-S5,
accidental.sharp-L4, digit.2-L4, digit.4-L2, note.quarter-L3, note.eighth-S2,
dot-S2, note.sixteenth-L2, barline-L1
```
(b) Agnostic encoding of the first bar.

```
**skern ↵ *clefG2 ↵ *k[f#c#g#d#] *M2/4 ↵ 4b ↵ 8.a ↵ 16g#y ↵ = ↵
```
(c) Semantic encoding of the first bar ('↵' represents the end-of-line token).

Fig. 3. Selected example from PrIMuS: Incipit RISM ID no. 000102547, Incipit 1.1.1 *Peace troubled soul whose plaintive moan.* Anonymous.

(a) Staff-section image.

```
clef.G:L2, accidental.flat:L3, ⋯ , note.beamedRight2_up-S2,
note.beamedBoth2_up-L3, note.beamedLeft2_up-S2, ⋯
```
(b) First tokens of the agnostic encoding.

```
**skern ↵ *clefG2 ↵ *k[] ↵ 4b- ↵ ⋯ 24aL ↵ 24b ↵ 24aJ ⋯
```
(c) First tokens of the semantic encoding.

Fig. 4. Selected example from FMT (Canción de Trilla) [9]

For the above, we measured the performance of the proposed models with the Sequence Error Rate (SER). Let H be the predicted sequence and R the reference one, this metric computes the edit distance between H and R and divides it by the length (in tokens) of R. We chose this metric as it both represents accurately the performance of the model in recognition tasks and correlates with the effort a user would have to invest to manually correct the results.

To obtain a more robust approximation of the generalization error, we follow a 5-fold cross-validation process, where the resultant SER is the average of the produced test error within the five data partitions.

5 Results

The experimentation results are given in Table 2, comparing the proposed two-step approach with a direct encoding, that acts as a baseline. We also report the intermediate results of the former, to provide more insights. In the case of the translation process, the intermediate results show the SER obtained starting from a ground-truth agnostic sequence.

Concerning the intermediate results, it can be observed that the graphical recognition step performs well on the printed dataset and gets much worse results in the handwritten one, as might be expected in terms of their training set size and complexity. In the translation task, the tendency is similar but this time we observe a more reasonable SER in both cases. The Transformer is the best only-translation option when there is enough training data, while the Seq2Seq-Attn results better in the case of limited training data. As discussed next, this fact does not extrapolate to the complete pipeline.

If we analyse the complete pipeline, the results obtained using the combination of CRNN and NMT models outperform the direct encoding approach, both in the PrIMuS and the FMT dataset. The difference is especially significant in the handwritten small-sized corpus FMT, where the SER of the CRNN+Seq2Seq-Attn outperforms the direct encoding approach by a wide margin (around 20% of SER). One interesting fact from these results is that the NMT models can deal reasonably well with the inconsistencies introduced during the graphics recognition, as we observe that the final SER figures are much more correlated to the graphical recognition than to the translation process.

Furthermore, it is interesting to note that the Transformer is the most NMT accurate model when translating from ground-truth data. However, if we pay attention to the complete pipeline, it does not produce a model as robust to inconsistencies as the Seq2Seq-Attn model does. This scenario, in practical terms, is the most frequent in OMR, where the graphical recognition step tends to make mistakes. Therefore, the Seq2Seq-Attn approach is, as far as our results generalize, the most suitable alternative for the translation process in the two-step pipeline.

Table 2. Average SER (%) over the test set. The table shows the error amount produced in the recognition and encoding steps (as they have been trained separately) and the final error done by the complete pipeline, which receives an image as input and a semantic sequence as output. We highlighted in bold typeface the results that show better performance in the complete pipeline.

	PrIMuS	FMT
Intermediate results		
Graphical recognition (CRNN)	3.52	34.88
Translation process w/ Seq2Seq-Attn	2.04	9.53
Translation process w/ Transformer	0.53	15.43
Complete pipeline		
CRNN + Seq2Seq-Attn	**4.28**	**36.76**
CRNN + Transformer	6.35	38.88
CRNN Direct encoding (baseline)	4.66	52.24

Despite the aforementioned evidence, some doubts may appear referring to the error fluctuation between the presented pipelines, as we observe a drastic change in the performance between the two datasets. To further analyze the situation, we repeated the same experimentation in reduced versions of the PriMUS dataset, where we try to find an intermediate point between FMT and PrIMuS complexities. This resulted in three new corpora with 10, 000, 5, 000 and 1,000 *samples*, (the FMT corpus has nearly 900 samples). The obtained results are graphically shown in Fig. 5. It can be observed that the tendency described from the original PrIMUS results, where the CRNN+Transformer performed the worst, is maintained until dropping to 5, 000 samples, where the direct approach is then outperformed by it. In all cases, however, the CRNN+Seq2Seq-Attn is postulated as the best option by different margins, depending on the complexity of the dataset.

This new experiment summarized the behaviour of all alternatives. On the one hand, a direct encoding pipeline—which acted as baseline—depends highly on the amount of training data, attaining competitive results in such case. On the other hand, the two-step process, especially when using the Seq2Seq-Attn as translation mechanisms, clearly stands for the best option when training data is limited, also reaching the best performance when the training set is of sufficient size.

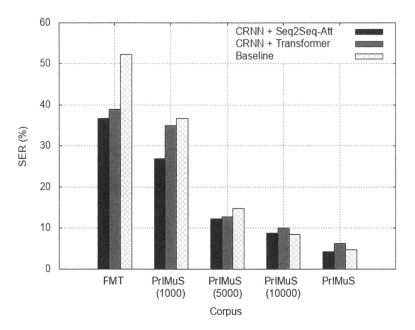

Fig. 5. Graphic bar plot comparison of the average SER produced by the proposed pipelines with the different corpora, which consists in the initially proposed datasets and two reductions on PrIMUS size in order to establish intermediate points between the handwritten and the printed corpus. The Baseline results refer to the Direct Encoding Encoding approach described in other sections.

6   Conclusions

We studied in this paper the development of complete OMR pipelines, which receive a music staff image as input and produce a standard music encoding sequence as output. We discussed how to approach this task by proposing a two-step pipeline based on a state-of-the-art image recognition model in OMR combined with NMT solutions for the encoding step. We also included a direct encoding pipeline that outputs directly the final encoding from the image. To evaluate these approaches, we experimented in two corpora of varying characteristics. After the experimentation, we observed different aspects about how these approaches perform over different corpora, where we obtained a relevant idea that outlines this work: the two-step pipeline with NMT is a good option when the target corpus to digitize does not have enough labeled data for learning the inherent complexity of the OMR, which is, in fact, an interest aim of this paper.

From a practical perspective, specifically in the case of early music heritage, it is common to find scenarios where manual data labeling is required in order to constitute a corpus before using OMR tools. As we saw in our experimentation, the OMR processes that include NMT models to perform the encoding step behave reasonably well in this case. This feature involves a great practical advantage for these scenarios, as there is no need to label a vast amount of data to start using this tool. However, the two-step pipeline also has a considerable drawback: the corpus has to be labeled in two encoding languages (agnostic and semantic) in order to make it work. Despite this issue, there are possible ways of mitigation because the translation process does not depend on a specific manuscript; therefore, just one pretrained translation model, relieving the effort of manually creating the semantic annotation.

Despite the advances presented in this paper, we consider that further research is required to maximize the benefits this approach might bring, as this paper only proves that it is a feasible option for cases where the corpus does not provide enough data. This future research may focus on different topics such as improving the robustness to input inconsistencies of the NMT models (especially the Transformer) with data augmentation, the modelling of cohesive vocabularies to obtain more profit from the encoding models, or the study on how to integrate these systems to produce a single-step OMR pipeline with a dual training process.

Acknowledgments. This work was supported by the Generalitat Valenciana through project GV/2020/030.

References

1. Répertoire International des Sources Musicales (RISM) Series A/II: Music manuscripts after 1600 on CD-ROM. Technical report (2005)
2. Baró, A., Badal, C., Fornés, A.: Handwritten historical music recognition by sequence-to-sequence with attention mechanism. In: 2020 17th International Conference on Frontiers in Handwriting Recognition (ICFHR), pp. 205–210 (2020)
3. Burgoyne, J.A., Devaney, J., Pugin, L., Fujinaga, I.: Enhanced bleedthrough correction for early music documents with recto-verso registration. In: Bello, J.P., Chew, E., Turnbull, D. (eds.) ISMIR 2008, 9th International Conference on Music Information Retrieval, Drexel University, Philadelphia, PA, USA, 14–18 September 2008, pp. 407–412 (2008)
4. Burnard, L., Bauman, S. (eds.): A gentle introduction to XML. Text encoding initiative consortium. In: TEI P5: Guidelines for Electronic Text Encoding and Interchange (2007). http://www.tei-c.org/release/doc/tei-p5-doc/en/html/SG.html
5. Byrd, D., Simonsen, J.: Towards a standard testbed for optical music recognition: definitions, metrics, and page images. J. New Music Res. **44**, 169–195 (2015). https://doi.org/10.1080/09298215.2015.1045424
6. Calvo-Zaragoza, J., Hajič Jr., J., Pacha, A.: Understanding optical music recognition. ACM Comput. Surv. **53**(4) (2020)
7. Calvo-Zaragoza, J., Rizo, D.: Camera-PrIMuS: neural end-to-end optical music recognition on realistic monophonic scores. In: Proceedings of the 19th International Society for Music Information Retrieval Conference, ISMIR 2018, Paris, France, 23–27 September 2018, pp. 248–255 (2018)
8. Calvo-Zaragoza, J., Toselli, A.H., Vidal, E.: Handwritten music recognition for mensural notation with convolutional recurrent neural networks. Pattern Recogn. Lett. **128**, 115–121 (2019)

9. Clares Clares, E.: Canción de trilla. Fondo de música tradicional IMF-CSIC. https://musicatradicional.eu/es/piece/12551. Accessed 01 Feb 2021

10. Dalitz, C., Droettboom, M., Pranzas, B., Fujinaga, I.: A comparative study of staff removal algorithms. IEEE Trans. Pattern Anal. Mach. Intell. **30**(5), 753–766 (2008)

11. Good, M., Actor, G.: Using MusicXML for file interchange. In: International Conference on Web Delivering of Music 0, 153 (2003)

12. Graves, A., Fernández, S., Gomez, F.J., Schmidhuber, J.: Connectionist temporal classification: labelling unsegmented sequence data with recurrent neural networks. In: Proceedings of the Twenty-Third International Conference on Machine Learning, (ICML 2006), Pittsburgh, Pennsylvania, USA, 25–29 June 2006, pp. 369–376 (2006)

13. Hajic, J., Pecina, P.: The MUSCIMA++ dataset for handwritten optical music recognition. In: ICDAR (2017)

14. Hankinson, A., Roland, P., Fujinaga, I.: The music encoding initiative as a document-encoding framework. In: Proceedings of the 12th International Society for Music Information Retrieval Conference (2011)

15. Huron, D.: Humdrum and Kern: Selective Feature Encoding, pp. 375–401. MIT Press, Cambridge (1997)

16. Koehn, P.: Statistical Machine Translation. Cambridge University Press, Cambridge (2009)

17. Koehn, P., et al.: Moses: open source toolkit for statistical machine translation. In: Proceedings of the 45th Annual Meeting of the Association for Computational Linguistics Companion Volume Proceedings of the Demo and Poster Sessions, pp. 177–180. Association for Computational Linguistics, Prague (2007)

18. Luong, T., Pham, H., Manning, C.D.: Effective approaches to attention-based neural machine translation. In: Màrquez, L., Callison-Burch, C., Su, J., Pighin, D., Marton, Y. (eds.) Proceedings of the 2015 Conference on Empirical Methods in Natural Language Processing, EMNLP 2015, Lisbon, Portugal, 17–21 September 2015, pp. 1412–1421. The Association for Computational Linguistics (2015)

19. Pacha, A., Calvo-Zaragoza, J., Hajič jr., J.: Learning notation graph construction for full-pipeline optical music recognition. In: 20th International Society for Music Information Retrieval Conference, pp. 75–82 (2019)

20. Pacha, A., Eidenberger, H.: Towards a universal music symbol classifier. In: 14th International Conference on Document Analysis and Recognition, pp. 35–36. IAPR TC10 (Technical Committee on Graphics Recognition), IEEE Computer Society, Kyoto (2017)

21. Pacha, A., Hajič, J., Calvo-Zaragoza, J.: A baseline for general music object detection with deep learning. Appl. Sci. **8**(9), 1488 (2018)

22. Parada-Cabaleiro, E., Batliner, A., Schuller, B.W.: A diplomatic edition of il lauro secco: ground truth for OMR of white mensural notation. In: Flexer, A., Peeters, G., Urbano, J., Volk, A. (eds.) Proceedings of the 20th International Society for Music Information Retrieval Conference, ISMIR 2019, Delft, The Netherlands, 4–8 November 2019, pp. 557–564 (2019)

23. Pugin, L., Zitellini, R., Roland, P.: Verovio: a library for engraving MEI music notation into SVG. In: Proceedings of the 15th International Society for Music Information Retrieval Conference, pp. 107–112. ISMIR, October 2014

24. Quirós, L., Toselli, A.H., Vidal, E.: Multi-task layout analysis of handwritten musical scores. In: Morales, A., Fierrez, J., Sánchez, J.S., Ribeiro, B. (eds.) IbPRIA 2019. LNCS, vol. 11868, pp. 123–134. Springer, Cham (2019). https://doi.org/10.1007/978-3-030-31321-0_11

25. Ríos-Vila, A., Calvo-Zaragoza, J., Rizo, D.: Evaluating simultaneous recognition and encoding for optical music recognition. In: 7th International Conference on Digital Libraries for Musicology, DLfM 2020, pp. 10–17. Association for Computing Machinery, New York (2020)
26. Ríos-Vila, A., Esplà-Gomis, M., Rizo, D., Ponce de León, P.J., Iñesta, J.M.: Applying automatic translation for optical music recognition's encoding step. Appl. Sci. **11**(9) (2021)
27. Sánchez, J., Romero, V., Toselli, A.H., Villegas, M., Vidal, E.: A set of benchmarks for handwritten text recognition on historical documents. Pattern Recognit. **94**, 122–134 (2019)
28. Sapp, C.S.: Verovio humdrum viewer. In: Proceedings of Music Encoding Conference (MEC), Tours, France (2017)
29. Tuggener, L., Elezi, I., Schmidhuber, J., Stadelmann, T.: Deep watershed detector for music object recognition. In: 19th International Society for Music Information Retrieval Conference, Paris, 23–27 September 2018 (2018)
30. Vaswani, A., et al.: Attention is all you need (2017)

Improving Machine Understanding of Human Intent in Charts

Sihang Wu[1], Canyu Xie[1], Yuhao Huang[1], Guozhi Tang[1], Qianying Liao[1],
Jiapeng Wang[1], Bangdong Chen[1], Hongliang Li[1], Xinfeng Chang[3], Hui Li[3],
Kai Ding[4], Yichao Huang[4], and Lianwen Jin[1,2(✉)]

[1] South China University of Technology, Guangzhou, China
`eelwjin@scut.edu.cn`
[2] Guangdong Artificial Intelligence and Digital Economy Laboratory (Pazhou Lab),
Guangzhou, China
[3] Lenovo Reserach, Beijing, China
[4] IntSig Information Co., Ltd., Shanghai, China

Abstract. Charts composed of images and text are a classical and compact method of displaying and comparing various data. Automated data extraction from charts is critical in the machine understanding and recognition of human intent and the meaning inherent in charts. Complex processes of automated chart data extraction have been divided into multiple basic tasks. However, these problems have not been solved well. In this paper, we principally focus on three key sub-tasks, including chart image classification (Task-1), text detection and recognition (Task-2), and text role classification (Task-3). For these tasks, we design and propose a set of effective methods. The experiments on the Adobe Synthetic and PubMedCentral datasets successfully demonstrate the effectiveness of our proposed systems. Notably, our proposed method outperforms competing systems from the ICDAR 2019 and ICPR 2020 CHART-Infographics competitions and achieved state-of-the-art performance. We hope this work will serve as a step toward enhancing the machine understanding of charts and inspiring new avenues for further research in the field of document analysis and recognition.

Keywords: Document analysis and recognition · Chart image classification · Text detection and recognition · Text role classification

1 Introduction

Charts have long been a classically compact method of displaying and comparing data. In scientific publications, charts and graphs are often used to summarize results, compare methodologies, and design guidance frameworks, among various other applications. In recent years, chart recognition and understanding have gained increasing attention.

This research is supported in part by NSFC (Grant No.: 61936003, 61771199), GD-NSF (no. 2017A030312006).
S. Wu, C. Xie, Y. Huang and G. Tang—Equal contribution.

J. Lladós et al. (Eds.): ICDAR 2021, LNCS 12823, pp. 676–691, 2021.
https://doi.org/10.1007/978-3-030-86334-0_44

The ICDAR 2019 CHART-Infographics [7] and ICPR 2020 CHART Info-graphics[1] [9], provide benchmarks for comparing the effectiveness of different methods on chart data extraction. The benchmarks provide two independent datasets, namely Adobe Synthetic and PubMedCentral (PMC). Each dataset represents a different type of challenges and includes training and testing datasets. The Adobe Synthetic dataset is based on several synthetic chart images (created with Matplotlib by Adobe) with corresponding automatically derived annotations. The PMC dataset is curated from real charts in scientific publications from PubMedCentral[2] with manual annotations. Some representative examples are shown in Fig. 1.

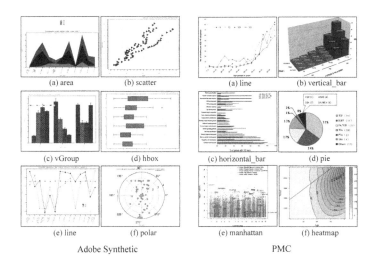

Fig. 1. Some examples of the Adobe Synthetic and PMC datasets.

According to the ICDAR 2019 [7] and ICPR 2020 CHART-Infographics (see Footnote 1) results, there remains considerable room for improvement on the benchmarks, especially the detection and recognition results under high intersec-tion over union (IOU) metric on the PMC dataset. In this paper, we principally focus on three key sub-tasks outlined as chart image classification (Task-1), text detection and recognition (Task-2) and text role classification (Task-3).

Task-1 classifies given chart images with various categories on the Adobe Synthetic and PMC datasets.

Task-2 performs text detection and recognition at the logical element level; that is, multi-line and multi-word titles or axis tick labels were detected as a single element.

Task-3 identifies the semantic role of text in chart interpretation. Taking text bounding boxes and transcripts as input, the model aims to accurately classify

---

[1] https://chartinfo.github.io.
[2] https://www.ncbi.nlm.nih.gov/pmc/.

each bounding box into one of the nine given roles (2020 PMC Datasets) or five (2020 Adobe Synth Datasets). More details can be found in [7,9].

These sub-tasks are key steps of automatically extracting data from charts, which has the potential to lead toward better machine understanding and recognition of the meaning and embedded context of a document.

In this paper, we design and propose a set of effective methods with targeted improvements for the aforementioned three sub-tasks. The experiments on the Adobe Synthetic and PMC chart datasets demonstrate the good performance of our systems. Critically, our proposed method outperforms the competing systems from the ICDAR 2019 and ICPR 2020 CHART-Infographics competitions.

The rest of the paper is organized as follows. Section 2 presents the recent developments of these three tasks in the relevant literature. Section 3 introduces our methods in detail, and Sect. 4 reports the experimental results and analysis on the ICDAR 2019 and ICPR 2020 chart datasets. Finally, we present our conclusions in Sect. 5.

2   Related Work

Classifying types of charts and processing charts to extract data have been of considerable recent interest in the field of Document Analysis and Recognition (DAR) [3,19,20,22]. Among them, several have been dedicated to chart mining, which refers to the process of automatic detection, information extraction, and analysis of charts to reproduce the tabular data that was originally used to create them. A recent survey on chart mining [8] presented comprehensive information on approaches across all components of the automated chart mining pipeline.

In parallel with the recent acceleration of the development of deep learning, several studies have been conducted to improve performance on chart classification tasks via deep neural networks. By focusing on the representation scalability and stability of charts, Tang et al. [28] proposed a novel framework to classify charts by combining convolutional networks and deep belief networks. Bajić et al. [2] used a simplified VGG model to implement chart classification. Araújo T et al. [1] decomposed chart classification tasks into three sub-tasks of classification, detection, and perspective correction, and then realized chart classification at the level of entire images. However, these studies can not handle well with the complex and diverse layout of chart images, especially on the PMC datasets.

Viewing the text as a specific type of object, Chen et al. [7] adopted Faster R-CNN [23] network with suitably modified anchor boxes and connected component analysis for detection, and attention-based modules [25] were used to handle text in multiple orientations for recognition. Their system performed best on the Adobe Synthetic dataset in ICDAR 2019 CHART-Infographics competition. However, Faster R-CNN can not meet the need of tighter bounding boxes under high IOU. Moreover, it can not handle the multi-oriented text well. Joao Pinheiro et al. [7] used PixelLink text detection system [10], pretrained on ICDAR 2015 scene text dataset [18], and was finetuned with the training data provided. For text recognition, Tesseract OCR [26] was used. They won the first place on the

PMC dataset in ICDAR 2019 CHART-Infographics competition. However, they only got the average IOU scores of 48.48%, which means there is still considerable room for improving tighter bounding boxes under high IOU.

For text role classification, Chen et al. [7] used gradient boosting decision tree trained on 20 features, where direction of text, relative position between text and axis/legend was used. Their system showed great performance on the Adobe Synthetic dataset in the ICDAR 2019 CHART-Infographics competition. However, this method can not accurately classify the semantic role of text in complex and diverse layout of PMC chart images. Joao Pinheiro et al. trained SVM on geometric features extracted from text bounding boxes [21], which worked better on the PMC dataset. However, it is easy to be misclassified categories by only using geometric features due to their similarity in layout organization. Therefore, multi-modal information is beneficial for accurately classifying semantic roles. In recent years, LayoutLM [29] model took the advantages of large-scale pretrained models, and integrated visual, semantic, layout, and other information, which showed superior performance on text classification and document image understanding.

3 Method

3.1 Task-1: Chart Image Classification

Approach. As a classification problem, we choose DenseNet121 [16] as our backbone. To improve the performance, we add the SE module [15] to the network. Moreover, a series of training strategies are used to train our model, including auto augmentation [6], label smoothing [27], and avoidance of bias decay [17]. An ablation study and its details are presented in Sect. 4.1.

We carefully analyze the aspect ratio of charts and their category distribution. An appropriate input size of the image is 450×450. Notably, there is a class imbalance problem on PMC dataset, as shown in Fig. 2.

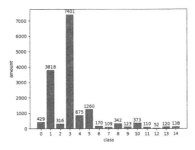

Fig. 2. Category distribution of the 2020 PMC dataset.

Considering this serious class imbalance problem, we adopt an oversampling strategy as a solution, similar to [12]. First, we calculate the frequency of each category, followed by the repeat factor of each category according to Eq. 1

$$r(c) = max(1, \sqrt{\frac{t}{f(c)}}) \qquad (1)$$

Here, $r(c)$ represents the repeat factor of category c, t represents the threshold, which is selected as the frequency of rare categories, and $f(c)$ is the frequency of a given category c. Subsequently, we sample each image according to the repeat factor $r(c)$.

3.2   Task-2: Text Detection and Recognition

Approach. The entire process flow pipeline of our proposed text detection and recognition model is shown in Fig. 3. We divide Task-2 in two, considering the sub-tasks of text detection and text recognition.

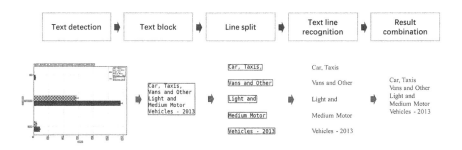

Fig. 3. Pipeline of Task-2.

Text Detection. The layout of Adobe Synthetic dataset is relatively clean and concise. Moreover, texts are all annotated in rectangle boxes. Therefore, it allows us to design a simple and effective detector by utilizing a general object detection framework called Faster R-CNN. We adopt Faster R-CNN with ResNet-101 [14] as a backbone and suitably modified anchor boxes to handle the text's significant variation in aspect ratios. Further, to enable high quality detection of extremely small text; such as axis values, we adopt the Cascade R-CNN [4] architecture with five detector heads trained with IOU thresholds of 0.5-0.9, as shown in Fig. 4. Each detector head aims to determine a good set of close false positives for training the next stage. This enables higher detection accuracies as the quality of the detection hypotheses increases gradually.

In the PMC dataset, charts have various layouts, whereas they are more complicated in the real scenarios, especially when including multi-oriented text. However, we can slightly modify our method using in the Adobe Synthetic dataset to handle the multi-oriented text. We add a mask head in each detector head

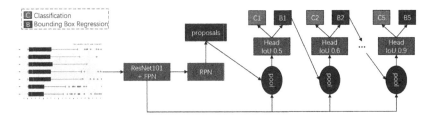

Fig. 4. Modified cascade R-CNN with five detector heads.

stage that trained with different IOU thresholds, also known as Cascade Mask R-CNN [4]. By visualizing the annotation of the PMC dataset, the bounding boxes are observed to not as tight as the Adobe Synthetic dataset. Therefore, we only adopt two detector heads trained with IOU thresholds of 0.5 and 0.6.

Text Recognition. For both the Adobe Synthetic and PMC datasets, the detected results of the text regions are text blocks, simplifying that the detected text regions may include more than one text line. Therefore, before the text line recognition, we first use the text line split operation to extract the text lines from text blocks, and then recognize each of the text lines.

For the text line split, we first obtain the inclination angle of the text blocks by detecting the minimum enclosing rectangle of the inclined text blocks via the Hough line detection algorithm [11]. Then, we can rotate the text blocks to the horizontal direction. Subsequently, we only need to project the text blocks horizontally to split the text line, as shown in Fig. 5.

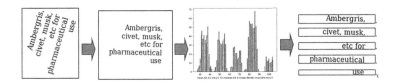

Fig. 5. Pipeline of text line extraction.

For the text line recognition, we adopt an ensemble including Convolutional Recurrent Neural Network with Connectionist Temporal Classification [24] (CRNN+CTC) based model and a CRNN+Attention based model. As in [5], a ResNet-based backbone extras features in both models. In addition, a Bidirectional Long Short-Term Memory (BiLSTM) is selected for the sequence modeling stage.

3.3 Task-3: Text Role Classification

Approach. The pipeline is shown in Fig. 6. Specifically, we first embed each word and its position with given OCR results. Then we extract word information and multi-modal information (including position, chart type) features through LayoutLM, and integrate them in each text block to bring richer context semantic representations. At the same time, a ResNet backbone with Roi Align [13] is used to extract the visual features of each text block in an image. Subsequently, the visual features are fused with the context features at text block level. Next, a standard self-attention module is used to learn the global local and non-local features. Finally, fully-connected layers output text role classification results.

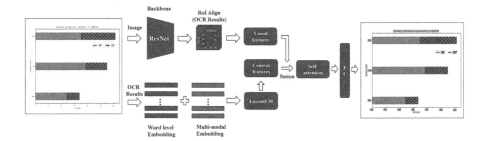

Fig. 6. Text role classification pipeline.

4   Experiments

4.1   Task-1

Evaluation Metric. Based on the class confusion matrix, we can calculate the precision, recall and F-measure of each class. The total score is the average of the F-measure for each class.

Experimental Details. The 2020 Adobe Synthetic training dataset contains 14412 chart images of 12 categories, while the 2020 PMC training dataset contains 15651 chart images of 15 categories. Similar to [21], we collected 980 chart images from [21] to improve the model performance of the PMC dataset. To evaluate our proposed method, we divided the training dataset into training and validation sets by a ratio of 7:3. Finally, we compared our method with the competing teams on the ICDAR 2019 Adobe Synthetic and 2020 PMC test sets.

We use the Adam optimizer with a learning rate and weight decay of 1e-4, and a step decay schedule to adjust the learning rate, which decays by 0.8 every 5 epochs. The batch size is set to 16 with 450×450 image resolution input, and we train our model with a single NVIDIA 2080Ti GPU. Table 1 shows the results.

Table 1. Results of the chart classification on the 2020 Adobe Synthetic and 2020 PMC validation sets.

Validation set	F-measure
2020 Adobe Synthetic	99.90
2020 PMC	92.89

Table 2. Improvement of training strategies on the 2020 PMC validation set.

Method	F-measure
Baseline	87.18
+auto augmentation	89.70
+avoidance of bias decay	91.02
+label smoothing	91.42
+SE module	91.59
+extra data	92.22
+class balance	**92.89**

We can easily find that the classification task of the PMC dataset is more challenging than Adobe Synthetic dataset, because the layout is more complex and diverse. To evaluate the effectiveness of our proposed network and training strategy, we conduct a series of ablation studies on the PMC dataset. As seen in Table 2, auto augmentation highly improves the performance by 2.52%. Furthermore, the class balance method enhances the performance to 92.89%.

Table 3 shows the promising results of the ICDAR 2019 Adobe Synthetic and ICPR 2020 PMC test sets. Note that in the ICDAR 2019 Adobe Synthetic test set, only 10 categories are considered, i.e., pie, donut, vertical box, horizontal box, grouped vertical bar, grouped horizontal bar, stacked vertical bar, stacked horizontal bar, line and scatter. It is worth to note that, there is no public training set of the ICDAR 2019 PMC; moreover, its categories are inconsistent with the 2020 PMC dataset. Thus, we cannot conduct experiments to compare our method with the ICDAR 2019 competition teams fairly.

As for the results of the 2020 PMC dataset, our proposed model achieves higher performance than other models. Furthermore, we integrate three pre-trained models, and average their prediction scores to calculate the final result. The model ensemble result shows that, compared with another model having the best performance, our proposed approach achieves a higher performance by a margin of 0.24%. Finally, we visualize our prediction result using a confusion matrix, as is illustrated in Fig. 7. Notable, our method can correctly predict every sample in the 2019 Adobe Synthetic test set. However, as for the confusion matrix of the PMC 2020 test set, the area, scatter-line and vertical interval plots are easily misclassified as line plots. This may be because all of these plots have lines, which do not represent their categories.

Table 3. Results of Task-1 on the 2019 Adobe Synthetic and PMC test set.

2019 synthetic test set		2020 PMC test set	
Team	F-measure	Team	F-measure
ABC* (1st)	99.81	Lenovo-SCUT-Intsig[†] (1st)	92.85
A-Team* (2nd)	94.82	Magic[†] (2nd)	90.48
ANU-Team* (3rd)	89.78	DeepBlueAI[†] (3rd)	90.43
Boomerang*	9.59	IPSA[†]	86.33
Ours	**100.00**	Ours	92.42
		Ours (ensemble)	**93.09**

* from [7].

[†] from https://chartinfo.github.io/leaderboards_2020.html.

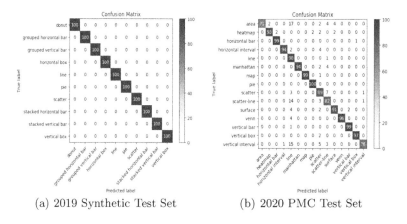

(a) 2019 Synthetic Test Set   (b) 2020 PMC Test Set

Fig. 7. Confusion matrices of 2019 Synthetic and 2020 PMC test sets.

4.2 Task-2

Evaluation Metric. For detection, predicted bounding boxes are compared to a ground truth set, and considered a sample as a match if the IOU exceeded a threshold of 0.5. Tighter detection bounding boxes means higher scores. Per image, the detection scores are averaged by maximum of number of ground truth boxes or number of predicted boxes, i.e., *max(gts, predictions)*.

For recognition, matched pairs are scored by IOU for detection as well as *max(1 - NCER, 0)*, where, NCER is the normalized character error rate measured as the edit distance between ground truth transcripts and predicted transcripts normalized with respect to ground truth string length.

Finally, the harmonic mean of the detection and recognition scores averaged across the entire test data set is presented as an final evaluation metric for Task-2.

Text Detection. In the 2020 Adobe Synthetic dataset, all images have 1280×960 resolution. Therefore, we train the modified Cascade R-CNN with 1280×960 fixed size as input. Some helpful data augmentations are applied with a probability of 0.5, such as color jittering and horizontal flip. In the inference phase, a test image is fed into the network and rectangular text predictions are directly obtained. Then, Non-Maximum Suppression (NMS) post-processing with a threshold of 0.5 is used to acquire the final results.

We divide the ICPR 2020 Adobe Synthetic training dataset of 9600 images, into training and validation sets with a 4:1 ratio. We attempt some targeted improvements on the Cascade R-CNN, such as well designed anchors, more cascade detection heads, a deeper backbone, and random rotation augmentation within $\pm 45°$. Specifically, we add 16:1 and 32:1 extreme aspect radio anchors. We trained the model for 12 epochs with a batch size of 4 on a single NVIDIA 2080Ti GPU. Table 4 shows the ablation experiments to verify the model performance. Because the 2020 Adobe Synthetic test set has not been released, we use the 2019 Adobe Synthetic test set to further verify the effectiveness of targeted improvements. Note that, we did not use the corresponding 2019 Adobe Synthetic data for training. Thus, the result is lower than Table 8 owing to the different data distribution.

From Table 4, we can see that the most powerful strategy is well designed anchors, adding 8% on 2020 validation set and 1.16% on 2019 test set. However, there is only a small gap in performance between other targeted improvements. It seems to reach a bottleneck to some extent, partly because of labeling bias and a slight pixel precision deviation of bounding boxes under the IOU of 0.9. Here, we tend to use group 5's targeted improvements for the trade-off between model capacity and robustness.

Table 4. Ablation experiments of targeted improvements results on the 2020 Adobe Synthetic validation set and 2019 Adobe Synthetic test set (Average IOU %).

Group	Strategy				2020 validation set	2019 test set
	Designed anchors	Cascade heads	Backbone	Rotate transform		
Baseline	(1:2, 1:1, 2:1)	3	ResNet50		73.89	86.94
1	✓	3	ResNet50		85.89	88.10
2	✓	5	ResNet50		86.00	**88.73**
3	✓	5	ResNet18		85.77	88.24
4	✓	5	ResNet101		**86.01**	88.65
5	✓	5	ResNet101	✓	85.95	88.51

In the 2020 PMC dataset, following group 5's targeted improvements in Table 4, we adopt multi-scale training and testing for the Cascade Mask R-CNN as well. In addition, color jittering and horizontal flip data augmentations are applied with a probability of 0.5 as default setting in baseline. In the inference phase, the final results are obtained from the mask head, with connected component analysis and minimum bounding box post-processing.

Table 5. Ablation experiments of targeted improvements results on the 2020 PMC validation set (Average IOU %).

Group	Strategy							2020
	Designed anchors	Cascade heads	Backbone	Rotate transform	Multi-scale training	Multi-scale testing	Others	validation set
Baseline	(1:2, 1:1, 2:1)	3	ResNet50				Confidence > 0.5 padding = 0	85.76
1	✓	3	ResNet50					85.78
2	✓	1	ResNet50					85.76
3	✓	2	ResNet50					86.13
4	✓	2	ResNet101					86.34
5	✓	2	ResNet101	✓				85.35
6	✓	2	ResNet101	✓	✓			85.41
7	✓	2	ResNet101	✓	✓	✓		85.40
8	✓	2	ResNet101	✓	✓	✓	Confidence > 0.8 padding = 1	**90.42**

Similarly, we divide the 2020 PMC training dataset, comprising 3416 images, into a training set and validation set with a 4:1 ratio. Comparing the experiments of groups 1 to 7 and a baseline in Table 5, the results are not much different, and the reason is similar to synthetic dataset. Therefore, we carefully adjust the detection result boundary, with a higher confidence of 0.8 (0.5 of baseline) and padding 1 pixel to expand the boundary. Finally, we obtain 90.42% average IOU on the 2020 PMC validation set.

Table 6. Results of Task-2 on the 2020 Adobe Synthetic and PMC validation sets and 2020 PMC test set.

2020 validation sets			
Dataset	IOU (%)	OCR (%)	F-measure
Adobe Synthetic	85.95	94.50 (CTC)	90.02
	85.95	95.59 (Attention)	90.51
	85.95	**95.78** (Ensemble)	**90.59**
PMC	90.42	95.43 (CTC)	92.86
	90.42	**95.69** (Attention)	**92.97**
	90.42	95.43 (Ensemble)	92.85

Table 7. Results of Task-2 on the 2020 PMC test set.

2020 PMC test set			
Team	IOU (%)	OCR (%)	F-measure
Lenovo-SCUT-Intsig[†] (1st)	74.05	76.45	75.23
Magic[†] (2nd)	72.21	73.51	72.85
DeepBlueAI[†] (3rd)	73.71	58.42	65.18
PY[†]	67.56	60.60	63.89
IPSA[†]	27.47	32.02	29.57
Ours	73.77	76.25 (Ensemble)	74.99
Ours (+extra data)	**74.24**	**76.50 (Ensemble)**	**75.35**

[†] from https://chartinfo.github.io/leaderboards_2020.html.

Text Recognition. On the Adobe Synthetic dataset, we use the text line image cropped from the ground truth for text line recognition model training. The training set includes 2019 and 2020 Adobe Synthetic datasets. We evaluate our recognition method based on the detection result (group 5 in Table 4).

Table 8. Results of Task-2 on the 2019 chart datasets. (* from [7])

Team	IOU (%)	OCR (%)	F-measure
2019 Synthetic test set			
A-Team* (1st)	70.96	78.97	74.75
ABC* (2nd)	86.73	97.61	91.85
Ours	**98.38**	**98.18**	**98.28**
2019 PMC test set			
A-Team* (1st)	48.48	58.81	53.15
Ours	**79.57**	**94.32**	**86.32**

In the PMC dataset, the training set only includes the training set divided from 2020 PMC dataset. We evaluate our recognition method based on group 8's targeted improvements, as in Table 5.

As aforementioned, it seems to reach the model and dataset bottleneck to some extent, partly because the labeling bias from manul annotation and slight pixel precision deviation of bounding boxes under the IOU of 0.9. Therefore, to improve labeling accuracy and data diversity, we carefully annotated an extra set of 1041 images as a subset for the PMC dataset[3]. Results from Table 7 shows that the model gains about 0.36% improvement. Notably, because the 2020 Adobe Synthetic official test set is not released yet, we can not directly compare with its 2020 ICPR leaderboard. The end-to-end detection and recognition experiment results are shown in Table 6 and Table 7.

Similarly, we evaluated our method using the ICDAR 2019 Chart datasets, and achieved promising results compared to the CHART-Infographics competition participants [7], as shown in Table 8. In the Synthetic dataset, we only use the 2019 official training set. In the PMC dataset, we use the same model as Table 7 with extra data. Both the results of Table 7 and Table 8 show that our method performs the best compared with prior approaches.

4.3   Task-3

Evaluation Metric. Like Task-1, the evaluation metric is the average per-class F-measure.

---

[3] https://github.com/HCIILAB/CHART-Infographics_PMC_Extended_Dataset.

Experimental Details. We regard Task-3 as a classification task, and divide the training and validation set with a 4:1 ratio. We first calculate the overall accuracy of the model on the two datasets, as shown in the Table 9.

Table 9. Results of text role classification on the 2020 Adobe Synthetic and PMC validation set.

Datasets	F-measure
Adobe Synthetic	100.00
PMC	86.20

It can be seen that our model has achieved gratifying results on the Adobe Synthetic dataset and good results on the PMC dataset. To analyze the accuracy of the model in each category in the PMC dataset, we present the statistics on the accuracy of the nine categories, as shown in Table 10.

It can be seen from the results that the accuracy of the model in most categories is already high. The main categories with low accuracy are chart title, mark label, other and tick grouping. These four categories are very easy to confuse, and the data is insufficient.

Table 10. Results of text role classification on the 2020 PMC validation set.

Text role	F-measure	Text role	F-measure
Tick label	99.43	Chart title	87.05
Axis title	98.27	Mark label	75.76
Legend label	97.12	Other	74.16
Value label	93.74	Tick grouping	56.48
Legend title	93.74		

To study the importance of the modal information of each part, we design an ablation experiment on the 2020 Adobe Synthetic dataset to analyze the accuracy of each part. The results verify that multi-modal information is beneficial for accurately classifying semantic roles (Table 11).

Table 11. Ablation studies of modules on the 2020 Adobe Synthetic validation set.

Module	F-measure
Baseline (only Semantics)	65.29
+ Position	90.85
+ Chart type	94.12
+ Visual	**100.0**

To verify our method, we conduct experiments on the 2019 Synthetic and PMC test sets as well as the 2020 PMC test set. Table 12 shows the promising results of these experiments. Note that it only considers four categories, i.e., chart title, axis title, legend label, and tick label in the 2019 datasets.

Table 12. Results of Task-3 on the 2019 and 2020 test sets. (* from [7])

2019 synthetic test set		2020 synthetic test set	
Team	F-measure	Team	F-measure
IITB-Team* (1st)	60.25	Data not release	
A-Team* (2nd)	99.95		
ABC* (3rd)	100.00		
Ours	**100.00**		
2019 PMC test set		2020 PMC test set	
Team	F-measure	Team	F-measure
IITB-Team* (1st)	35.58	Lenovo-SCUT-Intsig† (1st)	85.85
A-Team* (2nd)	84.38	Magic† (2nd)	81.71
Ours	**92.02**	DeepBlueAI†(3rd)	77.19
		PY†	65.38
		Ours	83.72
		Ours (ensemble)	**85.85**

† from https://chartinfo.github.io/leaderboards_2020.html

5   Conclusion

In this paper, we propose effective methods to handle chart classification, chart text detection and recognition, and chart text role classification. We verify our system on the Synthetic and PMC chart datasets and provide promising results, while summarizing some useful investigation and targeted improvements. Moreover, our proposed method won all the first places on the six test sets of three tasks in the ICPR 2020 competition.

In the future, we will make an attempt to handle other sub-tasks of chart data extraction and construct an end-to-end system. We hope this work will serve as an inspiration to take further steps toward improving the machine understanding of charts and the DAR field.

References

1. Araujo, T., Chagas, P., Alves, J., Santos, C., Santos, B., Meiguins, B.: A real-world approach on the problem of chart recognition using classification, detection and perspective correction. Sensors **20**, 4370 (2020)
2. Bajić, F., Job, J., Nenadić, K.: Chart classification using simplified VGG model. In: IWSSIP, pp. 229–233 (2019)

3. Böschen, F., Scherp, A.: A comparison of approaches for automated text extraction from scholarly figures. In: Amsaleg, L., Guðmundsson, G.Þ, Gurrin, C., Jónsson, B.Þ, Satoh, S. (eds.) MMM 2017. LNCS, vol. 10132, pp. 15–27. Springer, Cham (2017). https://doi.org/10.1007/978-3-319-51811-4_2

4. Cai, Z., Vasconcelos, N.: Cascade R-CNN: high quality object detection and instance segmentation. TPAMI, 1 (2019)

5. Cheng, Z., Bai, F., Xu, Y., Zheng, G., Pu, S., Zhou, S.: Focusing attention: towards accurate text recognition in natural images. In: ICCV, pp. 5076–5084 (2017)

6. Cubuk, E.D., Zoph, B., Mané, D., Vasudevan, V., Le, Q.V.: AutoAugment: learning augmentation strategies from data. In: CVPR, pp. 113–123 (2019)

7. Davila, K., et al.: ICDAR 2019 competition on harvesting raw tables from infographics (chart-infographics). In: ICDAR, pp. 1594–1599 (2019)

8. Davila, K., Setlur, S., Doermann, D., Bhargava, U.K., Govindaraju, V.: Chart mining: a survey of methods for automated chart analysis. TPAMI, 1 (2020)

9. Davila, K., Tensmeyer, C., Shekhar, S., Singhand, H., Setlur, S., Govindaraju, V.: ICPR 2020 - competition on harvesting raw tables from infographics (chart-infographics). In: Pattern Recognition. ICPR International Workshops and Challenges, pp. 361–380 (2021)

10. Deng, D., Liu, H., Li, X., Cai, D.: PixelLink: detecting scene text via instance segmentation. In: AAAI, vol. 32 (2018)

11. Duda, R.O., Hart, P.E.: Use of the Hough transformation to detect lines and curves in pictures. Commun. ACM **15**(1), 11–15 (1972)

12. Gupta, A., Dollár, P., Girshick, R.: Lvis: a dataset for large vocabulary instance segmentation. In: CVPR, pp. 5351–5359 (2019)

13. He, K., Gkioxari, G., Dollár, P., Girshick, R.: Mask R-CNN. In: ICCV, pp. 2961–2969 (2017)

14. He, K., Zhang, X., Ren, S., Sun, J.: Deep residual learning for image recognition. In: CVPR, pp. 770–778 (2016)

15. Hu, J., Shen, L., Sun, G.: Squeeze-and-excitation networks. In: CVPR, pp. 7132–7141 (2018)

16. Huang, G., Liu, Z., Van Der Maaten, L., Weinberger, K.Q.: Densely connected convolutional networks. In: CVPR, pp. 2261–2269 (2017)

17. Jia, X., et al.: Highly scalable deep learning training system with mixed-precision: training imagenet in four minutes. arXiv preprint arXiv:1807.11205 (2018)

18. Karatzas, D., et al.: ICDAR 2015 competition on robust reading. In: ICDAR, pp. 1156–1160. IEEE (2015)

19. Liu, Y., Lu, X., Qin, Y., Tang, Z., Xu, J.: Review of chart recognition in document images. Opt. Eng. **8654**, 865410 (2013)

20. Mei, H., Ma, Y., Wei, Y., Chen, W.: The design space of construction tools for information visualization: a survey. Int. J. Comput. Vis. **44**, 120–132 (2018)

21. Poco, J., Heer, J.: Reverse-engineering visualizations: recovering visual encodings from chart images. Comput. Graph Forum. **36**, 353–363 (2017)

22. Purchase, H.C.: Twelve years of diagrams research. Int. J. Comput. Vis. **25**(2), 57–75 (2014)

23. Ren, S., He, K., Girshick, R.B., Sun, J.: Faster R-CNN: towards real-time object detection with region proposal networks. In: Cortes, C., Lawrence, N.D., Lee, D.D., Sugiyama, M., Garnett, R. (eds.) NIPS, pp. 91–99 (2015)

24. Shi, B., Bai, X., Yao, C.: An end-to-end trainable neural network for image-based sequence recognition and its application to scene text recognition. TPAMI **39**(11), 2298–2304 (2016)

25. Shi, B., Wang, X., Lyu, P., Yao, C., Bai, X.: Robust scene text recognition with automatic rectification. In: CVPR, pp. 4168–4176 (2016)
26. Smith, R.: An overview of the tesseract OCR engine. In: ICDAR, vol. 2, pp. 629–633. IEEE (2007)
27. Szegedy, C., Vanhoucke, V., Ioffe, S., Shlens, J., Wojna, Z.: Rethinking the inception architecture for computer vision. In: CVPR, pp. 2818–2826 (2016)
28. Tang, B., et al.: DeepChart: combining deep convolutional networks and deep belief networks in chart classification. Signal Process. **124**, 156–161 (2016)
29. Xu, Y., Li, M., Cui, L., Huang, S., Wei, F., Zhou, M.: LayoutLM: pre-training of text and layout for document image understanding. In: KDD, pp. 1192–1200 (2020)

DeMatch: Towards Understanding the Panel of Chart Documents

Hesuo Zhang[1], Weihong Ma[1], Lianwen Jin[1,2(✉)], Yichao Huang[3], Kai Ding[3], and Yaqiang Wu[4]

[1] South China University of Technology, Guangzhou, China
eelwjin@scut.edu.cn
[2] Guangdong Artificial Intelligence and Digital Economy Laboratory (Pazhou Lab), Guangzhou, China
[3] IntSig Information Co., Ltd., Shanghai, China
[4] Lenovo Research, Beijing, China

Abstract. Chart document understanding is a challenging task in document analysis because of the complex format and specific semantics it contains. In this paper, we first define the chart panel analysis problem and propose a complete framework that can be performed on various types of chart. Generally, a chart document contains multiple types of elements, such as title, axis, legend, plot elements and comments. To solve the problem of chart panel analysis, we developed our system focusing on two aspects: axis analysis and legend analysis. For axis analysis, we first design a fully convolutional network to detect tick marks and then use the clustering method to distinguish the X-axis and Y-axis. A rectangle-growing matching rule is proposed to associate the predicted tick mark with its corresponding text. For legend analysis, we use a cascaded head detector to determine the accurate location of the legend marks and mark-text pairs; and then we design the highest IoU matching rule to determine the legend label text and its corresponding legend mark. Experimental results on synthetic and real data sets demonstrate the effectiveness of the proposed method. Specifically, we obtained the best result on the test set of the ICDAR2019 competition on harvesting raw tables from infographics, and state-of-the-art performance on UB PMC2020 data set of the ICPR2020 competition. Code will be at https://github.com/iiiHunter/CHART-DeMatch.

Keywords: Chart · Document understanding · Detection

1 Introduction

As a distinctive document with rich semantic information, a chart can represent data visually. As a useful tools for the presentation of data in a visually appealing format, a chart is usually used to summarize experimental results or

This research is supported in part by NSFC (Grant No.: 61936003, 61771199), GD-NSF (no. 2017A030312006).

J. Lladós et al. (Eds.): ICDAR 2021, LNCS 12823, pp. 692–707, 2021.
https://doi.org/10.1007/978-3-030-86334-0_45

conclusions [7]. Although the field of document analysis and recognition (DAR) has been developed for decades [1, 8, 21], chart document analysis and recognition (CHART-DAR) remains an open problem. Charts have their own characteristics, diversity, and semantics, which make the CHART-DAR task significantly different from other DAR tasks. The understanding of and automatic data extraction from chart documents are of great challenge and academic value.

Some existing works related to chart documents mainly treat this issue as a question-answer problem [11, 12, 19]. However, in some cases, the information we want to obtain from the chart is uncertain, and the question-answer model cannot meet our requirements. A more common task is table reconstruction and data extraction from charts. The existing works tend to solve this problem of any certain types of chart, such as scatter [4], bar or pie chart [16]. And the proposed methods do not work for the complex chart. For the data extraction of complex and various types of charts, a common solution is to break it down into several sub-problems, such as chart classification, text spotting, panel analysis, and data conversion. At present, there is a lack of systematic research on these sub-problems in the field of CHART-DAR. In this study, we concentrated on the chart panel analysis (CPA) problem. The CPA task includes two subtasks: axis and legend analysis.

Axis Analysis: The axis analysis system is expected to output the location of tick mark and associated corresponding value text on both the X-axis and Y-axis. In general, X-axis refers to the axis that represents the independent variable, rather than the axis that is visually horizontal. For example, horizontal bar and horizontal boxplot have an X-axis that is vertical. Similarly, the Y-axis is not always the axis that is vertical.

Legend Analysis: In a chart, each legend mark is associated with a data series name. For the understanding of chart with multiple data series, it is critical to analyze the legend first. The legend analysis system is responsible for finding all the legend marks area, and link them to the associated label text.

In this work, we explored the CPA task for chart documents systemically. To our knowledge, DeMatch is the only complete chart panel analysis system that can be performed on various types of chart document, including bar, line, boxplot, and scatter. Here, to simplify the CPA task, we assumed that all the text in chart image was obtained and developed a system to perform the CPA task. The system consists of detection and matching modules and is called DeMatch. In DeMatch, we design segmentation-/detection-based networks to detect the key elements of a chart panel and two robust matching modules to match the detected elements with their associated text: tick value text or legend label text. The novel contribution of this study is that DeMatch is the first proposed complete system that can perform the CPA task for various types of chart documents. A rectangle-growing matching rule for axis analysis and the highest IoU-based matching rule for legend analysis are proposed. These rules were proved to deliver better performance than other matching rules, such as distance and containing rules. We achieved state-of-the-art performance on the data set of the

ICDAR 2019 competition on harvesting raw tables from infographics (Chart-Infographics) and ICPR 2020 Chart-Infographics. Compared with the competitors' complex models with ensemble learning, DeMatch is simple yet effective. Finally, we performed some ablation experiments to verify the effectiveness of the proposed matching modules.

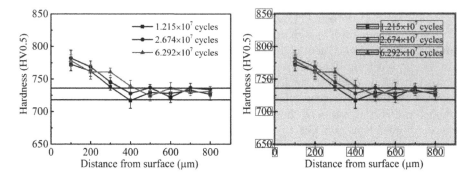

Fig. 1. Left: source image; right: corresponding label. CPA task aims to analyze the panel of a chart and can determine the precise position of the plot region, tick mark and legend marks that are associated with any text.

2   Related Work

Problems related to document analysis and understanding have been studied for decades, but due to the diversity and complexity of documents, these issues have not yet been completely addressed. Common documents include magazines, newspapers, historical books, receipts, chart documents, etc. Document analysis studies have been conducted on some types of such documents, such as historical documents [26], magazines [17,25], and receipts [15,18].

For chart documents, some existing studies have focused on reasoning tasks and chart document object detection tasks [11,12,19]. For these methods, text or plot element detection modules are particularly essential for the chart visual reasoning task. In a scientific plot, even minor localization errors can lead to large errors in subsequent numerical inferences. To address this issue, Ganguly et al. [9] proposed minor modifications to existing detection models by combining ideas from different networks, which helped to significantly improve the performance. In terms of work related to CPA, Kataris et al. [13] detected tick marks using the Fast Fourier Transform over pixel profiles of axes, under the assumption that tick marks are located at fixed intervals. With regard to deep neural methods, Cliche et al. [4] used an object detection network to find the bounding boxes of the tick mark, and associate it to a tick value text nearest to it. The clustering method was used to assign each pair of tick mark and tick value text to the X-axis or Y-axis. However, this method can only apply to synthetic scatter plots, which

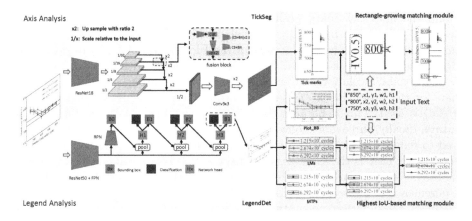

Fig. 2. Pipeline of DeMatch. It consists an axis analysis module (pink) and a legend analysis module (green). The TickSeg and LegendDet network are responsible for determining the position of tick mark and legend mark, and the rectangle-growing matching module and highest IoU-based matching module can match the detected elements with the associated text. (Color figure online)

are relative simple and only contain one series data. Liu et al. [16] used Faster-RCNN [22] to detect variable types of elements in bar and pie chart. RNs [23] were used to match the tick value text with the plot element directly, rather than detecting the tick value text. In this paper, we propose a more robust system towards complete analysis of chart elements and certain types.

3 Methodology

DeMatch includes two subsystems of axis analysis and legend analysis, which aim to associate the tick value text with the tick mark on the axes and the legend label text with the corresponding legend mark. The bounding boxes and transcriptions of all texts in the image are given as inputs. The pipeline of DeMatch is shown in Fig. 2. Both subsystems of axis analysis and legend analysis consist of two stages: detection and matching. We introduce them in detail in the following sections.

3.1 Axis Analysis

As shown in pink area in Fig. 2, the axis analysis system consists of two stages. We first designed a fully convolutional network (FCN) named TickSeg to determine the position of the tick mark; and then proposed a matching module to match the detected marks with the value text.

Tick Mark Detection. In the phase of tick mark detection, inspired by [24], we use FCN to predict a probability map, and post-processing is adopted to obtain the final tick mark position. The backbone architecture is based on ResNet18 [10], and a feature pyramid network (FPN) is used to enhance feature representation. The architecture of the FPN fusion block is illustrated in Fig. 2. The fusion block can fuse the higher level feature with that of lower level by additional convolutional layers and up-sample operation. It should be noted that, for the fusion block of the top feature, only one convolutional network (green) is used. This is because there is no concatenate operation for the top feature. Finally, three more convolutional layers and an up-sample operation are used to reduce the feature channel to 1; and obtain the final prediction output with a scale that is the same as that of the input. When training the network, the MSE loss function is used. Before training, the ground truth that meets the requirements of the segmentation network should be generated first.

Unlike the binary segmentation map, which labels each pixel discretely, we encode the tick mark center with a Gaussian heatmap. Heatmap representation has been used in many applications. For example, it is widely used in pose estimation [3,20] because of its high flexibility when dealing with ground truth regions that are not rigidly bounded. As shown in Fig. 3, for each ground truth tick mark, we split all ground truth tick marks onto a heatmap $Y[0,1]$ using a Gaussian kernel. The Gaussian value Y_{xy} is calculated as follows:

$$Y_{xy} = e^{-\frac{(x-p_x)^2+(y-p_y)^2}{2\sigma^2}} \tag{1}$$

where p_x, p_y denote the center of the tick mark, and σ is the object size-adaptive standard deviation. If two Gaussian maps overlap, we take the element-wise maximum [3].

Algorithm 1. Rectangle-growing Matching Rule

Input: P: position of detected tick marks, T: bounding box of all text
Output: $result$: pairs of tick mark and associated text

1: **function** MATCH(P,T)
2:     $T^* = [T_j$ if T_j not in Plot_BB for T_j in $T]$
3:     $result = \{\}$
4:     **for** $P_i(x_i, y_i)$ in P **do**
5:         $r \leftarrow r_0$
6:         **while** P_i not in $result$ **do**
7:             **if** P_i in vertical axis **then**
8:                 $w \leftarrow 2r, h \leftarrow r$
9:             **else**
10:                $w \leftarrow r, h \leftarrow 2r$
11:            $B_i \leftarrow (x_i, y_i, w, h)$
12:            **for** T_j in T^* **do**
13:                **if** $overlap$ (T_j, B_i) **then**
14:                    $result[P_i].$append(T_j)
15:            $r \mathrel{+}= 1$
16:     **return** $result$

Rectangle-Growing Matching Rule. Rectangle-growing matching rule as Algorithm 1 is proposed to associate a tick mark with its corresponding text. Specifically, to distinguish the detected tick marks of different axes, we first perform clustering based on the position of the marks. Second, we take the detected tick mark as the center; and draw a rectangle with a size of $r \times 2r$. For the detection mark on the horizontal axis, the longer side is in the vertical direction, while for the mark on the vertical axis, the longer side is in the horizontal direction. Third, we gradually increase the r value until the rectangle touches a text. Then, the text is regarded as the matched text. The initial $r0$ value is set to 30 and 10 for the synthetic and PMC data, respectively. It should be noted that if a tick mark associated with more than one text, heuristic post processing is adopted to remove the redundant text. Specifically, we will pick the text that's closest in the X and Y direction for the tick mark on horizontal and vertical axis.

In our experiments, we used an additional plot region's bounding box (Plot_BB) to remove some redundant text candidates. The plot region represents the main plot area of the chart panel. Usually, this is represented by drawing a rectangle, as shown in blue in Fig. 1. In DeMatch, we obtain the bounding box of the plot region from the detector of the legend analysis module.

3.2 Legend Analysis

The legend analysis system also contains two modules: detection and matching. They are responsible for detecting the position of the legend mark (LM) and matching LM with the corresponding legend label text (LT). Actually, To help the following matching module determine the corresponding LT, we detect the mark-text pair (MTP) simultaneously.

Legend Mark and Mark-Label Pair Detection. The legend mark is usually displayed as a small object in a chart document, which makes it challenging to determine its precise location. On the contrary, determining the precise position of LM is of great significance for the subsequent data conversion module, because different series of data have to be interpreted based on the graphical style presented in the small legend marks. Therefore, in the detection phase, it is critical to detect the LM with high precision. Therefore, We use a R-CNN-based network [2] as our detection network, with three regression heads of positive candidate IoU of 0.5, 0.6, and 0.7. The backbone is ResNet50 [10] with an FPN [14]. Considering the small size of the LMs, the RPN scales are set to (4, 8, 16, 32, 64). Furthermore, as mentioned above, the detector predicts the Plot_BB, which is the main plot area of the chart panel and is useful to the axis analysis system in DeMatch.

Because MTPs will be detected as well, their ground truth of the bounding box will be generated first. In our implementation, the bounding box of the MTP is drawn around the annotated box of the LM and its corresponding LT, as shown in Fig. 3.

Algorithm 2. Highest IoU Matching Rule
___
Input: B_{LM}: bounding box of LM, B_{MTP}: position of MTP, T: position of all text
Output: *result*: pairs of LM and associated text
1: **function** MATCH(B_{LM}, B_{MTP}, T)
2:     *result, cache* = [ ], [ ]
3:     **for** MTP_i in $MTPs$ and T_j in T **do**
4:         $IOU_p^{(i,j)} \leftarrow iou(MTP_i, T_j)$
5:     Sort IOU_p from large to small, yielding
6:     $Index_p = \{(p^{(k)}, q^{(k)}) | IOU_p^{(p^{(k)}, q^{(k)})} \leq IOU_p^{(p^{(k+1)}, q^{(k+1)})}\}$, $(p^{(k)}, q^{(k)})$ is a pair
        of indexes to the list of IOU_P
7:     **for** $(p^{(k)}, q^{(k)})$ in $Index_p$ **do**
8:         **if** both $MTP_{(p^{(k)})}$ and $T_{q^{(k)}}$ unmatched **then**
9:             cache.append($(MTP_{p^{(k)}}, T_{q^{(k)}})$)
10:     **for** (MTP_i, T_i) in *cache* and LM_j in LMs **do**
11:         $IOU_m^{(i,j)} \leftarrow iou(MTP_i, LM_j)$
12:     Sort IOU_m from large to small, yielding
13:     $Index_m = \{(p^{(k)}, q^{(k)}) | IOU_m^{(p^{(k)}, q^{(k)})} \leq IOU_m^{(p^{(k+1)}, q^{(k+1)})}\}$
14:     **for** $(p^{(k)}, q^{(k)})$ in $Index_m$ **do**
15:         **if** both $MTP^{(p^{(k)})}$ and $LM^{q^{(k)}}$ unmatched **then**
16:             result.append($(T^{(p^{(k)})}, LM^{q^{(k)}})$)
17:     **return** *result*
___

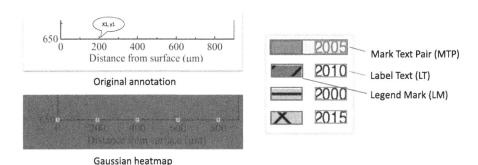

Fig. 3. Generation of ground truth. Gaussian heat maps of tick marks and MTPs of legend marks are generated for the axis analysis and legend analysis systems, respectively.

Table 1. Statistics of three data sets used.

Data set	Training	Validation	Test	Total
ICDAR Synth2019	20,000	–	528	20,528
Adobe Synth2020	8,640	960	518	10,118
UB PMC2020	2,388	279	726	3,393

Highest IoU Matching Rule. We use the highest IoU score-based rule as Algorithm 2 to match the predicted LM and MTP results with the given text. It should be noted that, when the IoU score is too low, for example, lower than 0.1, the pairs of $(MTP_{p^{(k)}}, T_{q^{(k)}})$ and $(T^{(p^{(k)})}, LM^{q^{(k)}})$ will be ignored. Thus, the number of the final matched pairs may be smaller than the number of LMs.

4 Experiments

4.1 Data Sets

The three data sets used for the evaluation of DeMatch include ICDAR Synth2019 of the ICDAR 2019 competition of Chart-Infographics [6], Adobe Synth2020 and UB PMC2020 of the ICPR 2020 competition of Chart-Infographics [5]. The synthetic data sets of Synth2019/2020 were created using the Matplotlib library, which contain multiple types of chart images, including bar (horizontal/vertical, stacked/grouped), boxplot (horizontal/vertical), scatter, and line. The synthetic chart images have a clean background and fixed resolution. The PMC2020 data set was created from real charts on scientific publications from PubMed Central[1]. It also contains multiple types of chart images but the images have different scales and resolutions. We divide Adobe's data into training and validation sets in a proportion of 4:1. Note that, for the UB PMC2020 data, the number of images in the original data set for the axis analysis and legend analysis tasks are slightly different. We take their intersection as our experimental data, which provides the same number of images as used for axis analysis. The statistics of the data split for these three data sets are shown in Table 1. In addition, we analyzed the distribution of the height, width, and ratio (height divided by width) of legend marks to help to set the hyperparameters of the legend analysis system, as shown in Fig. 4.

4.2 Evaluation Metric

F-Measure for Axis Analysis The metric for axis analysis is a weighted F-measure (F_m) [6], where predicted tick marks receive partial true positive scores between 0 and 1, based on the location accuracy. The evaluation mechanism defines two distance thresholds, a = 1.0% and b = 2.0% of the image diagonal, and the score is calculated based on the distance between the predicted tick mark and the ground truth tick location as:

$$score_{ij}(d_{ij}, a, b) = \begin{cases} 1, & d_{ij} \leqslant a \\ \frac{b - d_{ij}}{b - a}, & a \leqslant d_{ij} \leqslant b \\ 0, & d_{ij} \geqslant b \end{cases} \quad (2)$$

where d_{ij} is the Euclidean distance between the ground truth mark P_i and prediction mark P_j. Recall (R) is computed as the sum of the scores divided by the number of GT ticks, and precision (P) is the sum of scores divided by the number of predicted ticks. The F-measure is their harmonic average.

---

[1] https://www.ncbi.nlm.nih.gov/pmc/.

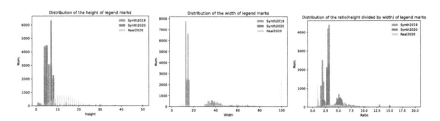

Fig. 4. Distribution of the height, width and ratio of legend marks on three data sets. The distributions of synthetic and PMC data sets are significantly different. The distribution of synthetic data is mainly concentrated in a relatively fixed interval, while the PMC data are scattered and irregular.

mIoU and Recall Score for Legend Analysis. For each ground truth LT, if there is an associated predicted LM, we compute the average IoU of the predicted LM's bounding box to that of the ground truth [6]. Assume that the predicted LM and ground truth LM of a ground truth LT are LM_j and $LM_{gt(i)}$, then the mIoU and recall score are

$$mIoU = \frac{1}{max(\#pred, \#gt)} \sum_{(i,j)} iou(LM_{gt(i)}, LM_j) \tag{3}$$

$$R = \frac{1}{max(\#pred, \#gt)} \sum_{(i,j)} \frac{intersection(LM_{gt(i)}, LM_j)}{area(LM_{gt(i)})} \tag{4}$$

where $\#pred$ is the number of predicted LMs, $\#gt$ is the number in the ground truth, $area$ and $intersection$ are the operation to obtain the area of a LM and the overlapping area of two LMs.

4.3   Implementation Details

In our experiments, we trained and evaluated DeMatch on Synth2020 and PMC2020 data sets, respectively. The ICDAR Synth2019 data was additionally used when training on Synth2020. In DeMatch, the axis analysis and legend analysis systems were trained severally. For the axis analysis system, the optimizer is Adam with an initial learning rate of 3e−3, and the learning rate decays to 3e−4 and 3e−5 at epochs 30 and 40, respectively. In the rectangle-growing matching module, the r value is set to 30 and 10 for the synthetic and PMC data, respectively. The total number of training epochs was 50. For the legend analysis system, we used SGD optimizer with an initial learning rate of 2e−2. The learning rate decays to 2e−3, 2e−4 and 2e−5 at epochs of 10, 16, and 24, respectively. The total number of training epochs was 30.

4.4   Results and Analysis

Results on Synthetic Data Set. In the first experiment, we compared DeMatch with two high-performance systems on the top of ranking list in the ICDAR2019 CHART-Infographics competition. The ICDAR Synth2019 test set of the competition contains 528 images. As shown in Table 2, our system outperforms the Rank1 result of the competition. For synthetic chart images with a fixed resolution, clean background, and relatively consistent style, the system can achieve a relatively high performance.

Table 2. Comparison of the results on different test sets.

Systems	ICDAR Synth2019			UB PMC2020		
	Axis analy.	Legend analy.		Axis analy.	Legend analy.	
	F_m	R	$mIoU$	F_m	R	$mIoU$
Rank1*	99.76	–	87.13	**95.69**	92.00	**86.00**
Rank2*	96.49	–	78.14	93.85	93.19	84.92
Ours	**99.94**	**99.88**	**92.48**	94.30	**93.63**	85.70

* latest results from https://chartinfo.github.io/leaderboards_2020.html, while the results in [5] are the old version.

Results on Real Data Set. In the second experiment, we conducted experiments on real data set of UB PMC2020, the test set of the ICPR2020 CHART-Infographics competition. The results are shown in Table 2. It is worth noting that the systems on the competition ranking usually use some complex techniques such as model ensemble, while our result is produced by only one DeMatch model. Compared with the performance on synthetic data, DeMatch achieved relatively poor performance on the PMC2020 test set. Real chart images are more challenging because of the variable styles and resolution, noisy background and some intrusive elements. Figure 5 shows the visualization results. To illustrate the findings clearly, we present the results of the two tasks of axis analysis and legend analysis separately. As mentioned above, the CPA task is more challenging for real chart images. Therefore, we present more visualization of the failure samples in Fig. 6. For the axis analysis system, the failure cases include redundant or missed detection of some tick marks, which will result in cumulative errors in the subsequent matching modules. When matching, irrelevant text near the tick mark may affect the matching operation. If semantic information is combined during matching, the interference caused by irrelevant text can be mitigated. Errors due to redundant or missed detection exist for legend analysis as well. It is difficult to detect and locate legends due to the wide variety of the style. Moreover, cumulative errors still exist in our two-stage system.

Table 3. Results for axis system with different matching rules.

Matching rules	ICDAR Synth2019			Adobe Synth2020			UB PMC2020		
	R	P	F_m	R	P	F_m	R	P	F_m
Distance [4]	99.93	99.63	99.58	97.60	95.69	96.64	89.77	93.58	91.63
Growing	99.92	99.95	99.94	99.11	97.59	98.34	93.81	94.42	94.12
Growing +Plot_BB	**99.92**	**99.95**	**99.94**	**99.73**	**99.93**	**99.83**	**93.81**	**94.42**	**94.30**

Table 4. Results for legend system with different matching rules.

Matching rules	ICDAR Synth2019		Adobe Synth2020		UB PMC2020	
	R	$mIoU$	R	$mIoU$	R	$mIoU$
Distance	99.11	91.78	98.86	94.10	82.66	76.64
Containing (margin = 7)	99.88	92.48	99.71	94.88	93.42	85.59
IoU	**99.88**	**92.48**	**99.86**	**95.02**	**93.63**	**85.70**

Ablation Study. As mentioned above, DeMatch contains detection and matching stages. To verify the superiority and effectiveness of the proposed matching module, we perform more ablation experiments. For addressing the axis analysis problem, we compared our rectangle-growing matching rule with the distance matching rule, under which a detected tick mark is linked to a text nearest to it [4]. In addition, we explored performance improvement of the system by using an additional Plot_BB to aid matching. The results are presented in Table 3. We observed that our rectangle-growing matching rule outperforms the distance matching rule. Moreover, with the help of Plot_BB, the system exhibits a certain performance improvement. This is because Plot_BB can help remove some text that might cause confusion. Similarly, for the legend analysis task, we compared the effects of different matching rules on the system performance, as shown in Table 4. The distance matching rule indicates that we match the LM with the text nearest to it. The containing rule means that an LM will match with the text contained in the same MTP. Before matching, we add margins of a certain width to the bounding box of the MTP in four directions. The width was set to 7 in the experiments. The IoU matching rule is introduced in Sect. 3.2. As shown in Table 4, the experiment results verify the superiority of the proposed highest IoU matching rules in DeMatch.

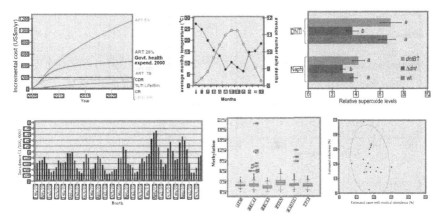

(a) Visualization of the axis analysis system (the blue bounding boxes indicate Plot_BB of the chart, and the scale value texts are connected with the associated tick marks on the axes with small lines).

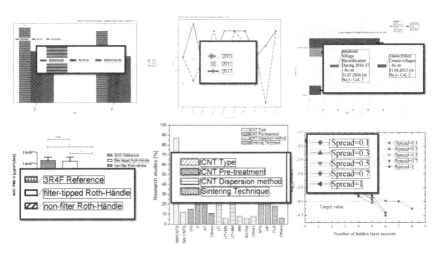

(b) Visualization of the legend analysis system (the label texts are connected with the associated legend marks with small lines).

Fig. 5. Some visualization results of DeMatch. (Color figure online)

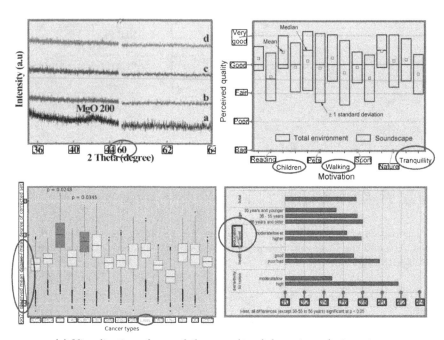

(a) Visualization of some failure results of the axis analysis system.

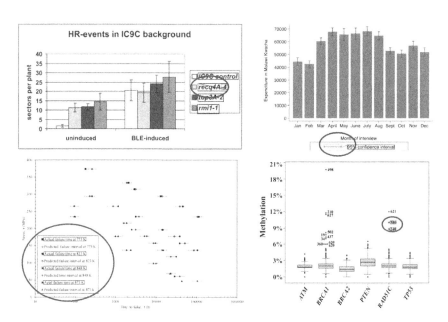

(b) Visualization of some failure results of the legend analysis system.

Fig. 6. Some failure samples of DeMatch, including missed or redundant detection and matching error.

5   Conclusion

In this study, we formally defined the CPA task and developed a complete framework DeMatch to solve the problem. In DeMatch, we focused on two subtasks: axis analysis and legend analysis. For addressing the axis analysis problem, we designed an FCN to find the tick mark and proposed rectangle-growing matching rule to find the value text associated with the mark. For addressing the legend analysis problem, we designed a multi-head object detection network to find the legend elements and the highest IoU-based method to match them. In addition, to locate the chart panel, the bounding box of the main plot area is predicted as well, which can help find the tick marks on the axes. We conducted experiments on synthetic and real chart document analysis data sets, and achieved state-of-the-art performance. To summarize, we have proposed a simple yet effective framework to solve the CPA problem of various types of chart systematically. We hope the solving of CPA problem contributes to the development of chart document analysis.

References

1. Binmakhashen, G.M., Mahmoud, S.A.: Document layout analysis: a comprehensive survey. ACM Comput. Surv. (CSUR) **52**(6), 1–36 (2019)
2. Cai, Z., Vasconcelos, N.: Cascade R-CNN: high quality object detection and instance segmentation. IEEE Trans. Pattern Anal. Mach. Intell. (2019)
3. Cao, Z., Simon, T., Wei, S.E., Sheikh, Y.: Realtime multi-person 2D pose estimation using part affinity fields. In: Proceedings of the IEEE Conference on Computer Vision and Pattern Recognition (CVPR), pp. 7291–7299 (2017)
4. Cliche, M., Rosenberg, D., Madeka, D., Yee, C.: Scatteract: automated extraction of data from scatter plots. In: Ceci, M., Hollmén, J., Todorovski, L., Vens, C., Džeroski, S. (eds.) ECML PKDD 2017. LNCS (LNAI), vol. 10534, pp. 135–150. Springer, Cham (2017). https://doi.org/10.1007/978-3-319-71249-9_9
5. Davila, K., Tensmeyer, C., Shekhar, S., Singh, H., Setlur, S., Govindaraju, V.: ICPR 2020 - competition on harvesting raw tables from infographics. In: Proceedings of the International Conference on Pattern Recognition (ICPR) Workshops (2020)
6. Davila, K., et al.: ICDAR 2019 competition on harvesting raw tables from infographics (chart-infographics). In: Proceedings of the International Conference on Document Analysis and Recognition (ICDAR), pp. 1594–1599. IEEE (2019)
7. Davila, K., Setlur, S., Doermann, D., Bhargava, U.K., Govindaraju, V.: Chart mining: a survey of methods for automated chart analysis. IEEE Trans. Pattern Anal. Mach. Intell. (2020)
8. Eskenazi, S., Gomez-Krämer, P., Ogier, J.M.: A comprehensive survey of mostly textual document segmentation algorithms since 2008. Pattern Recogn. **64**, 1–14 (2017)
9. Ganguly, P., Methani, N., Khapra, M.M., Kumar, P.: A systematic evaluation of object detection networks for scientific plots. arXiv preprint arXiv:2007.02240 (2020)

10. He, K., Zhang, X., Ren, S., Sun, J.: Deep residual learning for image recognition. In: Proceedings of the IEEE Conference on Computer Vision and Pattern Recognition (CVPR). pp. 770–778 (2016)
11. Kafle, K., Price, B., Cohen, S., Kanan, C.: DVQA: understanding data visualizations via question answering. In: Proceedings of the Conference on Computer Vision and Pattern Recognition, pp. 5648–5656 (2018)
12. Kahou, S.E., Michalski, V., Atkinson, A., Kádár, Á., Trischler, A., Bengio, Y.: FigureQA: an annotated figure dataset for visual reasoning. arXiv preprint arXiv:1710.07300 (2017)
13. Kataria, S., Browuer, W., Mitra, P., Giles, C.L.: Automatic extraction of data points and text blocks from 2-dimensional plots in digital documents. In: Proceedings of the Twenty-Third AAAI Conference on Artificial Intelligence (AAAI), vol. 8, pp. 1169–1174 (2008)
14. Lin, T.Y., Dollár, P., Girshick, R., He, K., Hariharan, B., Belongie, S.: Feature pyramid networks for object detection. In: Proceedings of the IEEE Conference on Computer Vision and Pattern Recognition (CVPR), pp. 2117–2125 (2017)
15. Liu, X., Gao, F., Zhang, Q., Zhao, H.: Graph convolution for multimodal information extraction from visually rich documents. In: Proceedings of the Conference of the North American Chapter of the Association for Computational Linguistics (NAACL), pp. 32–39 (2019)
16. Liu, X., Klabjan, D., NBless, P.: Data extraction from charts via single deep neural network. arXiv preprint arXiv:1906.11906 (2019)
17. Lu, T., Dooms, A.: Probabilistic homogeneity for document image segmentation. Pattern Recogn. **109**, 107591 (2020)
18. Majumder, B.P., Potti, N., Tata, S., Wendt, J.B., Zhao, Q., Najork, M.: Representation learning for information extraction from form-like documents. In: Proceedings of the meeting of the Association for Computational Linguistics (ACL), pp. 6495–6504 (2020)
19. Methani, N., Gauguly, P., Khapra, M.M., Kumar, P.: PlotQA: reasoning over scientific plots. In: Proceedings of the IEEE Winter Conference on Applications of Computer Vision (WACV), pp. 1527–1536 (2020)
20. Newell, A., Yang, K., Deng, J.: Stacked hourglass networks for human pose estimation. In: Leibe, B., Matas, J., Sebe, N., Welling, M. (eds.) ECCV 2016. LNCS, vol. 9912, pp. 483–499. Springer, Cham (2016). https://doi.org/10.1007/978-3-319-46484-8_29
21. Prasad, D., Gadpal, A., Kapadni, K., Visave, M., Sultanpure, K.: CascadeTabNet: An approach for end to end table detection and structure recognition from image-based documents. In: Proceedings of the IEEE Conference on Computer Vision and Pattern Recognition (CVPR) Workshops, pp. 572–573 (2020)
22. Ren, S., He, K., Girshick, R., Sun, J.: Faster R-CNN: towards real-time object detection with region proposal networks. IEEE Trans. Pattern Anal. Mach. Intell. **39**(6), 1137–1149 (2016)
23. Santoro, A., et al.: A simple neural network module for relational reasoning. In: Proceedings of the International Conference on Neural Information Processing Systems (NIPS), pp. 4974–4983 (2017)
24. Sun, K., Xiao, B., Liu, D., Wang, J.: Deep high-resolution representation learning for human pose estimation. In: Proceedings of the IEEE Conference on Computer Vision and Pattern Recognition (CVPR). pp. 5693–5703 (2019)

25. Tran, T.A., Na, I.S., Kim, S.H.: Page segmentation using minimum homogeneity algorithm and adaptive mathematical morphology. Int. J. Doc. Ana. Recogn. (IJDAR) **19**(3), 191–209 (2016). https://doi.org/10.1007/s10032-016-0265-3
26. Xu, Y., Yin, F., Zhang, Z., Liu, C.L.: Multi-task layout analysis for historical handwritten documents using fully convolutional networks. In: Proceedings of the International Joint Conference on Artificial Intelligence (IJCAI), pp. 1057–1063 (2018)

Sequential Next-Symbol Prediction for Optical Music Recognition

Enrique Mas-Candela[✉], María Alfaro-Contreras, and Jorge Calvo-Zaragoza

U.I. for Computer Research, University of Alicante, Alicante, Spain
emc89@alu.ua.es, {m.alfaro,jorge.calvo}@ua.es

Abstract. Optical Music Recognition is the research field that investigates how to computationally read music notation from document images. State-of-the-art technologies, based on Convolutional Recurrent Neural Networks, typically follow an end-to-end approach that operates at the staff level; i.e., a single stage for completely processing the image of a single staff and retrieving the series of symbols that appear therein. This type of models demands a training set of sufficient size; however, the existence of many music manuscripts of reduced size questions the usefulness of this framework. In order to address such a drawback, we propose a sequential classification-based approach for music documents that processes sequentially the staff image. This is achieved by predicting, in the proper reading order, the symbol locations and their corresponding music-notation labels. Our experimental results report a noticeable improvement over previous attempts in scenarios of limited ground truth (for instance, decreasing the Symbol Error Rate from 70% to 37% with just 80 training staves), while still attaining a competitive performance as the training set size increases.

Keywords: Optical Music Recognition · Handwritten music recognition · Deep learning · Reading order

1 Introduction

One of the main tools for preserving music compositions over time is their engraving—by handwriting or typesetting—in the so-called music scores. We find that a significant body of musical heritage is only available as physical documents. Most of them have never been stored in a structured digital format that allows their indexing, editing, or critical publication. Given the considerable cost and time required for manual transcription, this scenario invites automation much in the same way as other technologies enable the processing of written texts from document images.

Optical Music Recognition (OMR) refers to the field of research that studies how to computationally read music notation in documents [5]. As in many other fields, modern Machine Learning techniques, namely Deep Neural Networks (DNN), brought new successful approaches. End-to-end systems that

© Springer Nature Switzerland AG 2021
J. Lladós et al. (Eds.): ICDAR 2021, LNCS 12823, pp. 708–722, 2021.
https://doi.org/10.1007/978-3-030-86334-0_46

retrieve the series of symbols that appear in a single-section staff image, can be considered the current state of the art [3,7,20]. To develop these approaches, only training pairs are needed, consisting of problem images, together with their corresponding transcript solutions. As long as there is sufficient and adequate training data, the results achieved can be considered effective for transcribing music notation.

However, the size of the training set might become an issue, especially when transcribing small music manuscripts. In these cases, which are quite common in historical music heritage, the amount of data needed to train an accurate system might be close to the total amount of data to be transcribed. This could lead to a scenario in which the use of automatic technology is not useful at all.

In this paper, we propose an alternative classification-based system that is inspired by a sliding window approach. Instead of performing an exhaustive search over the target image, the points of interest over which the window should move are computed iteratively, starting from a previous location. Then, each location is classified in terms of the appearing music notation symbol.

We introduce a Next-Symbol Prediction (NSP) model, which aims at predicting the sequence of locations of each symbol of a staff in their reading order. For that, a Convolutional Neural Network (CNN) is trained to predict the location of the next symbol in the staff, conditioned to a current symbol location. Once a symbol is located, we retrieve both its shape and its vertical position in the staff, which are features related to the rhythm and the tone, respectively. This is achieved by simple and common classification CNN, widely used in symbol classification [13].

Our results show that the proposed approach yields good results with a small amount of training data, as opposed to the state of the art, while still providing a competitive performance as the ground-truth size increases. In addition, the present work opens up new avenues for research that will be discussed below.

The rest of the paper is organized as follows: Sect. 2 summarizes related attempts to OMR; Sect. 3 thoroughly develops our recognition framework; Sect. 4 describes the experimental setup; Sect. 5 reports the obtained results and their main outcomes; and finally, Sect. 6 concludes the present work, along with some avenues for future research.

2   Background

Traditional OMR attempts have been systematized in a multi-stage workflow. Bainbridge and Bell [2] properly described and formalized the *de facto* standard pipeline, which was later reviewed thoroughly by Rebelo et al. [14].

The recent paradigm shift towards the use of DNN has diversified the way OMR is addressed. The ongoing research is focused on an end-to-end or holistic approach, that operates at the staff level; i.e., a single step that fully processes the image of a single-staff section and retrieves the symbols therein.

In this context, we find in the literature multiple approaches based on Convolutional Recurrent Neural Networks (CRNN). A common one is that trained using the Connectionist Temporal Classification (CTC) loss function [10]. The work by Calvo-Zaragoza et al. [7] was the first one to use the CRNN-CTC architecture to deal with handwritten music documents. Recent advances on this line of research focus on how to complement the mentioned methodology. On the one hand, language models and dictionaries are used to enhance the information obtained from recognized symbols [19,20]. On the other hand, new configurations and output representations are being proposed to take into account the intrinsic two-dimensional nature of music notation [1,16]. Furthermore, CRNN-based sequence-to-sequence models with attention mechanisms have also been postulated as successful alternatives to the CRNN-CTC [3].

The CRNN architecture represents an approach that fits perfectly well with the task at issue; however, it is highly dependent on the amount of data available, which in some cases might not be enough to properly train the model due to its inherent complexity.

Our paper fills a gap in the existing literature and presents a staff-section sequential classification-based OMR system. Our basic premise is that instead of using complex end-to-end models, we can devote two model schemes to detecting and classifying music symbols sequentially. As will be shown, the proposed system provides a remarkable improvement in terms of recognition performance with limited training data, while still competing with the CRNN-CTC in cases of larger ground-truth data.

3   Framework

We define the OMR problem here as the task of retrieving the music-notation symbols that appear in a given staff-section image. As in the state-of-the-art works referenced above, we also assume that a previous process isolated each staff of the page, much in the same way as most Handwritten Text Recognition systems assume a previous line-level segmentation.

Typical end-to-end approaches process the input image as a whole. Given an input $\mathbf{x}$ representing a single staff-section, they seek for $\hat{\sigma} = \arg\max_{\sigma \in \Sigma^*} P(\sigma \mid \mathbf{x})$, where Σ is the alphabet of symbols. To properly estimate this probability through DNN, a large number of ground-truth pairs $(\mathbf{x}, \sigma)$ is necessary. Despite this drawback, this approach is ideal for the computational reading of text—especially handwritten—where characters are often linked to form words and it is difficult to establish individual bounds for each one. However, in much of the body of historical written music, engraved in different notations than the one used nowadays, this continuous approach is not that necessary, so its goodness does not always outweigh the disadvantage of the need for data.

For humans, reading is a process where we decode a written collection of symbols. To properly retrieve the information present in the document, we follow a sequential reading order, which tells us where we must focus our attention next after reading a specific symbol. All of this entails an iterative process: look for the position of the next attention focus, decode it in terms of its type—that depends on the application—and do it all over again until reaching the end of the document. Encouraged by the idea of this reading behavior, we propose an OMR system that moves over a single-staff section image, focusing each time on the *next* symbol and predicting its shape and position. By iteratively repeating the process, we are able to decode the full staff.

Formally, let us assume that a staff section $\mathbf{x} = (\sigma_1, \sigma_2, \ldots, \sigma_n)$ is a collection of symbols whose reading order is sequential. Each symbol is, in turn, modeled by a pair $\sigma_i = (c_i, l_i)$ where $c_i \in \mathbb{R}^2$ represents the center of the symbol in the image and l_i represents its label. In addition, given that music symbols are defined by both their shape and their vertical position within the staff (height), the label l_i consists of a pair (s_i, h_i), where $s_i \in \Sigma_s$ and $h_i \in \Sigma_h$ represent the shape and height components, respectively, from fixed alphabets. This is illustrated in Fig. 1.

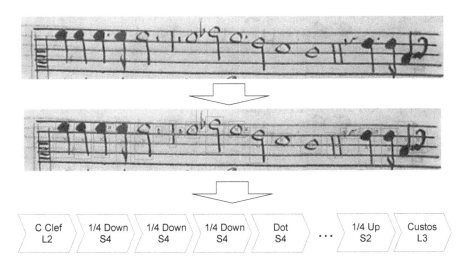

Fig. 1. Illustration of our reading music process. From top to bottom: the single staff image, the location of the symbols as the center (red crosses) of hypothetical bounding boxes, and the decoded sequence (comprising two features: shape and vertical position of the symbol in the staff). (Color figure online)

Broadly speaking, our approach seeks to estimate two functions, both conditioned to a symbol center c_i. The first function must predict the label l_i of the symbol whose center is given. A second function must predict the center c_{i+1} of the next symbol to be read. In this work, we resort to simple classification schemes for the former, while the latter is performed by means of the aforementioned NSP module. Finally, it must be remarked that the choice of using symbols' centers instead of their bounding boxes is due to the poor results obtained in preliminary experiments for the latter option.

A graphical overview of our proposal is illustrated in Fig. 2. We below delve into the architectures performing each function.

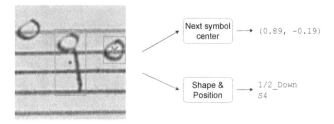

Fig. 2. Graphical description of the proposed approach. Given the blue ground truth symbol center, the shape and position models properly recognize the symbol in terms of its graphical meaning while the NSP module predicts the location of the next symbol in the sequence, illustrated in red. In light gray are represented the labeled bounding boxes which are used to compute the center of each symbol. The predicted coordinates are in the range $[-1, 1]$, as the NSP module works in this range. The labels *1/2 down* and *S4* correspond to the shape and height of the symbol respectively—the first one means the symbol is a down-oriented note that lasts half of the duration of a whole note, and the second one means that the note is in the fourth space between lines. (Color figure online)

3.1   Shape and Height Prediction

We consider that every music symbol can be completely recognized by two graphics components: shape and height. These typically condition the duration and the pitch, respectively, of the symbol. However, note that even those that do not represent any duration (clefs or alterations) or any pitch (rests), still have a specific shape and are placed in a specific location within the staff lines. Therefore, a double classification is always necessary.

This double classification process can be approached in several ways. A recent study by Nuñez-Alcover et al. [12] proposed different CNN architectures to classify both the shape and position of handwritten symbols. Two out of these different approaches—one based on using a CNN with an input-output pair associated with each symbol feature, and another based on using an independent CNN for each feature—performed comparably well and similar. In this paper, we decided to use the latter approach for its simplicity.

Therefore, the label prediction module is here implemented as two different classifiers, one CNN for each component. Given a symbol center $c_i = (x_i, y_i)$, we take a neighboring region from the top-left corner $\left(x_i - C_h^L, y_i - C_w^L\right)$ to the bottom-right corner $\left(x_i + C_h^L, y_i + C_w^L\right)$, where C_h^L and C_w^L represent the *context* height and width, respectively, of the corresponding label $L \in \{\text{shape}, \text{height}\}$ classifier. The specific values for these neighboring regions will be determined empirically. We then input the corresponding region to each CNN so that they predict the label, s_i or h_i, as appropriate.

3.2 Next-Symbol Prediction Module

The NSP module is intended to predict, in the proper reading order, a sequence of locations corresponding with each of the symbols on the given staff image.

It works as follows: given a current symbol location $c_i = (x_i, y_i)$, we follow the same procedure as in the previous section and take a neighboring region from the top-left corner $(x_i - C_h^{NSP}, y_i - C_w^{NSP})$ to the bottom-right corner $(x_i + C_h^{NSP}, y_i + C_w^{NSP})$, where C_h^{NSP} and C_w^{NSP} represent the *context* height and width of the NSP module, respectively—whose values will be empirically studied. We then feed a model with such region to estimate the center of the next symbol $c_{i+1} = (x_{i+1}, y_{i+1})$.

In our framework, the coordinates of the next symbol c_{i+1} are always within the range $[-1, 1]$, in normalized relation to the current center c_i. Therefore, we map the neighboring region, also referred to as *context image*, onto the $[-1, 1]^2$ space.

When starting the process, the first context image is centered horizontally in the first column of the staff image, with zero padding on the negative side, and vertically centered in the staff region. Similarly, zero padding will be included at the end of the staff, so that when the location of the next symbol predicted by NSP falls in that area (that is, outside the actual staff), the reading of the section will be finished.

To implement the NSP, we consider a CNN, which is formed by an initial convolutional stage, that acts as a *backbone*, followed by a regression layer, that acts as an *output block*. Given an RGB image of the context of the current symbol, the backbone performs a feature extraction of that image that the output block later synthesizes into a numerical coordinate representation, (x, y), indicating the estimated location of the next symbol center.

4 Experiments

In this section, we describe the data set, the evaluation protocol, and the architectures of the involved models.

4.1 Corpus

We consider the *Captain* corpus [6], which contains a manuscript from the 17th century of a *missa* (sacred music) in the so-called Mensural notation.[1] This corpus has been used in previous works and therefore serves as an excellent benchmark for comparison with the state of the art. An example of a staff section from this corpus is depicted in Fig. 3.

The ground-truth data already provides the segmentation of the pages into staves. This leads to a total of 99 pages, 704 single-staff sections, that amount to 17,112 running symbols, belonging to 53 and 16 different classes for shape and height, respectively. The annotation of this corpus includes the bounding

---

[1] Music notation system used for the most of the XVI and XVII centuries in Europe.

Fig. 3. Handwritten music staff-section from *Capitan* dataset.

boxes of the symbols and, therefore, we assume the center of those boxes as their locations.

4.2 Evaluation Protocol

Taking into account the different modules of the proposed framework, we consider several metrics to measure the performance of each of them individually, and also as a whole, namely:

- The shape and height classifiers are evaluated individually with simple categorical accuracy (cACC).
- The performance of the NSP module is evaluated with the Euclidean Distance between the ground-truth position of a symbol and the location predicted under two scenarios: (i) when given the ground truth location of the previous symbol, referred to as Euclidean Error Rate (EER), and (ii) when taking the previous predicted location as the initial location, so that evaluation is more accurate to a real scenario, referred to as Continuous Euclidean Error Rate (CEER).
- The NSP module together with the shape and height classifiers constitute a complete OMR model. The performance of such model is evaluated with the Symbol Error Rate (SER) computed as the average number of elementary editing operations (insertions, deletions, or substitutions) necessary to match the predicted sequence with the ground truth one, normalized by the length of the latter. It must be noted that this figure of merit is applied when both labels, s_i and h_i, are treated as a unique category, i.e., the label space considered is $\Sigma_s \times \Sigma_h$. This is the common metric of evaluation in state-of-the-art OMR.

For the experiments, we follow a 5-fold cross-validation, each fold containing its corresponding training (61 pages), validation (19 pages), and test (19 pages) partitions. We make use of the training partitions to adjust the parameters of the involved neural networks through gradient descent. The validation partition will be considered for the preliminary experiments (NSP tuning). The final evaluation, along with the comparison with the state of the art, will be performed over the test set. In all cases, the average result over the 5 folds will be reported.

4.3 NSP Configuration

We will empirically study different architectures for both the backbone and the output layer of the NSP module.

Concerning the backbone, we experiment over two well-known architectures: the convolutional layers of the VGG16 [17]—pretrained with *ImageNet* [8]—a common reference in computer vision, and a U-Net architecture [15], as it has been demonstrated to perform well on detection tasks.

Furthermore, the output block can be implemented either as a fully-connected (FC) regression layer, that receives the flattened features extracted by the backbone and transforms them into a pair of numerical coordinates or as a *differentiable spatial to numerical transform* (DSNT) layer [11]. The DSNT is of special interest for our approach since it was specifically designed for predicting image coordinates. This layer is fed with a single-channel normalized heatmap, $\hat{Z}$, of width m and height n. The term "normalized" indicates that all values of $\hat{Z}$ are non-negative and sum to one. Such normalized heatmap is obtained after applying a two-dimensional softmax activation to the feature map predicted by the backbone. The DSNT output is computed using the formula described in Eq. 1, which converts the normalized spatial heatmap to numerical coordinates of range $[-1, 1]$, as mentioned in Sect. 3.2.

$$\text{DSNT}(\hat{Z}) = \left[\left\langle \hat{Z}, X \right\rangle_F \left\langle \hat{Z}, Y \right\rangle_F \right] = \left[x, y \right] \tag{1}$$

where X and Y are $m \times n$ matrices with the x- and y-coordinates, respectively, of the input image.

As a last remark, locations are predicted from a context image of size $\left(2\,C_h^{\text{NSP}} \times 2\,C_w^{\text{NSP}} \right)$ as described in Sect. 3.2. By means of informal experimentation, we found that (192×192) was an appropriate size as it was large enough to encompass sufficient information without causing difficulties in the learning process.

4.4 Label Classifiers Configuration

The two CNN models for retrieving the shape and height features of music notation symbols, respectively, will be set as an equally-configured CNN.

The CNN will consist of five convolutional blocks. Each block consists, in turn, of a 3×3 convolution layer with $32 \times 2^{n-1}$ filters, followed by 2×2 max-pooling for downsampling and a dropout of 30%, where n represents the layer (from 1 to 5). After the 5th convolutional block, we set a fully-connected layer with 512 units followed by a batch normalization layer and a dropout of 50%. Then, the last layer of the model consists of a fully-connected layer with the number of units equals the number of classes for each classifier, along with a softmax function that converts the activations into probabilities.

As mentioned in Sect. 3.1, given a context image of $(2\,C_h^L \times 2\,C_w^L)$, the corresponding label, s_i or h_i, is predicted. We empirically found that (200×200) and (250×125) were appropriate for $L = shape$ and $L = height$, respectively. These sizes provide sufficient relevant information for the classification process. However, to reduce the complexity of the learning process, we resized these context images to (96 × 96) and (100 × 50), respectively, before feeding the information to the CNN. It must be noted that, when trained, the classifiers are augmented by adding small offsets to the ground-truth centers, making them more robust to slightly incorrect locations.

5   Results

In this section, we present the results obtained in our experiments in the following order: first, those obtained for the fine-tuning of the NSP module; and second, those concerning the complete OMR task, along with a comparison with the state of the art. Afterward, we will discuss the main outcomes.

5.1   Preliminary Experiments

We perform a series of preliminary experiments to study the impact of the different architectures for the backbone and output block of the NSP module defined in Sect. 4.3.

Table 1. Average EER and CEER figures attained by each backbone-output pair over the 5-fold validation partitions.

Backbone	Output			
	FC		DSNT	
	EER	CEER	EER	CEER
VGG16	13.28	58.59	**8.54**	**18.01**
U-Net	23.62	112.98	10.32	30.40

Table 1 shows the EER and CEER obtained in these experiments. It reports that the best configuration is the VGG16 architecture for the backbone and the DSNT layer for the output block. According to these results, the choice of the output layers seems much more relevant than the specific backbone. The use of DSNT decreases the error metrics by a wide margin—especially the CEER—compared to the use of a simple FC regression layer. For the following experiments, we adopt an NSP configuration consisting of a VGG16 architecture for the backbone and a DSNT layer for the output block of the module.

5.2   OMR Experiments

We carried out the final evaluation over the test set using the best configuration determined in Sect. 5.1.

These experiments are divided into two parts: (i) a performance analysis of each part of our method and also all of them as a whole, and (ii) a comparison between our method and the baseline. Unlike the previous preliminary experiment, in this case, the NSP is always conditioned by a previous predicted location, and not to ground-truth locations, so as to deal with the challenges of a real OMR scenario.

Performance Analysis. To properly assess the performance of our proposal, we also evaluate the contribution of each module. Table 2 provides the average results of this analysis over the test set. As it may be checked, the classifiers perform reasonably well, considering a general neighboring region and non-perfect localization. Whereas the shape classifier is closer to perfect performance, the height one reports an accuracy of around 90%. This is because, despite having a smaller number of classes, height is usually harder to learn than shape as there is greater variability between samples of the same class. Furthermore, we can observe that the NSP generalizes well, as obtains a performance similar to the one attained in the best case of our preliminary experiments. For a qualitative analysis, Fig. 4 illustrates the NSP performance over a selected part of one test image. Note that most of the centers are not perfectly located but close enough for assuming a correct detection. As reported in the table, all these modules yield a 14.2% SER when combined to perform a complete pipeline.

Table 2. Average over a 5-fold cross-validation over the test set attained by each module of the proposal and the method as a whole (Complete pipeline). Note that each part depicts its own metrics of interest.

Modules	
Label classifier (shape)	98.3 % cACC
Label classifier (height)	89.8 % cACC
NSP	15.9 % CEER
Complete pipeline	14.2 % SER

In order to gain more insights into which module of the proposed method has a greater impact on the complete pipeline, we compute a histogram of the edit operations (insertions, deletions, and substitutions) when measuring the edit distance between the ground-truth symbol sequences and the predicted ones. We consider that each insertion or deletion is caused by an error on the NSP module, which means predicting a sequence shorter or longer than the ground truth one—and therefore an incorrect symbol location—and each substitution is attributed to the shape or the height classifiers, which are assigning an incorrect symbol feature. The computation report that out of the total editing operations performed to match the predicted sequences to the ground truth ones, 21% correspond to the NSP module (6.5% and 14.5% of insertions and deletions, respectively) and the remaining 79.0% to substitutions (that is, classifiers). Hence, the

Fig. 4. Illustration of a qualitative evaluation for a selected part of a single test sample. The blue squared marks represent the ground truth symbol locations, computed from the light gray bounding boxes (ground-truth annotations of the corpus); and the red crosses denote the symbol locations predicted by NSP. (Color figure online)

shape and height prediction models are most likely to be causing most of the errors of the complete pipeline. This fact suggests two possible reasons: (i) the neighboring regions and/or the CNN considered might not be very appropriate for these classifications, or (ii) the NSP subtle errors might cause difficulties in the classification processes.

Comparison with the State of the Art. In a final evaluation, we want to observe how the amount of training data impacts the learning process. For that, we carried out an incremental training experiment. As a starting point, we evaluated the model using only 11 pages out of the total 61 that make up each training partition. We then repeat the process by increasing the number of training pages by 10 and so on, until we reach the scenario where all training pages are used. In addition, the performance of the considered approach is compared with the baseline algorithm. Specifically, a CTC-trained CRNN has been implemented following the details provided in the work by Calvo-Zaragoza et al. [7].

The curves of this experiment are shown in Fig. 5. An inspection of the reported figures reveals two relevant conclusions. On one end, a sequential classification-based approach, based on the combined use of our modules (NSP and label classifiers), allows the retrieval of the series of symbols that appear in the image of a single staff successfully. On the other end, the aforementioned approach drastically improves all the results obtained with CTC-trained CRNN in the cases when using a limited amount of training data. For sufficient-data cases, our approach shows a competitive performance, since the differences in the figure errors with respect to those of the baseline are relatively low.

5.3   Discussion

In this section we discuss the main outcomes drawn from our experiments.

As an initial remark, our approach can be considered to be competitive against the state of the art (CRNN-CTC) in cases of enough data. More importantly, it is postulated as an excellent alternative to the state of the art in

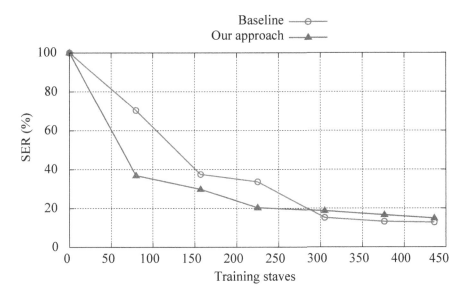

Fig. 5. SER (%) attained by the compared methods with respect to the number of training staves.

scenarios of limited data, thus validating the initial premise of our work. As mentioned above, this is a very common situation in music heritage. Considering a training set of only 80 single-staff section images, the method presented in this paper manages to halve the SER of the CRNN-CTC, from 70% to 37%. In spite of this benefit, the main disadvantage of our proposal is that it involves fine-labeling symbol positions, which requires more effort to create labeled corpora than in the CRNN-CTC approach that only needs sequence transcripts.

From another point of view, we believe that this work opens up several opportunities for further research. It must be noted that the proposed framework breaks with the limitation of a strict left-to-right order. By having a module devoted to predicting coordinates, we no longer assume a specific and always-the-same reading order. The order is set by the ground truth data: it could be from right to left, from top to bottom, in circles, or even in spirals. This could be of great interest when dealing with irregular writing styles, even in the non-music context.

Furthermore, due to the particularities of musical cultural heritage, there are many differences among music manuscripts. This fact negatively influences the OMR task because the model fails to transfer their knowledge from one corpus to another, thereby requiring learning from almost scratch each time a corpus—that might not even be labeled—needs to be transcribed. The existing labeling data is not being exploited to the full, and forcing to label a portion of each existing collection of documents is not feasible in practice. This issue could be solved by applying *transfer learning* techniques, such as *domain adaptation* techniques. While these techniques have provided good results in the literature for classification models [4, 9]—such as the ones considered independently in our modules—it might be difficult to make them work for the CRNN-CTC approach.

Finally, this formulation perfectly fits within other learning paradigms, as for instance that of *reinforcement learning* [18]. An agent could try to learn the NSP module by only providing a delayed reward correlating the final SER achieved in the single-staff section. This avoids the need to provide fine-labeling of symbol positions, only needing the same transcripts required by the CRNN-CTC approach. This is indeed an idea worth exploring.

6   Conclusions

Further improvements of state-of-the-art OMR systems are limited by the size of the available ground truth. To address such shortcoming, we propose a sequential classification-based OMR system, that retrieves the series of symbols that appear in the image of a single staff by performing symbol detection and classification as two consecutive but dependent tasks.

For that, we present the Next-Symbol Prediction module, implemented as a CNN that predicts the location of the next symbol in the staff as a pair of cartesian coordinates. This prediction is conditioned to a current symbol location, characterized in terms of a neighboring context image. We use it along with two CNN that classify the located symbol by its two-dimensional nature: shape and height (vertical position within the staff lines), respectively.

In our experiments over a handwritten 17th-century manuscript, we empirically evaluate which is the most suitable neural network configuration for the NSP module. Then, the recognition results prove that our proposed approach can be appropriate to solve the OMR task, as the learning process is successful and lower figure error rates are attained. Most importantly, when given a small amount of training data, the presented system improves considerably the state-of-the art results (for instance, it decreases the symbol error rate from 70% to 37% with 80 training images), while still reporting a competitive approach as the training set size increases.

Given the variability of music notation and the relative scarcity of existing labeled data, we aim at exploring *transfer learning* techniques to study different strategies to properly exploit the knowledge gathered from a given corpus on a different one. We also consider that *data augmentation* could be exploited to make the neural models much more robust. The use of a classification-based approach, rather than a holistic approach, suggests that these future research avenues may yield new insights on how to improve the overall performance of our model.

Acknowledgments. This work was supported by the Generalitat Valenciana through project GV/2020/030. Second author acknowledges the support from the Spanish Ministerio de Universidades through grant FPU19/04957.

References

1. Alfaro-Contreras, M., Valero-Mas, J.J.: Exploiting the two-dimensional nature of agnostic music notation for neural optical music recognition. Appl. Sci. **11**(8), 3621 (2021)
2. Bainbridge, D., Bell, T.: The challenge of optical music recognition. Comput. Humanit. **35**(2), 95–121 (2001)
3. Baró, A., Badal, C., Fornés, A.: Handwritten historical music recognition by sequence-to-sequence with attention mechanism. In: 2020 17th International Conference on Frontiers in Handwriting Recognition (ICFHR), pp. 205–210 (2020)
4. Bousmalis, K., Silberman, N., Dohan, D., Erhan, D., Krishnan, D.: Unsupervised pixel-level domain adaptation with generative adversarial networks. In: Proceedings of the IEEE Conference on Computer Vision and Pattern Recognition, pp. 3722–3731 (2017)
5. Calvo-Zaragoza, J., Jr, J.H., Pacha, A.: Understanding optical music recognition. ACM Comput. Surv. (CSUR) **53**(4), 1–35 (2020)
6. Calvo-Zaragoza, J., Toselli, A.H., Vidal, E.: Handwritten music recognition for mensural notation: formulation, data and baseline results. In: 2017 14th IAPR International Conference on Document Analysis and Recognition (ICDAR), vol. 1, pp. 1081–1086. IEEE (2017)
7. Calvo-Zaragoza, J., Toselli, A.H., Vidal, E.: Handwritten music recognition for mensural notation with convolutional recurrent neural networks. Pattern Recogn. Lett. **128**, 115–121 (2019)
8. Deng, J., Dong, W., Socher, R., Li, L.J., Li, K., Fei-Fei, L.: ImageNet: a large-scale hierarchical image database. In: 2009 IEEE Conference on Computer Vision and Pattern recognition, pp. 248–255. IEEE (2009)
9. Ganin, Y., et al.: Domain-adversarial training of neural networks. J. Mach. Learn. Res. **17**(1), 2030–2096 (2016)
10. Graves, A., Fernández, S., Gomez, F., Schmidhuber, J.: Connectionist temporal classification: labelling unsegmented sequence data with recurrent neural networks. In: Proceedings of the 23rd International Conference on Machine Learning, ICML 2006, New York, NY, USA, pp. 369–376. ACM (2006)
11. Nibali, A., He, Z., Morgan, S., Prendergast, L.: Numerical coordinate regression with convolutional neural networks. Computer research repository abs/1801.07372 (2018). http://arxiv.org/abs/1801.07372
12. Nuñez-Alcover, A., de León, P.J.P., Calvo-Zaragoza, J.: Glyph and position classification of music symbols in early music manuscripts. In: Morales, A., Fierrez, J., Sánchez, J.S., Ribeiro, B. (eds.) IbPRIA 2019. LNCS, vol. 11868, pp. 159–168. Springer, Cham (2019). https://doi.org/10.1007/978-3-030-31321-0_14
13. Pacha, A., Eidenberger, H.: Towards a universal music symbol classifier. In: 2017 14th IAPR International Conference on Document Analysis and Recognition (ICDAR), vol. 2, pp. 35–36. IEEE (2017)
14. Rebelo, A., Fujinaga, I., Paszkiewicz, F., Marçal, A., Guedes, C., Cardoso, J.: Optical music recognition: state-of-the-art and open issues. Int. J. Multimed. Inf. Retr. **1**, 173–190 (2012)
15. Ronneberger, O., Fischer, P., Brox, T.: U-Net: convolutional networks for biomedical image segmentation. In: Navab, N., Hornegger, J., Wells, W.M., Frangi, A.F. (eds.) MICCAI 2015. LNCS, vol. 9351, pp. 234–241. Springer, Cham (2015). https://doi.org/10.1007/978-3-319-24574-4_28

16. Ríos-Vila, A., Calvo-Zaragoza, J., Iñesta, J.M.: Exploring the two-dimensional nature of music notation for score recognition with end-to-end approaches. In: 2020 17th International Conference on Frontiers in Handwriting Recognition (ICFHR), pp. 193–198 (2020)
17. Simonyan, K., Zisserman, A.: Very deep convolutional networks for large-scale image recognition. In: Bengio, Y., LeCun, Y. (eds.) 3rd International Conference on Learning Representations, ICLR 2015, San Diego, CA, USA, 7–9 May 2015, Conference Track Proceedings (2015)
18. Sutton, R.S., Barto, A.G.: Reinforcement Learning: An Introduction. MIT Press, Cambridge (2018)
19. Villarreal, M., Sánchez, J.A.: Handwritten music recognition improvement through language model re-interpretation for mensural notation. In: 2020 17th International Conference on Frontiers in Handwriting Recognition (ICFHR), pp. 199–204 (2020)
20. Wick, C., Puppe, F.: Experiments and detailed error-analysis of automatic square notation transcription of medieval music manuscripts using CNN/LSTM-networks and a neume dictionary. J. New Music Res. 1–19 (2021)

Which Parts Determine the Impression of the Font?

Masaya Ueda[1]([⊠]), Akisato Kimura[2], and Seiichi Uchida[1] [iD]

[1] Kyushu University, Fukuoka, Japan
masaya.ueda@human.ait.kyushu-u.ac.jp
[2] NTT Communication Science Laboratories, NTT Corporation, Sapporo, Japan

Abstract. Various fonts give different impressions, such as legible, rough, and comic-text. This paper aims to analyze the correlation between the local shapes, or parts, and the impression of fonts. By focusing on local shapes instead of the whole letter shape, we can realize more general analysis independent from letter shapes. The analysis is performed by newly combining SIFT and DeepSets, to extract an arbitrary number of essential parts from a particular font and aggregate them to infer the font impressions by nonlinear regression. Our qualitative and quantitative analyses prove that (1) fonts with similar parts have similar impressions, (2) many impressions, such as legible and rough, largely depend on specific parts, and (3) several impressions are very irrelevant to parts.

Keywords: Font shape · Impression analysis · Part-based analysis

1 Introduction

Different font shapes will give different impressions. Figure 1 shows several font examples and their impressions attached by human annotators [5]. A font (`Garamond`) gives a *traditional* impression and another font (`Ruthie`) an *elegant* impression. Fig. 1 also shows that multiple impressions are given to a single font. Note that the meaning of the term "impression" is broader than usual in this paper; it refers to not only (more subjective) actual impressions, such as *elegant*, but also (less subjective) words of font shape description, such as *sans-serif*.

The relationship between fonts and their impressions is not well explained yet, despite many attempts from 100 years ago (e.g., [17] in 1923). This is because the past attempts were subjective and small-scale. Moreover, we need to deal with the strongly nonlinear relationship between fonts and impressions and find image features that are useful for explaining the relationship.

Fortunately, recent image analysis and machine learning techniques provide a reliable and objective analysis of complex nonlinear relationships. In addition, a large font image dataset with impression annotations is available now by Chen et al. [5]; their dataset contains 18,815 fonts, and several impression words (from

© Springer Nature Switzerland AG 2021
J. Lladós et al. (Eds.): ICDAR 2021, LNCS 12823, pp. 723–738, 2021.
https://doi.org/10.1007/978-3-030-86334-0_47

Garamond

ABCDabcd

garalde, swash, transitional

Lucida-sans

ABCDabcd

*legible, old-style, sans-serif,
1980s, american, humanist*

Ruthie

𝒜ℬ𝒞𝒟𝒶𝒷𝒸𝒹

fashionable, elegant, script

Copa-sharp-btn

ABCDabcd

*decorative, funny, energetic, point, script,
dynamic, sharp, clean, lively, handwrite*

Fig. 1. Fonts and their impressions (from [5]).

1,824 vocabularies) are attached to each font. The examples of Fig. 1 are taken from this dataset.

The purpose of this paper is to analyze the relationship between fonts and their impressions as objectively as possible by using the large font image dataset [5] with a machine learning-based approach. The relationship revealed by our analysis will help design a new font with a specific impression and judge the font appropriateness in a specific situation. Moreover, it will give hints to understand the psychological correlation between shape and impression.

We focus on local shapes, or *parts*, formed by character strokes for the relationship analysis. Typical examples of parts are the endpoints, corners, curves, loops (i.e., holes) and intersections. These parts will explain the relationship more clearly and appropriately than the whole character shape because of the following three reasons. First, the whole character shape is strongly affected by the character class, such as 'A' and 'Z,' whereas the parts are far less. Second, the decorations and stylizations are often attached to parts, such as serif, rounded corners, and uneven stroke thickness. Third, part-based analysis is expected to provide more explainability because it can localize the part that causes the impression. Although it is also true that the parts cannot represent some font properties, such as the whole character width and the aspect ratio, they are still suitable for analyzing the relationship between shape and impression, as proved by this paper.

Computer vision research in the early 2000s often uses parts of an image, called *keypoints*, as features for generic object recognition. Using some operators, such as the Difference of Gaussian (DoG) operator, the L informative parts (e.g., corners) are first detected in the input image. Then, each detected part is described as a feature vector that represents the local shape around the part. Finally, the image is represented as a set of L feature vectors. As we will see in Sect. 2, many methods, such as SIFT [14] and SURF [2], have been proposed to detect and describe the parts.

This paper proposes a new approach where this well-known local descriptor is integrated into a recent deep learning-based framework, called DeepSets [23].

DeepSets can accept an arbitrary number of D-dimensional input vectors. Let $\mathbf{x}_1, \ldots, \mathbf{x}_L$ denote the input vectors (where L is variable). In DeepSets, the input vectors are converted into another vector representation, $\mathbf{y}_1, \ldots, \mathbf{y}_L$ by a certain neural network g, that is, $\mathbf{y}_l = g(\mathbf{x}_l)$. Then, they are *summarized* as a single vector by taking their sum[1], i.e., $\tilde{\mathbf{y}} = \sum_l \mathbf{y}_l$. Finally, the vector $\tilde{\mathbf{y}}$ is fed to another neural network f that gives the final output $f(\tilde{\mathbf{y}})$.

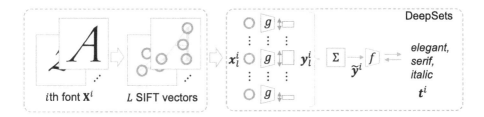

Fig. 2. Overview of the part-based impression estimation by DeepSets.

As shown in Fig. 2, we use DeepSets to estimate a font's impressions by a set of its parts. Precisely, each input vector $\mathbf{x}_l^i$ of DeepSets corresponds to the D-dimensional SIFT descriptor representing a part detected in the ith font. The output $f(\tilde{\mathbf{y}}^i)$ is the vector showing the K-dimensional impression vector, where K is the vocabulary size of the impression words. The entire network of DeepSets, i.e., f and g, is trained to output the K-dimensional m^i-hot vector for the font with m^i impression words in an end-to-end manner.

The above simple framework of DeepSets is very suitable for our relation analysis task due to the following two properties. First, DeepSets is invariant to the order of the L input vectors. Since there is no essential order of the parts detected in a font image, this property allows us to feed the input vectors to DeepSets without any special consideration.

The second and more important property of DeepSets is that it can learn the *importance of the individual parts* in the impression estimation task. If a part $\mathbf{x}_l^i$ is not important for giving an appropriate estimation result, its effect will be weakened by minimizing the norm of $\mathbf{y}_l^i$. Ultimately, if $\|\mathbf{y}_l^i\|$ becomes zero by the representation network g, the part $\mathbf{x}_l^i$ is totally ignored in the estimation process. Therefore, we can understand which part is important for a specific impression by observing the norm $\|\mathbf{y}_l^i\|$. This will give a far more explicit explanation of the relationship between shape and impression than, for example, the style feature that is extracted from the whole character shape by disentanglement (e.g., [13]).

The main contributions of this paper are summarized as follows:

- This paper proves that a specific impression of a font largely correlates to its local shapes, such as corners and endpoints of strokes. This is the first proof

---

[1] As noted in [23], it is also possible to use another operation than the summation, such as element-wise max operation.

, of this correlation by an objective and statistical analysis with a large-scale dataset to the authors' best knowledge.
- The well-known local descriptor called SIFT is integrated into a recent deep learning-based framework called DeepSets for the part-based impression analysis. DeepSets has appropriate properties for the task; especially, it allows us to evaluate the importance of each part for a specific impression.
- The analysis results provide various findings of the correlation between local shapes and impressions. For example, we could give an answer to the well-known open-problem; *what is the legibility?* Our analysis results show that constant stroke width, round corners, and wide (partially or entirely) enclosed areas are important elements (parts) for gaining legibility.

2   Related Work

2.1   Font Shape and Impression

As noted in Sect. 1, the relationship between font shape and impression has been a topic in psychology research since the 1920's [7,17]. In those studies, a limited number of people provide their impression about fonts. A similar subjective analysis approach is still used even in the recent psychological studies [1,3,8,9,11,15,16,21]. For an example of the recent trials, Shaikh and Chaparro [18] measured the impression of 40 fonts (10 for each of serif, sans-serif, script/handwriting, and display) by collecting the answers from 379 subjects for 16 semantic differential scales (SDS), where each SDS is a pair of antonyms, such as calm-exciting, old-young, and soft-hard. Their results clearly show the dependency of the impressions on font shape variations but do not detail what shapes are really relevant to raise an impression.

Computer science research has tried to realize a large-scale collection of font-impression pairs. O'Donovan et al. [16] use crowd-sourcing service for evaluating the grades of 37 attributes ($\sim$ impressions), such as *friendly*, of 200 different fonts. Based on this dataset, Wang et al. [22] realize a font generator called Attribute2Font and Choi et al. [6] realize a font recommendation system a called FontMatcher. Shinahara et al. [19] use about 200,000 book cover images to understand the relationship between the genre and each book's title font.

More recently, Chen et al. [5] realize a dataset that contains 18,815 fonts and 1,824 impression words, which are collected from `MyFonts.com` with a cleansing process by crowd-sourcing. Since our experiment uses this dataset, we will detail it in Sect. 3. Note that the primary purpose of [5] is font image retrieval by an impression query and not analyze the relationship between shape and impression.

Disentanglement is a technique to decompose the sample into several factors and has been applied to font images [4,13,20] to decompose each font image to style information and global structure information (showing the shape of each letter 'A'). Although their results are promising and useful for few-shot font generation and font style transfer, their style feature has no apparent relationship neither font shape nor impression.

All of the above trials use the *whole character shape* to understand the relationship between shape and impression. Compared to parts, the whole character shape can only give a rough relationship between the shape and the impression. The whole character shape is a complex composition of strokes with local variations. We, instead, focus on parts and analyze their relationship to specific impressions more directly.

2.2 Local Descriptors

Parts of an image sample have been utilized via local descriptors, such as SIFT [14] and SURF [2], to realize generic object recognition, image retrieval, and image matching. Local descriptors are derived through two steps: the keypoint detection step and the description step. In the former step, parts with more geometric information, such as corner and intersection, is detected. In the later step, geometric information of each part is represented as a fixed-dimensional feature vector, which is the so-called local descriptor. In the context of generic object recognition, a set of N local descriptors from an image is summarized as Bag-of-Visual Words (BoVW), where each feature vector is quantized into one of Q representative feature vectors, called visual words. Then all the L feature vectors are summarized as a histogram with Q-bins.

There are many attempts to combine local descriptors and CNNs as surveyed in [24]. However, (as shown in Table 5 of [24],) they still use BoVW summarization of the local descriptors. To the authors' best knowledge, this is the first attempt to combine the local descriptors (SIFT vectors) and DeepSets [23] and learn the appropriate representation for the estimation task in an end-to-end manner.

3 Font-Impression Dataset

As shown in Fig. 1, we use the font-impression dataset by Chen et al. [5]. The dataset is comprised of 18,815 fonts collected from Myfonts.com. From each font, we use 52 letter binary images of 'A' to 'z.' Their image size varies from 50 × 12 to 2,185 × 720.

For each font, $0 - 184$ impression words are attached. The average number of impression words per font is 15.5. The vocabulary size of the impression words is 1,824. Some of them are frequently attached to multiple fonts. For example, *decorative*, *display*, and *headline* are the most frequent words and attached to 6,387, 5,325, and 5,170 fonts, respectively. In contrast, some of them are rarely attached. The least frequent words (*web-design*, *harmony*, *jolly*, and other 13 words) are attached to only 10 fonts.

In the following experiment, we discarded minor impression words; specifically, if an impression word is attached to less than 100 fonts, it is discarded because it is a minor impression word with less worth for the analysis and insufficient to train our system. As a result, we consider $K = 483$ impressions words. The number of fonts was slightly decreased to 18,579 because 236 fonts have only minor impression words. The dataset was then divided into the train, validation, and test sets while following the same random and disjoint division of [5]. The numbers of fonts in these sets are 14,876, 1,856, and 1,847, respectively.

4   Part-Based Impression Estimation with DeepSets

4.1   Extracting Local Shapes by SIFT

From each image, we extract local descriptors using SIFT [14]. Consequently, we have a set of L_i local descriptors, $\mathbf{X}^i = \{\mathbf{x}_1^i, \ldots, \mathbf{x}_{L_i}^i\}$, from the 52 images of the ith font, where $\mathbf{x}_l^i$ is a D-dimensional SIFT vector. The number of SIFT descriptors L_i is different for each font. The average number of L over all the 18,579 fonts is about $2,505$ (i.e., about 48 descriptors per letter on average)[2]. Note that every SIFT vector is normalized as a unit vector, i.e., $\|\mathbf{x}_l^i\| = 1$.

SIFT is well-known for its invariance against rotations and scales. This property might affect our analysis positively and negatively. A positive example of the rotation invariance is that it can deal with the horizontal and the vertical serifs as the same serif; a negative example is that it cannot distinguish the oblique and the upright strokes. Fortunately, our experimental results show that the negative effect is not large and the positive effect is much larger.

4.2   Impression Estimation with DeepSets

To understand the relationship between parts and impressions of fonts, we consider a task of estimating the impression from the set of local descriptors $\mathbf{X}^i$. The ground truth of the estimation result is represented as a K-dimensional m^i-hot vector $\mathbf{t}^i$, where $K = 483$ is the vocabulary size of the impression words and m^i is the number of impression words attached to the i-th font.

As shown in Fig. 2, we solve this estimation task by DeepSets [23]. DeepSets first converts each $\mathbf{x}_l^i$ into another vector representation $\mathbf{y}_l^i$ by a neural network g; that is, $\mathbf{y}_l^i = g(\mathbf{x}_l^i)$. Then, a single sum vector $\tilde{\mathbf{y}}^i = \sum_l \mathbf{y}_l^i$ is fed to another neural network f to have the K-dimensional impression vector $f(\tilde{\mathbf{y}}^i)$. By the permutation-free property of the summation operation, DeepSets can deal with the set $\mathbf{X}_i$ as its input. In addition, it can deal with different L, without changing the network structure.

The networks f and g are trained in an end-to-end manner. The loss function is the binary cross-entropy between the output $f(\tilde{\mathbf{y}}^i)$ and the m^i-hot ground-truth $\mathbf{t}^i$. We use a fixed-sized minibatch during training DeepSets just for computational efficiency. Each minibatch contains 64 SIFT vectors which are randomly selected from L_i. In learning and inference phases, a mini-batch is created

---

[2] Fonts whose stroke is filled with textures such as "cross-hatching" give a huge number of SIFT descriptors because they have many corners. They inflate the average number of L; in fact, the median of L is $1,223$. In the later histogram-based analysis, we try to reduce the effect of such an extreme case by using the median-based aggregation instead of the average.

in the same way and thus gives the same computation time. Multi-Layer Perceptrons (MLPs) are used as the first and the second neural networks in the later experiment. More specifically, g: (128)-FC-R-(128)-FC-R-(128)-FC-(128) and f: (128)-T-FC-R-(256)-FC-R-(256)-FC-S(483), where FC, R, T, and S stand for fully-connected, ReLU, tanh, and sigmoid, respectively, and the parenthesized number is the dimension.

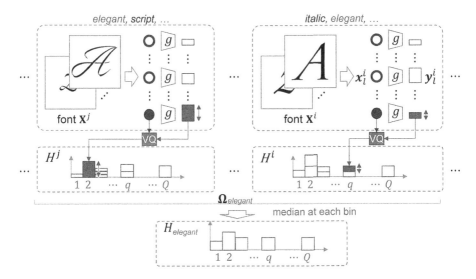

Fig. 3. Understanding the important parts for a specific impression by quantization and accumulation. VQ is a quantization module where x_l^i is quantized to the closest one among Q representative vectors. If x_l^i is quantized to qth vector, $\|y_l^i\|$ is added to the q-th bin of the histogram H^i.

4.3   Which Parts Determine a Specific Impression?

An important property of DeepSets for our impression estimation task is that we can obtain the importance of individual parts by observing the intermediate outputs $\{y_l^i\}$. If the l-th part of the i-th font is significant for giving the impression of the font, we can expect that the norm $\|y_l^i\|$ will become relatively larger than the norm of unimportant parts.[3] In an extreme case where the l-th part does not affect the impression at all, the norm $\|y_l^i\|$ will become zero.

Figure 3 illustrates its process to understand which parts are important for a specific impression $k \in [1, K]$, by using the above property. First, for each font i, we create a weighted histogram H^i of its SIFT feature vectors $\mathbf{X}^i$. The histogram has Q bins, and each bin corresponds to a representative SIFT vector derived by k-means clustering of all SIFT vectors $\bigcup_i \mathbf{X}^i$. This is similar to the classical

---

[3] Recall that the original SIFT vector is a unit vector, i.e., $\|x_l^i\| = 1$.

BoVW representation but different in the use of weighted votes. If a SIFT vector $\mathbf{x}_l^i$ is quantized to the q-th representative vector, the norm $\|\mathbf{y}_l^i\| = \|g(\mathbf{x}_l^i)\|$ is added to the q-th bin. This is because different parts have different importance as indicated by the norm, which is determined automatically by DeepSets.

Second, for each impression k, we create a histogram H_k by aggregating the histograms $\{H^i\}$ for $i \in \Omega_k$, where Ω_k denote the set of fonts annotated with the k-th impression. More specifically, the histogram H_k is given by $H_k = \mathrm{Med}_{i \in \Omega_k} H^i$, where Med is the bin-wise median operation. The median-based aggregation is employed because we sometimes have a histogram H^i with an impulsive peak at the q-th bin, and thus the effect of the single i-th font is overestimated in H_k.

The representative SIFT vectors with larger weighted votes in H_k are evaluated as important for the k-th impression. This is because (1) such SIFT vectors are evaluated as important ones with larger norms and/or (2) such SIFT vectors frequently appear in the fonts with the impression k. In the later experiment, we will observe the weighted histogram H_k to understand the important parts.

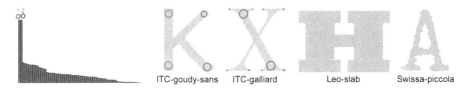

Fig. 4. The average histogram $\bar{H}$. The Q representative vectors are arranged in descending order of frequency. The two example images on the left show the parts that correspond to the two most frequent representative vectors.

5   Experimental Results

5.1   Important Parts for a Specific Impression

Figure 4 shows the average histogram $\bar{H} = \sum_{\kappa=1}^{K} H_\kappa / K$, which is equivalent to the frequency of Q representative vectors for all local descriptors $\bigcup_i \mathbf{X}^i$. The representative vectors are arranged in the descending order of frequency.

This histogram shows that there two very frequent local shapes; the two example images in the figure show the parts that correspond to them. Those parts represented by $q = 1$ and 2 often appear near the end of straight strokes with parallel contours (i.e., strokes with a constant width), which are very common for sans-serif fonts, such as ITC-goundy-sans, and even serif fonts, such as ITC-galliard. In contrast, they appear neither around extremely thick stokes even though they have a constant width, such as Leo-slab nor the font whose stroke width varies, like Swissa-piccola.

Figure 5 visualizes the important parts for eight impressions, such as *serif* and *legible*. The parenthesized number below the impression word (such as 244

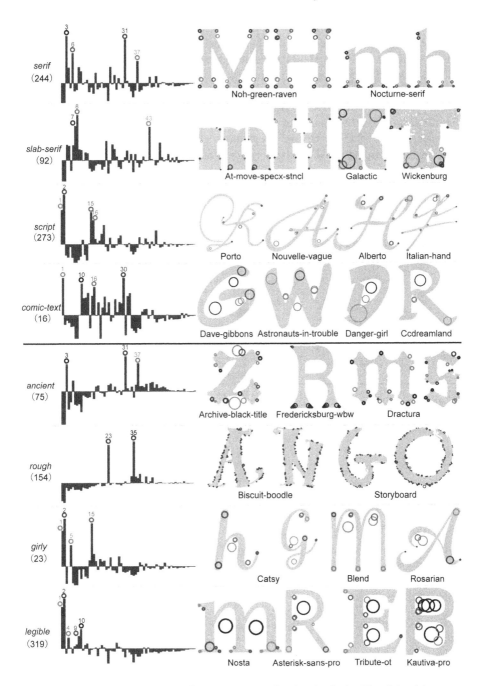

Fig. 5. Which parts determine the impression of a font?—Left: The delta-histogram ΔH_k and its peaks (marked by circles). Right: Several font images with the impression k and the location of the important parts that correspond to the peaks. The same color is used for the peak (i.e., a representative local vector) and the corresponding parts.

for *serif*) is the number of fonts with the impression, i.e., $\|\Omega_k\|$. The font name, such as Noh-green-raven, is shown below each font image.

The left-side of Fig. 5 shows a "delta"-histogram $\Delta H_k = H_k - \bar{H}$, which indicates the difference from the average histogram. The positive peaks of ΔH_k are marked by tiny circles and their number (i.e., q). Each peak corresponds to an important representative vector for the k-th impression because the vectors have far larger weights than the average for k. The delta-histogram can have a negative value at the q-th bin when parts that give the q-th representative vector are less important than the average. The delta histograms ΔH_k suggests the different local shapes (i.e., parts) are important for different impressions because the locations of their peaks are often different. We will see later that similar impression words have similar peaks.

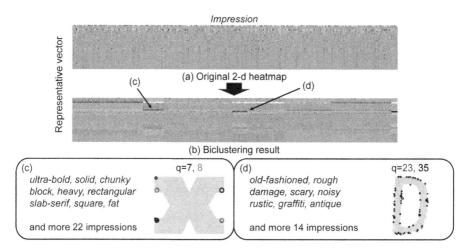

Fig. 6. A biclustering result to understand the strongly-correlated pairs between parts and impressions.

The right-side of Fig. 5 shows the parts corresponding to the peaks of the delta histogram on several font images—in other words, they indicate the important parts for the impression. The top four rows are the results of less subjective impressions:

- *Serif* has peaks at $q = 3, 6, 31, 37$ that shape serifs. Precisely, the pair of $q = 3$ and 6 form the serif tips, and the other pair of $q = 31$ and 37 form the neck of the serif. It should be noted that *serif* has less strokes with "parallel ending parts" represented by $q = 1$ and 2.
- *slab-serif* has extremely thick rectangular serifs and has peaks at $q = 7, 8$, which cannot find for *serif*.
- *Script* has four peaks; $q = 1$ and 2 correspond ∪-shaped round and parallel stroke ends. The part of $q = 15$ forms an 'ℓ'-shaped curve, unique to the

pen-drawn styles. $q = 16$ forms an oblique stroke intersection, which can also be found in 'ℓ'-shaped stroke.

- *Comic-text* has peaks at $q = 10$, 16, and 30. The part of $q = 10$ corresponds to a large partially (or entirely) enclosed area (called "counter" by font-designers) formed by a circular stroke. $q = 16$ is also found in *script* and suggests that *comic-text* also has an atmosphere of some handwriting styles.

The four bottom rows of Fig. 5 show the results of more subjective impressions:

- *Ancient* is has a similar ΔH_k to *Serif*; however, *ancient* does not have a peak at $q = 6$, which often appears at a tip of a serif. This suggests the serif in *ancient* fonts are different from standard serifs.
- *Rough* has two clear peaks that indicate fine jaggies of stroke contours.
- *Girly* is similar to *script*; however, *Girly* has an extra peak at $q = 5$. Comparing the parts corresponding to $q = 5$ and 15, the parts of $q = 5$ show a wider curve than 15. The parts of $q = 5$ also appear at counters.
- *Legible* shows its peaks at $q = 1$ and 2 and thus has parallel stroke ends. Other peaks at $q = 4$ and 9 are rounded corners. As noted at *comic-text*, the peak of $q = 10$ shows a wider "counter" (i.e., enclosed area). These peaks prove the common elements in more legible fonts; strokes with constant widths, rounded corners, and wider counters.

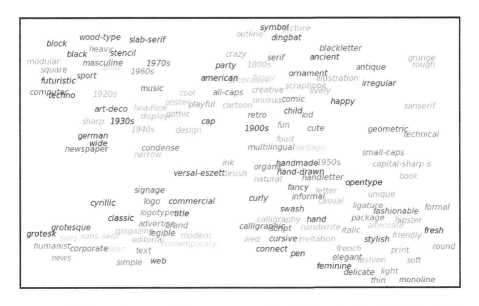

Fig. 7. t-SNE visualization of the impression distributions by the similarity of the weighted histograms $\{H_k\}$. Only top-150 frequent impressions words are plotted for better visibility.

The highlight of the above observation is that it shows that we can *explain* the font impressions using local shapes (i.e., parts) with DeepSets and local descriptors. This fact is confirmed in the later sections where we show that similarity between two histograms H_k and $H_{k'}$ reflects the similarity between the corresponding impressions. Note that we have also confirmed that scale invariance and rotation invariance brought by SIFT greatly contributed to the analysis; the scale invariance allows us to catch the detailed and tiny local shapes. The rotation invariance allows us to identify the local shapes that are horizontally or vertically symmetrical, such as both tips of a serif.

5.2 Parts and Impressions Pairs with a Strong Correlation

A biclustering analysis is conducted to understand the correlation between parts and impressions more comprehensively. Figure 6 shows its process and results. As shown in (a), we first prepare a matrix each of which column corresponds to ΔH_k, and thus each of which row corresponds to the representative vector q. Then, the column (and the row) are re-ordered so that similar column (row) vectors come closer. Consequently, as shown in (b), the original matrix is converted as a block-like matrix, and some blocks will have larger values than the others. These blocks correspond to a strongly-correlated pair of an impression word subset and a representative vector (i.e., local shape) subset. We used the `scikit-learn` implementation of spectral biclustering [12].

Among several highly correlating blocks in Fig. 6 (b), two very prominent blocks are presented as (c) and (d). The block (c) suggests that the impressions showing solid and square atmospheres strongly correlate to the local shapes represented by $q = 7$ and 8. The block (d) suggests that the impressions showing jaggy and old-fashioned atmospheres strongly correlate to $q = 23$ and 35. This analysis result shows that our strategy using parts has a very high explainability for understanding the correlations between the part and impressions.

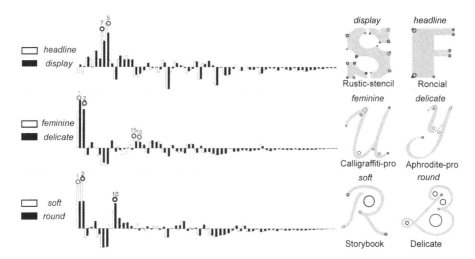

Fig. 8. The delta-histograms ΔH_k for similar impression words.

5.3 Similarity Among Impression Words by Parts

Figure 7 shows the t-SNE visualization of the distributions of the weighted histograms $\{H_k\}$ (or, equivalently, $\{\Delta H_k\}$). For better visibility and higher reliability, the top 150 frequent impression words are selected and plotted. This plot clearly shows that similar impression words are often close to each other. For example, *calligraphic*, *script* and *cursive* are neighboring because they imply a font with curvy strokes. (*bold*, *heavy*) and (*soft*, *round*) are also neighboring. In fact, by looking only less subjective words (i.e., words of typography), this distribution are roughly divided into four regions: bold sans-serif fonts (top-left), regular sans-serif fonts (bottom-left), script and handwriting font (bottom-right), and serif fonts (top-right).

Table 1. The top 20 impression words in average precision (AP). For each impression word, 5 font examples are shown.

Rank	Impression	AP(%)	font examples	Rank	Impression	AP(%)	font examples
1	*sans-serif*	68.66	AAAAA	11	*bitmap*	50.70	AAAAA
2	*handwrite*	68.55	AAAAA	12	*stencil*	49.90	AAAAA
3	*script*	67.80	AAAAA	13	*comic-text*	49.10	AAAAA
4	*black-letter*	66.71	AAAAA	14	*elegant*	48.91	AAAAA
5	*slab-serif*	63.95	AAAAA	15	*magazine*	48.79	AAAAA
6	*serif*	60.00	AAAAA	16	*decorative*	48.78	AAAAA
7	*text*	57.30	AAAAA	17	*legible*	48.78	AAAAA
8	*round*	55.62	AAAAA	18	*headline*	47.68	AAAAA
9	*grunge*	54.16	AAAAA	19	*didone*	46.80	AAAAA
10	*rough*	52.82	AAAAA	20	*brush*	45.54	AAAAA

Table 2. The bottom 20 impression words in average precision (AP).

Rank	Impression	AP(%)	font examples	Rank	Impression	AP(%)	font examples
464	*2000s*	2.04	AAAAA	474	*graphic*	1.71	AAAAA
465	*heart*	2.03	AAAAA	475	*sassy*	1.64	AAAAA
466	*random*	2.01	AAAAA	476	*distinctive*	1.58	AAAAA
467	*curve*	1.95	AAAAA	477	*style*	1.57	AAAAA
468	*fantasy*	1.92	AAAAA	478	*magic*	1.45	AAAAA
469	*arrow*	1.91	AAAAA	479	*chic*	1.44	AAAAA
470	*girl*	1.90	AAAAA	480	*revival*	1.34	AAAAA
471	*new*	1.86	AAAAA	481	*oblique*	1.18	AAAAA
472	*circle*	1.81	AAAAA	482	*thick*	1.13	AAAAA
473	*package*	1.76	AAAAA	483	*travel*	0.64	AAAAA

More subjective impression words also form a cluster. For example, *delicate*, *elegant*, *french*, and *feminine* are close to each other, and belong to the "script and handwriting font" region. *Legible, modern,* and *brand* are also very close and belong to the "regular sans-serif" region. *Cute, kid,* and *fun* belong to an intermediate region. Note that we can also observe the history of font trends by watching the transition from *1800s* to *1970s*. According to those observations, Fig. 7 is confirmed as the first large evidence-based proof that local shapes and impressions clearly correlate.

Figure 8 shows H_ks of a pair of similar words for confirming the above observation. Each pair has very similar histograms by sharing the same peak locations.

5.4   Evaluating Impression Stability by Estimation Accuracy

As we see in Fig. 2, DeepSets is trained to estimate the impression words for a given set of SIFT vectors. By using the estimation accuracy on the *test set*, we can understand the stability of the impressions. When an impression word k is estimated accurately, the impression k is stably correlating to the parts of the fonts with the impression k. For the quantitative evaluation of the estimation accuracy, we use average precision (AP) for each impression k. AP for k is given as $(\sum_{h=1}^{|\Omega_k|} h/r_h)/|\Omega_k|$, where r_h is the rank of the font $h \in \Omega_k$ in the list of the likelihoods of the impression of k; AP for k becomes larger when the fonts with the impression k gets higher likelihoods of k by DeepSets.

Table 1 shows the impression words with the 20 highest AP values and five font images with the impressions. This table indicates that the impression words for describing font shapes less subjectively are more stable. Especially, top-6 words (from *sans-serif* to *serif*) are technical terms for font designs and corresponding to specific font shapes. This fact proves that our analysis with DeepSets is reasonable.

Table 1 also shows that more subjective impressions can have a stable correspondence with parts. For example, *grunge* (9th), *rough* (10th), *elegant* (14th) and *legible* (17th) have a high AP and thus those impressions are clearly determined by the set of parts of a font image. The high AP of *legible* indicates that better legibility is common for many people. We also share similar elegant impressions from fonts with specific parts. It is also shown that parts can specify certain styles. For example, *comic-text* (13th), *magazine* (15th) , and *headline* (18th) are styles that are determined by parts. This means that we can imagine similar fonts that are suitable for comics and headlines.

Table 2 shows 20 impression words with the lowest AP values. Those impression words are least stable in their local parts. In other words, it is not easy to estimate the impression from parts. Instability of several impression words in the table is intuitively understandable; for example, it is difficult to imagine some valid font shapes from vague impressions, such as *random* (466th), *new* (471th), *graphic* (474th), and *style* (477th). It should be noted that the words relating to some specific shapes, such as *curve, circle, oblique,* and *thick*, are also listed in this table. This means that those shapes have large variations and are not stable, at least, in their parts. In other words, these shape-related impression words are not suitable as a query for searching fonts because the search results show too large diversities.

It would be interesting to see how much accuracy we can get if we use a simple CNN-based model for the impression word classification rather than using SIFT. It would tell us the expressive ability of the SIFT-based method. We conducted a experiment to recognize the impression by inputting the whole font image to a standard CNN (ResNet50[10]). Since the whole image conveys more information than a set of its local parts (more specifically, the whole image can use the absolute and relative location of its local parts), the use of the whole image gave a better mean Average Precision(mAP) as expected. More precisely, the standard CNN approach with the whole image achieved about 5.95-point higher mAP than our part-based method. However, it should be emphasized that the use of the whole font image cannot answer our main question – where the impression comes from. In other words, our SIFT-based approach can explain the relevant local parts very explicitly with the cost of 5.95-point degradation.

6   Conclusion

This paper analyzed the correlation between the parts (local shapes) and the impressions of fonts by newly combining SIFT and DeepSets. SIFT is used to extract an arbitrary number of essential parts from a particular font. DeepSets are used to summarize the parts into a single impression vector with appropriate weights. The weights are used for an index for the importance of the part for an impression. Various correlation analyses with 18,579 fonts and 483 impression words from the dataset [5] prove that our part-based analysis strategy gives clear explanations about the correlations, even though it still utilizes representation learning in DeepSets for dealing with the nonlinear correlations. Our results will be useful to generate new fonts with specific impressions and solve open problems, such as what legibility is and what elegance is in fonts.

Acknowledgment. This work was supported by JSPS KAKENHI Grant Number JP17H06100.

References

1. Azadi, S., Fisher, M., Kim, V.G., Wang, Z., Shechtman, E., Darrell, T.: Multi-content GAN for few-shot font style transfer. In: CVPR, pp. 7564–7573 (2018)
2. Bay, H., Tuytelaars, T., Van Gool, L.: SURF: Speeded up robust features. In: Leonardis, Aleš, Bischof, Horst, Pinz, Axel (eds.) ECCV 2006. LNCS, vol. 3951, pp. 404–417. Springer, Heidelberg (2006). https://doi.org/10.1007/11744023_32
3. Brumberger, E.R.: The rhetoric of typography: the awareness and impact of type-face appropriateness. Tech. Commun. **50**(2), 224–231 (2003)
4. Cha, J., Chun, S., Lee, G., Lee, B., Kim, S., Lee, H.: Few-shot compositional font generation with dual memory. In: Vedaldi, Andrea, Bischof, Horst, Brox, Thomas, Frahm, Jan-Michael. (eds.) ECCV 2020. LNCS, vol. 12364, pp. 735–751. Springer, Cham (2020). https://doi.org/10.1007/978-3-030-58529-7_43
5. Chen, T., Wang, Z., Xu, N., Jin, H., Luo, J.: Large-scale tag-based font retrieval with generative feature learning. In: ICCV (2019)

6. Choi, S., Aizawa, K., Sebe, N.: FontMatcher: font image paring for harmonious digital graphic design. In: ACM IUI (2018)
7. Davis, R.C., Smith, H.J.: Determinants of feeling tone in type faces. J. Appl. Psychol. **17**(6), 742–764 (1933)
8. Doyle, J.R., Bottomley, P.A.: Dressed for the occasion: font-product congruity in the perception of logotype. J. Consum. Psychol. **16**(2), 112–123 (2006)
9. Grohmann, B., Giese, J.L., Parkman, I.D.: Using type font characteristics to communicate brand personality of new brands. J. Brand Manage. **20**(5), 389–403 (2013)
10. He, K., Zhang, X., Ren, S., Sun, J.: Deep residual learning for image recognition. In: CVPR, pp. 770–778 (2016)
11. Henderson, P.W., Giese, J.L., Cote, J.A.: Impression management using typeface design. J. Market. **68**(4), 60–72 (2004)
12. Kluger, Y., Basri, R., Chang, J.T., Gerstein, M.: Spectral biclustering of microarray data: coclustering genes and conditions. Genome Res. **13**(4), 703–716 (2003)
13. Liu, Y., Wang, Z., Jin, H., Wassell, I.: Multi-task adversarial network for disentangled feature learning. In: CVPR (2018)
14. Lowe, D.G.: Distinctive image features from scale-invariant keypoints. Int. J. Comp. Vis. **60**(2), 91–110 (2004)
15. Mackiewicz, J.: Audience perceptions of fonts in projected powerpoint text slides. Tech. Commun. **54**(3), 295–307 (2007)
16. O'Donovan, P., Lībeks, J., Agarwala, A., Hertzmann, A.: Exploratory font selection using crowdsourced attributes. ACM Trans. Graphics **33**(4), 92 (2014)
17. Poffenberger, A.T., Franken, R.: A study of the appropriateness of type faces. J. Appl. Psychol. **7**(4), 312–329 (1923)
18. Shaikh, D., Chaparro, B.: Perception of fonts: perceived personality traits and appropriate uses. In: Digital Fonts and Reading, chap. 13. World Scientific (2016)
19. Shinahara, Y., Karamatsu, T., Harada, D., Yamaguchi, K., Uchida, S.: Serif or Sans: visual font analytics on book covers and online advertisements. In: ICDAR (2019)
20. Srivatsan, A., Barron, J., Klein, D., Berg-Kirkpatrick, T.: A deep factorization of style and structure in fonts. In: EMNLP-IJCNLP (2019)
21. Velasco, C., Woods, A.T., Hyndman, S., Spence, C.: The taste of typeface. i-Perception **6**(4), 1–10 (2015)
22. Wang, Y., Gao, Y., Lian, Z.: Attribute2Font. ACM Trans. Graphics **39**(4) (2020)
23. Zaheer, M., Kottur, S., Ravanbhakhsh, S., Póczos, B., Salakhutdinov, R., Smola, A.J.: Deep sets. NIPS (2017)
24. Zheng, L., Yang, Y., Tian, Q.: SIFT meets CNN: A decade survey of instance retrieval. IEEE Trans. Patt. Anal. Mach. Intell. **40**(5), 1224–1244 (2018)

Impressions2Font: Generating Fonts by Specifying Impressions

Seiya Matsuda[1]([✉]), Akisato Kimura[2], and Seiichi Uchida[1]

[1] Kyushu University, Fukuoka, Japan
seiya.matsuda@human.ait.kyushu-u.ac.jp
[2] NTT Communication Science Laboratories, NTT Corporation, Tokyo, Japan

Abstract. Various fonts give us various impressions, which are often represented by words. This paper proposes Impressions2Font (Imp2Font) that generates font images with specific impressions. Imp2Font is an extended version of conditional generative adversarial networks (GANs). More precisely, Imp2Font accepts an arbitrary number of impression words as the condition to generate the font images. These impression words are converted into a soft-constraint vector by an impression embedding module built on a word embedding technique. Qualitative and quantitative evaluations prove that Imp2Font generates font images with higher quality than comparative methods by providing multiple impression words or even unlearned words.

Keywords: Font impression · Conditional GAN · Impression embedding

1 Introduction

One of the reasons we have huge font design variations is that each design gives a different impression. For example, fonts with serif often give a *formal* and *elegant* impression. Fonts without serif (called sans-serif) often give a *frank* impression. Decorative fonts show very different impressions according to their decoration style. Figure 1 shows several fonts and their impression words. These examples are extracted from MyFonts dataset [3], which contains 18,815 fonts and 1,824 impression words (or tags) from MyFonts.com.

The ultimate goal of our research is to provide font images well suited for any given impression. One possible solution is tag-based font image retrieval; we can retrieve any existing font images by impression queries if they are appropriately annotated with impression tags. However, such annotations will require a substantial amount of effort. Even with full impression annotations, we can obtain neither any font images with new text tags nor completely new font images.

Based on the above discussion, one of the best ways to the goal is to generate font images with specific impressions[1]. For example, we aim to generate a font

---

[1] In this paper, we use the term "impression" in a broader meaning; some impression is described by words that relate more to font shapes, such as sans-serif, rather than subjective impression.

© Springer Nature Switzerland AG 2021
J. Lladós et al. (Eds.): ICDAR 2021, LNCS 12823, pp. 739–754, 2021.
https://doi.org/10.1007/978-3-030-86334-0_48

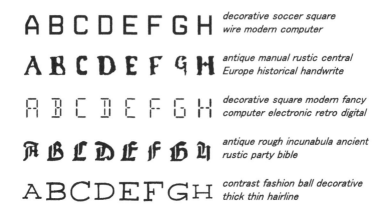

Fig. 1. Fonts and their impression words from MyFonts dataset [3].

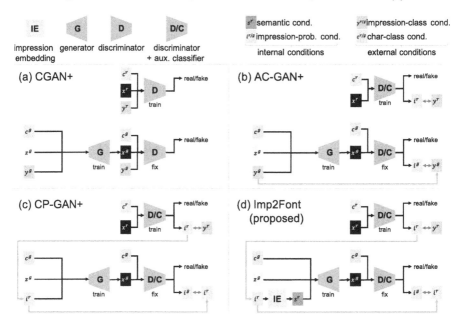

Fig. 2. (a)–(c) three representative conditional GANs, CGAN [14], AC-GAN [15], and CP-GAN [10], and (d) the proposed Impressions2Font (Imp2Font, for short). For (a)–(c), we add '+' to their name, like CGAN+, to indicate that they are an extended version from their original model for dealing with the character class condition, $c^{r/g}$.

image with an *elegant* impression or a font with a *scary* impression. Such a font image generator will be very useful to various typographic applications. In addition, it is also meaningful to understand the relationship between font style and impression.

Figure 2 (a) shows a naive GAN model for generating a font image with a specific impression. It is based on Conditional Generative Adversarial Networks (CGAN) [14]. Unlike the traditional GAN [5], CGAN can control the output images by using an external condition $y^{r/g}$, where the superscripts r and g suggest that the condition is prepared for the real data and the generated data, respectively. For our purpose, we accept a one-hot representation with K impression word vocabulary as an external condition $y^{r/g}$ and try to generate a font image with an impression specified by the one-hot representation. Here, we note that the CGAN in Fig. 2 (a) is slightly extended from its original model to incorporate another condition $c^{r/g}$ to specify the character class ('A'–'Z'). We thus call this extended CGAN as CGAN+ hereafter, for clarifying that it is not an original CGAN.

This naive realization font image generation by CGAN+ still has a large room for improvement. The most significant problem is that it treats different impression words just as different conditions. For example, a pair of synonyms, such as *large* and *big*, must work as almost the same condition for generating font images. However, they are treated as two totally different one-hot vectors in CGAN+, like the conditions of totally different words, *heavy* and *elegant*.

In this paper, we propose a novel network model, called Impressions2Font (Imp2Font, for short), for generating font images by specifying impression words. Figure 2 (d) shows the overall structure of Imp2Font, whose components are detailed in the later section. Although its structure is inspired by AC-GAN [15] (Fig. 2 (b)) and CP-GAN [10] (Fig. 2 (c)), Imp2Font has a novel module, called *impression embedding module*, to deal with various impressions more efficiently and more flexibly.

The impression embedding module enables us to deal with the mutual relationships among impression words by using their semantic similarity. More specifically, this new module uses a word embedding technique to convert each word to its real-valued semantic vector, and it is trained with external large-scale text corpora. It is well-known that word-embedding modules trained with large-scale text corpora will give similar semantic vectors to words with similar meanings. Consequently, using this semantic vector representation as real-valued conditions, mutual relationships among multiple impression words will be naturally incorporated into the CGAN framework.

Based on the above discussion, the advantages of the impression embedding module for our font generation task are summarized as follows.

1. We can specify *multiple* impression words for generating font images.
2. We can generate font images for *unlearned* impression words that are not included in the vocabulary used in the training phase.
3. The impression embedding module relaxes the data imbalance problem. Several impression words are rarely attached to fonts, and therefore difficult to associate them with font images. However, with the impression embedding module, similar and major impression words will help training for the rare impression words by sharing examples automatically.
4. The impression embedding module relaxes the noisy-label problem. Impression words attached to fonts are often very subjective and noisy. For example,

two rather inconsistent impressions, "modern" and "retro", are attached to the third font of Fig. 1. Since the proposed module converts the hard one-hot constraint (showing the labeled impression) into a soft real-valued constraint internally, it can weaken the effect of noisy labels.

In the experiment, we train Imp2Font by font images and their impressions from the dataset (hereafter called MyFonts dataset) published by Chen et al. [3] and then generate font images with various impressions. We also conduct qualitative and quantitative evaluations and comparisons with other CGANs. We also generate font images by specifying multiple impression words and unlearned impression words. The main contributions of this paper are summarized as follows:

- This paper proposed a novel GAN-based font generation method, called Impressions2Font, which can generate font images by specifying the expected impression of the font by possibly multiple words.
- Different from [20] where 37 impression words are pre-specified, this is the first attempt to cope with arbitrary impression words, even unlearned words.
- The experimental results and their quantitative and qualitative evaluations show the flexibility of the proposed method, as well as its robustness to the data imbalance problem and the noisy-label problem.

2   Related Work

2.1   Font Style and Impression

Research to reveal the relationship between font style and its impression has an about 100-year history, from the pioneering work by Poffenebrger [17]. As indicated by the fact that this work and its succeeding work [4] were published in *American Psychological Association*, such research was mainly in the scope of psychology for a long period. From around the 1980s, fonts have also been analyzed by computer science research. After tackling a simple bold/italic detection task (for OCRs), font identification, such as Zramdini et al. [23], becomes one of the important tasks. DeepFont [21] is the first deep neural network-based attempt for the font identification task. Shirani et al. [18] analyze the relationship between font styles and linguistic contents.

O'Donovan et al. [16] is one of the pioneering works on the relationship between font style and its impression in computer science research. They used a crowd-sourcing service to collect impression data for 200 fonts. The degrees of 37 attributes (i.e., impressions, such as dramatic and legible) are attached to each font by the crowd-workers. They have provided several experimental results, such as the attribute estimation by a gradient boosted regression.

Recently, Chen et al. [3] publish the MyFonts dataset by using the data from `Myfonts.com` and use it for a font retrieval system. They also use a crowd-sourcing service to clean up this noisy dataset as much as possible. It contains 18,815 fonts annotated with 1,824 text tags (i.e., the vocabulary size of impression words). We will use this dataset for training our Imp2Font; this large vocabulary helps us to generate font images by specifying arbitrary impression words.

2.2 GAN-Based Font Generation

Recent developments in generative adversarial networks (GANs) have contributed significantly to font generation. Lyu et al. [11] propose a GAN-based style transfer method for generating Chinese letters with various calligraphic styles. Hayashi et al. [6] have proposed GlyphGAN, which is capable of generating a variety of fonts by controlling the characters and styles. As a related task, font completion has also been investigated [1,2,9,22], where methods try to generate letter images of the whole alphabet in a font by using a limited number of letter images of the font (i.e., so-called few-shot composition). All of the previous methods for font completion employ GAN-based models as well.

Recently, Wang et al. [20] proposed a GAN-based font generation model that can specify font impressions. In this work, 37 impressions used in the dataset by O'Donovan et al. [16] are prepared to control the font shape. Although their purpose is similar to our trial, there exist several differences as follows: First, their impressions are limited to 37, but ours are infinite in principle, thanks to our proposed impression embedding module. Second, their method needs to specify the strength values of all the 37 impressions. In contrast, ours does not need[2]. Third, their method does not consider the semantics of each impression word, whereas ours does—this leads to our main contributions, as listed in Sect. 1.

2.3 Conditional GANs

Conditional GANs have been proposed to generate images that fit a specific condition, such as class labels. The original Conditional GAN (hereafter, called CGAN) was proposed by Mizra et al. [14] for controlling the generated image by giving a class label to the generator and the discriminator. Figure 2 (a) is CGAN with a slight modification that can deal with the character class condition $c^{r/g}$, in addition to the main condition $y^{r/g}$ related to the semantics of impression words. In this paper, a plus symbol '+' is attached to the GAN's name (e.g., CGAN+, AC-GAN+, CP-GAN+) to clarify that the character class condition augments them.

Odena et al. [15] proposed Auxiliary Classifier GAN (AC-GAN), where output images are classified by an auxiliary classifier sharing the feature extraction layers with the discriminator. This auxiliary classifier makes it possible to evaluate how a generated image fits a given condition $y^{r/g}$. Figure 2 (b) shows AC-GAN plus character class conditions. In our font generation scenario, the auxiliary classifier will output a K-dimensional vector $i^{r/g}$ whose element corresponds to the likelihood of an impression, and it will be trained so that the one-hot impression condition $y^{r/g}$ is as close to $i^{r/g}$ as possible.

---

[2] These two differences make it very difficult to fairly compare Wang et al. [20] and our proposed method.

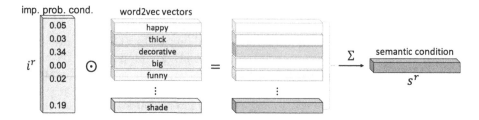

Fig. 3. The architecture of impression embedding module.

Kaneko et al. [10] proposed Classifier's Posterior GAN (CP-GAN), which is an improved version of AC-GAN. CP-GAN also has an auxiliary classifier; however, the usage of its outputs is unique. As shown in Fig. 2 (c), an output i^r of the auxiliary classifier is also used as an *internal* condition for a generator in CP-GAN+ along with *external* condition c^g. Since i^r is a real-valued vector representing the likelihood of K impression words, it works as a softer constraint than the one-hot condition y^g. It thus can realize a more flexible control of the generator during its training and testing. In fact, the vector i^r can be seen as a multi-hot vector representing the strength of all K conditions, and therefore we can generate images satisfying multiple conditions with certain strengths.

Although we developed our Imp2Font with inspiration from CP-GAN, the introduction of the impression embedding module realizes several essential differences between CP-GAN and ours. This new module works as an appropriate condition mechanism for our font generation task; especially, it can consider the mutual relationship among conditions and thus help to generate the fonts with arbitrary conditions even by the training data with imbalance and noisy labels. We will compare our method with CP-GAN+, AC-GAN+, and CGAN+, experimentally in the later section.

3   MyFonts Dataset

As noted above, we use the MyFonts dataset published by Chen et al. [3], which is comprised of 18,815 fonts. Since the dataset contains the dingbat fonts and the circled fonts, we removed them by manual inspection; as the result, we use 26 letter glyph images of 'A' to 'Z' from 17,202 fonts for training Imp2Font.

The vocabulary size of the impression words (see Fig. 1 for their examples) is 1,824[3] and at most 184 impression words are attached to each font. The examples in Fig. 1 are collected from the MyFonts dataset.

---

[3] As noted later, each impression word is converted to a semantic vector by word2vec [13]. Therefore, we remove too rare impression words that are not included even in the 3-million English vocabulary for training word2vec. This results in $K = 1,574$ impression words that we used in the following. Note that an impression word with hyphenation is split into sub-words, and then its semantic vector is derived by taking the sum of the semantic vectors of the sub-words.

The MyFonts dataset has several problems for training a GAN-based font image generator. First, it has a data imbalance problem. For example, the impression word *decorative* is attached to 6,387 fonts, whereas 16 impression words (including *web-design*, *harmony*, etc.) is associated with only 10 fonts. Second, it contains noisy labels. Since the impression words are attached by many people (including non-experts) and the impression of a font is often person-dependent, the impression words of a font are not very consistent. For example, two contradicting impressions, *thin* and *thick*, are attached to the bottom font of Fig. 1. Another interesting example is *modern* and *retro* of the second font; it might seem modern for an older person but retro for a young person. Third, the number of impression words, K, is still limited to 1,574, although it is far larger than 37 of [16]. Since it is known that English word vocabulary for common use is more than 10,000, we might encounter out-of-vocabulary words in several applications, such as the font generation with specifying its impression by a "sentence" or a "text." Fortunately, as noted in the later sections, our Imp2Font is robust to these problems by its word embedding module.

4   Impressions2Font—Conditional GAN with Impression Embedding Module

4.1   Overview

This section details our proposed Imp2Font, which can generate font images by specifying an arbitrary number of arbitrary impression words. As already seen in Fig. 2(d), Imp2Font accepts two external conditions, namely a character class $c^{r/g}$ and impression words $y^{r/g}$. The former is a 26-dimensional one-hot vector (for generating 26 English capitals, 'A'-'Z'), and the latter is a K-dimensional one-hot vector, where randomly selected an impression word from the impression words annotated in the font at each epoch. We do not use the K-hot representation to satisfy $\sum i^r = 1$ in the impression embedding module. The auxiliary classifier is trained so that its real-valued output i^r becomes more similar to a one-hot ground-truth y^r, just like AC-GAN.

Inspired by CP-GAN, we use this i^r also as an *internal condition* for training the generator. We call i^r an impression probability condition. We then optimize the generator so that this impression probability condition i^g of generated data corresponds with that i^r of real data. This formulation allows our model to capture between-impression relationships in a data-driven manner and generate a font image conditioned on the impression specificity.

The impression probability condition i^r relates to the highlight of the proposed method, i.e., the impression embedding module. More specifically, the condition i^r is converted to another internal condition, called *semantic condition s^r*, by the module, as detailed in the next Sect. 4.2. The semantic condition is a D-dimensional real-valued vector and specifies the impression to be generated. During the training phase of the generator, the semantic condition is fed to the generator.

4.2   Impression Embedding Module

Figure 3 shows the architecture of our impression embedding module. This module gives the D-dimensional semantic condition s^r, which is expected to be an integrated vector of the impressions specified by the impression probability condition, i^r. In the example of this figure, the probabilities of the impressions, *decorative* and *happy* are high (0.34) and low (0.05), respectively. Therefore, the semantic condition s^r should represent more *decorative* and *happy* as the integrated impression.

In the impression embedding module, the semantic condition is derived as

$$s^r = \sum_{k=1}^{K} i_k^r w_k, \tag{1}$$

where i_k^r is the k-th element of i^k and w_k is a semantic vector of the k-th impression word. The semantic vector w_k is expected to represent the meaning of the k-th impression word as a real-valued D-dimensional vector. Also, the k-th and k'-th words have similar semantic meanings, w_k and $w_{k'}$ are also expected to be similar.

Although an arbitrary word embedding method can be employed for extracting semantic vectors, we introduce word2vec [13], which is a traditional but still useful method. We use a word2vec model pre-trained by the large-scale Google News dataset (about 100 billion words). The model outputs a $D = 300$-dimensional semantic vector for each word in the 3-million English vocabulary.

Our formulation Eq. (1) of the impression embedding module offers several advantages for generating font images from impression words.

(1) Robustness against data imbalance: Let us assume that the impression word *fat* is rarely attached to font images, whereas a similar impression word *bold* is frequent. If a typical K-dimensional one-hot vector represents the semantic condition, it is hard to generate font images with *fat* impression due to the lack of its training samples. However, our semantic condition by Eq. (1) allows to transfer knowledge from font styles the *bold* impression to those with the *fat* impression, because we can expect $w_{\text{bold}} \sim w_{\text{fat}}$.

(2) Robustness to noisy labels: Let us assume that three impression words, *bold*, *heavy*, and *thin*, are attached to a font. The impression *thin* seems to be inconsistent with the others, and thus it is expected to be a noisy label. Note that, as shown in Fig. 1, the MyFonts dataset contains such inconsistent annotations. The impression class condition y^r will become a one-hot vector which represents any of the three impression words; however, the resulting semantic condition s^r is less affected by the noisy-label *thin* than the other two words, because the meanings of *bold* and *heavy* are similar, that is, $w_{\text{bold}} \sim w_{\text{heavy}}$, and have a double impact in s^r.

4.3 Generating Font Images from the Trained Generator

After the adversarial training of the generator and the discriminator (plus the auxiliary classifier), we can generate character images with specific impressions by giving the character-class condition c^g, the noise vector z^g, and an arbitrary semantic condition s^g. By changing z^g, we have various fonts even with the same impression s^g and the same character class c^g.

The condition s^g is fed to the generator like s^r; however, different from the internal condition s^r, the condition s^g is an external condition and specified by directly using arbitrary semantic vectors given by word2vec. For example, if we want to generate a font with the impression *elegant*, we can use w_{elegant} as s^g.

The flexibility of setting the condition s^g realizes Imp2Font's strength that we can specify an arbitrary number of arbitrary impression words. For example, a font with multiple impressions is generated by setting s^g as the sum of the corresponding semantic vectors. Moreover, it is possible to use the semantic vector of *unlearned* impression words as s^g; even if the impression word *gigantic* is unlearned, Imp2Font will generate a gigantic font with the help of some similar and learned words, such as *big*.

4.4 Implementation Details

Table 1 shows the network structure of the generator and the discriminator in the proposed method. The structure of the generator and discriminator is based on commonly known DCGANs. To avoid mode-collapse, the mode seeking regularization term by Mao et al. [12] is employed for the generator. The Kullback-Leibler (KL) divergence is employed as the loss function of the auxiliary classifier. During the 100-epoch training, ADAM optimization was used with the learning rate at 0.0002. The batch size is 512, and the generator and discriminator are updated alternately in a 1:5 ratio.

In the experiments, CGAN+, and AC-GAN+, and CP-GAN+ are used. For a fair comparison, we used the same setup for them, including the model seeking regularization. Since Imp2Font is a super-set of those methods (as indicated by Fig. 2), the following experiments can be seen as ablation studies.

Table 1. Network structure.

Generator	Discriminator /Aux. Classifier
$z^g \in \mathbb{R}^{300} \sim N(0, I)$, $c^g \in \{0,1\}^{26}$, $s^r \in \mathbb{R}^{300}$	$x^{r/g} \in \mathbb{R}^{64 \times 64 \times 1}$, Expand$(c^{r/g}) \in \{0,1\}^{64 \times 64 \times 26}$
FC→1500 for Concat(z^g, c^g), FC→1500 for s^r	4 × 4 str.=2 pad.=1 Conv 64, SN, LReLU
FC→32768, BN, LReLU	4 × 4 str.=2 pad.=1 Conv 128, SN, LReLU
Reshape 16 × 16 × 128	FC → 1 for Disc.. / FC → 1574 for Cl.
4 × 4 str.=2 pad.=1 Deconv 64, BN, LReLU	
4 × 4 str.=2 pad.=1 Deconv 1, Tanh	

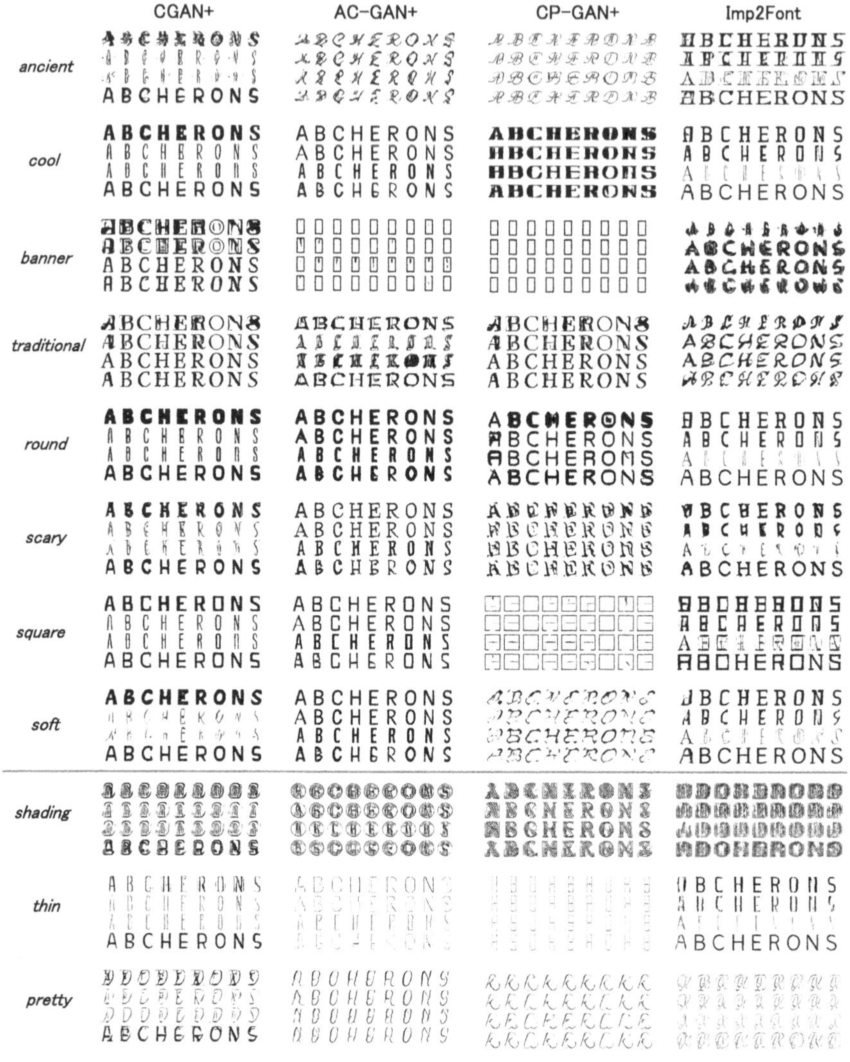

Fig. 4. Font images generated by specifying a single impression word.

5   Experimental Results

5.1   Font Generation Specifying a Single Impression Word

Figure 4 shows the fonts generated by the CGAN+, AC-GAN+, CP-GAN+, and our proposed Imp2Font, conditioned on a single impression word in the $K = 1,574$ vocabulary. For each impression, four images are generated for "ABC" and "HERONS"[4], by changing the noise input z^g. As noted in Sect. 4.4, the same setup was used for all the models for a fair comparison. In addition, since each of the competitors can be seen as a subset of Imp2Font, as shown in Fig. 2, their experimental results can be regarded as ablation studies.

The result shown in Fig. 4 indicates that Imp2Font could generate font images with the impression specified by the word. For example, all the generated images for *square* and *round* had more squared and round shapes than others, respectively, while maintaining the diversity of the fonts. In contrast, the other methods sometimes failed to capture the style specified by the word, as shown in *round* and *square* for CGAN+. The result also indicates that images generated by the proposed method maintained the readability of letters in most cases. Meanwhile, the other methods sometimes failed even in the cases where the proposed method succeeds, for example *banner* for AC-GAN+ and CP-GAN+ and *square* for CP-GAN+. Although the proposed method failed in *shading* and *pretty*, the result of the other methods implies that those words seem to be difficult to reproduce fonts.

5.2　Font Generation Specifying Multiple Impression Words

Figure 5 shows the font images generated by giving two words as conditions. Our proposed method is very suitable for the generation from multiple impression words; meanwhile the previous methods are not good at this task. This result indicates that Imp2Font could generate font images more visually readable than the ones generated from a single label shown in Fig. 4. For example, only the generated images by the proposed method for *lcd* and *ancient* look like dot representations, but they have serif typefaces.

Figure 6 shows the evolution of generated images for increasing numbers of impression words. In those examples, we chose real pairs of an image and the corresponding set of impression words as references and employed each reference set of impression words for generating font images. This result indicates that generated fonts by the proposed method for a reference set of impression words were gradually approaching to the corresponding reference font images; meanwhile the results by the existing methods were far from the reference fonts.

Imp2Font can change the strength of impression words as inputs, which means that we can interpolate between different fonts. For example, interpolation $i_{italic \leftrightarrow normal}$ between two impression words *italic* and *normal* can be implemented by $i_{italic \leftrightarrow normal} = (1 - \lambda)i_{italic} + \lambda i_{normal}$, where $\lambda \in [0, 1]$ represents an interpolation coefficient. Figure 7 shows examples of fonts generated with certain interpolation coefficients. In all these examples, the noise z^g is fixed. Imp2Font allows for smooth transitions between two impression words. This result implies that our proposed method could generate a font for a word with multiple impressions, specifying each percentage.

---

[4] "HERONS" is a common word to check the font style since it contains sufficient variations of stroke shapes.

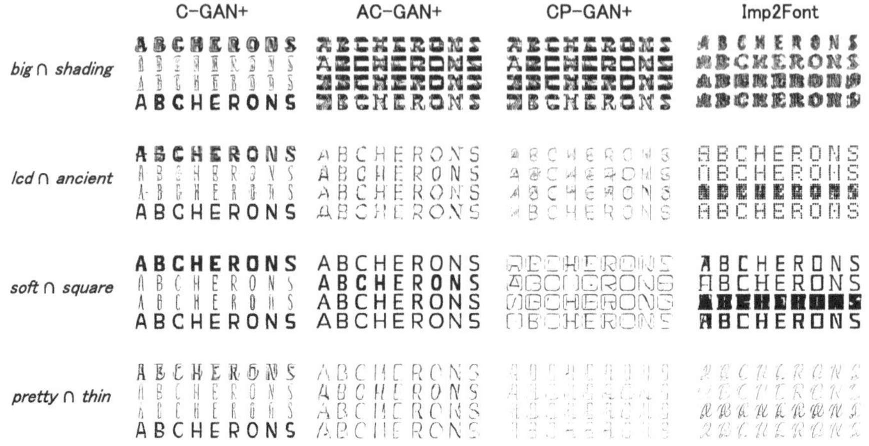

Fig. 5. Font images generated by specifying multiple impression words.

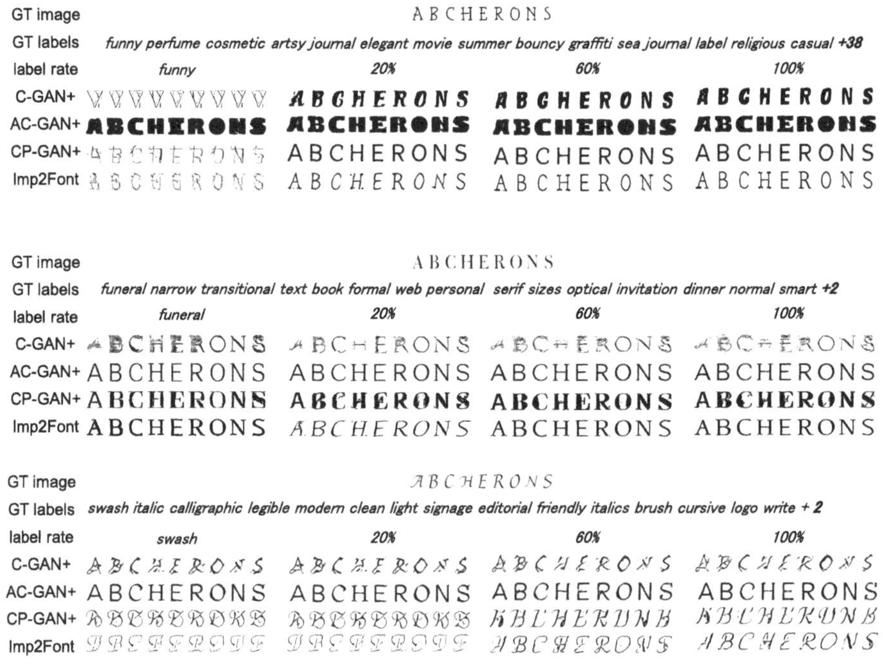

Fig. 6. Image generated conditional on all impression words in ground truth.

Similarly, Imp2Font can control the diversity of fonts even for a fixed set of impression words. We can implement this scheme by interpolating between different noises, i.e., $z^g = (1 - \lambda)z_1 + \lambda z_2$. Figure 8 shows examples of fonts generated with certain interpolation coefficients for a fixed single impression

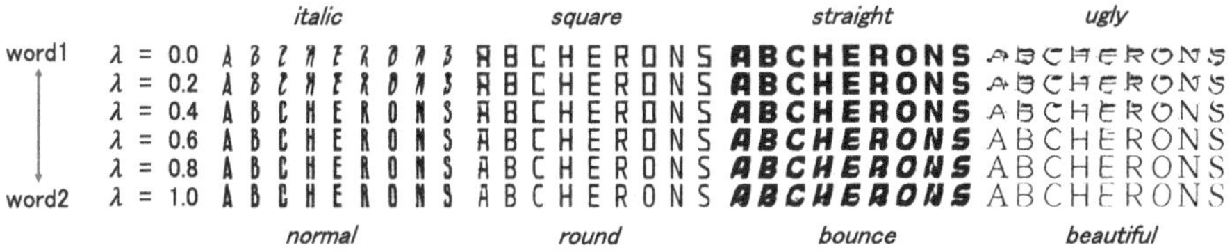

		italic				square			straight		ugly

word1
$\lambda = 0.0$ A B Z H I X D H Z   H B C H E R D N S   **ABCHERONS**   ABCHERONS

$\lambda = 0.2$ A B L H I X D H Z   H B C H E R D N S   **ABCHERONS**   ABCHERONS

$\lambda = 0.4$ A B L H E R D N Z   H B C H E R D N S   **ABCHERONS**   ABCHERONS

$\lambda = 0.6$ A B L H E R D N Z   H B C H E R D N S   **ABCHERONS**   ABCHERONS

$\lambda = 0.8$ A B L H E R D N Z   A B C H E R D N S   **ABCHERONS**   ABCHERONS

word2
$\lambda = 1.0$ A B L H E R D N Z   A B C H E R O N S   **ABCHERONS**   ABCHERONS

	normal		round		bounce		beautiful

Fig. 7. Generated glyph images by interpolation between two impression words.

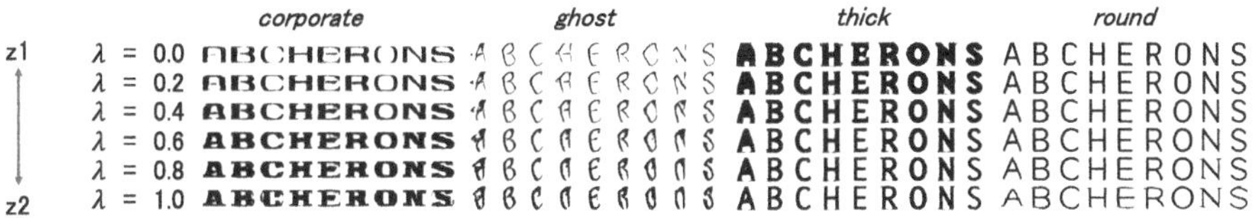

		corporate		ghost		thick		round

z1
$\lambda = 0.0$ ABCHERONS   ABCHERONS   **ABCHERONS**   ABCHERONS

$\lambda = 0.2$ ABCHERONS   ABCHERONS   **ABCHERONS**   ABCHERONS

$\lambda = 0.4$ ABCHERONS   ABCHERONS   **ABCHERONS**   ABCHERONS

$\lambda = 0.6$ **ABCHERONS**   ABCHERONS   **ABCHERONS**   ABCHERONS

$\lambda = 0.8$ **ABCHERONS**   ABCHERONS   **ABCHERONS**   ABCHERONS

z2
$\lambda = 1.0$ **ABCHERONS**   ABCHERONS   **ABCHERONS**   ABCHERONS

Fig. 8. Generated glyph images by interpolation between two noise.

word. It can be seen that the change in noise affected the stroke width and jump in the font style.

5.3   Font Image Generation for Unlearned Impressions

Imp2Font can also generate font images even for impression words not included in the vocabulary for training. This is because our impression embedding module effectively exploits the knowledge of large-scale external text corpora containing 3 million words through word2vec. This is a unique function of the proposed method that the previous method lacks. Figure 9 shows the generated font conditioned on the impression words (1) in the vocabulary for learning and (2) those not used for learning but having a similar meaning to the one in the vocabulary for training. The result shown in this figure indicates that the proposed method could generate fonts even for unlearned impressions, and they were similar to the fonts for similar learned impressions in terms of shapes and styles.

5.4   Quantitative Evaluations

We also compared the proposed method with the previous work in a quantitative manner. We applied two metrics to the evaluation: (1) Frechet inception distance (FID) [7] for measuring the distance between distributions of real and generated images. (2) Mean average precision (mAP) that is a standard metric for multi-label prediction. For computing mAPs, we built two additional multi-label ResNet-50 predictors, where one was trained with real images, and their impression words (called *pred-real*) and the other was trained with generated images and impression images given for generation (called *pred-gen*). In the following, we call mAP for the *pred-real* model and generated test images *mAP-train*, and the one for the *pred-gen* and real test images *mAP-test*, respectively.

Fig. 9. Font image generation for unlearned impressions.

Table 2. Quantitative evaluation results.

	CGAN+	AC-GAN+	CP-GAN+	Imp2Font
↓FID	39.634	39.302	33.667	**24.903**
↑ mAP-train	1.524	1.157	**1.823**	1.765
↑ mAP-test	1.155	1.158	1.600	**1.708**

Table 2 shows the quantitative results. The results indicate that the proposed method greatly outperformed the others for almost all the cases. More specifically, the proposed method and CP-GAN+ improved mAPs against CGAN+ and AC-GAN+. This implies that the introduction of the impression embedding module that provides soft constraints for generators was effective for our task. Also, we can see that the proposed method greatly improved FID, which implies that the proposed method can generate high-quality font images while maintaining the diversity of fonts. Note that the mAP-train of Imp2Font is slightly lower than CP-GAN+; one possible reason is that the word embedding performance is still not perfect. Currently, Imp2Font employs word2vec, which is based on the distributional hypothesis and thus has a limitation that words with very different meanings (even antonyms) have a risk of similar embedding results. Using advanced embedding methods will easily enhance the performance of Imp2Font.

6   Conclusion

This paper proposes Impressions2Font (Imp2Font), a novel conditional GAN, and enables generating font images with specific impressions. Imp2Font accepts an arbitrary number of impression words as its condition to generate the font. Internally, by an impression embedding module, the impression words are converted to a set of semantic vectors by a word embedding method and then unified into a single vector using likelihood values of individual impressions as weights.

This single vector is used as a soft condition for the generator. Qualitative and quantitative evaluations prove the high quality of the generated images. Especially, it is proved that giving more impression words will help to generate the expected font shape accurately. Moreover, it is also proved that Imp2Font can accept even *unlearned* impression words by using the flexibility of the impression embedding module.

Future work will focus on incorporating different word embedding techniques instead of word2vec. Especially if we can realize a new word embedding technique grounded by incorporating the shape-semantic relationship between fonts and their impressions, like sound-word2vec [19] and color word2vec [8], it is more appropriate for our framework. We also plan to conduct evaluation experiments for understanding not only the legibility but also the impression of the generated font images.

Acknowledgment. This work was supported by JSPS KAKENHI Grant Number JP17H06100.

References

1. Azadi, S., Fisher, M., Kim, V.G., Wang, Z., Shechtman, E., Darrell, T.: Multi-content GAN for few-shot font style transfer. In: CVPR (2018)
2. Cha, J., Chun, S., Lee, G., Lee, B., Kim, S., Lee, H.: Few-shot compositional font generation with dual memory. In: Vedaldi, A., Bischof, H., Brox, T., Frahm, J.-M. (eds.) ECCV 2020. LNCS, vol. 12364, pp. 735–751. Springer, Cham (2020). https://doi.org/10.1007/978-3-030-58529-7_43
3. Chen, T., Wang, Z., Xu, N., Jin, H., Luo, J.: Large-scale tag-based font retrieval with generative feature learning. In: ICCV (2019)
4. Davis, R.C., Smith, H.J.: Determinants of feeling tone in type faces. J. Appl. Psychol. **17**(6), 742–764 (1933)
5. Goodfellow, I.J., et al.: Generative adversarial networks. arXiv preprint arXiv: 1406.2661 (2014)
6. Hayashi, H., Abe, K., Uchida, S.: GlyphGAN: style-consistent font generation based on generative adversarial networks. Knowledge-Based Syst. **186**, 104927 (2019)
7. Heusel, M., Ramsauer, H., Unterthiner, T., Nessler, B., Hochreiter, S.: GANs trained by a two time-scale update rule converge to a local nash equilibrium. In: NIPS (2017)
8. Ikoma, M., Iwana Brian, K., Uchida, S.: Effect of text color on word embeddings. In: DAS (2020)
9. Jiang, Y., Lian, Z., Tang, Y., Xiao, J.: DCFont: an end-to-end deep Chinese font generation system. In: SIGGRAPH Asia (2017)
10. Kaneko, T., Ushiku, Y., Harada, T.: Class-distinct and class-mutual image generation with GANs. In: BMVC (2019)
11. Lyu, P., Bai, X., Yao, C., Zhu, Z., Huang, T., Liu, W.: Auto-encoder guided GAN for Chinese calligraphy synthesis. In: ICDAR, vol. 1, pp. 1095–1100 (2017)
12. Mao, Q., Lee, H.Y., Tseng, H.Y., Ma, S., Yang, M.H.: Mode seeking generative adversarial networks for diverse image synthesis. In: CVPR (2019)

13. Mikolov, T., Sutskever, I., Chen, K., Corrado, G.S., Dean, J.: Distributed representations of words and phrases and their compositionality. In: NIPS (2013)
14. Mirza, M., Osindero, S.: Conditional generative adversarial nets. arXiv preprint arXiv:1411.1784 (2014)
15. Odena, A., Olah, C., Shlens, J.: Conditional image synthesis with auxiliary classifier GANs. In: ICML (2017)
16. O'Donovan, P., Lībeks, J., Agarwala, A., Hertzmann, A.: Exploratory font selection using crowdsourced attributes. ACM Trans. Graph. **33**(4), 92 (2014)
17. Poffenberger, A.T., Franken, R.: A study of the appropriateness of type faces. J. Appl. Psychol. **7**(4), 312–329 (1923)
18. Shirani, A., Dernoncourt, F., Echevarria, J., Asente, P., Lipka, N., Solorio, T.: Let me choose: from verbal context to font selection. In: ACL (2020)
19. Vijayakumar, A., Vedantam, R., Parikh, D.: Sound-Word2Vec: learning word representations grounded in sounds. In: EMNLP (2017)
20. Wang, Y., Gao, Y., Lian, Z.: Attribute2font: creating fonts you want from attributes. ACM Trans. Graph. **39**(4), 69 (2020)
21. Wang, Z., et al.: DeepFont: identify your font from an image. In: ACM Multimedia (2015)
22. Zhu, A., Lu, X., Bai, X., Uchida, S., Iwana, B.K., Xiong, S.: Few-shot text style transfer via deep feature similarity. IEEE Trans. Image Proc. **29**, 6932–6946 (2020)
23. Zramdini, A., Ingold, R.: Optical font recognition using typographical features. IEEE Trans. Patt. Anal. Mach. Intell. **20**(8), 877–882 (1998)

Author Index

Printed in the United States
by Baker & Taylor Publisher Services